Electrode Materials for Energy Storage and Conversion

Electrode Materials for Energy Storage and Conversion

Edited by
Mesfin A. Kebede and Fabian I. Ezema

CRC Press
Taylor & Francis Group
Boca Raton London New York

CRC Press is an imprint of the
Taylor & Francis Group, an **informa** business

First edition published 2022
by CRC Press
6000 Broken Sound Parkway NW, Suite 300, Boca Raton, FL 33487-2742

and by CRC Press
2 Park Square, Milton Park, Abingdon, Oxon, OX14 4RN

Library of Congress Cataloging-in-Publication Data
Names: Kebede, Mesfin A., editor. | Ezema, Fabian I., editor.
Title: Electrode materials for energy storage and conversion / edited by Mesfin A. Kebede and
Fabian I. Ezema.
Description: First edition. | Boca Raton : CRC Press, 2022. | Includes bibliographical references and index.
Identifiers: LCCN 2021023334 | ISBN 9780367697907 (hardback) | ISBN 9780367703042 (paperback) |
ISBN 9781003145585 (ebook)
Subjects: LCSH: Electric batteries--Electrodes. | Energy storage--Materials. | Electric power production from chemical action--Materials.
Classification: LCC TK2945.E44 E44 2022 | DDC 621.31/2420284--dc23
LC record available at https://lccn.loc.gov/2021023334

ISBN: 978-0-367-69790-7 (hbk)
ISBN: 978-0-367-70304-2 (pbk)
ISBN: 978-1-003-14558-5 (ebk)

DOI: 10.1201/9781003145585

Typeset in Times
by MPS Limited, Dehradun

Contents

Foreword

Contemporary era is viewed as the era of globalization, which is principally portrayed as the era of modern technology, with high energy demand. Due to the fast-growing economy and human population worldwide, the global energy consumption has been rising steadily, necessitating other substitutes or means for satiating the growing population's energy demand. Thus, the issue of the sustainability of energy supplies has become a matter of concern to the entire world, with rising energy crisis due to the rapid depletion of fossil fuel energy resources; this is accompanied by serious environmental pollution problems arising from the use of fossil fuels. The pressing demand to introduce alternative energy sources to fossil fuels and minimize carbon(IV) oxide emissions breeds substantial research attentiveness in the expansion of renewables, and conversion and storage systems for various types of energy. Research has shown that reducing fossil fuel use is necessary because of their negative environmental impacts, such as the water and air pollution that lead to global warming.

Reliable and cost-effective expertise associated with energy conversion and storage is needed to solve the energy problems devastating the world. The critical requirements for a good storage device are affordability, availability, high energy and power densities, which the recent nanotechnology advancements using various electrode material gear up to tackle.

This book is certainly a welcome exposition to the much-required methodical studies of topical advancement of energy conversion and storage applications, responding confidently to the global energy requirement and revealing the need for environment-friendly energy sources, and conversion and storage mechanisms for a healthy environment with essential approaches to appreciate the diversity and complexity associated with high energy demand in our present time.

Prof. Charles A. Igwe
Vice-Chancellor
University of Nigeria, Nsukka
June 2021

Preface

Climate change has become an issue of great concern to all of us, by affecting our planet earth negatively. We are witnessing natural disasters across the globe on a large scale because of this problem. Thus, there is the need to tackle the issues arising from the use of traditional fossil energy sources, which produce global warming due to pollution. A clean environment, followed by sustainable development, can only be attained using clean energy technologies. Energy storage and conversion are considered to play a vital role in addressing the situation by producing clean energy sources to replace traditional fossil energy and coal-based energy sources. Presently, it is realized that the combination of solar, wind, and battery energy is cost-competitive to the traditional energy sources. Persistent R&D efforts for a couple of decades have started to pay off in finding competitive green energy alternatives that cause lesser pollution of the environment.

This book presents comprehensive knowledge on energy conversion and energy storage systems. Chapters 1–12 look at various aspects of energy storage systems such as lithium-ion battery, zinc-ion battery, supercapacitor, etc. Chapters 13–24 discuss the energy conversion systems such as fuel cell, solar cell, etc. The chapters are written in such a way as to assist beginners in the field of electrochemical science, while at the same time providing experts greater understanding of the field.

More specifically, Chapters 1–6 present various battery systems as energy storage applications. Chapters 7–9 discuss different types of supercapacitors and their electrodes. Chapters 10–11 explain the application of the biomass-based energy storage.

Chapters 12 and 13 review the working principle of fuel cells and their electrodes, while chapters 14–24 discuss mainly the solar cell systems and nanomaterials for optoelectronics applications in general.

On the whole, this book is designed to provide insight into a wide range of energy storage and conversion devices to the readers.

Mesfin A. Kebede
Pretoria, South Africa
Fabian I. Ezema
Nsukka, Nigeria
June 2021

Editors

Dr. Mesfin A. Kebede, obtained his PhD in Metallurgical Engineering from Inha University, South Korea. He is now a Principal Research Scientist at the Energy Centre of Council for Scientific and Industrial Research (CSIR), South Africa. He worked previously as Assistant Professor in the Department of Applied Physics and Materials Science at Hawassa University, Ethiopia. His extensive research experience covers the use of electrode materials for energy storage and energy conversion.

Prof. Fabian I. Ezema, is a Professor at the University of Nigeria, Nsukka. He obtained a PhD in Physics and Astronomy from University of Nigeria, Nsukka. His research focuses on several areas of Materials Science with an emphasis on energy applications, specifically electrode materials for energy conversion and storage.

Contributors

Adeolu A. Adediran
Department of Mechanical Engineering
Landmark University
Omu-Aran, Kwara State, Nigeria

Ishaq Ahmad
National Center for Physics
Islamabad, Pakistan
and
NPU-NCP Joint International Research Center on
    Advanced Nanomaterials and Defects
    Engineering
Northwestern Polytechnical University
Xi'an, China
and
Nanosciences African Network (NANOAFNET)
    iThemba LABS-National Research Foundation
Somerset West, Western Cape Province, South
    Africa

Dinsefa Mensur Andoshe
Adama Science and Technology University
Department of Materials Engineering
Adama, Ethiopia

M. Anusuya
Indra Ganesan College of Engineering
Trichy, Tamilnadu
India

Ebubechukwu N. Dim
Department of Science Laboratory Technology
University of Nigeria
Nsukka, Nigeria

Farai Dziike
Research and Postgraduate Support Directorate
Durban University of Technology
Durban, South Africa

Mkpamdi N Eke
Department of Mechanical Engineering
University of Nigeria
Nsukka, Nigeria

A. B. C. Ekwealor
Department of Physics and Astronomy
University of Nigeria
Nsukka, Enugu State, Nigeria

Blessing N. Ezealigo
Dipartimento di Ingegneria Meccanica, Chimica e
    dei Materiali
Università degli Studi di Cagliari
Cagliari, Italy

Calister N. Eze
Department of Physics
Federal University of Technology
Minna, Niger State, Nigeria

Chinenye Ezeokoye
Department of Industrial Physics
Ebonyi State University
Abakaliki, Nigeria

Fabian I. Ezema
Department of Physics and Astronomy
University of Nigeria
Nsukka, Enugu State, Nigeria
and
Nanosciences African Network (NANOAFNET),
    iThemba Labs
National Research Foundation
Somerset West, Western Cape, South Africa
and
UNESCO-UNISA Africa Chair in Nanosciences/
    Nanotechnology
University of South Africa (UNISA)
Pretoria, South Africa
and
National Center for Physics
Islamabad, Pakistan

Sabastine Ezugwu
Department of Physics and Astronomy
The University of Western Ontario
Ontario, Canada

Xolile Fuku
Energy Center, Smart Places
Council for Scientific and Industrial Research
   (CSIR)
Pretoria, South Africa

Fekadu Gashaw Hone
Addis Ababa University
Department of Physics
Addis Ababa, Ethiopia

Innocent S. Ike
Department of Chemical Engineering
Federal University of Technology
Owerri, Nigeria
and
African Centre of Excellence in Future Energies
   and Electrochemical Systems (ACE-FUELS)
Federal University of Technology
Owerri, Nigeria

Mesfin A. Kebede
Energy Centre
Council for Scientific & Industrial Research
Pretoria, South Africa
and
Department of Physics
Sefako Makgatho Health Science University
Medunsa, South Africa

Lehlohonolo F. Koao
Department of Physics
University of the Free State (Qwa Qwa campus)
Phuthaditjhaba, South Africa

Linda Z. Linganiso
Research and Postgraduate Support Directorate
Durban University of Technology
Durban, South Africa

Sylvester M. Mbam
Department of Physics and Astronomy
University of Nigeria
Nsukka, Enugu State, Nigeria

S. L. Mammah
Department of Science Laboratory Technology
School of Applied Sciences, Rivers State
   Polytechnic
Bori, Nigeria

Mandira Majumder
Department of Applied Physics
Indian Institute of Technology (Indian School of
   Mines)
Dhanbad, India

Mmalewane Modibedi
Energy Center, Smart Places
Council for Scientific and Industrial Research
   (CSIR)
Pretoria, South Africa

Chioma E. Njoku
Department of Materials and Metallurgical
   Engineering
Federal University of Technology
Owerri, Nigeria

Agnes Chinecherem Nkele
Department of Physics and Astronomy
University of Nigeria
Nsukka, Enugu State, Nigeria
and
Nanosciences African Network (NANOAFNET)
   iThemba LABS-National Research Foundation
Somerset West, Western Cape Province, South
   Africa

Paul Sunday Nnamchi
Department of Metallurgical & Materials
   Engineering
University of Nigeria
Nsukka, Enugu State, Nigeria

Assumpta C. Nwanya
Department of Physics and Astronomy
University of Nigeria
Nsukka, Enugu State, Nigeria
and
Nanosciences African Network (NANOAFNET)
   iThemba LABS-National Research Foundation
Somerset West, Western Cape Province, South Africa
and
UNESCO-UNISA Africa Chair in Nanosciences/
   Nanotechnology
University of South Africa (UNISA)
Pretoria, South Africa

Cynthia C. Nwaeju
Department of Mechanical Engineering
Nigerian Maritime University
Okerenkoko, Warri, Nigeria

Mary O. Nwodo
Department of Physics and Astronomy
University of Nigeria
Nsukka, Enugu State, Nigeria

Patrick A. Nwofe
Department of Industrial Physics
Ebonyi State University
Abakaliki, Nigeria

Camillus Sunday Obayi
Department of Metallurgical & Materials Engineering
University of Nigeria
Nsukka, Enugu State, Nigeria

Henry Uchenna Obetta
Science and Engineering Unit
Nigerian Young Researchers Academy
Anambra State, Nigeria
and
River State University – Nkpolu-oroworukwu
Port-Harcourt, River State, Nigeria

Raphael M. Obodo
Department of Physics and Astronomy
University of Nigeria
Nsukka, Enugu State, Nigeria
and
National Center for Physics
Islamabad, Pakistan
and
NPU-NCP Joint International Research Center on
    Advanced Nanomaterials and Defects
    Engineering
Northwestern Polytechnical University
Xi'an, China

Chukwujekwu Augustine Okaro
Science and Engineering Unit, Nigerian Young
    Researchers Academy
Anambra State, Nigeria
and
Department of Physics and Astronomy
University of Nigeria
Nsukka, Enugu State, Nigeria

Onyeka Stanislaus Okwundu
Science and Engineering Unit
Nigerian Young Researchers Academy
Anambra State, Nigeria
and
Department of Physics and Astronomy
University of Nigeria
Nsukka, Enugu State, Nigeria

Nithyadharseni Palaniyandy
Energy Centre
Council for Scientific & Industrial Research
Pretoria, South Africa

Kittessa Roro
Smart Places Cluster: Energy Centre
Council for Scientific and Industrial Research
    (CSIR)
Brummeria, Pretoria

V. Saravanan
Department of Physics
Sri Meenatchi Vidiyal Arts and Science College
Valanadu, Trichy, India

Mutsumi Suguyima
Department of Electrical Engineering
Tokyo University of Science
Yamazaki, Japan

Philips Chidubem Tagbo
Science and Engineering Unit
Nigerian Young Researchers Academy
Anambra State, Nigeria
and
Department of Physics and Astronomy
University of Nigeria
Nsukka, Enugu State, Nigeria

Newayemedhin A. Tegegne
Addis Ababa University
Department of Physics
Addis Ababa, Ethiopia

Zikhona Tshemese
Department of Chemistry
University of Zululand
KwaDlangezwa, KwaZulu Natal, South Africa

Anukul K. Thakur
Department of Advanced Components and
    Materials Engineering
Sunchon National University
Chonnam, Republic of Korea

Nqobile Xaba
Energy Center, Smart Places, Council for
    Scientific and Industrial Research (CSIR)
Pretoria, South Africa
and
Chemistry Department
University of the Western Cape Bellville
Cape Town, South Africa

Jude N. Udeh
Department of Physics and Astronomy
University of Nigeria
Nsukka, Enugu State, Nigeria

Cyril Oluchukwu Ugwuoke
Science and Engineering Unit
Nigerian Young Researchers Academy
Anambra State, Nigeria
and
Department of Physics and Astronomy
University of Nigeria
Nsukka, Enugu State, Nigeria

1

Lithium-Ion Batteries: From the Materials' Perspective

Camillus Sunday Obayi[1], Paul Sunday Nnamchi[1], and Fabian I. Ezema[2]
[1]*Department of Metallurgical & Materials Engineering, University of Nigeria, Nsukka, Enugu State, Nigeria*
[2]*Department of Physics & Astronomy, University of Nigeria, Nsukka, Enugu State, Nigeria*

1.1 Introduction

Currently, rechargeable lithium-ion battery (LIB) is the fastest growing energy storage device. It dominates portable and smart electronic devices such as cellular phones, laptop computers, and digital cameras due to its light weight, high energy density, good capacity retention, and long lifespan (Jiantie Xu et al. 2017, 1–14; Qi Wen et al. 2017, 19521–19540; K. Liu et al. 2018, eaas 9820). Tremendous research efforts led to its commercialization by SONY 29 years ago (1991), and these material-driven research efforts have led to significant improvement in the electrochemical performances of LIB. Lithium-ion batteries are now safer, have higher energy capacity, better cycling stability, and highest energy density compared to other common secondary batteries such as lead-acid, nickel-cadmium, and nickel-metal hydride. A greater part of this progress can be ascribed to the search and introduction of new materials with improved properties. LIBs are still the most promising option for large-scale energy storage for electric vehicles (EVs), hybrid electric vehicles (HEVs), grid-scale energy storage, and critical space and aeronautical applications (Jiantie Xu et al. 2017, 1–14; Qi Wen et al. 2017, 19521–19540; X. Lou et al. 2019, 6089–6096). The quest to use LIB for these higher energy storage applications will continue to challenge performance of LIBs and to drive the search for improved LIB materials.

The electrochemical performance of LIB is critically dependent on the intrinsic nature of electrode materials and chemistry associated with electrode–electrolyte interface. Besides the inherent nature of electrode materials which include physical and chemical properties, performance depends also on the design and synthesis routes for the electrode materials, which can enable modification of composition or structures, reduction in defects or creation of performance-enhancing architectures (Qi Wen et al. 2017, 19521–19540). The early development of rechargeable LIBs had material challenges associated with intrinsic reactivity of lithium metal, which posed safety and rechargeability problems. This initial problem was reduced by incorporating pure lithium metal into weakly bonded compounds such as graphite, to form intercalated lithium ion electrode that is slightly lower in energy density than pure lithium metal, but less reactive and safe.

Over time due to increasing demand for better performance, there has been intense and progressive research effort to overcome many of the LIB material problems by designing and developing a variety of new anode, cathode, and electrolyte materials. Materials that have been studied as anode materials include metallic lithium (K. Xu 2004, 4303–4418; B. Scrosati B 2011, 1623–1630), various forms of carbon (N. Nitta et al. 2015, 252–264; Z. Yan et al. 2017, 495–501; H. Zheng et al. 2012, 4904–4912; C. Ma et al. 2013, 553–556; K. Wang et al. 2017, 1687–1695), and other materials based on silicon (Si)

DOI: 10.1201/9781003145585-1

(N. Nitta et al. 2015, 252–264; A. Casimir et al. 2016, 359–376), tin (Sn) (O. Crosnier et al. 1999, 311–315), titanium (Ti) (N. Nitta et al. 2015, 252–264; S. Chauque et al. 2017, 142–155), and conversion reaction-type materials (J. Cabana et al. 2010, E170–E192; Scrosati and Garche 2010, 2419–2430; Y. Lu et al. 2018, 972–996). The major cathode materials include layered lithium transition metal oxides of the types $LiMO_2$ (M = V, Cr, Mn, Fe, Co, N) [Y. Yang et al. 1997, 227–230, J. Shu et al. 2010, 3323–3328], $LiFePO_4$ (L.H. Hu et al. 2013, 1687; C.T. Hsieh et al. 2014, 1501–1508), V_2O_5 (Y. Yang et al. 2014, 9590–9594), non-oxide FeS_2 (D.T. Tran et al. 2015, 87847–87854), fluoride-based compounds, and poly-anionic compounds (electronic conducting polymers) (Sngodu, Deshmukh 2015, 42109–42130; T.M. Higgins et al. 2016, 3702–3713).

Though these research efforts have resulted in the production of higher capacity and safer LIBs, but the realization of electrode materials with optimum electrochemical performances, ease of synthesis, and manufacture for massive energy storage is still pending. This chapter reviews the research efforts and challenges in the progressing development of new anode and cathode electrode materials in response to the increasing demand for cheaper, safer, lighter, higher energy storage capacity, and environmentally friendly LIBs.

1.2 Brief History of Lithium-Ion Battery Materials

The use of lithium metal as battery anode electrode material started much earlier than now due to its attractive properties such as low density ($0.534 \ gcm^{-3}$), high specific capacity ($3860 \ mAhg^{-1}$), low redox potential (−3.04 V vs. SHE), and low resistance (Krivik and Baca 2013; Mogalahalli 2020, 1884). These properties result in significantly high energy density and high operating voltage in lithium- based batteries compared to traditional batteries such as lead-acid, nickel-cadmium, and nickel-metal hydride batteries. The design of non-rechargeable or disposable lithium-metal battery started around 1912, and the first primary lithium batteries were commercially available in the early 1970s (Lewis and Keyes 1913, 340–344; Krivik and Baca 2013). Such primary non-aqueous LIBs included lithium sulphur dioxide $Li//SO_2$ which was introduced in 1969 (Meyers and Simmons 1969); lithium-carbon monofluoride (Li//CFx) primary cell introduced in 1973; lithium-manganese dioxide primary cells ($Li//MnO_2$) which came into being in 1975, and lithium-copper oxide Li//CuO batteries (Goodenough 2013, 51–92). These early primary Lithium cells found applications in LED fishing floats, cameras, calculators, and memory backup applications.

The increasing demand for batteries of higher performance and the drive to harness the attractive properties of lithium metal attracted great research efforts towards converting lithium primary cells into rechargeable cells with high energy density. These attempts to develop rechargeable lithium batteries had challenges such as safety problems associated with high chemical reactivity of lithium metal and the tendency for lithium to precipitate on the negative electrode during charging forming dendrites which easily cause short circuiting (Akira Yoshino 2012, 2–5). Lithium is a very reactive rare-earth metal, and this characteristic poses problems not only in manufacture but also in the selection of other battery components.

The difficulties of working with pure lithium metal were reduced with the discovery of intercalation compounds or non-stoichiometric compounds or weakly bonded compounds by inorganic and solid-state chemists in the 1970s (Guerard and Herold 1975, 337–345; Whittingham 1978, 41–99), which gave birth to rechargeable LIBs. This enabled intercalation reactions of an ion, atom, or molecule into a crystal lattice of a host material without destroying the crystal structure (Akira Yoshino 2012, 2–5). Lithium atoms were incorporated in the weakly bonded layers of these compounds forming intercalated compounds (non-metallic Li-based compounds). Although, intercalated lithium ion is slightly lower in energy density than pure lithium metal, it is safer. Both anode and cathode materials were made of lithium intercalated compounds such that lithium ions could intercalate and de-intercalate in the cathode and anode reversibly when the cell is in operation without compromising their electronic and ion conductivities (Armand and Touzain 1977, 319–329; Armand 1980, 145–161).

The concept to fabricate a LIB based on cathode and anode materials made of two different intercalation compounds was made by Armand in the 1970s (Akira Yoshino 2012, 2–5) and was demonstrated in 1980 by Lazzari and Scrosati (Lazzari and Scrosati 1980, 773–774). This battery was called the rocking-chair battery due to the movement of lithium ions from cathode to anode and vice versa during the charge–discharge process (Armand and Duclot 1978; Mauger 2019, 070507). This concept was commercialized twenty-nine years ago (1991) by the Sony Corporation as the first rechargeable LIB, consisting of the anode (Li_xC_6) and cathode ($Li_{1-x}CoO_2$) materials in which Li atoms are intercalated into a carbonaceous material and into cobalt oxide, respectively (Broussely 1999, 3–22; Tarascon and Armand 2001, 359–367). Since then much research has been and is still ongoing in improving the anode and cathode materials in Li-ion cells and developing new ones based on conversion reactions instead of intercalation reactions (Jordi Cabana 2010, E170–E192).

1.3 Lithium-Ion Battery and Its Principle of Operation

A Lithium ion battery consists mainly of four components: the anode (negative electrode), cathode (positive electrode), and an electronic-insulated separator immersed in Li^+ conductive electrolyte that supports external movement of electrons and lithium ions from the cathode to anode and vice versa (Y. Wang et al. 2012, 770–779; Islam and Fisher 2014, 185–204; Hannan et al. 2018, 19362–19378;). The anode is usually an electropositive electron donor such as lithium metal inserted in a graphite lattice, while the cathode is normally an electron acceptor which is strongly electronegative such as $LiMO_2$ (M = Co, Ni, Mn, etc.) compounds. The electrolyte which is typically non-aqueous is composed of a lithium salt in polar organic solvent, which allows the movement of Li ions only. The anode enables electric current to flow through the external circuit and reversible absorption/release of lithium ions from the cathode.

The cathode material is the source of Li ions and determines the working voltage, capacity/energy density, and rate capability of a LIB according to its limited theoretical capacity and thermodynamics (J. Xu et al. 2013, 439–442). The separator functions as a physical barrier between the anode and cathode, preventing direct flow of electrons and letting only flow of lithium ions through its microscopic pores. While overall performance of a LIB battery depends on the structure and properties of anode and cathode materials, the electrolyte and separator determine the safety of the battery.

The structure of a LIB cell is shown in Figure 1.1 and is used to explain the principle of operation of LIB with $LiCoO_2$ as the cathode and layered carbon as the anode. During charging process lithium ions (Li^+) move from the cathode ($LiCoO_2$ lattice) through electrolyte/separator to the anode and get inserted into the layered graphite structure and are stored, driven by an external power source, providing an electron flow in the same direction (Figure 1.1a). Reversely, during discharging, Li^+ ions spontaneously migrate back from the anode to the cathode through the electrolyte, causing an electron flow in the same direction via the external circuit to power the external device (Figure 1.1b). During discharging, oxidation reaction takes place at the anode while reduction reaction occurs at the cathode. The spontaneous Li^+ migration is driven by the difference of the electromotive force between the two half-cell reactions. Thus, the movement of Li ions reversibly between the anode and cathode led to the reversible conversion between electric energy and chemical energy in LIBs (Guyomard and Tarascon 1992, 937).

The two half-cell reactions at the anode (oxidation, electron generation) and cathode (reduction, electron consumption) are represented by equations 1.1 and 1.2, while the overall reaction is shown in equation 1.3.

$$Li_xC_6(s) \xrightarrow[\leftarrow \text{charging}]{\text{discharging} \rightarrow} 6C(s) + xLi^+(s) + xe^- \tag{1.1}$$

$$Li_{1-x}CoO_2(s) + Li^+(s) + xe^- \xrightarrow[\leftarrow \text{charging}]{\text{discharging} \rightarrow} LiCoO_2(s) \tag{1.2}$$

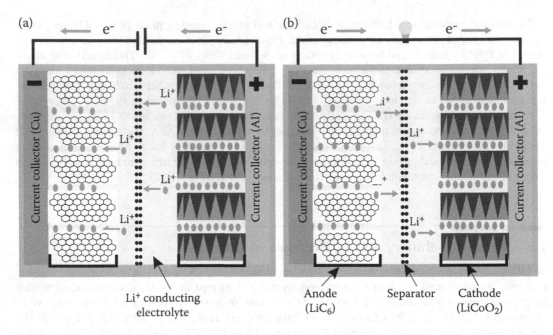

FIGURE 1.1 Principle of operation of a Li-ion battery, (a) charging process, (b) discharging process. Adapted from Ref. Yeru Liang et al. 2019, 1–27.

$$\text{Li}_x\text{C}_6(s) + \text{Li}_{1-x}\text{CoO}_2(s) \xrightarrow[\leftarrow\text{charging}]{\text{discharging}\rightarrow} 6\text{C}(s) + \text{Li}_{1-x}\text{CoO}_2(s) \tag{1.3}$$

It has been revealed that the Li-ion cells can operate at high potentials (up to 3.7 V) and demonstrate high energy density (from 150 Wh kg^{-1} to 200 Wh kg^{-1}) (Tarascon and Armand 2001, 359–367). Other very important properties of lithium-ion cell are good charge-discharge characteristics, with possible life cycles greater than 500 and rapid recharging (2 h), low self discharge (<10% per month), absence of a memory effect, and much safer than equivalent cells which use lithium metal.

1.4 Li-Ion Battery Component Materials

An LIB consists mainly of four component materials: the anode, cathode, a separator, and Li$^+$ conductive electrolyte. Others are conductive additives, current collector, and binder.

1.4.1 Li-Ion Battery Anode Materials, Characteristics, Advantages, and Limitations

The anode material is the electrochemical active material in the anode electrode, which serves as a lithium reservoir, donating lithium ions and electrons during discharging. An anode material plays a pivotal role in LIB performance. The battery behaviour depends not only on the intrinsic properties of the anode material such as the physical and chemical properties and energy storage capacity, but equally on the degree of crystallinity or amorphousness of structure of the anode material as well as on its shape, size, and state (Qi Wen et al. 2017, 19521–19540). To harness optimum performance in LIBs, anode materials should exhibit relatively low reaction potentials in order to obtain a high full-cell voltage, possess high energy and power density, long cycling life, and low cost.

The anode materials that have been studied for LIB application can be classified in four types: Li metal, intercalative materials, alloying materials, and conversion materials. Intercalative anode materials are materials possessing layered structures, which can store lithium ions in the interlayers.

They include materials based on carbon (C) and titanium (Ti), while alloying materials are those based on silicon (Si) and tin (Sn). Among these, carbon-based materials are the most studied and most successful.

1.4.1.1 Lithium Metal

Lithium metal would have been an ideal LIB anode material due to its low molecular weight, very high theoretical specific capacity (3860 mAh g^{-1}), and very low reduction potential (–3.04 V vs. standard hydrogen electrode (SHE), except for the observed safety problems of lithium metal batteries associated with its reactivity and rechargeability (Julien et al. 2016). During cycling lithium dendrites grow on Li metal anode and can penetrate the separator causing internal short circuit, consuming active Li and electrolyte and irreversibly reducing Coulombic efficiency. The presence of the dendrites leads to short cycling life due to electrode polarization and volume expansion. This shortcoming led to extensive search for non-metallic LIB anode materials, which was heightened by Armand's new concept in late 1970s that suggested that safer LIBs could be produced with lithium ions moving reversely between two intercalative electrodes (Reddy 2011). Currently, researchers are looking back to the Li metal anode with a view to harnessing Li high energy density by analyzing the factors that affect Li deposition and searching for methods to suppress dendrite growth (Cheng et al. 2017, 10403–10473).

1.4.1.2 Intercalative Anode Materials

1.4.1.2.1 Carbon-Based Anode Materials

Carbonaceous materials such as natural and artificial graphite and disordered hard carbons are the earlier and most popular anode material due to their high reversibility, in addition to being readily available, cheap, low weight, and non-toxic (Xu K 2004, 4303–4418; Krivik and Baca 2013). They also exhibit low reaction potential, long cycle life, and moderate specific theoretical capacity (372 mAh g^{-1}) and low operating potential (~0.1 V) (Blomgren 2017, A5019–A5025; Ayodele 2020, 21–39).

Carbon has the ability to absorb a good quantity of lithium ions and electrons into its crystal structures with little volume variation (Qi Wen et al. 2017, 19521–19540). Lithium ion inserts into the vacant sites of carbon forming lithiated carbon during polarization and de-inserts from the lithiated carbon during reverse polarization as shown in equation 1.4.

$$6C + Li^+ + e^- \leftrightarrow LiC_c \tag{1.4}$$

For higher power applications demanding high reversible capacity and stable cycling life, moderate specific theoretical capacity and low operating potential of carbon-based anode materials cannot suffice. Low operating potential would inevitably cause lithium dendrite growth during the charge-discharge process, thereby raising concerns on safety (Arico et al. 2005, 366–377; Y.M. Chen et al. 2016; Y.M. Chen et al. 2016, 5990–5993; Liu Z et al. 2018, 636–642).

The crystallinity, microstructure, and morphology of the carbonaceous host determine the extent of lithium accommodation in the carbon. Carbons are generally classified into two – graphitic and non-graphitic. Graphitic carbon or graphite consists of orderly arranged graphene layers bonded together by weak Van der Waals forces. It is highly ordered and has large number of vacant sites to intercalate both anions and cations. The intercalation mechanism in graphitic carbon occurs in steps and consists of insertion of lithium into the vacant space between the graphene layers by opening the Van der Waals gap between the layers (Shukla and Kumar 2008, 314–331).

Non-graphitic carbons are highly disordered structurally and can be further divided into graphitizing and non-graphitizing carbon (hard or soft carbon). Lithium intercalation mechanism is different from that of graphitic carbon and is more complex. Other carbon-based materials with special structure such as fullerenes, carbon nanotube (CNT), graphene, and graphene-based composites have also been studied as anode materials (Kaskhedikar and Maier 2009, 2664–2680). A combination of this lithium impregnated carbon anode and a lithium metal oxide cathode gives a Li-ion cell a relatively high voltage from 4 V in the fully charged state to 3 V in discharged state (Dell and Rand 2001). Recently, the most frequently used anode materials have been one-dimensional (1D), 2D, 3D carbon-based materials, as well as porous and core-shell structures.

1.4.1.2.2 Titanium-Based Anodes

Oxides of titanium such as $Li_4Ti_5O_{12}$ and TiO_2 have also been used as anodes for LIBs (Poizot et al. 2000, 496–499). Li intercalates in titanium oxides at a higher potential (>1 V) than in carbon-based anodes. This results in reduced operating voltage (<2.5 V) when combined with cathodes like $LiCoO_2$, $LiMn_2O_4$, or $LiFePO_4$. However, a battery with a higher operating voltage of ~3 V can be realized by combining these oxides with high-voltage cathodes such as $LiMn_{1.5}Ni_{0.5}O_4$ or $LiMn_{0.5}Ni_{0.5}O_2$. $Li_4Ti_5O_{12}$ that crystallize in spinel phase.

One big advantage of $Li_4Ti_5O_{12}$ over Si- and Sn-based anodes is that it does not induce volume change and is regarded as zero strain anode material. A capacity of ~175 mAh g^{-1} can be sustained for thousands of cycles without any capacity loss. However, one major shortcoming is lesser electronic conductivity, which thereby requires carbon coating and other doping methods in order to increase the conductivity (L. Cheng et al. 2007, A692).

Titanium IV oxide (TiO_2), which is a candidate transition metal oxide in the fields of catalysis, photocatalysis, hydrogen storage, etc., has also been studied as an LIB anode material. TiO_2 exhibits three different crystallographic forms namely anatase, rutile, and brookite. All the three forms are electrochemically active, and the ability to accommodate lithium in the lattices depends on the crystallography and microstructure of the material (Y.-K. Zhou 2003, A1246–A1249; Dambournet 2011, 3085–3090). The anatase phase is the most electrochemically active form of TiO_2, and it can reversibly accommodate 0.6 lithium exhibiting a stable capacity of ~170 mAh g^{-1}. However the lithium insertion potential is even higher (1.7 V) than $Li_4Ti_5O_{12}$, necessitating its combination with high-voltage cathode (Yoon and Manthiram 2011, 9410–9416).

1.4.1.3 Alloying Anode Materials

Due to the growing demand for LIBs with high energy density and relative low capacity of intercalative anode materials, many alloys and intermetallic compounds of group III, IV, and V elements (Sn, Sb, Al, and Mg) have been investigated as potential and next-generation anodes due to their extremely high capacity (W.-J. Zhang 2011, 13–24).

1.4.1.3.1 Si Alloy Anode Material

Among the most promising alloy anode materials, Si is the most studied due to its highest theoretical capacity and moderate working potential (W.-J. Zhang 2011, 13–24). In addition to its low molecular weight, relatively low discharge potential of ~0.4 V, low cost, low toxicity, earth abundance, stable chemical property, and outstanding environmental compatibility (K. Feng 2018, 1702737), Si has a theoretical capacity of ~4000 mAh.g^{-1}, which corresponds to that of a popularly studied anode material $Li_{4.4}Si$ (H Li et al. 2009, 4593–4607). The major shortcoming of $Li_{4.4}Si$ anode is volume change during the discharge-charge process, which is about 300% (J. Cho 2010, 4009–4014). This swelling and shrinking of the volume lead to quick loss of capacity on few cycles because the anode disintegrates, losing contact between each particle and disturbs Li^+ insertion.

Various strategies that have been applied to overcome the volume change problem in Si-based anodes include the following:

a. Creating stable oxide/carbon/polymer matrix or by using nanocrystalline materials for alloy-based anodes, which has reduced the volume expansion to certain extent (Ji and Zhang 2010, 124–129),

b. Using composite of Si with various carbonaceous materials which can absorb the stress associated with volume change (H. Wu et al. 2012, 904–909; T. Hwang et al. 2012, 802–807). Carbonaceous materials such as soft carbon, hard carbon, graphite, CNT, and graphene are used as carbon matrix for Si anode (H. Liu et al. 2010, 10055–10057; Wang and Kumta 2010,

2233–2241; X Zhou et al. 2012, 2198–2200), and their large surface areas enable them to accommodate the volume expansion exerted by silicon particles.

c. Using inactive-active composites formed by adding an inactive metal like Fe, Co, Ni, Mn, etc., and active metals like Sn, Si, or Al. Inactive metal provides a stable matrix that restrains volume expansion and maintains the contact and thereby preserves stable capacity for long cycling (Thackeray et al. 1999, 60–66).

1.4.1.3.2 Tin-Based Alloy Anodes

Another alloying element of group III, IV, and V that has been studied as an anode material is tin (Sn). Metallic tin reacts with Li to form a $Li_{4.4}Sn$ alloy that exhibits a capacity of ~1000 mAh g^{-1}. However, similar to $Li_{4.4}Si$, the formation of $Li_{4.4}Sn$ is associated with large volume expansion. Effective methods that have been used to reduce the problem of volume expansion include nanostructuring of Sn (H.B. Wu et al. 2011, 24605–24610), making Sn-C composites (R. Liang 2011, 17654–17657) or forming inactive-active material matrix (Y. Xia et al. 2001, A471).

Alloys that enable formation of $Li_{4.4}Sn$ in a stable matrix include Cu-Sn, Ni-Sn, Co-Sn. Cu forms an alloy with Sn (Cu_6Sn_5) and reacts with Li to form $Li_{4.4}Sn$ alloy in a stable Cu matrix (Y. Xia et al. 2001, A471). Cu is electrochemically inactive and forms a nanocrystalline matrix to accommodate the volume expansion. Also, Ni and Co form alloys with Sn (Ni_3Sn_2) and Co_3Sn_2, respectively, and react with Li similarly as above (H. Groult et al. 2011, 2656–2664; J. Xie et al. 2007, 386–389) and after effect is restriction on volume expansion.

1.3.1.4 Conversion-Type Anode Materials

Conversion-type anode materials are also considered to be the next-generation anode materials like alloying anode materials for LIBs owing to their high theoretical capacities and abundant resources (Y. Lu et al. 2018, 972–996; J Cabana et al. 2010, 170–192). They are a group of transition metal compounds (M_xN_y) that do not have vacant sites to accommodate lithium, but can be reduced electrochemically into metal nanoparticles in lithium matrix (Li_nX) (Y. Lu et al. 2018, 972–996; J. Cabana et al. 2010, 170–192; J Lu et al. 2018, 35–53; P. Poizot et al. 2000, 496–499), where M = transition metal; X = anion like O, S, P, N, Se, F, and n = oxidation state of X (Y Lu et al. 2018, 972–996). The nanocomposite (metal in Li_nX matrix) so formed is highly reactive and can decompose back to Li and M_aX_y through a process known as *conversion reaction* when reverse polarization is applied.

Among the metal oxides/sulphides/nitrides/phosphides and even hydrides that have been studied as conversion anode materials, the oxides are the most promising ones and have been intensively researched because of their easier preparatory routes and more stable chemical properties (Reddy et al. 2013, 5364–5457). It has been reported that this type of compound delivers stable capacities two to three times greater than that of conventional graphite anode. However, conversion-type anode materials have some shortcomings such as poor electronic and ionic conductivity, relatively large volume expansion (<200%), and continuous electrolyte decomposition during cycling, as well as large voltage hysteresis (Y. Lu et al. 2018, 972–996). Furthermore, the conversion reaction mechanism is complex involving electrochemical oxidation/reduction reaction and electrodes based on conversion mechanism are yet to be commercialized.

Overall, there have been significant efforts to improve the performance of anode materials but finding an alternative to carbon anode still remains as a daunting task.

1.5 Li-ion Battery Cathode Materials, Characteristics, Advantages, and Limitations

The cathode material to a great extent determines the performance of LIB such as cell potential, energy density, power density, lifetime, and safety. The cathode material also accounts for the larger part of cost of electrode materials. Out of about 44% attributed to the cost of electrode materials in a typical

LIB, the cathode material accounts for about 30% while anode material accounts for about 14%. The major studied cathode materials include layered transition metal oxides of the types $LiMO_2$ (M = Mn, Co, Ni), spinel-structured $LiMn_2O_4$ along with its derivatives and olivine-structured $LiFePO_4$, fluoride-based compounds (tavorites), polyanionic compounds, and other transition metal oxides such as MnO_2 and V_2O_5) (Whittingham 2004, 4271–4302; Bensalah and Dawood 2016, 258).

1.5.1 Layered Transition Metal Oxides Cathode Material

Layered transition metal oxides of the form $LiMO_2$, where M = Mn, Co, Ni, or a combination of two or more have been studied extensively and are the most successful as Li-ion battery cathode materials (Bensalah and Dawood 2016, 258). They exhibit superior electrochemical behaviour due to their layered structure which allows for a large number of diffusion paths for lithium ions (Bensalah and Dawood 2016, 258). The following transition metal oxides (TMOs) or mixed or combined transition metal oxides (MTMOs) have been studied as LIB cathode materials.

1.5.1.1 Lithium Cobalt Oxides (LiCoO₂)

$LiCoO_2$ is the earliest and the dominant cathode material for LIBs (Bensalah and Dawood 2016, 258). Mizushima and coworkers were first to recognize the reversible lithium reactivity in $LiCoO_2$ in 1980 (Mizushima et al. 1980, 783–789). Since then $LiCoO_2$ has remained the most exploited commercial cathode material for Li ion due to its low molecular weight, high discharge potential, high stability during electrochemical cycling, high energy capacity, and ease of preparation in bulk quantities (Wakihara 2001, 109–134). Depending on the synthesis route, $LiCoO_2$ crystallizes in either α-$NaFeO_2$ structure with hexagonally closely packed oxygen array forming O-Li-O-M-O chains at high temperatures or cubic spinel-like structure at low-temperatures (Antolini 2004, 159–171). The hexagonal arrangement enables Li to diffuse into $LiCoO_2$ structure through two channels with the concurrent oxidation/reduction of cobalt ion.

One of the challenges in the practical use of $LiCoO_2$ is the difficulty of extracting active Li/Co material, which reduces expected specific capacity of $LiCoO_2$ from ~280 mAh g^{-1} to a low capacity of 140 mAh g^{-1} (Y. Gan et al. 2008, 81–84). Another challenge is that charging $LiCoO_2$ to higher voltages leads to higher oxygen evolution and raises safety concerns (Manthiram 2011, 176–184). Furthermore, cobalt is expensive and toxic, and this has pushed the research towards alternative cathodes based on nickel and manganese-based lithium oxides. In order to improve the ionic conductivity and cycling performance of the $LiCoO_2$ cathode, some strategies that have been attempted include carbon coating (Q. Cao et al. 2007, 1228–1232; C.Z. Lu et al. 2008, 392–401), oxide coating with Al_2O_3, ZnO, and $LiTiO_{12}$, and cationic doping on aluminium (S. Myung et al. 2001, 47–56), chromium, (Madhavi and Rao 2002, 219–226) and silver (Huang et al. 2005, 72–77).

Due to their relative abundance, lower cost, and less toxicity (Armstrong et al. 2002, 710–719), $LiMnO_2$ and $LiNiO_2$ have been studied extensively. However, $LiMnO_2$ has performance limitations associated with its low capacity, difficulty of mass production, and power charge/discharge performance, especially at high temperatures. When compared to $LiCoO_2$, $LiMnO_2$ has advantages of high safety and low cost, which make it a promising substitute in the future. Although, $LiNiO_2$ has a layered structure and charge-discharge behaviour comparable to $LiCoO_2$, it has challenges of poor solubility in organic electrolyte solutions, particularly at high temperature, and difficulty of synthesis. Other oxides that have been studied are $LiVO_2$, $LiCrO_2$, $LiTiO_2$, and $LiFeO_2$. These oxides have challenges like difficulty of synthesis and instability during cycling.

1.5.1.2 LiMn₂O₄ Cathode Material

$LiMn_2O_4$ belongs to oxides that exhibit spinel crystal structure, a thermodynamically stable structure with cubic symmetry and 3D channels for Li diffusion. $LiMn_2O_4$ is among the most studied transition metal oxide for LIB cathode material due to its low cost, non-toxicity, environmental-friendliness, and high natural abundance of Mn (Guyomard and Tarascon 1992, 937; Kim et al. 2008, 3948–3952; Park et al. 2011, 1621–1633).

The movement of Li+ through octahedral sites in $LiMn_2O_4$ is difficult and occurs at a higher energy. At high-voltage region (4 V) $LiMn_2O_4$ can insert ~0.4 lithium without destabilizing cubic symmetry, and discharge capacity in the range of 100–120 mAh g^{-1} is achievable compared to its theoretical capacity of 148 mAh g^{-1} (Santiago et al. 2001, 447–449). At low-voltage region (3 V), $LiMn_2O_4$ is capable of inserting ~2 lithium into the structure, but more absorption leads to distortion of the structure and large variation in the volume of a unit cell. Thus, at lower voltage there is distortion of structural integrity accompanied by reduction in capacity.

Methods that have been employed to increase the structural stability and electrochemical performance of the $LiMn_2O_4$ cathode include cationic substitution, surface modification, and nanostructuring (Bensalah and Dawood 2016, 258). Cationic substitution which involves replacing manganese ions with metal ions like Cr, Co, Ni, etc., are found to improve the electrode properties of $LiMn_2O_4$ (Yoon et al. 2007, 780–784; Hassoun et al. 2011, 3139–3143). Surface modification entails coating $LiMn_2O_4$ with salts such as Al_2O_3, $AlPO_4$, Cr_2O_3 (Sun et al. 2008, 2390–2395) or doping with metal ions like Al^{3+}, Zn^{2+} (Ouyang et al. 2010, 4756–4759; Ryu et al. 2011, 15337–15342; Chan et al. 2003, 110–118; Ellis et al. 2010, 691–714). The nanostructuring approach attempted includes the encapsulation of $LiMn_2O_4$ nanowires in ZnO nanotubes (Liu et al. 2006, 248–253) and homogenous dispersion of $LiMn_2O_4$ nanoparticles in CNT composites (Ding et al. 2012, 197–200). Other materials that belong to this group that have been tested to replace $LiMn_2O_4$ are $LiFe_2O_4$, $LiCr_2O_4$, $LiCo_2O_4$, and LiV_2O_4.

1.5.2 Olivine Transition Metal Phosphates (LiFePO₄) Cathode Material

$LiFePO_4$ was introduced to solve the observed shortcomings of layered compounds and spinels such as lowering of capacity on long cycling (>1000 cycles) and poor stability of fully charged material. $LiFePO_4$, an olivine type of compound, has been found to be strongest alternative material for LIB cathode due to its low cost, non-toxicity, long cyclability, and good thermal and chemical stability (Guo et al. 2009, 166–169; Park et al. 2011, 1621–1633).

In 1997, Padhi et al. (1997, 1188–1194) recognized and demonstrated that Li can be electrochemically extracted from $LiFePO_4$, leaving behind $FePO_4$ that belongs to same space group as LiFePO4. Li extraction and re-insertion proceed by a two-phase process without distortion of the olivine structure. A capacity value of 170 mAh.g^{-1} is obtainable upon electrochemical extraction of approximately one lithium from $LiFePO_4$.

$LiFePO_4$ has some shortcomings. One is very low electronic conductivity (10^{-9}–10^{-10} S.cm^{-1}) at room temperature due to strong covalent nature of bonds, which restricts lithium diffusion (Morgan et al. 2004, A30–A32; Guo et al. 2009, 166–169). In order to realize full capacity charge-discharge process at much slower rate is required because of the inherent poor electronic conductivity and slow Li diffusion. Attempts made to increase intrinsic electronic conductivity and rate capability of $LiFePO_4$ include making nanoparticles of LiFePO4, nanocoating with conducting carbon, and solid solution doping with other metal ions (Chung et al. 2002, 123–128; Gibot et al. 2008, 741–747; Li and Zhou 2012, 1201–1217; Bensalah and Dawood 2016, 258).

Another drawback of $LiFePO_4$ is lower energy density caused by its low redox potential (3.5 V). Other alternative materials with higher redox potentials such as $LiNiPO_4$ (5.2 V) (Minakshi et al. 2011, 4356–4360), $LiMnPO_4$ (4.1 V) (Kim et al. 2010, 1305–1307), LiCoPO4 (4.8 V) (Liu et al. 2011, 9984–9987) have also been studied.

1.5.3 Fluoride-Based Compounds

The low redox potential shortcoming of $LiFePO_4$ as a cathode material ignited research interest on fluoride-based compounds that exhibit higher redox voltages. Some fluorides such as (FeF_3) (T. Li et al. 2010, 3190–3195), fluorophosphates ($LiFePO_4F$) (Recham et al. 2010, 1142–1148), and oxyfluorides have been studied. These fluorosulphates have tavorite structures (Sebastian et al. 2002, 374–377) consisting of 3D framework that enables faster lithium diffusion. A stable capacity of >130 mAh.g^{-1} has been obtained with LiFeSO$_4$F at higher rates. LiFeSO$_4$F has a drawback of being soluble in water and unstable at high temperatures. Further, it is difficult to synthesize LiMSO$_4$F in bulk quantities.

1.5.4 Polyanionic Compound Cathode Material

Polyanionic compounds such as lithium-rich Li$_2$M$_2$ (XO$_4$)$_3$ (M= Ni, Co, Mn, Fe, Ti, & V and X= P, S, As, Mo, W) and lithium less-rich Li$_x$M$_2$ (XO$_4$)$_3$ compounds have also been investigated as cathode materials. They crystallize in open type of 3D framework, which provides multiple spaces to accommodate a variety of guest species. As such, these compounds have interesting lithium insertion properties.

Polyanionic compounds like Fe$_2$ (WO$_4$)$_3$ and Fe$_2$ (SO$_4$)$_3$ have also been found to exhibit reversible lithium insertion property (Manthiram and Goodenough 1987, 349–360; Manthiram and Goodenough 1989, 403–408). Fe$_2$ (SO$_4$)$_3$ crystallizes in two different structures and both forms could reversibly insert ~2 lithium at 3.6 V (Manthiram and Goodenough 1989, 403–408). Also lithium vanadium phosphate Li$_3$V$_2$(PO$_4$)$_3$ exists in number of phases and have the capability to insert reversibly 1.3 to 3 lithium to give capacity up to 197 mAh.g^{-1} (Tang et al. 2008, 1646–1648).

1.5.5 Other Transition Metal Oxide Cathode Materials

1.5.5.1 Vanadium-Based Cathode Materials

Other oxides that have been studied as cathode materials include oxides of vanadium and manganese. The strong interest of researchers in vanadium-based oxides is a result of their good electronic conductivity, excellent chemical stability in polymeric electrolytes, and high energy density (X Ren et al. 2012, 929–934). In view of the fact that Vanadium exists in a number of oxidation states (from 2$^+$ in VO to 5$^+$ in V$_2$O$_5$), its oxides could provide a wide range of capacities as cathode materials (Kannan and Manthiram 2006, 1405–1408). Several vanadium oxides have been studied as potential cathode materials, but among them V$_2$O$_5$ (Z. Li 2007, 720–726; Y. Chen Y et al. 2008, 372–375; Pomerantseva et al. 2012, 282–287; X Ren et al. 2012, 929–934), LiV$_3$O$_8$ (H.Y. Xu et al. 2004, 349–353; H Ma et al. 2011, 6030–6035; Q. Shi et al. 2011, 9329–9336; X. Ren et al. 2012, 929–934), and Li$_3$V$_2$ (PO$_4$)$_3$ (L. Wang et al. 2010, 52–55; X.H. Rui et al. 2010, 2384–2390; Y.Q. Qiao et al. 2012, 132–137) have shown the most promising electrochemical performance as demonstrated by their high discharge capacities and good capacity retention. Similarly some manganese-based amorphous oxides also show lithium insertion property with huge capacity of about 300 mAh.g^{-1}.

1.5.5.2 Advanced/Green Cathode Materials

Major lithium ion cathode materials are made from non-renewable and expensive inorganic compounds (Amatucci and Pereira 2007, 243–262; T. Li et al. 2009, 3190–3195; Y. Yang et al. 2012, 15387–15394; F. Gschwind et al. 2016, 76–90). As a result, researchers are now searching for alternative green materials or material combinations or synthesis routes that are not only cheap but could improve the power, energy density, and safety of LIBs. One such alternative is the use of organic electrode material due to its inherent flexibility, non-toxicity, cheapness, and natural abundance (Goriparti et al. 2013, 7234–7236). Another alternative is the use of cheap sulphur compounds to produce cheap Li$_2$S that has a theoretical capacity density of 1672 mA/g (J. Wang et al. 2003, 487–492). Others are the use of conversion reactions that can enable the harvesting of oxidation phases of a transition metal (Choi and Manthiram 2007, 1541–1545; P. Poizot et al. 2000, 496–499) as fluorides (metal fluorides, oxyfluorides, carbon fluorides), fluorophosphates, and fluorosulphates.

1.6 Li-Ion Battery Electrolyte and Separator Materials

1.6.1 Li-Ion Battery Electrolyte Materials

A LIB electrolyte material acts as a medium for the transfer of ions or ionic conductivity pathways between the anode and cathode. The electrolyte could be a solid or a liquid medium that allows the Li ions to move between the electrodes and drive the electrons to move through the external circuit. Thus, electrolyte should exhibit high ionic and low electronic conductivity. Other desirable properties include low viscosity; broad chemical-electrochemical stability; wide working temperature range; proper passivation of the electrode surfaces; and high safety (Aurbach et al. 2004, 247–254).

The usual LIB liquid electrolyte consists of lithium salts dissolved in polar organic solvents. The popular organic solvents can be grouped into three types: ethers, esters, and alkyl carbonates (K. Xu et al. 2014a, 11503–11618). The alkyl carbonates are the most extensively utilized due to their high polarity and solubility. Examples of alkyl carbonates are ethylene carbonate (EC), dimethyl carbonate (DMC), diethyl carbonate (DEC), and ethyl-methyl carbonate (EMC). The potential Li salts comprise non-coordinating anionic complexes of Li such as $LiPF_6$, $LiClO_4$, $LiBF_4$, $LiN(SO_2CF_2CF_3)_2$ (LiBETI), $LiB(C_2O_4)_2$ (LiBOB), $LiPF_3(CF_2CF_3)_3$ (LiFAP), and $LiN(SO_2CF_3)_2$ (LiTFSI) (K. Xu et al. 2014b, 11503–11618).

However, many of these lithium salts have shortcomings. While BF_4 can interfere with the anode surface passivation, the LiTFSI can lead to corrosion of the aluminium current collector in the cathode. The $LiPF_6$ can hydrolyze and decompose to produce highly toxic and corrosive species such as LiF, PF_5, HF, and PF_3O (Aurbach et al. 2000, 1322; Kanamura et al. 2002, A339). Due to these shortcomings, Li salts and solvents are selected and combined to obtain a compromise. For example, EC which has a relatively high melting point is mixed with DMC or DEC and DMC to form EC-DMC and EC-DEC-DMC solvents, in order to obtain a wide working temperature range and enable passivation of the anode surface (Smart et al. 2002, A361). The popular electrolyte used in commercial batteries consists of 1M $LiPF_6$ dissolved in EC-DMC in the ratio of 1:1 by volume.

1.6.2 Li-Ion Battery Separator Materials

The separator is usually a thin porous film material inserted between the electrodes and permits ionic transport through the pores, and at the same time acts as an electronic insulator. It prevents short circuit induced by direct contact of the cathode and anode. Separators can be one of these types: commercially, a separator can be made of microporous polyolefin polymers such as polyethylene (PE), polypropylene (PP), and their blends (PE-PP) (Krivik and Baca 2013; Leijonmarck 2013). This type of separator is extremely thin and the multilayered sheet such as bilayer PE-PP and trilayer PP-PE-PP.

Another type of separator is a non-woven mat fabricated by bonding fibrous materials, including celluloses, polyvinylidene fluoride (PVDF), poly(vinylidene fluoride-co-hexafluoropropylene) (PVDF-HFP), and others (Kim JR et al. 2005, A295). They are mostly characterized by high porosity (60~80 %) and low cost (S.S. Zhang 2007, 351–364). The third type is a ceramic composite separator, which has excellent wettability and thermal stability (zero shrinkage). It is made of nanosized inorganic particles (e.g. Al_2O_3, SiO_2, MgO) bonded by a small amount of binder (e.g. PVDF, PVDF-HFP) (S.S. Zhang 2007, 351–364).

1.6.3 Other Li-Ion Battery Materials – Conductive Additives, Current Collector, and Binder

Conductive additive materials are materials added to electro-active material of the electrode during formulation in order to improve its inherent low electronic conductivity. The most frequently used conductive additive is acetylene carbon black. Sometimes conductive graphite, graphene, and carbon nanotubes (CNTs) are also used (Y. Shi et al. 2019, 19–26).

An electrode requires current collector in order to enhance its mechanical strength and electronic conductivity. A current collector should also be chemically and electrochemically compatible and stable with the active material of electrode and the electrolyte at operating conditions. Al and Cu foils are most suitable for positive and negative electrodes, respectively (Iwakura et al. 1997, 301–303). For anodes, current collectors comprising some carbonaceous materials (like carbon cloth, carbon fibre, and flexible graphite) are also used with the attendant benefit from their light weight and extra Li+ storage capacity (Yazici et al. 2005, 171–176; J. Guo et al. 2010, 981–984). Three-dimensional porous structure (like Ni-foam and porous Cu) is also employed as current collector in anodes enabling the increase in the contact area of active materials with electrolyte (Q. Li et al. 2017, 1606422; Sa and Wang 2012, 46–51).

The binder is a very important, but inactive material in LIB that maintains the connection between the electrode materials and current collector and imparts mechanical flexibility. A good binder should have high adhesive strength, flexibility, in addition to high electrochemical stability to maintain the integrity of the electrode during cycling and therefore ensuring a high cycling stability. Polyvinylidene fluoride (PVDF) is the binder used in the first commercial LIBs and still dominant in LIB industries (H. Chen et al. 2018, 8936-8982).

However, PVDF has limitations which prompted investigations of other binders. It is not so stable with the lithiated graphite, and the C–F bonds can decompose during charge-discharge cycle (K.Y. Cho et al. 2013, 2044–2050). Moreso, PVDF is only soluble in organic solvents, where the toxic N-methyl-2-pyrrolidone (NMP) is usually employed. The other binders which have been investigated and demonstrated better performance than PVDF include sodium alginate (SA), carboxyl methyl cellulose (CMC), CMC-SBR (styrene butadiene rubber), poly(acrylic acid) (PAA), and bio-derived polymers. These binders show higher tensile strength and flexibility than PVDF, due to the hydrogen bonding triggered by their abundant hydroxyl and/or carboxylate groups. They can provide stronger interfacial forces than the weak van der Waals force of PVDF (Chen H et al. 2018; 8936–8982). Moreover, these binders are water-soluble and more ecofriendly than PVDF (Bresser et al. 2018, 3096–3127).

1.7 Synthesis and Characterization of Li-Ion Battery Electrode Materials

Prior and very important step in LIB construction is the synthesis of the electrode materials. The method of synthesis greatly influences chemical and physical nature of the electrode material (its chemistry, size and morphology), which in turn determines a battery's electrochemical performance. Furthermore, the synthesis route influences desirable properties of the electrode materials such as lithium ion and electron transport, shortening of Li^+ diffusion path, electronic and ionic conductivities. Despite the fact that inherent property of the electrode material is very important for lithium ion storage, the appropriate synthesis method is of greater importance than the selected material (Goodenough et al. 1979). Thus, synthesis routes play a key role in devising better battery electrode materials (Schougaard et al. 2006, 905–909).

Various routes of synthesis of electrode materials have been studied, and they include solid-state method; coprecipitation; microwave synthesis; sol gel synthesis; pechini method; combustion method; hydrothermal synthesis; spray pyrolysis; oil emulsion method; freeze drying; mechanochemical process, solvo-thermal process, etc. (Bensalah and Dawood 2016, 258). Most of these methods are based on nanotechnology processing routes and result in nanosized electrode materials with high surface areas, which confer improved electrochemical performance to the LIB by providing short diffusion paths for Li^+ and electron pathway (Liu and Cao 2010, 1218–1237; Qi Wen et al. 2017, 19521). Although, each synthetic method has its own merits and demerits, the best method of synthesis should enable low cost of production; bulk production with minimal phase impurity; production of homogeneous materials with proper and suitable physical and chemical characteristics for battery application.

An accompanying aspect to synthesis is characterization which entails elucidating physical, chemical, and electrochemical characteristics of the electrode material and quantifying them. The common characterization techniques for ascertaining the electrochemical performance of electrode materials and in turn LIB include scanning electron microscopy (SEM); transmission electron microscopy (TEM); X-ray diffractometry (XRD); X-ray photoelectron spectroscopy (XPS), X-ray absorption spectroscopy

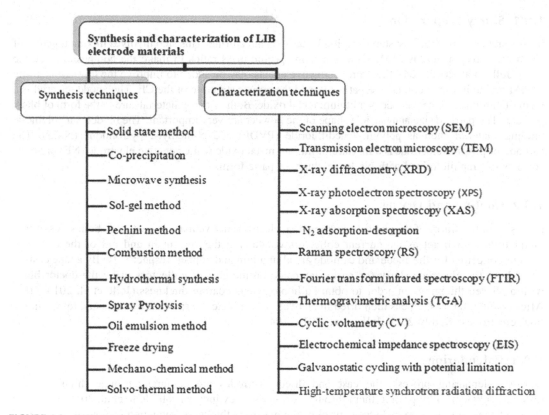

FIGURE 1.2 Different synthesis and characterization techniques used to produce Li-ion electrode materials.

(XAS); N_2 adsorption-desorption; Raman spectroscopy (RS), Fourier transform infrared spectroscopy (FTIR), Thermogravimetric analysis (TGA); cyclic voltammetry (CV); electrochemical impedance spectroscopy (EIS); galvanostatic cycling with potential limitation (GCPL), etc. The different synthesis and characterization techniques used to produce Li-ion electrode materials are shown in Figure 1.2.

1.8 Li-Ion Battery Manufacturing

The major task in Li-ion battery construction or any battery construction is confining all the materials including cathode and anode active materials, electrolytes, separators, and current collectors in a limited and enclosed space, enabling energy as high as possible to be delivered without any safety problems (Tagawa and Brodd 2009, 181–194). The steps in LIB manufacturing are shown in Figure 1.3.

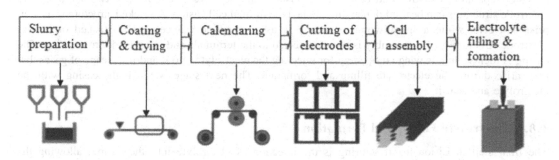

FIGURE 1.3 Lithium-ion battery manufacturing steps adapted from Jelle et al. (2016, 104).

1.8.1 Slurry Preparation

Slurry preparation is the first step in Li-ion battery manufacturing. This step involves mixing together of electrode active materials (AM) with binder in specific mass ratios to make the composite electrode (CE) (Jelle et al. 2016, 104). Electron-conductive agents may be added to improve the low conductivity of AM and to improve conductive pathways and the porous structure of the CE. The anode material is a form of carbon, and the cathode is a lithium metal oxide. Both of these materials are in the form of black powder. The particle size and particle shape of the powder are very important. The binder is a polymeric binding material such as polyvinylenedifluoride (PVDF) and N-methyl-2-pyrrolidone (NMP). The cathode or positive electrode is a mixture of lithium metal oxide while negative electrode (NE) or anode consists of graphite mixed with binder in slurry or paste form.

1.8.2 Coating and Drying

In this step, the slurry of the positive and negative electrode active mass is coated on both sides of thin metal foils, which act as the current collectors conducting the current in and out of the cell. The current collectors for the anode and cathode are aluminium and copper, respectively. By a tape casting procedure, the electrode slurry (or paste) is coated on the current collectors using the doctor blade with a chosen thickness, in order to obtain a homogenous coating thickness (Jelle et al. 2016, 104). After coating, the cast tape is then dried in an oven to eliminate the residual solvents and to obtain the final electrodes (Reddy 2011; Jelle et al. 2016, 104).

1.8.3 Calendaring

During calendaring process, the cast and dried electrodes are compressed under high pressure ($300–2000$ kg·cm^{-2}) by passing them through two massive cylindrical rolls Jelle et al. 2016, 104). The effect of the high pressure includes control of the electrode thickness, reduction in porosity, improved adhesion, and increase in density of electrode materials Jelle et al. 2016, 104). In order to remove all water contamination, the electrodes can be dried for a second time after calendaring process.

1.8.4 Cutting of Electrodes

At this stage the calendared electrodes are cleanly cut into strips of a desired shape so as to avoid burrs on the edges. Burrs can cause internal short circuit in the cells. Contact tabs are also fixed to the electrodes at this stage. The anodes are usually made slightly oversize than cathodes to avoid lithium deposition and dendride formation on the edge of the anode (Jelle et al. 2016, 104). However, large oversizing of the anode results in a loss of energy density of the complete cell (Jelle et al. 2016, 104).

1.8.5 Cell Assembly

The first step in the assembly process is building electrode sub-assembly in which a microporous polymer separator is sandwiched between the positive and negative electrode. The two sub-assembly electrode structures can be used depending on the type of final cell casing. A stacked structure is used in prismatic cells while a spiral wound structure is used in cylindrical cells. The stacked or wound electrode structure or sub-assembly is then connected to the terminals and integrated into a casing. The casing is then sealed, leaving an opening for injecting the electrolyte into it and for escape of gases that may arise during the electrolyte filling and formation. The next stage is to fill the casing with the electrolyte and seal it.

1.8.6 Electrolyte Filling and Formation

The final step in Li-ion manufacturing is the injection of electrolyte into the casing, allowing the electrolyte to adequately permeate/wet the electrode, sealing, and formation. Complete permeation of the electrolyte into the pores in the electrode and separator is very important (Whittingham 2004, 4271–4302). Wettability affects specific surface area of active material available for reaction and the formation of uniform SEI layer. Poor wettability will reduce the specific surface area of active material available for reaction during battery operation leading to increase in cell impedance.

Once the cell assembly is complete, the cell must be put through a formation process or subjected to a precise controlled charge-discharge cycle to activate the working materials and to create a solid electrolyte interface (SEI) on the anode, a passivating layer which is essential for moderating the charging process during normal use. The formation process enables data on the cell performance such as capacity and impedance to be gathered and recorded for quality analysis. Afterwards the cell is given an identification label by printing a batch or serial number on the case.

1.9 Conclusion and Future Trends

This chapter discussed a brief history of Li-ion batteries, their electrochemical performance assessment indices, and charge-discharge principles. The advantages and shortcomings of several commonly studied anode and cathode materials and attempts made by researchers to improve the properties of LIB materials were highlighted. The anode materials include carbon, alloys, transition metal oxides, and silicon while cathode materials considered are layered transition metal oxides of the type $LiMO_2$ (M = Mn, Co, Ni), spinel-structured $LiMn_2O_4$ along with its derivatives, and olivine-structured $LiFePO_4$, fluoride-based compounds (tavorites), polyanionic compounds, and other transition metal oxides. Despite the fact that there have been significant efforts to improve the performance of LIBs via various material-driven routes, finding the optimum combination of material properties for LIBs still remains a daunting task.

The synthesis routes and characterization techniques, which were mainly based on nanotechnology strategies to enhance and quantify battery performance, were examined. Since one of the most important thing in LIB systems is to confine all the component materials (cathode and anode active materials, electrolytes, separators, current collectors, and binder) in a limited and enclosed space that enable energy as large as possible to be brought out without any safety problems, this chapter also considered the LIB manufacturing steps.

Since commercialization of rechargeable LIB by SONY 29 ago (1991), it has shown rapid growth in the sales more than all the earlier rechargeable battery systems. This competitive edge of LIBs is a result of continued improvement in its component materials and processing techniques. The quest to use LIB for higher energy storage applications will continue to drive research to improve LIB component materials. Recent works on developing new LIB materials indicates strongly that LIB will continue to improve in cost, energy, safety, and power capability and will remain a formidable competitor for many years to come.

Acknowledgements

This work was supported by the University of Nigeria Nano Research Group and African Centre of Excellence for Sustainable Power and Energy Development (ACE-SPED) University of Nigeria, Nsukka, Nigeria.

REFERENCES

Amatucci, G.G., and N. Pereira. 2007. "Fluoride Based Electrode Materials for Advanced Energy Storage Devices". *Journal of Fluorine Chemistry* 128: 243–262.

Antolini, E. 2004. "LiCoO2: Formation, Structure, Lithium and Oxygen Nonstoichiometry, Electrochemical Behaviour and Transport Properties". *Solid State Ionics* 170(3–4): 159–171.

Arico, A.S., P. Bruce, B. Scrosati, J.M. Tarascon, and W. van Schalkwijk. 2005. "Nanostructured Materials for Advanced Energy Conversion and Storage Devices". *Nature Materials* 4: 366–377.

Armand, M., and M. Duclot. 1978. Ionically and Pref. electronically conductive electrode-comprising ag-glomerate of active electrode material and solid Solution of ionic Cpd. in polymer pref. polyoxyalk-ylene. French Patent 7,832,976.

Armand, M., and P. Touzain. 1977. "Graphite Intercalation Compounds as Cathode Materials". *Materials Science Engineering* 31: 319–329.

Armand, M.B. 1980. "Intercalation Electrodes". In *Materials for Advanced Batteries*, edited by D.W. Murphy, J. Broadhead, and B.C.H. Steele, 145–161. New York: Springer.

Armstrong, A.R., A.J. Paterson, A.D. Robertson, and P.G. Bruce. 2002. "Nonstoichiometric Layered $Li_xMn_yO_2$ with a High Capacity for Lithium Intercalation/Deintercalation". *Chemical Materials* 14: 710–719.

Aurbach, D., K. Gamolsky, B. Markovsky, G. Salitra, Y. Gofer, U. Heider, R. Oesten, and M. Schmidt. 2000. "The Study of Surface Phenomena Related to Electrochemical Lithium Intercalation into Li_xMO_y Host Materials (M = Ni, Mn)". *Journal of Electrochemical Society* 147: 1322.

Aurbach, D., Y. Talyosef, B. Markovsky, E. Markevich, E. Zinigrad, L. Asraf, J.S. Gnanaraj, and H.J. Kim. 2004. "Design of Electrolyte Solutions for Li and Li-Ion Batteries: A Review". *Electrochimica Acta* 50: 247–254.

Bensalah, N., and H. Dawood. 2016. "Review on Synthesis, Characterizations, and Electrochemical Properties of Cathode Materials for Lithium Ion Batteries. *Journal of Materials Science & Engineering* 5(4): 258.

Blomgren, G.E. 2017. "The Development and Future of Lithium Ion Batteries". *Journal of Electrochemical Society* 164: A5019–A5025.

Bresser, D., D. Buchholz, A. Moretti, A. Varzi, and S. Passerini. 2018. "Alternative Binders for Sustainable Electrochemical Energy Storage-the Transition to Aqueous Electrode Processing and Bio-Derived Polymers". *Energy & Environmental Science* 11: 3096–3127.

Broussely, M., P. Biensan, and B. Simon. 1999. "Lithium Insertion into Host Materials: The Key to Success for Li Ion Batteries". *Electrochimica Acta* 45: 3–22.

Cabana, J., L. Monconduit, D. Larcher, and M.R. Palacin. 2010. "Beyond Intercalation-Based Li-Ion Batteries: The State of the Art and Challenges of Electrode Materials Reacting through Conversion Reactions". *Advanced Materials* 22: E170–E192.

Cao, Q., H.P. Zhang, G.J. Wang, Q. Xia, Y.P. Wu, and H.Q. Wu. 2007. "A Novel Carbon Coated $LiCoO_2$ as Cathode Material for Lithium Ion Battery". *Electrochemistry Communications* 9(5): 1228–1232.

Casimir, A., H. Zhang, O.J. Ogoke, C. Amine, J. Lu, and G. Wu. 2016. "Silicon-Based Anodes for Lithium-Ion Batteries: Effectiveness of Materials Synthesis and Electrode Preparation". *Nano Energy* 27: 359–376.

Chan, H.W., J.G. Duh, and S.R. Sheen. 2003. "$LiMn_2O_4$ Cathode Doped with Excess Lithium and Synthesized by Co-Precipitation for Li-Ion Batteries". *Journal of Power Sources* 115(1): 110–118.

Chauque, S., F. Oliva, A. Visintin, D. Barraco, E. Leiva, and O. Cámara. 2017. "Lithium Titanate as Anode Material for Lithium Ion Batteries: Synthesis, Post-Treatment and Its Electrochemical Response". *Journal of Electroanalytical Chemistry* 799:142–155.

Chen, Y., H. Liu, and W. Ye. 2008. "Preparation and Electrochemical Properties of Submicron Spherical V_2O_5 as Cathode Material for Lithium Ion Batteries". *Scripta Materialia* 59: 372–375.

Chen, Y.M., X.Y. Yu, Z. Li, U. Paik, and X.W. Lou. 2016. "Hierarchical MoS_2 Tubular Structures Internally Wired by Carbon Nanotubes as a Highly Stable Anode Material for Lithium-Ion Batteries". *Scienctific Advances* 2:1600021.

Chen, Y.M., L. Yu, and X.W. Lou. 2016. "Hierarchical Tubular Structures Composed of Co_3O_4 Hollow Nanoparticles and Carbon Nanotubes for Lithium Storage". *Angew Chemistry International Education* 55: 5990–5993.

Cheng, L., X.-L. Li, H.-J. Liu, H.-M. Xiong, P.-W. Zhang., and Y.-Y. Xia. 2007. "Carbon-Coated $Li_4Ti_5O_{12}$ as a High Rate Electrode Material for Li-Ion Intercalation. *Electrochemical Society* 154(7): A692.

Cheng, X.-B., R. Zhang, C.-Z. Zhao, and Q. Zhang. 2017. "Toward Safe Lithium Metal Anode in Rechargeable Batteries: A Review". *Chemical Reviews* 117: 10403–10473.

Cho, J. 2010. "Porous Si Anode Materials for Lithium Rechargeable Batteries". *Journal of Materials Chemistry* 20: 4009–4014.

Cho, K.Y., Y.I.L. Kwon, J.R. Youn, and Y.S. Song. 2013. "Interaction Analysis between Binder and Particles in Multiphase Slurries". *Analyst* 138: 2044–2050.

Choi, W. and A. Manthiram. 2007. "Influence of Fluorine Substitution on the Electrochemical Performance of 3??V Spinel $Li_4Mn_5O_{12-2x}F_x$ Cathodes". *Journal of the Electrochemical Society* 154: A614–A618.

Chung, S.-Y., J.T. Bloking, and Y.-T. Chiang. 2002. "Electronically Conductive Phospho-Olivines as Lithium Storage Electrodes". *Nature Materials* 1: 123–128.

Crosnier, O., T. Brousse, and D. Schleich. 1999. "Tin Based Alloys for Lithium Ion Batteries". *Ionics* 5: 311–315.

Chen, H., M. Ling, L. Hencz, H.Y. Ling, G. Li, and S. Zhang. 2018. "Exploring Chemical, Mechanical, and Electrical Functionalities of Binders for Advanced Energy-Storage Devices". *Chemical Reviews* 118: 8936–8982.

Dambournet, D., L. Belharouak, J. Ma, and K. Amine. 2011. "Toward High Surface Area TiO_2 Brookite with Morphology Control". *Journal of Materials Chemistry* 21: 3085–3090.

Dell, R.M. and D.A.J. Rand. 2001. *Understanding Batteries*. Cambridge, UK: The Royal Society of Chemistry.

Ding, Y., J. Li, Y. Zhao, and L. Guan. 2012. "Direct Growth of $LiMn_2O_4$ on Carbon Nanotubes as Cathode Materials for Lithium Ion Batteries". *Materials Letters* 68: 197–200.

Ellis, B.L., K.T. Lee, and L.F. Nazar. 2010. "Positive Electrode Materials for Li-Ion and Li-Batteries". *Chemical Materials* 22: 691–714.

Feng, K., M. Li, W. Liu, A.G. Kashkooli, X. Xiao, M. Cai, and Z. Chen. 2018. "Silicon-Based Anodes for Lithium-Ion Batteries: From Fundamentals to Practical Applications". *Small* 14(8): 1702737.

Gan, Y., L. Zhang, Y. Wen, F. Wang, and H. Su. 2008. "Carbon Combustion Synthesis of Lithium Cobalt Oxide as Cathode Material for Lithium Ion Battery". *Particuology* 6: 81–84.

Gibot, P., M. C- Cabanas, L. Laffont, S.P. Levasseur Carlach, S. Hamelet, J.M. Tarascon, and C. Masquelier. 2008. "Room-Temperature Single-Phase Li Insertion/Extraction in Nanoscale Li x $FePO_4$". *Nature Materials* 7: 741–747.

Goodenough, J.B. 2013. "Battery Components, Active Materials for". In *Batteries for Sustainability: Selected Entries from the Encyclopedia of Sustainability Science and Technology*, edited by R.J. Brodd, 51–92. New York: Springer Science.

Goodenough, J.B., K. Mizushima, and P.J. Wiseman. 1979. Electrochemical cell and method of making ion conductors for said cell. Eur. Patent EP0017400A1.

Goriparti, S., M.N.K. Harish, and S. Sampath. 2013. "Ellagic Acid-A Novel Organic Electrode Material for High Capacity Lithium Ion Batteries". *Chemical Communications* 49: 7234–7236.

Groult, H., H.E. Ghallali, A. Barhoun, E. Briot, C.M. Julien, F. Lantelme, and S. Borensztjan. 2011. "Study of Co–Sn and Ni–Sn Alloys Prepared in Molten Chlorides and Used as Negative Electrode in Rechargeable Lithium Battery". *Electrochimica Acta* 56(6): 2656–2664.

Gschwind, F., G. Rodriguez-Garcia, D.J.S. Sandbeck, A. Gross, M. Weil, M. Fichtner, and N. Hörmann. 2016. "Fluoride Ion Batteries: Theoretical Performance, Safety, Toxicity, and a Combinatorial Screening of New Electrodes". *Journal of Fluorine Chemistry* 182: 76–90.

Guerard, D., and A. Herold. 1975. "Intercalation of Lithium into Graphite and Other Carbons". *Carbon* 13(4): 337–345.

Guo, H.J., K.X. Xiang, X. Cao, X.-H. Li, X. Cao, Z. Wang, and L.-M. Li. 2009. "Preparation and Characteristics of Li2FeSiO4/C Composite for Cathode of Lithium Ion Batteries". *Transactions of Nonferrous Metals Society of China* 19(1): 166–169.

Guo, J., A. Sun, and C. Wang. 2010. "A Porous Silicon-Carbon Anode with High Overall Capacity on Carbon Fiber Current Collector". *Electrochemical Communications* 12: 981–984.

Guyomard, D., and J.M. Tarascon. 1992. "Li Metal-Free Rechargeable $LiMn_2O_4$/Carbon Cells: Their Understanding and Optimization". *Journal of Electrochemical Society* 139: 937.

Hannan, M.A., M.M. Hoque, A. Hussain, Y. Yusof, and P.J. Ker. 2018. "State-of-the-Art and Energy Management System of Lithium-Ion Batteries in Electric Vehicle Applications: Issues and Recommendations". *IEEE Access* 6: 19362–19378.

Hassoun, J., K.-S. Lee, Y.-K. Sun, and B. Scrosati. 2011. "An Advanced Lithium Ion Battery Based on High Performance Electrode Materials". *Journal of American Chemical Society* 133(9): 3139–3143.

Higgins, T.M., S.H. Park, P.J. King, C. Zhang, N. McEvoy, N.C. Berner, D. Daly, A. Shmeliov, U. Khan, and G. Duesberg. 2016. "A Commercial Conducting Polymer as both Binder and Conductive Additive for Silicon Nanoparticle-Based Lithium-Ion Battery Negative Electrodes". *ACS Nano* 10: 3702–3713.

Hsieh, C.T., C.T. Pai, Y.F. Chen, I.L. Chen, and W.Y. Chen. 2014. "Preparation of Lithium Iron Phosphate Cathode Materials with Different Carbon Contents Using Glucose Additive for Li-Ion Batteries". *Journal of Taiwan Institute of Chemical Engineers* 45(4): 1501–1508.

Hu, L.H., F.Y. Wu, C.T. Lin, A.N. Khlobystov, and L.J. Li. 2013. "Graphene-Modified $LiFePO_4$ Cathode for Lithium Ion Battery beyond Theoretical Capacity". *Nature Communications* 4: 1687.

Huang, S., Z. Wen, X. Yang, Z. Gu, and X. Xu. 2005. "Improvement of the High-Rate Discharge Properties of $LiCoO_2$ with the Ag Additives". *Journal of Power Sources* 148: 72–77.

Hwang, T.H., Y.M. Lee, B.S. Kong, J.-S. Seo, and J.W. Choi. 2012. "Electrospun Core–Shell Fibers for Robust Silicon Nanoparticle-Based Lithium Ion Battery Anodes". *Nano Letters* 12(2): 802–807.

Islam, M.S., and C.A. Fisher. 2014. "Lithium and Sodium Battery Cathode Materials: Computational Insights into Voltage, Diffusion and Nanostructural Properties". *Chemical Society Reviews* 43: 185–204.

Iwakura, C., Y. Fukumoto, H. Inoue, S. Ohashi, S. Kobayashi, H. Tada, and M. Abe. 1997. "Electrochemical Characterization of Various Metal Foils as a Current Collector of Positive Electrode for Rechargeable Lithium Batteries". *Journal of Power Sources* 68: 301–303.

Ji, L. and X. Zhang. 2010. "Evaluation of Si/Carbon Composite Nanofiber-Based Insertion Anodes for New-Generation Rechargeable Lithium-Ion Batteries". *Energy Environmental Science* 3: 124–129.

Julien, C., A. Mauger, A. Vijh, and K. Zaghib. 2016. "Electrolytes and Separators for Lithium Batteries". In *Lithium Batteries*. Cham: Springer.

Kanamura, K., W. Hoshikawa, and T. Umegaki. 2002. "Electrochemical Characteristics of $LiNi_{0.5}Mn_{1.5}O_4$ cathodes with Ti or Al Current Collectors". *Journal of Electrochemical Society* 149: A339.

Kannan, A.M., and A. Manthiram. 2006. "Low Temperature Synthesis and Electrochemical Behavior of LiV_3O_8 Cathode". *Journal of Power Sources* 159: 1405–1408.

Kaskhedikar, N.A., and J. Maier. 2009. "Lithium Storage in Carbon Nanostructures". *Advanced Materials* 21: 2664–2680.

Kim, D.K., P. Muralidharan, H.W. Lee, R. Ruffo, Y. Yang, Candace K. Chan, Hailin Peng, Robert A. Huggins, and Yi Cui. 2008. "Spinel $LiMn_2O_4$ Nanorods as Lithium Ion Battery Cathodes". *Nano Letters* 8(11): 3948–3952.

Kim, J., D.-H. Seo, S.-W. Kim, Y.-U. Park, and K. Kang. 2010. "Mn Based Olivine Electrode Material with High Power and Energy". *Chemical Communications* 46: 1305–1307.

Kim, J.R., S.W. Choi, S.M. Jo, W.S. Lee, and B.C. Kim. 2005. "Characterization and Properties of P(VdF-HFP)-Based Fibrous Polymer Electrolyte Membrane Prepared by Electrospinning". *Journal of Electrochemical Society* 152: A295.

Krivik, P., and P. Baca. 2013. "Electrochemical Energy Storage". In *Energy Storage: Technologies and Applications*, edited by A.F. Zobaa, 79–101. United Kingdom: INTECHOPEN.

Lazzari, M., and B. Scrosati. 1980. "A Cyclable Lithium Organic Electrolyte Cell Based On Two Intercalation Electrodes". *Journal of Electrochemica Society* 127: 773–774.

Leijonmarck, S. 2013. "Preparation and Characterization of Electrochemical Devices for Energy Storage and Debonding. PhD diss., School of Chemical Science and Engineering, Stockholm, ISBN 978-91-7501-685-6.

Lewis, G.N. and F.G. Keyes. 1913. "The Potential of the Lithium Electrode". *Journal of American Chemcal Society* 35: 340–344.

Li, H., Z. Wang, L. Chen, and H. Huang. 2009. "Research on Advanced Materials for Li-Ion Batteries". *Advanced Materials* 21(45): 4593–4607.

Li, H., and H. Zhou. 2012. "Enhancing the Performances of Li-Ion Batteries by Carbon-Coating: Present and Future". *Chemical Communications* 48: 1201–1217.

Li, Q., S. Zhu, and Y. Lu. 2017. "3D Porous Cu Current Collector/Li-Metal Composite Anode for Stable Lithium-Metal Batteries". *Advanced Functional Materials* 27: 1606422.

Li, T., L. Li, Y. Cao, A. Xing, and H.X. Yang. 2009. "Reversible 3-Electron Redox Behaviors of FeF_3 Nanocrystals as High Capacity Cathode-Active Materials for Li-Ion Batteries". *Journal of Physical Chemistry* 114(7): 3190–3195.

Li, T., L. Li, Y. Cao, A. Xing, and H.X. Yang. 2010. "Reversible Three-Electron Redox Behaviors of FeF_3 Nanocrystals as High-Capacity Cathode-Active Materials for Li-Ion Batteries". *Journal of Physical Chemistry* 114: 3190–3195.

Li, Z. 2007. "Synthesis of LiV_2O_5/VO_2 Mixture by Thermal Lithiation of Vanadium (+ 4, + 5) Oxides". *Transactions of Nonferrous Metals Society of China* 17: 720–726.

Liang, R., H. Cao, D. Qian, J. Zhang, and M. Qu. 2011. "Designed Synthesis of SnO_2-Polyaniline-Reduced Graphene Oxide Nanocomposites as an Anode Material for Lithium-Ion Batteries". *Journal of Materials Chemistry* 21: 17654–17657.

Liang, Yeru, Chen-Zi Zhao, Hong Yuan, Yuan Chen, Weicai Zhang, Jia-Qi Huang, Dingshan Yu, et al. 2019. "A Review of Rechargeable Batteries for Portable Electronic Devices". *InfoMaterials* 1(1): 1–27.

Liu, D., and G. Cao. 2010. "Engineering Nanostructured Electrodes and Fabrication of Film Electrodes for Efficient Lithium Ion Intercalation". *Energy & Environmental Science* 3: 1218–1237.

Liu, H.K., Z.P. Guo, J.Z. Wang, and K. Konstantinov. 2010. "Si-Based Anode Materials for Lithium Rechargeable Batteries". *Journal of Materials Chemistry* 20: 10055–10057.

Liu, J., T.E. Conry, X. Song, L. Yang, M.M., Doeff, and T.J. Richardson. 2011. "Spherical Nanoporous $LiCoPO_4$/C Composites as High Performance Cathode Materials for Rechargeable Lithium-Ion Batteries". *Journal of Materials Chemistry* 21: 9984–9987.

Liu, K., Y. Liu, D. Lin, A. Pei, and Y. Cui. 2018. "Materials for Lithium-Ion Battery Safety". *Science Advances* 4(6): eaas9820.

Liu, X., J. Wang, J. Zhang, and S. Yang. 2006. "Sol–Gel-Template Synthesis of ZnO Nanotubes and Its Coaxial Nanocomposites of $LiMn_2O_4$/ZnO". *Materials Science and Engineering A* 430(1-2): 248–253.

Liu, Z., J. Yu, X. Li, L. Zhang, D. Luo, X. Liu, Xioawei Liu, et al. 2018. "Facile Synthesis of N-Doped Carbon Layer Encapsulated Fe2N as an Efficient Catalyst for Oxygen Reduction Reaction". *Carbon* 127: 636–642.

Lou X., R. Li, X. Zhu, L. Luo, Y. Chen, C. Lin, H. Li, and X.S. Zhao. 2019. "New Anode Material for Lithium-Ion Batteries: Aluminum Niobate (AlNb11O29)". *ACS Applied Materials & Interfaces* 11(6): 6089–6096.

Lu, C.Z., J.M. Chen, Y.D. Cho, W.H. Hsu, P. Muralidharan, and G. Ting-KuoFey. 2008 "Electrochemical Performance of $LiCoO_2$ Cathodes by Surface Modification Using Lanthanum Aluminum Garnet". Journal of Power Sources 184(2): 392–401.

Lu, J., Z. Chen, F. Pan, Y. Cui, and K. Amine. 2018. "High-Performance Anode Materials for Rechargeable Lithium-Ion Batteries". *Electrochemical Energy Reviews* 1: 35–53.

Lu, Y., L. Yu, and X.W. Lou. 2018. "Nanostructured Conversion-Type Anode Materials for Advanced Lithium-Ion Batteries". *Chem* 4(5): 972–996.

Ma, C., Y. Zhao, J. Li, Y. Song, J. Shi, Q. Guo, and L. Liu. 2013. Synthesis and Electrochemical Properties of Artificial Graphite as an Anode for High-Performance Lithium-Ion Batteries". Carbon 64(2013) 553–556.

Ma, H., Z. Yuan, F. Cheng, J. Liang, Z. Tao, and J. Chen. 2011. "Synthesis and Electrochemical Properties of Porous LiV_3O_8 as Cathode Materials for Lithium Ion Batteries". *Journal of Alloys and Compounds* 509: 6030–6035.

Madhavi, S. and G.S. Rao. 2002. "Effect of Cr Dopant on the Cathodic Behavior of $LiCoO_2$". *Electrochimica Acta* 48(3): 219–226.

Manthiram, A. 2011. "Materials Challenges and Opportunities of Lithium Ion Batteries". *Journal of Physical Chemistry Letters* 2: 176–184.

Manthiram, A., and J.B. Goodenough. 1987. "Lithium Insertion into $Fe_2(MO_4)_3$ Frameworks: Comparison of M = W with M = Mo". *Journal of Solid State Chemistry* 71(2): 349–360.

Manthiram, A. and J.B. Goodenough. 1989. "Lithium Insertion into Fe_2 $(SO_4)_3$ Frameworks". *Journal of Power Sources* 26(3-4): 403–408.

Mauger, A., C.M. Julien, J.B. Goodenough, and K. Zaghib. 2019. "Tribute to Michel Armand: From Rocking Chair-Li-Ion to Solid-State Lithium Batteries". *Journal of Electrochemical Society* 167: 070507.

Meyers,W.F., and J.W. Simmons. 1969. Electric current-producing cell with anhydrous organic liquid electrolyte. U.S. Patent 3,423,242.

Minakshi, M., P. Singh, D. Appadoo, and D.E. Martin. 2011. "Synthesis and Characterization of Olivine $LiNiPO_4$ for Aqueous Rechargeable Battery". *Electrochimica Acta* 56(11): 4356–4360.

Mizushima, K., P.C. Jones, P.J. Wiseman, and J.B. Goodenough. 1980. "Li_xCoO_2 (0<x<-1): A New Cathode Material for Batteries of High Energy Density". *Materials Research Bulletin* 15(6): 783–789.

Morgan, D., A. Van der Ven, and G. Ceder. 2004. "Li Conductivity in Li_xMPO$_4$ (M=Mn, Fe, Co, Ni) Olivine Materials. Electrochemical". *Solid State* 7(2): A30–A32.

Myung, S., N. Kumagai, S. Komaba, and H. Chung. 2001. "Effects of Al Doping on the Microstructure of $LiCoO_2$ Cathode Materials". *Solid State Ionics* 139(1–2): 47–56.

Nitta, N., F. Wu, J.T. Lee, and G. Yushin. 2015. "Li-Ion Battery Materials: Present and Future". *Materials Today* 18: 252–264.

Ouyang, C.Y., X.M. Zeng, Z. Sljivancanin, and A. Baldereschi. 2010. "Oxidation States of Mn Atoms at Clean and Al_2O_3-Covered $LiMn_2O_4(001)$ Surfaces". *Journal of Physical Chemistry C* 114(10): 4756–4759.

Padhi, A.K., K.S. Nanjundaswamy, and J.B. Goodenough. 1997. Phospho-Olivines as Positive-Electrode Materials for Rechargeable Lithium Batteries". *Journal of Electrochemical Society* 144(4): 1188–1194.

Park, O.K., Y. Cho, S. Lee, H.-C. Yoo, H.-K. Song, and J. Cho. 2011. "Who Will Drive Electric Vehicles, Olivine or Spinel?" *Energy Environmental Science* 4: 1621–1633.

Poizot, P., S. Laruelle, S. Grugeon, L. Dupont, L., and J.-M. Tarascon. 2000. "Nano-Sized Transition-Metal Oxides as Negative-Electrode Materials for Lithium-Ion Batteries". *Nature* 407: 496–499.

Pomerantseva, E., K. Gerasopoulos, X. Chen, G. Rubloff, and R. Ghodssi. 2012. "Electrochemical Performance of the Nanostructured Bio-Templated V_2O_5 Cathode for Lithium-Ion Batteries". *Journal of Power Sources* 206: 282–287.

Qiao, Y.Q., X.L. Wang, Y.J. Mai, X.H. Xia, J. Zhang, C.D. Gu, and J.P. Tu. 2012. "Freeze-Drying Synthesis of $Li_3V_2(PO_4)3/C$ Cathode Material for Lithium-Ion Batteries". *Journal of Alloys and Compounds* 536: 132–137.

Recham, N., J.-N. Chotard, J.-C. Jumas, L. Laffont, M. Armand, and J.-M. Tarascon. 2010. "Ionothermal Synthesis of Li-Based Fluorophosphates Electrodes". *Chem. Mater.* 22: 1142–1148.

Reddy, Mogalahalli V., Alain Mauger, Christian M. Julien, Andrea Paolella and Karim Zaghib. 2020. "Brief History of Early Lithium-Battery Development". Materials 13: 1884.

Reddy, Mogalahalli V., G.V.S. Rao, and B.V.R. Chowdari. 2013. "Metal Oxides and Oxysalts as Anode Materials for Li Ion Batteries". *Chemical Reviews* 113: 5364–5457.

Reddy, T.B. 2011. *Linden's Handbook of Batteries*. New York: McGraw-Hill Education.

Ren, X., C. Shi, P. Zhang, Y. Jiang, J. Liu, and Q. Zhang. 2012. "An Investigation of V_2O_5/Polypyrrole Composite Cathode Materials for Lithium-Ion Batteries Synthesized by Sol–Gel". *Materials Science and Engineering B* 177: 929–934.

Rui, X.H., N. Ding, J. Liu, C. Li, and C.H. Chen. 2010. "Analysis of the Chemical Diffusion Coefficient of Lithium Ions in Li3V2(PO4)(3) cathode materia". *Electrochimica Acta* 55: 2384–2390.

Ryu, W.-H., J.-Y. Eom, R.-Z. Yin, D.-W. Han, W.-K. Kim, and H.-S. Kwon. 2011. "Synergistic Effects of Various Morphologies and Al Doping of Spinel $LiMn_2O_4$ Nanostructures on the Electrochemical Performance of Lithium-Rechargeable Batteries". *Journal of Materials Chemistry* 21: 5337–15342.

Sa, Q., and Y. Wang. 2012. "Ni Foam as the Current Collector for High Capacity C–Si Composite Electrode". *Journal of Power Sources* 208: 46–51.

Santiago, E.I., S.T. Amancio-Filho, P.R. Bueno, and L.O.S. Bulhões. 2001. "Electrochemical Performance of Cathodes Based on LiMn2O4 Spinel Obtained by Combustion Synthesis". *Journal of Power Sources* 98: 447–449.

Schougaard, S.B., J. Breger, M. Jiang, C.P. Grey, and J.B. Goodenough. 2006. "$LiNi_{0.5+\delta}Mn_{0.5-\delta}O_2$ – a High-Rate, High-Capacity Cathode for Lithium Rechargeable Batteries". *Advanced Materials* 18: 905–909.

Scrosati, B. 2011. "History of Lithium Batteries". *Journal of Solid State Electrochemistry* 15: 1623–1630.

Scrosati, B., and J. Garche. 2010. "Lithium Batteries: Status, Prospects and Future". *Journal of Power Sources* 195(9): 2419–2430.

Sebastian, L., J. Gopalakrishnan, and Y. Piffard. 2002. "Synthesis, Crystal Structure and Lithium Ion Conductivity of $LiMgFSO_4$". *Journal of Material Chemistry* 12, 374–377.

Shi, Q., R. Hu, M. Zeng, M. Dai, and M. Zhu. 2011. "The Cycle Performance and Capacity Fading Mechanism of a LiV_3O_8 Thin-Film Electrode with a Mixed Amorphous Nanocrystalline Microstructure". *Electrochimica Acta* 56: 9329–9336.

Shi, Y., L. Wen, S. Pei, M. Wu, and F. Li. 2019. "Choice for Graphene as Conductive Additive for Cathode of Lithium-Ion Batteries". *Journal Energy Chemistry* 30: 19–26.

Shu, J., M. Shui, F. Huang, Y. Ren, Q. Wang, D. Xu, and L. Hou. 2010. "A New Look at Lithium Cobalt Oxide in a Broad Voltage Range for Lithium-Ion Batteries". *Journal of Physical Chemistry C* 114: 3323–3328.

Shukla, A.K., and T.P. Kumar. 2008. "Materials for Next-Generation Lithium Ion Batteries". *Current Science* 94(3): 314–331.

Smart, M.C., B.V. Ratnakumar, and S. Surampudi. 2002. "Use of Organic Esters as Cosolvents in Electrolytes for Lithium-Ion Batteries with Improved Low Temperature Performance". *Journal of Electrochemical Society* 149: A361.

Smekens, Jelle, Rahul Gopalakrishnan, Nils Van den Steen, Noshin Omar, Omar Hegazy, Annick Hubin, and Joeri Van Mierlo. 2016. "Influence of Electrode Density on the Performance of Li-Ion Batteries: Experimental and Simulation Results". *Energies* 9(2): 104.

Sngodu, P. and A.D. Deshmukh. 2015. "Conducting Polymers and Their Inorganic Composites for Advanced Li-Ion Batteries: A Review". *RSC Advances* 5: 42109–42130.

Soge, Ayodele O. 2020. "Anode Materials for Lithium-Based Batteries: A Review". *Journal of Materials Science Research and Reviews* 5(3): 21–39.

Sun, J., Tang K., Yu X., Hu J., Li H., and Xuejie Huang. 2008. "Overpotential and Electrochemical Impedance Analysis on Cr_2O_3 Thin Film and Powder Electrode in Rechargeable Lithium Batteries". *Solid State Ionics* 179: 2390–2395.

Tagawa, K. and R.J. Brodd. 2009. "Production Processes for Fabrication of Lithium-Ion Batteries". In *Lithium-Ion Batteries Science and Technologies*, edited by. M. Yoshio, R.J. Brodd, and A. Kozawa, 181–194. New York: Springer.

Tang, A., X. Wang, and Z. Liu. 2008. "Electrochemical Behavior of $Li_3V_2(PO_4)_3$/C Composite Cathode Material for Lithium-Ion Batteries". *Materials Letters* 62: 1646–1648.

Tarascon, J.M., and M. Armand. 2001. "Issues and Challenges Facing Rechargeable Lithium Batteries". *Nature* 414: 359–367.

Thackeray, M.M., C.S. Johnson, A.J. Kahaian, K.D. Kepler, J.T. Vaughey, S.H. Yang, and S.A. Hackney. 1999. "Stabilization of Insertion Electrodes for Lithium Batteries". *Journal of Power Sources* 81–82: 60–66.

Tran, D.T., H. Dong, S.D. Walck, and S.S. Zhang. 2015. "Pyrite FeS_2-C Composite as a High Capacity Cathode Material of Rechargeable Lithium Batteries". *RSC Advances* 5: 87847–87854.

Wakihara, M. 2001. "Recent Developments in Lithium Ion Batteries". *Materials Science and Engineering R: Reports* 33(4): 109–134.

Wang, J., J. Yang, C. Wan, K. Du, J. Xie, and N. Xu. 2003. "Sulfur Composite Cathode Materials for Rechargeable Lithium Batteries". *Advanced Functional Materials* 13: 487–492.

Wang, K., Y. Jin, S. Sun, Y. Huang, J. Peng, J. Luo, Q. Zhang, Y. Qiu, C. Fang, and J. Han. 2017. "Low-Cost and High-Performance Hard Carbon Anode Materials for Sodium-Ion Batteries". *ACS Omega* 2: 1687–1695.

Wang, L., L.C. Zhang, I. Lieberwirth, H.W. Xu, and C.H. Chen. 2010. "A $Li_3V_2(PO_4)_3$/C Thin Film with High Rate Capability as a Cathode Material for Lithium-Ion Batteries". *Electrochemistry Communications* 12: 52–55.

Wang, W. and P.N. Kumta. 2010. "Nanostructured Hybrid Silicon/Carbon Nanotube Heterostructures: Reversible High-Capacity Lithium-Ion Anodes". *ACS Nano* 4(4): 2233–2241.

Wang, Y., P. He, and H. Zhou. 2012. "Li-Redox Flow Batteries Based on Hybrid Electrolytes: At the Cross Road between Li-Ion and Redox Flow Batteries. *Advanced Energy Materials* 2: 770–779.

Wen, Qui, Joseph G. Shapter, Qian Wu, Ting Yin, Guo Gao, and Daxiang Cui. 2017. "Nanostructured Anode Materials for Lithium-Ion Batteries: Principle, Recent Progress and Future Perspectives". *Journal of Materials Chemistry A* 5: 19521–19540.

Whittingham, M.S. 1978. "Chemistry of Intercalation Compounds: Metal Guests in Chalcogenide Hosts". *Progress in Solid State Chemistry* 12(1): 41–99.

Whittingham, M.S. 2004. "Lithium Batteries and Cathode Materials". *Chemical Reviews* 104(10): 271–4302.

Wu, H., G. Zheng, N. Liu, T.J. Carney, Y. Yang, and Y. Cui. 2012. "Engineering Empty Space between Si Nanoparticles for Lithium-Lon Battery Anodes". *Nano Letters* 12(2): 904–909.

Wu, H.B., J.S. Chen, X.W. Lou, and H.H. Hng. 2011. "Synthesis of SnO_2 Hierarchical Structures Assembled from Nanosheets and Their Lithium Storage Properties". *Journal of Physical Chemistry C* 115(50): 24605–24610.

Xia, Y., T. Sakai, T. Fujieda, M., Wada, and H. Yoshinaga. 2001. "Flake Cu-Sn Alloys as Negative Electrode Materials for Rechargeable Lithium Batteries". *Journal of Electrochemical Society* 148(5): A471.

Xie, J., X.B. Zhao, G.S. Cao, and J.P. Tu. 2007. "Electrochemical Performance of Nanostructured Amorphous Co_3Sn_2 Intermetallic Compound Prepared by a Solvothermal Route". *Journal of Power Sources* 164: 386–389.

Xu, H.Y., H. Wang, Z.Q. Song, Y.W. Wang, H. Yan, and Masahiro Yoshimura. 2004. "Novel Chemical Method for Synthesis of LiV_3O_8 Nanorods as Cathode Materials for Lithium Ion Batteries". *Electrochimica Acta* 49: 349–353.

Xu, J., S. Dou, H. Liu, and L. Dai. 2013. "Cathode Materials for Next Generation Lithium Ion Batteries". *Nano Energy*: 439–442.

Xu, Jiantie, Yuhai Dou, Zengxi Wei, Jianmin Ma, Yonghong Deng, Yutao Li, Huakun Liu, and Shixue Dou. 2017. "Recent Progress in Graphite Intercalation Compounds for Rechargeable Metal (Li, Na, K, Al)-Ion Batteries". *Advanced Science* 4: 1–14.

Xu, K. 2014a. "Electrolytes and Interphases in Li-Ion Batteries and Beyond". *Chemical Reviews* 114: 11503–11618.

Xu, K. 2004b. "Nonaqueous Liquid Electrolytes for Lithium-Based Rechargeable Batteries". *Chemical Reviews* 104(10): 4303–4418.

Yan, Z., Q. Hu, G. Yan, H. Li, K. Shih, Z. Yang, X. Li, Z. Wang, and J. Wang. 2017. "Co_3O_4/Co Nanoparticles Enclosed Graphitic Carbon as Anode Material for High Performance Li-Ion Batteries". *Chemical Engineering* 321: 495–501.

Yang, Y., L. Li, H. Fei, Z. Peng, G. Ruan, and J.M. Tour. 2014. "Graphen Nanoribbon/V_2O_5 Cathodes in Lithium-Ion Batteries". *ACS Applied Mater & Interfaces* 6(12): 9590–9594.

Yang, Y., D. Shu, H. Yu, X. Xia, and Z. Lin. 1997. "Investigations of Lithium Manganese Oxide Materials for Lithium-Ion Batteries". *Journal of Power Sources* 65: 227–230.

Yang, Yuan, Guangyuan Zheng, Sumohan Misra, Johanna Nelson, Michael F. Toney, and Yi Cui. 2012. "High-Capacity Micrometer-Sized Li_2S Particles as Cathode Materials for Advanced Rechargeable Lithium-Ion Batteries". *Journal of American Chemical Society* 134: 15387–15394.

Yazici, M., D. Krassowski, and J. Prakash. 2005. "Flexible Graphite as Battery Anode and Current Collector". *Journal of Power Sources* 141: 171–176.

Yoon, S. and A. Manthiram. 2011. "Hollow Core–Shell Mesoporous TiO_2 Spheres for Lithium Ion Storage". *Journal of Physical Chemistry C* 115: 9410–9416.

Yoon, Y.K., C.W. Park, H.Y. Ahn, D.H. Kim, Y.S. Lee, and J. Kim. 2007. "Synthesis and Characterization of Spinel Type High-Power Cathode Materials Li M_xMn_{2-x} O_4 (M=Ni, Co, Cr)". *Journal of Physical & Chemical Solids* 68: 780–784.

Yoshino, Akira. 2012. "The Birth of the Lithium-Ion Battery". *Angew Chemical International Education* 51(24): 5798–5800.

Zhang, S.S. 2007. "A Review on the Separators of Liquid Electrolyte Li-Ion Batteries". *Journal of Power Sources* 164: 351–364.

Zhang, W.-J. 2011. "A Review of the Electrochemical Performance of Alloy Anodes for Lithium-Ion Batteries". *Journal of Power Sources* 196(1): 13–24.

Zheng, H., Q. Qu, L. Zhang, G. Liu, and V.S. Battaglia. 2012. "Hard Carbon: A Promising Lithium-Ion Battery Anode for High Temperature Applications with Ionic Electrolyte". *RSC Advances* 2: 4904–4912.

Zhou, X., Y.-X. Yin, L.-J. Wan., and Y.-G. Guo. 2012. "Facile Synthesis of Silicon Nanoparticles Inserted into Graphene Sheets as Improved Anode Materials for Lithium-Ion Batteries". *Chemical Communications* 48: 2198–2200.

Zhou, Y-K., L. Cao, F-B. Zhang, B-L. He, and H.-I. Li. 2003. "Lithium Insertion into TiO2 Nanotube Prepared by the Hydrothermal Process". *Journal of Electrochemical Society* 150(9): A1246–A1249.

2

Carbon Derivatives in Performance Improvement of Lithium-Ion Battery Electrodes

Raphael M. Obodo[1,2,3], **Assumpta C. Nwanya**[1,4,5], **Ishaq Ahmad**[2,3,4], **Mesfin A. Kebede**[6,7], **and Fabian I. Ezema**[1,4,5]
[1]*Department of Physics and Astronomy, University of Nigeria, Nsukka, Enugu State, Nigeria*
[2]*National Center for Physics, Islamabad, Pakistan*
[3]*NPU-NCP Joint International Research Center on Advanced Nanomaterials and Defects Engineering, Northwestern Polytechnical University, Xi'an, China*
[4]*Nanosciences African Network (NANOAFNET) iThemba LABS-National Research Foundation, Somerset West, Western Cape Province, South Africa*
[5]*UNESCO-UNISA Africa Chair in Nanosciences/Nanotechnology, College of Graduate Studies, University of South Africa (UNISA), Pretoria, South Africa*
[6]*Energy Centre, Council for Scientific & Industrial Research, Pretoria, South Africa*
[7]*Department of Physics, Sefako Makgatho Health Science University, Medunsa, South Africa*

2.1 Introduction

With the world economy developing and the population increasing tremendously, the total worldwide energy demand has drastically increased, requiring effective energy fabrication and storage systems (Obodo et al. 2020a). The problem of the sustainability of energy production, storage, and delivery has received serious attention worldwide due to rapid depletion of fossil fuel resources, as well as problems associated with their environmental pollutions (Obodo et al. 2019a). The high rise in number of portable electronic and hybrid devices in the world for usages in homes, industries, hospitals, worship centers, automobiles, etc., promoted a tremendous growth in demand for high-capacity electrode materials for batteries (Feng et al. 2015). Nowadays, batteries are the most used energy storage devices for powering these electronic devices because of their high-energy storage capacity. Systems for electrochemical energy storage convert chemical energy into electrical energy. Energy storage systems are valued in terms of energy density also known as *specific energy* and power density also called *specific power*. In evaluating performances of energy storage devices using Ragone plot as shown in Figure 2.1 (Obodo et al. 2019a), energy densities are plotted against power densities. Moreover, as demonstrated in the Ragone plot in Figure 2.1, it reveals that batteries suffer from low power density but display high energy density. The problem of the low power density of batteries prompted scientists to research means of improving their energy densities. Lithium-ion battery (LIB) electrodes exhibit some electrochemical potential – a characteristic potential difference, which exists between their electrodes. LIBs possess an awesome property compared to other energy storage devices because they operate with a constant voltage through charge or discharge process while working

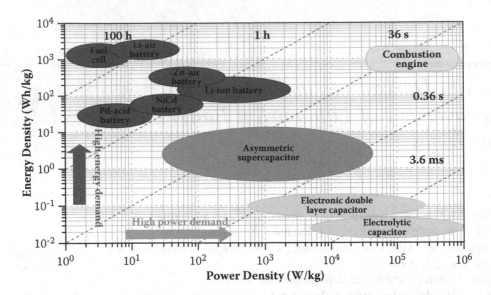

FIGURE 2.1 Energy Ragone plot of various energy storage devices. Reproduced with permission from Ref. Obodo et al. (2019).

voltage of other energy storage devices like supercapacitor declines linearly in their cycling process (Obodo et al. 2019b). LIBs' storage mechanisms occur using faradic reactions (oxidation/reduction) at the interface of their electrodes, which is controlled by slow diffusion rate and bring about low power density of LIBs.

The extensive usage of batteries seems constrained because of their low power density, which prompted various researches to enhance their energy density as well as power density. Various interest groups have launched enormous and passionate research determinations towards improving LIB performance in terms of energy and power densities, well-being, stability, availability, affordability, environmental friendliness, etc. (Obodo et al. 2019a).

Carbon derivatives such as graphene oxide (GO), reduced graphene oxide (rGO), carbon nanotubes (CNTs), activated carbon (AC), carbide-derived carbon (CDC), carbon aerogel (CA), etc. (Obodo et al., 2020b), have attracted fantastic interest in major researches and industrial usages, especially in energy storing and transformation devices like batteries, supercapacitor, etc. These carbon derivatives gained tremendous attention in energy storage applications because of their anticipated qualities such as low cost, accessibility, high surface area and environmental friendliness (Obodo et al. 2020c). Carbon derivatives possess high specific surface area, which results in a high capability for charge accumulation at the interface of electrode and electrolyte; hence, they are effectively used in energy LIBs (Obodo et al. 2020d). Carbon derivative–based LIBs' electrode architectures exhibit variable physical, chemical, and mechanical features, which enhance the performance of LIBs. In this chapter, we study the contribution and role of different carbon derivatives in enhancing LIB electrodes' performance. Various carbon derivatives' qualities that contributed to enhancing and producing LIB electrodes delivering higher energy and power densities were also studied.

In recent time, $LiCoO_2$, $LiFePO_4$, and $LiMn_2O_4$ cathode materials are most widely used in constructing LIB electrodes (Nyté n et al. 2005). Cobalt-centered LIBs' electrode material progresses in a pseudo tetrahedral feature, which permits two-dimensional lithium-ion diffusion (Nytén et al. 2005). LIB cobalt-based electrode is widely employed in fabrication of LIBs, because they possess high theoretical specific heat capacity, great volume capability, little self-discharge, high discharge voltage, and good cycle stability (Obodo et al. 2020e). However, LIB manganese-based electrode materials embrace a cubic crystal lattice system, which permits them to adopt three-dimensional lithium-ion diffusion (Nyte´n et al. 2005). Manganese cathodes are attractive because manganese is cheaper due to its abundance and because it could be theoretically used to make a more efficient, longer-lasting battery if its limitations could be overcome (Obodo et al. 2020f). These materials have their limitations; however, many new materials are being researched on towards reducing the cost of LIBs and increasing LIBs life span.

2.2 Battery

Conventionally battery' denotes a device comprising many cells connected in parallel or series. Batteries convert chemical energy stored in their system directly to electrical form of energy, which can be transformed into various forms of energy (Zimmerman 2004). Batteries normally consist of one or more electrochemical cells connected to an external circuit for operating various electrical devices like torchlights, computers, vehicles, mobile phones, machineries, and cars. If a battery is directly linked to external circuit (load), it triggers reduction/oxidation reactions within the battery, which converts the energy of reactants to electrical energy. Batteries are fabricated in a way that actively and satisfactory redox reactions can only occur if electrons migrate to the external circuit or load (Zimmerman 2004). Figure 2.2 denotes symbols of battery in a circuit and electronics systems.

There are majorly two types of batteries namely (a) primary (single use, disposable, or non-rechargeable) battery and (b) secondary (rechargeable) battery. Primary batteries can only be used once and discarded; the redox reactions in them are irreversible. Secondary batteries undergo several discharges and charges; this is achieved numerous times while maintaining the original composition of their electrodes. Figure 2.3 presents categories of batteries and mechanism of operation.

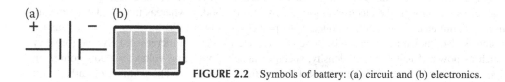

FIGURE 2.2 Symbols of battery: (a) circuit and (b) electronics.

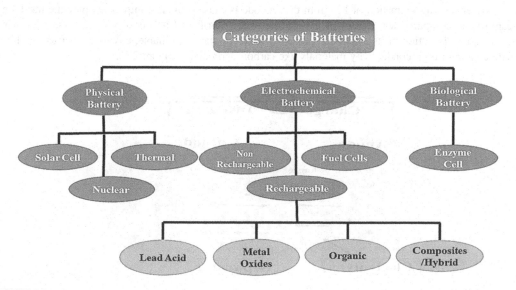

FIGURE 2.3 Categories of batteries.

2.2.1 LIB Components and Mechanisms of Operation

LIBs are electrochemical cells with two electrodes called *cathode* (positive electrode) and *anode* (negative electrode) alienated using a membrane called separator. The LIB configurations are filled with liquefied electrolyte to establish interactions between the cathode and anode. During the course of discharging, cations move from the anode to the cathode; instantaneously, electrons produced in the cathode migrate to the anode, thus moving to the external circuit to power electronics and electrical devices (Goodenough 2013). Typically, anodes of LIBs are made up of graphitic carbon and lithium-based coated metal oxide [e.g., $LiXXO_2$] cathodes, where XX represent transition metal oxides (TMOs), ions like (Co, Mn, or Ni), or composites of transition metal oxides and carbon derivatives like (GO, rGO, CNTs, CNFs, AC, etc.). $LiXXO_2$ configuration performs as the cathode, and an insulating membrane known as *separator*, which is electrolyte-permeable, separates the cathode and anode. Customarily, electrolyte in the LIBs is liquid organic solvent comprising lithium salt as source of lithium ions (Goodenough 2013). Electrodes of LIBs possess various inert current collectors that help in transporting electrons between the two electrodes. Copper current collectors are usually used for LIB anodes while aluminium current collectors are used for LIB cathodes (Chung, Bloking, Yet-Ming 2002).

$$LiXXO_2 + C \underset{\text{Discharge}}{\overset{\text{Charge}}{\rightleftharpoons}} Li_{1-x}XXO_2 + Li_xC \qquad (2.1)$$

where C is graphitic carbon and XX are various combining elements or composites like cobalt, manganese or cobalt/carbon derivatives, etc.

The processes of charge and discharge mechanisms in LIBs are chemical reaction as shown in equation (2.1). As soon as LIB is charged using an outer current source, Li^+ from $LiXXO_2$ cathode migrate to the anode by dispersing in the entire electrolyte and moving across the separator. The migration of Li^+ in the $LiXXO_2$ makes the cathode Li^+-deficient and the anode lithium-sufficient. This procedure is known as *delithiation* because lithium is removed from the $LiXXO_2$ (cathode). However, in discharging stage, Li^+ vacate the anode, move through the electrolyte to the $LiXXO_2$ (cathode), whereas the allied electrons are simultaneously accumulated using current collector to power an outward electrical or electronic device. This progression is also known as *lithiation* because the cathode Li^+ content increased. LIBs' performance parameters such as power density, energy density, specific capacity, operating voltage window commonly depend on the nature and structure of cathode used during their fabrication. The addition of carbon derivatives was incorporated in fabrication of LIBs' cathode in order to increase the storage capacity and power density by enhancing extraction of Li^+ from the cathode because cathode components play the most significant role as compared with other LIB parts. Moreover, majority of LIB's cost is mainly associated with synthesis and fabrication of its electrode; hence the need for cheap, available, environment-friendly, high surface area, and high conductivity materials like carbon derivatives (Figure 2.4).

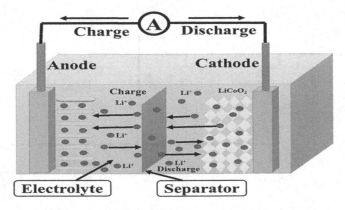

FIGURE 2.4 LIBs charge and discharge mechanisms.

2.3 LIB Electrodes Materials

The appropriate investigation of electrochemical changes or reactions that give optimum or less performance within LIB electrodes during cycling leads to better understanding of appropriate materials or conditions to use in synthesis or fabrication of LIB electrodes. In addressing these challenges, strategies adopted or considered to improve the performances should also consider availability, cost, and toxic nature of materials and methods. Several strategies such as controlled morphology, introduction of artificial defects in the nanostructured materials, nanostructured doping, surface coating and functionalization, etc., have been employed towards enhancing electrode performance. In addition, abundant interests have also be channelled in detecting new cathode materials that can accomplish the desired equilibrium among performance, affordability, availability, and environmental requests.

2.4 Anode Materials

Anode is one of the essential and vital components in LIB; it plays a very essential role in determination of general performance of LIBs. Recently, most LIB anodes are commonly fabricated using graphitic carbon; this graphite material possesses limited theoretical capacity and diffusion path for Li^+ dispersion is very long (Zhang 2011), which results in small energy and power densities of LIBs. However, researchers are looking for ways of enhancing anode material in order to meet the ever-expanding demand for next-generation LIBs. To solve problems associated with small theoretical capacity and long diffusion path length, properties of anode materials, varieties of nanostructured carbon derivatives like GO, rGO, CNTs, CNFs and their composites materials are receiving serious attention (Zhang et al. 2008; Fan et al. 2011; Zhang et al. 2015). Carbon derivatives' morphology can be controlled to produce porous, hollow, needle, rods, etc. – nanostructured nanocomposite materials that provide an enhanced surface-to-volume ratio, electrode/electrolyte interactions, short ionic diffusion length, and increased conductivity (Kim et al. 2006).

2.4.1 Carbonaceous Materials

Carbon derivatives interact with lithium ions (Li^+) at small potential difference, approximately 0.11 V, and hence are suitable for application in LIBs' negative electrode material. However, the quantity of Li^+ reversibly assimilated in the carbon lattice during the faradaic processes exhibit exceptional performance in LIB electrodes (Flandrois and Simon 1999). Certainly, great faradaic processes of carbon derivatives and their composites are intensely prejudiced by several macro- and microstructural properties such as high specific surface area, crystallinity, conductivities, enhanced morphologies, and well-ordered orientation of their crystallites (Markovsky et al. 1999; Endo et al. 2000). Carbon derivatives deliver high-capacity LIB electrodes because their morphologies provide substantial porosity capable of accommodating lithium ions (Zheng and Dahn 1996). Some examples of carbon derivatives include graphene oxide (GO), reduced graphene oxide (rGO), carbon nanotubes (CNTs), activated carbon (AC), carbide-derived carbon (CDC), carbon aerogel (CA), etc. These carbonaceous materials gained a lot of interest in energy storage applications due to their desirable qualities such as low price, availability, large surface area, and non-toxic nature. Carbon derivatives possess high specific surface area, which results in a high capability for charge accumulation at the border of electrode and electrolyte, hence, their effective usage in energy storage devices. Various examples of carbon derivatives are shown in Figure 2.5.

2.4.2 Transition Metal Oxides

Several transition metal oxides such as cobalt (Co), manganese (Mn), nickel (Ni), etc., have received wide attention as anode materials for LIBs because they are abundant in nature, non-toxic, affordable, and possess high specific capacitance (Kim et al. 2013; Deng et al. 2018). Moreover, these transition

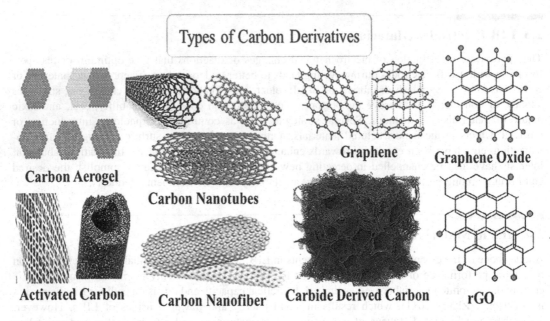

FIGURE 2.5 Typical examples of carbon derivatives.

metal oxides experience phase alteration during charge/discharge cycle, which results in huge volume expansion, resulting in destructive effect on the reliability of the transition metal oxide–based anodes (Zimmerman 2004). The amalgamation of carbon derivatives with transition metal oxides creates porous hybrid nanostructures, which help in solving problems associated with phase transformation witnessed in transition metal oxide anodes (Dsoke et al. 2015; Xu et al. 2015; Yang et al. 2017). These hybrid nanostructured materials conglomerate excellent qualities of nanostructured transition metal oxides with that of carbon derivative–based materials to enhance the performance of recently fabricated LIB anodes (Dong et al. 2017; Xie and Gao 2017). Transition metal oxides provide short ion diffusion path and large interfacial surfaces while carbon derivatives provide fast diffusion route for the electrons (ions). The combination of qualities of transition metal oxides and carbon derivatives enable scholars to synthesize and fabricate anodes with desirable qualities.

2.4.3 Polyanions

Polyanions can be defined as compounds with multiple negative ions. The relative numerous voids or spaces of polyanion compounds provide various lithium ions with numerous sites, thus enhancing quick diffusion and reactions between the electrolyte and anodes. Majority of these polyanion compounds are amphoteric in nature. These compounds can either be reduced by the use of lithium ion attachment or oxidized using lithium ion disappearance. The energy densities offered by polyanion compounds in anode materials of LIBs are very low, prompting scientists to device a means of formation of composite of carbon derivatives and polyanion compounds. This strategy serves as an alternative and effective method to increase polyanions' anode conductivities as well as their energy densities.

2.4.4 Metalloid/Metal Materials

Metals and metalloid materials have low lithiation potential and great theoretical specific capacitance, which make them encouraging materials for LIB anodes (Chan et al. 2008; Shen et al. 2018). Metals and metalloids experience high volume expansion, which causes pulverization of the anode materials. Furthermore, due to metals'/metalloids' intrinsic semiconductive features, they possess low conductivities, which restrict their charge/discharge efficiency especially at high current density

(Wang et al. 2015; Feng et al. 2018). The integration of carbon derivative to fabricate hybrid nanocomposites of metal/metalloid-based compounds reduces volume deviation in various composite anodes; produces firm electrode/electrolyte interfaces; and improves the conductivities, ions diffusion rate, length, and general performance of the anode (Yi et al. 2014).

2.5 Cathode Materials

Electrochemical potential of cathode material describes the positive terminal performance of LIBs. Conventionally, cathode material is the energy controlling or voltage regulating part of LIBs. Advancement of the cathode material for LIBs faces several challenges because present constituent materials like TMOs show several shortcomings. TMO cathode materials like $LiCoO_2$ suffer from instability, which affects their potential window and specific capacity. Furthermore, toxicity level and high fabrication cost of TMO-based cathode materials are still very high and unattractive. Moreover, Li^+-rich TMO composites like $Li_{1+x}M_{1-x}O_2$, given that M stands for mixture of TMOs like Mn, Co, Ni, etc., are encouraging cathode materials for LIBs because they can operate at high discharge voltage range and provide high specific capacitance. Several studies have shown that TMOs suffer from high voltage deterioration for the duration of cycles and extraordinary irreversible capacity loss during first cycle, which limit their use. On the other hand, TMOs' small electronic and ionic conductivities are other impediments towards expansion of their usage in LIB cathode material. Numerous strategies have been taken towards improvement of cathode material for LIBs to overcome contemporary shortcomings. Various approaches include fabrication of materials with organized morphologies, introduction of artificial defects in the material, doping, surface coatings and modifications, as well as functionalization to reduce unwanted reactions at electrolyte/cathode boundaries. A recent method widely adopted is the use of carbon derivatives to modify nanostructured cathode materials for efficient energy and power densities of LIBs.

2.5.1 Spinel Oxides

Spinel oxides are compounds with chemical formula LiX_2O_4 (where X stands for Mn, Co, Ni, etc., or their composites). They were first suggested as LIB cathode material by Michael Thackeray at the beginning of 1980s (Thackeray et al. 1983). Spinel oxides were considered because of their low cost, non-toxicity, and availability. The exceptional XO_2 structure within the framework of LiX_2O_4 delivers three-dimensional (3D) transmission path for Li^+. In LiX_2O_4, X cations continually reside in one-fourth of the octahedral locations of the lithium coating; therefore, one-fourth of octahedral positions remain unoccupied. Li^+ continuously dwell in the tetrahedral positions, sharing faces with vacant octahedral sites in the transition metal surfaces. Three-dimensional diffusion path length associated with LiX_2O_4 offers capabilities of exceptional rate, but undergoes huge capacity loss while cycling at high temperatures. Various reactions that prompt capacity fading include: (1) disbanding of X^{2+} within the electrolyte by the oxidization of hydrogen ions and (2) permanent fundamental conversion from spinel to tetragonal structure because of Jahn-Teller (J-T) energetic X^{3+} ions presence (Gummow et al. 1994; Aurbach et al. 1999). This disbanding of X ions is motivated by disparity of reactions shown in equation (2.2).

$$2X^{3+} \rightarrow X^{4+} + X^{2+} \tag{2.2}$$

where X stands for Mn, Co, Ni, etc., or their composites.

Various approaches have been employed to reduce spinel oxide cations' dissolution in order to lessen capacity fading of LIB cathode materials. Considering these approaches in use recently, doping (composite formation) raised the electrochemical performance of spinel oxide cathodes. Recent studies have shown that the addition of carbon derivatives in formation of composite cathode materials for LIBs helps to subdue the development of J-T active X^{3+} ions, which enhances the electrochemical performance of spinel oxide cathode materials (Myung et al. 2002; Takahashi et al. 2004).

2.5.2 Phosphates

Characteristic phosphate LIB cathode material is designated as LiXPO$_4$, where X represents Co, Mn, Ni, Fe, Cu, etc., or their various composites. The use of phosphate LIB cathode material was earliest reported by Padhi et al. (1997). Since their discovery, they are being studied as a favourable candidate for LIB cathode materials because they are cheap, abundantly available, and non-toxic, and exhibit small volume expansion during cycling, moderate specific capacitance, and reduced capacity fading. The departure of XO$_6$ octahedral by PO$_4$ polyanions considerably reduces the electronic conductivities of LiXPO$_4$ at room temperature (Chung et al. 2002; Whittingham 2004). LiXPO$_4$ low electrical conductivities cause low migration of Li$^+$ because of closely stacked hexagonal oxygen atoms. Techniques adopted in enhancing conductivities of LiXPO$_4$ include (a) reduction in particles/grains sizes; (b) applying conductive surfaces on LiXPO$_4$ grains or particles; and (c) doping LiXPO$_4$ compound with other elements, carbon derivatives, or other organic compounds. The use of reduction in particles or grains sizes has been reported as the best method of enhancing conductivity as well as increasing the performance of cathodes because it reduces the diffusion path for both electrons and ions (Herle et al. 2004). Studies have also shown that nanostructured synthesized LiXPO$_4$ improves cathodes' rate capability of increased surface area, which increases the rate capabilities of LiXPO$_4$ cathodes (Xia et al. 2006). Fabrication of LiXPO$_4$ through traditional solution reactions frequently results in the production of large particles/grain size, which causes low surface as a result of particle growth and agglomeration (Liu and Cao 2010). Moreover, increased surface area assists Li$^+$ reactions with electrolytes, consuming additional lithium ions while creating the solid electrolyte interphase (SEI) (Goodenough and Kim 2010) on the LiXPO$_4$ surface leading to an increased capacity stability. Combination of carbon derivatives with composites of metal oxides nanostructure is the most efficient and widely used material used in increasing the electronic conductivities of LiXPO$_4$ cathodes (Huang et al. 2001; Zaghib et al. 2008). The enhancement of LiXPO$_4$ electronic conductivities by addition of carbon derivatives allows LiXPO$_4$ to provide exceptional rate capability at room temperature (Chung et al. 2002) by partially substituting Li$^+$ with other elements easily and reducing diffusion path (Morgan et al. 2004).

2.5.3 Silicates

Researchers have identified a new class of intercalation composites known as silicates (Li$_2$XSiO$_4$) where X represents Co, Mn, Ni, Fe, Cu, etc., or various composites, proposed by Nyten in 2005 (Nyte´n et al. 2005). Silicates and TMO surfaces possess tetrahedral lattice structure and share common edges, which enable Li ions to diffuse easily through the Li$_2$XSiO$_4$ but in a meandering pathway (Muraliganth et al. 2010).

Li$_2$MSiO$_4$ and Li$_2$FeSiO$_4$ are recently and commonly used silicate compounds employed for LIB cathode materials. Li$_2$MnSiO$_4$ silicates show reduced electrochemical performance and unembellished capacity dwindling complications on the first cycle because of the presence of J-T active Mn^{3+} ions (Xu et al. 2012). A Li$_2$FeSiO$_4$ silicate shows multifaceted structural deficiencies, which make it hard to examine their structure and prompt voltage drop in the course of cycling (Xu et al. 2012).

These various challenges faced by silicate compounds demand for urgent attention towards enhancing their electrochemical behaviour; hence, the usage of carbon derivatives is now adopted to eradicate various problems associated with capacity fading and has shown remarkable results (Xu et al. 2012).

2.5.4 Borates and Tavorites

Borates are new generation cathode materials for LIBs with the chemical formula of LiXBO$_3$ where X represents Fe, Co, Mn, Ni, etc., composites of TMOs, or carbon/TMO composites. Yamada et al. (2010) disclosed that experimental capacity of borate-based cathode materials suffers from moisture adulteration on the surfaces, which depletes specific capacity and brings about fading during cycling.

Tavorite belongs to new-generation cathode materials with a chemical formula of LiXPO$_4$F, where X stands for Fe, Co, Mn, Ni, etc., composites of TMOs, or carbon/TMO composites and F is fluorine. Tavorites compared with other LIB cathode materials have better thermal strength because of the durable bond between phosphorous and oxygen. Lithium ions within tavorite compounds are situated in the TMO octahedral and phosphate tetrahedral sites while fluorine atom delivers 1D ionic path used in lithium diffusion (Gover et al. 2006). The Li^+ cannot be completely extracted from this compound because of the high redox potential of Fe(III) and Fe(IV).

2.6 Conclusion

Recent growth and extraordinary capacities in electrode materials as well as problems recorded in LIBs are receiving serious attention worldwide. Researchers are giving considerable attention towards providing advanced voltage and greater specific capacity electrode material for high-performance LIBs. The struggle will continue until high-capacity intercalation LIBs are obtained that will offer high energy and power densities to effectively cushion the effect of fossil fuel reduction and environmental problems associated with it worldwide, especially for future generations. Next-generation LIB fabrication will focus on decreasing activation energies, cycle stability, and efficiency by employing enhanced reaction paths. Recent progress recorded in LIB development and performances are attributed to advanced carbon-based electrode synthesis. New development in various carbon derivative–electrode composites offered high energy and power densities, performance, and cycle stability. Carbon derivatives display an exciting opportunity into the future for humankind in terms of LIB electrode materials. Several fabrication techniques have shown that solution methods like chemical bath (CBD), hydrothermal, and solvothermal are the cheapest and easiest techniques of fabricating LIB electrodes.

Acknowledgements

RMO and IA humbly acknowledge NCP for their PhD fellowship (NCP-CAAD/PhD-132/EPD) award and COMSATS for a travel grant for the fellowship.

RMO also acknowledge PPSMB Enugu State, Nigeria, for study leave permission granted.

FIE (90407830) affectionately acknowledge UNISA for VRSP Fellowship award and also graciously acknowledge the grant by TETFUND under contract number TETF/DESS/UNN/NSUKKA/STI/VOL.I/B4.33. We thank Engr. Meek Okwuosa for the generous sponsorship of April 2014, July 2016, and July 2018 conferences/workshops on applications of nanotechnology to energy, health, and environment, and for providing some research facilities.

REFERENCES

Aurbach, D., M.D. Levi, Gamulski, K., Markovsky, B., Salitra, G., Levi, E., Heider, U., Heider, L., and R. Oesten. 1999. "Review on Electrode–Electrolyte Solution Interactions, Related to Cathode Materials for Li-ion Batteries". *Journal of Power Sources* 81: 472.

Chan, C.K., H. Peng, G. Liu, K. McIlwrath, X.F. Zhang, R.A. Huggins, and Y. Cui. 2008. High-performance lithium battery anodes using silicon nanowires, *Nature Nanotechnology* 3: 31–35.

Chung, S.Y., J.T. Bloking and C. Yet-Ming. 2002. "Electronically Conductive Phospho-Olivines as Lithium Storage Electrodes". *Nature Materials* 1: 123–128.

Deng, B., T. Lei, W. Zhu, L. Xiao, and J. Liu. 2018. "In-Plane Assembled Orthorhombic Nb2O5 Nanorod Films with High-Rate Li+ Intercalation for High-Performance Flexible Li-Ion Capacitors". *Advanced Functional Materials* 28: 1704330.

Dong, S., H. Li, J. Wang, X. Zhang, and X. Ji. 2017. "Improved Flexible Li-Ion Hybrid Capacitors: Techniques for Superior Stability". *Nano Research* 10: 4448–4456.

Dsoke, S., B. Fuchs, E. Gucciardi, and M. Wohlfahrt-Mehrens. 2015. "The Importance of the Electrode Mass Ratio in a Li-Ion Capacitor based on Activated Carbon and Li4Ti5O12". *Journal of Power Sources* 282: 385–393.

Endo, M., C. Kim, K. Nishimura, T. Fujino, and K. Miyashita. 2000. Recent Development of Carbon Materials for Li-Ion Batteries. *Carbon* 38(2): 183–197.

Fan, Z.J., J. Yan, T. Wei, G.Q. Ning, L.J. Zhi, J.C. Liu, D.X. Cao, G.L. Wang, and Fei Wai. 2011. "Nanographene-Constructed Carbon Nanofibers Grown on Graphene Sheets by Chemical Vapor Deposition: High-Performance Anode Materials for Lithium Ion Batteries". *ACS Nano* 5: 2787.

Feng, K., M. Li, W. Liu, A.G. Kashkooli, X. Xiao, M. Cai, and Z. Chen. 2018. Silicon-Based Anodes for Lithium-Ion Batteries: From Fundamentals to Practical Applications, *Small*: 14: 1702737.

Flandrois, S. and B. Simon. 1999. "Carbon Materials for Lithium-Ion Rechargeable Batteries". *Carbon* 37: 165–180.

Feng, L.L., G.D. Li, Y.P. Liu, P.Y. Wu, H. Chen, Y. Wang, Y.C. Zou, D.J. Wang, and X.X. Zou. 2015. "Mesoporous Nitrogen, Sulfur Co-Doped Carbon Dots/CoS Hybrid as an Efficient Electrocatalyst for Hydrogen Evolution". *ACS Applied Materials & Interfaces* 7: 980–988.

Goodenough, J.B. 2013. "The Li-Ion Rechargeable Battery: A Perspective". *Journal of American Chemical Society* 135: 1167–1176.

Goodenough, J.B. and Y. Kim. 2010. "Challenges for Rechargeable Li Batterie". *Chemistry of Materials* 22: 587–603.

Gover, R.K.B., P. Burns, A.S.M.Y. Bryan, J.L. Swoyer, and J. Barker. 2006. "Li VPO4F: A new active material for safe lithium-ion batteries". *Solid State Ionics* 177: 2635–2638.

Gummow, R.J., A. Dekock, and M.M. Thackeray. 1994. "Improved capacity retention in rechargeable 4 V lithium/lithium manganese oxide (spinel) cells". *Solid State Ionics* 69: 59–67.

Herle, P.S., B. Ellis, N. Coombs, and L.F. Nazar. 2004. "Nano-Network Electronic Conduction in Iron a Phosphates". *Nature Materials* 3: 147–152.

Huang, H., S.C. Yin, and L.F. Nazar. 2001. "Approaching Theoretical Capacity of LiFePO4 at Room Temperature at High Rates". *Electrochemical and Solid-State Letters* 4: A170–A172.

Kim, C., K.S. Yang, M. Kojima, K. Yoshida, Y.J. Kim, Y.A. Kim, and M. Endo. 2006. "Fabrication of Electrospinning-Derived Carbon Nanofiber Webs for the Anode Material of Lithium Ion Secondary Batteries". *Advanced Functional Materials* 16: 2393.

Kim, H., M.Y. Cho, M.H. Kim, K.Y. Park, H. Gwon, Y. Lee, K.C. Roh, and K. Kang. 2013. A novel high-energy hybrid supercapacitor with an anatase TiO_2-reduced graphene oxide anode and an activated carbon cathode, *Advanced Energy Materials* 3: 1500–1506.

Liu, D. and G. Cao. 2010. "Engineering Nanostructured Electrodes and Fabrication of Film Electrodes for Efficient Lithium Ion Intercalation". *Energy & Environmental Science* 3: 1218–1237.

Aurbach, D., B. Markovsky, and Y. Ein-Eli. 1999. "On the correlation between surface chemistry and performance of graphite negative electrodes for Li-ion batteries". *Electrochimica Acta* 45: 67.

Morgan, D., A. Van der Ven and G. Ceder. 2004. "Li conductivity in LixMPO4 M = Mn, Fe, Co, Ni. Olivine Materials". *Electrochimica Solid State Letter* 7: A30–A32.

Myung, S.T., S. Komaba, N. Kumagai, H. Yashiro, H.T. Chung, and T.H. Cho. 2002. Nano-crystalline $LiNi_{0.5}Mn_{1.5}O_4$ synthesized by emulsion drying method, *Electrochimica Acta* 47: 2543–2549.

Muraliganth, T., K.R. Stroukoff and A. Manthiram. 2010. "Microwave-Solvothermal Synthesis of Nanostructured Li_2MSiO_4/C (M = Mn and Fe) Cathodes for Lithium-Ion Batteries", *Chemistry of Materials* 22: 5754–5761.

Nytén, A., A. Abouimrane, M. Armand, T. Gustafsson, and J. Thomas. 2005. "Electrochemical performance of Li2FeSiO4 as a new Li-battery cathode material." *Electrochemistry Communications* 7: 156–160.

Obodo, R.M., A. Ahmad, G.H. Jain, I. Ahmad, M. Maaza, and F.I. Ezema. 2020a. "8.0 MeV Copper Ion (Cu++) Irradiation-Induced Effects on Structural, Electrical, Optical and Electrochemical Properties of Co3O4-NiO-ZnO/GO Nanowires". *Materials Science for Energy Technologies* 3: 193–200.

Obodo, R.M., A.C. Nwanya, M. Arshad, C. Iroegbu, I. Ahmad, R. Osuji, M. Maaza and F.I. Ezema. 2020b. "Conjugated NiO-ZnO/GO Nanocomposite Powder for Applications in Supercapacitor Electrodes Material". *International Journal of Energy Research* 44: 3192–3202.

Obodo, R.M., A.C. Nwanya, A.B.C. Ekwealor, I. Ahmad, T. Zhao, M. Maaza, and F. Ezema. 2019a. "Influence of pH and Annealing on the Optical and Electrochemical Properties of Cobalt (III) Oxide (Co_3O_4) Thin Films". *Surfaces and Interfaces* 16: 114–119.

Obodo, R.M., A.C. Nwanya, Tabassum Hassina, Mesfin Kebede, Ishaq Ahmad, M. Maaza, and Fabian I. Ezema. 2019b. Transition Metal Oxide-Based Nanomaterials for High Energy and Power Density

Supercapacitor. In *Electrochemical Devices for Energy Storage Applications*, edited by F.I. Ezema and M. Kebede. United Kingdom: Taylor & Francis Group, CRC Press, 131–150.

Obodo, R.M., A.C. Nwanya, C. Iroegbu, B.A. Ezekoye, A.B.C. Ekwealor, I. Ahmad, M. Maaza, and F.I. Ezema. 2020c. "Effects of Swift Copper (Cu^{2+}) Ion Irradiation on Structural, Optical and Electrochemical Properties of Co_3O_4-CuO-MnO2/GO Nanocomposites Powder". *Advanced Powder Technology* 31 (4): 1728–1735.

Obodo, R.M., E.O. Onah, H.E. Nsude, A. Agbogu, A.C. Nwanya, I. Ahmad, T. Zhao, P.M. Ejikeme, M. Maaza, and F.I. Ezema. 2020d. "Performance Evaluation of Graphene Oxide Based Co_3O_4@GO, MnO_2@GO and Co_3O_4/MnO_2@GO Electrodes for Supercapacitors". *Electroanalysis* 32: 1–10.

Obodo, R.M., N.M. Shinde, U.K. Chime, S. Ezugwu, A.C. Nwanya, I. Ahmad, M. Maaza, and F.I. Ezema. 2020e. "Recent Advances in Metal Oxide/Hydroxide on Three-Dimensional Nickel Foam Substrate for High Performance Pseudocapacitive Electrodes". *Current Opinion in Electrochemistry* 21: 242–249.

Padhi, A.K., K.S. Nanjundaswamy. and J.B. Goodenough. 1997. "Phospho-Olivines as Positive-Electrode Materials for Rechargeable Lithium Batteries". *Journal of the Electrochemical Society* 144: 1188–1194.

Shen, T., Z. Yao, X. Xia, X. Wang, C. Gu, and J. Tu. 2018. Rationally Designed Silicon Nanostructures as Anode Material for Lithium-Ion Batteries, *Advanced Engineering Materials* 20: 1700591.

Takahashi, K., M. Saitoh, M. Sano, M. Fujita, and K. Kifune. 2004. "Electrochemical and Structural Properties of a 4.7 V-Class LiNi0.5Mn1.5 O 4 Positive Electrode Material Prepared with a Self-Reaction Method". *Journal of the Electrochemical Society* 151: A173.

Thackeray, M.M., W.I.F. David, P.G. Bruce, and J.B. Goodenough. 1983. Lithium insertion into manganese spinels, *Materials Research Bulletin* 18: 461–472.

Wang, B., X. Li, B. Luo, L. Hao, M. Zhou, X. Zhang, Z. Fan, and L. Zhi. 2015. "Approaching the downsizing limit of silico n for surface-controlled lithium storage". *Advanced Materials* 27: 1526–1532.

Whittingham, M.S. 2004. "Lithium Batteries and Cathode Materials". *Chemical Reviews*, vol. 104, pp. 4271–4302.

Xia, Y.G., M. Yoshio, and H. Noguchi. 2006. "Improved Electrochemical Performance of LiFePO4 by Increasing Its Specific Surface Area". *Electrochimica Acta* 52: 240–245.

Xie, Y. and R. Gao. 2017. "Electrochemical capacitance of titanium nitride modified lithium titanate nanotube array". *Journal of Alloys and Compounds* 725: 1–13.

Xu, B., D. Qian, Z. Wang, and Y.S. Meng. 2012. "Recent Progress in Cathode Materials Research for Advanced Lithium Ion Batteries". *Materials Science and Engineering R* 73: 51.

Xu, N., X. Sun, X. Zhang, K. Wang, and Y. Ma. 2015. "A Two-step Method For Preparing Li 4 Ti 5 O 12–Graphene as an Anode Material For Lithium-ion Hybrid Capacitors". *RSC Adv.* 5: 94361.

Yamada, A., N. Iwane, Y. Harada, S.I. Nishimura, Y. Koyama, and I. Tanaka. 2010. "Lithium Iron Borates as High-Capacity Battery Electrodes". *Advanced Materials* 22: 3583.

Yang, C., J.L. Lan, W.X. Liu, Y. Liu, Y.H. Yu, and X.P. Yang. 2017. High-Performance Li-Ion Capacitor Based on an Activated Carbon Cathode and Well-Dispersed Ultrafine TiO, *ACS Advanced Materials Interfaces* 9: 18710–18719.

Yi, R., S. Chen, J. Song, M.L. Gordin, A. Manivannan, and D. Wang. 2014. High-Performance Hybrid Supercapacitor Enabled by a High-Rate Si-based Anode, *Advanced Function Materials* 24: 7433–7439.

Zaghib, K., A. Mauger, F. Gendron, and C.M. Julien. 2008. "Surface Effects on the Physical and Electrochemical Properties of Thin LiFePO4 Particles". *Chemistry of Materials* 20: 462–469.

Zhang, J., Y.S. Hu, J.P. Tessonnier, G. Weinberg, J. Maier, R. Schlgl, and D.S. Su. 2008. "CNFs@CNTs: superior carbon for electrochemical energy storage". *Advanced Materials* 20: 1450.

Zhang, X., S. Han, C. Fan, L. Li and W. Zhang. 2015. "Hard Carbon Enveloped with Graphene Networks as Lithium Ion Battery Anode". *Materials Letters* 138: 259.

Zhang, W.J. 2011. "A Review of the Electrochemical Performance of Alloy Anodes for Lithium Ion Batteries". Journal of Power Sources 196: 13.

Zheng, Sue and. R. Dahn. 1996. "Carbons Prepared From Coals for Anodes of Lithium-ion Cells. *Carbon* 34: 1501–1507.

Zimmerman, A.H. 2004. "Self-Discharge Losses in Lithium-Ion Cells". *IEEE Aerospace and Electronic Systems Magazine* 19 (2): 19–24.

3

Current Status and Trends in Spinel Cathode Materials for Lithium-Ion Battery

Mesfin A. Kebede[1,3], Nithyadharseni Palaniyandy[1], Lehlohonolo F. Koao[2], Fabian I. Ezema[4], and Motlalepula R. Mhlongo[3]
[1]*Energy Centre, Council for Scientific & Industrial Research, Pretoria, South Africa*
[2]*Department of Physics, University of the Free State (Qwa Qwa campus), Phuthaditjhaba, South Africa*
[3]*Department of Physics, Sefako Makgatho Health Science University Medunsa, South Africa*
[4]*Department of Physics and Astronomy, University of Nigeria, Nsukka, Nigeria*

3.1 Introduction

Lithium-ion batteries (LIBs) are the preferred and important components in our society as power sources for diverse portable electronic devices and electric vehicles (Nitta et al. 2015; Zubi et al. 2018). LIBs are being widely customized and intensively used to run devices such as tablets, smartphones, cameras, etc., as energy storage. LIBs are well-developed, matured, and explored technologies and currently the most suitable energy storage system for powering electric vehicles (EVs), hybrid vehicles, military, and aerospace owing to their attractive properties including high energy efficiency, lack of memory effect, long cycle life, high energy density, and high power density (Ding et al. 2019).

The spinel-structured $LiMn_2O_4$ and $LiMn_{1.5}Ni_{0.5}O_4$ cathodes are attracting research interest among various cathode materials for LIBs because they being cheap, safe, and rich in resources (Kebede et al. 2015; Zhu et al. 2016; Wang et al. 2018). Particularly, $LiMn_2O_4$ cathodes has been considered as a very popular cathode material for commercialized LIBs and have been applied in mobile phones and Nissan Leaf pure electric vehicles as LIB energy source (Ding et al. 2019).

The main challenge for the application of $LiMn_2O_4$ facing is the capacity loss due to manganese dissolution and Jahn-Teller structural distortion in the 3 V region as the battery charge-discharge cycles repeatedly. LMNO is a high voltage and good rate performance cathode material with good cycling stability. However LMNO faces the challenge, that is, the high charging voltage (~4.7 V) exceeds the voltage window that is higher than the stable voltage of conventional carbonate-based $LiPF_6$ electrolytes, resulting in rapid oxidation decomposition of the electrolyte and unnecessary secondary reactions at the LMNO/electrolyte interface. The decomposition of electrolyte causes the formation of an insulating solid electrolyte interphase (SEI) layer, thereby impeding Li+ diffusion during the charge/discharge process, leading to serious capacity fading. Furthermore, dissolution of Mn is another serious issue of LMNO, which causes destroyed material structure and reduced cycle life of LMNO. Therefore, different strategies have been implemented/developed to solve the capacity fading through including morphology control, element doping, and surface modification, which will be discussed. This chapter presents the recent research achievements regarding the two spinel LMO and LMNO cathode materials. It also discusses the future research trends being developed to optimize the capacity and cyclability of these cathode materials.

DOI: 10.1201/9781003145585-3

Electrode Materials for Energy Storage

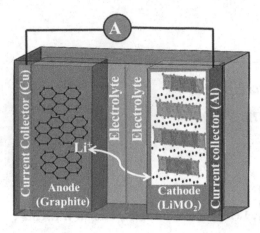

FIGURE 3.1 Schematic depiction of a lithium-ion battery (Kebede, Palaniyandy, and Koao 2019). Reprinted with permission copyright 2019, Taylor and Francis.

LIB consists of a negative (anode) electrode, a positive (cathode) electrode, and a separator, which is soaked in electrolyte, to ensure the charge transfer within the battery. The separator serves as an electrical insulator, prevents a short circuit between both electrodes, and at the same time allows rapid transport of ionic charge carriers that are needed to complete the circuit during the passage of current within the cell. The electrode materials are coated on current collectors, whereas copper is used for the negative electrode and aluminium for the positive electrode. The functional principle goes as, during discharge, Li ions move from the negative (anode) electrode to the positive (cathode) electrode, and reversely when charging. The schematic depiction of LIB is shown in Figure 3.1.

In the electrochemical redox reaction for lithium-ion battery, the oxidation occurs at the cathode and reduction occurs at the anode while the charging takes place. In the case of the discharge process, the redox reactions are inverted, the oxidation occurs at the anode and reduction occurs at the cathode. The electrochemical roles of the electrodes change between anode and cathode.

3.2 Spinel $LiMn_2O_4$ and $LiMn_{1.5}Ni_{0.5}O_4$ Cathode Materials

The commonly recognized spinel cathode materials $LiMn_2O_4$ and $LiMn_{1.5}Ni_{0.5}O_4$ have high operating voltage 4.1V, 4.7V (Li/Li+) and deliver reasonable experimental specific capacity of 120 mAh g^{-1} and 130 mAh g^{-1}, respectively. These manganese-based spinel cathode materials have attracted great research attention since they are cheap, safe, environmentally friendly, and rich in resources as compared to layered LCO by removing cobalt (Co) which is economically costly and environmentally toxic.

3.2.1 Spinel $LiMn_2O_4$ (LMO)

The spinel LMO is one of the commercially available cathode materials for LIB besides $LiCoO_2$ (LCO) cathode, which was first commercialized as cathode for LIB by SONY in 1990 and Olivine $LiFePO_4$ (LFP) (Logan et al. 2020). At CSIR, South Africa, Michael Thackeray first demonstrated the insertion of lithium into magnetite (Fe_3O_4) crystallizing in the spinel structure and using the same principle he further developed LMO cathode by inserting Li into spinel Mn_3O_4 framework (Thackeray, David, and Goodenough 1982; Manthiram 2020).

The conventional unit cell of the spinel $LiMn_2O_4$ has 8 Li, 16 Mn, and 32 O atoms and in a total of 56 atoms as shown in Figure 3.2. The Mn ions that are found in two oxidation states – Mn^{3+}/Mn^{4+} – occupy the 16d octahedral sites, the Li^+ occupy the 8a tetrahedral sites of the spinel framework with a cubic close-packed array of oxide ions. In addition, the O atoms sit at the 32e sites to host Li^+ during the charging/discharging (Kebede et al. 2014; Kebede et al. 2015). The stable $[Mn_2]_{16d}O_4$ framework with edge-shared octahedra offers a three-dimensional lithium-ion diffusion pathway with fast lithium-ion

(a)

(b)

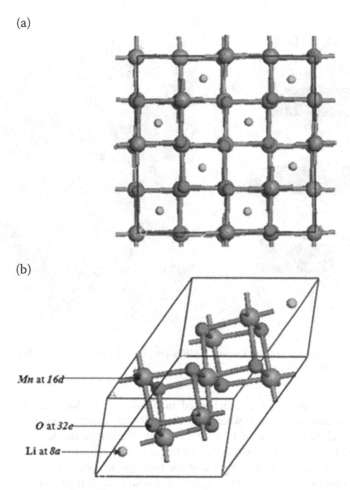

Mn at *16d*

O at *32e*

Li at *8a*

FIGURE 3.2 The crystal structure of LiMn$_2$O$_4$ (see Ref. Kebede et al. 2014). Reprinted with permission copyright 2014, Elsevier.

conductivity. The insertion/extraction of lithium into/from the tetrahedral sites with a deep site energy in Li$_{1-x}$Mn$_2$O$_4$ offers a high operating voltage of 4 V with a practical capacity of <130 mAh/g, as close to one lithium per two Mn ions can be reversibly extracted from the tetrahedral sites.

The synthesis approach followed to synthesize cathode materials contributes significant effect on the electrochemical performance of the material; recently, several researchers have enhanced the electrochemical performance of materials by improving the synthetic methods (Yin et al. 2018). The spinel LMO and LMNO cathode have been synthesized by various synthesis techniques such as solid-state reaction (Yin et al. 2018), solution combustion (Kebede et al. 2015), thermo-polymerization (Kebede et al. 2017), aqueous reduction (Kunjuzwa et al. 2016), molten salt (Kebede and Ozoemena 2017), etc. To get well-crystalized LMO cathode it normally requires heat treatment of above 700°C. Among the synthesis methods, co-precipitation is commonly considered a suitable/preferred procedure for synthesizing materials from mixed metal oxides and solid solutions. Cui et al. (2017) successfully synthesized electrochemically enhanced LMNO cathode using co-precipitation method and reported impressive high specific capacity of 123.8 mAh g^{-1} at a rate of 0.25 C.

Concerning the Li-ion diffusion kinetics, the Mn-based spinel (Fd3m); specifically LiMn$_2$O$_4$, has a three-dimensional diffusion channel for Li-ions formed by a 3D network of MnO6 octahedra (Lee et al. 2020). It is noteworthy to indicate the existence of one-dimensional olivine LFP and two-dimensional

(a)   (b)

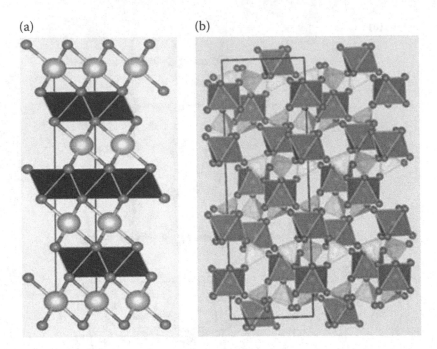

FIGURE 3.3 (a) The 2D layered (b) 1D Olivine Li diffusion (from Manthiram 2020). Reprinted with permission copyright 2020, Springer Nature.

layered LCO for Li ions diffusion as represented in Figure 3.3a,b (Manthiram 2020). In LCO, the Li-ions travel in two dimension ways, whereas in the LFP, they travel in one direction.

The electrochemical performance of LMO is seriously challenged in various aspects; its capacity loss is severe during repetitive charge/discharge cycling, especially at elevated temperature above 60°C. The Mn dissolution into electrolytes is the cause for the capacity degradation and poor cycle performance of cathode materials, which severely restrict the commercial application of $LiMn_2O_4$. Various mechanisms have been proposed and revealed in regards to capacity fading: the John Teller effect, manganese dissolution, and disproportion reaction. Different strategies are designed to tackle the capacity fading challenge; among them are cation substitution (Kebede et al. 2014; Kebede et al. 2015; Kunjuzwa et al. 2016), nanosizing of the cathodes (Kebede and Ozoemena 2017), microwave irradiation treatment (Kebede et al. 2017), etc.

3.2.1.1 Substitution of Mn-Ion by Transition Metal Ions

To improve the electrochemical properties and tackle the shortcomings of the LMO cathode material, other elements are introduced into the Mn-based spinel framework. Some of the numerous doping elements employed in the spinel $LiM_xMn_{2-x}O_4$ (M = Al, Mg, Co, Ni, Cr, Cu, Zn, etc.) to achieve moderate capacity (100–110 mAh/g)-type $Li_{1+x}Mn_{2-x}M_yO_4$ with better cycling performance and rate capability can be used as a power-type LIB for automotive and other electric tools (Mao, Dai, and Zhai 2012). However, the Ni-substituted spinel material denoted as $LiNi_{0.5}Mn_{1.5}O_4$ has been considered as the most promising cathode in high voltage spinel materials (Lee et al. 2020).

3.2.1.2 The Control of Morphology

Various commonly used strategies to improve the electrochemical performance of spinel $LiMn_2O_4$ cathode materials do include surface coating, bulk doping, and morphology control (downsizing, porous structure, etc.). Among these improvement strategies, downsizing of $LiMn_2O_4$ particles can significantly shorten the transport distance of lithium ions in solid, thus helping to improve their rate performance. In recent years, various morphologies of nanostructured $LiMn_2O_4$ have been extensively prepared such as a nanochain of $LiMn_2O_4$ (Tang et al. 2012) by a sol gel method with a very good rate capability, porous $LiMn_2O_4$ micro-/nanohollow spheres (Liu, Zhou, and Song 2018), nanorods $LiMn_2O_4$ (Kebede and Ozoemena 2017) by molten salt evaporation techniques (Hashem et al. 2019).

The downsizing of LMO to nanostructured is not always the preferred and perfect structure, since it usually makes the tap density generally low due to its irregular shape, high surface area, and high porosity, which results in low volumetric energy density of the LIB electrodes (Guo et al. 2014). Among the morphological structures, the preferred and currently focused structure is a porous microsphere. Particularly, nanocrystallites tightly compacted with three-dimensional channels are attractive which is convenient for ion diffusion in consideration of electron transport distance as reported by Hai et al. The schematic of synthesis technique and the X-ray diffraction (XRD) and scanning electron microscopy (SEM) of the samples is shown in Figure 3.4a–c (Y. Hai et al. 2019).

The electrochemical properties characterization data, cyclic voltammetry, the voltage versus capacity profile, and the galvanostatic charge/discharge capacity for repetitive cycles of the as-synthesized porous microspheres are shown in Figure 3.5.

Figure 3.5b reveals the current charge/discharge measurement by different rates over a voltage range of 3.0–4.5 V. The discharge capacities at rates of 0.1 C, 0.2 C, 0.5 C, 1.0 C, and 2.0 C were 103.18 mAh g^{-1}, 102.33 mAh g^{-1}, 101.50 mAh g^{-1}, 100.51 mAh g^{-1}, and 94.23 mAh g^{-1}, respectively. The rate performance is shown in Figure 3.5c. As can be seen, the rate performance of $LiMn_2O_4$ synthesized at 650 °C is quite good, which demonstrates clearly slower capacity decay with increasing discharge rates. For example, the porous $LiMn_2O_4$ microspheres retain a capacity of 94.23 mAh g^{-1}, which is 91.3% of the initial capacity at rate of 0.1 C. This is much higher than that (76%) of the commercial $LiMn_2O_4$ powders at the same rate. When the current went back to a rate of 0.1 C, a capacity of 102.27 mAh g^{-1} was resumed. The cycle stability of $LiMn_2O_4$ synthesized at 650°C at 1.0 C is shown in Figure 3.5d. The capacity of synthesized $LiMn_2O_4$ remains at 96.42 mAh g^{-1} after 100 cycles and drops by only 3.24% compared to that of the first cycle. For comparison, the commercial $LiMn_2O_4$ exhibits a discharge capacity of 85.15 mAh g^{-1} after 100 cycles, which is much lower than that of the synthesized porous $LiMn_2O_4$ microspheres sample. The good rate and cycling performance of the samples prepared are ascribed to a well-defined structure such as uniform size and high porosity, which is effective in increasing contact area, shortening the transport distance of lithium ions, and enhancing the structural stability of electrode material.

3.2.2 $LiMn_{1.5}Ni_{0.5}O_4$ (LMNO)

$LiMn_{1.5}Ni_{0.5}O_4$ (LMNO) is considered as a potential cathode material for a new generation of LIBs, to meet the continuous development of electric vehicles (EVs) and plug-in hybrid-electric vehicles (PHEVs), due to its ability to offer outstanding energy density and high operating voltage (4.7 V vs. Li/Li+) and theoretical capacity 147 mAh g^{-1} (Yin et al. 2018; Mao et al. 2020). The gravimetric energy density of LMNO is as high as ~700 Wh kg^{-1}. It is well known that synthesis of $LiMn_{1.5}Ni_{0.5}O_4$ is frequently hampered by the formation of $Li_xNi_{1-x}O$ impurities, which lowers capacity due to parasitic loss of active material.

The high voltage LMNO cathode thermodynamically favourably exists in two crystalline phases namely the ordered and disordered based on the heat treatment process. The ordered polymorph, which is cubic has the symmetry of $P4_332$, Mn (12b) and Ni (4a) ions are located in two separate octahedral sites, normally occurs at lower temperature (less than 800°C). Whereas the disordered polymorph is face-centered cubic with an $Fd\overline{3}m$ symmetry (favoured at high temperature, greater than 800°C), in which Mn and Ni ions are randomly distributed in the 16d octahedral sites. The ordered $P4_332$ structure is stoichiometric with a Mn valence of +4 in which Mn^{4+} is inactive electrochemically. Thus, the capacity is entirely due to the $Ni^{2+/4+}$ redox reaction. The $Fd\overline{3}m$ structure is nonstoichiometrically expressed as $LiNi_{0.5}Mn_{1.5}O_{4-\delta}$ (where δ represents the oxygen vacancy) and contains trace amounts of

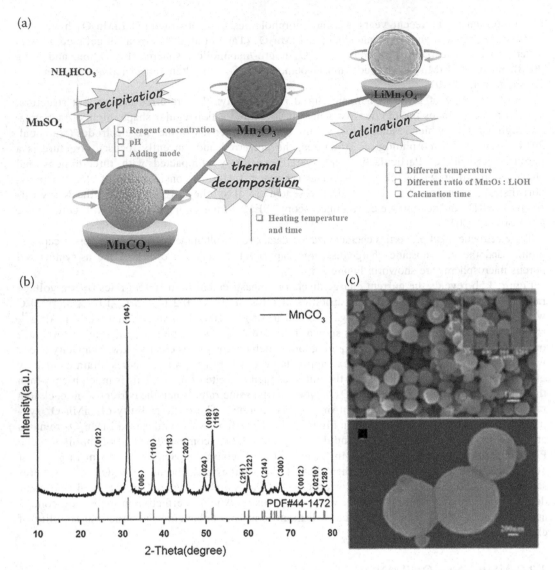

FIGURE 3.4 (a) Schematic of the preparation of the spinel $LiMn_2O_4$ porous microspheres, (b) & (c) the XRD and SEM data of the samples (Y. Hai et al. 2019). Reprinted with permission copyright 2019, Europe PMC.

Mn^{3+} in which Mn^{3+} contributes towards the capacity as it is active electrochemically (Kebede et al. 2017). The Mn^{3+} in disordered $Fd\overline{3}m$ structure is generated when oxygen escapes from the LMNO lattice at a high temperature (>800°C) and is not replaced by inhalation during rapid cooling. The disordered $Fd\overline{3}m$ structure is known to have a higher lithium-ion diffusion coefficient, greater conductivity, and stronger electrochemical performance in comparison to those of $P4_332$ (Wang et al. 2011), thus it is the favoured structure for LIBs.

In terms of phase and electrochemical properties, the ordered LMNO exhibited excellent rate capability, whereas detrimental effect on capacity fading and less specific capacity. For ordered high-voltage spinel $LiMn_{1.5}Ni_{0.5}O_4$ (LMNO) with the $P4_332$ symmetry, the two consecutive two-phase transformations at ~4.7 V (vs. Li+/Li), involving three cubic phases of LMNO, $Li_{0.5}Mn_{1.5}Ni_{0.5}O_4$ ($L_{0.5}MNO$), and $Mn_{1.5}Ni_{0.5}O_4$ (MNO), have been well established. Such a mechanism is traditionally associated with poor kinetics due to the slow movement of the phase boundaries and the large mechanical strain resulting from the volume changes among the phases; yet ordered LMNO has been shown to have excellent rate capability (Kan et al. 2017).

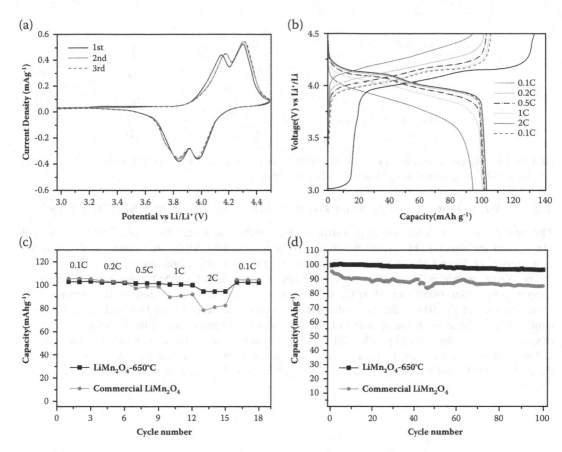

FIGURE 3.5 Electrochemical performances of the synthetic porous LiMn$_2$O$_4$ spheres. (a) CV curves at a scan rate of 0.1 mV s^{-1} in the voltage range of 3.0–4.5 V (vs. Li/Li+); (b) discharge-charge curves at different rate from 0.1 to 2.0C (c) variation of discharge capacity vs. cycle number of synthetic porous LiMn$_2$O$_4$ electrodes at 0.1–2.0C rate; (d) cycle performance of synthetic porous LiMn$_2$O$_4$ at 1.0C rate (Hai et al. 2019). Reprinted with permission copyright 2019, Europe PMC.

There is a risk of impurity formation as the temperature gets higher (900 C). According to work by Samarasingha et al. (2016), phase analysis by X-ray diffraction shows that extended heat treatment at 900°C causes precipitation of Li$_x$Ni$_{1-x}$O with consequent reduction in specific energy capacity.

The two phases – the ordered and disordered – within LMNO cannot be identified by X-ray diffractometer (XRD), as they provide similar XRD reflection, whereas it is possible to identify using neutron diffraction (Samarasingha et al. 2016), since it is a powerful analytical system. The other way to confirm is the use of Fourier Transfer Infrared (FTIR). In the disordered the Ni and Mn ions are randomly positioned as shown in Figure 3.6a, while in the ordered the Ni ions have their specific sites to sit as shown in Figure 3.6b. In the disordered Fd3m (#227) LMNO the Mn ion and Ni ion will be randomly distributed in the 16d octahedral sites, the Li-ion occupy 8a and O will occupy 32e sites. In the ordered P4332 (#212) LMNO, Mn is assigned to 12b and Ni to 4a octahedral sites (Kebede et al. 2017; Kebede et al. 2019 2019).

P.B. Samarasingha et al. (2016) reported a route to enhance performance of pure LiMn$_{1.5}$Ni$_{0.5}$O$_4$ spinel materials for LIB cathodes by the addition of propylene carbonate. Propylene carbonate significantly improves cathode performance when followed by a leaching step that yields a porous cathode tape. The ordered material calcined at 900°C and thereafter at 700°C for 10 h shows a specific capacity of ~137 mAh g^{-1} at 0.04 C and ~130 mAh g^{-1} at 2 C discharge rates – the highest capacities so far reported for submicron particles of the ordered phase.

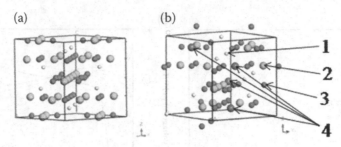

FIGURE 3.6   The crystal structures for (a) disordered Fd-3m (#227) (b) ordered P4₃43 (#212), where (1) -Li, (2) - Mn, (3) - O, and (4) - Ni generated using Materials Studio 2019 software.

3.2.2.1 X-Ray Powder and Neutron Powder Diffraction for LiMn$_{1.5}$Ni$_{0.5}$O$_4$ Cathodes

The neutron powder diffraction is powerful analysis technique to get the arrangements of Ni and Mn ions in the structure of LMNO which cannot be achieved by XRD. The ordered and disordered arrangement of Ni and Mn ions can be identified using neutron diffraction but not by XRD.

The XRD pattern in Figure 3.7a shows the crystal structure of LMNO synthesized at different temperatures from 600°C to 1000°C; this work was reported by Pushpaka B. Samarasingha (Samarasingha et al. 2016). There is no difference between ordered and disordered; all the patterns are similar which confirms the incapability of XRD technique to identify the ordered and disordered arrangements. Thus, the neutron powder diffraction needs to be used for the identification of such nature.

The crystal plan reflection in Figure 3.7b is the neutron diffraction data for LMNO for low temperature – 600°C and high temperature – 900°C. There is a clear difference in the reflection pattern,

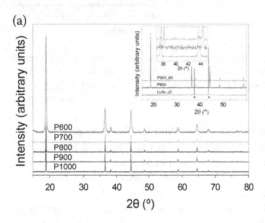

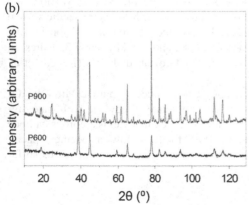

FIGURE 3.7   (a) Powder X-ray diffractograms of LiMn$_{1.5}$Ni$_{0.5}$O$_4$ heat treated at various temperatures. Inset represents the powder X-ray diffractograms for clarifying phase purity of LiMn$_{1.5}$Ni$_{0.5}$O$_4$ samples and emphasizes the presence of Li$_x$Ni$_{1-x}$O in pattern P900_60; (b) Powder neutron diffraction (PND) profiles for the P600 and P900 samples of LiMn$_{1.5}$Ni$_{0.5}$O$_4$ (Samarasingha et al. 2016). Reprinted with permission copyright 2016, Elsevier.

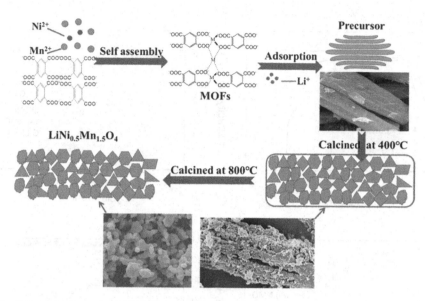

FIGURE 3.8 Schematic illustration of the preparation of M-LMNO (Yin et al. 2018). Reprinted with permission copyright 2018, ACS.

with additional peaks as evidence for the ordered LMNO. This is evidence that PND is the powerful technique for phase identification.

The PXRD (Cu Kα1) patterns are almost identical for ordered and disordered $LiMn_{1.5}Ni_{0.5}O_4$. The PND patterns for P600 and P900 are strikingly different, Figure 3.7b. The simple pattern for P600 corresponds to a disordered Fd-3m spinel.

Yin et al. (2018) in their recent work synthesized $LiNi_{0.5}Mn_{1.5}O_4$ cathode material with high surface orientation via a complexing reaction coupled with the elevated-temperature solid-state method; the as-synthesized $LiNi_{0.5}Mn_{1.5}O_4$ cathode material exhibited overwhelming high capacity and high stability. The schematic for the synthesis protocol of M-LMNO is presented in Figure 3.8. The as-synthesized M-LMNO was highly crystalline with low impurity, uniform grain size, and a preferred orientation in the (111) and (110) planes (Kebede et al. 2017). The (111) facets possess the lowest surface energy, and increasing the number of 3D Li+ transmission channels can improve the rate capacity and stability. However, the (100) surface facets also accelerate the Mn dissolution, causing specific-capacity attenuation, oxidation decomposition of the electrolyte, and structural collapse (Hai et al. 2013; Leung 2013; Huang et al. 2017).

Owing to these advantages, the M-LMNO cathode material exhibited overwhelmingly high cyclic stability and rate capability, and M-LMNO delivered a capacity of 145 mAh g^{-1} at a discharge rate of 0.1 C and a discharge capacity retention of 86.6% at 5 C after 1000 cycles. Even at an extremely high discharge rate (10 C), the specific capacity was 112.7 mAh g^{-1}, and 78.7% of its initial capacity was retained over 500 cycles. The superior electrochemical performance, particularly during a low-rate operation, was conferred by improved crystallinity and the crystal orientation of the particles.

The initial galvanostatic charge/discharge profiles of the samples M-LMNO, C-LMNO, and S-LMNO cathode materials is presented in Figure 3.9a. The initial discharge capacities of the three samples are 145 mAh g^{-1}, 106 mAh g^{-1}, and 119 mAh g^{-1}, respectively. The large specific surface area of M-LMNO provides a highly electrochemically active surface that is in contact with the electrolyte.

The cyclic performances of the obtained LMNO materials are shown in Figure 3.9c. After 500 charge/discharge cycles at a discharge rate of 5 C, the M-LMNO, C-LMNO, and SLMNO samples delivered specific discharge capacities of 136 mAh g^{-1}, 83 mAh g^{-1}, and 76 mAh g^{-1}, respectively, and retained 92.8%, 64.7%, and 31.6% of their largest capacities, respectively. After 500 cycles at 10 C, the specific-capacity retention rate of M-LMNO was 78.7%. The reasons for this excellent cyclic performance are

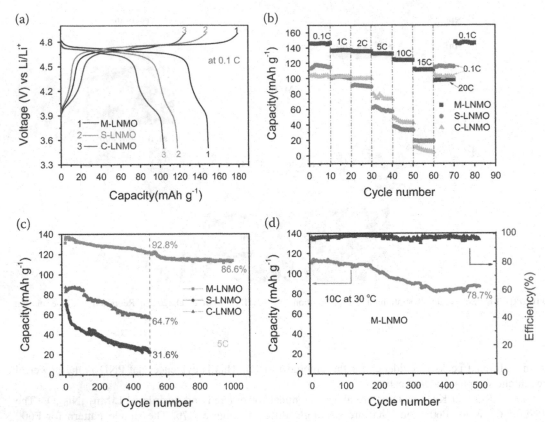

FIGURE 3.9 (a) Initial charge/discharge profiles measured at 0.1 C in the voltage range of 3.5–4.9 V; (b) rate capability investigation with a constant charge at 1 C followed by discharging at various C-rates ranging from 0.1 C to 20 C; (c) cycling performances of LMNO-MF and LMNO-NP at a 5 C discharge rate at 30°C; (d) long-term cycling stability of the M-LMNO material measured at a high current density of 10 C between 3.5 V and 4.9 V at 30°C (Yin et al. 2018). Reprinted with permission copyright 2018, ACS.

threefold. First, the cathode materials are obtained with uniform distribution of elements, high stoichiometry, high crystallinity, and few impurities because of the strong complexation between the organic ligands and metal ions. Second, the organic ligands provide a molecular guiding template with a stable (111) crystal surface that optimizes the crystal growth, while hindering the formation of the cyclical (100) crystal surface. Finally, the uniformly coated Li_2CO_3 on the M-LMNO surface effectively prevents the dissolution of Mn^{2+} during the charging/ discharging process. These three factors cooperate to enhance the performance of M-LMNO.

3.3 Conclusion

This chapter reviewed the current state of the progress in the field of high energy density LMO and LMNO cathode materials for LIB technology. The $LiMn_2O_4$ and $LiMn_{1.5}Ni_{0.5}O_4$ spinel cathode materials are well explored in applications of mobile technologies, stationary utilities, and electrifying of vehicles. Both manganese-based cathode materials are amongst the most attractive cathode materials for LIB technology due to low price and environmental benignity.

Currently, the research trend of LMO and LMNO is the use of optimized morphology being adopted to achieve enhanced capacity and acceptable cyclic stability. In the case of LMO, porous microspheres have found to offer high capacity and stable cyclic performance. The research trend regarding LMNO, there is synthesis of highly crystalline grains with uniform size and with preferred orientation in the (111) and (110) planes, unlike (100) plane which is prone to capacity fading; the (100) surface facets also accelerate the Mn dissolution, causing specific-capacity attenuation, oxidation decomposition of the electrolyte, and structural collapse. These facets orientations possess the lowest surface energy, and increasing the number of 3D Li transmission channels can improve the rate capacity and stability.

REFERENCES

Cui, Xiaoling, Hongliang Li, Shiyou Li, Shan Geng, and Jiachen Zhao. 2017. "Effects of Different Precipitants on $LiNi_{0.5}Mn_{1.5}O_4$ for Lithium Ion Batteries Prepared by Modified Co-Precipitation Method". *Ionics* 23: 2993–2999.

Ding, Yuanli, Zachary P. Cano, Aiping Yu, Jun Lu, and Zhongwei Chen. 2019. "Automotive Li-Ion Batteries: Current Status and Future Perspectives". *Electrochemical Energy Reviews* 2(1): 1–28.

Guo, Donglei, Zhaorong Chang, Hongwei Tang, Bao Li, Xinhong Xu, Xiao Zi Yuan, and Haijiang Wang. 2014. "Electrochemical Performance of Solid Sphere Spinel $LiMn_2O_4$ with High Tap Density Synthesized by Porous Spherical Mn_3O_4". *Electrochimica Acta* 123: 254–259.

Hai, Bin, Alpesh K. Shukla, Hugues Duncan, and Guoying Chen. 2013. "The Effect of Particle Surface Facets on the Kinetic Properties of $LiMn_{1.5}Ni_{0.5}O_4$ Cathode Materials". *Journal of Materials Chemistry A* 1:759–769.

Hai, Yun, Ziwei Zhang, Hao Liu, Libing Liao, Peng Fan, Yuanyuan Wu, Guocheng Lv, and Lefu Mei. 2019. "Facile Controlled Synthesis of Spinel $LiMn_2O_4$ Porous Microspheres as Cathode Material for Lithium Ion Batteries". *Frontiers in Chemistry* 7:437.

Hashem, Ahmed M., Somia M. Abbas, Xu Hou, Ali E. Eid, and Ashraf E. Abdel-Ghany, 2019. "Facile One Step Synthesis Method of Spinel $LiMn_2O_4$ Cathode Material for Lithium Batteries". *Heliyon* 5: e02027.

Huang, Jiajia, Haodong Liu, Naixie Zhou, Ke An, Ying Shirley Meng, and Jian Luo. 2017. "Enhancing the Ion Transport in $LiMn_{1.5}Ni_{0.5}O_4$ by Altering the Particle Wulff Shape via Anisotropic Surface Segregation". *ACS Applied Materials and Interfaces* 9(42): 36745–36754.

Kan, Wang Hay, Saravanan Kuppan, Lei Cheng, Marca Doeff, Jagjit Nanda, Ashfia Huq, and Guoying Chen. 2017. "Crystal Chemistry and Electrochemistry of $Li_xMn_{1.5}Ni_{0.5}O_4$ Solid Solution Cathode Materials". *Chemistry of Materials* 29, 16: 6818–6828.

Kebede, Mesfin A., and Kenneth I. Ozoemena. 2017. "Molten Salt-Directed Synthesis Method from LiMn2O4 Nanorods as a Cathode Material from a Lithium-Ion Battery with Superior Cyclability". *Materials Research Express* 4(2): 1–7.

Kebede, Mesfin A., Nithyadharseni Palaniyandy, and Lehlohonolo F. Koao. 2020. "Layered, Spinel, Olivine, and Silicate as Cathode Materials for Lithium-Ion Battery". In *Electrochemical Devices for Energy Storage Applications*. UK: Taylor and Francis Group.

Kebede, Mesfin A., Maje J. Phasha, Niki Kunjuzwa, Mkhulu K. Mathe, and Kenneth I. Ozoemena. 2015. "Solution-Combustion Synthesized Aluminium-Doped Spinel (LiAlxMn2–xO4) as a High-Performance Lithium-Ion Battery Cathode Material". *Applied Physics A: Materials Science and Processing* 121(1): 51–57.

Kebede, Mesfin A., Maje J. Phasha, Niki Kunjuzwa, Lukas J. le Roux, Donald Mkhonto, Kenneth I. Ozoemena, and Mkhulu K. Mathe. 2014. "Structural and Electrochemical Properties of Aluminium Doped LiMn2O4 Cathode Materials for Li Battery: Experimental and Ab Initio Calculations". *Sustainable Energy Technologies and Assessments* 5: 44–49.

Kebede, Mesfin A., Spyros N. Yannopoulos, Labrini Sygellou, and Kenneth I. Ozoemena. 2017. "High-Voltage $LiNi_{0.5}Mn_{1.5}O_{4-\delta}$ Spinel Material Synthesized by Microwave-Assisted Thermo-Polymerization: Some Insights into the Microwave-Enhancing Physico-Chemistry". *Journal of The Electrochemical Society* 164(13): A3259–A3265.

Kunjuzwa, Niki, Mesfin A. Kebede, Kenneth I. Ozoemena, and Mkhulu K. Mathe. 2016. "Stable Nickel-Substituted Spinel Cathode Material ($LiMn_{1.9}Ni_{0.1}O_4$) for Lithium-Ion Batteries Obtained by Using a Low Temperature Aqueous Reduction Technique". *RSC Advances* 6(113): 111882–111888.

Lee, Wontae, Shoaib Muhammad, Chernov Sergey, Hayeon Lee, Jaesang Yoon, Yong Mook Kang, and Won

Sub Yoon. 2020. "Advances in the Cathode Materials for Lithium Rechargeable Batteries". *Angewandte Chemie - International Edition* 59(7): 2578–2605.

Leung, Kevin. 2013. "Electronic Structure Modeling of Electrochemical Reactions at Electrode/Electrolyte Interfaces in Lithium Ion Batteries". *Journal of Physical Chemistry C* 117(4): 1539–1547.

Liu, Haowen, Yining Zhou, and Wenchuan Song. 2018. "Facile Synthesis of Porous $LiMn_2O_4$ Micro-/Nano-Hollow Spheres with Extremely Excellent Cycle Stability as Cathode of Lithium-Ion Batteries". *Journal of Solid State Electrochemistry* 22(8): 2617–2622.

Logan, E.R., Helena Hebecker, A. Eldesoky, Aidan Luscombe, Michel B. Johnson, and J.R. Dahn. 2020. "Performance and Degradation of $LiFePO_4$/Graphite Cells: The Impact of Water Contamination and an Evaluation of Common Electrolyte Additives". *Journal of The Electrochemical Society* 167: 130543.

Manthiram, Arumugam. 2020. "A Reflection on Lithium-Ion Battery Cathode Chemistry". *Nature Communications* 11: 1550.

Mao, Jing, Kehua Dai, and Yuchun Zhai. 2012. "Electrochemical Studies of Spinel $LiNi_{0.5}Mn_{1.5}O_4$ Cathodes with Different Particle Morphologies". *Electrochimica Acta* 63: 381–390.

Mao, Jing, Peng Zhang, Xin Liu, Yanxia Liu, Guosheng Shao, and Kehua Dai. 2020. "Entropy Change Characteristics of the $LiNi_{0.5}Mn_{1.5}O_4$ Cathode Material for Lithium-Ion Batteries". *ACS Omega* 5: 4108–4114.

Nitta, Naoki, Feixiang Wu, Jung Tae Lee, and Gleb Yushin. 2015. "Li-Ion Battery Materials: Present and Future". *Materials Today* 18(5): 252–264.

Samarasingha, Pushpaka B., Niels H. Andersen, Magnus H. Sørby, Susmit Kumar, Ola Nilsen, and Helmer Fjellvåg. 2016. "Neutron Diffraction and Raman Analysis of $LiMn_{1.5}Ni_{0.5}O_4$ Spinel Type Oxides for Use as Lithium Ion Battery Cathode and Their Capacity Enhancements". *Solid State Ionics* 284: 28–36.

Tang, W., X.J. Wang, Y.Y. Hou, L.L. Li, H. Sun, Y.S. Zhu, Y. Bai, Y.P. Wu, K. Zhu, and T. Van Ree. 2012. "Nano $LiMn_2O_4$ as Cathode Material of High Rate Capability for Lithium Ion Batteries". *Journal of Power Sources* 198: 308–311.

Thackeray, M.M., W.I.F. David, and J.B. Goodenough. 1982. "Structural Characterization of the Lithiated Iron Oxides $LixFe_3O_4$ and $LixFe_2O_3$ (0<x<2)". *Materials Research Bulletin* 17: 785–793.

Wang, Liping, Hong Li, Xuejie Huang, and Emmanuel Baudrin. 2011. "A Comparative Study of Fd-3m and $P4_332$ '$LiNi_{0.5}Mn_{1.5}O_4$'". *Solid State Ionics* 193: 32–38.

Wang, Peng Bo, Ming Zeng Luo, Jun Chao Zheng, Zhen Jiang He, Hui Tong, and Wan Jing Yu. 2018. "Comparative Investigation of $0.5Li_2MnO_3 \cdot 0.5LiNi_{0.5}Co_{0.2}Mn_{0.3}O_2$ Cathode Materials Synthesized by Using Different Lithium Sources". *Frontiers in Chemistry* 6: 1–9.

Yin, Chengjie, Hongming Zhou, Zhaohui Yang, and Jian Li. 2018. "Synthesis and Electrochemical Properties of $LiNi_{0.5}Mn_{1.5}O_4$ for Li-Ion Batteries by the Metal-Organic Framework Method". *ACS Applied Materials and Interfaces* 10(16): 13624–13634.

Zhu, Qing, Shuai Zheng, Xuwu Lu, Yi Wan, Quanqi Chen, Jianwen Yang, Ling Zhi Zhang, and Zhouguang Lu. 2016. "Improved Cycle Performance of $LiMn_2O_4$ Cathode Material for Aqueous Rechargeable Lithium Battery by LaF3 Coating". *Journal of Alloys and Compounds* 654: 384–391.

Zubi, Ghassan, Rodolfo Dufo-López, Monica Carvalho, and Guzay Pasaoglu. 2018. "The Lithium-Ion Battery: State of the Art and Future Perspectives". *Renewable and Sustainable Energy Reviews* 89(March): 292–308.

4

Zinc Anode in Hydrodynamically Enhanced Aqueous Battery Systems

Philips Chidubem Tagbo[1,2], Chukwujekwu Augustine Okaro[1,2], Cyril Oluchukwu Ugwuoke[1,2], Henry Uchenna Obetta[1,3], Onyeka Stanislaus Okwundu[1,2], Sabastine Ezugwu[4], and Fabian I. Ezema[2,5,6]

[1]*Science and Engineering Unit, Nigerian Young Researchers Academy, Onitsha, Anambra State, Nigeria*
[2]*Department of Physics and Astronomy, University of Nigeria, Nsukka, Enugu State, Nigeria*
[3]*River State University – Nkpolu-oroworukwu, Port-Harcourt, River State, Nigeria*
[4]*Department of Physics and Astronomy, The University of Western Ontario, London, Ontario, Canada*
[5]*Nanosciences African Network (NANOAFNET), iThemba Labs, National Research Foundation, Somerset West, Western Cape, South Africa*
[6]*UNESCO-UNISA Africa Chair in Nanosciences/Nanotechnology, College of Graduate Studies, University of South Africa (UNISA), Pretoria, South Africa*

4.1 Introduction

As evidenced by the vast area of application, Electrochemical Energy Storage (EES) devices such as fuel cells, supercapacitors, and batteries are more than capable of accommodating the intermittent nature of renewable energy sources. Of all, the battery is the most popularly employed due to its high energy capacity, long cycle life, and commendable stability (Chunlin Xie et al. 2020; Du et al. 2020). Rechargeable lithium-ion batteries (LIBs) have been the most widely used both in consumer electronics, small-scale power grids, and electric cars, as a result of their high energy efficiency, high energy density, and long cycle life. However, the large-scale utilization of this battery type (LIBs) has been constrained by the limited lithium reservoir, high cost of assembling, and the safety challenges posed by the usage of LIBs (Changgang Li et al. 2020a; Hao et al. 2020; N. Zhang et al. 2020a; Zhu et al. 2020). Therefore, there is a need for a safer, cheaper, more abundant, and high-performing battery system.

Monovalent alternatives such as sodium-ion (SIBs) and potassium-ion batteries (KIBs) gained the interest of researchers due to the abundance of the elements and low cost. But the high activity of Na and K, low operating voltage, and difficulty in finding suitable cathode materials due to the large radii of Na and K impede further development of such battery systems (Ming et al. 2019; Ponrouch and Palacín 2019). However, earth-abundant multivalent metal-ion batteries (such as Ca, Mg, Al, and Zn) are more promising alternatives not only as a result of their abundant nature but also because of their low cost, safety, high operating voltages, high stability, and high volumetric capacity (D. Wang et al. 2018; Y. Li et al. 2019; Biemolt et al. 2020). The unavailability of suitable electrolytes for the high-efficiency reversible metal plating/stripping, and slow intercalation kinetics of Ca, Mg, and Al have been the drawbacks towards the development of these battery systems (Son et al. 2018; D. Wang et al. 2018; Ponrouch and Palacín 2019). Additionally, the strong electrostatic interaction between these multivalent metals and the host materials has resulted in the inability of many of the discovered cathode materials to accommodate the mobile ions and consequently the dissolution of the active cathode materials (Ming et al. 2019; Ponrouch and Palacín 2019). In contrast to the above-mentioned challenges faced by multivalent metal-ion batteries, zinc-ion batteries (ZIBs) have shown to possess outstanding battery performances which have led to the growing interests of researchers in ZIBs in recent years.

This vast attraction of researchers can be attributed to many attractive features of ZIB and its components. The components of ZIB include zinc metal as the anode, cathode materials, and neutral/ slightly acidic aqueous electrolyte. The use of Zn metal as the anode is due to its high theoretical capacity, high abundance, low cost, and high stability. Also, Zn has low redox potential and commendable stability in water due to the high over-potential Zn has for hydrogen evolution, and hence the suitability of aqueous electrolytes for ZIBs (Song et al. 2018; Selvakumarana et al. 2019). Aside from the high stability of Zn metal in aqueous electrolytes, the use of aqueous electrolytes is desirable because they are cheaper, safer, and have higher ionic conductivity than non-aqueous ones (Zhao et al. 2019; Canpeng Li et al. 2020b). The other important component of ZIB is the cathode material which is made of tunnel- and layer-type structures. Amidst these desirable features, ZIBs still face some challenges that have resulted in the inability of ZIBs to attain their theoretical capacity and hence limit their practical applications – challenges such as dendrite formation, self-corrosion, passivation, and hydrogen evolution reaction. The effect of these challenges on the performance of ZIBs ranges from an increased rate of self-discharge to short circuiting of the battery system. Though these challenges are rampant when an alkaline electrolyte is used, they still exist in neutral/slightly acidic electrolytes but under certain conditions(Ma et al. 2018; Canpeng Li et al. 2020c).

To effectively mitigate these challenges, strategies such as surface modification, structural design modification of the anode, and electrolyte optimization have been employed by many researchers, and significant results have been achieved (Jia et al. 2020; Q. Zhang et al. 2020). The surface modification strategy entails the use of effective surface coating for the interfacial modification between the Zn anode and the electrolyte, while the structural design strategy involves the development of different structural appearances for the anode. Furthermore, the electrolyte optimization involves both the use of electrolyte additives and the employment of flowing electrolytes. The effect of flowing electrolytes on curbing the challenges faced by Zn anode hasn't had much attention from researchers and is therefore the concept of this chapter.

In this chapter, ZIBs in still-electrolytes, the various components, storage mechanisms, and challenges are discussed in Section 4.2. The various strategies for mitigating the challenges of ZIBs in still-electrolytes with a focus on the employment of flowing electrolytes are discussed in Section 4.3. In Section 4.4, some ZIBs utilizing flowing electrolytes and their performances are highlighted. In Section 4.5, the areas of applications of zinc flow batteries are briefly outlined. Finally, Section 4.6 offers a summary and a future outlook on zinc flow batteries.

4.2 Zinc Anode in Still-Aqueous Electrolyte: The Modus Operandi

4.2.1 The Conventional Zinc-Ion Batteries (ZIBs)

Besides the high power and energy density required of a battery system as an energy storage device, factors such as cost of assembling, environmental friendliness, safety of operation, and availability of materials are also considered vital when assembling battery systems. For example, the large-scale utilization of LIBs has witnessed some drawbacks even though they possess high energy and power densities. This is due to high cost, safety challenges, and limited lithium reservoirs (Fang et al. 2018; Ming et al. 2019; Jia et al. 2020). In this regard, zinc-ion batteries have proven to be cost-effective, readily available, high performing, and safer alternatives to LIBs for large-scale applications (Ming et al. 2019; Jia et al. 2020), as illustrated in the multi-angle comparison of ZIBs and LIBs in Figure 4.2(a). ZIBs are mainly comprised of zinc metal as the anode, a cathode material for the insertion of Zn ion, and an electrolyte. The overall performance of this battery system depends on its components and hence certain properties are required of each component. Below, some of these properties are briefly discussed.

4.2.1.1 Zinc Anode

The negative electrode of an electrochemical cell is known as the anode. The anode functions as the reducing electrode that issues electrons to the positive electrode (cathode) through an external circuit and oxidizes in the course of an electrochemical reaction (https://depts.washington.edu/matseed/batteries/MSE/components.html). Properties such as stability, high coulombic output, low toxicity, large specific area, low cost, good conductivity, and availability are considered during material selection for anode materials. The employment of Zn metal as an anode material especially in aqueous electrolytes is due to its high chemical stability in water, low toxicity, high conductivity, high abundance, ease of fabrication, high theoretical capacity of 820 $mAhg^{-1}$, and low redox potential of -0.76 V (vs. standard hydrogen electrode) (Song et al. 2018; Selvakumarana et al. 2019; N. Zhang et al. 2020a). Another advantage of Zn over Li as anode material is its divalent nature, which results in a higher capacitive performance of ZIBs than LIBs. This is because multivalent ions can transfer more than one electron per ion in contrast to monovalent ones (Selvakumarana et al. 2019). The above-mentioned theoretical capacity of Zn has not been completely attained due to some drawbacks faced by Zn anode in aqueous electrolytes as would be discussed in Section 4.2.3.

Various forms of zinc have been used as anode materials depending on what they have to offer, in terms of improving the performance of ZIBs. For industrial purposes, the most commonly used is the Zn foil, basically because of its wide availability. However, the limited specific area and the smooth planar structure of Zn foil serve as disadvantages to their utilization (Caramia and Bozzini 2014; X. Zeng et al. 2019; Du et al. 2020). As stated earlier, large specific areas of Zn anode is a desirable property, as more active sites will be utilized during the redox process. Also, the smooth planar structure of Zn foil encourages surface cracks/defects, which results in an uneven distribution of Zn^{2+} – a favourable condition for dendrite growth.

Other forms of zinc that can be used as anode materials, as can be seen in Figure 4.1 are porous zinc, zinc powder, and specific substrates supported zinc (Du et al. 2020). For the case of zinc powder, the zinc anode is made by slurry coating a combination of zinc powder, conductive carbon, and binder on a current collector.

4.2.1.2 Cathode

The cathode is a positive electrode that accepts electrons from the anode via an external circuit and is reduced during an electrochemical reaction. The cathode serves as the host material for the storage of anions or cations. The following properties are desirable of a cathode material: high stability, good conductivity, high porosity, large active surface area, chemical and corrosion resistance, low cost, suitable mechanical strength, and low energy pathways (Esan et al. 2020; Zhu et al. 2020). The cathode materials employed in ZIBs are tunnel-type and/or layer-type redox compounds that permit the intercalation/

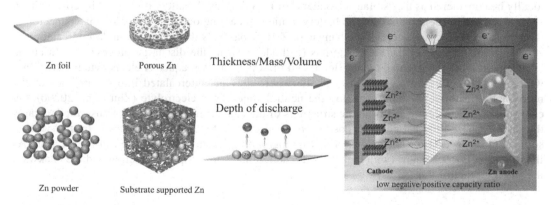

FIGURE 4.1 Various applicable forms of zinc metal as anode materials in zinc-ion batteries (Du et al. 2020). Reprinted with permission from Ref. Du et al. 2020, copyright 2020. The Royal Society of Chemistry.

de-intercalation of Zn^{2+} into/from the crystals. The search for suitable cathode materials with excellent reversible capacity, high cyclability, and suitable operating voltage has been the factor slowing the commercialization of ZIBs (Changgang Li et al. 2020). Due to the strong electrostatic interaction and high steric hindrance effects Zn^{2+} has on the host materials, available cathode materials often suffer from poor cyclability and sluggish intercalation kinetics (Changgang Li et al. 2020). Materials such as manganese-based oxides (e.g. MnO_2 with α, β, γ structures, $ZnMn_2O_4$, etc.), vanadium-based oxides (such as VO_2, V_2O_5, V_6O_{15}, etc.), Prussian blue analogues, olivine-based phosphates, and organic materials (such as Poly (benzoquinonylsulphide) (PBQS), polyaniline, 4-benzoquinone-(pchloranil), etc. (Fang et al. 2018; Song et al. 2018; N. Zhang et al. 2020a; Zhu et al. 2020b).

4.2.1.3 Electrolyte

The electrolyte (composed of electrolyte salt and solvent), which serves as the medium for the transportation of ions between the anode and cathode, has a tangible influence on the overall performance of ZIBs. Hence, when selecting an electrolyte, properties such as composition, thermal stability, ionic conductivity, solubility, concentration, cost, electrochemical potential window, and safety (Song et al. 2018; Esan et al. 2020; N. Zhang et al. 2020b; T. Zhang et al. 2020) are considered. Depending on the solvent, we have aqueous and non-aqueous electrolytes. For ZIBs, aqueous electrolytes are the most promising ones due to their good ionic conductivity, high safety, low cost, and good compatibility with Zn ions. In this system, the pH of the electrolyte plays a key role in the performance of Zn anode. For instance, the use of alkaline electrolytes results in severe capacity fading and low coulombic efficiency due to the formation of dendrites and zinc oxide (ZnO) (Song et al. 2018; N. Zhang et al. 2020a). Neutral or slightly acidic aqueous electrolytes (such as $ZnSO_4$, $ZnCl_2$, $Zn(CF_3SO_3)_2$, etc.) are the best option, as the above-mentioned challenges of alkaline electrolytes are to some extent mitigated. Non-aqueous electrolytes like ionic liquid electrolytes show traits of high operating potential windows and high thermal stability but their low ionic conductivity, high viscosity, and high cost limit their application in ZIBs (N. Zhang et al. 2020a; T. Zhang et al. 2020). Organic electrolytes though are very compatible with Zn ion but suffer from high flammability, high cost, low ionic conductivity, and high toxicity, hence are not readily employed.

In this section, the active components of ZIBs, their desirable properties, and example of materials that can be utilized have been discussed. These components all together contribute to the overall performance of ZIBs and therefore require more attention in materials selection for maximum performance of ZIBs.

4.2.2 Storage Mechanisms of Aqueous Zinc-Ion Batteries

From the meaning and functions of the electrodes discussed above, the anode and the cathode can ideally be considered as the "storage chambers" of ESS devices. A battery is charged by connecting a voltage source to the terminals of the battery while discharging occurs when a load is connected to the terminals. The charge storage mechanism of ZIBs is basically the intercalation/de-intercalation of Zn^{2+} into the crystals of the host materials (cathode). During the discharge process, the Zn anode loses electrons to the external circuit (load) and dissolves into the aqueous electrolyte to form Zn^{2+} which is then conveyed towards the cathode where they get intercalated into the crystals of the redox-active cathode, without changing the neutral state of the electrolyte (Zhu et al. 2020b). On charging, Zn^{2+} gets extracted from the structures of the cathode materials and is transported towards the anode where they get plated/deposited in the Zn anode by accepting electrons from a voltage source. This reaction is known as the Zn stripping/plating reaction as illustrated in Figure 4.2b. As illustrated in Eq. 9.1 below (with the forward reaction being the stripping reaction and the backward the plating reaction):

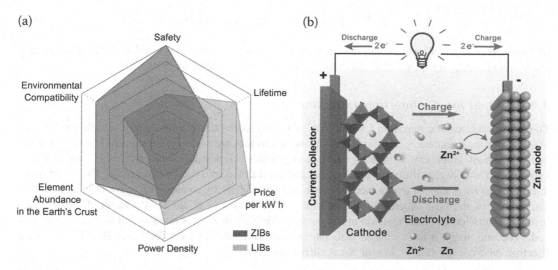

FIGURE 4.2 (a) performance comparison of zinc-ion batteries and lithium-ion batteries, (b) charge storage mechanism of secondary zinc-ion batteries (N. Zhang et al. 2020). Reprinted with permission from Ref. N. Zhang et al. 2020, copyright 2020. The Royal Society of Chemistry.

- Anode:

$$Zn(s) \leftrightarrow Zn^{2+}(aq) + 2e^- \qquad (4.1)$$

- Cathode:

$$Zn^{2+}(aq) + MO(s) + 2e^- \leftrightarrow ZnMO(s) \qquad (4.2)$$

- Overall:

$$Zn^{2+}(aq) + MO(s) \leftrightarrow ZnMO(s) \text{ (where MO is the cathode material)} \qquad (4.3)$$

It is worth mentioning that the electrochemical reaction that happens at the cathode is different for different cathode materials due to their different structures and morphologies (N. Zhang et al. 2020a). According to studies, the storage mechanism of cathode materials in aqueous ZIBs can be classified into insertion/extraction of Zn^{2+}; chemical conversion reaction; and dual ion co-insertion/extraction (Fang et al. 2018; Selvakumarana et al. 2019; Duo Chen et al. 2020).

4.2.2.1 Insertion/Extraction of Zn^{2+} Reaction

The small ionic radius of Zn^{2+} (0.74 Å) encourages the insertion/extraction of Zn^{2+} into the crystals of host materials of tunnel- and layer-type structures. During the discharge process, Zn^{2+} is inserted into the cathode, with the acceptation of electrons which leads to a reduction in the oxidation state. On charging, the Zn^{2+} is extracted from the cathode leading to the oxidation of the cathode (Duo Chen et al. 2020). This reaction is similar to that given in equation (4.3). This insertion/extraction reaction involves the phase transition of the cathode material and is the most studied storage mechanism of ZIBs (Hao 2020). Examples of such materials are α-MnO_2 (2 × 2 tunnels), β-MnO_2 (1 × 1 tunnels), VO_2 (B) (2 × 2 tunnels), etc. For the case of α-MnO_2, upon insertion of Zn^{2+} into the cathode, spinel $ZnMn_2O_4$ was reportedly formed. On charging, it reverses to the initial α-MnO_2 as demonstrated below:

$$Zn^{2+} + 2\alpha - MnO_2 + 2e^- \leftrightarrow ZnMn_2O_4 \tag{4.4}$$

In some other reported studies, the insertion of Zn^{2+} into the tunnel-structured α-MnO_2 resulted in a phase transformation of layered Zn-burserite or birnessite (B. Lee et al. 2014, 2015).

4.2.2.2 Dual Ion Co-Insertion/Extraction

The strong electrostatic repulsion force of Zn^{2+} is known to contribute to the sluggish diffusion kinetics of Zn^{2+} into host materials. This sluggish diffusion kinetics creates a favourable environment for the intercalation of other ions with higher diffusion kinetics. Monovalent ions such as H^+ and Na^+ are known to exhibit higher diffusion kinetics and are intercalated into the host materials. The insertion of these other ions is known to enhance the performance of ZIBs by ensuring effective utilization of the active sites of the host structures (W. Liu et al. 2019; P. Gao et al. 2020; X. Gao et al. 2020). This reversible dual ion insertion reaction of Zn^{2+} and H^+ (or Na^+) in ZIBs is investigated by electrochemical and structural analysis tools such as Galvanostatic Intermittent Titration Technique (GITT), Scanning Electron Microscope (SEM), in-situ X-ray diffraction, etc.

4.2.2.3 Chemical Conversion Reaction

In contrast to most storage mechanisms in ZIBs in which the capacity is as a result of the reversible (de) intercalation of Zn^{2+}, conversion reaction of the cathode material upon the acceptation of protons (H^+) from water has been reported by some researchers (J. Liu et al. 2016; Ma et al. 2018). For example, a reversible conversion reaction between α-MnO_2 and $MnOOH$ was demonstrated by J. Liu et al. (2016). Upon the insertion (discharge process) of H^+ from the dissolution of water, the α-MnO_2 was converted to $MnOOH$. To utilize the OH^- (created as a result of the dissolution of water) and Zn^{2+}, zinc hydroxide sulphate (ZHS) was formed, as can be seen below.

- Cathode:

$$H_2O \leftrightarrow H^+ + OH^- \tag{4.5}$$

$$MnO_2 + H^+ + e^- \leftrightarrow MnOOH \tag{4.6}$$

$$\tfrac{1}{2}\ Zn^{2+} + OH^- + 1/6\ ZnSO_4 + x/6\ H_2O \leftrightarrow 1/6\ ZnSO_4[Zn(OH)_2]_3 \cdot xH_2O \tag{4.7}$$

- Anode:

$$\tfrac{1}{2}\ Zn^{2+} \leftrightarrow \tfrac{1}{2}\ Zn^{2+} + e^- \tag{4.8}$$

It is worthy to note that in the above scenario, $MnSO_4$ was used as an electrolyte additive to prevent the dissolution of the cathode material (MnO_2). On charging, the reverse reaction takes place.

4.2.3 Challenges Facing Batteries Utilizing Zinc Anode in Still-Electrolytes

As mentioned above, the theoretical capacity of Zn metal has not been fully attained due to some challenges faced by the utilization of Zn metal in battery systems – challenges such as dendrite formation, zinc corrosion, passivation, and hydrogen evolution reactions, as can be seen in Figure 4.3. In this section, the fundamentals and effects of these challenges on the performance of ZIBs are discussed.

4.2.3.1 Dendrite Formation

The electrochemistry of Zn metal varies for different environments. The oxidation of Zn metal in aqueous media can lead to the formation of the dissolved form of Zn^2, $Zn(OH)_2$, $Zn(OH)_2^-$, and Zn

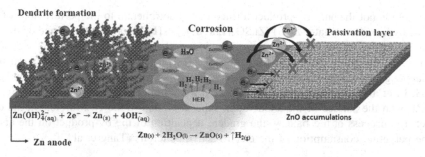

FIGURE 4.3 Major challenges facing zinc anode in aqueous still-electrolytes.

$(OH)_4^{2-}$ depending on the pH values (Shin et al. 2020b). In alkaline medium, on discharging, the Zn metal oxidizes to form $Zn(OH)_4^{2-}$, while during charging, the reverse occurs as illustrated in the equation below

$$Zn + 4OH^- \leftrightarrow Zn(OH)_4^{2-} + 2e^- \tag{4.9}$$

Dendrite is a leaf-like morphology that commonly occurs in alkaline ZIBs during the charging process. Dendrite formation is a result of uneven distribution of Zn^{2+} during zinc plating on the anode (the backward reaction of equation (4.9)). At the beginning of the Zn plating, little Zn deposits form on the surface of the anode, which disrupt the stable electric field of the system (Zhu et al. 2020b). This charge imbalance attracts subsequent charges on the Zn deposits and therefore leads to the gradual growth of dendrites from repeated charge/discharge cycles (Canpeng Li et al. 2020). The attraction of subsequent charges on the earlier Zn deposits occurs because Zn^{2+} preferentially deposits on the already formed nucleus rather than producing a new one. This is basically because the energy barrier of Zn nucleation is higher than the energy barrier of Zn growth on an already existing nucleus (L. Guo et al. 2020). It is worth stating that dendrites do not start developing in ZIBs until a critical over-potential is attained (Diggle, Despic, and Bockris 1969). The time required for the attainment of this potential value depends on factors such as current density, concentration, and temperature of the system (Yufit et al. 2019).

In a neutral or slightly acidic medium, the degree of dendrite formation is very much lesser than in an alkaline medium but increases with increasing current density (Canpeng Li et al. 2020). As Zn deposition continues, dendrites appear at different locations on the anode at different initiating times. This variance in the initiation time is attributed to the uneven surface finish of the Zn anode, which is responsible for the non-uniform local density that affects the critical potential (Yufit et al. 2019). The effects of dendrite formation on ZIBs include gradual capacity failure, the formation of dead zinc, increased rate of self-discharge, and finally short circuit (when the dendrites extend to the separator).

4.2.3.2 Zinc Corrosion

Corrosion of zinc anode is another drawback towards the full attainment of the theoretical capacity of zinc in a practical sense. The formation of $Zn(OH)_4^{2-}$ does not only favour dendrite growth but also leads to the gradual consumption of the electro-active Zn materials. This process of dissolution/corrosion in alkaline media is widely accepted to be a two-step reaction (Hao et al. 2020). The first step is the oxidation of Zn metal to form $Zn(OH)_4^{2-}$, as demonstrated in equation 4.9. The second reaction occurs when the solution near the surface is saturated with the dissolved $Zn(OH)_4^{2-}$ and hence the precipitation of the irreversible inert by-product shown in the equation below.

$$Zn(OH)_4^{2-} \rightarrow ZnO + 2OH^- + H_2O \tag{4.10}$$

The ZnO forms at the expense of the continuous consumption of the Zn^{2+}, leading to low plating/stripping efficiency (CE), increased rate of self-discharge, and decreased life cycles of zinc. It is worthy to note that ZnO is not the only by-product formed that is detrimental to the performance of ZIBs. For instance, zinc hydroxide sulphate (ZHS) ($ZnSO_4[Zn(OH)_2]_3.xH_2O$) commonly formed on the cathode as demonstrated above in the chemical conversion reaction mechanism can also form on the anode but now detrimental to the battery system. The formation of ZHS on the anode is different from that on the cathode (Canpeng Li et al. 2020). The difference is a result of the electrostatic interaction between the reactants (H. Li et al. 2015; Tang et al. 2019). The low solubility of the product usually results in the plating of ZHS on the surface of the anode upon supersaturation in the electrolyte. On the continuous charge/discharge process of the battery, the gradual deposition of the by-product on the anode would result in the perpetual consumption of the electro-active materials (Tang et al. 2019).

4.2.3.3 Passivation

Passivation is an impression originating from a corrosion reaction, which defines the formation of a protective layer on a metal surface, commonly composed of metal oxides (Ingelsson, Yasri, and Roberts 2020). The protective layer prevents the kinetics of the thermodynamically favourable metal oxidation, thereby increasing the resistance at the electrode-electrolyte interface (Schmuki 2002). In the case of Zn, ZnO product formed in equation (4.10) becomes supersaturated in the electrolyte and gets deposited on the surface of the Zn anode to form a protective/insulating layer that prevents further reaction on the surface. According to reports, this layer is of two types; type I and type II (Powers and Breiter 1969; Hao et al. 2020). The type I layer is the loose and porous layer which forms first on the anode surface in the absence of convention and can be effectively avoided by introducing flowing electrolyte (to be discussed in Section 4.3.4) while the type II is a more compacted layer, and can be prevented by reducing the depth of discharge (DoD). The effect of the insulating ZnO layer formed on the anode includes reduced capacity, reduced discharge and power capability, and increased impedance (Mainar et al. 2018; Tang et al. 2019).

4.2.3.4 Hydrogen Evolution Reaction (HER)

Hydrogen evolution reaction is a parasitic side reaction that occurs as a result of the corrosion of Zn metal (Mainar et al. 2018). In alkaline electrolytes, the standard reduction of Zn/ZnO (−1.22 V vs. SHE) is less than the hydrogen evolution potential (−0.83 V vs. SHE) (S. Lee et al. 2013), which means that HER is thermodynamically favoured rather than Zn deposition. However, in the neutral electrolyte, the standard reduction potential of Zn/Zn^{2+} is higher than that for hydrogen evolution and hence Zn ion reduction is thermodynamically favoured. In a slightly acidic medium, HER is thermodynamically favoured (as in the case of alkaline medium) (Hao et al. 2020). But in the actual sense, HER in an alkaline medium is averted as a result of the higher over-potential of Zn metal over hydrogen evolution. Therefore, Zn deposition is thermodynamically favoured as a result of the reduction in the actual potential value for HER (Shin et al. 2020b). The hydrogen evolution reaction is illustrated in Eq. (4.11).

$$Zn + 2H_2O \rightarrow Zn(OH)_2 + \uparrow H_2O \qquad (4.11)$$

The evolution of hydrogen gas in a sealed battery system leads to an increase in the internal pressure of the battery, a common cause of electrolyte leakage (resulting from internal swelling). Aside from this, HER results in an increase in the OH^- concentration, which encourages the formation of irreversible by-products like ZHS (Canpeng Li et al. 2020).

4.3 Optimization of the Performances of Zinc Anode Battery Systems

In this section, we aim at conversing on the solutions to the underlying setbacks imposed by the use of Zn anodes in aqueous ZIBs. The low cyclability and coulombic efficiency emanating from undesirable activities on the anode side such as irreversible by-products formation, corrosion, and harsh dendrite have become a big deal to tackle in the course of Zn anode enhancement. The strategies discussed in this

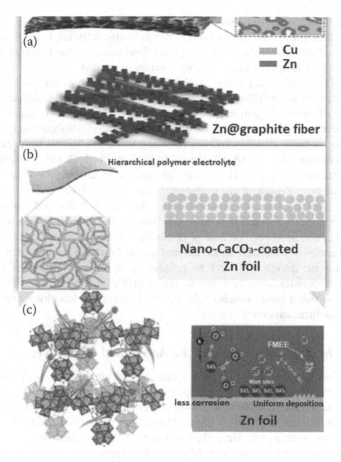

FIGURE 4.4 Optimization strategies for Zn anode in aqueous ZIBs (a) Structural architecture (b) Interfacial modification (c) Electrolyte engineering (Jia et al. 2020). Reprinted with permission from Ref. Jia et al. 2020, copyright 2020. Elsevier.

composition have been explored and geared towards the development of Zn anode to hamper all these undesirable effects. Effective ways of mitigating the effects of these issues have earlier been proposed as depicted in Figure 4.4, which include structural design of Zn anode, anode-electrolyte interfacial modification, and electrolyte optimization (Yi et al. 2020).

4.3.1 Structural Design towards High-Performing Zinc Anode

This is an effective strategy that has been studied and implemented towards improving the electro-chemical performance of Zn anode in aqueous battery systems. This approach has tolerated the growth of dendrite on Zn anode and curtailed the possible by-product formation which is detrimental to the electrochemical process within the system (P. Chen et al. 2018). Three-dimensional (3D) MOF-like nanostructured architecture has been developed to act as a sustainer to promote the uniform distribution of Zn ions, further leading to the even distribution of current. The structural optimization in this aspect could be interpreted as increasing sufficiently the surface area of the anode material and the current collector so as to encourage a uniform distribution of zinc during the electrodeposition as revealed in LIBs studies (X. B. Cheng et al. 2017). Structural design is generally categorized into 3D zinc-sponge electrode and hierarchical zinc-plated electrode. Both architectures are characterized by high surface area such that current and electrolyte ions can interact with the electrode surface in all directions. The 3D zinc–sponge when made with pure zinc faces the challenges of structural degradation when there is less or no sustaining framework in the electrode and this mostly occurs in a severe condition as reported by Fu et al. (2017). The 3D zinc-sponge when used in alkaline Zn-based rechargeable batteries has proven to stand dendrite growth and passivation better than originally used two-dimensional (2D) architectures (Ko et al. 2019). However, 3D zinc-sponge fabrication is characterized by so many complexities, and yet suffers structural instability after a series of cycling.

Various kinds of 3D current collectors and MOF-based substrates have been used as substrates for 3D hierarchical zinc-plated electrodes (Shi et al. 2019; Z. Wang et al. 2019). By engineering a hierarchical structure of the Zn/CNT anode, the current distribution has been reported to be improved (Y. Zeng et al. 2019). Some of these composites have been used in a mild acidic electrolyte copper foam (Shin et al. 2020a), the 3D porous copper skeleton on zinc (Z. Kang et al. 2019), and some other carbon-based architectures such as double-helix Zn/CNT yam (H. Li, Liu, et al. 2018), Zn/carbon flexible fiber (Y. Zeng et al. 2017), Zn/CNT paper (H. Li et al. 2018a), etc. it has been observed from studies that CNT matrix and some other Zn-plated electrodes made from carbon-based current collectors generally promote the surface electric current distribution, hence maintain stability and reversibility when employed for modification of Zn anode. These materials have shown excellent when used as an anode in modern electronic devices, however, the cost of obtaining carbon paper and carbon cloth poses a big challenge in its application commercialization. Current research on Zinc-plated electrode has shown that the bare substrate for the design will have to possess some intrinsic properties to be suitably incorporated with Zn. Properties such as hardness and ductility should be at the frontline when considering a substrate so that it could suitably tolerate the uniform Zn deposition and stripping thereby retaining its 3D hierarchical structure.

4.3.2 Interfacial Modification between the Anode and Electrolyte

While searching for lasting solutions to the current challenges facing the Zn anode, proper modification of the electrode-electrolyte interface with inorganic coatings can bring about quality electrochemical performances on the anode side. This is another sound approach towards achieving a dendrite-resistance Zn anode material during electrochemical charge/discharge cycles and could be actualized through surface coating and functionalization. Here, the coating material depending on the porous architecture (which serves as an ion filter) is used to mildly overlap the surface of the bare Zn anode thereby preventing direct interactions between the anode and the electrolyte. The thin layer coatings (usually a nanoshell) can only allow the desired ions to pass through it uniformly, thereby enabling a leveled Zn stripping/plating. When this strategy is properly employed, it effectively increases the cycling stability and lifespan of ZIBs by protecting the Zn metal from self-corrosion, dendrite growth, and hydrogen evolution reaction (Zheng et al. 2019; 2020). For instance, the coating of the inorganic $CaCO_3$ on Zn foil reportedly enhanced the lifespan of the ZIBs (L. Kang et al. 2018). The high porosity coating readily facilitates the infiltration of electrolyte on the electrode surface and allows uniform distribution of zinc nucleation. Furthermore, the use of an artificial solid electrolyte interface (SEI) is an effective way of optimizing the hydrogen evolution potential of electrodes, suppressing interfacial alkalization and undesired side reactions. The in-situ fabrication of SIE nanoparticles as a protective layer on the surface of the Zn anode has also been proposed to have suitably inhibited dendrite formation and self-corrosion of the Zn metal. Ultrathin TiO_2 deposited on zinc surface has been reported as an efficient anode shield against the detrimental activities on the battery system by preventing direct interaction between the anode and the electrolyte, hence enhancing the overall electrochemical performance of the cell (Shin et al. 2020a). Reduced graphene oxide coated on Zn foil has been effectively used in mitigating the effects of corrosion and harsh dendrite growth during the plating/stripping process (Shen et al. 2018; Xia et al. 2019). The engineered interface of the anode is characterized by high electroactive sites, lowered overpotential, high power and energy densities, and subsequently exhibiting high capacity retention after several cycles. Heterogeneous gold nanoparticles (Au-NPs) sputtered on Zn metal have shown excellent cyclability and long lifespan when compared to bare Zn anode (Cui et al. 2019). Researches are still ongoing on novel materials of high potentials including MOF-derived architectures for their suitability to act as very active protective layers on bare Zn anode as well as their tolerance to unnecessary challenges faced by Zn anode in ZIBs.

4.3.3 The Use of Electrolyte Additives

Being another focal component of ZIBs, modifying the electrolyte is an effective way of tackling the major challenges of ZIBs. One way of modifying the electrolyte and improving the performance of ZIBs is the use of electrolyte additives. The effects of electrolyte additives on Zn anode include smoother and denser Zn anode, the formation of a protective layer on the surface of the anode, and increased deposition overpotential of Zn (Naveed et al. 2019; S. Guo et al. 2020). A smooth, dense, and surface-protected Zn anode prevents the growth of dendrites while an increased overpotential leads to higher reversibility. Aside from the improvement of the Zn anode, the introduction of electrolyte additives also increases the ionic conductivity of the electrolyte, broadens the electrochemical window, and reduces the dissolution of the electroactive cathode materials (S. Guo et al. 2020).

Based on applications, additives can be grouped into organic additives, ionic additives, and inorganic additives. The organic additives are the most studied additives for ZIBs. They can be classified into five categories, with each affecting the anode performance in distinctive ways. Some serve as hydrogen inhibitors, some as crystal growth modifiers, and some as dendrite, passivation and corrosion inhibitors, and so on. These five categories are quaternary ammonium salts (such as tetraethylammonium chloride (TEACl), tetrapropylammonium chloride (TPrACl), tetrabutylammonium chloride (TBACl), etc.), polymers (such as polyethylene glycol (PEG), polyacrylamide (PAM), and polyethyleneimine (PEI), etc.), surfactants (like tetrabutylammonium sulphate (TBA_2SO_4), Sodium lignin sulphonate, etc.), natural compounds (such as chitin), organic acids (e.g. perfluoro carboxylic acids, malonic acids, etc.), and ionic liquid salts (like 1-butyl-3-methylimidazolium hydrogen sulphate [BMIM]HSO_4, etc.) (Sorour et al. 2017; Bayaguud et al. 2020; S. Guo et al. 2020).

The ionic additives are practically employable due to their high versatility, abundant production, and easy preparation (S. Guo et al. 2020). The use of these additive types provides the ZIBs with regulated Zn anode dissolution/deposition, dendrite growth inhibition, and improved cathode and electrolyte performance. Examples are Cu^{2+}, Mg^{2+}, Co^{2+}, Al^{2+}, In^{3+}, etc. (Ma et al. 2018; S. Guo et al. 2020; N. Li et al. 2020). Furthermore, researchers have shown limited interest in inorganic additives due to their low solubility in aqueous electrolytes, though their effect on ZIB performance upon application is very significant. Examples are tin oxide (SnO), indium sulphate, boric acid, etc. (Kim and Shin 2015; Sun et al. 2017).

By employing these additives, the overall performance of ZIBs significantly improves. In some cases, their uses result in suppression of zinc corrosion, inhibition of dendrite growth, increased coulombic efficiency, capacity increment, higher retention capacity, and improved cyclability (Xiao 2018; Xu et al. 2019; Lin et al. 2020; P. Wang et al. 2020).

4.3.4 Incorporation of Hydrodynamics into Zinc-Ion Battery System

Aside from the use of electrolyte additives, another way of optimizing the electrolyte is the introduction of hydrodynamics (flow) into the battery system. The introduction of flowing electrolyte has been considered as an effective way of mitigating dendrite growth and improving the stability of the Zn electrode through fluctuating the zincate ion concentration distribution and shunning the buildup of zincate ions (Ito et al. 2012; Yu et al. 2019, 2020). The zincate mass transfer being diffusion-controlled in conventional ZIBs creates a favorable environment for dendrite formation, as it results in uneven distribution of Zn ions during the plating process. However, by introducing flowing electrolyte, the zincate mass transfer changes from the traditional diffusion-controlled to convention-controlled mass transfer, leading to uniform Zn ion distribution and hence, the suppression of uneven electrodeposition of Zn ions (Ito et al. 2012). Since the by-products formed during the electrochemical processes are brought outside the electrolyte chamber in Zinc flow battery, the occurrence of passivation is mitigated by introducing a flowing electrolyte in the ZIB system.

The effect of flowing electrolyte on the stability of Zn anode was investigated through experimental and numerical analysis by Yu et al. (2020). In their work, the relationships between flow rates and battery operation, and the impact of flow orientation were studied and the results were compared with that of a static battery system. According to their findings, by employing flowing electrolyte, the lifespan of the Zn anode increases significantly from 900 cycles in the static electrolyte to 2580 cycles at a current density of 10 mA cm^{-2} with flow rates of 30 rad min^{-1}. By further increasing the flow rate to 50 rad min^{-1}, the lifespan leaped to 4725 cycles (Yu et al. 2020). After short circuiting of the battery system, photographs of the Zn anode in both the static and flow system were taken. It was found that due to the partial utilization of the flowing electrolyte, the position of dendrite growth in the flow system was similar to that of the static system. However, changing the inlet flow orientation from the side to the bottom led to the full utilization of the flow system as evidenced by the absence of short circuiting of the system (Yu et al. 2020). Also, this resulted in a lifespan of 18,000 cycles at the same current density of 10 mA cm^{-2}. From the numerical investigation, it was found that the introduction of flow increases the speed of the transport of Zn ions and contributes to a more even distribution of zincate ions (Yu et al. 2020). To portray this effect, Ito et al. (2012) investigated the changes in the electrodeposited zinc morphology in flowing alkaline electrolytes. They found that the ratio of the effective current density to the limiting current density [which is related to the concentration of zincate ions at the electrode surface] is a determinant of zinc morphology. It was found that with a ratio of 0.4 and 0.9, the morphology of zinc electrode was mossy and crystalline. By further increasing the ratio to above 0.9, the electro-deposited zinc morphology became crystalline and compact (Ito et al. 2012).

According to the International Flow Battery Forum (IFBF), a flow battery is a rechargeable battery system in which electrolyte flows through one or more electrochemical cells from one or more tanks. A conventional flow battery is composed of the anode, cathode, two tanks of electrolyte species (containing the redox couples), with one flowing to the anode (called the anolyte) and the other flowing to the cathode (the catholyte), and an ion exchanging membrane as shown in Figure 4.5. It should be noted that the types of flow batteries are basically a description of the choice of the redox pairs used. Some of the eminent redox pairs are:

- Vanadium/Vanadium
- Iron/Chromium
- Zinc/Bromine, etc. Refer to Section 4.4.1 for more flow batteries utilizing Zn anode.

In some cases, this battery system is called "redox flow batteries". This is because of the redox couples used as the electroactive species. Conventionally, both the electroactive species in the redox pairs are soluble in aqueous electrolytes. Batteries of this type store electric energy in the two external electrolyte tanks containing the redox couples while the electrodes act as active sites only. During discharging, the pumps transfer electrolytes through the electrodes where the reduction and oxidation process converts

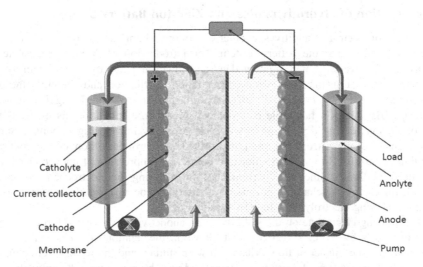

FIGURE 4.5 Conventional flow battery system.

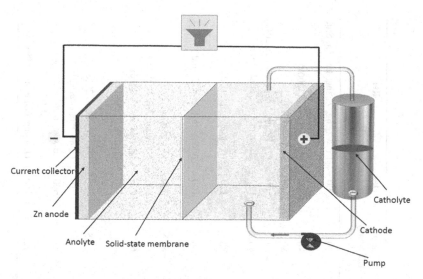

FIGURE 4.6   Conventional zinc hybrid flow battery.

chemical reaction to electric energy. During charging, the electrolytes gain chemical energy from the power source (Revankar 2019). A good example is the all vanadium redox flow battery (VRFB).

However, in the case of zinc flow batteries, the zinc metal is deposited on the electrode (rather than been dissolved in the electrolyte), and hence is commonly referred to as hybrid redox flow batteries, as can be seen in Figure 4.6. Unlike the VRFB, the zinc hybrid flow battery combines features of both conventional secondary battery and conventional redox flow battery. In some cases, this system comes in single-flow (one electrolyte tank), double-flow, and membraneless configurations depending on the electrolyte, the anode types, or the redox species used (Shah et al. 2018). For higher power, the battery can be constructed using multiple stacks of flow cells in a bipolar arrangement.

4.4 Types of Flow Batteries Utilizing Zn Anode and Their Performances

In this section, some of the major flow batteries utilizing zinc as the anode are briefly discussed. Table 4.1 shows the performances of the various zinc flow batteries.

4.4.1 Types of Zinc Flow Batteries

4.4.1.1 Zinc-Bromine Flow Battery

The zinc-bromine (Zn-Br) flow battery was first developed in the early 1970s by Exxon. Ever since then, the Zn-Br flow battery has been commercialized and considered to be the most promising among zinc-based flow batteries because it offers higher energy density and cell voltage (M. C. Wu et al. 2018). This battery system commonly utilizes carbon-based composite bipolar electrodes and two reservoirs of $ZnBr_2$ electrolyte. The use of carbon-based composite bipolar electrodes is due to the high surface area, high porosity, and good conductivity of carbon, which results in fast reaction rates (Arenas et al. 2018). During the charging process, Zn metal is plated on the negative carbon electrode while the bromide ions are oxidized to bromine at the positive electrode. The formed Br_2 is very corrosive, and hence requires complexing agents (such as N-methyl-ethyl-pyrrolidinium [MEP]) to help mitigate the effects of its corrosive nature (Arenas et al. 2018). While in the discharge process, the reverse reaction occurs.

TABLE 4.1

Performances of Flow Batteries Utilizing Zn Metal as the Anode Material

Flow Battery	Cell Configuration	Membrane	Electrolyte	Current Density ($mAcm^{-2}$)	Coulombic Efficiency (%)	Energy Efficiency (%)	Cycles	Ref
Zn-Ni	Single flow	Membraneless	KOH	80	97.3	80.1	200	(Y. Cheng, Zhang, Lai, Li, Shi, et al. 2013)
Zn-Ni	Single flow	Membraneless	KOH + ZnO	20	96	86	1000	(J. Cheng et al. 2007)
Zn-Ni	Single flow	Membraneless	KOH + ZnO	10	90	80	1500	(Ito et al. 2011)
Zn-Ni	Single flow	Membraneless	KOH+LiOH+ZnO	20	99.2	84.2	500	(T. Wang, Yang, and Pan 2017)
Zn-Fe	Double flow	Non-ionic poly(ether sulphone) (PES)	Catholyte: $Zn(OH)_4^{2-}$ + NaOH Anolyte: $Na_4Fe(CN)_6$ + KOH	80	99.3	86.8	120	(Dongju Chen et al. 2021)
Zn-Fe	Double flow	Polybenzimidazole	Catholyte: $Na_2Zn(OH)_4$ + NaOH Anolyte: $Na_2Fe(CN)_6$ + KOH	80	~99	~88	120	(X. Liu et al. 2020)
Zn-Fe	Double flow	Nafion 115	Catholyte: $FeCl_2$ + lycine Anolyte: $ZnBr_2$ + KCl	40	97.8	86.7	100	(Congxin Xie et al., n.d.)
Zn-Fe	Double flow	Polybenzimidazole	$Zn(OH)_4^{2-}$ + $K_4Fe(CN)_6$ + NaOH	160	99.5	82.8	500	(Yuan et al. 2018)
Zn-I	Single flow	Porous polyolefin	KI + $ZnBr_2$	40	97	81	500	(Congxin Xie et al. 2019)
Zn-Br	Single flow	Membraneless	$ZnBr_2$ + $ZnCl_2$	20	95	60	1000	(Biswas et al. 2016)
Zn-Br	Single flow	Daramic micro-porous membranes	$ZnBr_2$	20	92	82	70	(Lai et al. 2013)
Zn-Br	Double flow	—	HCl + NaBr	40	98	83.5	—	(M. Wu et al. 2018)
Zn-PbO	Single flow	Membraneless	H_2SO_4 + Zn(II)	150	90	80	10	(Junli Pan et al. 2016)
Zn-Air	Single flow	Membraneless	KOH + ZnO	50	90	54	600	(Bockelmann, Kunz, and Turek 2016)
Zn-Air	Single flow	Membraneless	MnO_2 + $NaBiO_3$	20	97.4	72.2	150	(Junqing Pan et al. 2009)

Zn-Br flow batteries are reliable, efficient, and offer energy efficiencies up to 90% as can be seen in the table above. The electroactive components of this battery system are cheap and highly abundant, which has led to the commercialization by several energy companies.

4.4.1.2 Zinc-Nickel Flow Battery

Though secondary zinc-nickel battery saw early commercialization in the 1970s and was applied in portable power tools and several vehicles, the introduction of flow system took about 37 years (Arenas et al. 2018). The main aim of introducing flow then is for the application in utility-scale energy storage (Turney et al. 2014). This is a non-toxic battery system that uses Zn metal as the anode and $Ni(OH)_2$/ NiO as the cathode. This cost-effective system not only makes use of cheap and abundant Zn and nickel metals but also employs the same electrolyte solution in both the anode and cathode with no membrane (Y. Cheng, Zhang, Lai, Li, Shi, et al. 2013). The rapid kinetics of the redox couple in Zn-Ni flow battery is one of its desirable features, as it provides fast charge/discharge capabilities. Zn-Ni single flow battery exhibits good properties but the accumulation of Zn when charged/discharged for a long cycle limits its employment as an energy storage system (T. Wang, Yang, and Pan 2017). To improve the performance of Zn-Ni flow batteries, many researchers have explored different structural designs of the anode. For instance, 3D nickel foam has successfully replaced the 2D sheet traditionally used. By doing this, the current density reportedly improved the energy efficiency to about 80% (Y. Cheng, Zhang, Lai, Li, and Shi 2013).

4.4.1.3 Zinc-Iron Flow Battery

At the early stage of development, this battery system was termed a *Zinc-ferricyanide flow battery* as a result of its employment of ferrocyanide/ ferricyanide as the positive electrode. This battery system is known for the abundant and cheap nature of its active materials, environmental friendliness, and high cell voltage (Shah et al. 2018). Though the Zn-Fe system has been shown little interest by researchers, ViZn energy has been the key player in the commercialization of Zn-Fe flow batteries for application in different areas of life. In an experimental setting, a high-performing Zn-Fe flow battery has been reported by Chen et al. (Dongju Chen et al. 2021), by employing poly (ether sulphone) (PES) as the battery membrane and PEG as the additive. A high coulombic and energy efficacies of 99.2% and 86.81% was reported, while a higher value of energy efficiency of 88% was reported by Yuan et al. (X. Liu et al. 2020) by using Na_2SO_4 electrolyte additives.

4.4.1.4 Zinc-Air Flow Battery

Zinc-air flow batteries are more compact and have a relatively higher energy density than conventional flow batteries (Yu et al. 2019); more compact in the sense that the active materials are just zinc metal and ambient air, hence only one storage tank is needed basically for enhancing mass transfer and bringing out the by-products unlike some conventional flow batteries with two redox couples, which require two storage tanks. The charge storage mechanism of this battery system is different from the conventional flow batteries, the reactants are the zinc metal and the gaseous oxygen on the air electrode. The zinc particles are added to the electrolyte to form a slurry and are stored in a tank. The negative electrode only provides sites for the stripping/plating reaction. On discharge, a combination of Zn particles and electrolyte is pumped into the negative electrode where the Zn particle oxidizes to Zn^{2+}, combines with the hydroxide ion to form $Zn(OH)_4^{2-}$, and then dissolves into the electrolyte (Yu et al. 2019). While on the air electrode, oxygen reduction reaction (ORR) occurs during the discharge process. On charging the reverse reaction occurs on the negative electrode, whereas on the air electrode, oxygen evolution reaction (OER) takes place. The ORR is a three-phase reaction involving the bifunctional solid catalyst-electrolyte-gaseous oxygen while the OER is a two-phase reaction involving only the catalyst and the electrolyte. In a practical sense, ZAFB offers a high energy density ranging between 350 to 11,000 Wh kg^{-1} (Arenas et al. 2018). Other advantages include safety, low cost, low self-discharge rate, and environmental friendliness.

Other types of zinc flow batteries include zinc-iodine flow battery (ZIFB), zinc-chlorine flow battery, zinc-cerium flow battery, zinc-lead-dioxide flow battery, and so on.

4.4.2 Performances of Zinc Flow Batteries

Fundamentally, the performance of electrode materials for battery systems and other energy storage devices have been hinged on capacity, rate capability, and cyclability. Which in a practical sense are measured in terms of the power and energy densities during charging/discharging. Ideally, the energy density for charging a battery system should be equal to that of discharge but in practice, energy is wasted. For long, the cost of energy waste has not been accounted for in battery systems because the measure of battery performance has been focused on higher capacity, good rate capability, and longer life cycles (Eftekhari 2017). The cost of this energy wastage is significantly small in small-scale energy systems. But in large-scale energy systems such as batteries in electric vehicles (EVs) and household storage systems, the cost of energy expended to charge the battery is considerably large and is incorporated into the cost of energy supplied by the storage system (Eftekhari 2017). From the above, it is evidenced that considering energy efficiency (EE) of battery systems is vital in battery research. Energy efficiency is defined as the ratio of the discharge energy density to the charge energy density as illustrated in the equation below (Eftekhari 2017).

$$\text{Energy efficiency} = \text{Energy density}_{(discharge)}/\text{Energy density}_{(charge)} \times 100 \qquad (4.12)$$

$$\text{Coulombic Efficiency} = \text{Total charge}_{(discharge)}/\text{Total charge}_{(charge)} \times 100 \qquad (4.13)$$

The comparison between the charging and the discharging process is usually made by calculating the coulombic efficiency as can be seen in equation. 4.13 above.

Table 4.1 gives a wider range of parameters (aside from the energy efficiency) for measuring the performances of zinc flow batteries.

4.5 Areas Where Zinc Flow Batteries Have Been Applied

The flow battery market is expected to grow with a compound annual growth rate (CAGR) of 12.41% within the forecast period of 2020–2025 (https://www.mordorintelligence.com/industry-reports/flow-battery-market n.d.). However, for zinc flow batteries, top global energy companies such as Redflow Ltd. (Australia), Primus Power (USA), ENSync Energy Systems (USA), LOTTE (Korea), MGX Renewable Inc. (Canada), Galion Technologies Pty Ltd. (Australia) ("Zinc-Bromine Battery Market to Reach US$ 23 Mn by 2027" n.d.), and many others are already developing zinc flow batteries by utilizing cathode materials such as bromine, iron, air, etc., with Redflow Ltd. being the major player. With the above CAGR of flow batteries in such a period, it is right saying that in the nearest future, flow batteries will overthrow the conventional secondary batteries and find applications in almost all areas of life. In the meantime, they have been used in applications such as power quality control, load levelling, coupling with renewable energy sources, and electric vehicles (Leung et al. 2012).

4.5.1 Power Quality Control

Disruption of an electric power system during power generation often results in the breakdown of the grid system. Therefore, there is a need for fast stabilization in order to avoid any complications. Desirable traits such as fast response time of <1 second to power demand and maximum short-time overload output that is numerous times that of the rated capacity of flow batteries make them

(a)   (b)

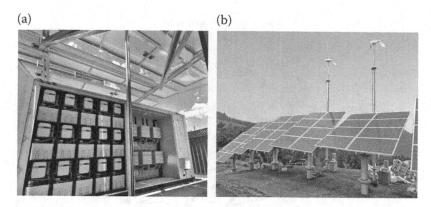

FIGURE 4.7 Applications of Zinc flow batteries (a) incorporation of a zinc-bromine flow battery in solar panels for commercial or industrial use by Redflow Ltd. (b) Redflow's installation of zinc-bromine flow battery as an energy supplier for Vodafone telecommunication tower. Reprinted from Ref. (https://redflow.com/applications/, 2020), copyright 2020. Redflow Ltd.

suitable for voltage and frequency control (Enomoto et al. 2002) and hence their application in battery-backed UPS (Uninterruptible power supply) system (Leung et al. 2012). The electricity provided by the UPS system during power disruption can basically be used to orderly shutdown available digital devices or to keep the system running while the backup generator is switched on. Therefore, the employment of UPS in such scenarios is for the protection of power utility, transmission, and distribution systems.

Though the use of flow batteries in UPS systems is ongoing for quite a while now for mainly vanadium-based flow batteries (Leung et al. 2012), zinc-bromine flow battery (ZBM2) has been developed by Redflow Ltd. for application in UPS systems (https://redflow.com/applications/, 2020). This battery with a dimension of 845 L × 23 H × 400 W (mm) is rated 10KWh with a roundtrip efficiency of about 80%, a voltage range of 40–50 V, and a minimum of 3650 cycles at 100% depth of discharge (DoD).

4.5.2 Incorporating with Renewable Energy Sources

The dependence of renewable energy sources (such as solar and wind) on weather, season, and time has created the need for the utilization of storage devices in order for the accommodation of their intermittent nature. Flow batteries are very suitable for this application due to their large capacity and long discharge time (Leung et al. 2012).

Many companies such as Primus Power, ViZn Energy, Redflow Ltd., etc., have developed zinc flow batteries for incorporation in both solar cells and wind turbines as shown in Figure 4.7. The energy from such battery systems like zinc-iron flow batteries, zinc-bromine, etc. can be employed for several purposes such as telecommunication towers, commercial and industrial use, residential uses, and grid-scale applications. For instance, Vodafone has been using Redflow's ZBM2 zinc-bromine flow battery to power their telecommunication tower in remote rural areas of New Zealand, where the battery system delivers more than 50 MWh of energy. The system uses both solar and wind as an energy source and a diesel generator as a backup in days of low radiation (https://redflow.com/applications/, 2020).

4.5.3 Electric Vehicles (EVs)

The use of zinc flow batteries in electric vehicles can be traced back to early 1980 when the Energy Development Associates (EDA) investigated the application of 50 kWh zinc-chlorine flow battery in the power system of vehicles (Leung et al. 2012). While in the 1990s, a zinc-bromine flow battery (rated 35

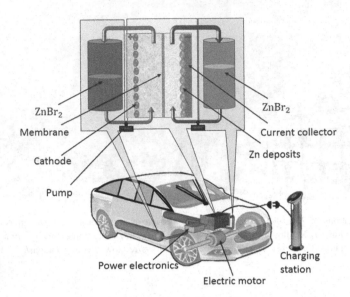

FIGURE 4.8 Early design of aqueous flow batteries for utilization in Electric Vehicles (EV) ("How GE Global Research Is Helping Shape the Future, GE News" n.d.). Reprinted from Ref. (https://www.ge.com/news/reports/how-ge-global-research-is-helping-shape-the-future n.d.), copyright 2020. GE Motors.

kWh) was tested in an electric vehicle by the University of California and was demonstrated by Toyota Motors (Japan) as model EV-3036 (Shah et al. 2018). Figure 4.8 below shows the early design of Zn-Br flow batteries for application in electric vehicles.

4.6 Summary and Future Perspectives

Dendrite formation, self-corrosion, passivation, hydrogen evolution reaction have been the drawbacks limiting the employment of zinc-ion batteries in large-scale energy storage systems such as power grid, household utility grid, and electric vehicles (EVs). Different strategies towards mitigating the effects of these challenges have been employed by many researchers. One of such is the introduction of hydrodynamics into the battery system. By introducing flowing electrolytes, these challenges are curbed and the performances of ZIBs significantly improved. Zinc hybrid flow batteries are embryonic battery technology exhibiting the advantageous features of the unique structural design of conventional redox flow batteries and excellent energy density of the traditional zinc-ion batteries. Hence, are promising candidates for large-scale energy storage systems. In this chapter, aside from the overview of the conventional ZIBs, different types of Zinc flow batteries, their applications, and suitable areas of applications have been discussed.

With the global concern on climate changes resulting in the consumption of fossil fuels growing and the commitment of the globe towards greener, safer, efficient, and sustainable energy supply, zinc flow batteries are reliable and more than capable of accommodating the intermittent nature of renewable energy sources [that are the only viable alternative to fossil fuels]. Furthermore, researchers will continue inventing ways of optimizing and improving the performance of this battery system. Based on the understanding so far, we suggest that the following areas be the focal points of future researches:

- **Cathode optimization and the invention of novel cathode materials:** Due to the nature of Zinc metal, varieties of materials are compatible and suitable for employment as cathode materials in zinc flow batteries. These materials such as bromine, air, iodine, nickel, iron, etc. have high electrochemical performances but their theoretical capacities have not yet been attained due to some challenges. For example, the air electrode used in zinc-air flow batteries faces sluggish redox kinetics, lack of high-performance catalysts for the OER/ORR. By micro/nanostructuring the air electrode can increase the surface area and mass transfer to improve intrinsic activities (Han et al. 2018), also optimizing and inventing novel bifunctional catalysts can result in improved performance of air cathode. Hence, there is a need to optimize the available cathode materials so that their high theoretical capacities can be attainable. Also, the invention of novel cathode materials with higher energy efficiencies, capacities, and high stability for zinc flow batteries is an efficient way of curbing the challenges of the already-utilized cathode materials.

- **Implementation of active-flow/mixing:** One of the major problems been faced by the convection-control mass transfer of flow battery active materials is the blockage of the pump and pipelines of the catholyte tank by solid active materials such as I_2 (Congxin Xie et al. 2019). By introducing active-mixing/flow, electrolyte circulation can be avoided. Hence innovative ways of implementing the introduction of active-mixing/flow of the electroactive materials are required.

- **Novel stack designs:** For higher energy density of zinc flow batteries, the battery system can be constructed using multiple stacks of flow cells in a bipolar arrangement. There is a need to explore novel stack designs that can improve the energy and power density, and the energy efficiency of zinc flow batteries more than the current existing ones.

REFERENCES

Arenas, Luis F, Adeline Loh, David P Trudgeon, Xiaohong Li, Carlos Ponce, De León, and Frank C Walsh. 2018. "The Characteristics and Performance of Hybrid Redox Flow Batteries with Zinc Negative Electrodes for Energy Storage". *Renewable and Sustainable Energy Reviews* 90 (October 2016): 992–1016. 10.1016/j.rser.2018.03.016.

Bayaguud, Aruuhan, Xiao Luo, Yanpeng Fu, and Changbao Zhu. 2020. "Cationic Surfactant-Type Electrolyte Additive Enables Three-Dimensional Dendrite-Free Zinc Anode for Stable Zinc-Ion Batteries". *ACS Energy Letters.* 10.1021/acsenergylett.0c01792.

Biemolt, Jasper, Peter Jungbacker, Tess van Teijlingen, Ning Yan, and Gadi Rothenberg. 2020. "Beyond Lithium-Based Batteries". *Materials* 13 (2): 1–31. 10.3390/ma13020425.

Biswas, Shaurjo, Aoi Senju, Robert Mohr, Thomas Hodson, Nivetha Karthikeyan, Kevin W. Knehr, Andrew G. Hsieh, et al. 2016. "Minimal Architecture Zinc-Bromine Battery for Low Cost Electrochemical Energy Storage". *Energy & Environmental Science*, 1–17. 10.1039/C6EE02782B.

Bockelmann, Marina, Ulrich Kunz, and Thomas Turek. 2016. "Electrically Rechargeable Zinc-Oxygen Flow Battery with High Power Density". *Electrochemistry Communications.* 10.1016/j.elecom.2016.05.013.

Caramia, Vincenzo, and Benedetto Bozzini. 2014. "Materials Science Aspects of Zinc – Air Batteries: A Review". *Materials for Renewable and Sustainable Energy.* 10.1007/s40243-014-0028-3.

Chen, Dongju, Chengzi Kang, Weiqi Duan, Zhizhang Yuan, and Xianfeng Li. 2021. "A Non-Ionic Membrane with High Performance for Alkaline Zinc-Iron Flow Battery". *Journal of Membrane Science* 618 (August 2020): 118585. 10.1016/j.memsci.2020.118585.

Chen, Duo, Mengjie Lu, Dong Cai, Hang Yang, and Wei Han. 2020. "Recent Advances in Energy Storage Mechanism of Aqueous Zinc-Ion Batteries". *Journal of Energy Chemistry.* 10.1016/j.jechem.2020.06.016.

Chen, Peng, Yutong Wu, Yamin Zhang, Tzu Ho Wu, Yao Ma, Chloe Pelkowski, Haochen Yang, Yi Zhang, Xianwei Hu, and Nian Liu. 2018. "A Deeply Rechargeable Zinc Anode with Pomegranate-Inspired Nanostructure for High-Energy Aqueous Batteries". *Journal of Materials Chemistry A* 6 (44): 21933–21940. 10.1039/C8TA07809B.

Cheng, Jie, Li Zhang, Yu Sheng Yang, Yue Hua Wen, Gao Ping Cao, and Xin Dong Wang. 2007. "Preliminary Study of Single Flow Zinc-Nickel Battery". *Electrochemistry Communications* 9 (11): 2639–2642. 10.1016/j.elecom.2007.08.016.

Cheng, Xin Bing, Rui Zhang, Chen Zi Zhao, and Qiang Zhang. 2017. "Toward Safe Lithium Metal Anode in Rechargeable Batteries: A Review". *Chemical Reviews.* American Chemical Society. 10.1021/acs.chemrev.7b00115.

Cheng, Yuanhui, Huamin Zhang, Qinzhi Lai, Xianfeng Li, and Dingqin Shi. 2013. "Performance Gains in Single Flow Zinc – Nickel Batteries through Novel Cell Configuration". *Electrochimica Acta* 105: 618–621. 10.1016/j.electacta.2013.05.024.

Cheng, Yuanhui, Huamin Zhang, Qinzhi Lai, Xianfeng Li, Dingqin Shi, and Liqun Zhang. 2013. "A High Power Density Single Flow Zinc-Nickel Battery with Three-Dimensional Porous Negative Electrode". *Journal of Power Sources* 241: 196–202. 10.1016/j.jpowsour.2013.04.121.

Cui, Mangwei, Yan Xiao, Litao Kang, Wei Du, Yanfeng Gao, Xueqin Sun, Yanli Zhou, et al. 2019. "Quasi-Isolated Au Particles as Heterogeneous Seeds to Guide Uniform Zn Deposition for Aqueous Zinc-Ion Batteries". *ACS Applied Energy Materials* 2 (9): 6490–6496. 10.1021/acsaem.9b01063.

https://depts.washington.edu/matseed/batteries/MSE/components.html. "Components of Cells and Batteries". n.d. Accessed October 27, 2020.

Diggle, J.W., A.R. Despic, and J.O.M. Bockris. 1969. "The Mechanism of the Dendritic Electrocrystallization of Zinc". *Journal of The Electrochemical Society* 116 (11): 1503. 10.1149/1.2411588.

Du, Wencheng, Edison Huixiang Ang, Yang Yang, Yufei Zhang, Minghui Yea, and Cheng Chao Li. 2020. "Challenges in Material and Structure Design of Zinc Anode toward High- Performance Aqueous Zinc-Ion Batteries". *Energy and Environmental Science.* 10.1039/D0EE02079F.

Eftekhari, Ali. 2017. "Energy Efficiency: A Critically Important but Neglected Factor in Battery Research". *Sustainable Energy & Fuels* 1: 2053–2060. 10.1039/C7SE00350A.

Enomoto, Kazuhiro, Tetsuo Sasaki, Toshio Shigematsu, and Hiroshige Deguchi. 2002. "Evaluation Study about Redox Flow Battery Response and Its Modeling". *IEEJ Transactions on Power and Energy* 122 (4): 554–560.

Esan, Oladapo Christopher, Xingyi Shi, Zhefei Pan, Xiaoyu Huo, Liang An, and T.S. Zhao. 2020. "Modeling and Simulation of Flow Batteries". *Advanced Energy Materials* 2000758: 1–42. 10.1002/aenm.202000758.

Fang, Guozhao, Jiang Zhou, Anqiang Pan, and Shuquan Liang. 2018. "Recent Advances in Aqueous Zinc-Ion Batteries". *ACS Energy Letters* 3 (10): 2480–2501. 10.1021/acsenergylett.8b01426.

Fu, Jing, Zachary Paul Cano, Moon Gyu Park, Aiping Yu, Michael Fowler, and Zhongwei Chen. 2017. "Electrically Rechargeable Zinc–Air Batteries: Progress, Challenges, and Perspectives". *Advanced Materials.* Wiley-VCH Verlag. 10.1002/adma.201604685.

Gao, Ping, Qiang Ru, Honglin Yan, Shikun Cheng, Yang Liu, Xianhua Hou, Li Wei, and Francis Chi-Chung Ling. 2020. "A Durable Na0.56V2O5 Nanobelt Cathode Material Assisted by Hybrid Cationic Electrolyte for High-Performance Aqueous Zinc-Ion Batteries". *ChemElectroChem* 7 (1): 283–288. 10.1002/celc.201901851.

Gao, Xu, Hanwen Wu, Wenjie Li, Ye Tian, Yun Zhang, Hao Wu, Li Yang, Guoqiang Zou, Hongshuai Hou, and Xiaobo Ji. 2020. "H+-Insertion Boosted α-MnO2 for an Aqueous Zn-Ion Battery". *Small* 16 (5): 1–10. 10.1002/smll.201905842.

https://www.ge.com/news/reports/how-ge-global-research-is-helping-shape-the-future. "How GE Global Research Is Helping Shape the Future | GE News". n.d. Accessed November 19, 2020.

Guo, Leibin, Hui Guo, Haili Huang, Shuo Tao, and Yuanhui Cheng. 2020. "Inhibition of Zinc Dendrites in Zinc-Based Flow Batteries". *Frontiers in Chemistry* 8 (July): 1–8. 10.3389/fchem.2020.00557.

Guo, Shan, Liping Qin, Tengsheng Zhang, Miao Zhou, Jiang Zhou, Guozhao Fang, and Shuquan Liang. 2020. "Fundamentals and Perspectives of Electrolyte Additives for Aqueous Zinc-Ion Batteries". *Energy Storage Materials.* 10.1016/j.ensm.2020.10.019.

Han, Xiaopeng, Xiaopeng Li, Jai White, Cheng Zhong, Yida Deng, and Wenbin Hu. 2018. "Metal–Air Batteries: From Static to Flow System". *Advanced Energy Materials* 1801396: 1–28. 10.1002/aenm.201801396.

Hao, Junnan. 2020. "Developing High-Performance Aqueous Zn-Based Batteries with Mild Electrolyte Developing High-Performance Aqueous Zn-Based Batteries with Mild Electrolyte". University of Wollongong.

Hao, Junnan, Xiaolong Li, Xiaohui Zeng, Dan Li, Jianfeng Mao, and Zaiping Guo. 2020. "Deeply Understanding the Zn Anode Behaviour and Corresponding Improvement Strategies in Different Aqueous Zn-Based Batteries". *Energy and Environmental Science.* 10.1039/D0EE02162H.

Ingelsson, Markus, Nael Yasri, and Edward P L Roberts. 2020. "Electrode Passivation, Faradaic Efficiency, and Performance Enhancement Strategies in Electrocoagulation – a Review". *Water Research* 187: 116433. 10.1016/j.watres.2020.116433.

Ito, Yasumasa, Michael Nyce, Robert Plivelich, Martin Klein, Daniel Steingart, and Sanjoy Banerjee. 2011. "Zinc Morphology in Zinc – Nickel Flow Assisted Batteries and Impact on Performance". *Journal of Power Sources* 196 (4): 2340–2345. 10.1016/j.jpowsour.2010.09.065.

Ito, Yasumasa, Xia Wei, Divyaraj Desai, Dan Steingart, and Sanjoy Banerjee. 2012. "An Indicator of Zinc Morphology Transition in Flowing Alkaline Electrolyte". *Journal of Power Sources* 211: 119–128. 10.1016/j.jpowsour.2012.03.056.

Jia, Hao, Ziqi Wang, Benjamin Tawiah, Yidi Wang, Cheuk Ying Chan, Bin Fei, and Feng Pan. 2020. "Recent Advances in Zinc Anodes for High-Performance Aqueous Zn-Ion Batteries". *Nano Energy* 70 (January): 104523. 10.1016/j.nanoen.2020.104523.

Kang, Litao, Mangwei Cui, Fuyi Jiang, Yanfeng Gao, Hongjie Luo, Jianjun Liu, Wei Liang, and Chunyi Zhi. 2018. "Nanoporous CaCO3 Coatings Enabled Uniform Zn Stripping/Plating for Long-Life Zinc Rechargeable Aqueous Batteries". *Advanced Energy Materials* 8 (25): 1801090. 10.1002/aenm. 201801090.

Kang, Zhuang, Changle Wu, Liubing Dong, Wenbao Liu, Jian Mou, Jingwen Zhang, Ziwen Chang, et al. 2019. "3D Porous Copper Skeleton Supported Zinc Anode toward High Capacity and Long Cycle Life Zinc Ion Batteries". *ACS Sustainable Chemistry and Engineering* 7 (3): 3364–3371. 10.1021/ acssuschemeng.8b05568.

Kim, Hong-ik, and Heon-Cheol Shin. 2015. "SnO Additive for Dendritic Growth Suppression of Electrolytic Zinc". *Journal of Alloys and Compounds* 645: 7–10. 10.1016/j.jallcom.2015.04.208.

Ko, Jesse S., Andrew B. Geltmacher, Brandon J. Hopkins, Debra R. Rolison, Jeffrey W. Long, and Joseph F. Parker. 2019. "Robust 3D Zn Sponges Enable High-Power, Energy-Dense Alkaline Batteries". *ACS Applied Energy Materials* 2 (1): 212–216. 10.1021/acsaem.8b01946.

Lai, Qinzhi, Huamin Zhang, Xianfeng Li, Liqun Zhang, and Yuanhui Cheng. 2013. "A Novel Single Flow Zinc-Bromine Battery with Improved Energy Density". *Journal of Power Sources* 235: 1–4. 10.1016/ j.jpowsour.2013.01.193.

Lee, Boeun, Hae Ri Lee, Haesik Kim, Kyung Yoon Chung, Byung Won Cho, and Si Hyoung Oh. 2015. "Elucidating the Intercalation Mechanism of Zinc Ions into α-MnO2 for Rechargeable Zinc Batteries". *Chemical Communications* 51 (45): 9265–9268. 10.1039/c5cc02585k.

Lee, Boeun, Chong Seung Yoon, Hae Ri Lee, Kyung Yoon Chung, Byung Won Cho, and Si Hyoung Oh. 2014. "Electrochemically-Induced Reversible Transition from the Tunneled to Layered Polymorphs of Manganese Dioxide". *Scientific Reports*. 10.1038/srep06066.

Lee, Sang-min, Yeon-Joo Kim, Seung-Wook Eom, Nam-Soon Choi, Ki-Won Kim, and Sung-Baek Cho. 2013. "Improvement in Self-Discharge of Zn Anode by Applying Surface Modi Fi Cation for Zn e Air Batteries with High Energy Density". *Journal of Power Sources* 227: 177–184. 10.1016/j.jpowsour. 2012.11.046.

Leung, Puiki, Xiaohong Li, Carlos Ponce de Leo, Leonard Berlouis, C. T. Low John, and Frank C. Walsh. 2012. "Progress in Redox Flow Batteries, Remaining Challenges and Their Applications in Energy Storage". *RSC Advances*: 10125–10156. 10.1039/c2ra21342g.

Li, Canpeng, Xuesong Xie, Shuquan Liang, and Jiang Zhou. 2020. "Issues and Future Perspective on Zinc Metal Anode for Rechargeable Aqueous Zinc-Ion Batteries". *Energy & Environmental Materials*: 146–159. 10.1002/eem2.12067.

Li, Changgang, Xudong Zhang, Wen He, Guogang Xu, and Rong Sun. 2020. "Cathode Materials for Rechargeable Zinc-Ion Batteries: From Synthesis to Mechanism and Applications". *Journal of Power Sources* 449 (December 2019).10.1016/j.jpowsour.2019.227596

Li, Hongfei, Cuiping Han, Yan Huang, Yang Huang, Minshen Zhu, Zengxia Pei, Qi Xue, et al. 2018. "An Extremely Safe and Wearable Solid-State Zinc Ion Battery Based on a Hierarchical Structured Polymer Electrolyte". *Energy and Environmental Science* 11 (4): 941–951. 10.1039/c7ee03232c.

Li, Hongfei, Zhuoxin Liu, Guojin Liang, Yang Huang, Yan Huang, Minshen Zhu, Zengxia Pei, et al. 2018. "Waterproof and Tailorable Elastic Rechargeable Yarn Zinc Ion Batteries by a Cross-Linked Polyacrylamide Electrolyte". *ACS Nano* 12 (4): 3140–3148. 10.1021/acsnano.7b09003.

Li, Hongfei, Chengjun Xu, Cuiping Han, Yanyi Chen, and Chunguang Wei. 2015. "Enhancement on Cycle Performance of Zn Anodes by Activated". *Journal of The Electrochemical Society* 162 (8): 1439–1444. 10.1149/2.0141508jes.

Li, Na, Guoqing Li, Changji Li, Huicong Yang, Gaowu Qin, Xudong Sun, Feng Li, and Hui Ming Cheng. 2020. "Bi-Cation Electrolyte for a 1.7 V Aqueous Zn Ion Battery". *ACS Applied Materials and Interfaces* 12 (12): 13790–13796. 10.1021/acsami.9b20531.

Li, Yun, Shanyu Wang, James R. Salvador, Jinpeng Wu, Bo Liu, Wanli Yang, Jiong Yang, Wenqing Zhang, Jun Liu, and Jihui Yang. 2019. "Reaction Mechanisms for Long-Life Rechargeable Zn/MnO_2 Batteries". *Chemistry of Materials* 31 (6): 2036–2047. 10.1021/acs.chemmater.8b05093.

Lin, Ming-Hsien, Chen-Jui Huang, Pai-Hsiang Cheng, Ju-Hsiang Cheng, and Chun-Chieh Wang. 2020. "Revealing the Effect of Polyethylenimine on Zinc Metal Anodes in Alkaline Electrolyte Solution for Zinc-Air Batteries: Mechanism Studies of Dendrite Suppression and Corrosion Inhibition". *Journal of Materials Chemistry A*. 10.1039/D0TA06929A.

Liu, Jun, Huilin Pan, Yuyan Shao, Pengfei Yan, Yingwen Cheng, Kee Sung Han, Zimin Nie, et al. 2016. "Reversible Aqueous Zinc/Manganese Oxide Energy Storage from Conversion Reactions". *Nature Energy* 1 (April): 1–7. 10.1038/NENERGY.2016.39.

Liu, Wenbao, Liubing Dong, Baozheng Jiang, Yongfeng Huang, Xianli Wang, Chengjun Xu, Zhuang Kang, Jian Mou, and Feiyu Kang. 2019. "Layered Vanadium Oxides with Proton and Zinc Ion Insertion for Zinc Ion Batteries". *Electrochimica Acta* 320: 134565. 10.1016/j.electacta.2019.134565.

Liu, Xiaoqi, Huamin Zhang, Yinqi Duan, Zhizhang Yuan, and Xianfeng Li. 2020. "Effect of Electrolyte Additives on the Water Transfer Behavior for Alkaline Zinc – Iron Flow Batteries". *Applied Materials & Interfaces*. 10.1021/acsami.0c16743.

Ma, Longtao, Shengmei Chen, Hongfei Li, Zhaoheng Ruan, Zijie Tang, Zhuoxin Liu, Zifeng Wang, et al. 2018. "Initiating A Mild Aqueous Electrolyte Co3O4/Zn Battery with 2.2 V-High Voltage and 5000-Cycle Lifespan by a Co(III) Rich-Electrode". *Energy and Environmental Science*: 0–13. 10.1039/C8EE01415A.

Mainar, Aroa R, Elena Iruin, Luis C Colmenares, Andriy Kvasha, Iratxe De Meatza, Miguel Bengoechea, Olatz Leonet, Iker Boyano, Zhengcheng Zhang, and J Alberto Blazquez. 2018. "An Overview of Progress in Electrolytes for Secondary Zinc-Air Batteries and Other Storage Systems Based on Zinc". *Journal of Energy Storage* 15: 304–328. 10.1016/j.est.2017.12.004.

Ming, Jun, Jing Guo, Chuan Xia, Wenxi Wang, and Husam N. Alshareef. 2019. "Zinc-Ion Batteries: Materials, Mechanisms, and Applications". *Materials Science and Engineering R: Reports* 135 (October 2018): 58–84. 10.1016/j.mser.2018.10.002.

https://www.mordorintelligence.com/industry-reports/flow-battery-market. "Flow Battery Market | Growth, Trends, and Forecasts (2020–2025)". n.d. Accessed November 17, 2020.

Naveed, Autoren Ahmad, Huijun Yang, Jun Yang, Yanna Nuli, and Jiulin Wang. 2019. "Highly Reversible and Safe Zn Rechargeable Batteries Based on Triethyl Phosphate Electrolyte Ahmad". *Angewandte Chemie*. 10.1002/ange.201813223.

Pan, Junli, Yuehua Wen, Jie Cheng, Junqing Pan, Shouli Bai, and Yusheng Yang. 2016. "Evaluation of Substrates for Zinc Negative Electrode in Acid PbO2 – Zn Single Flow Batteries". *Chinese Journal of Chemical Engineering* 24: 529–534. 10.1016/j.cjche.2016.01.001.

Pan, Junqing, Lizhong Ji, Yanzhi Sun, Pingyu Wan, Jie Cheng, Yusheng Yang, and Maohong Fan. 2009. "Preliminary Study of Alkaline Single Flowing Zn–O2 Battery". *Electrochemistry Communications* 11 (11): 2191–2194. 10.1016/j.elecom.2009.09.028.

Ponrouch, Alexandre, and M Rosa Palacín. 2019. "Post-Li Batteries: Promises and Challenges Subject Areas:" *Philosophical Transitions A*. 10.1098/rsta.2018.0297

Powers, R.W., and M.W. Breiter. 1969. "The Anodic Dissolution and Passivation of Zinc in Concentrated Potassium Hydroxide Solutions". *Journal of the Electrochemical Society* 116 (6): 719–729. 10.1149/1.2412040.

https://redflow.com/applications/ Applications – Redflow. n.d. Accessed November 18, 2020.

Revankar, Shripad T. 2019. "Chemical Energy Storage". In *Storage and Hybridization of Nuclear Energy*, 177–227. Elsevier Inc. 10.1016/B978-0-12-813975-2.00006-5.

Schmuki, Patrik. 2002. "From Bacon to Barriers: A Review on the Passivity of Metals and Alloys". *Journal of Solid State Electrochemistry*, 145–164. 10.1007/s100080100219.

Selvakumarana, Dinesh, Anqiang Pana, Shuquan Lianga, and Guozhong Cao. 2019. "A Review on Recent Developments and Challenges of Cathode Materials for Rechargeable Aqueous Zn-Ion Batteries". *Journal of Materials Chemistry A*. 10.1039/C9TA05053A.

Shah, A.A., A. Khor, P. Leung, M.R. Mohamed, C. Flox, Q. Xu, L. An, R.G.A. Wills, and J.R. Morante. 2018. "Review of Zinc-Based Hybrid Fl Ow Batteries: From Fundamentals to Applications". *Materisls Today Energy* 8: 80–108. 10.1016/j.mtener.2017.12.012.

Shen, Chao, Xin Li, Nan Li, Keyu Xie, Jian Gan Wang, Xingrui Liu, and Bingqing Wei. 2018. "Graphene-Boosted, High-Performance Aqueous Zn-Ion Battery". *ACS Applied Materials and Interfaces* 10 (30): 25446–25453. 10.1021/acsami.8b07781.

Shi, Xiaodong, Guofu Xu, Shuquan Liang, Canpeng Li, Shan Guo, Xuesong Xie, Xuemei Ma, and Jiang Zhou. 2019. "Homogeneous Deposition of Zinc on Three-Dimensional Porous Copper Foam as a Superior Zinc Metal Anode". *ACS Sustainable Chemistry and Engineering* 7 (21): 17737–17746. 10.1021/acssuschemeng.9b04085.

Shin, Jaeho, Jimin Lee, Youngbin Park, and Jang Wook Choi. 2020a. "Aqueous Zinc Ion Batteries: Focus on Zinc Metal Anodes". *Chemical Science.* 10.1039/D0SC00022A.

Shin, Jaeho, Jimin Lee, Youngbin Park, and Jang Wook Choi. 2020b. "Aqueous Zinc Ion Batteries: Focus on Zinc Metal Anodes View". *Chemical Science.* 10.1039/D0SC00022A.

Son, Seoung Bum, Tao Gao, Steve P. Harvey, K. Xerxes Steirer, Adam Stokes, Andrew Norman, Chunsheng Wang, Arthur Cresce, Kang Xu, and Chunmei Ban. 2018. "An Artificial Interphase Enables Reversible Magnesium Chemistry in Carbonate Electrolytes". *Nature Chemistry* 10 (5): 532–539. 10.1038/s41557-018-0019-6.

Song, Ming, Hua Tan, Dongliang Chao, and Hong Jin Fan. 2018. "Recent Advances in Zn-Ion Batteries". *Advanced Functional Materials* 1802564: 1–27. 10.1002/adfm.201802564.

Sorour, Nabil, Wei Zhang, Edward Ghali, and Georges Houlachi. 2017. "A Review of Organic Additives in Zinc Electrodeposition Process (Performance and Evaluation)". *Hydrometallurgy* 171 (April): 320–332. 10.1016/j.hydromet.2017.06.004.

Sun, Kyung Eun Kate, Tuan Hoang, The Nam Long Doan, Yan Yu, and Pu Chen. 2017. "Highly Sustainable Zinc Anodes for the Rechargeable Hybrid Aqueous Battery". *Chemistry - A European Journal.* 10.1002/chem.201704440.

Tang, Boya, Lutong Shan, Shuquan Liang, and Jiang Zhou. 2019. "Issues and Opportunities Facing Aqueous Zinc-Ion Batteries". *Energy and Environmental Science.* 3288–3304. 10.1039/C9EE02526J.

Turney, Damon E., Michael Shmukler, Kevin Galloway, Martin Klein, Yasumasa Ito, Tal Sholklapper, Joshua W. Gallaway, Michael Nyce, and Sanjoy Banerjee. 2014. "Development and Testing of an Economic Grid-Scale Flow-Assisted Zinc/Nickel-Hydroxide Alkaline Battery". *Journal of Power Sources* 264: 49–58. 10.1016/j.jpowsour.2014.04.067.

Wang, Da, Xiangwen Gao, Yuhui Chen, Liyu Jin, Christian Kuss, and Peter G. Bruce. 2018. "Plating and Stripping Calcium in an Organic Electrolyte". *Nature Materials* 17 (1): 16–20. 10.1038/NMAT5036.

Wang, Pengpeng, Fangxia Zhao, Hong Chang, Quan Sun, and Zhenzhong Zhang. 2020. "Effects of BTA and TBAB Electrolyte Additives on the Properties of Zinc Electrodes in Zinc – Air Batteries". *Journal of Materials Science: Materials in Electronics.* 10.1007/s10854-020-04347-x.

Wang, Tian, Meng Yang, and Junqing Pan. 2017. "A New Single Flow Zinc-Nickel Hybrid Battery Using A". *International Journal of Electrochemical Science* 12: 6022–6030. 10.20964/2017.07.84.

Wang, Zhuo, Jianhang Huang, Zhaowei Guo, Xiaoli Dong, Yao Liu, Yonggang Wang, and Yongyao Xia. 2019. "A Metal-Organic Framework Host for Highly Reversible Dendrite-Free Zinc Metal Anodes". *Joule* 3 (5): 1289–1300. 10.1016/j.joule.2019.02.012.

Wu, M.C., T.S. Zhao, L. Wei, H.R. Jiang, and R.H. Zhang. 2018. "Improved Electrolyte for Zinc-Bromine Flow Batteries". *Journal of Power Sources* 384: 232–239. 10.1016/j.jpowsour.2018.03.006.

Wu, Maochun, Tianshou Zhao, Ruihan Zhang, Haoran Jiang, and Lei Wei. 2018. "A Zinc – Bromine Flow Battery with Improved Design of Cell Structure and Electrodes". 333–339. 10.1002/ente.201700481.

Xia, Aolin, Xiaoming Pu, Yayuan Tao, Haimei Liu, and Yonggang Wang. 2019. "Graphene Oxide Spontaneous Reduction and Self-Assembly on the Zinc Metal Surface Enabling a Dendrite-Free Anode for Long-Life Zinc Rechargeable Aqueous Batteries". *Applied Surface Science* 481 (July): 852–859. 10.1016/j.apsusc.2019.03.197.

Xiao, You. 2018. "Effects of Electrolyte Additives on the Properties of Zinc-Bismuth Electrodes in Zinc-Air Batteries". *Journal of The Electrochemical Society* 165 (2): 47–54. 10.1149/2.0251802jes.

Xie, Chunlin, Yihu Li, Qi Wang, Dan Sun, Yougen Tang, and Haiyan Wang. 2020. "Issues and Solutions toward Zinc Anode in Aqueous Zinc-ion Batteries: A Mini Review". *Carbon Energy* June: 1–21. 10.1002/cey2.67.

Xie, Congxin, Yinqi Duan, Wenbin Xu, Huamin Zhang, and Xianfeng Li. n.d. "A Low Cost Neutral Zinc-Iron Flow Battery with High Energy Density for Stationary Energy Storage". *Journal of Thr Gesellschaft Deutscher Chemiker.* 10.1002/anie.201708664.

Xie, Congxin, Yun Liu, Wenjing Lu, Huamin Zhang, and Xianfeng Li. 2019. "Highly Stable Zinc-Iodine Single Flow Batteries with Super High Energy Density for Stationary Energy Storage". *Energy and Environmental Science* 12 (6): 1834–1839. 10.1039/c8ee02825g.

Xu, Weina, Kangning Zhao, Wangchen Huo, Yizhan Wang, Guang Yao, Xiao Gu, Hongwei Cheng, Liqiang Mai, Chenguo Hu, and Xudong Wang. 2019. "Diethyl Ether as Self-Healing Electrolyte Additive Enabled Long-Life Rechargeable Aqueous Zinc Ion Batteries". *Nano Energy* 62 (May): 275–281. 10.1016/j.nanoen.2019.05.042.

Yi, Zhehan, Guoyuan Chen, Feng Hou, Liqun Wang, and Ji Liang. 2020. "Strategies for the Stabilization of Zn Metal Anodes for Zn-Ion Batteries". *Advanced Energy Materials* (November): 2003065. 10.1002/aenm.202003065.

Yu, Wentao, Wenxu Shang, Peng Tan, Bin Chen, Zhen Wu, Haoran Xu, Zongping Shao, Meilin Liu, and Meng Ni. 2019. "Toward a New Generation of Low Cost, Efficient, and Durable Metal–Air Flow Batteries". *Journal of Materials Chemistry A*, 26744–26768. 10.1039/c9ta10658h.

Yu, Wentao, Wenxu Shang, Xu Xiao, Peng Tan, Bin Chen, Zhen Wu, Haoran Xu, and Meng Ni. 2020. "Achieving a Stable Zinc Electrode with Ultralong Cycle Life by Implementing a Flowing Electrolyte". *Journal of Power Sources* 453 (January): 227856. 10.1016/j.jpowsour.2020.227856.

Yuan, Zhizhang, Yinqi Duan, Tao Liu, Huamin Zhang, and Xianfeng Li. 2018. "Toward a Low Cost Alkaline Zinc-Iron Flow Battery with a Polybenzimidazole Custom Membrane for Stationary Energy Storage". *ISCIENCE*. 10.1016/j.isci.2018.04.006.

Yufit, Vladimir, Farid Tariq, S. Eastwood, Moshiel Biton, Peter D Lee, P. Nigel, Vladimir Yufit, et al. 2019. "Operando Visualization and Multi-Scale Tomography Studies of Dendrite Formation and Dissolution in Zinc Batteries Operando Visualization and Multi-Scale Tomography Studies of Dendrite Formation and Dissolution in Zinc Batteries". *Joule* 485–502. 10.1016/j.joule.2018.11.002.

Zeng, Xiaohui, Junnan Hao, Zhijie Wang, Jianfeng Mao, and Zaiping Guo. 2019. "Recent Progress and Perspectives on Aqueous Zn-Based Rechargeable Batteries with Mild Aqueous Electrolytes". *Energy Storage Materials* 20 (December 2018): 410–437. 10.1016/j.ensm.2019.04.022.

Zeng, Yinxiang, Xiyue Zhang, Yue Meng, Minghao Yu, Jianan Yi, Yiqiang Wu, Xihong Lu, and Yexiang Tong. 2017. "Achieving Ultrahigh Energy Density and Long Durability in a Flexible Rechargeable Quasi-Solid-State Zn–MnO2 Battery". *Advanced Materials* 29 (26). 10.1002/adma.201700274.

Zeng, Yinxiang, Xiyue Zhang, Ruofei Qin, Xiaoqing Liu, Pingping Fang, Dezhou Zheng, Yexiang Tong, and Xihong Lu. 2019. "Dendrite-Free Zinc Deposition Induced by Multifunctional CNT Frameworks for Stable Flexible Zn-Ion Batteries". *Advanced Materials* 31 (36). 10.1002/adma.201903675.

Zhang, Ning, Xuyong Chen, Meng Yu, Zhiqiang Niu, Fangyi Cheng, and Jun Chen. 2020a. "Materials Chemistry for Rechargeable Zinc-Ion Batteries". *Chemical Society Reviews* 49 (13): 4203–4219. 10.1039/c9cs00349e.

Zhang, Qi, Jingyi Luan, Yougen Tang, Xiaobo Ji, and Haiyan Wang. 2020b. "Interfacial Design of Dendrite-Free Zinc Anodes for Aqueous Zinc-Ion Batteries". *Angewandte Chemie International Edition in English*. 10.1002/ange.202000162.

Zhang, Tengsheng, Yan Tang, Shan Guo, Xinxin Cao, Anqiang Pan, Guozhao Fang, Jiang Zhou, and Shuquan Liang. 2020c. "Fundamentals and Perspectives in Developing Zinc-Ion Battery Electrolytes: A Comprehensive Review". *Energy and Environmental Science*. 10.1039/D0EE02620D.

Zhao, Qinghe, Xin Chen, Ziqi Wang, Luyi Yang, Runzhi Qin, Jinlong Yang, Yongli Song, et al. 2019. "Unravelling H + / Zn 2 + Synergistic Intercalation in a Novel Phase of Manganese Oxide for High-Performance Aqueous Rechargeable Battery". *Nano Micro Small* 1904545: 1–10. 10.1002/smll.201904545.

Zheng, Jingxu, Jiefu Yin, Duhan Zhang, Gaojin Li, David C. Bock, Tian Tang, Qing Zhao, et al. 2020. "Spontaneous and Field-Induced Crystallographic Reorientation of Metal Electrodeposits at Battery Anodes". *Science Advances* 6 (25). 10.1126/sciadv.abb1122.

Zheng, Jingxu, Qing Zhao, Tian Tang, Jiefu Yin, Calvin D. Quilty, Genesis D. Renderos, Xiaotun Liu, et al. 2019. "Reversible Epitaxial Electrodeposition of Metals in Battery Anodes". *Science* 366 (6465): 645–648. 10.1126/science.aax6873.

Zhu, Kaiyue, Tao Wu, Shichen Sun, Yeting Wen, and Kevin Huang. 2020. "Electrode Materials for Practical Rechargeable Aqueous Zn-Ion Batteries: Challenges and Opportunities". *ChemElectroChem*. 10.1002/celc.202000472.

"Zinc-Bromine Battery Market to Reach US\$ 23 Mn by 2027". n.d. Accessed November 14, 2020. https://www.transparencymarketresearch.com/zinc-bromine-battery-market.html.

5

Advanced Materials for Energy Storage Devices

Fekadu Gashaw Hone[1], Newayemedhin A. Tegegne[1], and Dinsefa Mensur Andoshe[2]
[1]*Addis Ababa University, Department of Physics, Addis Ababa, Ethiopia*
[2]*Adama Science and technology University, Department of Materials Engineering, Adama, Ethiopia*

5.1 General Introduction

In the day-to-day activities of human life, energy has become an important variable that determines the quality of life. Moreover, in modern day the need for energy utilization has been increasing and highly dependent on fossil fuel energy sources. However, depletion of fossil fuel energy sources and environmental concerns have become the biggest challenges to the human race. Basic human needs for all residential, commercial, transportation, and industrial activities are met by the support of energy. From manufacturing products to heating or cooling the buildings, all functions require energy (Zhang et al. 2013; Liu et al. 2014). In the past years, fossil fuels have been overexploited as the primary source of energy for industries and people's daily life activities around the world. Therefore, the fossil fuel resource shortage and environmental pollution due to the burning of fossil fuels have been motivating researchers to develop alternative renewable energy sources and storage devices and mechanisms (Fernandez et al. 2009). Nanomaterials offer many advantages in energy storage applications. Energy storage involves chemical and/or physical interaction reaction mostly at the surface or interface, so that the specific surface area, surface energy, and surface chemistry play a very important role (Zhang et al. 2013). The surface impacts are not limited to the kinetics and rate only; surface energy and surface chemistry can have significant influences on the thermodynamics of heterogeneous reactions occurring at the interface and the nucleation as well as subsequent growth when phase transitions are involved. The size and dimensions of nanomaterials may also offer more favourable heat, mass, and charge transfer, as well as accommodate dimensional changes associated with some phase transitions and chemical reactions. The development of nanotechnology has also contributed to the advance in energy storage mechanisms and devices, nanotechnology being a critical technology in this century (Liu et al. 2014). Nanomaterials also introduce new challenges in the application of energy storage. For example, smaller pores may limit the penetration of electrolyte ions in supercapacitors (Fernandez et al. 2009; Wang et al. 2020). The increasing demand for high energy-density and high power-density storage systems has motivated researchers to design and develop environmentally friendly energy-storing materials and devices. The Ragone plot (Figure 5.1) shows different energy storage technologies currently available and studied worldwide (Zou et al. 2020). Electrochemical energy storage systems with high comprehensive performance, such as lithium-ion batteries (LIBs), sodium-ion batteries (SIBs), hybrid supercapacitors (HSCs), metal-air batteries (MABs), and hydrogen storage, are very promising technologies to overcome the global energy challenge (Chiang 2010; Yu and Lou 2017).

Due to their inferior electrochemical performances and high cost, the large-scale application of most of the electrochemical energy technologies is still hindered. In this regard, numerous efforts have been made to design advanced and cost-effective electro-active materials so as to optimize the performance

DOI: 10.1201/9781003145585-5

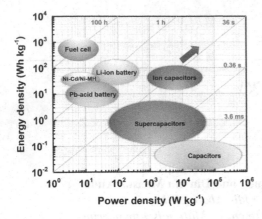

FIGURE 5.1   Ragone plots illustrate the performances of various energy storage devices and conversion in terms of specific power density vs. specific energy density. *Source*: (Aravindan et al. 2014)

of these electrochemical energy systems. To reach the goal of high energy density, three fundamental requirements must be satisfied by battery materials, which are often manifested on electrodes: (i) high specific charge (Ah kg^{-1}) and charge density (Ah l^{-1}), that is, a high number of available charge carriers per unit mass or volume of the material; (ii) high (for cathode electrode) and low (for anode electrode) standard redox potential for the respective electrode redox reaction, leading to high cell voltage; and (iii) highly reversible reactions at both anode and cathode electrodes to maintain the specific charge for hundreds of charge and discharge cycles (Kathy 2014). In addition, supercapacitors have gained significant attention due to their high power density; fast charge/discharge rate, and good cycle stability against galvanostatic charge/discharge (Yadav and Chavan 2017). The performance of supercapacitors/batters strongly depends on the electrode and electrolyte materials. So, active electrode and electrolyte materials with all the desired terms are still a question. Several books and review articles have been published dealing with the fundamentals and technical approaches for the design, fabrication, and characterization of nanomaterials for energy storage applications (Balbuena and Seminario 2006; Bruce et al. 2008). This chapter takes a few selected topics and focuses on the relatively recent progress, to highlight the most recent developments and the promise and limitations of nanomaterials in energy storage applications. It is also the intent of the authors to present the reader with basic information and slightly different perspectives on nanomaterials for energy storage applications. In this review, we will focus and discuss the applications of advanced nanomaterials for batteries and supercapacitors. One of the reasons these two technologies are chosen for this review is to identify research gaps, challenges, and possible future work for energy storage devices. Another reason is that these two fields deal with different fundamental and technical challenges, but are connected by the common use of nanostructured materials to form electrodes for electron/mass transport under an electrochemical environment. This chapter focuses on reviewing the most recent publications, with the majority being published in the past four years.

5.2 Supercapacitors

Supercapacitors (SCs) are one of the energy storage systems that are able to store and deliver energy at relatively high rates because the mechanism of energy storage is the simple charge separation at the electrochemical interface between the electrode and the electrolyte. SCs contain one positive electrode with electron deficiency and one negative electrode with electron excess, and the two electrodes can be built from the same or different materials (symmetric or asymmetric capacitors) (Kathy 2014). They are a special class of electrochemical devices of the capacitor type with a high density of electrical energy through ion adsorption ranging up to several joules per cm^3 (Kathy 2014). One particular advantage for supercapacitors is that they have several orders of magnitude higher energy density than that of conventional dielectric capacitors (Wang et al. 2009). Since charging-discharging occurred on the surface, which does not induce drastic structural changes upon electro-active materials, supercapacitors possess

TABLE 5.1

Basic Performance Comparison of Various Electrochemical Energy Storage Devices (Sarita Yadav 2020)

No.	Parameter	Capacitor	Supercapacitor	Battery
1	Specific Energy (Wh kg^{-1})	0.01 –0.1	1–10	10–20
2	Specific Power (W kg^{-1})	10^4 –10^7	10^2– 10^5	1–10^3
3	Discharge time	10^{-6}–10^{-3} s	Sec. to Min.	0.3–3 h
4	Charge time	10^{-6}–10^{-3} s	Sec. to Min.	1–4 h
5	Cycle time	Almost infinite	>100,000	500–2000

excellent cycling ability (Wang Huilin et al. 2020). Table 5.1 illustrates a comparison of basic performance parameters of capacitors, batteries, and supercapacitors. The characteristic features of supercapacitors are described based on their working principle, electrochemical behaviour, electrode materials, and electrolyte (Sarita Yadav 2020). Thus supercapacitor can be a serious alternative to batteries when the amount of energy to be stored is relatively small for a battery and the peak power requirement of the system is large compared to the average power (Burke 2014). In this section, the latest progress in supercapacitors, pseudocapacitive and lithium-ion capacitor charge storage mechanisms, electrode and electrolyte materials will be discussed. Moreover, the prospects and challenges associated with these energy storage devices are also discussed.

5.2.1 Classifications of Supercapacitors

The SCs can be classified into several classes owing to the energy storage mechanism and the active materials utilized. The first one is electrochemical double-layer capacitors (EDLCs), and the second one is pseudocapacitors or redox SCs, but there are other classes such as hybrid capacitors and battery-type capacitors (see Figure 5.2) (Utetiwabo et al. 2020). Each of them is defined by the charge storage mechanism. The capacitance of SCs mainly arises from surface reactions of electrode materials, including electrochemical adsorption /desorption of cations and anions at the electrode/electrolyte interface. In EDLCs the charge storage mechanism is like conventional capacitors. However, the EDLCs are characterized by large surface area, enabled by the porous nature of electrode materials, and the use of electrolyte that contains positive and negative ions dissolved in a solvent like water (Sarita Yadav 2020). Mathematically, the capacitance (C) of EDLCs is related to the surface area (A), the effective dielectric constant (ε), and the double layer thickness (d) by an inverse linear relationship ($C = \varepsilon A/d$). To achieve high capacitance, the combination of small charge separation and high surface area is crucial. For instance, a typical smooth surface will have a double-layer capacitance of about 10 to 20 μF cm^{-2}, however, if a high surface area electrode is used, the capacitance can be increased to 100 F g^{-1} for conducting materials with a specific surface area of 1000 m^2 g^{-1} (Kotz and Carlen 2000). Also, EDLCs can deliver ultra-high power density and an excellent lifespan due to the non-degrative processes between the electrode and the electrolyte (Utetiwabo et al. 2020).

In EDLCs the energy is stored via reversible adsorption and desorption of electrolyte ions at the electrode interface, with no net transfer of charges between the two mediums (Soc et al. 2012). When the cell is being charged, one electrode becomes positively charged and the other becomes negatively charged. At this point, the ions in the electrolyte are attracted to the surface of the electrodes (anions to the positive electrode and cations to the negative electrode). This creates a "double-layer" at the interface of the electrode surface and the electrolyte. Figure 5.3 shows a schematic of this process.

Because of the difference in electrochemical potential between the electrodes, the open circuit potential of the cell is varied, which responds to the passage of electric charges in the external circuit. The energy stored in this process is purely electrical, where no contribution occurs from the chemical reactions or faradaic reactions between electrode and electrolyte (Choi and Yoon 2015). EDLCs have large cycling stabilities due to negligible morphological or volumetric changes that occur in the electrode materials. To maximize the charge storage capacity, the electrode materials are usually made from

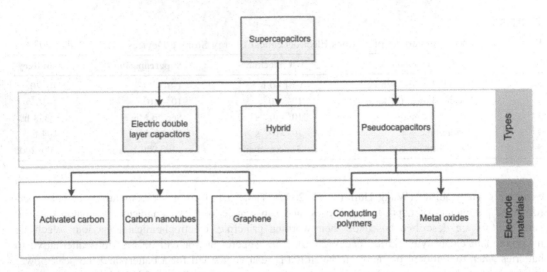

FIGURE 5.2 Classification of different supercapacitors with their electrode materials. *Source*: (González et al. 2016)

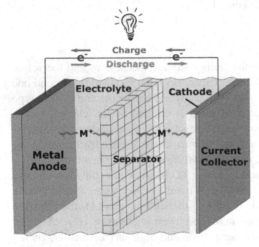

FIGURE 5.3 Schematic illustration of the charging/discharging process in a basic EDLC supercapacitor. *Source*: (Miller et al. 2018)

highly porous carbon materials. These materials possess unique structural and mechanical properties such as large surface area, high chemical and mechanical stability, excellent electrical conductivity, and low cost (Li and Wei 2013). Pseudocapacitors (PCs) are based on faradaic charge storage in which electrons are produced by redox reactions and transferred across the electrode–electrolyte interface. This mechanism is illustrated below in Figure 5.4. These reactions are in addition to the EDLC, and can significantly increase the capacitance and energy density of the SC. The charge storage attainable by this mechanism is approximately one order of magnitude higher than for the electric double-layer (EDL) mechanism (Conway and Pell 2003). However, the charge transfer and build up of charge on the surface produce electrostatic stresses, which likely leads to mechanical strain within the electrode resulting in a decrease in the charge storage capacity.

Pcs typically lose capacity with repeated charging and discharging much more rapidly than double-layer capacitors. This is often referred to as "fading". High charge/discharge rates also appear to produce extra stress, yet leading to shorter and effective device lifetimes (Yang and Ionescu 2017). The most used materials for PCs are conducting polymers and metal oxides like ruthenium dioxide (Ming Chen et al. 2014). Ruthenium dioxide is probably the most studied example and the highest capacitance is reported. It has low equivalent series resistance (ESR) and very high specific capacitance. But its high

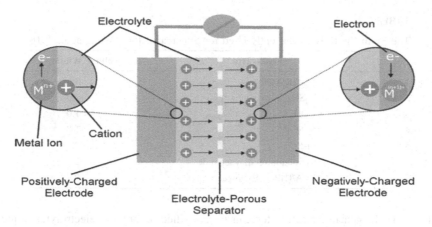

FIGURE 5.4 Schematic illustration of the charge storage mechanism of a basic pseudocapacitor. *Source*: (Kopczyński et al. 2016)

cost compared to other transition metal oxides has caused researches to diversify into alternate possibilities (Najib and Erdem 2019). Conducting polymer-based PCs have comparably high capacitance, conductivity, and low ESR than carbon-based materials (Ramya et al. 2013). On the other hand, transition metal oxides have high specific capacitance, conductivity, power density, energy density, and low resistance, which make them potential candidates for electrochemical energy storage applications (Augustyn and Dunn 2014). Capacitance contribution of PCs is 10 to 100 times larger than EDLCs, which is due to faradaic redox reactions. However, the power performance and cycling stability are much lower than EDLCs, due to slow faradaic mechanism as well as swelling and phase transformations of electrode materials during charge-discharge processes (Lu et al. 2011). Hybrid capacitor is one of the classes of supercapacitors with better energy and power densities with long-term cycling stability (Dubal et al. 2014). Two factors play a key role in deciding the energy density of SCs: cell voltage and electrode capacitance. With a correct electrode combination, it is possible to increase the cell voltage, which in turn leads to an improvement in energy and power densities. Several combinations have been tested in the past with both positive and negative electrodes in aqueous and inorganic electrolytes. Generally, the faradic electrode results in an increase of energy density at the cost of cyclic stability, which is the main drawback of hybrid devices compared to EDLCs. Currently, researchers have focused on the three different types of HScs, which can be distinguished by their electrode configurations: composite and asymmetric (Simon and Naoi 2008; Iro et al. 2016). There are several electrode configurations which resulted in different types of hybrid SCs. The first one is asymmetric hybrids, they comprised of EDLC and PC electrodes. These materials have higher electrochemical cycling stability, energy and power densities than EDLCs (Song et al. 2018). Composite hybrids integrate both carbonaceous and pseudocapacitive electrode materials within the same electrode is another approach. Porous carbon materials act as a backbone with a large surface area, to improve the conductivity and mechanical strength of conducting polymers or metal oxides, thus enhancing the electrochemical behaviour (Ates and El-Kady 2018). Lithium-ion capacitor is one of hybrid electrochemical energy storage systems and will be discussed in detail in the next section.

5.2.2 Electrolyte for Supercapacitor

It is well recognized that the working potential of the SCs is highly reliant on the electrochemical stability of the electrolytes. The electrolyte, including solvent and salt, is one of the most important constituents of electrochemical supercapacitors due to its advantages of ionic conductivity and charge compensation on both electrodes of the cell. Up to now, various types of electrolytes have been developed by researchers (Wang et al. 2016). In addition to the operating voltage, the electrolytes have substantially influenced the other parameters such as power density, cycling stability, operating temperature, equivalent series resistance, lifetime, and self-discharge rate of the capacitors

TABLE 5.2

The Main Electrolytes Currently Used for Supercapacitors (Wang et al. 2016)

No.	Electrolyte	Voltage Window (V)
1	TEABFA/PC	3.5
2	H_2SO_4	1.0
3	KOH	1.0
4	$NaSO_4$	1.8
5	Li_2SO_4	2.2
6	Pyrrolidinium dicyanamide	2.6
7	PVA/H_3PO_4 Hydrogel	0.8

(Yan et al. 2014). In general, electrolytes can be classified as organic electrolytes, aqueous electrolytes, ionic liquids, and solid-state polymer electrolytes (Wang et al. 2016). The main electrolytes currently used for SC devices are tabulated in Table 5.2. Organic electrolytes and ionic liquid (IL)-derived SCs can easily be handled at a large potential window of 2.5–2.7 V and 3.5–4.0 V, respectively (Jiang et al. 2012). There are still some drawbacks associated with organic electrolytes, such as higher cost, lower specific capacitance, and inferior conductivity, as well as safety concerns associated with flammability, volatility, and toxicity.

The typical organic electrolytes for commercial EDLC consist of conductive salts (e.g. tetraethyl-ammonium tetrafluoroborate [TEABF]) dissolved in acetonitrile or propylene carbonate (PC) solvent (Zhong et al. 2015). Although acetonitrile can dissolve much more salt than other solvents, it is harmful to the environment. In contrast, PC-based electrolytes are not only friendly to the environment but also provide a wide voltage window, a wide range of working temperatures, and good electrical conductivity. Therefore, the organic electrolyte, TEABF/PC, has been widely used in EDLC studies (Jung et al. 2013). However, two issues should be kept in mind. One is that the ion size of the selected electrolyte should be matched with the pore diameter of carbon materials to achieve the maximum specific capacitance. Otherwise, the accessibility for electrolyte ions will be limited. The other is that organic electrolytes must contain very little water, that is, below 3–5 ppm. Otherwise, the capacitor's voltage will be greatly reduced. Recently, ionic liquids have received significant interest as alternative electrolytes for supercapacitors because of their negligible volatility, high thermal, chemical, and electrochemical stability, low flammability, and wide electrochemical stability window of 4.5 V, as well as good conductivity of *ca.* 10 mS cm^{-1} (Zhong et al. 2015). They are composed of ions (cations and anions) with melting points below 100^0C (De Adhikari et al. 2020). Ionic liquids are classified into three major types: (1) protic (ionic liquid that comprises of a labile H^+), (2) aprotic (ionic liquid molecules do not have a hydrogen atom attached to an atom of an electronegative element), and (3) zwitterionic (molecules consisting of two or more functional groups in ionic form) (Aldama et al. 2017). The main ionic liquids investigated for SCs are pyrrolidinium, imidazolium, or aliphatic quaternary ammonium salts coupled with such anions as PF_6^-, BF_4^-, $TFSI^-$, or FSI^- (Mousavi et al. 2016). Solid polymer electrolyte-based SCs have attracted great interest in recent years due to the rapidly growing demand for power for various types of electronics. The solid polymer electrolytes can act as ionic conducting media and electrode separators. There are three types of polymer-based solid electrolytes for SCs: dry polymer electrolyte, gel polymer electrolyte, and polyelectrolyte (Yan et al. 2020). Among these, the gel polymer electrolyte has recently been the most extensively investigated electrolyte because of its high ionic conductivity, stable electrochemical performances, being leakage-free, and providing more opportunities for structure design (Jin et al. 2018). Additionally, the gel polymer electrolyte is known as a hydrogel polymer electrolyte when using water as the plasticizer. And this type of hydrogel polymer electrolyte generally possesses three-dimensional polymeric networks (Huang Yuan et al. 2019). Recently, redox couples have been added into the electrolyte to increase the charge storage capability via redox transformation and thus enhance the electrochemical performances (Su et al. 2009). According to the various reaction mechanisms, redox electrolytes are mainly classified into redox-additive electrolytes and redox-active electrolytes (Akinwolemiwa et al. 2015). It should be pointed that redox mediator is the key component of redox electrolytes. Normally, redox mediators are divided into inorganic and organic redox mediators, respectively. The inorganic mediators like transition metal ions, halide ions, and cyanide ions are the common redox additives for aqueous electrolyte, offering pseudocapacitance via chemical valence variation, while the organic redox mediators (e.g. methylene blue, indigo carmine, and *p*-phenylenedi-amine) can achieve capacitance improvement via the conjugated effects (Huang Yuan et al. 2019; Gou et al. 2020). Generally, to date, there has been no perfect electrolyte developed, meeting all the requirements discussed previously. Each electrolyte has its own advantages and disadvantages (Zhong et al. 2015). An ideal electrolyte of supercapacitor needs to have some fundamental requirements: (1) broad potential window, (2) a wide range of working temperature, (3) high ionic conductivity, (4) low viscosity, (5) high electrochemical stability, (6) environmentally friendly, (7) low cost, and (8) low flammability. Each electrolyte has its merits and drawbacks, and it is feasible to meet all the above specifications with one electrolyte. Nonstop and tremendous research efforts have been made in the present and will also keep going in the future for the electrolyte development investigation.

5.2.3 Advanced Electrode Materials for Supercapacitor

A SC stores energy using either ion adsorption (EDLCs) or fast and reversible faradic reactions (PSCs). These two mechanisms can function simultaneously depending on the nature of the electrode material (Zhang et al. 2016). The selection of electrode materials and their fabrication plays a crucial role in enhancing the capacitive performance of SCs. Electrodes of SCs must provide thermal stability, high SSA, corrosion resistance, high electrical conductivity, appropriate chemical stability, and suitable surface wettability. They should also be low-cost and environmentally benign. Besides, their capability of transferring the faradic charge is important to enhance the capacitance performance (Lai et al. 2012; Xie et al. 2016). The progress of SC technologies continues mostly towards the development of nanostructured electrode materials. From the morphology point of view, to develop high-performance electrode materials, scientists have designed and integrated one-dimensional (1D) (nanotubes and nanowires), two-dimensional (2D) (nanosheets and nanodiscs), and three-dimensional (3D) nanoarchitectures into electrode materials for the production of SCs and batteries as well (Najib and Erdem 2019). Therefore, both synthesis and the design of storage systems play a vital role in obtaining high performance from the device. In general, the electrode materials are categorized based on four main classes, including carbonaceous materials like activated carbons, carbon fibres, graphene, and carbon nanotubes (Han et al. 2013; Md Moniruzzaman et al. 2016). Transition-metal oxides, and conducting polymers (CPs) like polyaniline (PANI), polypyrrole (PPy), polythiophene (PTh)) (Shi et al. 2014; Forouzandeh et al. 2020), and their derivatives have been explored extensively as electrode materials. Most of SC research currently being conducted focuses on electrode materials based partly or entirely on carbon. The most common type of these materials is activated carbon (AC), which is carbon that has been put through chemical or physical processes in order to increase its porosity and adsorptive capabilities. Presently other carbon-based materials researchers use that do not exactly fall into the category of ACs increased rapidly, and their performance as specific capacitance (SC) electrodes often exceeds that of standard ACs. Transition metal oxides are excellent pseudocapacitive electrode materials. Those metal oxides typically have several redox states or structures and contribute to the charge storage in pseudocapacitors via fast redox reactions. The most active structure is that of hydrous ruthenium oxide because of its intrinsic reversibility for various surface redox couples and high conductivity (Zhang et al. 2009). Other simple metal oxides usually have some limitations such as poor electrical conduction, insufficient electrochemical cycling stability, limited voltage operating window, and low specific capacitance. Mixed binary or ternary metal oxide systems, such as Ni–Mn oxide, Mn–Co oxide, Mn–Fe oxide, Ni–Ti oxide, Sn–Al oxide, Mn–Ni–Co oxide, Co–Ni–Cu oxide, and Mn–Ni–Cu oxide have shown improved properties as electro-active materials for pseudocapacitors and have shed new lights in this area of research (Wanga et al. 2018). In addition, transition metal sulphides, such as cobalt sulphides and nickel sulfides, have been investigated as a new electrode material for pseudocapacitors with a good performance (Zhu et al. 2018). Particularly, $NiCo_2S_4$ exhibited an electrical conductivity about 100 times that of $NiCo_2O_4$ and approximately 10^4 times higher than conventional mono-metal counterparts (Wu et al. 2017). In general, transition metal oxides proved to achieve much higher SC than other types of electrode materials in regards to their high theoretical capacitance through the occurrence of Faradaic redox reactions (Md Moniruzzaman et al. 2016). Extensive research has been carried out on various metal oxide–based electrode materials like ruthenium oxide, iridium oxide, manganese oxide, cobalt oxide, nickel oxide, zinc oxide, and iron oxide (Qu et al. 2011; Dubal et al. 2013). Conducting polymers (CPs) are commonly found in composite materials used for SC electrodes. The energy density of redox capacitors consisting of conducting polymers is higher than that of double-layer capacitors composed of carbon materials because conducting polymers can store charge not just in the EDL but also through the rapid faradic charge transfer called *pseudocapacitance* (Zhang et al. 2016). Several research groups reported different approaches to come up with better electrode materials for SCs. For instance, Bidhan Pandit et al. synthesized Bi_2S_3:PbS solid-solution thin-film electrodes by using low-cost SILAR methods. The thin films exhibited excellent specific capacitance of 402.4 F g^{-1} at current density of 1 mA cm^{-2} with modest charge-discharge cycles. In terms of energy storage, it exhibited maximum specific power of 20.1 Wh kg^{-1} with accepting specific power of 1.2 kW kg^{-1} (Pandit et al. 2017). Ni foam supported Co_3O_4 nanoflakes reported by Shuying Kong et al. and the nanoflakes were synthesized through a simple three-step route. The unique architecture's morphology of the metal oxide electrodes with porosity and interconnected channels enhances not only the surface area but also the ion/electron diffusion. The electrochemical tests show that Co_3O_4 nanoflakes exhibit a high specific capacity up to 576.8 C g^{-1} at a current density of 1 A g^{-1} and remain at 283.7 C g^{-1} capacity at a high current density of 50 A g^{-1}, as well as 82% capacitance retained after 5000 cycles. The above results demonstrate the great potential of Co_3O_4 nanoflakes in the development of battery-type SCs (Kong et al. 2017). Recently, Chaojing Yan et al. prepared a novel flexible polyelectrolyte-based gel polymer electrolyte (PGPE) by using polymerization process assisted with UV light. The PGPE exhibits favourable mechanical strength and excellent ionic conductivity (66.8 mS cm^{-1} at 25 degree). In addition, the all-solid-state supercapacitor fabricated with a PGPE membrane and activated carbon electrodes shows outstanding electrochemical performance. The specific capacitance of the PGPE supercapacitor is 64.92 F g^{-1} at 1 A g^{-1}, and the device shows a maximum energy density of 13.26 Wh kg^{-1} and a maximum power density of 2.26 kW kg^{-1}. After 10,000 cycles at a current density of 2 A g^{-1}, the all-solid-state supercapacitor with PGPE reveals a capacitance retention of 94.63%. Furthermore, the authors pointed out that the specific capacitance and charge-discharge behaviours of the flexible PGPE device hardly change with the bending states. Porous $NiCo_2S_4$/CNTs nanocomposites were synthesized by Yunxia Huang et al. using a hydrothermal method followed by the sulphurization process using different sulphide sources. By comparing two different sulphur sources, the samples using thioacetamide as sulphide source delivered more remarkable electrochemical performance with a high specific capacitance of 1765 F g^{-1} at 1 A g^{-1} and good cycling stability with capacitance retention of 71.7% at a high current density of 10 A g^{-1} after 5000 cycles in 2M KOH aqueous electrolyte. Furthermore, an asymmetric supercapacitor (ASC) device was successfully fabricated with the $NiCo_2S_4$/CNTs electrode as the positive electrode and graphene as the negative electrode. The device provided a maximum energy density of 29.44 Wh kg^{-1} at a power density of 812 W kg^{-1}. Even at a high power density of 8006 W kg^{-1}, the energy density still reaches 16.68 Wh kg^{-1}. Moreover, the ASC presents 89.8% specific capacitance retention after 5000 cycles at 5 A g^{-1} (Huang et al. 2018). Rare earth elements have also been applied in SCs. An ultra-high specific capacitance of 2060 F g^{-1} could be obtained by the direct use of commercial $Ce(NO_3)_3$ as an electrode material in KOH electrolyte without any additional processing. Ce^{+3}/Ce^{+4} could deliver a high practical specific capacitance close to its theoretical value (Chen and Xue 2014). Graphene a one-atom-thick layer 2D structure has emerged as a unique carbon material that has potential for energy storage device applications because of its superb characteristics of high electrical conductivity, chemical stability, and large surface area (Liu et al. 2010; Iro et al. 2016). PANI/rGO composites with a macroscopically phase-separated structure were prepared through electrochemical deposition by J. Wu et al. The composite electrode exhibited a high specific capacitance of 783 F g^{-1} (8773 mF cm^{-2}) at a high current density of 27.3 A g^{-1} (305.7 mA cm^{-2}), which is 99% of the value at a current density of 1.14 A g^{-1}, demonstrating an excellent rate performance. The authors attributed that the phase-separated structure is beneficial to the diffusion of electrolyte, and thus can significantly improve the electrochemical performance of the composites

(Wu et al. 2016). Recently, S.J. Rowley and C.E. Banks have reported an asymmetrical SC device based on graphene oxide (GO). The device has been fabricated by a facile screen-printing technique, where the capacitive performance is found to increase from 0.82 F g^{-1} to 423 F g^{-1}, after incorporating GO. The device has shown power density of up to 13.9 kW kg^{-1} and an energy density of 11.6 Wh kg^{-1} (Rowley-Neale and Banks 2018).

5.3 Li-Ion Capacitors

Lithium-ion capacitors (LICs) were first developed by Amatucci et al. in 2001 (Amatucci et al. 2001), since then, the studies on LICs have increased rapidly. LICs are hybrid electrochemical energy storage systems that combine chemical reactions: faradaic at the anode where intercalation occurs and non-faradaic at the cathode where only surface adsorption-desorption occurs. The structure is made of lithium-ion battery anode materials (hard carbon) and electrochemical double-layer capacitor cathode (activated carbon) materials (Cao and Zheng 2013; Bolufawi et al. 2019). SCs have a high power density (10 kW kg^{-1}) and long cycle life but suffer from low energy density (only 5–10 Wh kg^{-1}). On the other hand, lithium-ion batteries can provide high energy densities (150–200 Wh kg^{-1}), but their power densities are relatively low (below 1 kW kg $^{-1}$) and their cycle life is quite poor (usually less than 1000 cycles). However, in the view of applications in electronic devices and hybrid electric vehicles, energy storage devices with both high energy and power density are in demand now (Dubal et al. 2015). Hybrid lithium-ion capacitors (LICs) have been recently proposed as a way to bridge the gap between LIBs and SCs by combining the merits of the two systems (Ma et al. 2015). Factors that affect the LIC performance are the pre-lithiation process of the negative battery electrode, the carbonaceous active electrode materials, safe operational limits of the electrolyte, and the configuration of the electrodes for hybrid cell design. Pre-lithiation of the negative battery electrode (e.g. hard carbon) for LICs is required and improves the working voltage, energy density, and cycling stability. The pre-lithiated battery electrodes exhibit better retention of the irreversible capacity and lower electrode resistance (Brandt et al. 2014). Based on the function of cathode and anode LICs are divided into two. In the first case, the battery-type electrode acts as the anode and the capacitor-type electrode serves as the cathode, such as a Li$_4$Ti$_5$O$_{12}$//AC system (see Figure 5.5). For the second case, the capacitor-type electrode acts as the anode and the battery-type electrode serves as the cathode, such as an AC//LiFePO$_4$ system (see Figure 5.5) (Shang et al. 2018; Wang Yumei et al. 2019). In Li-ion capacitors, Li$^+$ sources are included in the electrolyte in the form of LiPF$_6$ ethylene carbonate (EC) and dimethyl carbonate (DMC) (Cao and Zheng 2012). As a result, their energy densities are much larger than those of supercapacitors. Unlike rechargeable batteries, the electrodes of Li-ion capacitors function based on adsorption/desorption and thus facilitate fast kinetics for discharge/charge, which enables higher power densities than those of rechargeable batteries.

Furthermore, Li-ion capacitors do not undergo significant volume expansion for the electrodes and can therefore afford to run over a large number of cycles, which could be even comparable to the cycle lives of SCs (Kathy 2014). In LICs the two electrodes operate reversibly in different potential ranges,

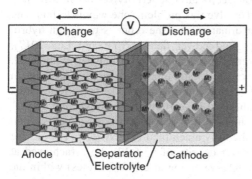

FIGURE 5.5 The charge storage mechanisms of LICs. *Source*: (Wang Yumei et al. 2019)

thus increasing the operation voltage and providing an opportunity to achieve fast charge capability, robust cycle life, and high energy density. The challenge to fabricate such a high-performance hybrid supercapacitor is to couple appropriately both high-performance capacitor-type and battery-type electrode materials in the devices.

5.3.1 Electrolyte for LICs

One of the three basic elements of LICs, the electrolyte plays an essential role in improving all aspects of the performance of LICs. The electrolyte in LICs serves as the medium for the transfer of charges between a pair of electrodes (Li Bing et al. 2018). Electrolytes constitute any solution that is ionically conducting yet electronically insulating and is required to facilitate the movement of ions from cathode to anode and vice versa. Organic solvents, lithium salts, and additives are the major aspects of interest in LIC electrolyte research (Xu 2010; Xu and Av 2011). In the case of LICs a stable electrolyte usually consists of a combination of alkyl carbonates, including organic solvents of ethylene carbonate (EC), propylene carbonate (PC), dimethyl carbonate (DMC), diethyl carbonate(DEC), ethyl-methyl carbonate (EMC), etc. The electrolyte solution must also contain lithium salts like $LiPF_6$ or $LiBF_4$ to facilitate free ions (Aurbach et al. 2004; Khomenko et al. 2008). Kwon and coworkers investigated the effect of lithium difluoro(oxalate) borate (LiDFOB) as an electrolyte additive on the electrochemical performance of an LIC of $LiMn_2O_4$/AC at 10 C within 0–2.5 V. The results showed that the additive not only led to retardation of the phase transition of $LiMn_2O_4$ during repeated charging-discharging cycles and reduced the charge-transfer resistance of the cell but also reduced the polarization of the electrode with a decreased over-potential. Moreover, the interfacial stability was increased to achieve an excellent cycle life of the LIC of AC/$LiMn_2O_4$ (Li Bing et al. 2018).

5.3.2 Recently Developed Electrode Materials for LICs

Among several types of hybrid capacitors proposed and developed, LICs are the most promising ones (Yan et al. 2013). One of the key difficulties in developing high-performance LICs is to increase the energy and power density simultaneously with enhanced safety benefit (Wang et al. 2016). For the high-capacity Li intercalation anodes, the biggest challenge is to control the cycling-induced electrode degradation like volume change, inactivation, and pulverization at high rates to match the high power characteristic of cathodes (Pan et al. 2015). Accordingly, the electrode materials play a crucial role in the electrochemical performance of LICs. Therefore, extensive studies have been undertaken to prepare excellent electrode materials. As for the battery-type anode of LICs, various intercalation materials have been discussed, such as carbonaceous materials, titanium-based compounds, and manganese oxides. Among these compounds, graphite is still widely and commercially used in LICs because of its low lithium-ion intercalation potential (Wang et al. 2015; Zhang et al. 2015). The cathode is an important part of the LIC. To ensure high power density and high energy density of the LICs, it is necessary to have a high operating voltage and good electrical conductivity. Cathode materials are divided into carbon materials, Li^+-intercalated compounds, and composite materials (Li Bing et al. 2018). As for the cathode, porous carbon is a promising candidate due to its large specific area and chemical stability. However, the poor electrical conductivity results in a limited power capacity of the hybrid system (Lei et al. 2013). To solve this problem, one way is to fabricate new electrode materials with high electrochemical performance, and the other is to design hybrid energy-storage devices that possess both high energy and power density (Yu and Lou 2017). Lithium-ion hybrid supercapacitors (LIHSs) have emerged as such a solution. Huanwen Wang et al. designed a new nanostructured LIC electrode that both exhibits a dominating capacitive mechanism for double-layer and pseudocapacitive with a diminished intercalation process. The electrodes are a 3D interconnected TiC nanoparticle chain anode, synthesized by carbothermal conversion of graphene/TiO_2 hybrid aerogels, and a pyridine-derived hierarchical porous nitrogen-doped carbon (PHPNC) cathode. Electrochemical study verified that the fully assembled PHPNC//TiC LIC device delivers an energy density of 101.5 Wh kg^{-1} and a power density of 67.5 kW kg^{-1} (achieved at 23.4 Wh kg^{-1}), and a reasonably good cycle stability (≈82% retention after 5000 cycles) within the voltage range of 0.0–4.5 V (Wang et al. 2016). An approach to further improve the performance of the lithium-ion capacitor has been demonstrated by Xiaoyu Gao et al. They used the graphitic porous carbon (GPC) and high-purity vein graphite (PVG) prepared from Sri Lanka graphite ore by KOH activation, and high-temperature purification. An electrochemical performance with a maximum energy density of 86 Wh kg^{-1} at 150 W kg^{-1} and 48 Wh kg^{-1} at a high-power density of 7.4 kW kg^{-1} was achieved at a relatively low cost. Moreover, 55.8% retention at 7.4 kW kg^{-1} was found (Gao et al. 2017). Fei Sun et al. designed an LIC system by integrating a rationally designed Sn-C anode with a biomass-derived activated carbon cathode. A new type of biomass-derived activated carbon featuring both high surface area and high carbon purity is also prepared to achieve high capacity for cathode. The assembled LIC (Sn-C//PAC) device delivers high energy densities of 195.7 Wh kg^{-1} and 84.6 Wh kg^{-1} at power densities of 731.25 W kg^{-1} and 24375 W kg^{-1}, respectively. This work offers a new strategy for designing high-performance hybrid system by tailoring the nanostructures of Li insertion anode and ion adsorption cathode (Sun et al. 2017). Various approaches have been reported to construct graphene-based architectures for LICs. For instance, X. Ma et al. fabricated micron-sized porous graphene belts (PGBs) by a chemical vapour deposition approach. According to the authors, the obtained PGBs had good structural integrity and exhibited good durability. Besides, the Li-ion capacitor fabricated with the Li$_4$Ti$_5$O$_{12}$ anode and PGBs cathode exhibits satisfactory Ragone performance, i.e. 8044 W kg^{-1} at 51 Wh kg^{-1} and 120 Wh kg^{-1} at 503 W kg^{-1} (Maa et al. 2018). Analogously, Shi et al. reported activated carbon nanofibre (a-PANF) with a hierarchical porous structure and a high degree of graphitization. Half-cell evaluation of the as-prepared a-PANF gave a discharge capacity of 80 mAh g^{-1} at 0.1 A g^{-1} within 2–4.5 V and no capacity fading after 1000 cycles at 2 A g^{-1}. Furthermore, an as-assembled LIC with a-PANF cathode and Fe$_3$O$_4$ anode showed a superior energy density of 124.6 Wh kg^{-1} at a specific power of 93.8 W kg^{-1}, which remained at 103.7 Wh kg^{-1} at 4687.5 W kg^{-1} (Shi et al. 2018a). Ruiying Shi et al. demonstrated the synthesis of hierarchical porous carbon with an extremely large specific surface area of 3898 m^2 g^{-1} and an improved graphitization degree by using egg white biomass as a precursor and NaCl as a template. The developed porous carbon exhibits a noticeably enhanced specific capacity of 118.8 mAh g^{-1} at 0.1 A g^{-1} with excellent rate capability and improved cycling stability over 4000 cycles in an organic Li-ion conducting electrolyte. Furthermore, the obtained porous carbon was employed as a cathode paired with a Fe$_3$O$_4$@C anode for LIC applications, which delivers an integrated high energy density of 124.7 Wh kg^{-1} and a power density of 16,984 W kg^{-1} as well as a superior capacity retention of 88.3% after 2000 cycles at 5 A g^{-1} (Shi et al. 2018a). Green hybrid lithium-ion capacitor electrodes were developed from Silica Rice Hull Ash (SRHA) an agricultural waste by Eleni Temeche et al. The electrochemical properties were assessed by assembling Li/SDRHA half-cells and LiNi$_{0.6}$Co$_{0.2}$Mn$_{0.2}$O$_2$ (NMC622)-SDRHA full-cells. The half-cell delivered a high specific capacity of 250 mAh g^{-1} at 0.5C and retained a capacity of 200 mAh/g at 2 C for 400 h. Moreover, the hybrid full-cell demonstrated a high specific capacitance of 200 F/g at 4 C. In addition, both the half and full hybrid cells demonstrate excellent Coulombic efficiencies about ~100% (Temeche et al. 2020). Similarly, J.L. Gomez-Urbano et al. reported an easy, eco-friendly, and cheap synthetic approach for the preparation of carbon composites from the pyrolysis and activation of coffee waste and graphene oxide for LIC electrode. Optimized electrodes are allowed to go one step beyond the state-of-the-art of biowaste-based dual carbon LICs in terms of energy, power, and cyclability. Assembled LICs show values of 100 Wh kg^{-1} at 9000 W kg^{-1} and retain above 80% of the initial capacitance after 3000 cycles, which is enhanced to 15,000 cycles by decreasing the voltage window (Gomez-Urbano et al. 2020). Graphite/copper oxide composite as the anode material and porous carbon as the positive electrode material were reported by Seong-Hun Lee et al. for LICs. The optimized hybrid cell shows a high specific energy density of 212.3 Wh kg^{-1} at a specific power density of 1.3 kW kg^{-1} and maintains 85% of its initial energy density after 500 cycles (Lee et al. 2020). High-energy LIC using the defect-rich and N-doped hard carbon (DNC) as anode has been successfully developed by J. Jiang et al. The DNC shows nanospherical structure with a diameter of about 100 nm. Owing to the two-pronged strategy of N-doping and defect engineering, it delivers a high specific capacity (580.3 mAh g^{-1} at 0.05 A g^{-1}), excellent rate capability, and long cycle stability (1000 cycles). The as-fabricated LIC delivers a remarkable energy density (101.7 Wh kg^{-1}), an outstanding rate capability 56.3 Wh kg^{-1} at 12.5 kW kg^{-1}. In addition, it exhibits a superior cycling life of 82.2% capacity retention after 3000 cycles, corresponding to 0.0059% fading per cycle (Jiang et al. 2020). Recently, Ruyi Bi et al. prepared triple-shelled (3S) Nb_2O_5 hollow multishelled structures (HoMSs) through the sequential tin-plating approach and then applied for the anode of LIC. The 3S-Nb_2O_5-HoMSs anode exhibited a high reversible capacity of 172.6 mA h g^{-1} after 2000 cycles at 1 C and high specific capacities of 135.2 and 108.1 mA h g^{-1} at the high rates of 10 C and 20C over 2000 cycles. Moreover, the fabricated 3S-Nb_2O_5-HoMSs//AC LICs showed outstanding energy and power densities (93.8 Wh kg^{-1} at 112.5 W kg^{-1} and 22.5 kW kg^{-1} at 19.6 Wh kg^{-1}) with good cycle stability (89 % of initial capacity after 10,000 cycles at 1 A g^{-1}) (Bi et al. 2020). Huanhuan Zhou et al. developed $Li_4Ti_5O_{12}$-TiO_2 (LTO-TO) composite and coated it on carbon foam (CF) for anode of LICs. The asymmetric LICs based on activated carbon cathode and CF@LTO-TO anode (AC//CF@LTO-TO) with 25 wt.% of LTO-TO delivers a specific capacity of 65 mA h g^{-1} at 300 C. In addition, slowest self-discharge was achieved from AC//CF@LTO-TO LIC with 45 wt.% of LTO-TO, which showed a low leakage current of 0.00575 Ag^{-1} and an OCV drop from 2.5 V to 1.1 V after one week. For CF@LTO-TO with 45 wt.% of LTO-TO, a high discharge capacity of 70 mAh g^{-1} at 300 C was attained (Zhou et al. 2020). Reduced graphene oxide (ErGO) with high porosity was synthesized by Yi Zhan et al. through high-voltage ECD to form a binder-free thin-film capacitor electrode for LICs. They found that increasing the film thickness did not linearly increase the areal capacitance, which was attributed to the resistive electrolyte diffusion through internal pores. Furthermore, a good capacitance as high as 168 F g^{-1} at 0.1 A g^{-1} was obtained by combining ErGO with V_2O_5 nanoparticles (Zhan et al. 2020).

5.4 Battery

The growing demand for electricity and depletion of fossil fuels have led to the increase in development of various non-conventional energy storage devices. Among those, batteries are the most significant energy storage devices (Wang et al. 2013). Electrochemical batteries are considered the most important devices for energy storage. It produces electricity by releasing the potential energy stored in the chemicals of the battery. Generally speaking, a battery consists of five major components – anode, cathode, the current collectors these may sit on, electrolyte, and separator, as shown in Figure 5.6. The purpose of the anode is to hold the active ions in a high energy state. The higher the energy state, the higher the eventual voltage of the cell. In principle, pure metal is the best anode material (Borah et al. 2020). The chemical reaction between the negative electrode and the positive electrode has two components: (i) electronic and (ii) ionic. The ionic component is driven by the electrolyte that forces the electronic component to circulate through an external circuit.

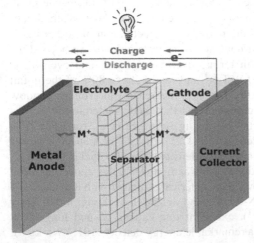

FIGURE 5.6 A typical cell format of battery. Charging processes are indicated in the left direction, and discharging processes are indicated in the right direction. On discharge, the high potential metal atoms oxidize, and the resultant ions move toward and interact with the reducing cathode. *Source*: (Borah et al. 2020)

Current collectors at the negative and positive electrodes deliver the electronic current to the external circuit (Martins et al. 2020).

During battery operation, some mechanisms behind the electrochemical process occur – the intercalation and the transformation processes. The intercalation process occurs by the insertion of metal ion during charging/discharging processes. During discharge, the ion moves within the electrolyte toward the positive electrode and interleaves into the material. During battery charging, the reverse process occurs, the ion now moves toward the negative electrode, which is the reason why this is known as a *reversible process*. In contrast to an intercalation material, transformation reactions can also take place in batteries, with the break and formation of chemical bonds during battery charge and discharge (Zhou et al. 2017; Martins et al. 2020). The increasing demand for portable electronic devices has led to drastic improvements in their performance. On this account, lithium-ion and sodium-ion batteries have gained considerable attention and have diverse applications (Iqbal et al. 2019).

5.4.1 Lithium-Ion Batteries (LIBs)

Lithium has long received much attention as a promising anode material. The interest in this alkali metal has arisen from the combination of its two unique properties: (1) it is the most electronegative metal (~−3.0 V vs. SHE) which yields a high output voltage when coupled with a cathode for high energy-density devices and (2) it is the lightest metal (0.534 g cm^{-3}) which can be translated into a very high theoretical specific capacity (3860 mAh g^{-1}) (Xu 2004; Jie et al. 2020). On the basis of the electrodes and electrolyte nature and physical state, several lithium-based battery classification schemes were proposed. Lithium battery (LB) is the common name given to primary (disposable) devices having lithium metal or a lithium compound as the anode. Lithium-ion battery (LIB) indicates a family of secondary (rechargeable) devices where both the electrodes are intercalation materials, and the electrolyte is commonly a lithium salt dissolved in a mixture of organic solvents (Li et al. 2016). LMBs use metallic lithium (Li) as anodes, which can be paired with a variety of cathode materials. LIBs are the most commonly used source of power for modern electronic devices and attracting much attention in the past few decades. LIBs can store a large amount of energy, but the slow kinetics in the electrochemical process restrain the rate of energy storing and releasing, or the charging current rate and output power density. There are plenty of applications that require high-power and high-rate energy storage with a much longer lifecycle where LIBs cannot meet the demand (Li and Fan 2020). LIBs are widely used as electrochemical sources in portable electronic applications including mobile phones, personal computers, and video cameras due to their favourable performance of electrochemical properties; they are also likely to play an important role in providing power for electric automobiles in the future (Deng et al. 2009; Sun and Qiu 2012). The main reason for such increasing the demand for LIBs is because of its very good electrochemical properties such as high energy density (120 Wh kg^{-1}), high voltage (up to 3–6 V), longevity (500–1000 cycles), wide temperature range (-200°C to 600°C) and minimum memory effect. Moreover, LIBs do not contain hazardous heavy metals such as cadmium and lead (Shin et al. 2015). LIBs are highly advanced as compared to other commercial rechargeable batteries, in terms of gravimetric and volumetric energy. Figure 5.7 compares the energy densities of different commercial rechargeable batteries, which clearly shows the superiority of the Li-ion batteries as compared to other batteries (Tarascon and Armand 2001). LIBs have occupied most of the growing market due to their outstanding merits in safety, operation lifespan, and energy density, which heavily eclipse other rechargeable batteries (such as lead-acid batteries). However, the rise of practical energy density of LIB devices is too sluggish to keep pace with the current requests by the portable electronics and electric vehicles. In 1990s, the best record of energy density for LIB devices was *ca.*90 Wh kg^{-1}. Currently, the energy density of LIB products mainly ranges from 120 Wh kg^{-1} to 220 Wh kg^{-1}. By optimization of electrode materials, electrolyte, separator, binder, and current collector, the practical energy density of LIBs is expected to reach its limit soon (Janek and Zeier 2016; Shen et al. 2018). The worldwide market in 2010 for rechargeable lithium batteries reached ~$11 billion per annum and continues to grow (Kam and Doeff 2012). They are the technology of choice for future hybrid electric vehicles, which are central to the reduction of CO_2 emissions arising from transportation (Bruce et al. 2008).

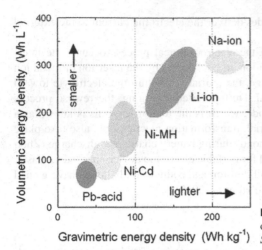

FIGURE 5.7 Comparison of energy densities and specific energy of different rechargeable batteries. Reproduced with permission. *Source*: (Tarascon and Armand 2001)

An LIB is constructed by connecting basic Li-ion cells in parallel (to increase current), in series (to increase voltage), or combined configurations. Multiple battery cells can be integrated into a module. Multiple modules can be integrated into a battery pack (Deng et al. 2009). The rechargeable lithium battery does not contain lithium metal. It is a lithium-ion device, consisting of a positively charged cathode, negatively charged anode, separator, electrolyte, and positive and negative current collectors. While discharging, the lithium ions travel from the anode to the cathode through the electrolyte, thus generating an electric current, and, while charging the device, lithium ions are released by the cathode and then go back to the anode. Figure 5.8 shows the basic working principle of an LIB. Since the electrolyte is the key component in batteries, it affects the electrochemical performance and safety of the batteries (Chawla et al. 2019). The rechargeable lithium battery is a supreme representation of solid-state chemistry in action (Whittingham 2008). Given its fundamental advantages, LIBs will in all likelihood continue to dominate portable electrochemical energy storage for many years to come. Since LIBs are the first choice source of portable electrochemical energy storage, improving their cost and performance can greatly expand their applications and enable new technologies which depend on energy storage (Nitta et al. 2015).

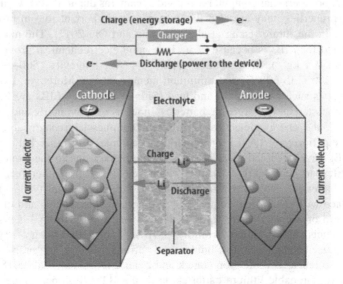

FIGURE 5.8 Illustration to show the basic components and operation principle of a Li-ion cell. *Source*: (Chawla et al. 2019)

5.4.1.1 Electrolyte for LIBs

In order to increase the battery operating voltage, electrolytes with higher electrochemical stability towards oxidation are necessary. Various approaches have been extensively investigated and discussed in the literatures to make Li metal anodes viable for practical applications. Among all the strategies, electrolytes play a key role in obtaining stable cycling of Li metal anode. In fact, the search for suitable electrolytes for Li metal plays a critical role in the development of LIBs. Thermodynamically, no liquid electrolyte is stable for Li metal anode because of very low electronegativity of the latter. However, it is possible to kinetically stabilize the Li metal anode when a passivation layer is formed at the interface between Li anode and the electrolyte. The extremely reductive Li metal anode inevitably reacts with electrolytes to form an SEI which was firstly named by Peled in the 1970s (Peled 1979). An electrolyte could be viewed as the inert component in the battery, and it must demonstrate stability against both cathode and anode surface. During operation, the electrolyte should undergo no net chemical changes of the battery, and all faradaic processes are expected to occur within the electrodes. LIBs have been used to power portable electrical devices for 30 years (Li Bing et al. 2018). However, the contemporary commercial LIBs with flammable liquid organic electrolytes cannot satisfy the requirements, especially regarding safety and power density, of the ever-increasing scale of battery applications (Nanda et al. 2018). Addressing most of these requirements passes through the development of new-generation electrolytes able to overcome the issues of the state-of-the-art liquid ones, which are based on highly volatile and flammable organic solvents (Quartarone and Mustarelli 2020). Given the nature of flammable organic electrolytes, a battery can be a fire hazard in case of over-charging or short circuiting. Solid electrolytes possess a much higher thermal stability, and this makes the solid-state battery one of the best choices for the next generation of batteries. Moreover, inorganic solid electrolytes can work in hostile environments, such as in the temperature range from −50°C to 200°C or even higher, in which organic electrolytes fail due to freezing, boiling, or decomposition (Wang Huilin et al. 2020). Recently, some new concepts were proposed for the design of novel liquid, quasi-solid, and solid electrolytes with improved safety, durability, and electrochemical performance. These new concepts include (i) biomimetic electrolytes (ii) ionogels/eutectogels; (iii) patterned membranes obtained by 3D printing and/or photo- or stereolithographic processes; and (iv) super-concentrated solutions, which in some cases also show self-extinguishing properties. Whereas several reviews appeared on solid-state electrolytes (e.g. ceramic, polymer-ceramic composites, and solvent-free polymer systems), to date scarce attention has been devoted to new liquid-like or quasi-solid ion conductors, which can indeed lead to technological breakthroughs in battery development. Here, "quasi-solid" electrolytes are defined as those systems where a liquid phase is chemically or physically entrapped into a solid matrix, which is generally nanostructured, e.g. in the case of metal-organic frameworks (MOFs) or ionogels (Manthiram et al. 2017; Joos et al. 2018; Quartarone and Mustarelli 2020). It is clear and widely accepted that the quest for new-generation electrolytes is closely connected to the effort of replacing the graphite-based anode with a lithium metal one. This will strongly improve the performance of LIBs, as well as help to make new chemistries, such as Li-air and Li-S, industrially available (Quartarone and Mustarelli 2020). Self-extinguishing electrolytes based on ionic liquids have reached such a degree of development as to imagine their use even in the short term. MOFs and, in case, COFs (Covalent Organic Frameworks) are also promising in the medium term, chiefly for increasing attainable power density by means of anion blocking. In future, it is expected that interesting opportunities will be assured by the development and exploitation of quasi-solid single-ion conductors with high ionic conductivity (Bertasi et al. 2019). Another strategy to be considered towards more stable batteries is the study of self-healing/self-repairing electrolytes. As stated before, this route has so far been undertaken chiefly for proton conductors and must be extended to lithium and sodium carriers. To date, polymers able of self-repairing through extended hydrogen-bond networks have been proposed as binders for Si anodes (Munaoka et al. 2018). Supramolecular chemistry concepts could help to extend this approach to separators (Kwon et al. 2018). Besides the improvement of the functional performance of electrolytes, it will be also important to concentrate efforts on aspects related to their sustainable production, also in the frame of the modern concepts of circular economy (Han et al. 2020). In principle, a good electrolyte for Li metal anode should meet the following minimal requirements: (1) it should be a good ionic conductor and electronic insulator, so that ion (Li^+) transport can be facile and self-discharge can be kept to a minimum; (2) it should have a high mechanical strength; (3) it should be flexible to some extent; (4) it should be homogeneous along the Li anode surface; (5) it should be stable with Li anode and electrolyte components; (6) It should be thermally stable; for liquid electrolytes both the melting and boiling points should be well outside the operation temperatures; and (7) It must have low toxicity and successfully meet also other measures of limited environmental hazard (Jie et al. 2020; Li et al. 2016).

5.4.1.2 Electrode Materials of Current Interest for LIBs

An LIB may be largely divided into three parts of anode, cathode, and electrolyte, where a variety of materials may be used. The electrode materials can be either positive or negative. The negative electrode material is the one that carries the electrons to the external circuit and oxidizes during the electrochemical reaction. The ideal negative electrode for lithium batteries is undoubtedly metallic lithium, since it presents enormous theoretical specific capacity (3860 mA h g^{-1}), low density (0.59 g cm^{-3}), and low working potential (-3.04 V vs. SHE). On the other hand, rechargeable batteries based on metallic lithium negative electrodes are not commercialized due to a practical problem: dendrites are formed during numerous charge/discharge cycles, which can result in loss of metallic lithium and generation of internal short circuits affecting battery safety and cycle life (Shen et al. 2018). Positive electrode materials in an LIB play an important role in determining capacity, rate performance, cost, and safety. The most used positive electrodes in LIBs are layered $LiCoO_2$ (LCO), $LiNiO_2$ (LNO), spinel $LiMn_2O_4$ (LMO), etc. (Ellis et al. 2010; Liu and Mukherjee 2015). Recently, several new types of lithium intercalation materials for positive electrodes with higher capacities have been developed. The materials that are typically used for fabricating the anode are metallic lithium, graphitic carbon, hard carbon, synthetic graphite, lithium titanate, tin-based alloys, and silicon-based materials (Yan et al. 2017; Mishra et al. 2018; Selis and Seminario 2018). The materials used for making cathode are an oxide of lithium and manganese, lithium cobalt oxide, FeS_2, V_2O_5, lithium nickel cobalt manganese oxide, lithium-ion phosphate, and electronic conducting polymers. The materials used as electrolytes include $LiPF_6$, $LiClO_4$, $LiAsF_6$, and $LiCF_3SO_3$. Apart from these main components, there are other components such as a binder, flame retardant, gel precursor, and electrolyte solvent (Elia et al. 2014; Li et al. 2016; Mishra et al. 2018). A great volume of research in LIBs has thus far been in electrode materials. Electrodes with higher rate capability, higher charge capacity, and (for cathodes) sufficiently high voltage can improve the energy and power densities of Li batteries and make them smaller and cheaper. However, this is only true assuming that the material itself is not too expensive or rare. In order to meet the demand for ever-increasing energy/power densities, enormous efforts have been devoted to searching for the advanced electrode materials with favourable kinetic property, high electric conductivity, and improved safety. Recently there has been impressive progress in the exploration of electrode materials for lithium-based batteries such as various metal oxides and polyanionic compounds as well as anode materials as shown in Figure 5.9 (Roy and Srivastava 2015; Li et al. 2016). Here, in this sub-topic we present the recent trends in electrode materials and some new strategies of electrode fabrication for Li-ion batteries. Some promising materials with better electrochemical performance have also been represented along with the traditional electrodes, which have been modified to enhance their performance and stability. Few years ago Wei Tao et al. synthesized a series of $BaLi_{2-x}Na_xTi_6O_{14}$ (0 = x = 2) compounds as lithium storage materials by a facile solid-state method. Compared with other samples, $BaLi_{0.5}Na_{1.5}Ti_6O_{14}$ exhibits higher reversible capacity, better rate capability, and superior cyclability. $BaLi_{0.5}Na_{1.5}Ti_6O_{14}$ delivers the delithiation capacities of 162.1 mAh g^{-1} at 50 mA g^{-1}, 158.1 mAh g^{-1} at 100 mA g^{-1}, 156.7 mAh g^{-1} at 150 mA g^{-1}, 152.2 mAh g^{-1} at 200 mA g^{-1}, 147.3 mAh g^{-1} at 250 mA g^{-1}, and 142 mAh g^{-1} at 300 mA g^{-1}, respectively. They also point out that $BaLi_{0.5}Na_{1.5}Ti_6O_{14}$ as anode electrode shows an acceptable electrochemical performance (Tao et al. 2017). A facile low-temperature approach was used to prepare graphitic carbons with a spherically shaped morphology (CO-CS) by V. Aravindan et al. The excellent Li-insertion properties were noted with high reversibility (~369 mAh g^{-1}) in the half-cell assembly. The LIB was assembled with olivine type $LiFePO_4$/CO-CS and delivered a maximum energy density of ~337 Wh kg^{-1} (Jayaraman et al. 2018). Liu et al. synthesized a material from ($Li_{1.2}Ni_{0.2}Mn_{0.6}O_2$) doping with Cr

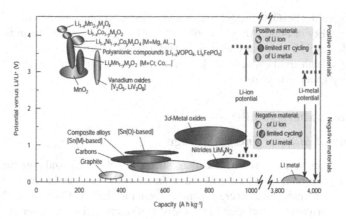

FIGURE 5.9 Electrode materials presently used or under serious considerations for rechargeable lithium-based batteries.
Source: (Li et al. 2016)

($Li_{1.2}Ni_{0.16}Mn_{0.56}Cr_{0.08}O_2$) and finally coating with $LiAlO_2$. The 3 wt.% $LiAlO_2$-coated LNMCr showed the highest discharge specific capacity of 268.8 mAh g^{-1} and the best cycling stability among different coating levels (1 wt.%, 3 wt.%, and 5 wt.%) compared with pristine $Li_{1.2}Ni_{0.2}Mn_{0.6}O_2$ (230.4 mAh g^{-1}) and Cr-doped $Li_{1.2}Ni_{0.16}Mn_{0.56}Cr_{0.08}O_2$ (248.6 mAh g^{-1}) (Liu et al. 2019).

Kim and his coworkers developed a hybrid positive electrode $Li[Ni_{0.886}Co_{0.049}Mn_{0.050}Al_{0.015}]O_2$ formed by a core of $Li[Ni_{0.934}Co_{0.043}Al_{0.015}]O_2$ encapsulated by $Li[Ni_{0.844}Co_{0.061}Mn_{0.080}Al_{0.015}]O_2$. This core@ shell structure provided an exceptionally high discharge capacity of 225 mAh g^{-1} at 4.3 V and 236 mAh g^{-1} at 4.5 V, which are better values than the separated electrodes. The authors believe that the ordering of Li ions in the new hybrid on a microscopic scale led to a stabilization of the host structure during the cycles and facilitated Li^+ intercalation (Kim et al. 2019). The performance of MXenes (Ti_2CT_x) combined with electrolytic manganese dioxide (EMD) in three different weight ratios (i.e. MXene: EMD = 20:80; 50:50; 80:20) were studied by Sharona A. Melchior et al. to examine as anode material for LIBs. The prepared materials were investigated for their electrochemical properties. The best ratio was found to be MXene: EMD = 80:20. The capacity obtained for this material after 200 cycles is 460 mAh g^{-1} at a current density of 100 mA g^{-1}. The Li-ion accessibility improved with cycling. The authors claimed that this study provides a first insight into the viability of using one of the lightest known MXenes and EMD composite for improved LIB anodes (Melchior et al. 2019). Y. Liu et al. demonstrated the synthesis and electrochemical performance of MAX phase (Ti_2AlC) derived $C-K_2Ti_4O_9$ nanocomposites by one-step KOH-assisted hydrothermal treatment. Due to the layered structure of potassium titanate and disordered carbon, the composites exhibit excellent electrochemical properties as anode materials in LIBs. The $C-K_2Ti_4O_9$ delivered high reversible capacity of 314 mAh g^{-1} at 100 mA g^{-1} and superior rate performance (Liu et al. 2019). Very recently, Abdelfettah Lallaoui et.al synthesized a new titanium(III) phosphite $Ti_2(HPO_3)_3$ under hydrothermal conditions. This compound exhibits electrochemical activity toward lithium insertion at very low potential, 0.24 V. Nevertheless, $Ti_2(HPO_3)_3$ suffers from a huge capacity loss and poor cycling life. This is the first study on developing a phosphite polyanion-based anode for LIBs, and it is intended to serve as a key driver in the area of phosphite chemistry for new anode materials for LIBs (Lallaoui et al. 2020). Three-dimensional (3D) TiO_2@nitrogen-doped carbon (NC)/Fe_7S_8 (TONCFS) composite was prepared by Xinlu Zhang et al. They fabricated the material by a facile and simple hybrid strategy via in situ polymerization of pyrrole monomer with alkalized $Ti_3C_2T_x$ 8 and subsequent vulcanization at 700°C. When evaluated as an anode material for LIBs, the TiO_2@NC/Fe_7S_8 demonstrates a high reversible capacity (516 mAh g^{-1} after 100 cycles at 0.1 A g^{-1}), excellent rate capability (337 mAh g^{-1} at 1 A g^{-1}), and robust long cycling stability (282 mAh g^{-1}after 1000 cycles at 4 A g^{-1}) (Zhang Xinlu et al. 2020). Da Xu et al. have developed novel Fe_3O_4@Ti_3C_2 hybrids, in which the Fe_3O_4 nanocrystals are well-anchored between Ti_3C_2 interlayers with the assistance of the synergistic effects of 2D physical confinement and Ti-O-Fe covalent bonds.

These advantages give the $Fe_3O_4@Ti_3C_2$ hybrids a very high specific capacity of 1172 mAh g^{-1} and a rapid charging capability of 366 mAh g^{-1} in 66 s. A 90% capacity retention can be maintained even through 1000 cycles at 5 A g^{-1}. They also achieved a free-standing electrode by a simple vacuum filtering, exhibiting a high areal capacity of 4.2 mAh cm^{-2} at 4.4 mg cm^{-2} almost without sacrificing gravimetric capacity (Xu et al. 2020).

5.4.2 Sodium-Ion Batteries (SIBs)

5.4.2.1 Rationale of SIBs for Energy Storage

Although Li-ion batteries have received tremendous success in electronic and portable devices, the high price of lithium due to its limited and unequally distributed resources may hinder its further application in novel large-scale electrochemical energy storages systems (ESSs) (Chen et al. 2020). Sodium-ion batteries (SIBs) were originally developed in the early 1980s, approximately over the same time period as LIBs (Newman and Klemann 1980). As promising alternatives to lithium-ion batteries (LIBs), in recent years, SIBs have attracted much attention due to the similarities between Na and Li in terms of the electrochemical/chemical properties. Besides, sodium is the fourth most abundant metal element with a suitable redox potential (–2.71 V vs. the standard hydrogen electrode). Recently, room-temperature stationary sodium-ion batteries (SIBs) have received extensive investigations for large-scale energy storage systems (ESSs) and smart grids due to the huge natural abundance and low cost of sodium (Chen et al. 2019). Moreover, the low melting point of sodium at 97.7°C also presents a safety hazard for devices using Na metal electrodes at ambient temperature (Hwang et al. 2017). Also, SIB systems have good performance in aqueous systems, unlike their LIB counterparts, and this greatly helps to bring down costs as inexpensive electrolytes and less complicated fabrication processes can be used (Kim Haegyeom et al. 2014). However, the larger radius, ~25% larger size of Na$^+$, and slower reaction kinetics of sodium ions compared to those of lithium ions would cause large volume variation and large polarization of anode materials, which lead to poor cycling performance and low reversible capacity (Lao et al. 2017; Wang et al. 2018). Along with this, the charge transportation and stability of crystal structure become jeopardized and sodium metal has a much lower specific capacity, only 1165 mAh g^{-1}, due to its higher mass in comparison to lithium (Zhang Xinlu et al. 2020). Thus, this challenge should be taken care to get better sodium storage properties. However, it is impossible to reduce the size of the Na$^+$ per se, but can improve the insertion efficiency of the Na$^+$ into the material structure (Rajagopalan et al. 2020). In addition, the larger volume shrinkage will have further negative effects on the long-term cycling performance. So, even though the reaction mechanism of SIBs is similar to that of LIBs, there is still an urgent demand for new strategies and novel material designs for SIBs, apart from the existing routes or experience acquired from LIBs. The current research outcomes have demonstrated many unique characteristics of Na-containing composites (Shi et al. 2018a; Chen et al. 2019). Recent studies have shown that rechargeable batteries that pair a Na anode with highly energetic O_2-based cathodes are intrinsically more stable during discharge than their Li analogues because the species generated electrochemically in the cathode, the metal superoxide, is more stable when the anode is Na, as opposed to Li (Yadegari et al. 2016).

5.4.2.2 Physical Principles of SIBs

A sodium-ion battery system is an energy storage system based on electrochemical charge/discharge reactions that occur between a positive electrode (cathode) composed of sodium-containing layered materials, and a negative electrode (anode) that is typically made of hard carbons or intercalation compounds (Durmus et al. 2020). The electrodes are separated by some porous material which allow ionic flow between them and are immersed in an electrolyte that can be made up of either aqueous solution (such as Na_2SO_4 solution) or non-aqueous solution (e.g. salts in propylene carbonate). When the battery is being charged, Na atoms in the cathode release electrons to the external circuit and become ions which migrate through the electrolyte toward the anode, where they combine with electrons from the external circuit while reacting with the layered anode material. This process is reversed during discharge (Liu et al. 2016). The battery components and the electrical storage mechanism of SIBs and LIBs are basically the same except for their ion carriers. In terms of cathode materials, the intercalation chemistry of sodium is very similar to that of lithium, making it possible to use similar compounds for both systems. However, there are some obvious differences between these systems. Na ions (1.02 Å) are larger compared to Li ions (0.76 Å), which affects the phase stability, transport properties, and inter-phase formation (Adelhelm et al. 2015). During insertion or extraction of Na ions, various changes occur at electrode and electrode/electrolyte interface including volume expansion/contraction, phase transition, morphological evolution, and surface reconstruction. Monitoring and understanding these physical and chemical changes of the electrodes and interfaces are vitally important to guide the improvement of the electrochemical performance of SIBs (Xu et al. 2018). The following scientific issues are the major concerns during SIBs investigation: (1) the relationship of crystal structure, local structure, and Na storage mechanism; (2) the origin of charge compensation and capacity; (3) the evolution of surface feature and solid electrolyte interphase (SEI) layer; (4) the relationship between the performance and reaction anisotropy; (5) thermal dynamics and kinetics of the electrode materials play a key role in understanding these scientific issues of SIBs' research (Shadike et al. 2018).

5.4.2.3 Electrolytes Materials for SIBs

In SIBs, the electrolyte that can provide rapid Na^+ transportation between the anode and cathode and simultaneously block the diffusion of electron is deemed as one of the most crucial components for high-performance rechargeable batteries (Qiao et al. 2020). The electrolyte is an essential and decisive component in all SIBs, and it plays a key role in balancing and transferring charges in the form of ions between the two electrodes. It basically determines the electrochemical window of a cell according to the energy levels of the mixture, which reflects the thermodynamic stability. Additionally, operational kinetics control of SIBs is closely connected to Na^+ transference numbers of the electrolyte and the solid-electrolyte interphase (SEI) on the anode, which is related to the electrolyte composition. Generally, the choice of electrolyte decides or affects energy density, safety, cycle life, storage performance, operating conditions, etc. (Che et al. 2017). For SIB systems, the electrolytes are generally composed of sodium salts with or without additives dissolved. In general, electrolytes in SIBs can be categorized as organic liquid electrolytes, inorganic solid electrolytes, and flexible polymer/plastic electrolytes (See Figure 5.10). The current SIBs derived from organic liquid electrolytes (OLEs) made from sodium salts and organic solvents usually suffer from several problems, including limited electrochemical window, flammability, and leakages to the potential safety hazard in large-scale applications (Che et al. 2017). Therefore, sodium metal batteries with higher energy density are limited and difficult to be improved for the current liquid electrolyte batteries, although some strategies have been used to effectively suppress the growth of dendrites by introducing additives in current OLEs or surface protection to avoid direct contact between sodium metal and electrolytes (Lin et al. 2017). Unlike liquid SIBs, solid-state SIBs can completely address the safety issue because they can employ non-flammable solid-state electrolytes, which eliminate the leakage or flammability problems for liquid SIBs. In addition, a solid-state electrolyte with a wide electrochemical window makes possible the use of a metal sodium anode and high-potential cathode for high energy-density batteries (Che et al. 2017; Zhao Chenglong et al. 2018). Solid-state electrolytes could be categorized into inorganic ceramic/glass-ceramic electrolyte, organic polymer electrolyte, and ceramic-polymer composite electrolyte. Inorganic electrolyte is more suitable for rigid battery design which could be operated under aggressive environment, as it possesses high elastic module, good thermal/chemical stability, wide electrochemical window, high ionic conductivity, and low electronic conductivity (Kim et al. 2017). Compared to the polymer-based electrolytes, ISEs with a high ionic conductivity and a high Na^+ ion transference number at RT can effectively improve battery performance for both long-term cycling and high power density. In addition, ISEs with a much higher mechanical strength can suppress the growth of sodium dendrites; however, the low chemical/electrochemical stability may cause an unavoidable side reaction between electrolyte and electrode materials, resulting in a large interfacial impedance (Khurana et al. 2014). Compared with liquid electrolytes, polymer electrolytes possess several inherent advantages, which

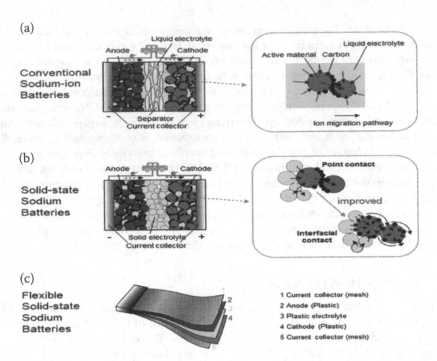

FIGURE 5.10 Schematics of representative (a) organic liquid electrolytes (OLEs), (b) inorganic solid electrolytes (ISEs), and (c) solid polymer/plastic electrolytes (SPEs) for conventional sodium-ion batteries (NIBs) and solid-state sodium batteries. *Source*: (Zhao Chenglong et al. 2018)

include high safety, suppression of sodium dendrite formation, and reduced electrolyte decomposition (Yang et al. 2019). The polymer electrolytes (PEs) used for sodium batteries are generally classified into three categories (Qiao et al. 2020), including (i) solid polymer electrolytes (SPEs) (ii) composite solid polymer electrolytes (CSPEs), and (iii) plasticized polymer electrolytes and gel polymer electrolytes (PPEs/GPEs). SPEs contain only sodium salts and polymer matrices. They are considered as appealing electrolytes for developing all-solid-state sodium batteries due to their low density, good flexibility, and process ability; inherent safety and reduced side-reactions; good interfacial contact with electrodes (Liu Lilu et al. 2019). Poly(ethylene oxide) (PEO) is a popular polymer host for solid polymer electrolyte, due to its good electrochemical stability, mechanical properties, and the capability to solvate different metal salts (Bitner-Michalska et al. 2017). CSPEs involve the addition of inorganic fillers in above SPEs to enhance the ionic conductivity of electrolyte (Liu et al. 2017). Plasticized polymer electrolytes and gel polymer electrolytes (PPEs/GPEs) contain liquid phases together with the components used in SPEs or CSPEs. The distinction between PPEs and GPEs lies in the fraction of liquid phases, where the former electrolytes contain less than 50 wt.% liquid plasticizers and the latter electrolytes involve more than 50 wt.% amount of liquid plasticizers. In order to further enhance the ionic conductivity of SPEs, several strategies have been implemented, including polymer cross-linking, blending, copolymerization, doping of additives such as plasticizers, ionic liquids, inorganic fillers, and liquid electrolytes (Zhao Chenglong et al. 2018). Passive ceramic fillers like SiO_2, TiO_2, ZrO_2 incorporated into the polymer hosts form the ceramic-polymer composite electrolyte. Taking the advantages of both inorganic ceramic electrolyte and the organic polymer electrolyte, enhanced ionic conductivity, good mechanical property, and high chemical/thermal stability could be achieved in the composite electrolyte simultaneously (Zhang et al. 2016; Wang Yumei et al. 2019).

Recently, chalcogenide solid electrolytes (CSEs) have become a very active area of all-solid-state SIB research due to their excellent room-temperature ionic conductivity (10^{-3}–10^{-2} S cm^{-1}), low activation energy (<0.6 eV), and easy cold-pressing consolidation. At present, CSEs are mainly divided into four categories and two of them are Na_3MS_4 (M = P, Sb) and Na_3MSe_4 (M = P, Sb). Chalcogenide-based (S, Se) chemistries offer the potential for higher ionic conductivities than oxides. Though it is likely that sulphide and selenide-based solid electrolytes may exhibit lower intrinsic electrochemical stability, the formation of passivating phases at the electrode–solid electrolyte interface can potentially mitigate further reactions. Sulphide electrolytes, particularly sodium sulphides, also tend to be softer than oxides, which allows intimate contact between electrode and solid electrolyte to be achieved via cold pressing instead of high-temperature sintering (Chu et al. 2016). Since the electronegativity of sulphur is lower than that of oxygen, sulphide electrolytes usually exhibit faster sodium ion conduction than oxide electrolytes because the electrostatic forces of sulphur and sodium ion are lesser than that of oxygen and sodium ion. Compared with oxide solid electrolyte, chalcogenide solid electrolytes could be synthesized at low temperature, which reduces the production cost. Due to the deformability of CSEs, it can achieve good interface contact with electrode interface via simple cold pressing to further reduce the actual production cost. Among all kinds of solid electrolytes, CSEs with good formability are a good choice for the preparation of bulk-type all-solid-state SIBs. Therefore, CSEs have an important application prospect in the field of bulk-type all-solid-state SIBs for large-scale energy storage. Despite these advantages, there are still some challenges before CSEs can be widely used. Initially, the composite electrodes of bulk-type all-solid-state SIBs are in poor contact. Although CSEs have deformability, investigation results demonstrate cold pressing alone is not enough to form compact components without voids. Secondly, CSEs are unstable in the air, which are easy to deliquesce, oxidize, and generate toxic gases, resulting in a sharp drop of conductivity. Therefore, it is very important to improve the performance of CSEs in bulk-type all-solid-state SIBs (Dai et al. 2020).

5.4.2.4 Electrode Materials for SIBs

The key in realizing the practical applications of SIBs lies in the development of high-performance electrode materials with acceptable characteristics such as specific capacity and operation voltage; therefore, the major challenge in advancing SIB technology lies in finding novel electrode materials with improved Na kinetics (Xiao et al. 2016). Further the electrode materials must also prove durable and provide a significantly longer cycle life, especially since SIBs are being considered for grid-scale storage, where reliability and longevity are two of the most important parameters (Mukherjee et al. 2019). Current electrode materials for SIBs include inorganic compounds (e.g. transition metal oxides, phosphates, fluorides, sulphides, phosphides, alloys, Prussian blue analogues, and carbonaceous materials) and organic materials (Zhao Chenglong et al. 2018). Each of them has its own advantages and disadvantages, for instance, many inorganic compounds contain toxic metals, causing further concerns over resources and environmental contamination. In addition, the inorganic compounds undergo the insertion-type sodium storage mechanism. During the intercalation and deintercalation of sodium ions (1.02 Å), large volume changes and irreversible phase transitions may occur, leading to low reversible capacity and poor cycling performance (Wang et al. 2016). Hence, different research groups have been adopting different approaches to overcome these challenges, and one of the most important avenues ahead for SIB systems is from the perspective of materials engineering, that is, developing newer and better electrodes with novel morphologies to enhance performance and longevity (Mukherjee et al. 2018). The main parameters to evaluate the high performance of an electrode material are broadly classified into the four categories: (a) energy density, (b) rate capability, (c) cyclability, and (d) thermodynamic stability (Bhandavat et al. 2012). Energy density is defined as the product of average operating potential (V) and the total amount of charge transfer (Ah) and is expressed in Wh g^{-1}/WH L^{-1}. In general, for a high-performance electrode, it is desired that the electrode materials undergo fast charge and discharge while maintaining high-energy density. On the other hand, the ability of the anode material to reversibly cycle Na ions with the least irreversible capacity (IRC) is the cyclability of the material. To achieve thermodynamic stability and prevent the possible change in anode structure, the addition of external matrix material or similar chemical modifications are desired for a high-performance electrode material (Mukherjee et al. 2019). Important performance characteristics of batteries, such as capacity, cycling stability, and operation voltage, are mainly determined by the electrochemical performance of the electrode materials. Therefore, the major challenge in advancing SIBs technology lies in finding high-performance electrode materials including anode and cathode materials (Cheng et al. 2018). Cathodes are a very important aspect of SIBs and should be able to reversibly intercalate the Na^+, preferably at voltages greater than 2 V. Ideally a cathode should exhibit low volume expansion upon intercalating Na^+ and structural stability (Mukherjee et al. 2019). A wide variety of cathode materials for SIBs have been explored up to now. Among them, layered sodium-containing transition-metal oxides have been the subject of the most extensive research, similar to the most commonly used layered-oxide cathodes in LIBs (Dai et al. 2017). The large Na ion also grants to the sodium-containing oxides tunnel-type structures, which are rarely seen in LIB electrode materials. In general, layered materials exhibit larger specific capacities but are less stable in long-term cycling owing to their structural instability upon Na-ion extraction. Most layered SIB cathode materials exhibit operating voltages of up to 1.5 V lower than those of their layered LIB counterparts, which is the main factor responsible for the lower energy densities of SIBs (Ping et al. 2011). Compared with sodium-free metal oxides, sodium-inserted layered metal oxides may present a desired electrochemical stability especially in the first two cycles, where an initial irreversible sodium insertion occurs for sodium-free metal oxides (Cheng et al. 2011). Sulfate and phosphate-based polyanionic compounds usually provide a more open pathway for Na-ion transport, (Ni et al. 2017) although the volumetric energy density would be sacrificed to some extent due to their non-close packed structure. Appropriate organic molecules could be used as cathode materials for SIBs as well. In addition, the Na containing Prussian-blue analogues have also been proven viable for sodium storage (Sun et al. 2018). Polyanion-type cathodes have also received much attention as promising cathodes for future Na-ion batteries on account of their structural stability, safety, and appropriate operating potential, which is owing to the strong inductive effect originating from the large high electronegativity anion groups (Xu et al. 2018). It can be noticed that the cathode of SIBs is currently the hot subject of the research into sodium-ion batteries. With a better understanding of how sodium transports and reacts at the atomic level, new solutions or structural modification of cathode materials can be developed, in the hope of better performance of sodium-ion batteries (Li et al. 2017). Hence, cathode materials should be given more attention, owing to their significant effect on the electrochemical performance of SIBs in terms of specific energy, cycling life, and specific power. In general ideal candidates for cathode materials should exhibit low volume expansion upon inter-calating/de-intercalating of the Na ion, in order to provide superior cycling performance for the cell (Wang et al. 2015). The voltages and capacities of representative SIB cathodes are summarized in Figure 5.11. Anode, also known as negative electrode, is an essential component of SIBs amounting for 14% of the total cell cost. It must possess an element in its composition with low atomic weight, low density, and be able to accommodate vast quantity of sodium ions per formula unit having good cyclability in order to yield stable and high volumetric ($mAhcm^{-3}$) and gravimetric ($mAhg^{-1}$) capacities (Perveen et al. 2020). Development of the anode materials is even slower. Graphite, the most widely used anode material in LIBs, is inactive towards Na ions. Na-ion diffusion in hard carbon occurs along channels and cavities with an irregular geometry, causing relatively poor rate performance (Choi and Aurbach 2016). Hence, preparing high-performance anode materials with a high reversible capacity, stable cycling performance, and high rate capability is still a challenge (Mukherjee et al. 2019; Cheng et al. 2018). Negative electrode is a necessary part of Na-ion batteries. The optimization of Na-ion technology urgently needs improvement for the anode materials (Kang et al. 2015). Unlike Li, Na metal cannot be directly used as an anode considering its safety hazard and unstable passivation layer in most organic electrolytes at room temperature.

Besides, graphite, the conventional anode for LIBs, is electrochemically less active in SIBs due to the mismatching of the interlayer distance to the larger Na^+ radius. Some organic materials with reasonably low operating voltages have also been introduced as SIB anodes, but their feasibility under practical conditions is yet to be verified. Although it is probable that only small molecules are competitive against other non-organic counterparts in terms of the specific capacity, factors such as their solubility in electrolytes at various potentials and poor adhesion to the current collector might create a non-trivial hurdle to overcome (Deng et al. 2013). Carbon-based materials are insertion-based

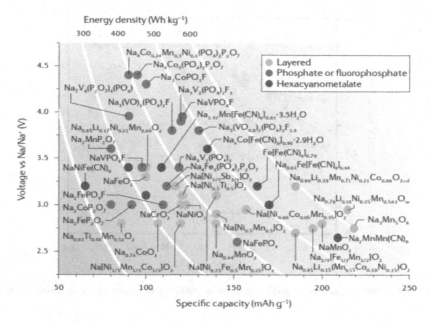

FIGURE 5.11 Operation voltages vs. specific capacities of sodium-ion battery cathode materials. *Source*: (Choi and Aurbach 2016)

materials which insert/extract a certain amount of Na during charge/discharge process. However, graphitic carbons hinder the intercalation of Na-ions due to their large radius. A minimum interlayer distance of ~0.37 nm is required for Na-ion insertion, whereas the interlayer spacing of graphite is ~0.335 nm (Cao et al. 2012). Hard carbons have also been utilized as anodes for Na-ion batteries due to their disordered structure and large interlayer distance. Further, hard carbon can deliver high capacity and low voltage plateau near 0.1 V (vs. Na/Na^+) (Cheng et al. 2018; Saurel et al. 2018). Phosphorus (P) has three allotropes: white, black, and red phosphorus. Among these allotropes, red phosphorus is relatively stable, commercially available, and most studied as anode for NIBs. P provides a highest theoretical capacity of 2596 mA h g^{-1} (Na_3P) and appropriate redox potential of about 0.4 V vs. Na+/Na (Kim et al. 2013). Due to its superior mechanical and electronic properties phosphorene, a single layer of black phosphorus, has recently emerged as a candidate for SIB anodes. It is also important to highlight that phosphorene interlayer spacing is 5.4 Å, which is greater than graphite (3.7 Å). As a result, the specific capacity of phosphorene for SIB is theoretically 2596 mAh g^{-1}, almost seven times hard carbon's specific capacity (Mukherjee et al. 2019; Huang et al. 2017). However, The energy density of phosphate-based SIBs is slightly lower, as a whole, than those of the metal oxide and prussian blue analogues: most are below 400 Wh kg^{-1} (Li et al. 2017). Because of their high energy density and low redox potential, alloys have garnered considerable interest as anode materials for Na-ion batteries. Alloy-based materials have been proved to be promising anode electrodes owing to their high theoretical capacity. They can alloy with sodium to form rich alloy phases, yielding a much higher capacity than carbon-based materials. For LIBs, Si is the most studied alloy material due to its ultra-high theoretical specific capacity of 4200 mAh g^{-1}; however, it is indeed Na inactive. Indium (In) is also explored as anode for SIBs; however, it just exhibits a low capacity of *ca.* 100 mAh g^{-1}. Elements in group 14 (Sn, Ge) and group 15 (P, Sb) are extensively studied owing to their excellent electrochemical performance, low cost, and environmental friendliness. The main challenge for alloy-based materials is the enormous volume expansion, which will result in continuous pulverization of these electrode materials and then a steep deterioration of electrochemical performance (Kang et al. 2015; Kundu et al. 2015). Metal sulphides (MSα) are also widely studied anode materials. In general, the mechanism in Na-MSα occurs through two steps. Na^+ first intercalates into MSα to form an intermediate ($Na_xMSα$), then $Na_xMSα$ decomposes to Na_2S and M through a conversion reaction (Hu et al. 2014). However, the latter conversion

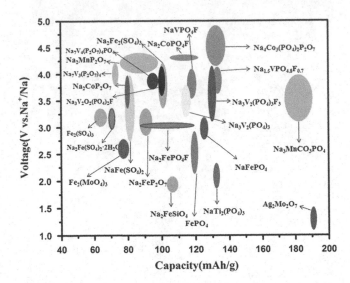

FIGURE 5.12 Operation voltages vs. specific capacities of sodium-ion battery anode materials. *Source*: (Cheng et al. 2018)

reaction usually causes severe volume expansion of the electrode materials and sluggish kinetics for Na^+ intercalation/de-intercalation. Many approaches, such as preparing MS /C composites, controlling the cut-off voltage, have been adopted to solve these problems. Generally speaking, metal sulphides are composed of layered metal disulphides and non-layered metal sulphides (Kang et al. 2015; Hu et al. 2015). Recently, various titanium-based compounds have been explored as promising sodium insertion host due to their low-cost, non-toxicity, low operation voltage, low strain, and excellent cyclability. However, the performances of them strongly depend on the electrolyte, the binder, and the morphology. In general, Ti-based composites provide low specific capacity due to the limited Na storage sites (Sun et al. 2013; Kim Haegyeom et al. 2014). The voltages and capacities of representative SIB anode materials are summarized in Figure 5.12. Based on reaction mechanism research on anode materials SIBs can be categorized into five types: (1) carbon-based materials, (2) alloy-based materials, (3) metal oxides and sulfides based on conversion reaction, (4) titanium-based composites with insertion mechanism, and (5) organic composites (Cheng et al. 2018; Kang et al. 2015; Xu et al. 2017; Liu et al. 2015). Currently, the rapid development of anode materials for SIBs, with various and numerous anodes quickly springing back up. Some recent articles have reviewed the research progress in electrode materials for SIBs. In this paper, we offer more detailed and updated research progress for SIBs electrode materials reported recently. Jiale Xia et al. synthesized a series of lanthanide (Ln = La, Ce, Nd, Sm, Gd, Er, and Yb)-doped microsized $Na_2Ti_3O_7$ anode materials and tested their electrochemical performance. Especially, the Yb^{3+}-doped sample not only delivers a high reversible capacity of 89.4 mAh g^{-1} at 30°C, but also maintains 71.6 mAh g^{-1} at 5 C after 1600 cycles, nearly twice that of pristine Na_2Ti_3O. It was found for the first time that the enhancement in doped samples is attributed to the introduction of lanthanides which induces lattice distortion and oxygen vacancies (Xia et al. 2018). Electrolyte mixtures based on 1-ethyl-3-methyl-imidazolium bis (trifluoromethanesulphonyl) (EMI-TFSI) and carbonate solvents (EC-PC) were prepared by T. D. Vo et al. for SIBs. The electrochemical compatibility in half-cell configuration with respect to sodium metal anode of various electrode materials, including SnS/C, hard carbon (HC), and $Na_{0.44}MnO_2$, was evaluated. Stable discharge capacity of SnS/C (up to 410 mAh·g^{-1}) and hard carbon (up to 300 mAh·g^{-1}) was obtained when they were cycled in half-cell using EC-PC (1:1) + 20 wt.% IL + 2 wt.% FEC. In case of $Na_{0.44}MnO_2$ cathode material, excellently stable discharge capacity was obtained even without using FEC additive (Vo et al. 2019). Zheng-Long Xu research group investigated the factors that influence the co-intercalation potential of graphite. They achieved as large as 0.38 V by adjusting the relative stability of ternary graphite intercalation compounds and the solvent activity in electrolytes. The feasibility of graphite anode in sodium-ion batteries was confirmed in conjunction with $Na_{1.5}VPO_{4.8}F_{0.7}$ cathodes by using the optimal electrolyte. The sodium-ion battery delivers an improved voltage of 3.1 V, a high power density of 3863 W kg^{-1} at both electrodes, negligible temperature dependency of energy/power densities, and an extremely low capacity fading rate of 0.007% per cycle over 1000 cycles (Xu et al. 2019).

Hard carbons are considered to be promising anode materials for sodium-ion batteries (SIBs) with their improved specific capacity and good long-term cycle life. However, a significant low initial Coulombic efficiency (ICE) and high cost consequently impede its use in future commercialization of SIBs. Ghulam Yasin et al. introduced a cost-effective strategy to prepare the large-scalable hard carbon (HC) anode material for SIBs. The prepared compound shows excellent initial reversible capacity of 308 mA h g^{-1} under the current rate of 50 mA g^{-1} and considerable higher ICE of 78%. After more than 300 cycles, the material exhibits a specific capacity of 160 mAh g^{-1} and presenting superior capacity retention of more than 94% under a current density of 500 mA g^{-1}. They claim that obtained results establish a new technique to construct low-cost hard carbon-based anode materials for next-generation high-performance SIBs (Yasina et al. 2019). $NaNi_xCo_{1-x}O_2$ and α-Fe_2O_3 compounds were successfully synthesized using aqueous extract of *Zea mays* L. dry silk by Assumpta Chinwe Nwanya et al. The electrochemical energy storage capabilities of the NPs were studied individually as well as a full device using the $NaNi_xCo_{1-x}O_2$ and the α-Fe_2O_3 as the positive and negative electrodes, respectively. A charge and discharge capacity of about 62 C g^{-1} and 25 C g^{-1}, respectively, were obtained at a current density of 50 mA g^{-1} for the full device. However, when they used activated carbon (AC) as the negative and $NaNi_xCo_{1-x}O_2$ as the positive electrode, they obtained charge and discharge capacities of 378 C g^{-1} (105 mAh g^{-1}), respectively, at a current density of 80 mA g^{-1} (105 mAh g^{-1}) and 283 C g^{-1} (79 mAh g^{-1}) (Nwanya et al. 2020). Michael Ruby Raj et al. reported for the first time, two 3,4:9,10-perylenetetracarboxylicdianhydride (PTCDA)-based polyimides, namely, perylenediimide-benzidine (PDI-Bz) and perylenediimide-urea (PDI-Ur), and utilized them as organic cathode materials for SIBs. The obtained organic metal-ion batteries employing PDI-Bz demonstrate a high discharge capacity of 120 mAh/g (with a reversible capacity of ~54 mAh/g) vs. Li+/Li and the second discharge capacity of 111 mAh/g (~74 mAh/g) vs. Na+/Na with two discharge voltage plateaus in the range of 1.9–2.4 V. The cells retained a capacity retention of 46% vs. Li+/Li and 55.2% vs Na+/Na over 50 cycles. PDI-Ur exhibits higher lithiation capacity of~119 mAh/g at the 14th cycling (increased discharge capacity of~118 mAh/g at the 25thcycling). In SIBs, PDI-Ur shows an initial discharge capacity of~119 mAh/g with a single discharge voltage plateau around 1.9Vvs Na^+/Na and the capacity retention of~78.7% (~93 mAh/g) over 50 cycles, both of which are suggesting a potential feasibility of these PTCDA-based polyimides as promising organic cathode materials for high-capacity metal-ions batteries (Raj et al. 2020). Owing to its unique properties, two-dimensional (2D) reduced graphene oxide (rGO) is often combined with metal oxides for energy-storage applications. Recently, Xianying Han et al. compared the electrochemical performance of Nb_2O_5-rGO and amorphous carbon-coated-Nb_2O_5 composites, synthesized in similar conditions. The composite materials made of Nb_2O_5 and amorphous carbon (using 1,3,5-triphenylbenzene as carbon source) outperforms the Nb_2O_5-rGO counterpart as a high rate anode electrode material in Li-ion and Na-ion half-cells. The hybrid supercapacitors, delivering specific capacities of 134 mAh g^{-1} at 25 C against 98 mAh g^{-1} for the rGO-based composite (in Li electrolyte) and 125 mAh g^{-1} at 20 C against 98 mAh g^{-1} (in Na electrolyte) (Zhang Xinlu et al. 2020). Ananta Sarkar and his co-workers prepared hydrogenated metal-doped sodium titanium oxide ($Na_2Ti_3O_7$) anode material for SIBs. The material exhibited high capacity (237 mAh g^{-1}) at a current rate of 200 mA g^{-1} and prolonged cycling stability up to 2500 cycles with an average 99.70 % Coulombic efficiency with excellent rate capability. The material first suffered irreversible loss (53.86%), however, 60 min chemical shorting was enough to minimize the irreversible loss to 4.11% from 53.6% and enhance the Coulombic efficiency to 84% from 36.22% respectively without compromising the specific capacity and cycling stability (Sarkar et al. 2020). High power/energy density (6750 W kg^{-1} /364.2 Wh kg^{-1}) $Na_{2.4}Fe_{1.8}(SO4)_3$ (NFS) composite material has been reported by Jie Hou et al. They applied surface chemistry assistant effective carbon coating at 350^0C and achieved highly improved electrochemical performance as cathode materials for sodium-ion batteries (SIBs). The prepared NFS/C composite shows excellent electrochemical performance of high capacity (100.2 mAh g^{-1} at 1 C), great cyclability (95.0 mAh g^{-1} after 100 cycles at 1 C), and superb rate performance (95.1 mAh g^{-1} at 5 C, 91.3 mAh g^{-1} at 10 C). The full battery of NFS/C coupled with TiO$_2$ anode material also exhibits high reversible capacity of 87.0 mAh g^{-1} at 10 C with an average discharge voltage over 3.0 V (Hou et al. 2020). Debanjana Pahari et al. prepared a novel electrode materials of P2-type layered oxides, Na$_{0.67}$Ni$_{0.33-x}$Ti$_x$ Mn$_{0.67}$O$_2$ (x = 0, 0.08, and 0.16) by using solid-state synthesis method and electrochemically characterized as cathode materials for rechargeable non-aqueous sodium-ion batteries. The Na$_{0.67}$Ni$_{0.17}$Ti$_{0.16}$Mn$_{0.67}$O$_2$ electrodes can deliver initial discharge capacity of 167 mAh g^{-1} at 0.1 C rate with an average voltage of 3.7 V. Na$_{0.67}$Ni$_{0.25}$Ti$_{0.08}$Mn$_{0.67}$O$_2$ electrodes lose only 12% of their initial capacity after 50 cycles when cycled to an upper cut-off voltage of 4.3 V. Controlling the P2-O2 transition at higher voltages through either optimized cut-off voltages or substitution is critical for improving the capacity retention. A complete reversible P2-OP4 transition occurs only in optimally substituted sample, Na$_{0.67}$Ni$_{0.25}$Ti$_{0.08}$Mn$_{0.67}$O$_2$ whereas P2-O2 transition reemerges with higher amount of substitution in Na$_{0.67}$Ni$_{0.17}$Ti$_{0.16}$Mn$_{0.67}$O$_2$ (Pahari and Puravankara 2020). NASICON-structured Na$_3$MnTi(PO$_4$)$_3$ (NMTP) is a high-energy SIB cathode material. However, the low rate capability and unsatisfactory cycle life limit its practical applications. Ting Zhu et al. proposed a dual carbon decoration strategy to tackle the above-mentioned issues of NMTP. The semi-graphitic carbon and reduced graphene oxide co-functionalized NMTP (NMTP/C@rGO) demonstrates a specific capacity of ~114 mAh g^{-1}, reaching a high energy density of ~410 Wh kg^{-1} with an average discharge potential of ~3.6 V. Besides, the NMTP/C@rGO also manifests long-term durability and high rate capability. More importantly, the NMTP/C@rGO also demonstrates ideal electrochemical properties in NMTP/C@rGO//soft carbon full cells. The impressive electrochemical performances make the NMTP/C@rGO a reliable cathode material for SIBs (Zhu et al. 2020).

5.5 Summary and Future Prospects

This review gives information about the basic working principles of energy storage devices, important criteria for the selection of electrode, and the recent development in the electrode and electrolyte materials. The basic charge storage mechanisms in SCs, LICs, LIBs, and SIBs have been also discussed. Recent trends in the three main groups of the electrode materials including carbon-based material, metal oxides/hydroxides, and conductive polymers have been reviewed well with special focus on the developments incurred within the last four years. Improving the energy density, cathode capacity, and operating voltage range remains the most pressing challenges in cathode material development. Similarly, the main development direction for anodes is to achieve high power, high energy density, good cycling performance, and low cost, still a challenging issue. With unique potentials to achieve high energy density and low cost, rechargeable batteries based on metal anodes are capable of storing more energy via an alloying/de-alloying process, in comparison to traditional graphite anodes via an intercalation/de-intercalation process. However, the drawbacks of metal anodes such as high initial capacity loss and short cycling life need to be solved before commercialization. In addition to aqueous electrolyte, in this chapter, we have also tried to highlight three types of solid electrolytes: inorganic, polymer, and HSEs. Inorganic-type electrolytes allow for the realization of high power density. However, inorganic-type electrolytes are brittle, have low adhesion to electrodes, and are difficult to manufacture, especially using current battery manufacturing processes, which limit their practical application. Polymer electrolytes are easy to process, are flexible, and have good adhesion to electrodes, while their ionic conductivity is far from satisfactory. Therefore, HSEs developed by the mixing of an inorganic filler with a polymer matrix have attracted much attention. Many HSEs demonstrate better ionic conductivity than that of the polymer electrolyte matrix itself. Nonetheless, the mixing of filler and matrix still needs improvement to increase ionic conductivity of composite electrolytes. Generally, for different electrolytes, it is still necessary to choose appropriate modification methods according to the actual requirement, so as to better augment the performance of electrolytes. In this chapter, a comprehensive discussion has been included to link the nature of electrode and electrolyte with their performance, for them to act as the electrode materials in SC, LIC,

LIB, and SIB applications. Furthermore, by using important schematic diagrams, figures, and tables, we created a comparative account on the state-of-the-art in SCs, LICs, LIBs and SIBs, so as to help readers achieve a deepened understanding and promote further thinking. By citing the most recent references in each section of this chapter, we have shown the ongoing cutting-edge research around the globe on supercapacitors and batteries, and also pointed out the potential directions where challenges remain for further exploration.

REFERENCES

Adelhelm, P., P. Hartmann, C.L. Bender, M. Busche, C. Eufinger, and J. Janek. 2015. "Materials for Sustainable Energy Production, Storage, and Conversion". *Journal of Nanotechnology* 6: 1016–1055.

Akinwolemiwa, B., C. Peng, and G.Z. Chen. 2015. "Redox Electrolytes in Supercapacitors". *Journal of the Electrochemical Society* 62: A5054–A5059.

Aldama, I., V. Barranco, M. Kunowsky, J. Ibañez, and J.M. Rojo. 2017. "Contribution of Cations and Anions of Aqueous Electrolytes to the Charge Stored at the Electric Electrolyte/Electrode Interface of Carbon-Based Supercapacitors". *The Journal of Physical Chemistry C* 121: 12053–12062.

Amatucci, G.G., F. Badway, A. Du Pasquier, and T. Zheng. 2001. "An Asymmetric Hybrid Nonaqueous Energy Storage Cell". *Journal of the Electrochemical Society* 148: A930–A939.

Aravindan, Vanchiappan, Joe Gnanaraj, Yun-Sung Lee, and S. Madhavi. 2014. "Insertion-Type Electrodes for Nonaqueous Li-Ion Capacitors". *Chemical Reviews* 114: 11619–11635.

Ates, M., and R.B.K. El-Kady. 2018. "Three-Dimensional Design and Fabrication of Reduced Graphene Oxide/Polyaniline Composite Hydrogel Electrodes for High Performance Electrochemical Supercapacitors". *Nanotechnology* 29: 175402.

Augustyn, V., and B. Dunn. 2014. "Environmental Science, Pseudocapacitive Oxide Materials for High-Rate Electrochemical Energy Storage". *Energy & Environmental Science* 7: 1597–1614.

Aurbach, Doron, Yosef Talyosef, Boris Markovsky, Elena Markevich, Ella Zinigrad, Liraz Asraf, Joseph S. Gnanaraj, and Hyeong-Jin Kim. 2004. "Design of Electrolyte Solutions for Li and Li-Ion Batteries: A Review". *Electrochimica Acta* 50: 247–254.

Balbuena, P., and J.M. Seminario. 2006. *Nanomaterials: Design and Simulation: Design and Simulation*. New York: Elsevier.

Bertasi, Federico, Gioele Pagot, Keti Vezzù, Angeloclaudio Nale, Giuseppe Pace, Yannick Herve Bang, Giovanni Crivellaro, Enrico Negro, and VitoDi Noto. 2019. "Lithiated Nanoparticles Doped with Ionic Liquids as Quasi-Solid Electrolytes for Lithium Batteries". *Electrochimica Acta* 307: 51–63.

Bhandavat, R., Z. Pei, and G. Singh. 2012. "Polymer-Derived Ceramics as Anode Material for Rechargeable Li-Ion Batteries: A Review". *Nanomater Energy* 1: 324–337.

Bi, Ruyi, Nan Xu, Hao Ren, Nailiang Yang, Yonggang Sun, Anmin Cao, Ranbo Yu, and Dan Wang. 2020. "Hollow Multi-Shelled Structure Benefited Charge Transport andActive Sites for Li-Ion Capacitors". *Angewandte Chemie International Edition* 59: 4865–4868.

Bitner-Michalska, A., G.M. Nolis, G. Zukowska, A. Zalewska, M. Poterała, T. Trzeciak, Maciej Dranka, et al. 2017. "Fluorine-Free Electrolytes for All-Solid Sodium-Ion Batteries Based on Percyano-Substituted Organic Salts". *Scientific Reports* 7: 40036.

Bolufawi, Omonayo, Annadanesh Shellikeri, and J.P. Zheng. 2019. "Lithium-Ion Capacitor Safety Testing for Commercial Application". *Batteries* 5: 1–13.

Borah, R., F.R. Hughson, J. Johnston, and T. Nann. 2020. "On Battery Materials and Methods". *Materials Today Advances* 6: 100046.

Brandt, A., A. Balducci, U. Rodehorst, S. Menne, M. Winter, and A. Bhaskar. 2014. "Investigations about the Use and the Degradation Mechanism of $LiNi_{0.5}Mn_{1.5}O_4$ in a High Power LIC". *Journal of the Electrochemical Society* 161 (6): A1139.

Bruce, Peter G, Bruno Scrosati, and J.-M. Tarascon. 2008. "Nanomaterials for Rechargeable Lithium Batteries". *Angewandte Chemie International Edition* 47: 2930–2946.

Burke, A.F. 2014. *Ultracapacitors in Hybrid and Plug-In Electric Vehicles*. New York: John Wiley & Sons, Ltd.

Cao, W.J., and J.P. Zheng. 2012. "Li-Ion Capacitors with Carbon Cathode and Hard Carbon/Stabilized Lithium Metal Powder Anode Electrodes". *Journal of Power Sources* 213: 180–185.

Cao, W.J., and J.P. Zheng. 2013. "The Effect of Cathode and Anode Potentials on the Cycling Performance of Li-Ion Capacitors". *Journal of the Electrochemical Society* 160: A1572–A1576.

Cao, Yuliang, Lifen Xiao, Maria L Sushko, Wei Wang, Birgit Schwenzer, Jie Xiao, Zimin Nie, Laxmikant V. Saraf, Zhengguo Yang, and Jun Liu. 2012. "Sodium Ion Insertion in Hollow Carbon Nanowires for Battery Applications". *Nano Letters* 12: 3783–3787.

Chawla, Neha, Neelam Bharti, and S. Singh. 2019. "Recent Advances in Non-Flammable Electrolytes for Safer Lithium-Ion Batteries". *Batteries* 5: 1–26.

Che, H., S. Chen, Y. Xie, H. Wang, K. Amine, X.-Z. Liao, and Zi-Feng Ma. 2017. "Electrolyte Design Strategies and Research Progress for Room-Temperature Sodium-Ion Batteries". *Energy and Environmental Science* 10: 1075–1101.

Chen, K.F. and D.F. Xue. 2014. "Water-Soluble Inorganic Salt with Ultrahigh Specific Capacitance: Ce $(NO_3)_3$ Can Be Designed as Excellent Pseudocapacitor Electrode". *Journal of Colloid and Interface Science* 416: 172.

Chen, M., Q. Liu, S.W. Wang, E. Wang, X. Guo, and S.L. Chou. 2019. "High-Abundance and Low-Cost Metal-Based Cathode Materials for Sodium-Ion Batteries: Problems, Progress, and Key Technologies". *Advanced Energy Materials* 9 (14): 1803609.

Chen, Mingzhe, Yanyan Zhang, Guichuan Xing, and Y. Tang. 2020. "Building High Power Density of Sodium-Ion Batteries: Importance of Multidimensional Diffusion Pathways in Cathode Materials". *Frontiers in Chemistry* 8: 152.

Cheng, De-Liang, Li-Chun Yang, and M. Zhu. 2018. "High-Performance Anode Materials for Na-Ion Batteries". *Rare Metals* 37: 167–180.

Cheng, F.Y., J. Liang, Z.L. Tao, and J. Chen. 2011. "Functional Materials for Rechargeable Batteries". *Advanced Materials* 23: 1695–1715.

Chiang, Y.-M. 2010. "Building a Better Battery". Science 330: 1485–1486.

Choi, H., and H. Yoon. 2015. "Nanostructured Electrode Materials for Electrochemical Capacitor Applications". *Nanomater* 5: 906–936.

Choi, J.W. and D. Aurbach. 2016. "Promise and Reality of Post-Lithium-Ion Batteries with High Energy Densities". *Nature Review Materials* 1: 1–16.

Chu, Lek-Heng, Christopher S. Kompella, Han Nguyen, Zhuoying Zhu, Sunny Hy, Zhi Deng, Ying Shirley Meng, and Shyue Ping Ong. 2016. "Room-Temperature All-Solid-State Rechargeable Sodium-Ion Batteries with a Cl-Doped Na_3PS_4 Superionic Conductor". *Scientific Reports* 6: 33733.

Conway, B.E., and W.G. Pell. 2003. "Double-Layer and Pseudocapacitance Types of Electrochemical Capacitors and Their Applications to the Development of Hybrid Devices". *Journal of Solid State Electrochemistry* 7: 637–644.

Dai, Hanqing, Wenqian Xu, Zhe Hu, Yuanyuan Chen, Xian Wei, Bobo Yang, Zhihao Chen, et al. 2020. "Effective Approaches of Improving the Performance of Chalcogenide Solid Electrolytes for All-Solid-State Sodium-Ion Batteries". *Frontiers in Energy Research* 8: 97.

De Adhikari, Amrita, Anukul K. Thakur, Santosh K. Tiwari, Nannan Wang, and Y. Zhu. 2020. "Current Research of Graphene-Based Nanocomposites and Their Application for Supercapacitors". *Nanomaterials* 10: 1–48.

Deng, D., M.G. Kim, J.Y. Lee, and J. Cho. 2009. "Green Energy Storage Materials: Nanostructured TiO_2 and Sn-Based Anodes for Lithium-Ion Batteries". *Energy and Environmental Science* 2: 818–837.

Deng, Wenwen, Hanxi Yang, Xinmiao Liang, Xianyong Wu, Jiangfeng Qian, Yuliang Cao, et al. 2013. "A Low Cost, All-Organic Na-Ion Battery Based on Polymeric Cathode and Anode". *Scientific Reports* 3: 2671.

Dubal, D.P., G.S. Gund, R. Holze, H.S. Jadhav, C.D. Lokhande, and C.-J. Park. 2013. "Solution-Based Binder-Free Synthetic Approach of RuO2 Thin Films for All Solid State Supercapacitors". *Electrochimica Acta* 103: 103–109.

Dubal, D.P., O. Ayyad, V. Ruiz, and P. Gomez-Romero. 2015. "Hybrid Energy Storage: The Merging of Battery and Supercapacitor Chemistries". *Chemical Society Reviews* 44: 1777–1790.

Dubal, D.P., R. Holze, and P. Gomez-Romero. 2014. "Development of Hybrid Materials Based on Sponge Supported Reduced Graphene Oxide and Transition Metal Hydroxides for Hybrid Energy Storage Devices". *Scientific Reports* 4: 1–10.

Durmus, Yasin Emre, Huang Zhang, Florian Baakes, Gauthier Desmaizieres, Hagay Hayun, Yang, Liangtao, M. Kolel, et al. 2020. "Side by Side Battery Technologies with Lithium-Ion Based Batteries". *Advanced Energy Materials* 10: 2000089.

Elia, G.A., J.B. Park, Y.K. Sun, B. Scrosati, and J. Hassoun. 2014. "Role of the Lithium Salt in the Performance of Lithium-Oxygen Batteries: A Comparative Study". *ChemElectroChem* 1: 47–50.

Ellis, Brian L., Kyu Tae Lee, and L.F. Nazar. 2010. "Positive Electrode Materials for Li-Ion and Li-Batteries". *Chemistry of Materials* 22: 691–714.

Fernandez, J.A., S. Tennison, and O. Kozynchenko. 2009. "Effect of Mesoporosity on Specific Capacitance of Carbons". *Carbon* 47: 1598–1604.

Forouzandeh, Parnia, Vignesh Kumaravel, and S.C. Pillai. 2020. "Electrode Materials for Supercapacitors: A Review of Recent Advances". *Catalysts* 10: 969.

Gao, Xiaoyu, Changzhen Zhan, Xiaoliang Yu, Qinghua Liang, Ruitao Lv, Guosheng Gai, Feiyu Kang, and Zheng-Hong Huang. 2017. "A High Performance Lithium-Ion Capacitor with Both Electrodes Prepared from Sri Lanka Graphite Ore". *Materials* 10: 1–11.

Gomez-Urbano, Juan Luis, Gelines Moreno-Fernandez, María Arnaiz Teofilo Rojo, Daniel Carriazo, and J. Ajuria. 2020. "Graphene-Coffee Waste Derived Carbon Composites as Electrodes for Optimized Lithium Ion Capacitors". *Carbon* 162: 273–282.

González, Ander, Eider Goikolea, Jon Andoni Barrena, and R. Mysyk. 2016. "Review on Supercapacitors: Technologies and Materials". *Renewable and Sustainable Energy Reviews* 58: 1189–1206.

Gou, Qianzhi, Shuang Zhao, Jiacheng Wang, and M. Li. 2020. "Recent Advances on Boosting the Cell Voltage of Aqueous Supercapacitors". *Nano-Micro Letters* 12: 98.

Han, Lu, Michelle L. Lehmann, Jiadeng Zhu, Tianyi Liu, Zhengping Zhou, Xiaomin Tang, C.-T. Heish, et al. 2020. "Recent Developments and Challenges in Hybrid Solid Electrolytes for Lithium-Ion Batteries". *Frontiers in Energy Research* 8: 202.

Han, Pengxian, Wen Ma, Shuping Pang, Qingshan Kong, Jianhua Yao, Caifeng Bib, and G. Cui. 2013. "Graphene Decorated with Molybdenum Dioxide Nanoparticles for Use in High Energy Lithium Ion Capacitors with an Organic Electrolyte". *Journal of Materials Chemistry A* 1: 5949–5954.

Hou, Jie, Wei Wang, Pingyuan Feng, Kangli Wang, and K. Jiang. 2020. "A Surface Chemistry Assistant Strategy to High Power/Energy Density and Cost-Effective Cathode for Sodium Ion Battery". *Journal of Power Sources* 453: 227879.

Hu, L., T. Zhai, H. Li, and Y. Wang. 2019. "Redox-Mediator-Enhanced Electrochemical Capacitors: Recent Advances and Future Perspectives". *ChemSusChem* 12: 1118–1132.

Hu, Z., L.X. Wang, K. Zhang, J.B. Wang, F.Y. Cheng, Z.L. Toa, and Jun Chen. 2014. "MoS_2 Nanoflowers with Expanded Interlayers as High-Performance Anodes for Sodium-Ion Batteries". *Angewandte Chemie International Edition* 53: 12794–12798.

Hu, Z., Z.Q. Zhu, F.Y. Cheng, K. Zhang, J.B. Wang, C.C. Chen, and J. Chen. 2015. "Pyrite FeS_2 for High-Rate and Long-Life Rechargeable Sodium Batteries". *Energy and Environmental Science* 8: 1309–1316.

Huang, Yunxia, Ming Cheng, Zhongcheng Xiang, and Y. Cui. 2018. "Facile Synthesis of $NiCo_2S_4$/CNTs Nanocomposites for High-Performance Supercapacitors". *Royal Society Open Science* 5: 180953.

Huang, Zhaodong, Hongshuai Hou, Yan Zhang, Chao Wang, Xiaoqing Qiu, and X. Ji. 2017. "Layer-Tunable Phosphorene Modulated by the Cation Insertion Rate as a Sodium-Storage Anode". *Advanced Materials* 29: 1702372.

Hwang, Jang-Yeon, Seung-Taek Myung, and Y.-K. Sun. 2017. "Sodium-Ion Batteries: Present and Future". *Chemical Society Reviews* 46: 3529–3614.

Iqbal, Sajid, Halima Khatoon, Ashiq Hussain Pandit, and S. Ahmad. 2019. "Recent Development of Carbon Based Materials for Energy Storage Devices". *Materials Science for Energy Technologies* 2: 417–428.

Iro, Zaharaddeen S., C. Subramani, and S.S. Dash. 2016 "A Brief Review on Electrode Materials for Supercapacitor". *International Journal of Electrochemical Science* 11: 10628–10643.

Janek, Jürgen, and W.G. Zeier. 2016. "A Solid Future for Battery Development". *Nature Energy* 1: 1–4.

Jayaraman, Sundaramurthy, Srinivasan Madhavi, and V.J. Aravindan. 2018. "High energy Li-ion capacitor and battery using graphitic carbon spheres as an insertion host from cooking oil". *Journal of Materials Chemistry A* 6: 3242–3248.

Jiang, D.E., Z. Jin, D. Henderson, and J. Wu. 2012. "Solvent Effect on the Pore-Size Dependence of an Organic Electrolyte Supercapacitor". *Journal of Physical Chemistry Letters* 3: 1727–1731.

Jiang, Jiangmin, Yadi Zhang, Zhiwei Li, Yufeng An, Qi Zhu, Yinghong Xu, Shuai Zang, Hui Dou, Xiaogang Zhang. 2020. "Defect-Rich and N-Doped Hard Carbon as a Sustainable Anode for High-Energy Lithium-Ion Capacitors". *Journal of Colloid and Interface Science* 567: 75–83.

Jie, Yulin, Xiaodi Ren, Ruiguo Cao, Wenbin Cai, and S. Jiao. 2020. "Advanced Liquid Electrolytes for Rechargeable Li Metal Batteries". *Advanced Functional Materials* 30 (25): 1910777.

Jin, M., Y. Zhang, C. Yan, Y. Fu, Y. Guo, and X. Ma. 2018. "High-Performance Ionic Liquid-Based Gel Polymer Electrolyte Incorporating Anion-Trapping Boron Sites for All-Solid-State Supercapacitor Application". *ACS Applied Materials & Interfaces* 10: 39570–39580.

Joos, Bjorn, Thomas Vranken, Wouter Marchal, Mohammadhosein Safari, Marlies K. Van Bael, and A.T. Hardy. 2018. "Eutectogels: A New Class of Solid Composite Electrolytes for Li/Li-Ion Batteries". *Chemistry of Materials* 30: 655–662.

Jung, N., S. Kwon, D. Lee, D.M. Yoon, Y.M. Park, A. Benayad, Jae-Young Choi, and Jong Se Park. 2013. "Synthesis of Chemically Bonded Graphene/Carbon Nanotube Composites and Their Application in Large Volumetric Capacitance Supercapacitors". *Advanced Materials* 25: 6854–6858.

Kam, Kinson C., and M.M. Doeff. 2012. "Electrode Materials for Lithium Ion Batteries". *Material Matters* 7: 182–187.

Kang, H., Y. Liu, K. Cao, Y. Zhao, L. Jiao, Y. Wang, and Huatang Yuana. 2015. "Update on Anode Materials for Na-Ion Batteries". *Journal of Materials Chemistry A* no. 35: 1–15.

Kathy, L. 2014. *Materials in Energy Conversion, Harvesting, and Storage*. New Jersey: John Wiley & Sons, Inc, 487.

Khomenko, V., E. Raymundo-Piñero, and F, Béguin. 2008. "High-Energy Density Graphite/AC Capacitor in Organic Electrolyte". *Journal of Power Sources* 177 (2): 643–651.

Khurana, R., J.L. Schaefer, L.A. Archer, and G.W. Coates. 2014. "Suppression of Lithium Dendrite Growth Using Cross-Linked Polyethylene/Poly(ethylene oxide) Electrolytes: A New Approach for Practical Lithium-Metal Polymer Batteries". *J Am Chem Soc* 136: 7395.

Kim, H., J. Hong, K.Y. Park, H. Kim, S.W. Kim, and K. Kang. 2014. "Aqueous Rechargeable Li and Na Ion Batteries". *Chemical Reviews* 114: 11788–11827.

Kim, J.J., K. Yoon, I. Park, and K. Kang. 2017. "Progress in the Development of Sodium-Ion Solid Electrolytes Small". *Methods* 1: 1700219.

Kim, Ki. T., G. Ali, K.Y. Chung, C.S. Yoon, H. Yashiro, Y.-K. Sun, Jun Lu, Khalil Amine, and Seung-Taek Myung. 2014. "Anatase Titania Nanorods as an Intercalation Anode Material for Rechargeable Sodium Batteries". *Nano Letters* 14: 416–422.

Kim, U.H., J. Hk, J.Y. Hwang, H.H. Ryu, C.S. Yoon, and Y.K. Sun. 2019. "Compositionally and Structurally Redesigned High-Energy Ni-Rich Layered Cathode for Next-Generation Lithium Batteries". *Materials Today* 23: 26–36.

Kim, Y., Y. Park, A. Choi, N.S. Choi, J. Kim, J. Lee, and Ji Heon Ryu, Seung M. Oh, and Kyu Tae Lee. 2013. "An Amorphous red Phosphorus/Carbon Composite as a Promising Anode Material for Sodium Ion Batteries". *Advanced Materials* 25: 3045–3049.

Kong, Shuying, Fan Yang, Kui Cheng, Tian Ouyang, Ke Ye, Guiling Wang, and D. Chao. 2017. "In-Situ Growth of Cobalt Oxide Nanoflakes from Cobalt Nanosheet on Nickel Foam for Battery-Type Supercapacitors with High Specific Capacity". *Journal of Electroanalytical Chemistry* 785: 103–108.

Kopczyński, K., L. Kolanowski, M. Baraniak, K. Lota, A. Sierczyńska, and Lota G. 2016. "Highly Amorphous PbO_2 as an Electrode in Hybrid Electrochemical Capacitors". *Current Applied Physics* 17: 66–71.

Kotz, R., and M. Carlen. 2000. "Principles and Applications of Electrochemical Capacitors". *Electrochimica Acta* 45: 2483–2498.

Kundu, Dipan, Elahe Talaie, Victor Duffort, and L.F. Nazar. 2015. "The Emerging Chemistry of Sodium Ion Batteries for Electrochemical Energy Storage". *Angewandte Chemie International Edition* 54: 3431–3448.

Kwon, T.-W., J.W. Choi, and A. Coskun. 2018. "The Emerging Era of Supramolecular Polymeric Binders in Silicon Anodes". *Chemical Society Reviews* 47: 2145.

Lai, Linfei, Huanping Yang, Liang Wang, Boon Kin Teh, Jianqiang Zhong, Harry Chou, Luwei Chen, et al. 2012. "Preparation of Supercapacitor Electrodes through Selection of Graphene Surface Functionalities". *ACS Nano* 6: 5941–5951.

Lallaoui, Abdelfettah, Zineb Edfouf, Omar Benabdallah, Siham Idrissi, Mohammed Abd-Lefdil, and F.C.E. Moursli. 2020. "New Titanium (III) Phosphite Structure and Its Application as Anode for Lithium Ion Batteries". *International Journal of Hydrogen Energy* 45: 11167–11175.

Lao, Mengmeng, Yu Zhang, Wenbin Luo, Qingyu Yan, Wenping Sun, and S.X. Dou. 2017. "Alloy-Based Anode Materials toward Advanced Sodium-Ion Batteries". *Advanced Materials* 29: 1700622.

Lee, Seong-Hun, Gayeong Yoo, Jinil Cho, Seokgyu Ryu, Youn Sang Kim, and J. Yoo. 2020. "Expanded Graphite/Copper Oxide Composite Electrodes for Cell Kinetic Balancing of Lithium-Ion Capacitor". *Journal of Alloys and Compounds* 829: 154566.

Lei, Yu, Zheng-Hong Huang, Ying Yang, Wanci Shen, Yongping Zheng, Hongyu Sun, and Feiyu Kang. 2013. "Porous Mesocarbon Microbeads with Graphitic Shells: Constructing a High-Rate, High-Capacity Cathode for hybrid Supercapacitor". *Scientific Reports* 3: 2477.

Li, Bing, Junsheng Zheng, Hongyou Zhang, Liming Jin, Daijun Yang, Hong Lv, C Shen, et al. 2018. "Electrode Materials, Electrolytes, and Challenges in Nonaqueous Lithium-Ion Capacitors". *Advanced Materials* 30 (17): 1705670

Li, Matthew, Jun Lu, Zhongwei Chen, and K. Amine. 2018. "30 Years of Lithium-Ion Batteries". *Advanced Materials* 30: 1800561.

Li, Q., J. Chen, L. Fan, X. Kong, and Y. Lu. 2016. "Progress in Electrolytes for Rechargeable Li-Based Batteries and Beyond". *Green Energy and Environment* 1: 18–42.

Li, Shiqi, and Z. Fan. 2020. "Special Issue: Advances in Electrochemical Energy Materials". *Materials* 13: 844.

Li, Wei-Jie, Chao Han, Wanlin Wang, Florian Gebert, Shu-Lei Chou, Hua-Kun Liu, Xinhe Zhang, and Shi-Xue Dou. 2017. "Commercial Prospects of Existing Cathode Materials for Sodium Ion Storage". *Advanced Energy Materials* 1700274.

Li, X., and B. Wei. 2013. "Supercapacitors Based on Nanostructured Carbon". *Nano Energy* 2: 159–173.

Lin, D., Y. Liu, and Y. Cui. 2017. "Reviving the Lithium Metal Anode for High-Energy Batteries". *Nature Nanotechnology* 12: 194–206.

Liu, C., Z. Yu, D. Neff, A. Zhamu, and B.Z. Jang. 2010. "Graphene-Based Supercapacitor with an Ultrahigh Energy Density". *Nano Letters* 8: 4863–4868.

Liu, Chaofeng, Zachary G. Neale, and G. Cao. 2016. "Understanding Electrochemical Potentials of Cathode Materials in Rechargeable Batteries". *Materials Today* 19: 1–32.

Liu, F., and P.P. Mukherjee. 2015. "Materials for Positive Electrodes in Rechargeable Lithium-Ion Batteries". In *Rechargeable Lithium Batteries*, edited by A.A. Franco, 21–39. Cambridge: Woodhead Publishing.

Liu, L., X. Qi, S. Yin, Q. Zhang, X. Liu, L. Suo, Hong Li, Liquan Chen, and Yong-Sheng Hu. 2019. "In Situ Formation of a Stable Interface in Solid-State Batteries". *ACS Energy Letters* 4: 1650–1657.

Liu, Nian, Li Wei yang, Mauro Pasta, and Y. Cui. 2014. "Nanomaterials for Electrochemical Energy Storage". *Frontiers in Physics* 9 (3): 323–350.

Liu, W., S.W. Lee, D. Lin, F. Shi, S. Wang, A.D. Sendek, and Y. Cui. 2017. "Enhancing Ionic Conductivity in Composite Polymer Electrolytes with Well-Aligned Ceramic Nanowires". *Nature Energy* 2: 17035.

Liu, Y., X. Fan, Z. Zhang, H.H. Wu, D. Liu, A. Dou, Mingru Su, Qiaobao Zhang, and Dewei Chu. 2019. "Enhanced Electrochemical Performance of Li-Rich Layered Cathode Materials by Combined Cr Doping and LiAlO$_2$ Coating". *ACS Sustainable Chemistry and Engineering* 7: 2225–2235.

Liu, Y.C., N. Zhang, L.F. Jiao, Z.L. Tao, and J. Chen. 2015. "Ultrasmall Sn Nanoparticles Embedded in Carbon as High-Performance Anode for Sodium-Ion Batteries". *Advanced Functional Materials* 25: 214.

Liu, Yi, Yingxin Li, Fan Li, Yizhuo Liu, Xiaoyan Yuan, Lifeng Zhang, and S. Guo. 2019. "Conversion of Ti$_2$AlC to C-K$_2$Ti$_4$O$_9$ via a KOH Assisted Hydrothermal Treatment and Its Application in Lithium-Ion Battery Anodes". *Electrochimica Acta* 295: 599–604.

Longwei, Liang, Wenheng Zhang, Denis Kionga, and Jinyang Zhang. 2019. Comparative investigations of high-rate NaCrO2 cathodes towards wide-temperature-tolerant pouch-type Na-ion batteries from −15 to 55 °C: nanowires vs. bulk. *Journal of Materials Chemistry A* 7: 11915.

Lu, Q., M.W. Lattanzi, Y. Chen, X. Kou, W. Li, X. Fan, Karl M. Unruh, et al. 2011. "Supercapacitor Electrodes with High-Energy and Power Densities Prepared from Monolithic NiO/Ni Nanocomposites". *Angewandte Chemie International Edition* 50: 6979–6982.

Ma, Yanfeng, Huicong Chang, Miao Zhang, and Y. Chen. 2015. "Graphene-Based Materials for Lithium-Ion Hybrid Supercapacitors". *Advanced Materials* 27: 5296–5308.

Maa, Xinlong, Lei Zhao, Xinyu Song, Zhiqing Yua, Lu Zhao, Yintao Yu, Zhihua Xiao, Guoqing Ning, and Jinsen Gao. 2018. "Superior Capacitive Behaviors of the Micron-Sized Porous Graphene Belts with High Ratio of Length to Diameter". *Carbon* 140: 314–323.

Manthiram, Arumugam, Xingwen Yu, and S. Wang. 2017. "Lithium Battery Chemistries Enabled by Solid-State Electrolytes". *Nature Reviews Materials* 2: 16103.

Martins, Vitor L., Herbert R. Neves, Ivonne E. Monje, Marina M. Leite, Paulo F.M. De Oliveira, Rodolfo M. Antoniassi, S. Chauque, et al. 2020. "An Overview on the Development of Electrochemical Capacitors and Batteries – part II". *An Acad Bras Cienc* 92: 1–29.

Md Moniruzzaman S.K., Chee Yoon Yue, Kalyan Ghosh, and R.K. Jena. 2016. "Review on Advances in Porous Nanostructured Nickel Oxides and Their Composite Electrodes for High-Performance Supercapacitors". *Journal of Power Sources* 308: 121–140.

Miller, Elizabeth Esther, Ye Hua, and F.H. Tezel. 2018. "Materials for Energy Storage: Review of Electrode Materials and Methods of Increasing Capacitance for Supercapacitors". *Journal of Energy Storage* 20: 30–40.

Ming Chen, S., and R. Ramachandran, V. Mani, and R. Saraswathi. 2014. "Recent Advancements in Electrode Materials for the High-Performance Electrochemical Supercapacitors: A Review". *International Journal of Electrochemistry and Science* 9: 4072–4085.

Mishra, Amit, Akansha Mehta, Soumen Basu, Shweta J. Malode, Nagaraj P. Shetti, Shyam S. Shukla, Mallikarjuna N. Nadagouda, and Tejraj M. Aminabhavi. 2018. "Electrode Materials for Lithium-Ion Batteries". *Materials Science for Energy Technologies* 1: 182–187.

Mousavi, Maral P.S., Benjamin E. Wilson, Sadra Kashefolgheta, Evan Anderson, Siyao He, Philippe Buhlmann, and A. Stein. 2016. "Ionic Liquids as Electrolytes for Electrochemical Double-Layer Capacitors: Structures that Optimize Specific Energy". ACS Applied Materials and Interfaces 1–41.

Mukherjee, S., Z. Ren, and G. Singh. 2018. "Molecular Polymer-Derived Ceramics for Applications in Electrochemical Energy Storage Devices". *Journal of Physics D: Applied Physics* 51: 463001.

Mukherjee, Santanu, Shakir Bin Mujib, Davi Soares, and G. Singh. 2019. "Electrode Materials for High-Performance Sodium-Ion Batteries". *Materials* 12: 1952.

Munaoka, Takatoshi, Xuzhou Yan, Jeffrey Lopez, John W. F. To, Jihye Park Jeffrey, B.H. Tok, Y. Cui, and Z. Bao. 2018. "Ionically Conductive Self-Healing Binder for Low Cost Si Microparticles Anodes in Li-Ion Batteries". *Advanced Energy Materials* 8 (14): 1703138.

Najib, Sumaiyah, and E. Erdem. 2019. "Current Progress Achieved in Novel Materials for Supercapacitor Electrodes: Mini Review". *Nanoscale Advances* 1: 2817–2827.

Nanda, Jagjit, Chongmin Wang, and P. Liu. 2018. *Frontiers of Solid-State Batteries*. Cambridge: Cambridge University Press, 740–745.

Newman, G.H., and L.P. Klemann. 1980. "Ambient Temperature Cycling of an Na-TiS$_2$ Cell". *Journal of Electrochemical Society* 127: 2097–2099.

Ni, Q., Y. Bai, F. Wu, and C. Wu. 2017. "Polyanion-Type Electrode Materials for Sodium-Ion Batteries". *Advanced Science* 4: 1600275.

Nitta, Naoki, Feixiang Wu, Jung Tae Lee, and G. Yushin. 2015. "Li-Ion Battery Materials: Present and Future". *Materials Today* 18: 252–264.

Nwanya, Assumpta Chinwe, Miranda M. Ndipingwi, Chinwe O. Ikpo, Fabian I. Ezema, Emmanuel I. Iwuoha, and M. Maaza. 2020. "Biomass Mediated Multi Layered NaNi$_x$Co$_{1-x}$O$_2$ (x = 0.4) and α-Fe$_2$O$_3$ Nanoparticles for Aqueous Sodium Ion Battery". *Journal of Electroanalytical Chemistry* 858: 113809.

Pahari, Debanjana, and S. Puravankara. 2020. "On Controlling the P2-O2 Phase Transition by Optimal Ti-Substitution on Nisite in P2-type Na$_{0.67}$Ni$_{0.33}$Mn$_{0.67}$O$_2$ (NNMO) Cathode for Na-Ion Batteries". *Journal of Power Sources* 455: 227957.

Pan, Long, Xiao-Dong Zhu, Xu-Ming Xie, and Y.-T. Liu. 2015. "Smart Hybridization of TiO$_2$ Nanorods and Fe$_3$O$_4$ Nanoparticles with Pristine Graphene Nanosheets: Hierarchically Nanoengineered Ternary Heterostructures for High-Rate Lithium Storage". *Advanced Functional Materials* 25 (22): 3341–3350.

Pandit, Bidhan, Gagan Kumar Sharma, and B.R. Sankapal. 2017. "Chemically Deposited Bi$_2$S$_3$:PbS Solid Solution Thin film as Supercapacitive Electrode". *Journal of Colloid and Interface Science* 505: 1011–1017.

Peled, E. 1979. "The Electrochemical Behavior of Alkali and Alkaline Earth Metals in Nonaqueous Battery Systems – the Solid Electrolyte Interphase Model". *Journal of Electrochemical Society* 126: 2047.

Perveen, Tahira, Muhammad Siddiq, Nadia Shahzad, Rida Ihsan, Abrar Ahmad, M.I. Shahzad. 2020. "Prospects in Anode Materials for Sodium Ion Batteries – a Review". *Renewable and Sustainable Energy Reviews* 119: 109549.

Ping, Ong Shyue, Chevrier Vincent, Hautier Geoffroy, Jain Anubhav, Moore Charles Jacob, Kim Sangtae, and Xiaohua Maa, and, Gerbrand Ceder. 2011. "Voltage, Stability and Diffusion Barrier Differences between Sodium-Ion and Lithium-Ion Intercalation Materials". *Energy and Environmental Science* 4: 3680–3688.

Qiao, Lixin, Xabier Judez, Teofilo Rojo, Michel Armand, and H. Zhang. 2020. "Review – Polymer Electrolytes for Sodium Batteries". *Journal of Electrochemical Society* 167: 070534.

Qu, Q., S. Yang, and X. Feng. 2011. "2D Sandwich-like Sheets of Iron Oxide Grown Ongraphene as High Energy Anode Material for Supercapacitors". *Advanced Materials* 23: 5574–5580.

Quartarone, E., and P. Mustarelli. 2020. "Review – Emerging Trends in the Design of Electrolytes forLithium and Post-Lithium Batteries". *Journal of Electrochemical Society* 167: 050508.

Raj, Michael Ruby, Ramalinga Viswanathan Mangalaraja, David Contreras, Kokkarachedu Varaprasad, Mogalahalli Venkatashamy Reddy, and S. Adams. 2020. "Perylenedianhydride-Based Polyimides as Organic Cathodes forRechargeable Lithium and Sodium Batteries". *ACS Applied Energy Materials* 3: 240–252.

Rajagopalan, Ranjusha, Yougen Tang, Chuankun Jia, Xiaobo Ji, and H. Wang. 2020. "Understanding the Sodium Storage Mechanisms of Organic Electrodes in Sodium Ion Batteries: Issues and Solutions". *Energy and Environmental Science* no. 6: 1–27.

Ramya, R., R. Sivasubramanian, and M.V. Sangaranarayanan. 2013. "Conducting Polymersbased Electrochemical Supercapacitors – Progress and Prospects". *Electrochimica Acta* 101: 109–129.

Rowley-Neale, S.J. and C.E. Banks. 2018. "Fabrication of Graphene Oxide Supercapacitor Devices". *ACS Applied Energy Materials* 1: 707–714.

Roy, P., and S.K. Srivastava. 2015. "Nanostructured Anode Materials for Lithium Ion Batteries". *Journal of Materials Chemistry A* 3: 2454–2484.

Sarita Yadav, Devi 2020 "A. Recent Advancements of Metal Oxides/Nitrogen-Doped Graphene Nanocomposites for Supercapacitor Electrode Materials". *Journal of Energy Storage* 30: 101486.

Sarkar, Ananta, C.V. Manohar, and S. Mitra. 2020. "A Simple Approach to Minimize the First Cycle Irreversible Loss of Sodium Titanate Anode towards the Development of Sodium-Ion Battery". *Nano Energy* 70: 104520.

Saurel, Damien, Brahim Orayech, Biwei Xiao, Daniel Carriazo, Xiaolin Li, and T. Rojo. 2018. "From Charge Storage Mechanism to Performance: A Roadmap toward High Specific Energy Sodium-Ion Batteries through Carbon Anode Optimization". *Advanced Energy Materials* 8: 1703268.

Selis, L.A., and J.M. Seminario. 2018. "Dendrite Formation in Silicon Anodes of Lithium-Ion Batteries". *RSC Advances* 8: 5255–5267.

Shadike, Zulipiya, Enyue Zhao, Yong-Ning Zhou, Xiqian Yu, Yong Yang, Enyuan Hu, Seongmin Bak, Lin Gu, and Xiao-Qing Yang. 2018. "Advanced Characterization Techniques for Sodium-Ion Battery Studies". *Advanced Energy Materials* 8: 1702588.

Shang, H., Z. Zuo, L. Yu, F. Wang, F. He, and Y. Li. 2018. "Low-Temperature Growth of All-Carbon Graphdiyne on a Silicon Anode for High-Performance Lithium-Ion Batteries". *Advanced Materials* 30: 1801459.

Sharona A. Melchior, Nithyadharseni Palaniyandy, Iakovos Sigalas, Sunny E. Iyuke, and K.I. Ozoemena. 2019. "Probing the Electrochemistry of MXene (Ti_2CT_x)/Electrolytic Manganese Dioxide (EMD) Composites as Anode Materials for Lithiumion Batteries". *Electrochimica Acta* 297: 961–973.

Shen, Xin, He Liu, Xin-Bing Cheng, Chong Yan, and J.-Q. Huang. 2018. "Beyond Lithium Ion Batteries: Higher Energy Density Battery Systems Based on Lithium Metal Anodes". *Energy Storage Materials* 12: 161–175.

Shi, J.L., D.D. Xiao, M. Ge, X. Yu, Y. Chu, X. Huang, and Xu-Dong Zhang, et al. 2018. "High-Capacity Cathode Material with High Voltage for Li-Ion Batteries". *Advanced Materials* 30: 1705575.

Shi, Ruiying, Cuiping Han, Xiaofu Xu, Xianying Qin, Lei Xu, Hongfei Li, Junqin Li, Ching-Ping Wong, and Baohua Li. 2018. "Electrospun N-Doped Hierarchical Porous Carbon Nanofiber with Improved Degree of Graphitization for High-Performance Lithium Ion Capacitor". *Chemistry – A European Journal* 24: 10460–10467.

Shi, Ruiying, Cuiping Han, Hongfei Li, Lei Xu, Tengfei Zhang, Junqin Li, Junqin Li, Ching-Ping Wong, and Baohua Li. 2018. "NaCl-Templated Synthesis of Hierarchical Porous Carbon with Extremely Large Specific Surface Area and Improved Graphitization Degree for High Energy Density Lithium Ion Capacitors". *Material Chemistry A* 6: 1–23.

Shi, Ye, Lijia Pan, Borui Liu, Yaqun Wang, Yi Cui, Zhenan Bao, and G. Yu. 2014. "Nanostructured Conductive Polypyrrole Hydrogelsas High-Performance, Flexible Supercapacitor Electrodes". *Journal of Materials Chemistry A* 2: 6086–6091.

Shin, S.M., G.J. Jung, Woo-Jin Lee, C.Y. Kang, and J.P. Wang. 2015. "Recovery of Electrodic Powder from Spent Lithium Ion Batteries (LIBs)". *Archives of Metallurgy and Materials* 60: 1145–1149.

Simon, P., and K. Naoi. 2008. "New Materials and New Configurations for Advanced Electrochemical Capacitors". *Electrochemical Society* 17 (1): 34.

Soc, C., G. Wang, and J. Zhang. 2012. "A Review of Electrode Materials for Electrochemical Supercapacitors". *Chemical Society Reviews* no. 2: 797–828.

Song, D., J. Zhu, L. Xuan, C. Zhao, L. Xie, and L. Chen. 2018. "Freestanding Two-Dimensional Ni(OH) Thin Sheets Assembled by 3D Nanoflake Array as Basic Building Units for Supercapacitor Electrode Materials". *Journal of Colloid and Interface Science* 509: 163–170.

Su, L.H., X.G. Zhang, C.H. Mi, B. Gao, and Y. Liu. 2009. "Improvement of the Capacitive Performances for Co–Al Layered Double Hydroxide by Adding Hexacyanoferrate into the Electrolyte". *Physical Chemistry Chemical Physics* 11: 2195–2202.

Sun, Fei, Jihui Gao, Yuwen Zhu, Xinxin Pi, Lijie Wang, Xin Liu, and Y. Qin. 2017. "A High Performance Lithium Ion Capacitor Achieved by the Integration of a Sn-C Anode and a Biomass-Derived Microporous Activated Carbon Cathode". *Scientific Reports* 7: 40990.

Sun, L., and K. Qiu. 2012. "Organic Oxalate as Leachant and Precipitant for the Recovery of Valuable Metals from Spent Lithium-Ion Batteries". *Waste Manage* 32: 1575–1582.

Sun, Y., L. Zhao, H.L. Pan, X. Lu, L. Gu, Y.S. Hu, Hong Li, et al. 2013. "Direct Atomic-Scale Confirmation of Three-Phase Storage Mechanism in $Li_4Ti_5O_{12}$ Anodes for Room-Temperature Sodium-Ion Batteries". *Nature Communications* 4: 1870.

Sun, Yang, Shaohua Guo, and H. Zhou. 2018. "Exploration of Advanced Electrode Materials for Rechargeable Sodium-Ion Batteries". *Advanced Energy Materials* 9 (3): 1800212.

Tao, Wei, Mao-Lian Xu, Yan-Rong Zhu, Qianyu Zhang, and T.-F. Yi. 2017. "Structure and Electrochemical Performance of $BaLi_{2-x}Na_xTi_6O_{14}$ (0≤x≤2) as Anode Materials for Lithium-Ion Battery". *Science China Materials* 60: 728–738.

Tarascon, J.M., and M. Armand. 2001. "Issues and Challenges Facing Rechargeable Lithium Batteries". *Nature* 414: 359–367.

Temeche, Eleni, Mengjie Yu, and R.M. Laine. 2020. "Silica Depleted Rice Hull Ash (SDRHA), an Agricultural Waste, as a High-Performance Hybrid Lithium-Ion Capacitor". *Green Chemistry* 1–13.

Utetiwabo, Wellars, Le Yang, Muhammad Khurram Tufail, Lei Zhou, Renjie Chen, Yimeng Lian, and Wen Yang. 2020. "Electrode Materials Derived from Plastic Wastes and Other Industrial Wastes for Supercapacitors". *Chinese Chemical Letters* 31: 1474–1489.

Vo, T.D., H.V. Nguyen, Q.D. Nguyen, Q. Phung, V.M. Tran, and P.L.M. Le. 2019. "Carbonate Solvents and Ionic Liquid Mixtures as an Electrolyte to Improve Cell Safety in Sodium-Ion Batteries". *Journal of Chemistry* 2019: 1–10.

Wang, H.W., C. Guan, X.F. Wang, and H.J. Fan. 2015. "A High Energy and Power Li-Ion Capacitor Based on a TiO_2 Nanobelt Array Anode and a Graphene Hydrogel Cathode". *Small* 11: 1470–1477.

Wang, Huanwen, Yu Zhang, Huixiang Ang, Yongqi Zhang, Hui Teng Tan, Yufei Zhang, Y. Guo, et al. 2016. "A High-Energy Lithium-Ion Capacitor by Integration of a 3D Interconnected Titanium Carbide Nanoparticle Chain Anode with a Pyridine-Derived Porous Nitrogen-Doped Carbon Cathode". *Advanced Functional Materials* 26: 3082–3093.

Wang, Huilin, Xitong Liang, Jiutian Wang, Shengjian Jiao, and D. Xue. 2020. "Multifunctional Inorganic Nanomaterials for Energy Applications". *Nanoscale* 12: 14–42.

Wang, L., X. Bi, and S. Yang. 2016. "Partially Singl-Crystalline Mesoporous Nb2O Nanosheets in between Graphene for Ultrafast Sodium Storage". *Advanced Materials* 28: 7672–7679.

Wang, L., X. He, J. Li, J. Gao, M. Fang, and G. Tian. 2013. "Graphene-Coated Plastic film as Current Collector for Lithium/Sulfur Batteries". *Journal of Power Sources* 239: 623–627.

Wang, Liguang, Jun Li, Guolong Lu, Wenyan Li, Qiqi Tao, Caihong Shi, Huile Jin, Guang Chen, and Shun Wang. 2020. "Fundamentals of Electrolytes for Solid-State Batteries: Challenges and Perspectives". *Frontiers in Materials* 7: 111.

Wang, Luyuan Paul, Linghui Yu, Xin Wang, Madhavi Srinivasan, and Z.J. Xu. 2015. "Recent Developments in Electrode Materials Forsodium-Ion Batteries". *Journal of Materials Chemistry A* 3: 9353–9378.

Wang, Xia, Xueying Li, Qiang Li, Hongsen Li, Jie Xu, Hong Wang, G. Zhao, et al. 2018. "Improved Electrochemical Performance Based on Nanostructured $SnS_2@CoS_2$–rGO Composite Anode for Sodium-Ion Batteries". *Nano-Micro Letters* 10: 46–58.

Wang, Xiaojun, Lili Liu, and Z. Niu. 2019. "Carbon-Based Materials for Lithium-Ion Capacitors". *Mater Chem Front* 3: 1265–1279.

Wang, Y., Z. Shi, Y. Huang, Y. Ma, C. Wang, M. Chen, and Y. Chen. 2009. "Supercapacitor Devices Based on Graphene Materials". *Journal of Physical Chemistry C* 113: 13103–13107.

Wang, Yonggang, Yanfang Song, and Y. Xia. 2016. "Electrochemical Capacitors: Mechanism, Materials, Systems, Characterization and Applications". *Chemical Society Reviews* 45: 5925–5950.

Wang, Yumei, Shufeng Song, Chaohe Xu, Ning Hu, Janina Molenda, and L. Lu. 2019. "Development of Solid-State Electrolytes for Sodium-Ion Battery – a Short Review". *Nano Materials Science* 1: 91–100.

Wanga, Teng, Hai Chao Chenb, Feng Yua, X.S. Zhaob, and H. Wang. 2018. "Boosting the Cycling Stability of Transition Metal Compounds-Based Supercapacitors". *Energy Storage Materials* 16: 545–573.

Whittingham, M.S. 2008. "Lithium Batteries and Cathode Materials". *Chemical Reviews* 104: 4271.

Wu, J., Q.E. Zhang, A.A. Zhou, Z. Huang, H. Bai, and L. Li. 2016. "Phase-Separated Polyaniline/Graphene Composite Electrodes for High-Rate Electrochemical Supercapacitors". *Advanced Materials* 28: 10211–10216.

Wu, X., S. Li, B. Wang, J. Liu, and M. Yu. 2017. "In Situ Template Synthesis of Hollow Nanospheres Assembled from $NiCo_2S_4@C$ Ultrathin Nanosheets with High Electrochemical Activities for Lithium Storage and ORR Catalysis". *Physical Chemistry* 19: 11554–11562.

Xia, Jiale, Hongyang Zhao, Wei Kong Pang, Zongyou Yin, Bo Zhou, Gang He, Zaiping Guo, and Yaping Du. 2018. "Lanthanide Doping Induced Electrochemical Enhancement of $Na_2Ti_3O_7$ Anodes for Sodium-Ion Batteries". *Chemical Science* 9: 3421–3425.

Xiao, Ying, Seon Hwa Lee, and Y.-K. Sun. 2016. "The Application of Metal Sulfides in Sodium Ion Batteries". *Advanced Energy Materials* 7 (3): 1601329.

Xie, Lijing, Guohua Sun, Fangyuan Su, Xiaoqian Guo, Qingqiang Kong, Xiaoming Li, X. Huang, et al. 2016. "Hierarchical Porous Carbon Microtubes Derived from Willow Catkins for Supercapacitor Applications". *Journal of Materials Chemistry A* 4: 1637–1646.

Xu, Da, Kun Ma, Ling Chen, Yanjie Hu, Hao Jiang, and C. Li. 2020. "MXene Interlayer Anchored Fe_3O_4 Nanocrystals for Ultrafast Li-Ion Batteries". *Chemical Engineering Science* 212: 115342.

Xu, Gui-Liang, Rachid Amine, Ali Abouimrane, Haiying Che, Mouad Dahbi, Zi-Feng Ma, Ismael Saadoune, et al. 2018. "Challenges in Developing Electrodes, Electrolytes, and Diagnostics Tools to Understand and Advance Sodium-Ion Batteries". *Advanced Energy Materials* 8 (14): 1702403.

Xu, K. 2004. "Nonaqueous Liquid Electrolytes for Lithium-Based Rechargeable Batteries". *Chemical Reviews* 104: 4303–4417.

Xu, K. 2010. "Electrolytes and Interphasial Chemistry in Li Ion Devices". *Energies* 3: 135–154.

Xu, Kang, and Cresce Av. 2011. "Interfacing Electrolytes with Electrodes in Li Ion Batteries". *Journal of Materials Chemistry* 21: 9849.

Xu, X.M., C.J. Niu, M.Y. Duan, X.P. Wang, L. Huang, J.H. Wang, L. Pu, et al. 2017. "Alkaline Earth Metal Vanadates as Sodium-Ion Battery Anodes". *Nature Communications* 8: 11.

Xu, Zheng-Long, Gabin Yoon, Kyu-Young Park, Hyeokjun Park, Orapa Tamwattana, Sung Joo Kim, et al. 2019. "Tailoring Sodium Intercalation in Graphite for High Energy and Power Sodium Ion Batteries". *Nature Communications* 10: 2598.

Yadav, A.A., and U.J. Chavan. 2017. "Electrochemical Supercapacitive Performance of Spray Deposited Co_3O_4 Thin film Nanostructures". *Electrochimica Acta* 232: 370–376.

Yadegari, H., Q. Sun, and X. Sun. 2016. "Sodium-Oxygen Batteries: A Comparative Review from Chemical and Electrochemical Fundamentals to Future Perspective". *Advanced Materials* 28: 7065–7093.

Yan, Chaojing, Mengyuan Jin, Xinxin Pan, Longli Ma, and X. Ma. 2020. "Aflexible Polyelectrolyte-Based Gel Polymerelectrolyte for High-Performance All-Solid-Statesupercapacitor Application". *RSC Advances* 10: 9299–9308.

Yan, J., Q. Wang, T. Wei, and Z. Fan. 2014. "Recent Advances in Design and Fabrication of Electrochemical Supercapacitors with High Energy Densities". *Advanced Energy Materials* 4: 1300816.

Yan, Jingwang, Yuanyuan Sun, Liang Jiang, Yiang Tian, Rong Xue, Lixing Hao, Wei Liu, and B. Yi. 2013. "Electrochemical Performance of Lithium Ion Capacitors Using Aqueous Electrolyte at High Temperature". *J Renewable Sustainable Energy* 5: 021404.

Yan, Z., Q. Hu, G. Yan, H. Li, K. Shih, Z. Yang, Xinhai Li, and Zhixing Wang, and Jiexi Wang. 2017. "Co_3O_4/Co Nanoparticles Enclosed Graphitic Carbon as Anode Material for High Performance Li-Ion Batteries". *Chemical Engineering Journal* 321: 495–501.

Yang, Dongfang, and M.I. Ionescu. 2017. "Metal Oxide-Carbon Hybrid Materials for Application in Supercapacitors". In *Metal Oxides Supercapacitors*, edited byDeepak P. Dubal and Pedro Gomez-Romero, 193–218. Amsterdam: Elsevier.

Yang, Jinfeng, Huanrui Zhang, Qian Zhou, Hongtao Qu, Tiantian Dong, Min Zhang, Ben Tang, Jianjun Zhang, and Guanglei Cui. 2019. "Safety-Enhanced Polymer Electrolytes for Sodium Batteries: Recent Progress and Perspectives". *ACS Applied Materials and Interfaces* 11: 17109–17127.

Yasina, Ghulam, Muhammad Abubaker Khanb, Waheed Qamar Khanc, Tahira Mehtabd, Rashid Mustafa Koraia, Xia Lua, et al. 2019. "Facile and Large-Scalable Synthesis of Low Cost Hard Carbon Anode for Sodium-Ion Batteries". *Results in Physics* 14: 102404.

Yu, Xin Yao, and X.W. Lou. 2017. "Mixed Metal Sulfides for Electrochemical Energy Storage and Conversion". *Advanced Energy Materials* 8 (3): 1701592.

Yuan, Huang, Jiuwei Liu, Jiyan Zhang, Shunyu Jin, Yixiang Jian, Shengdong Zhang, et al. 2019. Flexible quasi-solid-state zinc ion batteries enabled by highly conductive carrageenan bio-polymer electrolyte RSC Advances 9: 16313–16319.

Zhan, Yi, Eldho Edison, William Manalastas, Ming Rui Joel Tan, R. Satish, Andrea Buffa, Srinivasan Madhavi, and D. Mandler. 2020. "Electrochemical Deposition of Highly Porous Reduced Graphene Oxide Electrodes for Li-Ion Capacitors". *Electrochimica Acta* 337: 135861.

Zhang, Hongwei, Zhisheng Lv, Qinghua Liang, Huarong Xia, Zhiqiang Zhu, Wei Zhang, Xiang Ge, Pei Yuan, Qingyu Yan, and Xiaodong Chen. 2020. "Highly Elastic Binders Incorporated with Helical Molecules to Improve the Electrochemical Stability of Black Phosphorous Anodes for Sodium-Ion Batteries". *Batteries & Supercaps* 3: 1–7.

Zhang, J., Z. Shi, J. Wang, and J. Shi. 2015. "Composite of Mesocarbon Microbeads/Hard Carbon as Anode Material for Lithium Ion Capacitor with High Electrochemical Performance". *Journal of Electroanalytical Chemistry Electroanalytical Chemistry* 747: 20–28.

Zhang, Li Li, Zhibin Lei, Jintao Zhang, Xiaoning Tian, and X.S. Zhao. 2016. "Supercapacitors: Electrode Materials Aspects". In *Encyclopedia of Inorganic Chemistry* 1–23. New York: John Willey & Sons.

Zhang, Qifeng, Evan Uchaker, Stephanie L. Candelaria, and G. Cao. 2013. "Nanomaterials for Energy Conversion and Storage". *Chemical Society Reviews* 42: 3127–3171.

Zhang, Xinlu, Junfeng Li, Jiabao Li, Lu Han, Ting Lu, Xiaojie Zhang, and L. Pan. 2020. "3D TiO_2@ Nitrogen-Doped Carbon/Fe_7S_8 Composite Derived from Polypyrrole-Encapsulated Alkalized MXene as Anode Material for High-Performance Lithium-Ion Batteries". *Chemical Engineering Journal* 385: 123394.

Zhang, Y., H. Feng, X. Wu, L. Wang, A. Zhang, T. Xia, Huichao Dong, Xiaofeng Li, and Linsen Zhang. 2009. "Progress of Electrochemical Capacitor Electrode Materials: A Review". *International Journal of Hydrogen Energy* 34 (!1): 4889–4899.

Zhang, Z., Q. Zhang, C. Ren, F. Luo, Q. Ma, Y.-S. Hu, Zhibin Zhou, Hong Li, Xuejie Huang, and Liquan Chena. 2016. "A Ceramic/Polymer Composite Solid Electrolyte for Sodium Batteries". *Journal of Materials Chemistry* 4: 15823–15828.

Zhao, Chenglong, Lilu Liu, Xingguo Qi, Yaxiang Lu, Feixiang Wu, Junmei Zhao, Yan Yu, Yong-Sheng Hu, and Liquan Chen. 2018. "Solid-State Sodium Batteries". *Advanced Energy Materials* 8 (17): 1703012.

Zhao, Qinglan, Andrew K. Whittaker, and X.S. Zhao. 2018. "Polymer Electrode Materials for Sodium-ion Batteries". *Materials* 11: 2567.

Zhengfei Ulaganathan Mani, Hui Teng Tan, and Q. Yan. 2017. "Advanced Cathode Materials for Sodium-Ion Batteries: What Determines Our Choices?" *Small Methods* 1 (5): 1700098.

Zhong, Cheng, Yida Deng, Wenbin Hu, Jinli Qiao, Lei Zhangd, and J. Zhang. 2015. "A Review of Electrolyte Materials and Compositions for Electrochemical Supercapacitors". *Chemical Society Reviews* 44: 7484–7539.

Zhou, Huanhuan, Qun Ma, Wei Yang, and X. Lu. 2020. "$Li_4Ti_5O_{12}$-TiO_2 Composite Coated on Carbon Foam as Anode Material for Lithium Ion Capacitors: Evaluation of Rate Performance and Self-Discharge". *ChemNanoMat* 6: 280–284.

Zhou, Limin, Kai Zhang, Zhe Hu, Zhanliang Tao, Liqiang Mai, Yong-Mook Kang, and Shu-Lei Chou, and Jun Chen. 2017. "Recent Developments on and Prospects for Electrode Materials with Hierarchical Structures for Lithium-Ion Batteries". *Advanced Energy Materials* 8 (6): 1701415.

Zhu, Ting, Ping Hu, Congcong Cai, Ziang Liu, Guangwu Hu, Quan Kuang, Liqiang Mai, and Liang Zhou. 2020. "Dual Carbon Decorated $Na_3MnTi(PO_4)_3$: A High-Energy-Density Cathode Material for Sodium-Ion Batteries". *Nano Energy* 70: 104548.

Zhu, Y.R., Z.D. Huang, Z.L. Hu, L.J. Xi, X.B. Ji, and Y. Liu. 2018. "3D Interconnected Ultrathin Cobalt Selenide Nanosheets as Cathode Materials for Hybrid Supercapacitors". *Electrochimica Acta* 269: 30–37.

Zou, Kangyu, Peng Cai, Xiaoyu Cao, Guoqiang Zou, Hongshuai Hou, and X. Ji. 2020. "Carbon Materials for High-Performance Lithium-Ion Capacitor". *Current Opinion in Electrochemistry* 21: 31–39.

6

Li_6PS_5X (X = Cl, Br, or I): A Family of Li-Rich Inorganic Solid Electrolytes for All-Solid-State Battery

Anukul K. Thakur[1] and Mandira Majumder[2]
[1]*Department of Advanced Components and Materials Engineering, Sunchon National University, Chonnam, Republic of Korea*
[2]*Department of Applied Physics, Indian Institute of Technology (Indian School of Mines), Dhanbad, India*

6.1 Introduction

Inorganic sulphide materials have recently been studied as solid electrolytes (SEs) for all-solid-state batteries (ASSBs) attributed to high safety, broad temperature window, and improved mechanical properties as compared to those of liquid electrolytes. SEs find several applications in the field of solid-state Li-batteries (Mauger et al. 2019), Li-air (Li et al. 2013; Mauger et al. 2019; Liu et al. 2020) and Li-S (Liu et al. 2018) batteries, sensors (Fergus 2011; Vernoux et al. 2013), and fuel cells (Mauger et al. 2019). Although SEs can be used in widespread applications, this chapter specially emphasizes on SEs used in all-solid-state rechargeable batteries based on lithium. Early history of Li batteries has been summarized in various literatures (Reddy et al. 2020a). Briefly, a Li battery comprises of a positive electrode (cathode), a negative electrode (anode), an electrolyte (Li-ion conductor), and a separator. The cathode material comprises of $LiCoO_2$ (LCO), $LiFePO_4$ (LFP), $Li(Ni_xMn_yCo_z)O_2$ (NMC), or $LiMn_2O_4$ (LMO). Apart from all these compounds, intercalated binary oxides are also used as cathode. On the other hand, graphite, Li metal, Li-In alloys, Si, $Li_4Ti_5O_{12}$ (LTO), or Sn-Co-C mixed composites are employed as anode (Ellis and Nazar 2012). Li batteries can have liquid (Xu 2004), gel polymer (Song et al. 1999; Armand et al. 2009; W. Zhang et al. 2017) consisting of one or more polymers, and solid electrolytes. Also, the methods used for electrode preparation are different for all-solid-state lithium batteries (ASSLBs) and other commercial Li batteries. In general, to increase the conductivity of the electrolyte carbon is added during the designing of sulphide electrolytes. In addition, according to the mechanical characteristics of the sulphide electrolytes, a compatible stack pressure is necessary for successful assembly of ASSLBs. Figure 6.1 represents the general design of an ASSLB. There are certain characteristics that an ideal electrolyte material should meet for application in the ASSLBs, as mentioned: (i) Elevated ionic conductivity of at-least 10^{-3} S cm^{-1} at ambient temperature, (ii) less electronic conductivity of $<10^{-8}$ S cm^{-1} for refraining from self-discharge, (iii) broad electrochemical potential window, (iv) good chemical stability towards the electrodes and over the working potential range, (v) transference number near to 1, (vi) thermal expansion coefficients similar to cathode materials, (vii) substantial chemical stability; electrolyte should refrain from any structural phase transformation for the electrode active till their sintering temperatures, (viii) the sintering temperature for the electrode materials

DOI: 10.1201/9781003145585-6

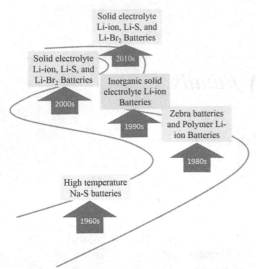

FIGURE 6.1 A road map demonstrating the development of solid-state electrolyte batteries.

should be similar to the sintering temperature of the electrode active materials, and (ix) electrolyte should be non-toxic and less expensive.

A significant amount of research is dedicated to investigate novel solid electrolytes with an aim to replace flammable liquid electrolytes and/or augment the performance of present solid electrolytes and clarify their basic properties and related technological developments. Many reviews have been published already (Huggins 1977; Weppner 1981; Kulkarni et al. 1984; Minami 1985; Pradel and Ribes 1989a; Adachi et al. 1996; Owens 2000; Thangadurai and Weppner 2002; Knauth 2009; Fergus 2010) over the past three decades on solid electrolytes. This has been considered a hot research topic worldwide and has resulted in many publications. To highlight the cutting-edge research on solid electrolyte and their fundamentals, including electrode/electrolyte interface interactions' analysis, their various applications have been reviewed by many researchers. In this chapter, we highlight the fundamental research on Li-based SEs for Li-based solid-state batteries.

Attributed to the safety features solid-state batteries saw a hike in their use in electric vehicle leading to increased research and numerous reviews on them in the past three decades. For instance, Sakuda et al. (2017) and Tatsumisago et al. (2013) reviewed sulphide electrolytes. On the other hand, Thangadurai et al. (2014) and Thangadurai et al. (2015) wrote a review on garnet electrolytes. Moreover, the basic principles of ASSBs were highlighted in several reviews (Goodenough and Singh 2015; Bachman et al. 2016; Mauger et al. 2017; Sun et al. 2017; Yang et al. 2017; Chen et al. 2018; Y. Gao et al. 2018; Wang et al. 2020aWang 2020). Expectedly, the number of reviews on the various attributes of mechanical properties, electrolytes, interface engineering, and cathodes has increased exponentially since 2018 (Reddy et al. 2020b). For instance, fundamentals of electrolytes were reviewed by X.-.Q Zhang et al. (2018) and Famprikis et al. (2019), while Julien and Mauger (2019), Oudenhoven et al. (2011), and Angalakurthi et al. (2018) highlighted the cutting edge technology concerning solid-state microbatteries. In addition, solid electrode/electrolyte interfaces interactions were explored both through in situ and ex situ techniques (Ohta et al. 2006; Takada et al. 2017; Liu et al. 2018; L. Xu et al. 2018Xu 2018; Y. Zhang et al. 2018; Wu et al. 2019; T. Liu et al. 2020; Krachkovskiy et al. 2020). The conduction mechanism of the sulphide electrolytes was studied through computational methods which were reviewed by Xiao et al. (2020).

In this chapter, we discuss the brief history of electrochemical batteries with special emphasis on the solid-state batteries. This is followed by a discussion about the emergence of sulphur-based solid electrolytes. Further, the recent advances in the field of sulphur-based solid electrolytes for ASSB applications have been discussed. A special emphasis has been placed on the importance of the cell designing technology and various process parameters on the mechanical properties, electrochemical charge storage performance, various interfacial mechanisms within the cells, and the challenges faced by the large-scale fabrication of ASSBs. Furthermore, we have summarized the recent reports on remarkable researches based on sulphur-based solid electrolytes.

6.2 History of Solid-State Batteries

The solid-state ionic conductors were discovered by Faraday in 1830s, when he discovered the outstanding property of conduction of heated solid Ag_2S and PbF_2 (Faraday 1833). Nevertheless, the 1960s is more conventionally considered as the turning point for the emergence of solid-state electrolytes and the coining of the term "solid-state ionics" (Figure 6.1 represents a timeline of developments regarding solid-state electrolytes) (Takahashi 1988). Back in 1960s, for the first time the idea of incorporating solid-state electrolytes into batteries was conceived, as a result of the discovery of a rapid 2D sodium-ion-transport phenomenon in β-alumina ($Na_2O \cdot 11Al_2O_3$). This was the material that was later used in designing high-temperature Na-S batteries (Knödler (1984). As a result of three consecutive positive demonstrations of energy storage with ionic conductive β-alumina, Ag_3SI, $RbAg_4I_5$ solid-state electrolytes in the 1960s and early 1970s (Reuter and Hardel 1960; ung-Fang Yu and Kummer 1967), the development in terms of practical applications of solid-state electrolytes accelerated. Following the discovery of ionic transport in poly(ethylene oxide) (PEO)-based solid polymer material in 1973, the research involving solid-state ionics no longer remained limited to inorganic materials (Fenton et al., (1973).

In the 1980s, Na-ion conductive β-alumina was implemented in the ZEBRA cell, which is another type of high-temperature battery "system" in South Africa (Coetzer 1986; Bones et al. 1987). The high-temperature Na-S battery has already been commercialized in Japan (Oshima et al. 2004) and the ZEBRA battery is still at its developmental state in the General Electric Corporation in the United States (Capasso and Veneri 2014). Since 1980, the term *solid-state ionics* has garnered wide coverage; as a result a journal having the same name was launched in the same year. Henceforth, both organic (that is, polymer) and inorganic solid-state electrolytes have garnered augmented attention. Together with the growth of materials and theories, solid-state electrolytes have also developed and have been implemented as essential components into a broad range of electrochemical devices, including sensors, fuel cells, supercapacitors, and batteries (Svensson and Granqvist 1985; Knauth and Tuller 2002; Funke 2013). At the beginning of the 21st century, much emphasis was given on the understanding of the ionic transport mechanism in the solid-state ionics research exploring new superionic conductors, implementing advanced characterization tools, augmenting the characteristics of electrochemical devices involving solid-state electrolytes, and incorporating them in various noble applications associated with ionic transport in solid materials (Gao et al. 2015; Li et al. 2009a). Implementation of solid-state electrolytes in ambient temperature batteries was actually instigated by safety concerns over the use of lithium-ion batteries. Conventionally used flammable organic electrolytes are challenged by various safety concerns including fire hazard and even explosion arising out of overcharging or short circuiting of a Li-based cell (Hanyu and Honma 2012); several cases of Li-ion battery explosions have been reported all over the world. The solid-state electrolytes implemented in Li-ion batteries can be broadly categorized into two classes of materials: inorganic lithium-ion-conductive ceramics and lithium-ion-conductive polymers. In the 1980s, after the discovery of lithium-ion conduction in a PEO-based system, attempts to implement solid-state polymer electrolytes were made in Li-based batteries (Fenton et al. 1973; Gorecki et al. 1986; Gray et al. 1986). Following this, various Li-ion conductive polymer materials, including poly(methyl methacrylate) (PMMA) (Appetecchi et al. 1995), poly(acrylonitrile) (PAN) (Abraham and Alamgir 1990; Wang et al. 1996), and poly(vinylidene fluoride) (PVDF) (Choe et al. 1995), have been much exploited for the designing of all solid-state polymer Li-ion batteries. Since 1990s, research on inorganic solid-state electrolytes has also gained pace for Li-ion battery, following the fabrication of a lithium phosphorus oxynitride (LiPON) material thin film by Oak Ridge National Laboratory (Bates et al. 1992; Dudney et al. 1992). Enthused by the discovery of LiPON, elevated effort has been put towards the designing of inorganic Li-ion conducting ceramics including sodium superionic conductor (NASICON)-type (Goodenough et al. 1976; Ubramanian st al. 1986), perovskite-type (Inaguma et al. 1993), garnet-type (Mazza 1988; Cussen 2006), and sulphide-type materials (Kennedy and Yang 1986; Mazza 1988). Year 2000 has seen the emergence of implementation of solid electrolytes in lithium batteries with gaseous or liquid cathodes, including lithium-sulphur batteries (Wang et al. 2015; Yu et al. 2015), lithium-air batteries (Li et al. 2009b; Lu and Goodenough 2011), and lithium-bromine batteries (Chang et al., 2014; Takemoto and Yamada 2015). Solid-electrolyte Na-ion batteries that operate at ambient temperatures have also been reported (Kim et al. 2015). Very recently, an exclusive "mediator-ion" battery concept has been highlighted where solid electrolytes are utilized for designing low-cost, high-energy, and aqueous electrochemical energy storage systems (Dong et al. 2014; Zhang et al. 2015; Manthiram et al. 2017; Fleischmann et al. 2019).

6.3 Mechanism of Ion Transport in Solid Electrolytes

In crystalline solid materials, ionic transport generally relies on the concentration and distribution of defects. Ion diffusion mechanisms based on Schottky and Frenkel point defects include the simple vacancy mechanism and relatively complicated diffusion mechanisms, such as the divacancy mechanism, interstitial mechanism, interstitial–substitutional exchange mechanism, and the collective mechanism (Mehrer 2007; Park et al. 2010; Wu et al. 2015). However, some materials with special structures can achieve high ionic conductivities without a high concentration of defects. Such structures normally consist of two sublattices, a crystalline framework composed of immobile ions, and a sublattice of mobile species. To achieve fast ionic conduction, three minimum criteria must be fulfilled for this kind of structure (Kumar and Yashonath 2006; Perram 2013): (1) the number of equivalent (or nearly equivalent) sites available for the mobile ions to occupy should be much larger than the number of mobile species; (2) the migration barrier energies between the adjacent available sites should be low enough for an ion to hop easily from one site to another; and (3) these available sites must be connected to form a continuous diffusion pathway. Similar to the diffusion process in a crystal structure, ionic transport in glassy materials starts with ions at local sites being excited to neighbouring sites and then collectively diffusing on a macroscopic scale (Hagenmuller and Van Gool 2015). For most glassy materials, short- and medium-range order still exists in the amorphous structure. The interaction between charge carriers and the structural skeleton cannot be neglected (Angell 1992).

In polymer electrolytes, microscopic ion transport is related to the segmental motion of polymer chains above the glass transition temperature (Berthier et al. 1983). The segmental motion of the chains can create free volumes for the hopping of lithium ions that coordinate with the polar groups. A lithium ion can hop from one coordinating site to another coordinating site, accompanying the segmental motion of polymer chains (Berthier et al. 1983; Nitzan and Ratner 1994; Forse et al. 2017). Under an electrical field, long-distance transport is realized by continuous hopping. The number of free ions depends on the dissociation ability of the lithium salt in the polymer (Manthiram et al. 2017). Table 2 includes the conductivity, advantages and disadvantages of various sulphide-based electrolytes.

6.4 Sulphide-Based Solid Electrolytes

Credited to high Li^+ conductivity at ambient temperature, sulphide-based electrolytes are more promising as compared to the oxide-based ones (Dzwonkowski et al. 1991). Furthermore, sulphide electrolytes are comparatively soft and deformable. In addition, sulphide electrolytes possess better polarizability as compared to that of their oxide counterparts, ultimately resulting in a weaker force of attraction between the sulphide framework and the Li^+ ions compared to that amongst the Li^+ ions and oxide framework. This results in better mobility for the sulphide electrolytes as compared to the oxide-based ones. In 1996, the conductivity of the glass system comprising of Li_2O –Li_2Cl_2–Li_2SO_4–SiO_2–B_2O_3 (35:10:30:12.5:12.5) was reported to be 9.7×10^{-2} S cm^{-1} and 3.3×10^{-6} at 350°C and 25°C by Otto, respectively (Dzwonkowski et al. 3rd, 20th, and 50th cycles. Reproduced). In 1997, ionic conductivities of 1.0×10^{-3} S cm^{-1} at 300°C for B_2O_3–Li_2O–Li_2SO_4 borate-based and the B_2O_3–Li_2O–LiX (X = F, Cl, Br, I) glassy electrolytes were reported by Mercier et al. (1981). These works highlighted the possibility of the sulphide-based electrolytes for various applications and triggered the pursuit of such new systems. In 1981, in a report by Mercier et al. (1981) the room-temperature conductivity was stated to be 10^{-3} S cm^{-1} for Li_2S–P_2S_5–LiI ($Li_4P_2S_7$–LiI) electrolyte. Further, in 1986, Pradel and Ribes (1986) and Pradel and Ribes (1989b) investigated

Li_2S–M (M = SiS_2, GeS_2, P_2S_5, B_2S_3, As_2S_3) and × Li_2S(1-x)SiS_2 (x ≤ 0.6) glasses. In 1986 and 1987, Kennedy and Yang (1986, 1987) demonstrated the melt quenching synthesis strategy and reported conductivity measurement results on Li_2S–SiS_2 LiX (X = Br, I); furthermore, in 1988 and 1989, Kennedy 1988) studied the SiS_2–P_2S_5–Li_2S–Li_2S–LiI system, in which LiX acted as an interstitial dopant leading to the augmentation of the ionic conductivity. Rao and Seshasayee (2006) investigated the molecular dynamics (MD) simulations of the × (0.5Li_2S–0.5P_2S_5)–(1 − x) LiI (x = 0.9, 0.75) and × (0.4Li_2S–0.6P_2S_5) − (1 − x) LiI superionic sulphide glasses ternary materials and related their large ionic conductivity at room temperature to the existence of non-bridging S atoms about the disseminating Li atoms. Moreover, the reduction in the glass transition temperature (T_g) of these systems was credited to the existence of iodine atoms, leading to the plasticization of the structure, making it less rigid, and reduction in P–P bonds resulted by altering the action of the Li atoms, also weakening the glass matrix and resulting in the reduction in T_g. From 1986 to 1989 there were several reports introducing thin-film electrolyte concept (Kennedy and Yang 1987; Balkanski et al. 1989; Creus et al. 1989; Meunier et al. 1989; Jones and Akridge 1992; Jones and Akridge 1993; Rao and Seshasayee 2006). In 1995, Takada et al. (1995) reported a capacity ranging from 80–90 mAh g^{-1} for ASSB based on thin-film electrolytes of $LiMO_2$ (M = Co, Ni)/Li_3PO_4 (LPO)–Li_2S–SiS_2/Li, at a current rate of 64 µA cm^{-2} and the voltage range of 2.0–3.8 V. All these studies triggered the exploration of several glassy and nanocrystalline sulphide electrolytes by the researchers worldwide. Many research groups explored argyrodites, Li–P–S-based glasses, glass-ceramics, thio-ISICONs, Li_6PS_5X (X = Cl, Br, I), and $Li_{(11-x)}M_{(2-x)}$P1+xS_{12} (M = Ge, Sn, and Si) as sulphide electrolytes (Takada 2018; Q. Zhang et al. 2019). Among all the reported electrolyte compositions, Li-based argyrodites Li_6PS_5 X (X=Cl, Br, I) have been the most studied, with due credit to their remarkable conductivity and extraordinary mechanical properties, facilitating the fabrication of high-performance ASSBs. A few reviews, including those published by Takada (2018) highlighted sulphide-based electrolytes. Herein, we highlight the most significant recent report and focus especially on the argyrodite electrolytes, synthesis, and implementation in the fabrication of high-performance ASSBs (Reddy et al. 2020). The following Table 6.1 enlists the performance of various Li-based argyrodites as reported in the previous literatures (Wada et al. 1983; Bron et al. 2016; Camacho-Forero and Balbuena 2019; Chen et al. 2016; Dietrwe broadly review and studyich et al. 2017; Hayashi et al. 2001; Hood et al. 2015; Kaib et al. 2012; Kanno and Murayama 2001; Kato et al. 2016; Kennedy et al. 1990; Kim and Martin 2019; Kuhn et al. 2013; Liu et al. 2013; Rangasamy et al. 2014; Rangasamy et al. 2015; Rao and Adams 2011; Sahu et al. 2014; Yamane et al. 2007; Y. Zhao et al. 2016; Zhou et al. 2016). Table 6.2 enlists the advantages and disadvantages of various solid argyrodite electrolytes.

TABLE 6.1

Room Temperature Ionic Conductivity of ($σ_{RT}$) and Activation Energy (E_a) of Sulphide-Based Solid Electrolytes

Material	$σ_{RT}$ (S cm^{-2})	E_a	Reference
Li_6PS_5Cl	3.3×10^{-5}	0.38 eV	(Rao and Adams 2011)
	1.9×10^{-9}		
Li_6PS_5Br	3.2×10^{-5}	0.32 eV	(Rao and Adams 2011)
	6.8×10^{-3}		
Li_6PS_5I	2.2×10^{-4}	0.26 eV	(Rao and Adams 2011)
	4.6×10^{-7}		
β-Li_3PS_4	2.8×10^{-4}	0.37 eV	(Dietrich et al. 2017)
β-Li_3PS_4	2.0×10^{-4}	0.34 eV	(Hayashi et al. 2001)
β-Li_3PS_4	1.6×10^{-4}	0.36 eV	(Liu et al. 2013)
	7.4×10^{-5}		
β-Li_3PS_4 -LZNO (2wt %)	2.4×10^{-4}	N/A	(Hood et al. 2015)
β-Li_3PS_4 -Al_2O_3(2wt %)	2.3×10^{-4}	N/A	(Hood et al. 2015)

(Continued)

TABLE 6.1 (Continued)

Room Temperature Ionic Conductivity of (σ_{RT}) and Activation Energy (E_a) of Sulphide-Based Solid Electrolytes

Material	σ_{RT} (S cm^{-2})	E_a	Reference
β-Li$_3$PS$_4$-SiO$_2$(2wt %)	1.8×10^{-4}	N/A	(Hood et al. 2015)
β-Li$_3$PS$_4$--LZNO (30 wt%)	1.8×10^{-4}	N/A	(Rangasamy et al. 2014)
Li$_7$P$_3$S11	$0.1\text{--}0.2 \times 10^{-3}$	0.2-0.4eV	(Yamane et al. 2007)
Li$_7$P$_2$S$_8$I	6.3×10^{-3}	0.31 eV	(Rangasamy et al. 2015)
Li$_7$P$_2$S$_8$I	6.07×10^{-3}	0.27 eV	(Choi et al. 2018)
Li$_{10}$GeP$_2$S$_{12}$	12×10^{-3}	0.24 eV	(Kamaya et al. 2011)
Li$_{10}$GeP$_2$S$_{12}$	9.0×10^{-3}	0.22 eV	(Kuhn et al. 2013)
Li$_{10}$GeP$_2$S$_{12}$	10×10^{-3}	0.30 eV	(Bron et al. 2016)
Li$_{10}$SiP$_2$S$_{12}$	2.0×10^{-3}	0.30 eV	(Bron et al. 2016)
Li$_{10}$SiP$_2$S$_{11.3}$O$_{0.7}$	3.1×10^{-3}	0.32 eV	(Kim and Martin 2019)
Li$_{10}$SnP$_2$S$_{12}$	6.0×10^{-3}	0.31 eV	(Bron et al. 2016)
Li$_{10}$Si$_{0.3}$Sn$_{0.7}$P$_2$S$_{12}$	8.0×10^{-3}	0.29 eV	(Bron et al. 2016)
Li$_{10.3}$ Al$_{0.3}$ Sn$_{0.7}$ P$_2$S$_{12}$	5.0×10^{-3}	0.29 eV	(Bron et al. 2016)
Li$_{9.42}$Si$_{1.02}$P$_{2.1}$S$_{9.96}$ O$_{2.04}$	1.1×10^{-4}	0.23 eV	(Kato et al. 2016)
Li$_{9.54}$ Si$_{1.74}$P$_{1.44}$S$_{11.7}$ Cl$_{0.3}$	2.53×10^{-2}	0.23 eV	(Kato et al. 2016)
Li$_{11}$AlP$_2$S$_{12}$	8.02×10^{-4}	0.25 eV	(Zhou et al. 2016)
60Li$_2$S.40P$_2$S$_5$ glass	3.2×10^{-6}	50.7 kJ mol^{-1}	(Dietrich et al. 2017)
67Li$_2$S.33P$_2$S$_5$ glass	3.8×10^{-5}	42.3 kJ mol^{-1}	(Dietrich et al. 2017)
70Li$_2$S.30P$_2$S$_5$ glass	3.7×10^{-5}	43.5 kJ mol^{-1}	(Dietrich et al. 2017)
40Li$_2$S.28SiS$_2$.30LiI glass	1.8×10^{-3}	N/A	(Kennedy et al. 1990)
30Li$_2$S26B$_2$S$_3$33LiI glass	1.7×10^{-3}	N/A	(Wada et al. 1983)
Thio-LiSICON Li$_{3.25}$ Ge$_{0.25}$P$_{0.75}$S$_4$	2.2×10^{-3}	20 kJ mol^{-1}	(Kanno and Murayama 2001)
Thio-LiSICON analogue Li$_4$SnS$_4$	7×10^{-5}	0.41 eV	(Kaib et al. 2012)
Thio-LiSICON analogue Li$_{3.833}$ Sn$_{0.833}$As$_{0.166}$S$_4$	1.4×10^{-3}	0.21 eV	(Sahu et al. 2014)
(PEO18-LiTFSI)-LGPS (LGPS 1 wt%)	1.2×10^{-5}	112 kJ mol^{-1}	(P. Zhao et al. 2016)
(PEO18-LiTFSI)–LGPS (LGPS 1 wt%)-SN (SN 10 wt %)	9.1×10^{-5}	82.77 kJ mol^{-1}	(Chen et al. 2016)

TABLE 6.2

The Conductivity, Advantages, and Disadvantages of Li$_6$PS$_5$ X (X = Cl, Br, I)

Sulfide Materials	Conductivity σ (RT) (S cm^{-1})	Advantages	Disadvantages
Li$_6$PS$_5$Cl	$10^{-7} - 10^{-3}$	High conductivity	Low oxidation stability
Li$_6$PS$_5$Br		Good mechanical strength and mechanical flexibility	Sensitive to moisture
Li$_6$PS$_5$I		Low grain-boundary resistance	Poor compatibility with cathode materials

6.5 Persisting Challenges Encountered and Possible Solution

Despite various advantages, the main challenges faced in the application of solid-state electrolytes in ASSBs are its meagre electrochemical and mechanical stabilities, resulting in a constant augmentation in interfacial resistance during operation (K. Xu et al. 2018; Yue et al. 2018). Moreover, sulphide-based electrolytes are chemically unstable against metallic lithium, leading to the creation of interphases

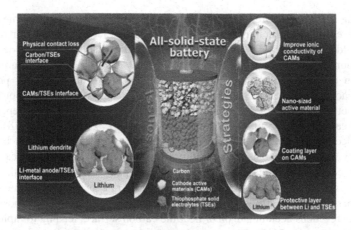

FIGURE 6.2 Summary of the interfacial issues in ASSBs with TSEs and the corresponding strategies. Reproduced from Ref. Wang et al. (2020) with permission. Copyright 2020 Elsevier.

causing lithium dendrite growth and an augmentation in the local current density at the interfaces of Li/TSEs (Cao et al. 2020). In this chapter, we broadly review and study the origin of these issues in ASSBs, as shown in Figure 6.2. Strategic approaches resulting in minimizing the instability of the interfaces that augment the performance of ASSBs with sulphide-based electrolytes have been discussed here. The interfacial glitches mainly include (1) loss of physical contact at the electrode/electrolyte interface, (2) unwanted side reactions at the cathode/electrolyte, carbon/electrolyte, and Li-metal anode/ electrolytes interfaces, and (3) rapid lithium dendrite creation across solid electrolytes enhancing the cathode active materials' ionic conductivity through developing nanosized cathode active materials. Li-metal shield at the Li/electrolyte interface and cathode active material surface coating are effective methods to address these issues (C. Wang et al. 2020b).

6.5.1 Physical Contact between Electrolyte and Electrodes

Chemo-mechanical failure is one of the main reasons for rapid degradation of capacity in ASSBs when the rigid solid-state electrolytes consequently lose physical contact with the electrodes during operation (Figure 6.2) (Koerver et al. 2017). Most of the intercalation-type cathodes undergo volume alteration of ~5% during repeated cycling, leading to secondary particle cracking associated with the active material; the newly created surface of the particles becomes contactless with solid electrolyte, resulting in an increase of the interfacial resistance and deterioration in the reversible capacity (Koerver et al. 2018). Conversion-type cathodes, including sulphur, also undergo detachment at the interface with the solid electrolyte and also the electronically conductive network as a result of repetitive cycling as a result of the large alteration (about 80%) of the active sulphur (Figure 6.2) (Ohno et al. 2019). The implications of cathode volume alteration can be minimized by (1) composition optimizing associated with the cathodes (Strauss et al. 2019); (2) selecting cathode materials with particles arranged in radial orientation (Jung et al. 2020); and (3) optimizing the particle sizes of the cathode active material (Ohno et al. 2019).

6.5.2 Electrochemical Interfacial Reactions

The range of potential with electrochemical stability of electrolytes is significantly determined by the difference in energy between the highest occupied molecular orbital (HOMO) and the lowest unoccupied molecular orbital (LUMO) of the electrolyte, where the reduction potential is determined by the LUMO and the oxidation potential is determined by HOMO (Goodenough (2013). Replacement of the inert blocking electrode with some composite comprising of solid electrolyte and carbon will result in augmentation of the total quantity of reaction and increase the measurement accuracy for the potential measurement of the solid electrolytes. This modified cyclic voltammetric measurement, in association with first-principle calculations, confirms a comparatively thin electrode width for the thiophosphate electrolytes (Dewald et al. 2019). The narrow electrode width of sulphide electrode is related to the small band gaps found for sulphide-based electrolytes (Mo et al. 2012). Li_2S or S formed as a result of reduction/oxidation of the sulphide-based solid electrolytes at low/high voltages possess reversible electrochemical reactivity, and some sulphide-based solid electrolytes possess potential as cathode materials (Hatzell et al. 2020). Li_6PS_5Cl (argyrodite) cathode material has been reported to exhibit a reversible specific capacity of 535 mAh g^{-1} at 0.176 mA cm^{-2} (Z. Zhang et al. 2018). Though oxygen doping in these electrolytes can enhance the working window by improving the band gap, oxygen doping significantly reduces their ionic conductivity (Sun et al. 2016). Conductive carbon provides the requisite electronic conductivity and establishes an electronic pathway within the cathode composite for all-solid-state batteries. However, this carbon increases the rate of decomposition reactions at the cathode/sulphide-based solid electrolyte interfaces (Walther et al. 2019). Decomposition of products from these parasitic reactions accumulate on or near the carbon surface, leading to rapid capacity loss of the batteries (H. Zhang et al. 2017). There are three fundamental strategies to alleviate the negative effects of aforementioned reactions at the sulphide-based solid electrolyte/carbon interface: (1) reduce exposed area of the electrolyte, (2) choose proper carbon additives possessing a relatively low specific surface area, which will provide electronic conduction up to a long distance, including multiwalled carbon nanotubes (Tan et al. 2019), and (3) integrate semiconductor to the composite cathodes, like poly (3,4-ethylenedioxythiophene) (PEDOT). A PEDOT integrated carbon-containing composite cathode will not only reduce the side reactions and minimize the interfacial resistance but also deliver effective electron transport at the interfaces (Deng et al. 2020).

6.5.3 Cathode Active Material/TSE Interface

6.5.3.1 Intercalation Cathode/TSE Interface

Elemental interdiffusion, space charge layers, and chemical reactions at the interface of cathode and sulphide based solid electrolyte can lead to fast capacity degradation, large interfacial resistance, and reduced rate performance of the all-solid-state batteries (Figure 6.2) (Sakuda et al. 2010; Otoyama et al. 2016; Zhang et al. 2020a). Approaches including doping the sulphide-based solid electrolyte coating cathode active materials with an electronic-insulating material/thin Li-ion-conduction can effectively minimize the degradation of the cathode/TSE interface (Culver et al. 2019). Owing to the much sturdier ionic bonding in between $Li^+-O_2^-$ as compared to that of $Li^+-S_2^-$, oxygen doping for sulphur atoms in sulphur-based solid-state electrolyte can enhance their elastic modulus, increase their chemical stability in moisture, and stabilize chemical/electrochemical properties against metallic lithium anode/ high-voltage cathode (Figure 6.2) (Culver et al. 2019).

6.5.3.2 Conversion Cathode/TSEs Interface

Elemental sulphur has garnered much attention as a low-cost cathode material possessing high theoretical specific capacity of 1672 mAh g^{-1} (Pang et al. 2016; Eng et al. 2017). The shuttle effect associated with the polysulphide species in lithium-sulphur batteries in presence of a liquid electrolyte can be eradicated with the sulphur-based solid electrolyte in an all-solid-state lithium-sulphur battery (Yamada et al. 2015). Nevertheless, Li-ion transfer at the sulphur/solid-state electrolyte interface is numerous orders of magnitude lesser than the conductivity of sulphur-based solid electrolytes, causing deprived rate-performance and cycle life in all-solid-state lithium-sulphur batteries (Yu et al. 2016). There is also further concern of sulphur reacting with the sulphur-based solid electrolyte as explained by theoretical calculations, showing that PS4(t) can be transformed to $PS3^{3-}$ and S^{2-}. $PS4^{3-}$ tetrahedron also possesses the possibility to react with S or other thiophosphates to lead to the formation of (PS4+n) 3- or P-(S)n-P complexes, respectively (Camacho-Forero and Balbuena 2019). Further, the lithiated sulphur- based solid electrolytes formed as a result of cycling are unable to provide enough Li-ion conductivity to the sulphur cathode (Nagai et al. 2019). Nanostructured active materials, architecting a robust composite cathode associated with good ionic and electronic channels, improving the active material ionic conductivity are feasible approaches leading to a reduction in the interfacial resistance between conversion-type cathodes and sulphur-based solid electrolytes.

6.5.4 Li-Metal Anode/TSEs Interface

In all solid-state batteries with sulphide-based solid electrolyte and a metallic lithium anode, Li/TSE interface stability and electronic conductivity are determined by the reaction products formed at the interface. This also decides the thickness and stability of the solid electrolyte interphase (SEI) (Wenzel et al. 2015). An SEI of selectively conducting Li-ions (restricting the flow of electrons) can block further reactions at the interface (Sun et al. 2020). An unstable SEI which will conduct electrons as well as Li-ions will lead to a continuous reaction between the Li metal and the TSEs, causing consumption of both materials and deteriorating the cycle life of the all-solid-state battery so fabricated (Wenzel et al. 2016). An artificial SEI layer compatible with Li metal and which will conduct Li-ions could be implemented into the Li/sulphide-based solid electrolyte interface to mitigate the interfacial issues.

6.5.5 Lithium Dendrites and Li-Metal Protection

Sulphide-based solid electrolytes exhibiting large Young's modulus are supposed to mitigate the growth of lithium dendrites (Monroe and Newman 2004). However, an irregular current density distribution at the surface of Li metal can cause non-uniform stripping and plating operation, resulting in short circuiting and the formation of lithium dendrites in all solid-state batteries (Cao et al. 2020; Hatzell et al. 2020).

Lithium dendrites grow in the defect-rich areas, including grain boundaries and pores in sulphide-based solid-state electrolytes (Kerman et al. 2017). Primarily, Li deposits in columns, as revealed by in situ SEM, and grows along the grain boundaries of the sulphide-based solid-state electrolytes with an appearance of local cracks (Nagao et al. 2013; Porz et al. 2017; Wang et al. 2020). Some effective approaches have been projected to mitigate the growth of lithium dendrite, like solid electrolyte doping, surface modification of the lithium metal or the solid electrolytes, and mechanical constriction (Z. Gao et al. 2018; Wang et al. 2019).

6.6 Fundamentals of Argyrodite Electrolyte

In 2008, Deiseroth et al. (2008) discovered a noble Li-argyrodite Li$_6$PS$_5$X (X = Cl, Br, I) which was a fast-ion conductor with preliminary room-temperature conductivity varying in the range 10^{-2}–10^{-3} S cm^{-1}. This work paved the path for understanding further the physical and structural properties of solid-state electrolytes leading to the development of ASSBs. Argyrodite exhibits large conductivity; furthermore, argyrodite-based batteries are facile to design as compared to those implementing oxide-based solid electrolytes. In the following section, we briefly explain certain important attributes of argyrodites (Hanghofer et al. 2019).

i. Li$_6$PS$_5$X (X = Cl, Br, I) compounds seem to be isostructural with Cu and Ag-argyrodites and cubic unit cells possessing materials with (F-43m space group) (Figure 6.3a–c) (Hanghofer et al. 2019). Li$^+$ ions are arbitrarily distributed all over the remaining tetrahedral interstices in the cubic structure, in which P atoms reside in the tetrahedral interstices, while 16e sites are completely occupied by sulphur forming a network of isolated PS$_4$ tetrahedra. X anions reside in the face centred cubic (fcc) lattice. In the case of Li$_6$PS$_5$Cl lattice Li resides in the 24g sites, whereas for Li$_6$PS$_5$Br framework they are present over the 24g and 48h sites (Kraft et al. 2017). Li$^+$ diffusion takes place through these partially occupied sites, which results in hexagonal cages joined to each other through the interstitial sites about the X$^-$ and S^{2-} ions for Li$_6$PS$_5$Cl and Li$_6$PS$_5$I argyrodites, respectively. The lattice parameters of the argyrodites, viz. Li$_6$PS$_5$Cl, Li$_6$PS$_5$Br, and Li$_6$PS$_5$I powders were reported to be a = 9.85 Å, 9.98 Å, and 10.142 Å,

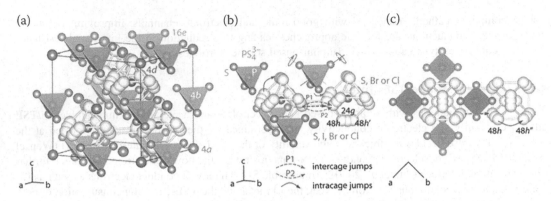

FIGURE 6.3 (a–c) Schematic representation of type Li_6PS_5X (X = Cl, Br, I) argyrodite crystal phase that crystallizes in cubic symmetry with representing the space group F43m. Reproduced with permission from Ref. Hanghofer et al. (2019) Copyright 2019, American Chemical Society.

respectively, by Rao and Adams (2011). The differences in the lattice parameter is mainly caused by the differences in the anions ionic radii (r) in Li_6PS_5X, i.e. radius (S^{2-}) = 1.84 Å, radius (Cl^{-1}) = 1.81 Å, radius (Br^-) = 1.95 Å, and radius (I^-) = 2.16 Å.

ii. In 2011, Rao and Adams (2011) and Rayavarapu et al. (2012) reported the synthesis of Li_6PS_5X (X = Cl, Br, I) and performed conductivity, neutron diffraction, and bond valence computational investigation on them. They reported the existence of a three-dimensional (3D) channel network that opted for long-path ion conduction of all Li_6PS_5X (X = Cl, Br, I) phases, consisting of interconnected low-energy local pathway cages (Kraft et al. (2017). The experimentally obtained ionic conductivity at the ambient temperature of Li_6PS_5Cl, Li_6PS_5Br, and Li_6PS_5I synthesized via ball milling and heat-treatment at 550°C were in the range 1.9 × 10^{-4}–7.0 × 10^{-3} S cm^{-1} and activation energies calculated were in the range 0.26–0.41 eV (Rao and Adams 2011; Rayavarapu et al. 2012; Kraft et al. 2 coworkers have reported solid017). Furthermore, Boulineau et al. (2012) showed the effect of increase in the conductivity of Li_6PS_5Cl from 2x10^{-4} S cm^{-1} to 1.33 × 10^{-3} S cm^{-1} with variation in the time of ball milling as 1 h to 10 h. Rao and Adams (2011) also drew a comparison between the values of E_a (activation energy) obtained by both computational method and experimental method for Li_6PS_5X (X = Cl, Br, I) in the range between 0.25 eV and 0.38 eV. Camacho-Forero and Balbuena (2019) carried out ab initio calculations and obtained the conductivity, E_a, and the diffusion coefficient of Li^+ at the room temperature, which were 6.07 × 10^{-3} S cm^{-1}, 0.27 eV, and 5.8 × 10^{-9} cm^2 s^{-1} for Li_6PS_5I, 0.17 × 10^{-3} S cm^{-1}, 0.37 eV, and 1.2 × 10^{-9} cm^2 s^{-1} for Li_6PS_5Cl, respectively. The reported diffusion coefficient of Li_6PS_5Cl was two orders of magnitude lesser than that obtained using ^7Li nuclear magnetic resonance (NMR) (7.7 × 1^{-8} cm^2 s^{-1} at 40 °C). According to the report by Camacho-Forero and Balbuena (2019) the ionic conductivity of Li_6PS_5I was significantly lesser as compared to Li_6PS_5Cl and Li_6PS_5Br.

iii. Solid electrolytes comprising of argyrodite can be synthesized using different methods (Akridge and Vourlis 1986; Deiseroth et al. 2008): solution-based methods (Miura et al. 2019; Arnold et al. 2020), solid-state reaction in a sealed tube, and ball milling (Rao and Adams 2011; Boulineau et al. 2013).

iv. The synthesis method, the grain boundary contributions, and the conductivity method are crucial factors in determining the conductivities of argyrodite electrolytes, and conductivity measurement method and fabrication technique of pelletized samples, considering sintering cold-pressed pellets influence the density of the specimens (Boulineau et al. 2012). Based on the literature survey, conductivity values also seem to be dependent on cooling rate (Boulineau et al. 2013), porosity, and pore distribution (Montes et al. 2008). Lesser Li^+ ionic conductivities, in the range 10^{-5}–10^{-4} mS cm^{-1}, were reported when the electrolytes were produced using the solution-based method, which was credited to the presence of extra impurity phases within the compounds (Miura et al. 2019).

v. Reports are available where NMR has been used to characterize the structure and dynamics of argyrodites. The chemical shifts from ^{31}P and ^{6}Li MAS NMR spectra (Hanghofer et al. 2019) are 93.9 ppm and 1.49 ppm for X = I, 85 and 1.6 ppm for X = Cl, and 96.3 and 1.3 ppm for X = Br nanostructured materials obtained by implementing solid-state and ball milling methods. The E_a conductivity and Li-jump rate values acquired from the NMR measurements were 0.2 eV, 10^{-3}–10^{-2} S cm^{-1}, and 109 s^{-1}, respectively, for Li₆PS₅Br and Li₆PS₅I argyrodites (Epp et al. 2013).

vi. The electrochemical stability potential span was reported in the range 0–7 V vs. Li^{+}/Li for Li₆PS₅X (X = Cl, Br, I) argyrodite (Takada et al. 2017; Takada 2018; Zhang et al. 2019).

vii. Kong et al. (2010) demonstrated that the O substitution at the S sites in Li₆PS₅X (X = Cl, Br) argyrodite results in the decrease in room-temperature conductivity by orders of magnitudes, to $\sim 10^{-9}$ S cm^{-1}. Furthermore, the E_a associated with the O-containing compound was obtained to be 0.66 eV. Rao and Adams (2011) further confirmed the low conduction implementing bond valence studies.

viii. Critical current density limits associated with Li plating on Li₆PS₅Cl argyrodite was investigated by Asemchainan et al. (2019) and Doux et al. (2020). They also studied the stack pressure limits of Li₆PS₅Cl.

ix. Yokokawa (2016) proposed a potential diagram approach by investigating the thermodynamic stability of sulphide electrolyte/oxide interface of all-solid-state batteries. In this approach, the phase relationships associated with the interfaces could be studied by a comparison of proper chemical potentials associated with the target devices. Fundamental understanding of the aforementioned parameters is significant for both theoretical and industrial applications.

x. In 2019, Prasada Rao et al. (2019) reported noble Li₁₅(PS₄)4Cl₃ and its Mg^{2+}-doped state, viz. Li₁₄.₈Mg₀.₁(PS₄)₄Cl₃ I-43d space group and lattice parameters a = 14.308 Å and 14.323 Å, respectively. This argyrodite was isostructural to Ag₁₅(PS₄)₄Cl₃ phases; furthermore, they reported that Mg^{2+} doping resulted in the increase of the ionic conductivity from 4 × 10^{-8} S cm^{-1} for to 10^{-7} S cm^{-1} for the Mg^{2+} doped state (Reddy et al. 2020).

6.7 Argyrodites for ASSBs

In this section, we will be discussing various Li-based argyrodites that have been reported recently for application in all solid-state batteries.

6.7.1 Argyrodite with X = Cl (Li₆PS₅Cl)

As reported by Ce-Wen Nan and coworkers Li-ion conducting Li₆PS₅Cl solid-state electrolytes (SSEs) were prepared by solid-state sintering method (Figure 6.4a). The dependence of ionic conductivity, phase formation, and E_a on the sintering temperature and duration for the argyrodite were also studied. Li₆PS₅Cl was investigated systematically. The Li₆PS₅Cl electrolyte exhibiting large ionic conductivity of 3.15 × 10^{-3} S cm^{-1} at ambient temperature was observed when sintered at 550°C for 10 min. All-solid-state lithium-sulphur batteries based on Li-In alloy as the anode and Li₆PS₅Cl solid-state argyrodite electrolyte integrated with nano-sulphur/multiwall carbon nanotube composite as the cathode showed good cycle life as shown using CV plots (Figure 6.4b). The assembled cell showed a large discharge capacity of 1850 mAh g^{-1} at ambient conditions associated with the first full cycle at 0.176 mA cm^{-2} corresponding to ~0.1 C. The discharge capacity exhibited by the assembled cell was 1393 mAh g^{-1} over 50 cycles (Figure 6.4c). In addition, the Coulombic efficiency exhibited was ~100% as observed from galvanostatic cycling (Figure 6.4d). The results indicated the aptness of Li₆PS₅Cl as solid-state electrolyte to be implemented in solid-state batteries (Wang et al. 2018).

Xiayin Yao and coworkers have reported solid electrolyte comprising of a flat-surface Li₆PS₅Cl nanorod pellet with large density, exhibiting an ionic conductivity of 6.11 mS cm^{-1} at ambient

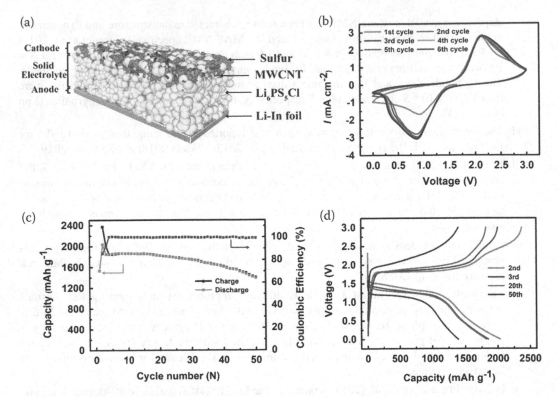

FIGURE 6.4   (a) Schematic diagram representing the layers of nano-sulphur/MWCNT composite cathode‖Li₆PS₅Cl‖Li-In all solid-state battery (ASSLSB); (b) Cycling life of nano-sulphur/MWCNT composite cathode‖Li₆PS₅Cl‖Li-In ASS cells. (c) Charge/discharge capacity variation, and Coulombic efficiency variation with cycle number at 0.176 mA cm⁻². (d) The charge/discharge plots of the assembled cells for 2nd, 3rd, 20th, and 50th cycles. Reproduced from Ref (Wang et al. 2018) with permission. Copyright 2018 American Chemical Society.

conditions (Figure 6.5a–d). For homogeneous lithium deposition flat surface of the pellet is necessary, and the suppression of lithium dendrites' growth is attributed to dense pellet microstructure, resulting in a significantly improved current density of 1.05 mA cm⁻² at 25 °C. The obtained dense Li_6PS_5Cl pellet is further used to fabricate a $LiCoO_2/Li_6PS_5Cl/Li$ all-solid-state lithium battery, exhibiting a discharge capacity of 115.3 mAh g⁻¹ at 1 C (0.35 mA cm⁻², 25 °C) associated with a capacity retention of 80.3% over 100 cycles (Figure 6.5e–g) (G. Liu et al. 2020).

Marnix Wagemaker and coworkers reported large lithium conductivity for Li_6PS_5Cl argyrodite projecting it as an attractive candidate to be implemented as solid-state electrolyte in all-solid-state Li-batteries. In this work, the mechanical milling times have been in order to reach an optimum preparation strategy for the structure and conductivity. Li_6PS_5Cl argyrodite with the ionic conductivity of 1.1×10^{-3} S cm⁻¹ was reached when the milling time was 8 hours at 550 rpm followed by heating the powder at 550°C. The assembled all-solid-state Li-S battery had a positive electrode comprising of a combination of Li_6PS_5Cl solid electrolyte with a carbon-sulphur mixture and Li-Al, Li-In, and Li as the negative electrode. Through CV experiments an optimum charge/discharge potential window ranging between 0.4 V and 3.0 V vs. Li-In was obtained. Galvanostatic cycling projected a capacity around 1400 mAh g⁻¹ at the initial cycles, reducing to a value lesser than 400 mAh g⁻¹ over 20 cycles. This capacity deterioration was attributed to increasing electrode–electrolyte interface which was confirmed by Impedance spectroscopy (Yu et al. 2016).

Wenkui Zhang and coworkers reported results obtained by in situ Raman spectroscopy and in situ electrochemical impedance spectroscopy measurements to reveal detailed interface evolutions in $LiNi_{0.8}Co_{0.1}Mn_{0.1}O_2$ (NCM)/Li_6PS_5Cl/Li cell (Figure 6.6a–b). This in combining with the ex situ characterizations such as X-ray photoelectron spectroscopy and scanning electron microscopy,

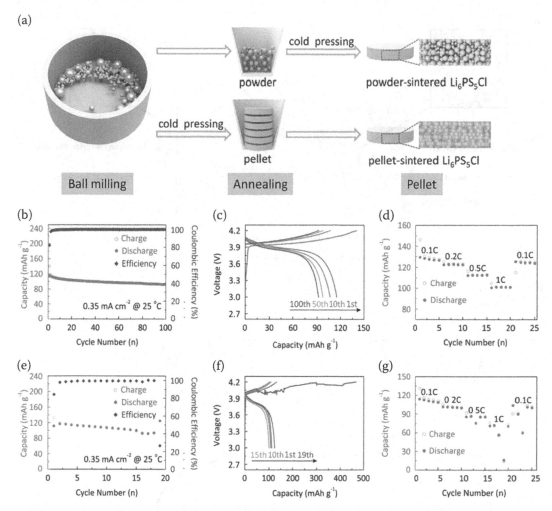

FIGURE 6.5 (a) The schematic representation for the synthesis method of Li₆PS₅Cl; (b) the cyclic life demonstration; (c) charge/discharge profiles, and (d) rate capability a LiCoO₂/pellet-sintered Li₆PS₅Cl/Li battery at the current density of 0.35 mA cm⁻², 25°C; (e) the cyclic stability performance; (f) charge/discharge profiles; and (g) rate performance of a LiCoO₂/powder-sintered Li₆PS₅Cl/Li battery at 0.35 mA cm⁻², 25°C. Reproduced from Ref. Liu et al. (2020) with permission. Copyright 2020 American Chemical Society.

the chemical bond vibration at NCM/Li₆PS₅Cl interface for the initial cycles has been elaborated. The results implied that Li⁺ ion migration, which is dependent on the value of the potential change, is a very crucial reason for these interface behaviours. In case of the long-term cycling, lithium dendrites, interfacial reactions, and chemo-mechanical failure result in an combined effect on interfaces, resulting in deterioration of the electrochemical performance and the interfacial structure. This work projects novel insight on interfacial evolution during intra- and inter-cycle of solid-state batteries (Zhang et al. 2020b).

Wen Nan and coworkers reported pioneering investigation on Li₆PS₅Cl solid-state electrolyte meant for use as a cathode active material in ASSLIB (Figure 6.7a–d). In the fabricated ASSLIBs, Li₆PS₅Cl was used both as the electrolyte and the active material in association with a Li-In alloy anode. The fabricated Li₆PS₅Cl-based cell having an active material loading of 3.1 mg cm⁻² showed a discharge capacity of ~535 mAh g⁻¹ over 650 cycles at 56 mA g⁻¹ (~0.1 C) at ambient conditions. From the literature survey, the discharge capacity value obtained was the largest value amongst all the ASSLIBs reported so far implementing sulphide SSEs as active materials. This value was even comparable with cells with Li₂S, pristine sulphur, or transition metal sulphides as the active material. This greater

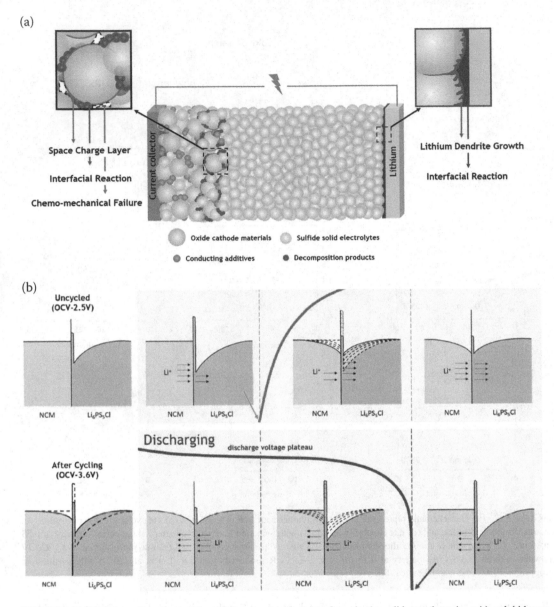

FIGURE 6.6 (a) Schematic representation of five drawbacks at interfaces in the solid-state batteries with sulphides as electrolytes and oxides as active materials; (b) schematic illustrations of the Li^+ concentration and migration at the interface of NCM/Li_6PS_5Cl: in uncycled state, while the first charge/discharge cycle, and after cycling. Reproduced from Ref. Zhang et al. (2020) with permission. Copyright 2020 Wiley-VCH.

long-term cycling performance is credited to the largely reversible electrochemical activity of Li_6PS_5Cl indicated by the ex situ XPS and XRD analyses (Wang et al. 2019).

Xueliang Sun and coworkers suggested introducing Cl to reduce the S/Cl defects in the argyrodite structure to improve the lithium ion conductivity. They synthesized $Li_{5.7}PS_{4.7}C_{11.3}$ exhibiting a lithium-ion conductivity of 6.4 mS cm^{-1}. The synthesis conditions for $Li_{7-x}PS_{6-x}Cl_x$ (x = ¼ 1.0, 1.1, 1.2, 1.3, 1.4, 1.5, 1.6, 1.7, 1.8, 1.9) are methodically investigated in order to receive pure lithium argyrodite phase associated with large ionic conductivity. ^7Li spin-lattice relaxation NMR and AC impedance spectroscopy results indicated significant augmentation in lithium-ion conductivity credited to the presence of Cl atom doping in the lattice of the argyrodite. Introduction of Cl also led to the reduction in the value energy barriers for Li-ion transfer in both long and short diffusion lengths as indicated by the ab initio

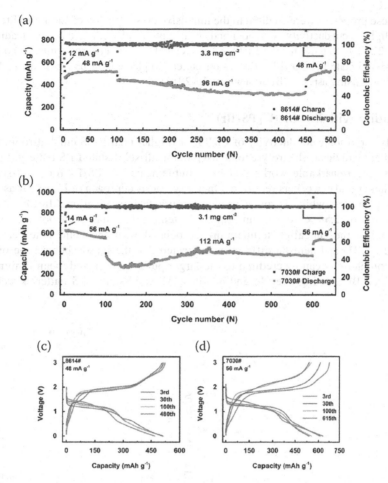

FIGURE 6.7 Cycling life of the Li₆PS₅Cl-MWCNT/Li₆PS₅Cl/Li–In solid-state battery with (a) 8614# and (b) 7030# cathodes at ambient temperature. Charge/discharge profiles of the cells using (c) 8614# and (d) 7030# cathodes at various cycles. Reproduced from Ref. Wang et al. (2019) with permission. Copyright 2019 Royal Society of Chemistry.

molecular dynamics simulations. All-solid-state lithium batteries implementing Li₅.₇PS₄.₇C₁₁.₃ solid electrolyte and LiNbO₃-coated LiNi₀.₈Mn₀.₁Co₀.₁O₂ cathode exhibited large discharge capacities and remarkable cycling performances at comparatively large current densities. Galvanostatic intermittent titration technique (GITT) together with EIS further affirmed that the augmented electrochemical performance credited to the minimization of voltage polarization and tailoring of the interfacial resistance at the cathode and solid electrolyte interface (Yu et al. 2020).

Maohua Chen and Stefan Adams reported fabrication of all-solid state lithium secondary batteries merging sulphur acting as the active cathode material together with argyrodite-type Li₆PS₅Br acting as the solid electrolyte. A uniformly distributed composite cathode powder of sulphur, Li₆PS₅Br, and super P carbon with particle size <100 was obtained through following two-step ball milling at rotating speed of 500 rpm. The obtained all-solid-state S/Li₆PS₅Br/In-Li batteries comprising of S contents altered over the range of 20–40 wt.% exhibited a maximum capacity of 1460 mAh g⁻¹ and a discharge capacity reaching a value of 1080 mAh g⁻¹ over 50 cycles at C/10 rate. No structural change of Li₆PS₅Br throughout the operating cycles was confirmed by ex situ X-ray diffraction (XRD) results (Chen associated with an activation energyand Adams 2015).

Virginie Viallet and coworkers reported ion-conductive Li₂PS₅X (X=Cl, Br, I) Li- argyrodites synthesized using high-energy ball milling. Ball milled compounds exhibited large conductivity ranging in between 2×10^{-4} S cm⁻¹ and 7×10^{-4} S cm⁻¹ associated with an activation energy of 0.3–0.4 eV for conduction. These properties were credited to the impulsive crystallization of Li-argyrodite as a result of ball milling. Highest conductivity was obtained at an optimized milling duration leading to a conductivity of 1.22×10^{-3} S cm^{-1} for the Li_6PS_5Cl associated with an electrochemical stability over 7 V vs. Li. The all-solid-state $LiCoO_2$/Elec./In Li-ion battery implementing ball-milled Li_6PS_5Cl as electrolyte showed promising aspects (Boulineau et al. (2012).

6.7.2 Argyrodite with X = Br (Li_6PS_5Br)

The remarkably rapid ionic conductivity of Li_6PS_5Br (higher than 1 mS cm^{-1} at room temperature) renders it as an apt candidate electrolyte to be used in the all-solid-state Li-S battery. In this section, we will discuss the remarkable works that have implemented Li_6PS_5Br for solid-state batteries. Marnix Wagemaker and coworkers reported a facile synthesis approach of Li_6PS_5Br associated with an easy scale-up process which is much needed the practical applications. In this work, an ionic conductivity value of 2.58×10^{-3} S cm^{-1} at room temperature has been obtained for Li_6PS_5Br through optimizing the annealing temperature associated with the solid-state reaction method (Figure 6.8a–d). XRD and neutron diffraction indicated that the reason behind the origin of augmented ionic conductivity can be credited to the larger purity, minimized mean lithium-ion jumps, and the optimized Br ordering over 4c and 4a sites. All-solid-state Li-S batteries were fabricated

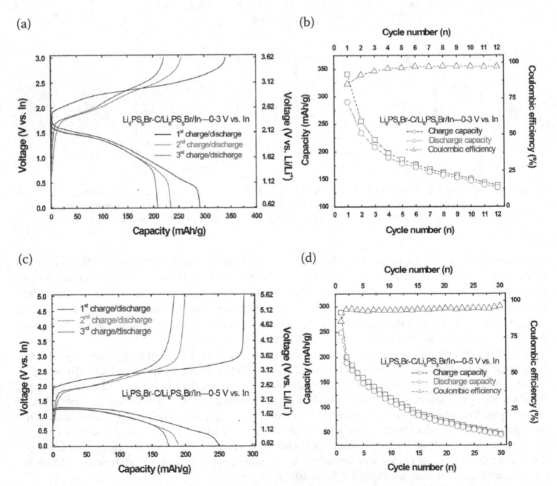

FIGURE 6.8 Charge/discharge profiles and cycling stability of Li_6PS_5Br–C/Li_6PS_5Br/In ASSBs at various voltage windows: (a and b) 0 and 3.0 V vs. In (c and d) 0 and 5.0 V vs. In at the current density of 0.32 mA cm^{-2}. Reproduced from Ref. Yu et al. (2019) with permission. Copyright 2019 Royal Society of Chemistry.

implementing an S-C composite cathode in association with the optimized Li_6PS_5Br solid-state electrolyte and Li-In anode which exhibited large discharge capacities. Various cycling modes (discharge-charge and charge-discharge) indicated that the capacity of the $S-C-Li_6PS_5Br/Li_6PS_5Br/$ Li-In battery had its origin from both the Li_6PS_5Br in the cathode mixture and active S-C composite. The contribution of the cathode mixture was verified from all-solid-state batteries implementing Li_6PS_5Br and its equivalent battery types as active materials. Ex situ XRD and electrochemical performance measurement results indicated that the influence of capacity from Li_6PS_5Br for the cathode mixture might be due to the decomposition product Li_2S, while the Li_6PS_5Br solid electrolyte layer in the bulk was steady during cycling (Yu et al. 2019).

In another report by Marnix Wagemaker and coworkers, the authors investigated the implication of the solid-state argyrodite electrolyte conductivity on the solid-state battery performance associated with the particle size, chemical structure, and the crystallinity depending on the synthesis conditions (Figure 6.9a–c). Neutron and X-ray diffraction had been used to study the detailed structure and impedance offered by the material. ^7Li solid-state NMR spectroscopic technique was used to investigate the Li-ion kinetics within the material. It is reported that the homogeneity of the distribution of Br dopant over the crystallographic sites in Li_6PS_5Br depends on the synthesis conditions. This can influence significantly the mobility of Li-ion fraction in the interfaces associated with the annealed argyrodite materials. It was proposed that solid-state battery property can be optimized by comparing the interfacial and bulk properties of the differently synthesized Li_6PS_5Br materials, requiring an unlike particle size in case of region where only solid electrolyte resides and

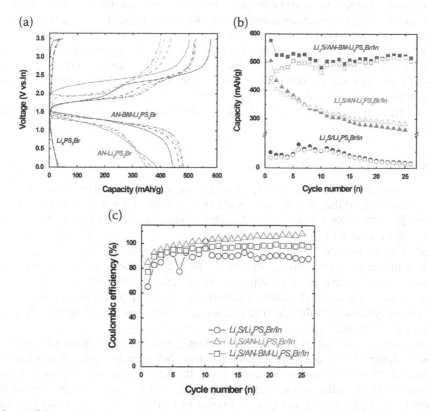

FIGURE 6.9 (a) Galvanostatic voltage curves corresponding to the first four cycles of the assembled nano-Li_2S/Li_6PS_5Br (Li_6PS_5Br or AN-Li_6PS_5Br or AN-BM-Li_6PS_5Br)/In solid-state batteries at 0.064 mA cm^{-2} applied in the range of 0 V to 3.5 V vs. In (0.62–4.12 V vs. Li$^+$/Li). (b) Cyclic stability of the solid-state batteries at 0.064 mA cm^{-2} applied in the range of 0 V and 3.5 V vs. In (0.62–4.12 V vs. Li$^+$/Li). (c) The associated Coulombic efficiency for the assembled nano-$Li_2S/$ Li_6PS_5Br (Li6PS5Br or AN-Li_6PS_5Br or AN-BM-Li_6PS_5Br)/In solid-state batteries. Reproduced from Ref. Yu et al. (2017) with permission. Copyright 2017 Royal Society of Chemistry.

solid electrolyte present in the cathode mixture. In the electrolyte region, the resistance offered by the grain boundaries is reduced by annealing the argyrodite leading to relatively large crystallites. Unlike the argyrodite residing in the electrolyte region, Li_6PS_5Br existing in the cathode mixture, however, needed extra reduction of the particle size when abundant Li_6PS_5Br–Li_2S interfaces responsible for reducing the resistance of this rate-limiting step in Li-ion transport are present. This work reveals the necessity of modifying the electrolyte structure in order to optimize the electrolyte performance (Yu et al. 2017).

Long Zhang and coworkers investigated O-doped Li_6PS_5Br as solid electrolyte obtained by solid-state sintering. Unlike other O-doped sulphides, the O atoms in $Li_6PS_{5-x}O_xBr$ selectively got substituted at the S atoms existing at free S^{2-} sites instead of those at the PS_4 tetrahedra. Remarkably, without truncating the ionic conductivity, this O-doped solid electrolyte exhibited significantly augmented properties such as superior chemical and electrochemical stability against Li-metal, outstanding dendrite limiting capability, as well as large voltage oxide cathodes and good air stability. $LiCoO_2$ and $Li(Ni_{0.8}Co_{0.1}Mn_{0.1})O_2$ -based ASSB implemented with $Li_6PS_{4.7}O_{0.3}Br$ electrolyte exhibited superior rate capability, large specific capacity, and remarkable cycling life together with less interfacial resistivity (Z. Zhang et al. 2019).

Wolfgang G. Zeier and coworkers have reported a detailed investigation on the effect of site-disorder linked with the anions S^{2-} and X^- on affecting the ionic transport and implications of the nature of halide present (Figure 6.10a–g). They have revealed the nature of dependency on such disorder in Li_6PS_5Br, which can further be engineered through altering synthesis approach. A comparison between quenching and slow cooling of the samples revealed larger than the anion site-disorder at elevated temperatures, and quenching could be implemented to kinetically trap the requisite disorder, resulting in larger ionic conductivities as demonstrated by impedance spectroscopy in association with ab initio molecular dynamics. Furthermore, it was seen that following milling, a crystalline lithium argyrodite phase formed within one minute of heating the sample. This quick crystallization projects the reactive nature of mechanically milled samples and indicates that large reaction times with large energy consumption were not necessary for these argyrodite samples. The observation indicated that site-disorder tempted by the means of quenching is beneficial for ionic transport and offers an extra strategy for the design and optimization of lithium superionic conductors (Gautam et al. 2019).

6.7.3 Argyrodite with X = I (Li_6PS_5I)

Wenkui Zhang and coworkers have reported a silicon-doped Li-based argyrodite solid electrolyte $Li_{6+x}P_{1-x}Si_xS_5I$, which exhibited an elevated ionic conductivity of 1.1×10^{-3} S cm^{-1} and truncated activation energy of 0.19 eV (Figure 6.11a–h). This modification could be possible as a result of altering the structural unit in the network of argyrodite. The $Li_{6+x}P_{1-x}S_5I$ solid electrolytes are implemented in all-solid-state Li-ion batteries with Li metal as anode and $Li(Ni_{0.8}Mn_{0.1}Co_{0.1})O_2$ (NCM-811) as the cathode material for evaluating their electrochemical performance. With × = 0.55, the battery exhibited discharge capacity for the first cycle of 105 mA h g^{-1} at a rate of 0.05 C and attained large Coulombic efficiency. Furthermore, chemical reactions on interfaces of the NCM/solid electrolyte and Li/solid electrolyte causing degradation of cell performance were also studied (Zhang et al. 2020).

Akitoshi Hayashi and coworkers reported an efficient approach for the preparation of argyrodite-like crystals implementing a liquid-phase technique using homogeneous ethanol solution to elevate cell performance through using an SE-coating on the active material (Figure 6.12a–d). The synthesis conditions including suitable alcohol solvents and halogen species, dissolution time, and temperature of drying were investigated to result in Li_6PS_5Br resulting in overestimation and with a Li-ion conductivity of 1.9×10^{-4} S cm^{-1}. It is noteworthy that in all the solid-state cells the obtained solution forms a promising solid–solid electrode–electrolyte interface associated with a large contact area, leading to an improved capacity as compared to the conventional techniques like hand mixing using a mortar (Yubuchi et al. 2018).

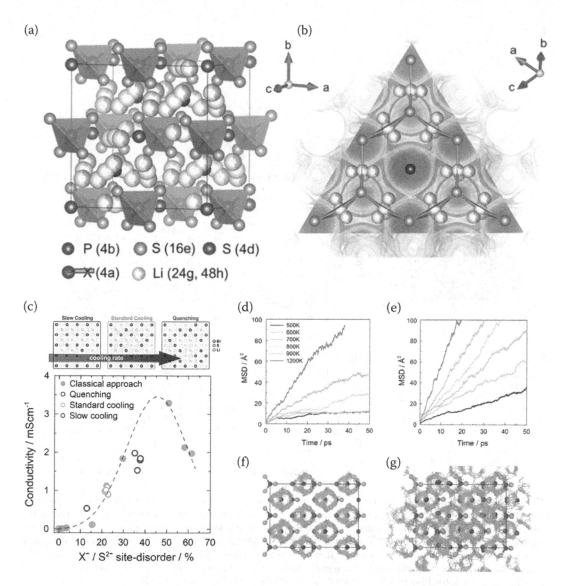

FIGURE 6.10 (a) Schematic representation of the crystal structure of Li₆PS₅Br without any site-disorder amongst the sites 4a and 4d; (b) obtained bond valence sum indicating that the long-range jump between the Li⁺ cages represents the bottleneck for diffusion; (c) comparison of conductivities vs. Br⁻/S²⁻ site-disorder of classically synthesized Li₆PS₅X; here, (d) and with 50% Br–/S2– site disorder; (e) representative Li+ trajectories at 700 K for 35 ps are shown for the fully ordered (f) and disordered (g) structure. Reproduced from Ref. Gautam et al. (2019) with permission. Copyright 2019 American Chemical Society.

Prasada Rao Rayavarapu and coworkers reported synthesis of argyrodite-type Li₆PS₅X (X = Cl, Br, I) by implementing mechanical milling and then by annealing the powdered mixed reactant. Creation and growth of Li₆PS₅X crystals in nature were revealed by X-ray diffraction characterization. Ionic conductivity values of 7×10^{-4} S cm⁻¹ in Li₆PS₅Cl and Li₆PS₅Br projects these phases apt for ASSLIBs. Joint structure refinements performed using laboratory X-ray diffraction and high-resolution neutron led to deep insight into the implication of the disorder on the rapid ionic conductivity. Apart from the disorder for distribution of Li-metal, the extent of disorder in the S²⁻/Cl⁻ or S²⁻/Br⁻ distribution promoted ion mobility, whereas the exchange of large I⁻ with S²⁻ was unsuccessful creating more ordered Li₆PS₅I resulting in moderate conductivity. Bond valence approach has been used to model the Li⁺

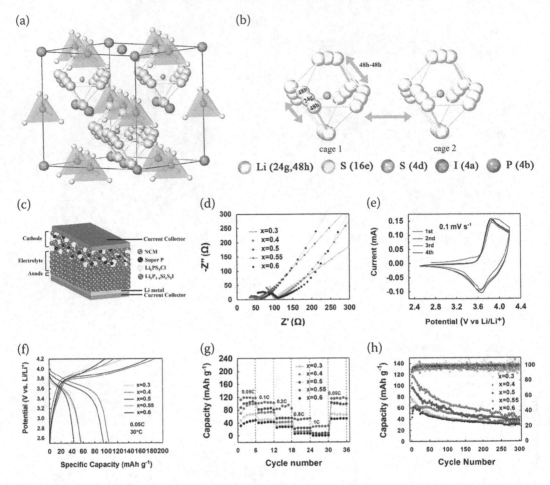

FIGURE 6.11 (a) Crystal structure unit of Li_6PS_5I; (b) The mobile Li^+ associated at 24g and 48h Wyckoff positions around the free S anions, forming a cage-like octahedral structure enable three possible jump paths: jumps between cages (intercage jump), the 48h-48h jump (intracage jump), the 48h-24g-48h jump (doublet jump); (c) The schematic representation of NCM-Li6PS5Cl/ $Li_{6+x}P_{1-x}Si_xS_5I$ (0.3 ≤ x ≤ 0.6)/ Li ASSLBs; (d) Arrhenius plots of temperature-dependent ionic conductivity values in the temperature range from 30°C to 100°C; (e) CV plots of NCM-Li_6PS_5Cl/ $Li_{6.55}P_{0.45}Si_{0.55}S_5I$ / Li ASSLBs at 0.1 mV s^{-1}; (f) initial cycle of charge-discharge curves, (g) cyclic performance; and (h) rate performance of the NCM-Li6PS5Cl/ Li6+xP1−xSixS5I (0.3 ≤ x ≤ 0.6)/ Li ASSLBs at RT. Reproduced from Ref. Zhang et al. (2020) with permission. Copyright 2020 American Chemical Society.

transportation pathways for the crystalline compounds to understand the variances between argyrodites containing diverse halide ions (Rayavarapu et al. 2012).

6.8 Conclusions and Perspectives

In a nutshell, we have described the essence of sulphide-based solid-state electrolytes with special emphasis on the Li-based argyrodites with chemical formula Li_6PS_5X (X = Cl, Br, and I). The emergence of solid-state batteries has been discussed in detail. Further, the ion conduction mechanism has been discussed briefly. The challenges still persisting with the solid-state batteries and their possible mitigation have been highlighted. Solid-state batteries have experienced rapidly growing interest in the past few years. However, the fundamental understanding of reactions occurring at the interfaces, cell

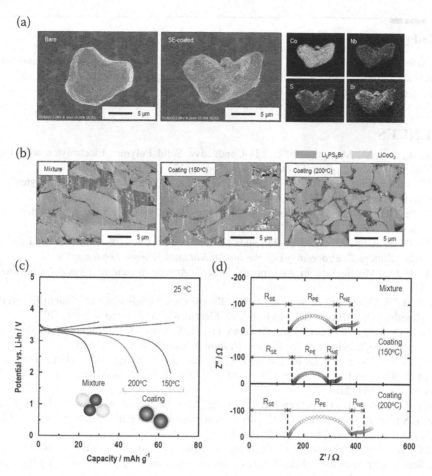

FIGURE 6.12 (a) FE-SEM images of the bare LCO and the SE-coated LCO particles and EDX maps of cobalt, niobium, sulphur, and bromine for the SE-coated LCO particle: (b) Cross-sectional FE-SEM images of the composite positive electrode layers using the conventional mixture of LCO and SE particles and the SE-coated LCO particles dried at 150°C or 200°C. (c) Charge-discharge curves of the cells In/Li₃PS₄/LCO with the conventional mixed electrode prepared by the hand mixing of LCO and SE particles using a mortar and the composite electrodes with the SE-coated LCO particles dried at 150°C or 200°C prepared via the liquid-phase technique. Specific capacities are calculated on the basis of the weight of LiNbO₃-coated LCO. (d) Nyquist plots for the cells after the initial charging process. Reproduced from Ref. Yubuchi et al. (2018) with permission. Copyright 2018 American Chemical Society.

design, and performance is still in its infancy. Three main points are highlighted which should be addressed: (1) Firstly, inefficient interfacial kinetics occurring between the solid electrolyte and active electrolyte material needs to be investigated more for optimization; (2) Secondly, mechanical pressure is necessary to assure stable working of a solid-state cell, and investigations related to quantitative results are requisite; and (3) Finally, apart from Li-metal anode other anodes need to be explored to cause large advantage to solid-state batteries for delivering high energy density. As solid electrolytes permit large current densities in absence of concentration polarization, large power densities can be achieved as well in these batteries by the means of introduction of suitable anode and cathode. Solid-state battery as a concept is connected with preconceptions and misjudgements—resulting in overestimation and pessimistic views. Several key issues need to be addressed yet and large efforts are requisite to cause an improved assessment of the technical future of all-solid-state batteries. In our view, high power densities and high voltages will be attained in the near future and the thick Li metal electrodes will be replaced by other efficient intercalation anodes to achieve superior kinetics in all-solid-state batteries.

Acknowledgements

AKT acknowledges Sunchon National University, Republic of Korea, for providing platform to conceive the idea and a thorough literature survey. MM acknowledges IIT (ISM), Dhanbad for providing a platform to compile the manuscript.

REFERENCES

Abraham, K.M., and M. Alamgir. 1990. "Li+-Conductive Solid Polymer Electrolytes with Liquid-Like Conductivity". *Journal of the Electrochemical Society* 137 (5): 1657.

Adachi, G.Y., N. Imanaka, and H. Aono. 1996. "Fast Li$^{\oplus}$ Conducting Ceramic Electrolytes". *Advanced Materials* 8 (2): 127–135.

Akridge, J.R., and H. Vourlis. 1986. "Solid State Batteries Using Vitreous Solid Electrolytes". *Solid State Ionics* 18: 1082–1087.

Angalakurthi, R., S. Krupanidhi, and P. Barpanda. 2018. "An Overview of Nanostructured Li-based Thin Film Micro-Batteries". *Proceedings of the Indian National Science Academy*, 98.

Angell, C.A. 1992. "Mobile Ions in Amorphous Solids". *Annual Review of Physical Chemistry* 43 (1): 693–717.

Appetecchi, G.B., F. Croce, and B. Scrosati. 1995. "Kinetics and Stability of the Lithium Electrode in Poly (Methylmethacrylate)-Based Gel Electrolytes". *Electrochimica Acta* 40 (8): 991–997.

Armand, M., F. Endres, D.R. MacFarlane, H. Ohno, and B. Scrosati. 2009. "Ionic-Liquid Materials for the Electrochemical Challenges of the Future". *Nature Materials* 8 (8): 621–629.

Arnold, W., D.A. Buchberger, Y. Li, M. Sunkara, T. Druffel, and H. Wang. 2020. "Halide Doping Effect on Solvent-Synthesized Lithium Argyrodites Li6PS5X (X= Cl, Br, I) Superionic Conductors". *Journal of Power Sources* 464: 228158.

Asemchainan, J., S. Zekoll, D. Spencer Jolly, Z. Ning, G.O. Hartley, J. Marrow, and P.G. Bruce. 2019. "Critical Stripping Current Leads to Dendrite Formation on Plating in Lithium Anode Solid Electrolyte Cells". *Nature Materials* 18 (10): 1105–1111.

Bachman, J.C., S. Muy, A. Grimaud, H.-H. Chang, N. Pour, S.F. Lux, O. Paschos, et al. 2016. "Inorganic Solid-State Electrolytes for Lithium Batteries: Mechanisms and Properties Governing Ion Conduction". *Chemical Reviews* 116 (1): 140–162.

Balkanski, M., Julien, C., and J.Y. Emery. 1989. "Integrable Lithium Solid-State Microbatteries". *Journal of Power Sources* 26 (3–4): 615–622.

Bates, J.B., N.J. Dudney, G.R. Gruzalski, R.A. Zuhr, A. Choudhury, C.F. Luck, and J.D. Robertson. 1992. "Electrical Properties of Amorphous Lithium Electrolyte Thin Films". *Solid State Ionics* 53: 647–654.

Berthier, C., W. Gorecki, M. Minier, M.B. Armand, J.M. Chabagno, and P. Rigaud. 1983. "Microscopic Investigation of Ionic Conductivity in Alkali Metal Salts-Poly(Ethylene Oxide) Adducts". *Solid State Ionics* 11 (1): 91–95.

Bones, R.J., J. Coetzer, R.C. Galloway, and D.A. Teagle. 1987. "A Sodium/Iron (II) Chloride Cell with a Beta Alumina Electrolyte". *Journal of the Electrochemical Society* 134 (10): 2379.

Boulineau, S., M. Courty, J.-M. Tarascon, and V. Viallet. 2012. "Mechanochemical Synthesis of Li-Argyrodite Li6PS5X (X = Cl, Br, I) as Sulfur-Based Solid Electrolytes for All Solid State Batteries Application". *Solid State Ionics* 221: 1–5.

Boulineau, S., J.-M. Tarascon, J.-B. Leriche, and V. Viallet. 2013. "Electrochemical Properties of All-Solid-State Lithium Secondary Batteries Using Li-Argyrodite Li6PS5Cl as Solid Electrolyte". *Solid State Ionics* 242: 45–48.

Bron, P., S. Dehnen, and B. Roling. 2016. "Li10Si0.3Sn0.7P2S12 – A Low-Cost and Low-Grain-Boundary-Resistance Lithium Superionic Conductor". *Journal of Power Sources* 329: 530–535.

Camacho-Forero, L.E., and P.B. Balbuena. 2019. "Elucidating Interfacial Phenomena between Solid-State Electrolytes and the Sulfur-Cathode of Lithium–Sulfur Batteries". *Chemistry of Materials* 32 (1): 360–373.

Cao, D., X. Sun, Q. Li, A. Natan, P. Xiang, and H. Zhu. 2020. "Lithium Dendrite in All-Solid-State Batteries: Growth Mechanisms, Suppression Strategies, and Characterizations". *Matter* 3(1): 57–94.

Capasso, C., and O. Veneri. 2014. "Experimental Analysis of a Zebra Battery Based Propulsion System for Urban Bus Under Dynamic Conditions". *Energy Procedia* 61, 1138–1141.

Chang, Z., X. Wang, Y. Yang, J. Gao, M. Li, L. Liu, and Y. Wu. 2014. "Rechargeable Li//Br Battery: A Promising Platform for Post Lithium Ion Batteries". *Journal of Materials Chemistry A* 2 (45): 19444–19450.

Chen, B., Z. Huang, X. Chen, Y. Zhao, Q. Xu, P. Long, S. Chen, and X. Xu. 2016. "A New Composite Solid Electrolyte PEO/Li10GeP2S12/SN for All-Solid-State Lithium Battery". *Electrochimica Acta* 210: 905–914.

Chen, M., and S. Adams. 2015. "High Performance All-Solid-State Lithium/Sulfur Batteries Using Lithium Argyrodite Electrolyte". *Journal of Solid State Electrochemistry* 19 (3): 697–702.

Chen, S., K. Wen, J. Fan, Y. Bando, and D. Golberg. 2018. "Progress and Future Prospects of High-Voltage and High-Safety Electrolytes in Advanced Lithium Batteries: From Liquid to Solid Electrolytes". *Journal of Materials Chemistry A* 6 (25): 11631–11663.

Choe, H.S., J. Giaccai, M Alamgir, and K.M. Abraham. 1995. "Preparation and Characterization of Poly (Vinyl Sulfone)-and Poly (Vinylidene Fluoride)-Based Electrolytes". *Electrochimica Acta* 40 (13–14): 2289–2293.

Choi, S.-J., S.-H. Lee, Y.-C. Ha, J.-H. Yu, C.-H. Doh, Y. Lee, J.-W. Park, S.-M. Lee, and H.-C. Shin. 2018. "Synthesis and Electrochemical Characterization of a Glass-Ceramic Li7P2S8I Solid Electrolyte for All-Solid-State Li-Ion Batteries". *Journal of The Electrochemical Society* 165 (5): A957.

Coetzer, J. 1986. "A New High Energy Density Battery System". *Journal of Power Sources* 18 (4): 377–380.

Creus, R., Sarradin, J., Astier, R., Pradel, A., and M. Ribes. 1989. "The Use of Ionic and Mixed Conductive Glasses in Microbatteries". *Materials Science and Engineering: B* 3 (1): 109–112.

Culver, S.P., R. Koerver, W.G. Zeier, and J. Janek. 2019. "On the Functionality of Coatings for Cathode Active Materials in Thiophosphate-Based All-Solid-State Batteries". *Advanced Energy Materials* 9 (24): 1900626.

Cussen, E.J. 2006. "The Structure of Lithium Garnets: Cation Disorder and Clustering in a New Family of Fast Li+ Conductors". *Chemical Communications* (4): 412–413.

Deiseroth, H.J., S.T. Kong, H. Eckert, J. Vannahme, C. Reiner, T. Zaiß, and M. Schlosser. 2008. "Li6PS5X: A Class of Crystalline Li-Rich Solids with an Unusually High Li+ Mobility". *Angewandte Chemie* 120 (4): 767–770.

Deng, S., Y. Sun, X. Li, Z. Ren, J. Liang, K. Doyle-Davis, J. Liang, et al. 2020. "Eliminating the Detrimental Effects of Conductive Agents in Sulfide-Based Solid-State Batteries". *ACS Energy Letters* 5 (4): 1243–1251.

Dewald, G.F., S. Ohno, M.A. Kraft, R. Koerver, P. Till, N.M. Vargas-Barbosa, J. R. Janek, and W.G. Zeier. 2019. "Experimental Assessment of the Practical Oxidative Stability of Lithium Thiophosphate Solid Electrolytes". *Chemistry of Materials* 31 (20): 8328–8337.

Dietrich, C., D.A. Weber, S.J. Sedlmaier, S. Indris, S.P. Culver, D. Walter, J. Janek, and W.G. Zeier. 2017. "Lithium Ion Conductivity in Li 2 S–P 2 S 5 Glasses–Building Units and Local Structure Evolution during the Crystallization of Superionic Conductors Li 3 PS 4, Li 7 P 3 S 11 and Li 4 P 2 S 7". *Journal of Materials Chemistry A* 5 (34): 18111–18119.

Dong, X., Y. Wang, and Y. Xia. 2014. "Re-building Daniell Cell with a Li-Ion Exchange Film". *Scientific Reports* 4: 6916.

Doux, J.M., H. Nguyen, D.H.S. Tan, A. Banerjee, X. Wang, E.A. Wu, C. Jo, H. Yang, and Y.S. Meng. 2020. "Stack Pressure Considerations for Room-Temperature All-Solid-State Lithium Metal Batteries". *Advanced Energy Materials* 10 (1): 1903253.

Dudney, N.J., J.B. Bates, R.A. Zuhr, C.F. Luck, and J.D. Robertson. 1992. "Sputtering of Lithium Compounds for Preparation of Electrolyte Thin Films". *Solid State Ionics* 53: 655–661.

Dzwonkowski, P., M. Eddrief, C. Julien, and M. Balkanski. 1991. "Electrical a.c. Conductivity of B$_2$O$_3$-xLi$_2$O Glass Thin Films and Analysis Using the Electric Modulus Formalism". *Materials Science and Engineering: B* 8 (3): 193–200.

Ellis, B.L., and L.F. Nazar. 2012. "Sodium and Sodium-Ion Energy Storage Batteries". *Current Opinion in Solid State and Materials Science* 16 (4): 168–177.

Eng, H.J., J.Q. Huang, X.B. Cheng, and Q. Zhang. 2017. "Review on High-Loading and High-Energy Lithium–Sulfur Batteries". *Advanced Energy Materials* 7 (24): 1700260.

Epp, V., O.Z.L. Gün, H.-J.R. Deiseroth, and M. Wilkening. 2013. "Highly Mobile Ions: Low-Temperature NMR Directly Probes Extremely Fast Li+ Hopping in Argyrodite-Type Li$_6$PS$_5$Br". *The Journal of Physical Chemistry Letters* 4 (13): 2118–2123.

Famprikis, T., P. Canepa, J.A. Dawson, M.S., Islam, and C. Masquelier. 2019. "Fundamentals of Inorganic Solid-State Electrolytes for Batteries". *Nature Materials* 18: 1278–1291.

Faraday, M., IV 1833. "Experimental Researches in Electricity – Third series". *Philosophical Transactions of the Royal Society of London* (123): 23–54.

Fenton, D.E., J.M. Parker, and P.V. Wright. 1973. "Complexes of Alkali Metal Ions with Poly(ethylene oxide)". *Polymer* 14 (11): 589.

Fergus, J.W. 2010. "Ceramic and Polymeric Solid Electrolytes for Lithium-Ion Batteries". *Journal of Power Sources* 195 (15): 4554–4569.

Fergus, J.W. 2011. "Sensing Mechanism of Non-equilibrium Solid-Electrolyte-Based Chemical Sensors". *Journal of Solid State Electrochemistry* 15 (5): 971–984.

Fleischmann, S., M. Widmaier, A. Schreiber, H. Shim, F.M. Stiemke, T.J.S. Schubert, and V. Presser. 2019. "High Voltage Asymmetric Hybrid Supercapacitors Using Lithium- and Sodium-Containing Ionic Liquids". *Energy Storage Materials* 16: 391–399.

Forse, Alexander C., John M. Griffin, C. Merlet, J. Carretero-Gonzalez, A. Raji, O. Rahman, Nicole M. Trease, and Clare P. Grey. 2017. "Direct Observation of Ion Dynamics in Supercapacitor Electrodes Using IN SITU DIFFUSION NMR SPECTROSCOPY". *Nature Energy* 2 (3): 16216.

Funke, K. 2013. "Solid State Ionics: From Michael Faraday to Green Energy – the European Dimension". *Science and Technology of Advanced Materials* 14 (4): 043502.

Gao, J., Shi, S.-Q., and H. Li. 2015. "Brief Overview of Electrochemical Potential in Lithium Ion Batteries". *Chinese Physics B* 25 (1): 018210.

Gao, Y., D. Wang, Y.C. Li, Z. Yu, T.E. Mallouk, and D. Wang. 2018. "Salt-Based Organic–Inorganic Nanocomposites: Towards A Stable Lithium Metal/Li10GeP2S12 Solid Electrolyte Interface". *Angewandte Chemie International Edition* 57 (41): 13608–13612.

Gao, Z., H. Sun, L. Fu, F. Ye, Y. Zhang, W. Luo, and Y. Huang. 2018. "Promises, Challenges, and Recent Progress of Inorganic Solid-State Electrolytes for All-Solid-State Lithium Batteries". *Advanced Materials* 30 (17): 1705702.

Gautam, A., M. Sadowski, N. Prinz, H. Eickhoff, N. Minafra, M. Ghidiu, S.P. Culver, et al. 2019. "Rapid Crystallization and Kinetic Freezing of Site-Disorder in the Lithium Superionic Argyrodite Li6PS5Br". *Chemistry of Materials* 31 (24): 10178–10185.

Goodenough, J.B. 2013. "Evolution of Strategies for Modern Rechargeable Batteries". *Accounts of Chemical Research* 46 (5): 1053–1061.

Goodenough, J.B., and P. Singh. 2015. "Review – Solid Electrolytes in Rechargeable Electrochemical Cells". *Journal of the Electrochemical Society* 162 (14): A2387–A2392.

Goodenough, J.B., H.-P. Hong, and J.A. Kafalas. 1976. "Fast Na+-Ion Transport in Skeleton Structures". *Materials Research Bulletin* 11 (2): 203–220.

Gorecki, W., R. Andreani, C. Berthier, M. Armand, M. Mali, J. Roos, and D. Brinkmann. 1986. "NMR, DSC, and Conductivity Study of a Poly (Ethylene Oxide) Complex Electrolyte: PEO (LiClO4) × ". *Solid State Ionics* 18: 295–299.

Gray, F.M., J.R. MacCallum, and C.A. Vincent. 1986. "Poly (Ethylene Oxide)-LiCF3SO3-Polystyrene Electrolyte Systems". *Solid State Ionics* 18: 282–286.

Hagenmuller, P., and W. Van Gool. 2015 *Solid Electrolytes: General Principles, Characterization, Materials, Applications*. New York: Elsevier.

Hanghofer, I., B. Gadermaier, and H.M.R. Wilkening. 2019. "Fast Rotational Dynamics in Argyrodite-Type Li6PS5X (X: Cl, Br, I) as Seen by 31P Nuclear Magnetic Relaxation – On Cation–Anion Coupled Transport in Thiophosphates". *Chemistry of Materials* 31 (12): 4591–4597.

Hanyu, Y., and I. Honma. 2012. "Rechargeable Quasi-Solid State Lithium Battery with Organic Crystalline Cathode". *Scientific Reports* 2: 453.

Hatzell, K.B., X.C. Chen, C.L. Cobb, N.P. Dasgupta, M.B. Dixit, L.E. Marbella, M.T. McDowell, P.P. Mukherjee, A. Verma, and V. Viswanathan. 2020. "Challenges in Lithium Metal Anodes for Solid-State Batteries". *ACS Energy Letters* 5 (3): 922–934.

Hayashi, A., S. Hama, H. Morimoto, M. Tatsumisago, and T. Minami. 2001. "Preparation of Li2S–P2S5 Amorphous Solid Electrolytes by Mechanical Milling". *Journal of the American Ceramic Society* 84 (2): 477–479.

Hood, Z.D., H. Wang, Y. Li, A.S. Pandian, M.P. Paranthaman, and C. Liang. 2015. "The "filler effect": A Study of Solid Oxide Fillers with β-Li3PS4 for Lithium Conducting Electrolytes". *Solid State Ionics* 283: 75–80.

Huggins, R.A. 1977. *Recent Results on Lithium Ion Conductors*, 773–781. Amsterdam: Elsevier.

Inaguma, Y., C. Liquan, M. Itoh, T. Nakamura, T. Uchida, H. Ikuta, and M. Wakihara. 1993. "High Ionic Conductivity in Lithium Lanthanum Titanate". *Solid State Communications* 86 (10): 689–693.

Jones, S.D., and J.R. Akridge. 1992. "A Thin Film Solid State Microbattery". *Solid State Ionics* 53-56: 628–634.

Jones, S.D., and J.R. Akridge. 1993. "A Thin-Film Solid-State Microbattery". *Journal of Power Sources* 44 (1): 505–513.

Julien, C.M., and A. Mauger. 2019. "Pulsed Laser Deposited Films for Microbatteries". *Coatings* 9 (6): 386.

Jung, S.H., U.H. Kim, J.H. Kim, S. Jun, C.S. Yoon, Y.S. Jung, and Y.K. Sun. 2020. "Ni-Rich Layered Cathode Materials with Electrochemo-Mechanically Compliant Microstructures for All-Solid-State Li Batteries". *Advanced Energy Materials* 10 (6): 1903360.

Kaib, T., S. Haddadpour, M. Kapitein, P. Bron, C. Schröder, H. Eckert, B. Roling, and S. Dehnen. 2012. "New Lithium Chalcogenidotetrelates, LiChT: Synthesis and Characterization of the Li+-Conducting Tetralithium Ortho-Sulfidostannate Li4SnS4". *Chemistry of Materials* 24 (11): 2211–2219.

Kamaya, N., K. Homma, Y. Yamakawa, M. Hirayama, R. Kanno, M. Yonemura, T. Kamiyama, Y. Kato, S. Hama, and K. Kawamoto. 2011. "A Lithium Superionic Conductor". *Nature Materials* 10 (9): 682–686.

Kanno, R., and M. Murayama. 2001. "Lithium Ionic Conductor Thio-LISICON: The Li2 S GeS2 P 2 S 5 System". *Journal of the Electrochemical Society* 148 (7): A742.

Kato, Y., S. Hori, T. Saito, K. Suzuki, M. Hirayama, A. Mitsui, M. Yonemura, H. Iba, and R. Kanno. 2016. "High-Power All-Solid-State Batteries Using Sulfide Superionic Conductors". *Nature Energy* 1 (4): 1–7.

Kennedy, J.H., and Y. Yang. 1986. "A Highly Conductive Li+-Glass System:(1-x)(0.4 SiS2-0.6 Li2S)-xLiI". *Journal of the Electrochemical Society* 133: 2437–2438.

Kennedy, J.H., and Y. Yang. 1987. "Glass-Forming Region and Structure in SiS2 Li2S LiX (X = Br, I)". *Journal of Solid State Chemistry* 69 (2): 252–257.

Kennedy, J.H., and Z. Zhang. 1988. "Improved Stability for the SiS2-P2S5-Li2S-LiI Glass System". *Solid State Ionics* 28: 726–728.

Kennedy, J.H. and Z. Zhang. 1989. "Preparation and Electrochemical Properties of the SiS2-P 2 S 5-Li2 S Glass Coformer System". *Journal of the Electrochemical Society* 136 (9): 2441.

Kennedy, J.H., Z. Zhang, and H. Eckert. 1990. "Ionically Conductive Sulfide-Based Lithium Glasses". *Journal of Non-Crystalline Solids* 123 (1–3): 328–338.

Kerman, K., A. Luntz, V. Viswanathan, Y.-M. Chiang, and Z. Chen. 2017. "practical Challenges Hindering the Development of Solid State Li Ion Batteries". *Journal of The Electrochemical Society* 164 (7): A1731.

Kim, J.K., E. Lee, H. Kim, C. Johnson, J. Cho, and Y. Kim. 2015. "Rechargeable Seawater Battery and Its Electrochemical Mechanism". *ChemElectroChem* 2 (3): 328–332.

Kim, K.-H., and S.W. Martin. 2019. "Structures and Properties of Oxygen-Substituted Li10SiP2S12–xOx Solid-State Electrolytes". *Chemistry of Materials* 31 (11): 3984–3991.

Knauth, P. 2009. "Inorganic Solid Li Ion Conductors: An Overview". *Solid State Ionics* 180 (14-16): 911–916.

Knauth, P., and H.L. Tuller. 2002. "Solid-State Ionics: Roots, Status, and Future Prospects". *Journal of the American Ceramic Society* 85 (7): 1654–1680.

Knödler, R. 1984. "Thermal Properties of Sodium-Sulphur Cells". *Journal of Applied Electrochemistry* 14 (1): 39–46.

Koerver, R., I. Aygün, T. Leichtweiß, C. Dietrich, W. Zhang, J.O. Binder, P. Hartmann, W.G. Zeier, and J.R. Janek. 2017. "Capacity Fade in Solid-State Batteries: Interphase Formation and Chemomechanical Processes in Nickel-Rich Layered Oxide Cathodes and Lithium Thiophosphate Solid Electrolytes". *Chemistry of Materials* 29 (13): 5574–5582.

Koerver, R., W. Zhang, L. de Biasi, S., Schweidler, A.O. Kondrakov, S. Kolling, T. Brezesinski, P. Hartmann, W.G. Zeier, and J. Janek. 2018. "Chemo-Mechanical Expansion of Lithium Electrode Materials – on the Route to Mechanically Optimized All-Solid-State Batteries". *Energy & Environmental Science* 11 (8): 2142–2158.

Kong, S.T., H.J. Deiseroth, J. Maier, V. Nickel, K. Weichert, and C. Reiner. 2010. "Li$_6$PO$_5$Br and Li$_6$PO$_5$Cl: The first Lithium-Oxide-Argyrodites". *Zeitschrift für anorganische und allgemeine Chemie* 636 (11): 1920–1924.

Krachkovskiy, S., M.L. Trudeau, and K. Zaghib. 2020. "Application of Magnetic Resonance Techniques to the In Situ Characterization of Li-Ion Batteries: A Review". *Materials* 13 (7): 1694.

Kraft, M.A., S.P. Culver, M. Calderon, F. Böcher, T. Krauskopf, A. Senyshyn, C. Dietrich, A. Zevalkink, J. Janek, and W.G. Zeier. 2017. "Influence of Lattice Polarizability on the Ionic Conductivity in the Lithium Superionic Argyrodites Li6PS5X (X = Cl, Br, I)". *Journal of the American Chemical Society* 139 (31): 10909–10918.

Kuhn, A., V. Duppel, and B. Lotsch. 2013. "Tetragonal Li10GeP2S12 and Li7GePS8 – Exploring the Li ion Dynamics in LGPS Li Electrolytes". *Energy & Environmental Science* 6: 3548.

Kulkarni, A.R., H.S. Maiti, and A. Paul. 1984. "Fast Ion Conducting Lithium Glasses–Review". *Bulletin of Materials Science* 6 (2): 201–221.

Kumar, P.P., and S. Yashonath. 2006. "Ionic Conduction in the Solid State". *Journal of Chemical Sciences* 118 (1): 135–154.

Li, F., H. Kitaura, and H. Zhou. 2013. "The pursuit of rechargeable solid-state Li–air batteries". *Energy & Environmental Science* 6 (8): 2302–2311.

Li, H., Z. Wang, L. Chen, and X. Huang. 2009a. "Research on Advanced Materials for Li-Ion Batteries". *Advanced Materials* 21 (45): 4593–4607.

Li, H., Y. Wang, H. Na, H. Liu, and H. Zhou. 2009b. "Rechargeable Ni-Li Battery Integrated Aqueous/Nonaqueous System". *Journal of the American Chemical Society* 131 (42): 15098–15099.

Liu, G., W. Weng, Z. Zhang, L. Wu, J. Yang, and X. Yao. 2020. "Densified Li6PS5Cl Nanorods with High Ionic Conductivity and Improved Critical Current Density for All-Solid-State Lithium Batteries". *Nano Letters* 20 (9): 6660–6665.

Liu, T., J.P. Vivek, E.W. Zhao, J. Lei, N. Garcia-Araez, and C.P. Grey. 2020. "Current Challenges and Routes Forward for Nonaqueous Lithium-Air Batteries". *Chemical Reviews* 120 (14): 6558–6625.

Liu, Y., B. Xu, W. Zhang, L. Li, Y. Lin, and C. Nan. 2020. "Composition Modulation and Structure Design of Inorganic-in-Polymer Composite Solid Electrolytes for Advanced Lithium Batteries". *Small* 16 (15): 1902813.

Liu, Y., P. He, and H. Zhou. 2018. "Rechargeable Solid-State Li–Air and Li–S Batteries: Materials, Construction, and Challenges". *Advanced Energy Materials* 8 (4): 1701602.

Liu, Z., W. Fu, E.A. Payzant, X. Yu, Z. Wu, N.J. Dudney, J. Kiggans, K. Hong, A.J. Rondinone, C. Liang. 2013. "Anomalous High Ionic Conductivity of Nanoporous β-Li3PS4". *Journal of the American Chemical Society* 135 (3): 975–978.

Lu, Y., and J.B. Goodenough. 2011. "Rechargeable Alkali-Ion Cathode-Flow Battery". *Journal of Materials Chemistry* 21 (27): 10113–10117.

Manthiram, A., X. Yu, and S. Wang. 2017. "Lithium Battery Chemistries Enabled by Solid-State Electrolytes". *Nature Reviews Materials* 2 (4): 1–16.

Mauger, A., C.M. Julien, A. Paolella, M. Armand, and K. Zaghib. 2019. "Building Better Batteries in the Solid State: A Review". *Materials (Basel)* 12 (23): 3892.

Mauger, A., M. Armand, C.M. Julien, and K. Zaghib. 2017. "Challenges and Issues Facing Lithium Metal for Solid-State Rechargeable Batteries". *Journal of Power Sources* 353: 333–342.

Mazza, D. 1988. "Remarks on a Ternary Phase in the La2O3-Me2O5-Li2O System (Me= Nb, Ta)". *Materials Letters* 7 (5–6): 205–207.

Mehrer, H. 2007. *Diffusion in Solids: Fundamentals, Methods, Materials, Diffusion-Controlled Processes*, Vol. 155. Cham: Springer Science & Business Media.

Mercier, R., J.-P. Malugani, B. Fahys, and G. Robert. 1981. "Superionic Conduction in Li2S-P2S5-LiI-Glasses". *Solid State Ionics* 5: 663–666.

Meunier, G., Dormoy, R., and A. Levasseur. 1989. "New Positive-Electrode Materials for Lithium Thin Film Secondary Batteries". *Materials Science and Engineering: B* 3 (1–2): 19–23.

Minami, T. 1985. "Fast Ion Conducting Glasses". *Journal of Non-crystalline Solids* 73 (1–3): 273–284.

Miura, A., N.C. Rosero-Navarro, A. Sakuda, K. Tadanaga, N.H.H. Phuc, A. Matsuda, N. Machida, A. Hayashi, and M. Tatsumisago. 2019. "Liquid-Phase Syntheses of Sulfide Electrolytes for All-Solid-State Lithium Battery". *Nature Reviews Chemistry* 3 (3): 189–198.

Mo, Y., S.P. Ong, and G. Ceder. 2012. "First Principles Study of the Li10GeP2S12 Lithium Super Ionic Conductor Material". *Chemistry of Materials* 24 (1): 15–17.

Monroe, C., and J. Newman. 2004. "The Effect of Interfacial Deformation on Electrodeposition Kinetics". *Journal of the Electrochemical Society* 151 (6): A880.

Montes, J.M., F.G. Cuevas, and J. Cintas. 2008. "Porosity Effect on the Electrical Conductivity of Sintered Powder Compacts". *Applied Physics A* 92 (2): 375–380.

Nagai, E., T.S. Arthur, P. Bonnick, K. Suto, and J. Muldoon. 2019. "The Discharge Mechanism for Solid-State Lithium-Sulfur Batteries". *MRS Advances* 4 (49): 2627–2634.

Nagao, M., A. Hayashi, M. Tatsumisago, T. Kanetsuku, T. Tsuda, and S. Kuwabata. 2013. "In Situ SEM Study of a Lithium Deposition and Dissolution Mechanism in a Bulk-Type Solid-State Cell with a Li2S–P2S5 Solid Electrolyte". *Physical Chemistry Chemical Physics* 15 (42): 18600–18606.

Nitzan, A., and M.A. Ratner. 1994. "Conduction in Polymers: Dynamic Disorder Transport". *The Journal of Physical Chemistry* 98 (7): 1765–1775.

Ohno, S., R. Koerver, G. Dewald, C. Rosenbach, P. Titscher, D. Steckermeier, A. Kwade, J.R. Janek, and W.G. Zeier. 2019. "Observation of Chemomechanical Failure and the Influence of Cutoff Potentials in All-Solid-State Li–S Batteries". *Chemistry of Materials* 31 (8): 2930–2940.

Ohta, N., K. Takada, L. Zhang, R. Ma, M. Osada, and T. Sasaki. 2006. "Enhancement of the High-Rate Capability of Solid-State Lithium Batteries by Nanoscale Interfacial Modification". *Advanced Materials* 18 (17): 2226–2229.

Oshima, T., M. Kajita, and A. Okuno. 2004. "Development of Sodium-Sulfur Batteries". *International Journal of Applied Ceramic Technology* 1 (3): 269–276.

Otoyama, M., Y. Ito, A. Hayashi, and M. Tatsumisago. 2016. "Raman Imaging for LiCoO2 Composite Positive Electrodes in All-Solid-State Lithium Batteries Using Li2S–P2S5 Solid Electrolytes". *Journal of Power Sources* 302: 419–425.

Oudenhoven, J.F.M., L. Baggetto, and P.H.L. Notten. 2011. "All-Solid-State Lithium-Ion Microbatteries: A Review of Various Three-Dimensional Concepts". *Advanced Energy Materials* 1 (1): 10–33.

Owens, B.B. 2000. "Solid State Electrolytes: Overview of Materials and Applications during the Last Third of the Twentieth Century". *Journal of Power Sources* 90 (1): 2–8.

Pang, Q., X. Liang, C.Y. Kwok, L.F. Nazar. 2016. "Advances in Lithium–Sulfur Batteries Based on Multifunctional Cathodes and Electrolytes". *Nature Energy* 1 (9): 1–11.

Park, M., X. Zhang, M. Chung, G.B. Less, and A.M. Sastry. 2010. "A Review of Conduction Phenomena in Li-Ion Batteries". *Journal of Power Sources* 195 (24): 7904–7929.

Perram, J.W. 2013. *The Physics of Superionic Conductors and Electrode Materials*, Vol. 92. Cham: Springer Science & Business Media.

Porz, L., T. Swamy, B.W. Sheldon, D. Rettenwander, T. Frömling, H.L. Thaman, S. Berendts, R. Uecker, W.C. Carter, and Y.M. Chiang. 2017. "Mechanism of Lithium Metal Penetration through Inorganic Solid Electrolytes". *Advanced Energy Materials* 7 (20): 1701003.

Pradel, A., and M. Ribes. 1986. "Electrical Properties of Lithium Conductive Silicon Sulfide Glasses Prepared by Twin Roller Quenching". *Solid State Ionics* 18: 351–355.

Pradel, A., and M. Ribes. 1989a. "Ionic Conductive Glasses". *Materials Science and Engineering: B* 3 (1–2): 45–56.

Pradel, A., and M. Ribes. 1989b. "Lithium Chalcogenide Conductive Glasses". *Materials chemistry and physics* 23 (1–2): 121–142.

Prasada Rao, R., H. Chen, and S. Adams. 2019. "Stable Lithium Ion Conducting Thiophosphate Solid Electrolytes Lix(PS4)yXz (X = Cl, Br, I)". *Chemistry of Materials* 31 (21): 8649–8662.

Rangasamy, E., G. Sahu, J.K. Keum, A.J. Rondinone, N.J. Dudney, and C. Liang. 2014. "A High Conductivity Oxide–Sulfide Composite Lithium Superionic Conductor". *Journal of Materials Chemistry A* 2 (12): 4111–4116.

Rangasamy, E., Z. Liu, M. Gobet, K. Pilar, G. Sahu, W. Zhou, H. Wu, S. Greenbaum, and C. Liang. 2015. "An Iodide-Based Li7P2S8I Superionic Conductor". *Journal of the American Chemical Society* 137 (4): 1384–1387.

Rao, R.P., and M. Seshasayee. 2006. "Molecular Dynamics Simulation of Ternary Glasses Li2S–P2S5–LiI". *Journal of Non-crystalline Solids* 352 (30–31): 3310–3314.

Rao, R.P., and S. Adams. 2011. "Studies of Lithium Argyrodite Solid Electrolytes for All-Solid-State Batteries". *Physica Status Solidi (A)* 208 (8): 1804–1807.

Rayavarapu, P.R., N. Sharma, V.K. Peterson, and S. Adams. 2012. "Variation in Structure and Li+-Ion Migration in Argyrodite-Type Li 6 PS 5 X (X= Cl, Br, I) Solid Electrolytes". *Journal of Solid State Electrochemistry* 16 (5): 1807–1813.

Reddy, M.V., A. Mauger, C.M. Julien, A. Paolella, and K. Zaghib. 2020a. "Brief History of Early Lithium-Battery Development". *Materials* 13 (8): 1884.

Reddy, M.V., C.M. Julien, A. Mauger, and K. Zaghib. 2020b. "Sulfide and Oxide Inorganic Solid Electrolytes for All-Solid-State Li Batteries: A Review". *Nanomaterials* 10 (8): 1606.

Reuter, B., and K. Hardel. 1960. "Silbersulfidbromid und silbersulfidjodid". *Angewandte Chemie* 72 (4): 138–139.

Sahu, G., Z. Lin, J. Li, Z. Liu, N. Dudney, and C. Liang. 2014. "Air-Stable, High-Conduction Solid Electrolytes of Arsenic-Substituted Li4 SnS4". *Energy & Environmental Science* 7 (3): 1053–1058.

Sakuda, A., A. Hayashi, and M. Tatsumisago. 2010. "Interfacial Observation between LiCoO2 Electrode and Li2S–P2S5 Solid Electrolytes of All-Solid-State Lithium Secondary Batteries Using Transmission Electron Microscopy". *Chemistry of Materials* 22 (3): 949–956.

Sakuda, A., A. Hayashi., and M. Tatsumisago. 2017. "Recent Progress on Interface Formation in All-Solid-State Batteries". *Current Opinion in Electrochemistry* 6 (1): 108–114.

Song, J.Y., Y.Y. Wang, and C.C. Wan. 1999. "Review of Gel-Type Polymer Electrolytes for Lithium-Ion Batteries". *Journal of Power Sources* 77 (2): 183–197.

Strauss, F., L. de Biasi, A.Y. Kim, J. Hertle, S. Schweidler, J.R. Janek, P. Hartmann, and T. Brezesinski. 2019. "Rational Design of Quasi-Zero-Strain NCM Cathode Materials for Minimizing Volume Change Effects in All-Solid-State Batteries". *ACS Materials Letters* 2 (1): 84–88.

Sun, C., J. Liu, Y. Gong, D.P. Wilkinson, and J. Zhang. 2017. "Recent Advances in All-Solid-State Rechargeable Lithium Batteries". *Nano Energy* 33: 363–386.

Sun, C., Y. Ruan, W. Zha, W. Li, M. Cai, and Z. Wen. 2020. "Recent Advances in Anodic Interface Engineering for Solid-State Lithium-Metal Batteries". *Materials Horizons* 7:1667–1696.

Sun, Y., K. Suzuki, K. Hara, S. Hori, T.-A. Yano, M. Hara, M. Hirayama, and R. Kanno. 2016. "Oxygen Substitution Effects in Li10GeP2S12 Solid Electrolyte". *Journal of Power Sources* 324: 798–803.

Svensson, J., and C.G. Granqvist. 1985. "Electrochromic Coatings for 'Smart Windows'". *Solar Energy Materials* 12 (6): 391–402.

Takada, K. 2018. "Progress in Solid Electrolytes toward Realizing Solid-State Lithium Batteries". *Journal of Power Sources* 394: 74–85.

Takada, K., N. Aotani, K. Iwamoto, and S. Kondo. 1995. "Electrochemical behavior of LixMO2 (M = Co, Ni) in All Solid State Cells Using a Glass Electrolyte". *Solid State Ionics* 79: 284–287.

Takada, K., T. Ohno, N. Ohta, T. Ohnishi, and Y. Tanaka. 2017. "Positive and Negative Aspects of Interfaces in Solid-State Batteries". *ACS Energy Letters* 3 (1): 98–103.

Takahashi, T. 1988. "Early History of Solid State Ionics". *MRS Online Proceedings Library Archive* 135: 3–13.

Takemoto, K., and H. Yamada. 2015. "Development of Rechargeable Lithium–Bromine Batteries with Lithium Ion Conducting Solid Electrolyte". *Journal of Power Sources* 281: 334–340.

Tan, D.H.S., E.A. Wu, H. Nguyen, Z. Chen, M.A.T. Marple, J.-M. Doux, X. Wang, H. Yang, A. Banerjee, and Y.S. Meng. 2019. "Elucidating Reversible Electrochemical Redox of Li6PS5Cl Solid Electrolyte". *ACS Energy Letters* 4 (10): 2418–2427.

Tatsumisago, M., M. Nagao, and A. Hayashi. 2013. "Recent Development of Sulfide Solid Electrolytes and Interfacial Modification for All-Solid-State Rechargeable Lithium Batteries". *Journal of Asian Ceramic Societies* 1 (1): 17–25.

Thangadurai, V., S. Narayanan, and D. Pinzaru. 2014. "Garnet-Type Solid-State Fast Li Ion Conductors for Li Batteries: Critical Review". *Chemical Society Reviews* 43 (13): 4714–4727.

Thangadurai, V., D. Pinzaru, S. Narayanan, and A.K. Baral. 2015. "Fast Solid-State Li Ion Conducting Garnet-Type Structure Metal Oxides for Energy Storage". *The Journal of Physical Chemistry Letters* 6 (2): 292–299.

Thangadurai, V., and W. Weppner. 2002. "Solid State Lithium Ion Conductors: Design Considerations by Thermodynamic Approach". *Ionics* 8 (3–4): 281–292.

Ubramanian, M.A., R. Subramanian, and A. Clearfield. 1986. "Lithium Ion Conductors in the System AB (IV) 2 (PO4) 3 (B= Ti, Zr and Hf)". *Solid State Ionics* 18: 562–569.

ung-Fang Yu, Y., and J.T. Kummer. 1967 "Ion Exchange Properties of and Rates of Ionic Diffusion in Beta-Alumina". *Journal of Inorganic and Nuclear Chemistry* 29 (9): 2453–2475.

Vernoux, P., L. Lizarraga, M.N. Tsampas, F.M. Sapountzi, A. De Lucas-Consuegra, J.-L. Valverde, S. Souentie, et al. 2013 "Ionically Conducting Ceramics as Active Catalyst Supports". *Chemical Reviews* 113 (10): 8192–8260.

Wada, H., M. Menetrier, A. Levasseur, P. Hagenmuller. 1983. "Preparation and Ionic Conductivity of New B2S3-Li2S-LiI Glasses". *Materials Research Bulletin* 18 (2): 189–193.

Walther, F., R. Koerver, T. Fuchs, S. Ohno, J. Sann, M. Rohnke, W.G. Zeier, and J.R. Janek. 2019. "Visualization of the Interfacial Decomposition of Composite Cathodes in Argyrodite-Based All-Solid-State Batteries Using Time-of-Flight Secondary-Ion Mass Spectrometry". *Chemistry of Materials* 31 (10): 3745–3755.

Wang, C., K. Fu, S.P. Kammampata, D.W. McOwen, A.J. Samson, L. Zhang, G.T. Hitz, A.M. Nolan, E.D. Wachsman, and Y. Mo. 2020. "Garnet-Type Solid-State Electrolytes: Materials, Interfaces, and Batteries". *Chemical Reviews* 120 (10): 4257–4300.

Wang, C., K.R. Adair, J. Liang, X. Li, Y. Sun, X. Li, J. Wang, Q. Sun, F. Zhao, and X. Lin. 2019. "Solid-State Plastic Crystal Electrolytes: Effective Protection Interlayers for Sulfide-Based All-Solid-State Lithium Metal Batteries". *Advanced Functional Materials* 29 (26): 1900392.

Wang, L., Y. Wang, and Y. Xia. 2015. "A High Performance Lithium-Ion Sulfur Battery Based on a Li2S Cathode Using a Dual-Phase Electrolyte". *Energy & Environmental Science* 8 (5): 1551–1558.

Wang, S., R. Fang, Y. Li, Y. Liu, C. Xin, F.H. Richter, and C.-W. Nan. 2020a. "Interfacial Challenges for All-Solid-State Batteries Based on Sulfide Solid Electrolytes". *Journal of Materiomics* 7(2): 209–218.

Wang, S., X. Xu, X. Zhang, C. Xin, B. Xu, L. Li, Y.-H. Lin, Y. Shen, B. Li, and, C.-W. Nan. 2019. "High-Performance Li 6 PS 5 Cl-Based All-Solid-State Lithium-Ion Batteries". *Journal of Materials Chemistry A* 7 (31): 18612–18618.

Wang, S., X. Zhang, S. Liu, C. Xin, C. Xue, F. Richter, L. Li, et al. 2020b. "High-Conductivity Free-Standing Li6PS5Cl/Poly(Vinylidene Difluoride) Composite Solid Electrolyte Membranes for Lithium-Ion Batteries". *Journal of Materiomics* 6 (1): 70–76.

Wang, S., Y. Zhang, X. Zhang, T. Liu, Y.-H. Lin, Y. Shen, L. Li, and C.-W. Nan. 2018. "High-Conductivity Argyrodite Li6PS5Cl Solid Electrolytes Prepared via Optimized Sintering Processes for All-Solid-State Lithium–Sulfur Batteries". *ACS Applied Materials & Interfaces* 10 (49): 42279–42285.

Wang, Z., B. Huang, H. Huang, L. Chen, R. Xue, and F. Wang. 1996. "Investigation of the Position of Li+ Ions in a Polyacrylonitrile-Based Electrolyte by Raman and Infrared Spectroscopy". *Electrochimica Acta* 41 (9): 1443–1446.

Wenzel, S., T. Leichtweiss, D. Krüger, J. Sann, and J. Janek. 2015. "Interphase Formation on Lithium Solid Electrolytes: An In Situ Approach to Study Interfacial Reactions by Photoelectron Spectroscopy". *Solid State Ionics* 278: 98–105.

Wenzel, S., S. Randau, T. Leichtweiß, D.A. Weber, J. Sann, W.G. Zeier, and J. Janek. 2016. "Direct Observation of the Interfacial Instability of the Fast Ionic Conductor Li10GeP2S12 at the Lithium Metal Anode". *Chemistry of Materials* 28 (7): 2400–2407.

Weppner, W. 1981. "Trends in New Materials for Solid Electrolytes and Electrodes". *Solid State Ionics* 5: 3–8.

Wu, M., B. Xu, and C. Ouyang. 2015. "Physics of Electron and Lithium-Ion Transport in Electrode Materials for Li-Ion Batteries". *Chinese Physics B* 25 (1): 018206.

Wu, Z., Z. Xie, A. Yoshida, Z. Wang, X. Hao, A. Abudula, and G. Guan. 2019. "Utmost Limits of Various Solid Electrolytes in All-Solid-State Lithium Batteries: A Critical Review". *Renewable and Sustainable Energy Reviews* 109: 367–385.

Xiao, Y., Y. Wang, S.-H. Bo, J.C. Kim, L.J., Miara, and G. Ceder. 2020. "Understanding Interface Stability in Solid-State Batteries". *Nature Reviews Materials* 5 (2): 105–126.

Xu, K. 2004. "Nonaqueous Liquid Electrolytes for Lithium-Based Rechargeable Batteries". *Chemical Reviews* 104 (10): 4303–4418.

Xu, L., S. Tang, Y. Cheng, K. Wang, J. Liang, C. Liu, Y.-C. Cao, F. Wei, and L. Mai. 2018. "Interfaces in Solid-State Lithium Batteries". *Joule* 2 (10): 1991–2015.

Xu, R.C., X.H. Xia, S.Z. Zhang, D. Xie, X.L. Wang, and J.P. Tu. 2018. "Interfacial Challenges and Progress for Inorganic All-Solid-State Lithium Batteries". *Electrochimica Acta* 284: 177–187.

Yamada, T., S. Ito, R. Omoda, T. Watanabe, Y. Aihara, M. Agostini, U. Ulissi, J. Hassoun, and B Scrosati. 2015. "All Solid-State Lithium–Sulfur Battery Using a Glass-Type P2S5–Li2S Electrolyte: Benefits on Anode Kinetics". *Journal of The Electrochemical Society* 162 (4): A646.

Yamane, H., M. Shibata, Y. Shimane, T. Junke, Y. Seino, S. Adams, K. Minami, A. Hayashi, and M. Tatsumisago. 2007. "Crystal Structure of a Superionic Conductor, Li7P3S11". *Solid State Ionics* 178: 1163–1167.

Yang, C., K. Fu, Y. Zhang, E. Hitz, and L. Hu. 2017. "Protected Lithium-Metal Anodes in Batteries: From Liquid to Solid". *Advanced Materials* 29 (36): 1701169.

Yokokawa, H. 2016. "Thermodynamic Stability of Sulfide Electrolyte/Oxide Electrode Interface in Solid-State Lithium Batteries". *Solid State Ionics* 285: 126–135.

Yu, C., S. Ganapathy, N.J.J. de Klerk, I. Roslon, E.R.H. van Eck, A.P.M. Kentgens, and M. Wagemaker. 2016. "Unravelling Li-Ion Transport from Picoseconds to Seconds: Bulk versus Interfaces in an Argyrodite Li6PS5Cl–Li2S All-Solid-State Li-Ion Battery". *Journal of the American Chemical Society* 138 (35): 11192–11201.

Yu, C., S. Ganapathy, E.R.H. van Eck, L. van Eijck, S. Basak, Y. Liu, L. Zhang, H. Zandbergen, W. Henny, and M. Wagemaker. 2017. "Revealing the Relation between the Structure, Li-Ion Conductivity and Solid-State Battery Performance of the Argyrodite Li6PS5Br Solid Electrolyte". *Journal of Materials Chemistry A* 5 (40): 21178–21188.

Yu, C., J. Hageman, S. Ganapathy, L. Eijck, L. Zhang, K. Adair, X. Sun, and M. Wagemaker. 2019. "Tailoring Li6PS5Br Ionic Conductivity and Understanding of Its Role in Cathode Mixtures for High Performance All-Solid-State Li-S Batteries". *Journal of Materials Chemistry A* 7 (17): 10412–10421.

Yu, C., Y. Li, M. Willans, Y. Zhao, K.R. Adair, F. Zhao, W. Li, S. Deng, J. Liang, and M.N. Banis 2020. "Superionic Conductivity in Lithium Argyrodite Solid-State Electrolyte by Controlled Cl-Doping". *Nano Energy* 69: 104396.

Yu, C., L. van Eijck, S. Ganapathy, and M. Wagemaker. 2016. "Synthesis, Structure and Electrochemical Performance of the Argyrodite Li6PS5Cl Solid Electrolyte for Li-Ion Solid State Batteries". *Electrochimica Acta* 215: 93–99.

Yu, X., Z. Bi, F. Zhao, and A. Manthiram. 2015. "Hybrid Lithium–Sulfur Batteries with a Solid Electrolyte Membrane and Lithium Polysulfide Catholyte". *ACS Applied Materials & Interfaces* 7 (30): 16625–16631.

Yubuchi, S., M. Uematsu, M. Deguchi, A. Hayashi, and M. Tatsumisago. 2018. "Lithium-Ion-Conducting Argyrodite-Type Li6PS5X (X = Cl, Br, I) Solid Electrolytes Prepared by a Liquid-Phase Technique Using Ethanol as a Solvent". *ACS Applied Energy Materials* 1 (8): 3622–3629.

Yue, J., M. Yan, Y.X. Yin, and Y.G. Guo. 2018. "Progress of the Interface Design in All-Solid-State Li–S Batteries". *Advanced Functional Materials* 28 (38): 1707533.

Zhang, H., C. Li, M. Piszcz, E. Coya, T. Rojo, L.M. Rodriguez-Martinez, M. Armand, and Z. Zhou. 2017. "Single Lithium-Ion Conducting Solid Polymer Electrolytes: Advances and Perspectives". *Chemical Society Reviews* 46 (3): 797–815.

Zhang, H., T. Yang, X. Wu, Y. Zhou, C. Yang, T. Zhu, and R. Dong. 2015. "Using Li+ as the Electrochemical Messenger to Fabricate an Aqueous Rechargeable Zn–Cu Battery". *Chemical Communications* 51 (34): 7294–7297.

Zhang, J., L. Li, C. Zheng, Y. Xia, Y. Gan, H. Huang, C. Liang, X. He, X. Tao, and W. Zhang. 2020a. "Silicon-Doped Argyrodite Solid Electrolyte Li6PS5I with Improved Ionic Conductivity and Interfacial Compatibility for High-Performance All-Solid-State Lithium Batteries". *ACS Applied Materials & Interfaces* 12 (37): 41538–41545.

Zhang, J., C. Zheng, L. Li, Y. Xia, H. Huang, Y. Gan, C. Liang, X. He, X. Tao, and W. Zhang. 2020b. "Unraveling the Intra and Intercycle Interfacial Evolution of Li6PS5Cl-Based All-Solid-State Lithium Batteries". *Advanced Energy Materials* 10 (4): 1903311.

Zhang, Q., D. Cao, Y. Ma, A. Natan, P. Aurora, and H. Zhu. 2019. "Sulfide-Based Solid-State Electrolytes: Synthesis, Stability, and Potential for All-Solid-State Batteries". *Advanced Materials* 31 (44): 1901131.

Zhang, W., T. Leichtweiß, S.P. Culver, R. Koerver, D. Das, D.A. Weber, W.G. Zeier, and J.R. Janek. 2017. "The Detrimental Effects of Carbon Additives in Li10GeP2S12-Based Solid-State Batteries". *ACS Applied Materials & Interfaces* 9 (41): 35888–35896.

Zhang, X.-Q., X.-B. Cheng, and Q. Zhang. 2018. "Advances in Interfaces between Li Metal Anode and Electrolyte". *Advanced Materials Interfaces* 5 (2): 1701097.

Zhang, Y., R. Chen, T. Liu, B. Xu, X. Zhang, L. Li, Y. Lin, C.-W. Nan, and Y. Shen. 2018. "High Capacity and Superior Cyclic Performances of All-Solid-State Lithium Batteries Enabled by a Glass-Ceramics Solo". *ACS Applied Materials & Interfaces* 10 (12): 10029–10035.

Zhang, Z., L. Zhang, X. Yan, H. Wang, Y. Liu, C. Yu, X. Cao, L. van Eijck, and B. Wen. 2019. "All-in-One Improvement toward Li6PS5Br-Based Solid Electrolytes Triggered by Compositional Tune". *Journal of Power Sources* 410-411: 162–170.

Zhang, Z., Y. Shao, B. Lotsch, Y.-S. Hu, H. Li, J. Janek, L.F. Nazar, C.-W. Nan, J. Maier, and M. Armand. 2018. "New Horizons for Inorganic Solid State Ion Conductors". *Energy & Environmental Science* 11 (8): 1945–1976.

Zhao, Y., C. Wu, G. Peng, X. Chen, X. Yao, Y. Bai, F. Wu, S. Chen, and X. Xu. 2016. "A new Solid Polymer Electrolyte Incorporating Li10GeP2S12 into a Polyethylene Oxide Matrix for All-Solid-State Lithium Batteries". *Journal of Power Sources* 301: 47–53.

Zhou, P., J. Wang, F. Cheng, F. Li, and J. Chen. 2016. "A Solid Lithium Superionic Conductor Li11AlP2S12 with a Thio-LISICON Analogous Structure". *Chemical Communications* 52 (36): 6091–6094.

7

Recent Advances in Usage of Cobalt Oxide Nanomaterials as Electrode Material for Supercapacitors

Jude N. Udeh[1], Raphael M. Obodo[1,2,3], Agnes C. Nkele[1], Assumpta C. Nwanya[1,4,5], Paul M. Ejikeme[6], and Fabian I. Ezema[1,4,5]

[1]*Department of Physics and Astronomy, University of Nigeria, Nsukka, Enugu State, Nigeria*
[2]*National Center for Physics, Islamabad, Pakistan*
[3]*NPU-NCP Joint International Research Center on Advanced Nanomaterials and Defects Engineering, Northwestern Polytechnical University, Xi'an, China*
[4]*Nanosciences African Network (NANOAFNET) iThemba LABS-National Research Foundation, Somerset West, Western Cape Province, South Africa*
[5]*UNESCO-UNISA Africa Chair in Nanosciences/Nanotechnology, College of Graduate Studies, University of South Africa (UNISA), Muckleneuk Ridge, Pretoria, South Africa*
[6]*Department of Pure and Industrial Chemistry, University of Nigeria, Nsukka, Enugu Stete, Nigeria*

7.1 Introduction

Recently, energy demands have been growing rapidly due to increasing population and industrial growth at large. With respect to research carried out by EIA, fossil fuels (natural gas, petroleum, coal, and others) establish the immensity of the recent global principal energy basis (Ho et al. 2014; Obodo et al. 2020a). Too much reliance on fossil fuels has brought about rapid greenhouse effect, escalating lubricant tariffs, and fossil oil reduction, which therefore have led to worldwide economic disasters and environmental problems. For that reason, developing renewable energy storage devices will be the need for the hour to achieve steady power and reduce the rate of reliance on fossil fuels. The starting of the 21st century saw an expansion within the developing areas of nanoscience and nanotechnology (Hulla et al. 2015) for provision of alternative energy storage devices for power consumption. Therefore, synthesizing nanoparticles, thin films, and developing energy capacity gadgets such as energy storage devices (photovoltaic cells, li-ion batteries, supercapacitors, etc.) are presently the leading ways to check the deficiency of control and vitality request for industrial, mechanical, and residential applications around the world. Energy is part of the largest economic expansion, development, and growth, including poverty abolition and security of any nation (Oyedepo 2012). The growing need for energy has motivated the search for energy storage devices with elevated energy densities and power (Deng et al. 2014) such as supercapacitors and other devices for energy storage. The design and amalgamation of nanoscale materials are the key development of the 21st century in a broad run of applications in various areas such as catalysis, vitality capacity and change gadgets, biosensors, and biomedical applications by utilizing metal oxides, metal carbon allotropes, and chalcogenides (Theerthagiri et al. 2019). Supercapacitor is an electronic device that stores large amount of electric charge. Supercapacitor is a modern energy-saving and conversion apparatus that has the potential of tall control thickness, good

DOI: 10.1201/9781003145585-7

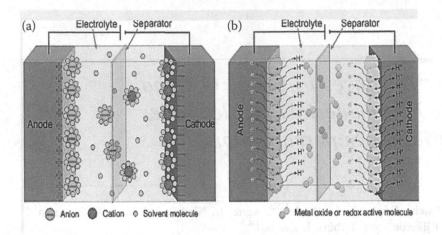

FIGURE 7.1 (a) shows the schematic representation of EDLC while that of (b) shows the schematic representation of pseudocapacitor (Chen et al. 2017). Reprinted with permission from Ref. Chen et al. (2017), copyright 2017, Oxford University Press.

circulation, quick discharge-charge, destitute self-discharging, secured working, and more fetched (Obodo et al. 2019). Supercapacitors store large amount of electric charge compared to the electrolytic capacitors and all other types of conventional capacitors. Current years have generated lots of progress in the practical and theoretical search for high production of supercapacitors for use both in industrial and domestic applications. Presently, supercapacitors are broadly used in industrial power systems, memory backup systems, consumer electronics, and energy management (Miller and Burke 2008; Zhang and Zhao 2009). The supercapacitor comprises of huge surface area electrodes and exceptionally clean dielectric, which makes it conccivable to attain exceptionally huge capacitance (large charge storage). Based on quite a few mechanisms of energy saving, supercapacitors are grouped into two classes together with pseudocapacitors and electric double-layer capacitors (EDLCs) (Zhang and Zhao 2009; Obodo et al. 2019a). In principle, EDLCs stockpile power by solutions of ions produced by electrostatic communications in excess of high surface specific area of the vigorous material electrode, while the energy storing device of pseudocapacitors is a rapid reversible electrochemical occurrence (Hu et al. 2020). Figure 7.1a–b shows the schematic diagram of supercapacitors (Chen et al. 2017). Furthermore, the pseudocapacitor stores electrical power or energy by the transfer of electrons within the terminals and electrolyte.

In this chapter, recent usage and development of Co_3O_4 nanomaterials as electrodc material for production of supercapacitors is summarized. The performance of supercapacitors in general, Co_3O_4 electrode materials, and the synthesis methods are presented, including the coprecipitation method, sol gel method, hydrothermal method, electrode position, and chemical bath deposition method. Also, cobalt oxide–containing electrode materials are summarized, as well as cobalt oxide/carbon, cobalt oxide/conducting polymer, and metal/cobalt oxide composite compounds.

7.2 Theoretical Overview of Supercapacitors

7.2.1 Supercapacitor Performance

Capacitor is an electrical module or component that stockpiles electric charge. In addition, It is also an instrument that stores electrical power or energy in the form of electrical charge and can distribute the stored charge on request within the frame of an electric current (Obodo et al. 2019b) for consumption purposes. In general, capacitor as an energy storage device is made of two close conductors (called *plates*) that are divided by an insulating (dielectric) material. The plates stores electric charges when connected to the power supply or source. One of the plates stores positive charge and the second stores negative charge. Capacitance of any capacitor is the measure of the quantity of charges that it stockpiles in it at voltage of 1 volt. In general, the capacitance of any capacitor can be obtained using

$$C = \frac{Q}{V} \qquad (7.1)$$

where C stands for the capacitance of capacitor which is measured in farad (F), Q is the quantity of charge that is measured in coulombs. The total capacitance of a capacitor is given as the permittivity (ε) times the plate area (A) over the gap amid the plates (d) as shown in equation (7.2):

$$C = \frac{\varepsilon A}{d} \qquad (7.2)$$

where V is the voltage, A stands for the area of the capacitor measured in meter square (m^2), d is known as the distance between the capacitor's plates and it is measured in meters (m). The two major features of many capacitor are their power and energy densities (Obodo et al. 2020b). The amount of energy stored in a capacitor can be calculated using equation (7.3)

$$E = \frac{1}{2}CV2 \qquad (7.3)$$

Conservative capacitors contain moderately lofty power density and moderately low energy density in contrast to electrochemical cells and fossil fuels. Specifically, batteries stock up extra energy when compared to capacitors, but cannot disseminate it rapidly because of its low power density. A supercapacitor is a sort of capacitor that stores huge amount of energy and power, regularly 10 to 100 times more energy when compared to electrolytic type of capacitors. Supercapacitors communicate with electric gadgets that work in the course of the electrostatic combinations of particles on the surface of electrodes (Wang et al. 2016).

7.3 Electrode Materials

The formulation of electrode material that has excellent electrical conductivity property has generated a lot of research interest, since the vital factor that has a very big effect on the functioning of the supercapacitor is the material used to fabricate the electrode. According to the electrode materials used, capacitors are classified as carbon-based capacitors, metal oxide capacitors, and conductive polymer capacitors. These electrode materials need to have good adsorption properties, elevated excellent electrical conductivity, and good specific surface area. Carbon-based electrode resources have the above advantages, and the raw materials are rich and cheap, the preparation process is simple, the pore state is easy to adjust, the chemical properties are stable, the thermal and electrical conductivity is good, the specific surface area is high, the cycle performance is excellent, and it can be used as the composite of the substrate, metal oxide, and conductive polymer, and it is the most widely used and the most mature commercial supercapacitor electrode material in the present century.

In the present century, carbon instruments are the main materials for research in industrial and other related applications of EDLCs (Xuelei Wang et al. 2020). Therefore, familiar electrode devices that are best for EDLCs are carbon materials (for example, activated carbon (Luo et al. 2015), carbon aerogel (Hao et al. 2014), carbon fibre (Chen et al. 2014), carbon nanotube (De Oliveira and De Oliveira 2014), porous carbon (Xiao et al. 2017), and grapheme (Li et al. 2014)), that demonstrate the rapid operation of the storage ability (Hu et al. 2020) and long life cycle. Carbon fabric is viewed as the best device or electrode material for EDLC sowing for good electrochemical operations, which comprise flexibility, good electrical and thermal conductivity, controllable porous structures, big specific surface area, and stretchability.

For example, Ji and coworkers (2014) worked on capacitance of carbon-based EDLCs. They discovered that the capacitance of the carbon-based EDLCs is overpowered close to impartiality, and is irregularly improved for thickness underneath very few films. It was found out that the bulky capacitance results obtained in their analysis show that the gravimetric power storage density within the solitary coated graphene limits are analogous to that of the battery. They anticipated that the findings obtained in their experimental analysis explained the idea of developing the latest theoretical prototypes in understanding the EDLCs of carbon devices and creating modern methodologies for enhancing energy and power densities of carbon-containing storage devices such as capacitors. A lot of preparation procedures of carbon-containing compounds for supercapacitors were reported Zhang et al. (2016). Zhang and his group reviewed the challenges of developing carbon-based SSCs and progress in the developing field. EDLCs have outstanding definite power densities, though the dreadful specific energy density is a serious shortcoming that restricts its practical application (Hu et al. 2020). Comparing with EDLCs, pseudocapacitors on the other hand have higher energy density and specific capacity. Mesoporous manganese dioxide/carbon (MnO_2/C) composites that have a lofty surface area of about $324.0 \text{ m}^2\text{g}^{-1}$ and spherical morphology were hydrothermally synthesized (Pan et al. 2014). The FESEM images of their result show a uniform sphere with average size in diameter of about 900 nm as in Figure 7.2a. A close look revealed that the surfaces of every single sphere are considerably uneven, which comprises enormous numbers of little particles as shown in Figure 7.2b.

It was discovered that the TEM images of manganese dioxide/carbon spheres which were in agreement with their FESEM results displayed solid sphere-shaped morphologies with homogeneous dimensions as shown in Figure 7.3.

Pen and his group revealed that the results of their electrochemical analysis show that the mesoporous manganese dioxide/carbon composites that have 72.860 wt.% of manganese dioxide demonstrate large specific capacitance of about 383.0 F g^{-1} at 2.0 mV s^{-1} which also display higher long-term recurring stability, with exact capacitance preservation 82.20% of the early assessment after 1000 cycles as shown in Figure 7.4 and Figure 7.5a and b, respectively.

In general, about 99% of electrode devices of pseudocapacitors are mostly transition metal oxides or hydrides (for example, RuO_2 (Ahn et al. 2007; Patake et al. 2009; Zhou et al. 2014), IRO_2 (Hu et al. 2002), MnO_2 (Jiang and Kucernak 2002; Yan and Fan 2010; Tadjer et al. 2014), NiO (Patil et al. 2008; Prasad and Miura 2013; Zheng et al. 2017), MoO (Patil et al. 2008; Zheng et al. 2017), Co_2O_3 (Tsukamoto et al. 2005; Kandalkar et al. 2008; Kandalkar et al. 2010; Al-Tuwirqi et al. 2011), $Ni(OH)_2$ (Ji et al. 2013), and V_2O_5 (Hu et al. 2008; Silva et al. 2008; Zhou et al. 2009), and electrically conducting polymers (for example, polyaniline (PANI) (Li et al. 2016) and polypyrrole (PPy) (Zhu et al. 2014)). Of the electrode materials itemized above, cobalt oxide (Co_3O_4) has the better appealing quality for electrode devices for supercapacitors attributed to its lofty outstanding electrochemical properties,

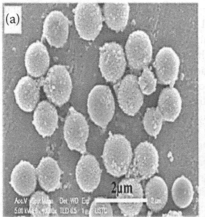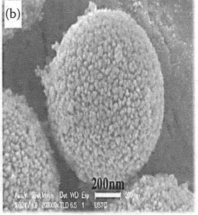

FIGURE 7.2 FESEM images of mesoporous manganese dioxide/carbon spheres (Pan et al. 2014). Reprinted with permission from Ref. Pan et al. (2014), copyright 2014, Elsevier.

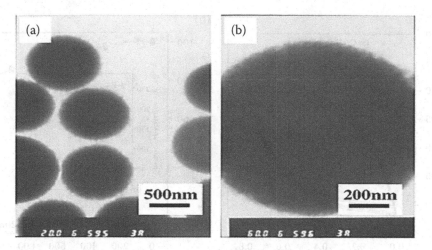

FIGURE 7.3 TEM images of mesoporous manganese dioxide/carbon spheres (Pan et al. 2014). Reprinted with permission from Ref. Pan et al. (2014), copyright 2014, Elsevier.

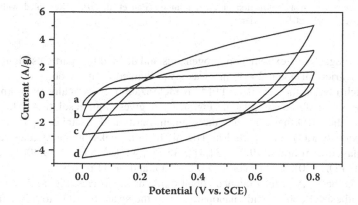

FIGURE 7.4 Cylic voltammetry curves of mesoporous manganese dioxide/carbon spheres of different scan rates (a) 2 mV/s; (b) 5 mV/s; (c) 10 mV/s; (d) 20 mV/s (Pan et al. 2014). Reprinted with permission from Ref. Pan et al., (2014), copyright 2014, Elsevier.

large hypothetical precise capacitance (3560 F/g), and high-quality invertibility. On the other hand, there are a number of disadvantages that restrict the view of supercapacitor performance, for example, laborious reaction kinetics, power density, and meagre cycle life (Wang et al. 2016; Obodo et al. 2019a). Current studies prove that Co_3O_4 could achieve excellent electrochemical performance by altering its morphology. Also, cobalt oxide (Co_3O_4) electrode material that possesses extraordinary morphology and microstructures has an excellent electrochemical capacitive behaviour (Guo et al. 2014). On the other hand, Co_3O_4 materials that serve as electrodes have shown low conductivity. Therefore, to eradicate the drawback of solitary electrode devices, the groundwork of cobalt oxide containing composite will attain a greater amalgamation functioning (Wang et al. 2014). In recent times, composites that are graphene based have been frequently researched (Lee et al. 2015), analyzed, and used as electrode devices for a variety of power storage apparatuses (Obodo et al. 2020b). However, 2D graphene porous framework of nanocomposites in supercapacitors is the hotbed of the current study (Yuan et al. 2015). When mixed with cobalt oxide nanomaterials, they display outstanding properties in asymmetric supercapacitors (Gao et al. 2018). Comparing to a particular part of graphene and cobalt oxide, cobalt oxide/graphene composites display superior exact capacitance, precise power, and definite energy (Xuelei Wang et al. 2020). For instance, cobalt oxide nanoparticles were obtained using low-temperature processes at a temperature range of 70°C (Obodo et al. 2020c). It was observed in their

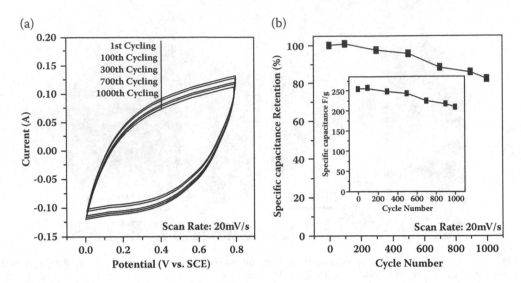

FIGURE 7.5   (a) Samples of MnO$_2$/C spheres that have various cycling numbers up to 100 at a scan of 20 mV/s. (b) Specific capacitance retention as a function of cycle number (Pan et al. 2014). Reprinted with permission from Ref. Pan et al. (2014), copyright 2014, Elsevier.

work that FESEM images show extremely homogeneous and defined few particles Figure 7.6a. Average particles volume invariety of about 20–30 nm was also obtained Figure 7.6b.

Figure 7.7a which reveals the details of TEM characterization gives analogous normal particles dimension in the series of about 20–30 nm. Figure 7.7b gives high resolution (HR) TEM images. Figure 7.8(a) shows the XRD that examined the segment purity and crystallographic formations of the obtained cobalt oxide (Co$_3$O$_4$) nanoparticles. The diffraction peaks of the synthesized cobalt oxide nanoparticles displayed as follows: 19.02°, 31.31°, 36.76°, 44.92°, 55.55°, 59.30°, and 65.22° which were allocated to (111), (220), (311), (400), (442), (511), and (411) planes.

Figure 7.8 reveals the TGA of the obtained cobalt oxide nanoparticles. Figure 7.8c reveals the FTIR spectrum of as-synthesized cobalt oxide nanoparticles in the space of 4000 to 400 cm^{-1}. Figure 7.8d displays absorbance bands as shown: 269 nm, 471 nm, and 767 nm, showing the two transition states. Synthesized cobalt oxide (Co$_3$O$_4$) nanoparticles–based electrode of their result displayed a good quantity of larger precise capacitance of 304 Fg^{-1} in one mole of potassium hydroxide solutions as

FIGURE 7.6   FESEM images at low (a) and high (b) magnification of the synthesized cobalt oxide (Co$_3$O$_4$) nanoparticles (Obodo et al. 2020a). Reprinted with permission from Ref. Obodo et al., (2020a), copyright 2020, Elsevier.

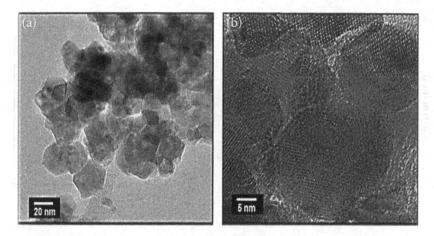

FIGURE 7.7 TEM (a) and HR TEM (b) images of the synthesized cobalt oxide (Co_3O_4) nanoparticles (Obodo et al. 2020a). Reprinted with permission from Ref. Obodo et al. (2020a), copyright 2020, Elsevier.

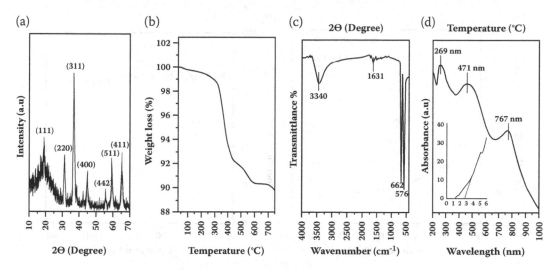

FIGURE 7.8 Shows (a) X-Ray Diffraction (XRD) Patterns, (b) TGA (FTIR spectrum of the synthesized cobalt oxide (Co_3O_4) nanoparticles) and (d) UV-vis spectroscopy of the synthesized cobalt oxide nanoparticles (Obodo et al. 2020a). Reprinted with permission from Ref. Obodo et al. (2020a), copyright 2020, Elsevier.

shown in Figure 7.9a–b. Their results revealed that the capacitance was significantly increased to 480.0 F/g at the time the working electrodes were achieved by combining cobalt oxide nanoparticles and super P carbon.

They reported that Co_3O_4 nanoparticles–based electrode materials reached the admirable cycling stability with the preservation proportion of about 88.6% immediately after 1000 cycles as shown in Figure 7.10a–b.

Li et al. (2018) synthesized superthin Co_3O_4 porous nanoflake via a simple natural biomembrance system. The result of their studies shows a remarkable electrochemical performance, which includes defined surface area of about 225 m^2g^{-1} and definite capacitance of 448 F g^{-1}. Deng (Obodo et al. 2020c) and his group reported 3D hierarchical flower and nanoneedle nanostructures of Co_3O_4 nano-materials by hydrothermal method. The results of their studies show that 3D hierarchical flower-like Co_3O_4 has large accurate capacitance (327.30 F/g within 0.50 A/g) in reference to nanoneedle-like Co_3O_4 with an outstanding preservation (96.070% within 5.0 A/g) for 10,000 cycles.

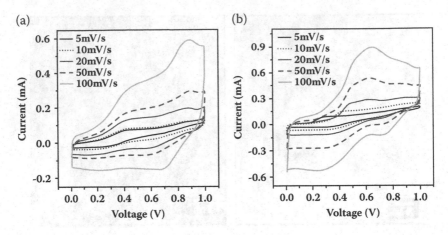

FIGURE 7.9 Cyclic voltammetry plots (a) of cobalt oxide (Co_3O_4) nanoparticles and (b) cobalt oxide (Co_3O_4) nano-particles with super-P carbon electrodes at different scan rates in 1M KOH electrode (Obodo et al. 2020a). Reprinted with permission from Ref. Obodo et al. (2020a), copyright 2020, Elsevier.

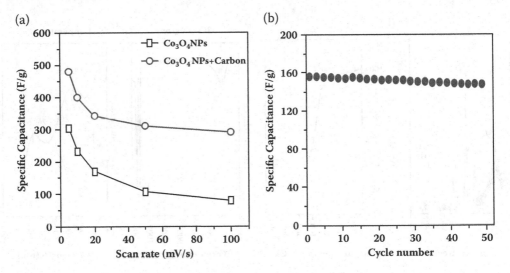

FIGURE 7.10 (a) shows the variation of specific capacitance of cobalt oxide (Co_3O_4) nanoparticles with super-P carbon at different scan rates while (b) shows the variation of specific capacitance over 50 cycles out of 1000 cycles (Obodo et al. 2020f). Reprinted with permission from Ref. Obodo et al., (2020d), copyright 2020, Elsevier.

7.4 Synthesis and Performance of Co_3O_4

7.4.1 Coprecipitation Method

Coprecipitaion is simply a means or a process by which a strong molecule (solid particle) is precipitated from a mixture or solution containing other ions (Leckie et al. 1980). Certainly, this process or method is typically comprehensible and inexpensive (Pudovkin et al. 2018). Coprecipitaion is an inexpensive, easy, and straightforward method, in which the size distribution and control of size are achieved by regulating the relative rates of formation of crystals and crystal growth during the amalgamation or synthesis procedures (Houshiar et al. 2014). Coprecipitaion reactions deal with the concurrent occurrence of crystal formation, coarsening, agglomeration, and growth processes (Rane et al. 2018). Typical coprecipitaion synthesis procedures or methods are formation of metals from aqueous mixtures or solutions, electrochemical reduction, reduction from non-aqueous solution, and decomposition of

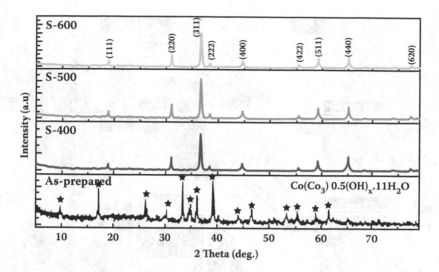

FIGURE 7.11 The results of X-ray diffraction (XRD) of as-prepared samples gotten at 90°C and annealed samples of S-600, S-500, and S-400 (Jadhav et al. 2014). Reprinted with permission from Ref. Jadhav et al. (2014), copyright 2012, Springer Science Business Media B.V.

metal-organic precursors. Many research works have been reported that coprecipitaion method is broadly used for synthesizing Co_3O_4 nanomaterials for supercapacitor applications as an energy storage device. For instance, spinel Co_3O_4 nanoparticles were obtained through chemical precipitation (Obodo et al. 2020f). Results obtained by Bashir et al. show that the screen-printed thin layer of p-type inorganic spinel of cobalt oxide in carbon PSCs presents a superior energy matching level, stability, and greater efficiency. Jadhav et al. (2014) used chemical coprecipitaion method to synthesize cobalt oxide, flower-like, and annealed at different temperatures of 400°C, 500°C, and 600°C for good two hours. The XRD pattern of their studies was obtained at 90°C from as-prepared sample and annealed samples of S-400, S-500, and S-600. It was revealed in their studies that all the diffraction peaks of the samples got at 90°C are well indexed with an intermediate phase of cobalt hydroxyl carbonate as shown in Figure 7.11.

Figure 7.12a, d, and g displayed the FESEM images of S-400, S-500, and S-600 samples correspondingly. The large magnification of FESEM images obtained reveals that the Co_3O_4 samples of different temperatures give flower-like structures that show a large number of porous nanorods. FE-TEM images of the samples S-400, S-500, and S-600 were also obtained, respectively, as shown in Figure 7.12b, e, and h. Figure 7.12 c, f, and i shows the HR-TEM images. The result of this analysis confirms unidirectional fringe patterns and gives lofty crystalline quality of all the samples.

Electrochemical investigation of their results showed that the cobalt oxide flower-like samples obtained at 500 C has an outstanding cycling stability (1536.9 mAh g^{-1} at 0.1 C up to 50 cycles and 1226.9 mAh g^{-1} at 0.5 C up to 350 cycles) and greater rate capability (845 mAh g^{-1} at 10 C) as shown in Figure 7.13a–b.

7.4.2 Hydrothermal Method

Hydrothermal method is simply defined as methods of synthesizing lone crystals which require solubility of raw materials in boiled water under a lofty force (Segal 1994; Pan et al. 2015). Nucleation is carried out in a steel pressure vessel also known as an *autoclave* in which the nutrient is made available along with water. This method that is broadly used to obtain nanostructured materials has the benefit of better and uncomplicated control of dimension, low-temperature reaction, crystal formation, and morphology. However, contraction of any molecules or particles undergoing reaction, and subjected to reaction temperature and pressure, will definitely have an effect on the size, type of the product, and morphology (Qiu et al. 2015) of the nanostructure. Furthermore, hydrothermal reaction method has been extensively used to obtain Co_3O_4 nanoparticles or thin films in many research works of the present

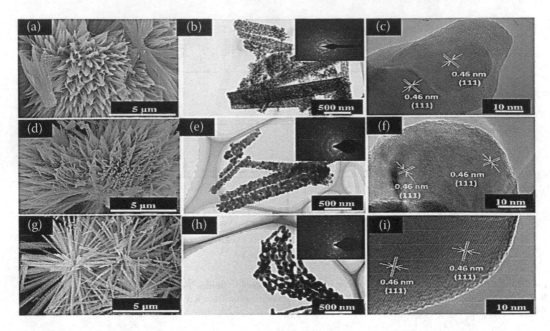

FIGURE 7.12 FE-SEM (a, d, and g), FE-EM (b, e, and h), and HR-TEM (c, f, and i) images of the samples corre-spondingly (Jadhav et al. 2014). Reprinted with permission from Ref. Jadhav et al. (2014), copyright 2012, Springer Science Business Media B.V.

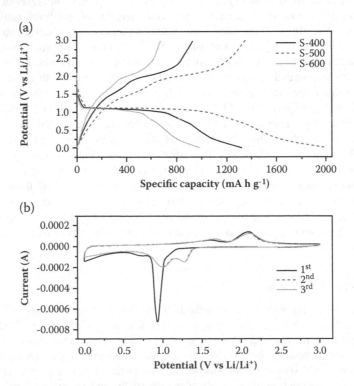

FIGURE 7.13 This figure reveals the electrochemical analysis of flower-like cobalt oxide (a) GCD voltages profiles for the samples S-600, S-500, and S-400, respectively. (b) CV for the three cycles (Jadhav et al. 2014). Reprinted with permission from Ref. Jadhav et al. (2014), copyright 2012, Springer Science Business Media B.V.

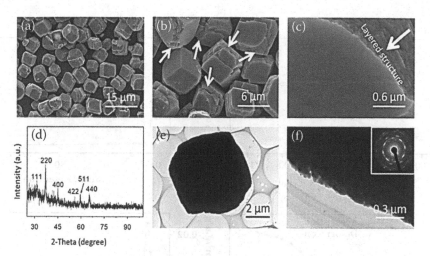

FIGURE 7.14 SEM images were shown in (a–c), (d) gives XRD pattern while (e) and (f) reveal the TEM images of porous cobalt oxide quasi-cubes. It is seen that (f) is for SAED pattern (Sun et al. 2019). Reprinted with permission from Ref. Sun et al., (2019), copyright 2012, Springer Nature.

century. Perhaps, novel cobalt oxide quasi- cubes alongside coated formation was achieved through hydrothermal method in company with egg albumin, and then the reactions were quickly annealed at 300°C to be transformed into uncontaminated cobalt oxide fine particles (Sun et al. 2019). Figure 7.14a gives SEM image. It reveals that the sample is composed mainly of big amount of cubic-like particles of sizes 5–6 μm. Figure 7.14b SEM image revealed that some edges of the cubes were not perfect and gives layered structures as indicted with the white arrows. The SEM image shown in Figure 7.14c gives the clarity of the layered structures. The XRD test which was carried out in their work as shown in Figure 7.14d gives the diffraction peaks. Figure 7.14e reveals cobalt oxide cube with a size of about 5 μm and the size shows nice agreement with the SEM data. Magnified TEM images were obtained as shown in Figure 7.14f.

Figure 7.15a gives the SEM image. Their work also revealed that when no egg albumin was used the end product gives a lot of cobalt oxide nanosheets and porous structures will be transparently seen in the TEM images which will be crystalline as shown in Figure 7.15b–c. Figure 7.15d, e, and f gives the structure of porous cobalt oxide with egg albumin.

FIGURE 7.15 (a) gives the SEM image, (b and c) give the TEM images of cobalt oxide nanosheets obtained without any egg albumin while (d, e, and f) reveal the TEM images of the samples prepared with egg albumin (Sun et al. 2019). Reprinted with permission from Ref. Sun et al., (2019), copyright 2012, Springer Nature.

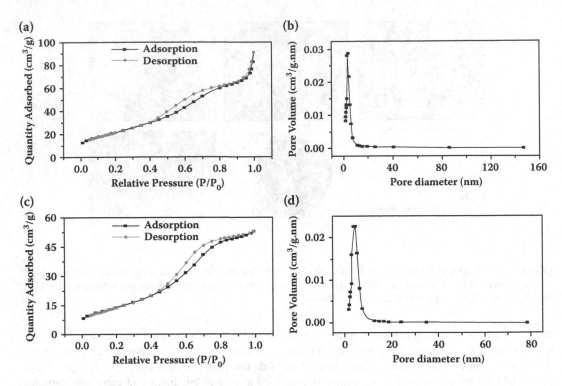

FIGURE 7.16 Above shows N_2 adsorption-desorption isotherms and the corresponding BJH pore size distribution for (a,b) porous Co_3O_4 curves and (c,d) porous Co_3O_4 nanosheets, respectively (Sun et al. 2019). Reprinted with permission from Ref. Sun et al. (2019), copyright 2012, Springer Nature.

The properties of cobalt oxide nanocubes were revealed by Sun and coworkers as shown in Figure 7.16a. Figure 7.16b shows the pore size distribution gotten by BJH method. The isotherms of cobalt oxide nanosheets were displayed in Figure 7.16c; it was analogous to the isotherms of nanocubes as shown in Figure 7.16a above. Figure 7.16d reveals that the pore diameter of cobalt oxide nanosheets is 4.44 nm.

Sun and his group revealed that layered cobalt oxide cubes have a mesoporous nature with specific surface area of 80.3m^2/g and a mean pore size of 5.58 nm (Figure 7.16). The results of their electrochemical analysis specified that precise capacitance of 754.0 F g^{-1} within 1 A g^{-1} was obtained as shown in Figure 7.17a–f.

Song et al. (2013) reported the synthesis of cobalt oxide/reduced composites of graphene oxide nanosheets (Co_3O_4/rGONS) through a simple hydrothermal process. Song et al. revealed that the electrochemical analysis of their experimental work were determined using cyclic voltammetry and GCD in 1 mole of potassium hydroxide in aqueous solution. The Co_3O_4/rGONS-II displays better and excellent coulomb efficiency and cyclic performance that have an exact capacitance of 400 F/g between current density of 0.5 A and 2.0 A. Synthesis of one-dimensional electroactive material of Co_3O_4 nanorods by hydrothermal method was reported by V. Venkatachalam and coworkers (2018). The results of their analysis shows that the pore diameter of rod-like Co_3O_4 is 3.45 nm and particular surface area is 38.6 m^2/g. It was also shown in their results that the definite capacitance of the modified electrode of Co_3O_4 is 655 F/g at 0.5 A/g and the retention rate of the capacitance reaches 82.7% after about 1000 cycles.

Wang and his group synthesized highly ultra-thin mesoporous Co_3O_4 nanosheets on Ni foam working electrode using hydrothermal method (Wang et al. 2016). Their results reveal the capacitance of Co_3O_4 nanosheets electrode materials gave 610 mF/cm^2 at a current density of 1 mA/cm^2. Their results also produce good cycling stability of about 95% immediately after 3000 cycles within the charge density of 4.0 mA/cm^2. Furthermore, they revealed that an asymmetric supercapacitor electrode material found on

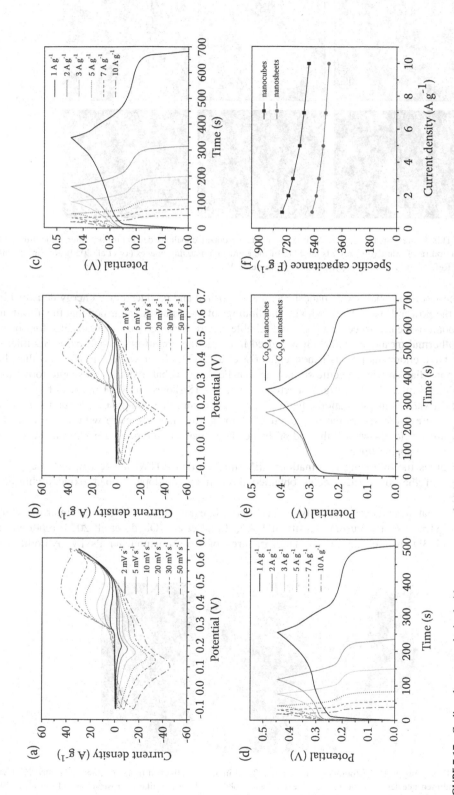

FIGURE 7.17 Cyclic voltammetry obtained with many scan rates for (a) porous cobalt oxide cubes and (b) for cobalt oxide nanosheets. (c and b) give CP curves for porous cobalt oxide cubes and porous cobalt oxide nanosheets. (e) shows the two electrodes CP curves obtained and (f) gives the specific capacitance (Sun et al. 2019). Reprinted with permission from Ref. Sun et al. (2019), copyright 2012, Springer Nature.

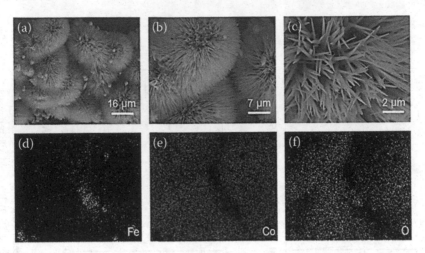

FIGURE 7.18 This figure displayed the SEM images of the Fe-doped cobalt oxide urchin-like microstructures with various magnification (a-c), while (d-e) reveals the EDS of the elemental mapping images (Li et al. 2020). Reprinted with permission from Ref. Li et al. (2020), copyright 2019, Elsevier B.V.

the ultrathin mesoporous cobalt oxide nanosheets was deposited and produced nice energy density 136 Wh/kg within the power density 0.750 Wh/kg. The findings of their studies confirmed that the ultra-thin mesoporous cobalt oxide nanosheet is a good and reliable electrode device for supercapacitor for power storage. Hydrothermal means of extraction was used in extracting urchin-like cobalt oxide 3D microstructures that have enormous porous nanoneedles (Li et al. 2020). Their work demonstrated that the BET average pore diameter and specific surface area of Fe-doped cobalt oxide microscopic formation were found to be 18 nm and 39.7 m^2/g, respectively. Figure 7.18 shows the SEM images of Fe-doped cobalt oxide with various magnifications. Figure 7.18a has 20 mm that comprises various urchin-like microspheres. Figure 7.18b shows the magnified SEM images of microstructure with urchin-like microspheres. Figure 7.18(c) shows that the end of the needle is nice and sharp. Figure 7.18d and e shows the EDS elemental mapping images.

Figure 7.19 gives the following information: TEM image (a), HRTEM image (b), and magnified HRTEM image of the combined needles obtained through the Fe-doped cobalt oxide urchin-like microstructures.

It was found that the precise capability of the urchin-like Fe-doped cobalt oxide microstructure electrode gives 315.8 C/g at a current density at 1 A/g. Deng ct al., (Obodo et al. 2020) reported on nanoneedles and 3D hierarchical flower nanostructures of Co_3O_4 nanomaterials by hydrothermal

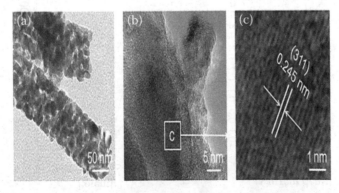

FIGURE 7.19 This figure gives the following information: TEM image (a), HRTEM image (b), and magnified HRTEM image of the combined needles obtained through the Fe-doped cobalt oxide urchin-like microstructures (Li et al. 2020). Reprinted with permission from Ref. Li et al. (2020), copyright 2019, Elsevier B.V.

method. Their work disclosed that 3D hierarchical cobalt oxide flower-like nanostructures have superior capacitance (327.3 F/g at 0.5 A/g) when compared with nanoneedle-like Co_3O_4. Figure 7.20a shows the cyclic voltammetry curves of Fe-doped cobalt oxide electrode with various scan rates, Figure 7.20b gives the cyclic voltammetry curve of pure cobalt oxide with various scan rates, Figure 7.20c reveals the galvanostatic charge/discharge of Fe-doped cobalt oxide electrode, Figure 7.20d GCD of pure cobalt oxide electrode at various current densities, Figure 7.20e gives the GCD of the combined Fe-doped and pure cobalt oxide electrodes at 1 A/G and Figure 7.20f reveals the specific capacitance of the combined electrodes at various current densities.

Bazrafshan and coworkers (2017) synthesized Co_3O_4 nanosheets via hydrothermal extraction by using $Co(NO_3)_2.6H_2O$. The results of their study indicated that Co_3O_4 nanosheets presented high efficiency of 0.92% for water electrolysis beneath replicated 1.5 worldwide sunshine air mass that further suggests the excellent potential of Co_3O_4 nanosheets for application in hydrogen generation.

7.4.3 Sol Gel Method

Sol gel method of extraction of nanoparticles or thin films is a significant technique that has been widely used for synthesizing transition metals oxide such as Co_3O_4 electrode materials and many others. In this method of extraction, the reagents are uniformly combined in the solution segment to produce clear and stable sol via hydrolysis and condensation means, and then aging it to produce gel (A, Georgea, and Eliasb 2015). Sol gel process is a wet chemical procedure often called *chemical solution deposition* that involves numerous stages, in the following sequential arrangements: densification, gelation, aging, drying crystallization, hydrolysis, and polycondensation (Neacşu et al. 2016) as shown in Figure 7.21. The chronological steps for synthesis using sol gel method comprise hydrolysis and condensation of molecular precursor, followed by gelation based on condensation of the aqueous solution of molecular metal chelates, aging time, drying, and crystallization as shown in Figure 7.3. Sol gel method of extraction has many advantages which includes smaller size, ease of operation, low temperature reaction, uniform molecular-level mixing, and miscellaneous (Pudukudy and Yaakob 2014).

Guo and coworkers synthesized mesoporous Co_3O_4 octahedra using sol gel process for electrode device for the application of energy storage devices such as lithum-iron battery (Guo et al. 2014). Guo and coworkers revealed that the electrochemical operations of their study exhibits remarkable lofty changeable capability (1195 mAh/g within the rate of 200 mA/g), excellent capability preservation, and outstanding charge ability (681 mAh/g within elevated charge of 2 A/g). Lima-Tenório and coworkers (2018) presented the synthesis of nanoparticles of Co_3O_4 using sol gel method with no secondary phase modification. The studies of composite compound capacitance and composite compound power of their work show that the capacitive and resistive uniqueness with high precise capacitance of 120 F/g. Devi et al. (2019) worked on lithium-ion battery and high operation of supercapattery storage device using nanostructured spinel cobalt oxide obtained through modified facile via sol gel process. Development of spinel cobalt oxide was obtained by materials characterization. They disclosed that the electrochemical studies of Co_3O_4 in potassium hydroxide disclosed superb redox features. Supercapattery apparatus of CR-2032 coincell comprising of the cobalt oxide as positive electrode material and reduced graphene-oxide as negative electrode produced superior energy density of 40 Whk/g at power density of 742 Wkg/g. Lakehal and his group (2018) worked on Co_3O_4 electrode materials using sol gel method. The results obtained from their experimental analysis demonstrated that the resistance of cobalt oxide polycrystalline reduced. Peterson and coworkers (2014) worked on the synthesis of Co_3O_4 using sol gel process and contrasted the operations of network-like, plate-like, sphere-like, and sponge-like forms of cobalt oxide. They revealed that the network-like material displayed superior capacitance of about 708.0 F/g within 5.0 mV/s and excellent charge capacity of 71.90% within 50 mV/s.

7.4.4 Chemical Bath Deposition Method (CBD)

Chemical bath deposition is simply the chemical reduction route and one of the easiest methods of extraction of nanoparticles and thin films. CBD method has been frequently engaged in fabricating thin films and nanoparticles of different materials (Ezekoye et al. 2012). Among the chemical methods of

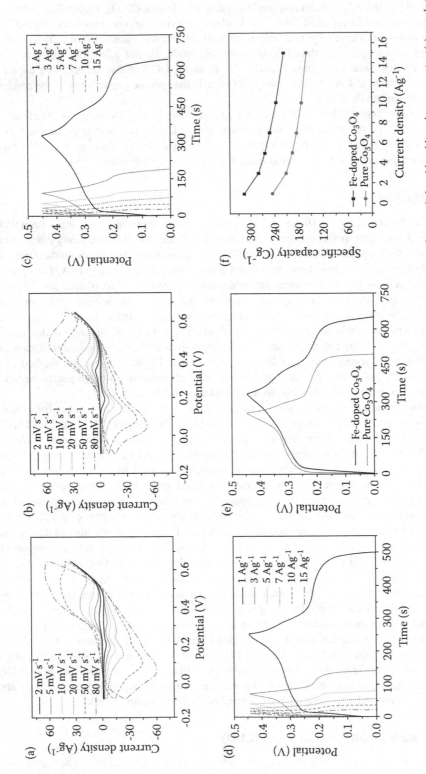

FIGURE 7.20 (a) shows the cyclic voltammetry curves of Fe-doped cobalt oxide electrode, (b) cyclic voltammetry curve of pure cobalt oxide with various scan rates, (c) reveals the galvanostatic charge/discharge of Fe-doped cobalt oxide electrode, (d) GCD of pure cobalt oxide electrode at various current densities, (e) gives the GCD of the combined Fe-doped and pure cobalt oxide electrodes at 1 A/G and (f) reveals the specific capacitance of the combined electrodes at various current densities (Li et al. 2020). Reprinted with permission from Ref. Li et al. (2020), copyright 2019, Elsevier B.V.

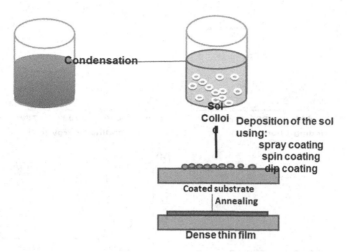

FIGURE 7.21 Sol gel techniques.

deposition of nanoparticles and thin films, CBD is the simplest (Mugle and Jadhav 2016) and best method to synthesize and fabricate cobalt oxide (Co_3O_4) for supercacitors and photovoltaic cell applications as energy storage devices. For example, Yadav et al. (2018) successfully prepared cobalt oxide (Co_3O_4) nanowire using CBD method as a means of fabricating nanoparticles on electrode materials. Figure 7.21a gives the results of XRD pattern analysis of Co_3O_4 films in their studies. The analysis shows the formation of polycrystalline Co_3O_4 with cubic structure and various peaks. Yadav and coworkers revealed that satellite peaks were obtained at 78.2, 798.1, and 803.3 in Figure 7.22a. Figure 7.23b also reveals the spectrum peaks of the various samples as seen below. Figure 7.24a shows various scan rates of cyclic voltammetry curves, Figure 7.24b shows variation of specific capacitance with various scan rates, while that of Figure 7.24c shows GCD curves at various current densities of cobalt oxide thin films.

Electrochemical analysis of their work revealed that Co_3O_4 nanowire delivers better specific surface area of about 66.33 m^2/g. Co_3O_4 nanowires displayed excellent electrochemical performance that gives lofty specific capacitance of 850 F/g with scan rate of 5 mV/s. Yadav and coworkers confirmed that cobalt oxide nanowires displayed an excellent long-term cycling stability of 86% immediately after 5000 cyclic voltammetry cycles. Furthermore, they fabricated symmetric supercapacitor by

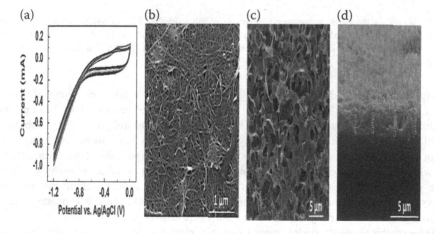

FIGURE 7.22 (a) is the cyclic voltammetry curves gotten during the deposition of cobalt hydroxide (Co(OH)2 on SWNT thin films, (b) is the SEM images of SWNT thin films, (c) is the cobalt oxide/SWNT thin films, while that of (d) is SEM crossectional image of cobalt oxide/SWNT thin films (Durukan et al. 2016). Reprinted with permission from Ref. Durukan et al. (2016), copyright 2016, Elsevier.

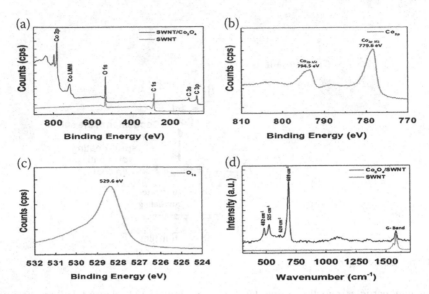

FIGURE 7.23 (a) displays the XPS survey of SWNT thin films and cobalt oxide/SWNT nanocomposite thin films, (b) shows the spin orbits of 2p Co, (c) shows spin orbits of O and (d) shows Raman spectra of cobalt oxide/SWNT and SWNT bare nanocomposites (Durukan et al. 2016). Reprinted with permission from Ref. Durukan et al. (2016), copyright 2016, Elsevier.

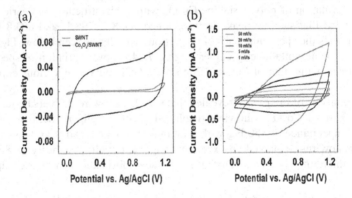

FIGURE 7.24 (a) shows the cyclic voltammetry curves of cobalt oxide/SWNT and SWNT bare nanocomposites at a single scan rate of 1 mV/s and (b) gives cyclic voltammetry curves of cobalt oxide/SWNT nanocomposite electrode material at various scan rates (Durukan et al. 2016). Reprinted with permission from Ref. Durukan et al. (2016), copyright 2016, Elsevier.

accumulating double electrodes of cobalt oxide nanowire, which gives excellent electrochemical properties, like high specific capacitance of 127 F/g, elevated power of 24.18 Wh/kg, and better cycling stability of 85% immediately after 3000 cycles. Zhang and his group (2016) obtained novel 3D hierarchical Co_3O_4/Ni(OH)$_2$ nanowire arrays (Co_3O_4/Ni(OH)$_2$ NAs) using hydrothermal method of treatment and CBD. The composite shows outstanding electrochemical performance, which is combined with the advantage of Co_3O_4 nanowires and nickel hydroxide nanowires. Their results displayed a considerable capacitance of about 912 F/g within 1 A/g, excellent multiplier stability which shows good retention rate of 81.10% immediately after 1500 cycles, and outstanding coulomb efficiency that is within the range of 100% in the cycle experimental analysis. Wang et al. 2014 reported 3D hierarchical cobalt oxide/copper oxide nanowire arrays on Ni foam mainly comprises of copper oxide nanowire stems and cobalt oxide nanosheets which were extracted and deposited using CBD method. An excellent reversible capacity that gives up to 1191 mAh/g with about 90.8% retention capacity immediately after 200 cycles at current density of 200 mA/g was obtained. Wang and his group further stated that reversible ability that gives up to 810 mAh/g immediately after 500 cycles within the region of the current density of 100 mA/g was also obtained in their experimental analysis. It was confirmed that the high electrochemical performance electrode materials in their work consist of cobalt oxide/copper oxide (Co_3O_4/CuO) nanowire array linked straightly on the Ni foam that made them possible anode equipment for excellent operations. Chen et al. (2019 fabricated Co_3O_4 nanorod arrays via the facile CBD method. Their experimental analysis displayed Co_3O_4 nanorod has the diameter of 450 nm. The specific capacitance of electrode delivered 387.25 F/g (154.9 c/g) at 1 A/g and gives cyclic stability of 88% after 1000 cycles. These striking outcomes made the exceptional cobalt oxide nanorod arrays an excellent electrode material for superior and excellent performance for supercapacitors and photovoltaic applications.

7.4.5 Electrodeposition

In electrodeposition, positively charged ions in a solution or mixture are deposited on the surface of the materials to be fabricated that is connected to the negatively charged electrode (the cathode) by simply applying or passing electric current through the circuit. Using the above method, nanomaterials are easily fabricated on the face of any electrode materials using electrochemical reduction reaction processes (Ren et al. 2019). Recently, electrodeposition technique has been widely used for fabricating metal coatings (Ebrahim-Ghajari et al. 2015; Aghazadeh et al. 2016) and metal nanostructures, as it is straightforward, cheap, and efficient compared to other methods. For example, Aghazadeh et al. (2016) prepared Co_3O_4 nanoplates using elecrodeposition process. They synthesized cobalt oxide hydroxide nanomaterials via free additive electrodeposition means, and later obtained Co_3O_4 nanoplates through calcination method. Their results show that Co_3O_4 nanoplates displayed high specific capacitance of 485 F/g and also gives capacitance retention of 84.1% after 3000 cycles at 5 A/g. Guo et al. (2016) prepared nano-Co_3O_4 films using one-step cathodic electrophoretic deposition. The electrochemical experiments of their work disclosed highest specific capacitance of 233.6 F/g at 0.5 A/g, 93.5% of which will be upheld immediately after 2000 GCD cycles. Aghazadeh (2012) prepared nanostructured Co_3O_4 via cathodic electrodeposition. Their findings displayed that the prepared metal oxide product has a surface area of 208.50 m^2/g and average pore diameter of 4.75 nm.

They evaluated the supercapacitive performance of the nanoplates with GCD and CV tests. Results of their electrochemical analysis displayed a very high specific capacitance of 393.6 F/g at current density of 1 A/g and gave better capacity retention of 96.5% after 500 charge/discharge cycles. Their experimental findings show that cobalt oxide nanoplates are good electrodes for high-quality performance applications. Razmjoo et al. (2014) prepared porous cobalt oxide nanoplates that give lofty surface area using cathodic deposition. The supercapacitive investigations of their results by electrochemical experiment of GCD and cyclic voltammetry give large capacitance of 420.50 F/g at an applied discharge current density of 5 A/g and also delivers better long-term cycling stability of 90% capacity retention immediately after 1000 cycles at current density of 5 A/g. Jagadale et al. (2014) worked on various electrodeposition processes such as GS (galvanostatic), PD (potentiodynamic), and PS (potentiodynamic) for fabrication of Co_3O_4 nanoparticles on the substrate. The electrochemical analysis of their experimental work was carried out via GCD, EIS, and CV. It was disclosed that Co_3O_4 nanoparticles deposited via potentiostatic have the highest value of exact energy, specific capacitance, and specific energy of 3.5 kW/kg, 248 F/g and 2.3 Wh/kg (Table 7.1).

7.5 Co_3O_4-Based Nanocomposites

7.5.1 Co_3O_4/Carbon Composites

Looking into the performance of EDLCs and pseudocapacitors, pseudocapacitors possess superior energy density and capacitance capacity; though poor electrochemical stability, elevated

TABLE 7.1

Shows Electrochemical Properties of Pure Cobalt Oxide (Co_3O_4) Electrode Reported in Earlier Literatures

Sample/Structure	Extraction Method	Capacitance (F/g)	Surface Area (m^2/g)	Cycle Stability (%)
Co_3O_4 porous nanoflake (Li et al. 2018)	Hydrothermal method	448 F/g	225 m^2/g	76.30%, 1000 cycles
Co_3O_4 nanosheet (Obodo et al. 2020)	Low temperature process method	304 F/g	336 m^2/g	88.6%, 1000 cycles
3D hierarchical flower and nanoneedles nanostructures of Co_3O_4 (Obodo et al. 2020)	Hydrothermal method	3327.3 F/g at 0.5 A/g	303 m^2/g	96.07% at 5A/g, 1000 cycles
Flower-like Co_3O_4 (Jadhav et al. 2014)	Coprecipitation method	328.4 F/g	228 m^2/g	92.0% 50 cycles
Layered Co_3O_4 cubes mesoporous (Sun et al. 2019)	Hydrothermal method	754 F/g	80.3 m^2/g	–
One-dimensional electroactive Co_3O_4 nanorods (Venkatachalam et al. 2018)	Hydrothermal method	655 F/g at 0.5 A/g	38.6 m^2/g	82.7% 1000 cycles
Mesoporous Co_3O_4 nanosheets (Wang et al. 2016)	Hydrothermal method	610 mF/cm^2 at 1 mA/cm^2	39.7 m^2/g	95% 3000 cycles
Nanoneedle and 3D hierarchial flower-like Co_3O_4 (Obodo et al. 2020e)	Hydrothermal	327.3 F/g	–	–
Mesoporous Co_3O_4 (Zhu et al. 2014)	Sol gel method	120 F/g		0.4% per cycle upon 60 cycles
Co_3O_4 network-like (Peterson et al. 2014)	Sol gel method	708 F/g		71.9%
Co_3O_4 nanowire (Yadav et al. 2018)	Chemical bath	850 F/g	66.33 m^2/g	86% over 5000 cycles
Two electrodes of Co_3O_4 nanowire (Yadav et al. 2018)	Chemical bath	127 F/g		85% over 3000 cycles
3D hierarchical Co_3O_4/CuO (Wang et al. 2014)	Chemical bath method	391 F/g	–	90.85% after 200 cycles
Co_3O_4 nanorod (Chen et al. 2019)	Chemical bath	387.25 F/g	–	88% after 1000 cycles
Co_3O_4 nanoplates (Aghazadeh et al. 2016)	Electrodeposition method	485 F/g	208.5 m^2/g	84.1% after 3000 cycles

resistance, and little power density deter its general performance. For that, 21st-century researchers have seen hybrid composites to have more advantages of carbon based for the application in EDLCs and transition metal oxide as more advantages in pseudocapacitors applications. They possess good electrical and thermal conductivity, outstanding mechanical properties, uniform pore structures and famous surface area operation showing they are very good electrode materials for supercapacitor application (Guan et al. 2015). For example, Kazazi et al. (2018) fabricated binder-free carbon nanotubes/cobalt oxide (CNT/CO) pseudocapacitive electrodes using a two-step procedure. The results of their findings shows that cobalt oxide thin films were homogeneously fastened on the surface of carbon nanotubes to produce porous structure of carbon nanotubes/cobalt composites electrode material that gives superior surface area of 144.90 m^2/g. On account of elevated thermal and electrical conductivity of CNTs, porous structure, and superior surface area, composites electrode apparatus displayed lofty specific capacitance of about 4.96 F/g within the current density of 2.0 mA/g. It also shows excellent rate performance of 64.70% capacitance preservation (2–5 mA/cm^2) and good cycling strength of about 92% after 2000 cycles. Durukan et al. (2016) fabricated cobalt oxide nanoflakes thin films for supercapacior electrodes.

The results of their study revealed that the capacitance gravimetric gives 313.90 F/g which corresponds to the capacitance of 70.50 F/g obtained from the deposited electrode material within the scan rate of 1 mV/s. Therefore, they further stated in their work that the capability preservation gives 80% immediately after 3000 cycles from deposited nanocomposite electrodes material. Figure 7.22a displays the cyclic voltammetry curves obtained during the deposition of cobalt hydroxide ($Co(OH)_2$ on SWNT thin films, Figure 7.22b is the SEM images of SWNT thin films. Figure 7.22c is the cobalt oxide/SWNT thin films obtained in their works, while that of Figure 7.22d is SEM cross-sectional image of cobalt oxide/SWNT thin films.

Also, Figure 7.23a displays the XPS survey of SWNT thin films and cobalt oxide/SWNT nanocomposite thin films as the experimental analysis of their studies revealed, Figure 7.23b shows the spin orbits of 2p Co, Figure 7.23c shows spin orbits of O and that of Figure 7.23d shows Raman spectra of cobalt oxide/SWNT and SWNT bare nanocomposites as analyzed in the experimental section of Durukan and coworkers.

Furthermore, Figure 7.24a shows the cyclic voltammetry curves of cobalt oxide/SWNT and SWNT bare nanocomposites at a single scan rate of 1 mV/s and Figure 7.24b gives cyclic voltammetry curves of cobalt oxide/SWNT nanocomposite electrode material at various scan rates.

Figure 7.25 displays the specific capacitance of SWNT/cobalt oxide nanocomposite electrode material in reference to the scan rates (Figure 7.25).

Furthermore, Kumar et al. (2016) synthesized carbon nanotubes and cobalt oxide ($CNTs/Co_3O_4$) nanomaterial via hydrothermal method. From their experimental results, the synthesized carbon nanotubes/cobalt oxide nanomaterial gives superior capacitance of 705.0 F/g at charging current of 3 A/g. Mondal and his group (2016) prepared multiwall CNT-nickel Co_3O_4 nanosheet mixture structure through microwave-assisted processes. The experimental analysis of their work revealed that multiwall CNT/Ni Co_3O_4 nanosheet displayed very high capacitance of 1395.0 F/g within current density of 1.0 A/g and good cycling stability after 5000 cycles (Table 7.2).

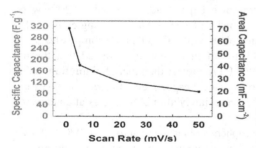

FIGURE 7.25 This figure displays the specific capacitance of SWNT/cobalt oxide nanocomposite electrode material in reference to the scan rates (Durukan et al. 2016). Reprinted with permission from Ref. Durukan et al. (2016), copyright 2016, Elsevier.

TABLE 7.2

Shows Electrochemical Properties of Cobalt Oxide (Co_3O_4)/Carbon Composite Electrode Reported in Earlier Literatures

Sample/Structure	Extraction Method	Capacitance (F/g)	Surface Area (m²/g)	Cycle Stability (%)
Carbon nanotubes/cobalt oxide (Kazazi et al. 2018)	Hydrothermal	496 F/g	144.9 m²/g	92% after 2000 cycles.
Co_3O_4 nano-flakes on a lone walled CNT (Durukan et al. 2016)	Hydrothermal method	313.9 F/g	-	80% after 3000 cycles
Carbon nanotubes and cobalt oxide (Kumar et al. 2016)	Hydothermal method	705.0 F/g	-	-
Multiwall CNT/Ni Co_3O_4 nanosheet amalgam structure (Mondal et al. 2016)	Microwave method	395.0 F/g	-	92.5% after 5000 cycles.

7.5.2 Co_3O_4/Graphene Composites

It has been shown that graphene is among the families of two-dimensional materials made of carbon that have incredible tensile strength and good mechanical property, especially on a small-scale value. Graphene is a lone film of sp^2carbon atoms organized hexagonally (Xu et al. 2013). Graphene has a very good intrinsic transport mobility (200,000 cm^2/v/s) (Geim and Novoselov 2007), outstanding electrical and thermal transmission ability (5000 W/m/k) (Calizo et al. 2007; Nsude et al. 2020; Nika and Balandin 2012), superior visible transmission of 97.70% (De and Coleman 2010), very good theoretical surface area of 2630.0 m^2/g, and high mechanical strength. In recent times, graphene has been one of the most promising nanomaterials due to its unique amalgamation of excellent properties. Apart from being thin, it is also very strong and conducts heat better than many other composite electrode materials. It is also an excellent material for electrical conductivity. Graphene as an electrode material has excellent mechanical properties, good specific surface area, and outstanding electrochemical properties. Graphene or metal oxide composite materials are among the promising or potential electrode materials for supercacitors (Yang et al. 2018) and solar cell applications that are currently in vogue. For instance, Du et al. (2016) fabricated cobalt oxide porous nanosheets attached to reduced graphene oxide (Co_3O_4atRGO) by using hyrothermal method. The experimental analysis of their work revealed that the supercacitor lined on Co_3O_4atRGO nanocomposite electrode gives a very high capacitance of about 894.0 F/g that is thrice greater when compared to pure cobalt oxide electrode material of 300 F/g at current density of 1 A/g. However, the electrodes displayed an excellent long-term life cycle, especially the electrode of Co_3O_4atRGO nanocomposite that gives no transparent attenuation immediately after 3000 cycles. Cobalt oxide nanoflake arrays anchored on nickel@graphene foam are fabricated using hydrothermal method (Tan et al. 2017). Tan and coworkers demonstrated that the result of electrochemical analysis shows that the hybrid electrode with the solvent nanoflakes displayed superior capacitance due to the complete utilization of the electroactive cobalt oxide (Co_3O_4). Furthermore, their studies show that cobalt oxide/nickel/graphene crossbreed electrode that has 12.59 nm in thickness nanoflakes gives good capacitance of 1.750 F/cm^2 at a current density of 1.0 mA/cm^2. There is an increase in capacitance of about 12.20% when compared to the initial one after 5000 cycles at a current density of 10.0 mA/cm^2. Zhang et al. (2019) fabricated aluminium oxide (Al_2O_3) doped with graphene/cobalt oxide (Co_3O_4) nanocomposites using calcination and hydrothermal method of extraction. Results of their analysis revealed that the nanocomposite materials displayed higher capacitance of 968.0 F/g greater than that of aluminium oxide doped with cobalt oxide nanocomposites and pure cobalt oxide, respectively. Besides, it gives excellent capability preservation retention of about 96.90% immediately after 2000 cycles at an elevated current density of 10.0 A/g.

Yan et al. (2017) prepared three-dimensional porous graphene/cobalt oxide aerogel (GA/Co_3O_4) via one-step hydrothermal method. Their work revealed that the amalgamation of extremely interrelated three-dimensional network framework of Co_3O_4 nanospheres and GA, GA/cobalt oxide (Co_3O_4) electrode displayed good electrical conductivity and pseudocapacitance that shows it is a better material for energy storage devices. It was observed in their work that GA/cobalt oxide composites displayed superior surface area of 127.0 m^2/g and large pore size distribution of 1.0–110.0 nm. They also revealed that the existence of porous Co_3O_4 is found effective in promoting capacitance of the composite aerogel of about 1512.70 F/g at 1.0 A/g (Table 7.3).

7.5.3 Cobalt Oxide (Co_3O_4)/Conducting Polymer

Conducting or conductive polymers are simply defined as organic material/polymers that are capable of conducting electricity. Conducting/conductive polymers are attractive due to the fact that they have excellent current density and are less expensive when compared to carbon electrode materials (Yang et al. 2017). Conducting composite polymers primarily comprise polypyrrole (PPy), polyaniline (PANI), polythiophene (PTh), and their derivatives. Conducting composite polymers have advantages of good stability, outstanding electrochemical properties, and ease of processing (Xuelei Wang et al. 2020). For instance, PANI is a good example. It has been reported in many works

TABLE 7.3

Shows Electrochemical Properties of Cobalt Oxide (Co_3O_4)/Graphene Composite Electrode Reported in Earlier Literatures

Sample/Structure	Extraction Method	Capacitance (F/g)	Surface Area (m^2/g)	Cycle Stability (%)
Cobalt oxide porous nanosheets attached on reduced graphene oxide (Co_3O_4 of RGO) (Du et al. 2016)	Hydrothermal	894 F/g	–	92.9% after 3000 cycles.
Cobalt oxide nanoflake anchored on nickel@graphene foam (Tan et al. 2017)	Hydrothermal method	1.750 F/cm^2	–	12.20% after 5000 cycles
Al_2O_3 doped with graphene/cobalt oxide nanocomposites (Zhang et al. 2019)	Hydrothermal/ calcination method	968 F/g	–	96.90% after 2000 cycles
3D porous graphene/Co_3O_4 aerogel (GA/Co_3O_4) (Yan et al. 2017)	Hydrothermal method	1512.70 F/g	127 m^2/g	92.5% after 5000 cycles

TABLE 7.4

Shows Electrochemical Properties of Cobalt Oxide (Co_3O_4)/Conducting Polymer Electrode Reported in Earlier Literature

Sample/Structure	Extraction Method	Capacitance (F/g)	Surface Area (m^2/g)	Cycle Stability (%)
Co_3O_4 conducting polyindole (Co_3O_4-Pind)	Cathodic electrodeposition	1805 F/g	–	–
Co_3O_4/Polyaniline (Co_3O_4/PANI)	Hydrothermal method	1625 F/g	–	–

of literature that many energy storage devices that are based on PANI such as supercapacitors give outstanding power of about 10 Wh/kg with infinitesimal specific power of about 2 kW/kg (Obodo et al. 2020e; Raj et al. 2015) whereas that of carbon-based supercapacitor apparatus can display a very high specific power of 3.0 kW/kg to 4.0 kW/kg and specific energy of about 3.0 Wh/kg to 5.0 Wh/kg (Obodo et al. 2020b; Obodo et al. 2020d). For example, Raj et al. (2015) synthesized Co_3O_4 conducting polyindole (Co_3O_4-Pind) composites via cathodic electrodeposition. They carried out electrochemical performance of Co_3O_4-Pind using GCD, CV, and EIS analysis. They revealed that the supercapacitors of Co_3O_4-decorated Pind displayed 1805 F/g at current density of 2 A/g with outstanding capability rate of 1625 F/g at current density of 25 A/g and excellent cycling stability. Obodo et al., (2020f) synthesized conductive cobalt oxide (Co_3O_4)/polyaniline (Co_3O_4/PANI) nanocomposites using in situ polymerization. XRD of their experimental analysis disclosed nanocrystalline cobalt oxide entrenched in polycrystalline PANI to obtain crystalline. They observed that the infrared Fourier transform and visible ultraviolet spectroscopy portrayed that Co_3O_4 nanoparticles acquired electrostatically bound to the specific location of polyaniline (Table 7.4).

7.6 Conclusion

In this critical review, we presented a broad overview of the current development and progress of cobalt oxide (Co_3O_4) based on electrode materials for superior performance of supercapacitors and conductive composite polymers as highly developed electrode materials for capacitors such as supercapacitors and many other devices for power storage. Capacitors like supercapacitors have a wide scope in the area of energy and power storage as the power storage devices in current century. In contrast to other metal oxide electrode materials, Co_3O_4 is the most fascinating material for supercapacitors owing to its superior electrochemical features, theoretical capacitance ratio, and excellent reversibility. The theoretical performance of supercapacitor was also discussed in detail. A variety of synthesis process or methods and the electrochemical features of Co_3O_4 plus Co_3O_4 containing electrode apparatus were also discussed in detail here. We compared the electrochemical features and the performance of Co_3O_4 electrode materials obtained through various synthesis methods. Of the various synthesis methods applied in extraction of cobalt oxide electrode materials, electrodeposition is most commonly used for synthesizing cobalt oxide and cobalt oxide containing electrode materials. Conversely, more researchers are paying attention to coprecipitation method in the recent times. However, this review sums up the properties, synthesis methods, and theoretical background of Co_3O_4 and Co_3O_4-containing supercapacitors. Finally, Co_3O_4 and Co_3O_4-containing electrodes materials have bigger usage and application in power storage areas due to their good electrochemical performance and reasonable structural and morphological design.

Acknowledgements

RMO and IA humbly acknowledge NCP for their PhD fellowship (NCP-CAAD/PhD-132/EPD) award and COMSATS for a travel grant for the fellowship.

RMO also acknowledge PPSMB Enugu State, Nigeria for study leave permission granted.

FIE (90407830) affectionately acknowledge UNISA for VRSP Fellowship award and also graciously acknowledge the grant by TETFUND under contract number TETF/DESS/UNN/NSUKKA/STI/VOL.I/ B4.33. We thank Engr. Meek Okwuosa for the generous sponsorship of April 2014, July 2016, and July 2018 conferences/workshops on applications of nanotechnology to energy, health & environment, and for providing some research facilities.

REFERENCES

Aghazadeh, M. 2012. "Electrochemical Preparation and Properties of Nanostructured Co_3O_4 as Supercapacitor Material". *J. Appl. Electrochem.* 42(2): 89–94. doi: 10.1007/s10800-011-0375-z.

Aghazadeh, M., R. Ahmadi, D. Gharailou, M.R. Ganjali, and P. Norouzi. 2016. "A Facile Route to Preparation of Co_3O_4 Nanoplates and Investigation of Their Charge Storage Ability as Electrode Material for Supercapacitors". *J. Mater. Sci. Mater. Electron.* 27 (8): 8623–8632. doi: 10.1007/s10854-016-4882-x.

Ahn, Y.R., C.R. Park, S.M. Jo, and D.Y. Kim. 2007. "Enhanced Charge-Discharge Characteristics of RuO_2 Supercapacitors on Heat-Treated TiO_2 Nanorods Enhanced Charge-Discharge Characteristics of RuO_2 Supercapacitors on Heat-Treated TiO_2 Nanorods". 122106. doi: 10.1063/1.2715038.

Al-Tuwirqi, R.M., A.A. Al-Ghamdi, Faten Al-Hazmi, Fowzia Alnowaiser, Attieh A. Al-Ghamdi, Nadia Abdel Aal, and Farid El-Tantawy. 2011. "Synthesis and Physical Properties of Mixed Co_3O_4/ CoO Nanorods by Microwave Hydrothermal Technique". *Superlattices Microstruct* 50(5): 437–448. doi: 10.1016/j.spmi.2011.06.007.

Babakhani, B. and D.G. Ivey. 2010. "Anodic Deposition of Manganese Oxide Electrodes with Rod-like Structures for Application as Electrochemical Capacitors". 195: 2110–2117. doi: 10.1016/j.jpowsour.2 009.10.045.

Bazrafshan, H., R. Shajareh Touba, Z. Alipour Tesieh, S. Dabirnia, and B. Nasernejad. 2017. "Hydrothermal Synthesis of Co_3O_4 Nanosheets and Its Application in Photoelectrochemical Water Splitting". *Chem. Eng. Commun.* 204(10): 1105–1112. doi: 10.1080/00986445.2017.1344651.

Calizo, I., A.A. Balandin, W. Bao, F. Miao, and C.N. Lau. 2007. "Temperature Dependence of the Raman Spectra of Graphene and Graphene Multilayers". *Nano Lett* 7(9): 2645–2649, doi: 10.1021/ nl071033g.

Chen, D., M.K. Song, S. Cheng, L. Huang, and M. Liu 2014. "Contribution of Carbon Fiber Paper (CFP) to the Capacitance of a CFP-Supported Manganese Oxide Supercapacitor". *J. Power Sources* 248: 1197–1200. doi: 10.1016/j.jpowsour.2013.09.068.

Chen, M., Q. Ge, M. Qi, X. Liang, F. Wang, and Q. Chen. 2019. "Cobalt Oxides Nanorods Arrays as Advanced Electrode for High Performance Supercapacitor". *Surf. Coatings Technol.* 360, December 2018: 73–77, doi: 10.1016/j.surfcoat.2018.12.128.

Chen, X., R. Paul, and L. Dai. 2017. "Carbon-Based Supercapacitors for Efficient Energy Storage". *Natl. Sci. Rev.* 4(3): 453–489. doi: 10.1093/nsr/nwx009.

Choudhary, R.K., P. Mishra, and V. Kain. 2017. "Pulse DC Electrodeposition of Zn–Ni–Co Coatings". *Surf. Eng.* 33(2): 90–93, doi: 10.1080/02670844.2015.1108072.

De Oliveira, A.H.P. and H.P. De Oliveira. 2014. "Carbon Nanotube/ Polypyrrole Nanofibers Core-Shell Composites Decorated with Titanium Dioxide Nanoparticles for Supercapacitor Electrodes". *J. Power Sources* 268: 45–49, doi: 10.1016/j.jpowsour.2014.06.027.

De, S. and J.N. Coleman. 2010. "Are There Fundamental Limitations on the Sheet Resistance and Transmittance of Thin Graphene Films?". *ACS Nano* 4(5): 2713–2720. doi: 10.1021/nn100343f.

Deng, J., Litao Kang, Gailing Bai, Ying Li, Peiyang Li, Xuguang Liu, Yongzhen Yang, Feng Gao, and Wei Liang. 2014. "Solution Combustion Synthesis of Cobalt Oxides (Co_3O_4 and Co_3O_4/CoO) Nanoparticles as Supercapacitor Electrode Materials". *Electrochim. Acta* 132: 127–135. doi: 10.1016/j.electacta.2014.03.158.

Devi, V. Sankar, M. Athika, E. Duraisamy, A. Prasath, A. Selva Sharma, and P. Elumalai. 2019. "Facile Sol-Gel Derived Nanostructured Spinel Co_3O_4 as Electrode Material for High-Performance Supercapattery and Lithium-Ion Storage". *J. Energy Storage* 25, no. May: 100815, doi: 10.1016/j.est.2019.100815.

Du, F., Xueqin Zuo, Qun Yang, Guang Li, Zongling Ding, Mingzai Wu, Yongqing Ma, Shaowei Jin, and Kerong Zhu. 2016. "Facile Hydrothermal Reduction Synthesis of Porous Co_3O_4 Nanosheets@ RGO Nanocomposite and Applied as a Supercapacitor Electrode with Enhanced Specific Capacitance and Excellent Cycle Stability". *Electrochim. Acta* 222: 976–982. doi: 10.1016/j.electacta.2016.11.065.

Durukan, M.B., R. Yuksel, and H.E. Unalan. 2016. "Cobalt Oxide Nanoflakes on Single Walled Carbon Nanotube Thin Films for Supercapacitor Electrodes". *Electrochim. Acta* 222: 1475–1482. doi: 10.1016/j.electacta.2016.11.126.

Ebrahim-Ghajari, M., S.R. Allahkaram, and S. Mahdavi. 2015. "Corrosion Behaviour of Electrodeposited Nanocrystalline Co and Co/ZrO_2 Nanocomposite Coatings". *Surf. Eng.* 31(3): 251–257, doi: 10.1179/1743294414Y.0000000355.

Ezekoye, B.A., P.O. Offor, V.A. Ezekoye, and F.I. Ezema. 2012. "Chemical Bath Deposition Technique of Thin Films: A Review". *Int. J. Sci. Res.* 2(8): 452–456. doi: 10.15373/22778179/aug2013/149.

Gao, Z. Chen Chena, Jiuli Chang, Liming Chen, Dapeng Wu, Fang Xu, and Kai Jiang. 2018. "Balanced Energy Density and Power Density: Asymmetric Supercapacitor Based on Activated Fullerene Carbon Soot Anode and Graphene-Co_3O_4 Composite Cathode". *Electrochim. Acta.* doi: 10.1016/j.electacta.2017.12.070.

Geim, A.K. and K.S. Novoselov. 2007. "The Rise of Graphene". *Nat. Mater.* 6(3): 183–191, doi: 10.1038/nmat1849.

George, G., Elias, L., Hegde, A.C., and Anandhan, S. 2015. "Morphological and Structural Characterisation of Sol-Gel Electrospun Co_3O_4 Nanofibres and Their Electro-Catalytic Behaviour Gibin". *RSC Adv.* doi: 10.1039/C5RA06368J.

Guan, C., Xu Qian, Xinghui Wang, Yanqiang Cao, Qing Zhang, Aidong Li, and John Wang. 2015. "Atomic Layer Deposition of Co_3O_4 on Carbon Nanotubes/Carbon Cloth for High-Capacitance and Ultrastable Supercapacitor Electrode". *Nanotechnology* 26(9): 94001. doi: 10.1088/0957-4484/26/9/094001.

Guo, J., L. Chen, X. Zhang, B. Jiang, and L. Ma. 2014. "Sol-Gel Synthesis of Mesoporous Co_3O_4 Octahedra toward High-Performance Anodes for Lithium-Ion Batteries". *Electrochim. Acta* 129: 410–415. doi: 10.1016/j.electacta.2014.02.104.

Guo, X., Xueming Li, Zhongshu Xiong, Chuan Lai, Yu Li, Xinyue Huang, Hebin Bao, Yanjun Yin, Yuhua Zhu, and Daixiong Zhang. 2016. "A Comprehensive Investigation on Electrophoretic Self-Assembled Nano-Co_3O_4 Films in Aqueous Solution as Electrode Materials for Supercapacitors". *J. Nanoparticle Res.* 18(6). doi: 10.1007/s11051-016-3456-4.

Hao, P., Zhenhuan Zhao, Jian Tian, Haidong Li, Yuanhua Sang, Guangwei Yu, and Huaqiang Cai. 2014. "Hierarchical Porous Carbon Aerogel Derived from Bagasse for High Performance Supercapacitor Electrode". *Nanoscale* 6(20): 12120–12129. doi: 10.1039/c4nr03574g.

Ho, M.Y., P.S. Khiew, D. Isa, T.K. Tan, W.S. Chiu, and C.H. Chia. 2014. "A Review of Metal Oxide Composite Electrode Materials for Electrochemical Capacitors". *Nano* 9(6): 1–25, doi: 10.1142/S1793292014300023.

Houshiar, M., F. Zebhi, Z.J. Razi, A. Alidoust, and Z. Askari. 2014. "Synthesis of Cobalt Ferrite (CoFe2O4) Nanoparticles Using Combustion, Coprecipitation, and Precipitation Methods: A Comparison Study of Size, Structural, and Magnetic Properties". *J. Magn. Magn. Mater.* 371: 43–48, doi: 10.1016/j.jmmm. 2014.06.059.

Hu, C., C. Huang, and K. Chang. 2008. "Anodic Deposition of Porous Vanadium Oxide Network with High Power Characteristics for Pseudocapacitors". 185, 1594–1597. doi: 10.1016/j.jpowsour.2008.08.017.

Hu, C., Y. Huang, and K. Chang. 2002. "Annealing Effects on the Physicochemical Characteristics of Hydrous Ruthenium and Ruthenium ± Iridium Oxides for Electrochemical Supercapacitors". *J. Pow. Sourc.* 108, 117–127.

Hu, X., L. Wei, R. Chen, Q. Wu, and J. Li. 2020. "Reviews and Prospectives of Co_3O_4-Based Nanomaterials for Supercapacitor Application". *ChemistrySelect* 5(17): 5268–5288. doi: 10.1002/slct.201904485.

Hulla, J.E., S.C. Sahu, and A.W. Hayes. 2015. "Nanotechnology: History and Future". *Hum. Exp. Toxicol.* 34(12): 1318–1321, doi: 10.1177/0960327115603588.

Jadhav, H.S., A.K. Rai, J.Y. Lee, J. Kim, and C. Park. 2014. "Enhanced Electrochemical Performance of Flower-like Co_3O_4 as an Anode Material for High Performance Lithium-Ion Batteries". *Electrochim. Acta* 146: 270–277. doi: 10.1016/j.electacta.2014.09.026.

Jagadale, A.D., V.S. Kumbhar, R.N. Bulakhe, and C.D. Lokhande. 2014. "Influence of Electrodeposition Modes on the Supercapacitive Performance of Co_3O_4 Electrodes". *Energy*, 64, 234–241. doi: 10.1016/ j.energy.2013.10.016.

Ji, H., Xin Zhao, Zhenhua Qiao, Jeil Jung, Yanwu Zhu, Yalin Lu, Li Li Zhang, Allan H. MacDonald, & Rodney S. Ruoff. 2014. "Capacitance of Carbon-Based Electrical Double-Layer Capacitors". *Nat. Commun.* 5(no. Cmcm): 1–7, doi: 10.1038/ncomms4317.

Ji, J., Li Li Zhang, Hengxing Ji, Yang Li, Xin Zhao, Xin Bai, Xiaobin Fan, Fengbao Zhang, and Rodney S. Ruoff. 2013. "Nanoporous Ni(OH)2 Thin Filmon 3D Ultrathin-Graphite Foam for Asymmetric Supercapacitor". 7: 6237–6243.

Jiang, J. and A. Kucernak. 2002. "Electrochemical Supercapacitor Material Based on Manganese Oxide: Preparation and Characterization". 47: 2381–2386.

Kandalkar, S.G., D.S. Dhawale, C. Kim, and C.D. Lokhande. 2010. "Chemical Synthesis of Cobalt Oxide Thin Film Electrode for Supercapacitor Application". *Synth. Met.* 160(11–12): 1299–1302. doi: 10.101 6/j.synthmet.2010.04.003.

Kandalkar, S.G., J.L. Gunjakar, and C.D. Lokhande. 2008. "Applied Surface Science Preparation of Cobalt Oxide Thin Films and Its Use in Supercapacitor Application". 254, 5540–5544. doi: 10.1016/j.apsusc. 2008.02.163.

Kazazi, M., A.R. Sedighi, and M.A. Mokhtari. 2018. "Pseudocapacitive Performance of Electrodeposited Porous CO_3O_4 Film on Electrophoretically Modified Graphite Electrodes with Carbon Nanotubes". *Appl. Surf. Sci.* 441, 251–257. doi: 10.1016/j.apsusc.2018.02.054.

Kumar, N., Y.C. Yu, Y.H. Lu, and T.Y. Tseng. 2016. "Fabrication of Carbon Nanotube/Cobalt Oxide Nanocomposites via Electrophoretic Deposition for Supercapacitor Electrodes". *J. Mater. Sci.* 51(5): 2320–2329. doi: 10.1007/s10853-015-9540-9.

Lakehal, A., B. Bedhiaf, A. Bouaza, H. Benhebal, A. Ammari, and C. Dalache. 2018. "Structural, Optical and Electrical Properties of Ni-Doped Co_3O_3 Prepared via Sol-Gel Technique". *Mater. Res.* 21(3): doi: 10.1590/1980-5373-MR-2017-0545.

Leckie, S., J.O. Benjamin, M.M., Hayes, K., Kaufman, G., Altman. 1980. "Adsorption/Coprecipitation of Trace Elements from Water with Iron Oxyhydroxide," 1–270.

Lee, M., B.H. Wee, and J.D. Hong. 2015. "High Performance Flexible Supercapacitor Electrodes Composed of Ultralarge Graphene Sheets and Vanadium Dioxide". *Adv. Energy Mater.* 5(7): 1–9, doi: 10.1002/ aenm.201401890.

Li, J., X. Hu, D. Chen, J. Gu, and Q. Wu. 2018. "Facile Synthesis of Superthin Co_3O_4 Porous Nanoflake for Stable Electrochemical Supercapacitor". *Inorganic Chemistry* 2: 9622–9626, doi: 10.1002/slct.201802131.

Li, L., G. Zhou, Z. Weng, X.Y. Shan, F. Li, and H.M. Cheng. 2014. "Monolithic Fe_2O_3/Graphene Hybrid for Highly Efficient Lithium Storage and Arsenic Removal". *Carbon N. Y.* 67: 500–507, doi: 10.1016/ j.carbon.2013.10.022.

Li, S., Y. Wang, J. Sun, Y. Zhang, C. Xu, and H. Chen. 2020. "Hydrothermal Synthesis of Fe-Doped Co_3O_4 Urchin-like Microstructures with Superior Electrochemical Performances". *J. Alloys Compd.* 821: 153507. doi: 10.1016/j.jallcom.2019.153507.

Li, W., F. Gao, X. Wang, N. Zhang, and M. Ma. 2016. "Strong and Robust Polyaniline-Based Supramolecular Hydrogels for Flexible Supercapacitors". *Angew. Chemie* 128(32): 9342–9347. doi: 10.1002/ange. 201603417.

Li, X. and B. Wei. 2013. "Supercapacitors Based on Nanostructured Carbon". *Nano Energy* 2(2): 159–173. doi: 10.1016/j.nanoen.2012.09.008.

Lima-Tenório, M.K., Carlos Sergio Ferreira, Querem Hapuque Felix Rebelo, Rodrigo Fernando Brambilla de Souza, Raimundo Ribeiro Passos, Edgardo Alfonso Gómes Pineda, and Leandro Aparecido Pocrifka. 2018. "Pseudocapacitance Properties of Co_3O_4 Nanoparticles Synthesized Using a Modified Sol-Gel Method". *Mater. Res.* 21(2): 1–7. doi: 10.1590/1980-5373-MR-2017-0521.

Luo, Q.P., Liang Huang, Xiang Gao, Yongliang Cheng, Bin Yao, Zhimi Hu, Jun Wan, Xu Xiao, and Jun Zhou *et al.* 2015. "Activated Carbon Derived from Melaleuca Barks for Outstanding High-Rate Supercapacitors". *Nanotechnology* 26(30): 304004. doi: 10.1088/0957-4484/26/30/304004.

Miller, J.R. and A.F. Burke. 2008. "Electrochemical Capacitors: Challenges and Opportunities for Real-World Applications". *Electrochem. Soc. Interface* 17(1): 53–57.

Mondal, A.K., H. Liu, Z.F. Li, and G. Wang. 2016. "Multiwall Carbon Nanotube-Nickel Cobalt Oxide Hybrid Structure as High Performance Electrodes for Supercapacitors and Lithium Ion Batteries". *Electrochim. Acta* 190: 346–353, doi: 10.1016/j.electacta.2015.12.132.

Mugle, D. and G. Jadhav. 2016. "Short Review on Chemical Bath Deposition of Thin Film and Characterization". *AIP Conf. Proc.* 1728, doi: 10.1063/1.4946648.

Nakayama, M., A. Tanaka, Y. Sato, T. Tonosaki, and K. Ogura. 2005. "Electrodeposition of Manganese and Molybdenum Mixed Oxide Thin Films and Their Charge Storage Properties". no. 17: 5907–5913.

Neacşu, I.A., A.I. Nicoară, O.R. Vasile, and B.Ş. Vasile. 2016. "Inorganic Micro- and Nanostructured Implants for Tissue Engineering". *Nanobiomaterials Hard Tissue Eng. Appl. Nanobiomaterials*, 271–295. doi: 10.1016/B978-0-323-42862-0.00009-2.

Nika, D.L. and A.A. Balandin. 2012. "Two-Dimensional Phonon Transport in Graphene". *J. Phys. Condens. Matter* 24(23): doi: 10.1088/0953-8984/24/23/233203.

Nsude, H.E., K.U. Nsude, G.M. Whyte, R.M. Obodo, C. Iroegbu, M. Maaza, and F.I. Ezema. 2020. Green Synthesis of $CuFeS_2$ Nanoparticles Using Mimosa Leaves Extract for Photocatalysis and Supercapacitor Applications. *J. Nanopart. Res.* 22: 352.

Obodo, R.M., A. Ahmad, G.H. Jain, I. Ahmad, M. Maaza, and F.I. Ezema. 2020a. "8.0 MeV Copper Ion (Cu^{++}) Irradiation-Induced Effects Structural, Electrical, Optical and Electrochemical Properties of Co_3O_4-NiO-ZnO/GO Nanowires". *Mater. Sci. Ener. Technol* 3: 193–200.

Obodo, R.M., A.C. Nwanya, M. Arshad, C. Iroegbu, I. Ahmad, R. Osuji, M. Maaza and F.I. Ezema. 2020b. "Conjugated NiO-ZnO/GO Nanocomposite Powder for Applications in Supercapacitor Electrodes Material". *Int. J. Energy Res.* 44: 3192–3202.

Obodo, R.M., A.C. Nwanya, A.B.C. Ekwealor, I. Ahmad, T. Zhao, M. Maaza, and F. Ezema. 2019a. "Influence of pH and Annealing on the Optical and Electrochemical Properties of Cobalt (III) Oxide (Co_3O_4) Thin Films". *Surfaces and Interfaces* 16: 114–119.

Obodo, R.M., A.C. Nwanya, Tabassum Hassina, Mesfin Kebede, Ishaq Ahmad, M. Maaza and Fabian I. Ezema. 2019b. "Transition Metal Oxide-Based Nanomaterials for High Energy and Power Density Supercapacitor". In *Electrochemical Devices for Energy Storage Applications*. United Kingdom: Taylor & Francis Group, CRC Press, 131–150.

Obodo, R.M., A.C. Nwanya, C. Iroegbu, I. Ahmad, A.B.C. Ekwealor, R.U. Osuji, M. Maaza, and F.I. Ezema. 2020c. "Transformation of GO to rGO due to 8.0 MeV Carbon (C^{++}) Ions Irradiation and Characteristics Performance on MnO_2–NiO–ZnO@GO Electrode". *Int. J. Energy Res.* 44: 6792–6803.

Obodo, R.M., A.C. Nwanya, C. Iroegbu, B.A. Ezekoye, A.B.C. Ekwealor, I. Ahmad, M. Maaza, and F.I. Ezema. 2020d. "Effects of Swift Copper (Cu^{2+}) Ion Irradiation on Structural, Optical and Electrochemical Properties of Co_3O_4-CuO-MnO_2/GO Nanocomposites Powder". *Adv. Pow. Technol.*

Obodo, R.M., E.O. Onah, H.E. Nsude, A. Agbogu, A.C. Nwanya, I. Ahmad, T. Zhao: M. Ejikeme, M. Maaza, and F.I. Ezema. 2020e. "Performance Evaluation of Graphene Oxide Based Co_3O_4@GO, MnO_2@GO and Co_3O_4/MnO_2@GO Electrodes for Supercapacitors". *Electroanalysis* 32: 1–10.

Obodo, R.M., N.M. Shinde, U.K. Chime, S. Ezugwu, A.C. Nwanya, I. Ahmad, M. Maaza and F.I. Ezema. 2020f. "Recent Advances in Metal Oxide/Hydroxide on Three-Dimensional Nickel Foam Substrate for High Performance Pseudocapacitive Electrodes". *Curr. Opinio. Electrochem.* 21: 242–249.

Oyedepo, S.O. 2012. "Energy and Sustainable Development in Nigeria: The Way Forward Energy Situation in Nigeria Energy Consumption Pattern in Nigeria". *Energy. Sustain. Soc.* 2 (15): 1–11, doi: 10.1186/2192.

Pan, Y., Z. Mei, Z. Yang, W. Zhang, B. Pei, and H. Yao. 2014. "Facile Synthesis of Mesoporous MnO_2/C Spheres for Supercapacitor Electrodes". *Chem. Eng. J.* 242: 397–403, doi: 10.1016/j.cej.2013.04.069.

Pan, Z., Y. Wang, H. Huang, Z. Ling, Y. Dai, and S. Ke. 2015. "Recent Development on Preparation of Ceramic Inks in Ink-Jet Printing". *Ceram. Int.* 41(10): 12515–12528. doi: 10.1016/j.ceramint.2015.06.124.

Patake, V.D., C.D. Lokhande, and O. Shim. 2009. "Applied Surface Science Electrodeposited Ruthenium Oxide Thin Films for Supercapacitor: Effect of Surface Treatments". 255: 4192–4196. doi: 10.1016/j.apsusc.2008.11.005.

Patil, U.M., R.R. Salunkhe, K.V. Gurav, and C.D. Lokhande. 2008. "Applied Surface Science Chemically Deposited Nanocrystalline NiO Thin Films for Supercapacitor Application". 255: 2603–2607. doi: 10.1016/j.apsusc.2008.07.192.

Peterson, G.R., F. Hung-Low, C. Gumeci, W.P. Bassett, C. Korzeniewski, and L.J. Hope-Weeks. 2014. "Preparation-Morphology-Performance Relationships in Cobalt Aerogels as Supercapacitors". *ACS Appl. Mater. Interfaces* 6 (3): 1796–1803. doi: 10.1021/am4047969.

Prasad, K.R. and N. Miura. 2013. "Electrochemically Deposited Nanowhiskers of Nickel Oxide as a High-Power Pseudocapacitive Electrode Electrochemically Deposited Nanowhiskers of Nickel Oxide as a High-Power Pseudocapacitive Electrode". 4199(2004): 1–4. doi: 10.1063/1.1814816.

Pudovkin, M.S., Pavel V. Zelenikhin, Victoria Shtyreva, Oleg A. Morozov, Darya A. Koryakovtseva, Vitaly V. Pavlov, and Yury N. Osin, et al. 2018. "Coprecipitation Method of Synthesis, Characterization, and Cytotoxicity of Pr3+:LaF3 (CPr = 3, 7, 12, 20, 30%) Nanoparticles". *J. Nanotechnol.* doi: 10.1155/2018/8516498.

Pudukudy, M. and Z. Yaakob. 2014. "Sol-Gel Synthesis, Characterisation, and Photocatalytic Activity of Porous Spinel Co_3O_4 Nanosheets". *Chem. Pap.* 68(8): 1087–1096. doi: 10.2478/s11696-014-0561-7.

Qiu, K. Yang Lu, Jinbing Cheng, Hailong Yan, Xiaoyi Hou, Deyang Zhang, Min Lu, Xianming Liu, and Yongsong Luo. 2015. "Ultrathin Mesoporous Co_3O_4 Nanosheets on Ni Foam for High-Performance Supercapacitors". *Electrochim. Acta* 157: 62–68, doi: 10.1016/j.electacta.2014.12.035.

Raj, R.P., P. Ragupathy, and S. Mohan. 2015. "Remarkable Capacitive Behavior of a Co_3O_4-Polyindole Composite as Electrode Material for Supercapacitor Applications". *J. Mater. Chem. A* 3(48): 24338–24348, doi: 10.1039/c5ta07046e.

Rane, A.V., K. Kanny, V.K. Abitha, and S. Thomas. 2018. *Methods for Synthesis of Nanoparticles and Fabrication of Nanocomposites.* New York: Elsevier Ltd.

Razmjoo, P., B. Sabour, S. Dalvand, M. Aghazadeh, and M.R. Ganjali. 2014. " Porous Co_3O_4 Nanoplates: Electrochemical Synthesis, Characterization and Investigation of Supercapacitive Performance". *J. Electrochem. Soc.* 161(5): D293–D300. doi: 10.1149/2.059405jes.

Ren, S., Y. Guo, L. Ju, H. Xiao, A. Hu, and M. Li. 2019. "Facile Synthesis of Petal-like Nanocrystalline Co_3O_4 Film Using Direct High-Temperature Oxidation". *J. Mater. Sci.* 54(10): 7922–7930, doi: 10.1007/s10853-019-03412-z.

Segal, D. 1994. "Chemical Synthesis of Advanced Ceramic Materials". *Cambridge Univ. Press. 1991* 9(10): 807–808.

Silva, D.L., Rafael G. Delatorre, Gyana Pattanaik, Giovanni Zangari, Wagner Figueiredo, Ralf-Peter Blum, Horst Niehus, and Andre A. Pasa. 2008. "Electrochemical Synthesis of Vanadium Oxide Nanofibers". 14–17. doi: 10.1149/1.2804856.

Song, Z., Yujuan Zhang, Wei Liu, Song Zhang, Guichang Liu, Huiying Chen, and Jieshan Qiu. 2013. "Hydrothermal Synthesis and Electrochemical Performance of Co_3O_4/Reduced Graphene Oxide Nanosheet Composites for Supercapacitors". *Electrochim. Acta* 112, 120–126. doi: 10.1016/j.electacta.2013.08.155.

Sun, J., Y. Wang, Y. Zhang, C. Xu, and H. Chen. 2019. "Egg Albumin-Assisted Hydrothermal Synthesis of Co_3O_4 Quasi-Cubes as Superior Electrode Material for Supercapacitors with Excellent Performances". *Nanoscale Res. Lett.* 14 (1). doi: 10.1186/s11671-019-3172-y.

Tadjer, M.J., Michael Mastro, José M. Rojo, Alberto Boscá Mojena, Madrid Fernando Calle, Francis J. Kub, and Charles R. Eddy. 2014. "MnO_2-Based Electrochemical Supercapacitors on Flexible Carbon Substrates". *J. Electron. Mater.* 43 (4): 1188–1193, doi: 10.1007/s11664-014-3047-z.

Talbi, H., P.E. Just, and L.H. Dao. 2003. "Electropolymerization of Aniline on Carbonized Polyacrylonitrile Aerogel Electrodes: Applications for Supercapacitors". *J. Appl. Electrochem.* 33 (6): 465–473. doi: 10.1023/A:1024439023251.

Tan, H.Y., Bao Zhi Yu, Lin Li Cao, Tao Cheng, Xin Liang Zheng, Xing Hua Li, Wei Long Li, and Zhao Yu Ren. 2017. "Layer-Dependent Growth of Two-Dimensional Co_3O_4 Nanostructure Arrays on Graphene for High Performance Supercapacitors". *J. Alloys Compd.* 696: 1180–1188. doi: 10.1016/j.jallcom.2016.12.050.

Theerthagiri, J., Sunitha Salla, R.A. Senthil, P. Nithyadharseni, A. Madankumar, Prabhakarn Arunachalam, T. Maiyalagan, and Hyun-Seok Kim. 2019. "A Review on ZnO Nanostructured Materials: Energy, Environmental and Biological Applications". *Nanotechnology* 30 (39): doi: 10.1088/1361-6528/ab268a.

Tsukamoto, R., K. Iwahori, M. Muraoka, and I. Yamashita. 2005. "Synthesis of Co_3O_4 Nanoparticles Using the Cage-Shaped Protein, Apoferritin". *Bull. Chem. Soc. Jpn.* 78(11): 2075–2081. doi: 10.1246/bcsj.78.2075.

Venkatachalam, V., A. Alsalme, A. Alswieleh, and R. Jayavel. 2018. "Shape Controlled Synthesis of Rod-like Co_3O_4 Nanostructures as High-Performance Electrodes for Supercapacitor Applications". *J. Mater. Sci. Mater. Electron.* 29 (7): 6059–6067. doi: 10.1007/s10854-018-8580-8.

Wang, J., Qiaobao Zhang, Xinhai Li, Daguo Xu, Zhixing Wang, Huajun Guo, and Kaili Zhang. 2014. "Three-Dimensional Hierarchical Co_3O_4/CuO Nanowire Heterostructure Arrays on Nickel Foam for High-Performance Lithium Ion Batteries". *Nano Energy*, 6: 19–26, doi: 10.1016/j.nanoen.2014.02.012.

Wang, X., H. Xia, X. Wang, J. Gao, B. Shi, and Y. Fang. 2016. "Facile Synthesis Ultrathin Mesoporous Co_3O_4 Nanosheets for High-Energy Asymmetric Supercapacitor". *J. Alloys Compd.* 686: 969–975. doi: 10.1016/j.jallcom.2016.06.156.

Wang, Y., Anqiang Pan, Qinyu Zhu, Zhiwei Nie, Yifang Zhang, Yan Tang, Shuquan Liang, and Guozhong Cao. 2014. "Facile Synthesis of Nanorod-Assembled Multi-shelled Co_3O_4 Hollow Microspheres for High-Performance Supercapacitors". *J. Power Sources* 272: 107–112, doi: 10.1016/j.jpowsour.2014.08.067.

Wang, Y., Y. Song, and Y. Xia. 2016. "Electrochemical Capacitors: Mechanism, Materials, Systems, Characterization and Applications". *Chem. Soc. Rev.* 45 (21): 5925–5950. doi: 10.1039/c5cs00580a.

Xiao, P.W., Q. Meng, L. Zhao, J.J. Li, Z. Wei, and B.H. Han 2017. "Biomass-Derived Flexible Porous Carbon Materials and Their Applications in Supercapacitor and Gas Adsorption". *Mater. Des.* 129 (May): 164–172. doi: 10.1016/j.matdes.2017.05.035.

Xu, C., B. Xu, Y. Gu, Z. Xiong, J. Sun, and X.S. Zhao. 2013. "Graphene-Based Electrodes for Electrochemical Energy Storage". *Energy Environ. Sci.* 6 (6): 1388–1414. doi: 10.1039/c3ee23870a.

Xuelei Wang, X.H., Anyu Hu, Chao Meng, Chun Wu, Shaobin Yang. 2020. "Recent Advance in Co_3O_4 and Co_3O_4-Containing Electrode Materials for High-Performance Supercapacitors". *J. Phys. Chem.*

Yadav, A.A., Y.M. Hunge, and S.B. Kulkarni. 2018. "Chemical Synthesis of Co_3O_4 Nanowires for Symmetric Supercapacitor Device". *J. Mater. Sci. Mater. Electron.* 29 (19): 16401–16409, doi: 10.1007/s10854-018-9731-7.

Yan, H., Jianwei Bai, Mingrui Liao, Yang He, Qi Liu, Jingyuan Liu, Hongsen Zhang, Zhanshuang Li, and Jun Wang. 2017. "One-Step Synthesis of Co_3O_4/Graphene Aerogels and Their All-Solid-State Asymmetric Supercapacitor". *Eur. J. Inorg. Chem.* 2017 (8): 1143–1152. doi: 10.1002/ejic.201601202.

Yan, J. and Z. Fan. 2010. "Fast and Reversible Surface Redox Reaction of Graphene – MnO_2 Composites as Supercapacitor Electrodes". 8. doi: 10.1016/j.carbon.2010.06.047.

Yang, J., Y. Liu, S. Liu, L. Li, C. Zhang, and T. Liu 2017. "Conducting Polymer Composites: Material Synthesis and Applications in Electrochemical Capacitive Energy Storage". *Mater. Chem. Front.* 1 (2): 251–268. doi: 10.1039/c6qm00150e.

Yang, S., Yuanyue Liu, Yufeng Hao, Xiaopeng Yang, William A. Goddard III, Xiao Li Zhang, and Bingqiang Cao. 2018. "Oxygen-Vacancy Abundant Ultrafine Co_3O_4/Graphene Composites for High-Rate Supercapacitor Electrodes". *Adv. Sci.* 5 (4): doi: 10.1002/advs.201700659.

Yuan, K., Yazhou Xu, Johannes Uihlein, Gunther Brunklaus, Lei Shi, Ralf Heiderhoff, and Mingming Que et al. 2015. "Straightforward Generation of Pillared, Microporous Graphene Frameworks for Use in Supercapacitors". *Adv. Mater.* 27 (42): 6714–6721, doi: 10.1002/adma.201503390.

Zhang, G.F., P. Qin, and J.M. Song. 2019. "Facile Fabrication of Al2O₃-Doped Co₃O₄/Graphene Nanocomposites for High Performance Asymmetric Supercapacitors". *Appl. Surf. Sci.* 493, no. June, 55–62. doi: 10.1016/j.apsusc.2019.06.288.

Zhang, L. and X.S. Zhao. 2009. "Carbon-Based Materials as Supercapacitor Electrodes". *Chem. Soc. Rev.* 38 (9): 2520–2531. doi: 10.1039/b813846j.

Zhang, R., Y. Xu, D. Harrison, J. Fyson, and D. Southee. 2016a."A Study of the Electrochemical Performance of Strip Supercapacitors under Bending Conditions". 11: 7922–7933, doi: 10.20964/2016.09.59.

Zhang, X., J. Xiao, X. Zhang, Y. Meng, and D. Xiao. 2016b. *Three-Dimensional Co₃O₄ Nanowires@ Amorphous Ni(OH)₂ Ultrathin Nanosheets Hierarchical Structure for Electrochemical Energy Storage.* New York: Elsevier Ltd, 191.

Zheng, M., Hanwu Dong, Yong Xiao, Hang Hu, Chenglong He, Yeru Liang, Bingfu Lei, Luyi Sun, and Yingliang Liu. 2017. "Hierarchical NiO Mesocrystals with Tuneable High-Energy Facets for Pseudocapacitive Charge Storage". *J. Mater. Chem. A* 5 (15): 6921–6927. doi: 10.1039/C7TA00978J.

Zhou, X., H. Chen, D. Shu, C. He, and J. Nan. 2009. "Study on the Electrochemical Behavior of Vanadium Nitride as a Promising Supercapacitor Material". *J. Phys. Chem. Solids* 70: 495–500, doi: 10.1016/j.jpcs.2008.12.004.

Zhou, Z., Y. Zhu, Z. Wu, F. Lu, M. Jing, and X. Ji. 2014. "Amorphous RuO₂ Coated on Carbon Spheres as Excellent Electrode Materials for Supercapacitors". *RSC Adv.* 4 (14): 6927–6932, doi: 10.1039/c3ra4 6641h.

Zhu, Y., K. Shi, and I. Zhitomirsky. 2014. "Polypyrrole Coated Carbon Nanotubes for Supercapacitor Devices with Enhanced Electrochemical Performance". *J. Power Sources* 268, 233–239. doi: 10.1016/j.jpowsour.2014.06.046.

8

Recent Developments in Metal Ferrite Materials for Supercapacitor Applications

Mary O. Nwodo[1], Raphael M. Obodo[1,2,3], Assumpta C. Nwanya[1,4,5], A. B. C. Ekwealor[1], and Fabian I. Ezema[1,4,5]

[1]*Department of Physics and Astronomy, University of Nigeria, Nsukka, Enugu State, Nigeria*
[2]*National Center for Physics, Islamabad, Pakistan*
[3]*NPU-NCP Joint International Research Center on Advanced Nanomaterials and Defects Engineering, Northwestern Polytechnical University, Xi'an, China*
[4]*Nanosciences African Network (NANOAFNET) iThemba LABS-National Research Foundation, Somerset West, Western Cape Province, South Africa*
[5]*UNESCO-UNISA Africa Chair in Nanosciences/Nanotechnology, College of Graduate Studies, University of South Africa (UNISA), Muckleneuk Ridge, Pretoria, South Africa*

8.1 Introduction

Non-renewable energy resources from fossil fuels such as petroleum and coal have been the basic sources of energy generation. More than 80% global energy consumption is derived from these sources of energy. The world has experienced rapid surge in energy and power usage as a result of rapid growth in technology and industrialization. This has facilitated the increase in the production of electrical gadgets in the society. The rapid development of industry and technology has been accompanied by an increase in power consumption and energy utilization (Theerthagiri et al. 2018; Aydin 2019; Fethi 2019). Over-dependence on fossil fuel has a disadvantage of them running out, which will pose a very big threat to the world if alternative sources of energy generating system are not developed. Nearly all the appliances we encounter in our daily activities require electricity (Das et al. 2016; Theerthagiri et al. 2018; Duraisamy et al. 2019). Usage of fossil fuel and combustion of coal also lead to pollution of the environment. Environmental degradation results when petroleum and coal are utilized. This happens when released gases in the form of soot and smoke enter the atmosphere by wind and react with some gases in the atmosphere. Thus, chlorofluorohydrocarbons (CFCs) formed here break down into chlorine and carbon dioxide. The chlorine from the CFCs hits the ozone layer and causes its depletion leading to rapid absorption of enormous ultraviolet radiation to the earth surface. This results in global warming and consequently green house effect, when this intense radiation is accumulated (Aydin 2019; Fethi 2019; Obodo et al. 2020a). In order to avert these problems of shortage of energy and pollution, there is vital need for alternative sources of energy.

Wind, tidal, and solar energy are utilized in electricity generation. These energy sources are irregular even though they can meet the global energy demand presently. Figure 8.1 shows the diagram of the renewable energy resources where it is shown that as against the traditional fossil fuel source, these energy resources can be found in all geographical locations, and they can be replaced when used since

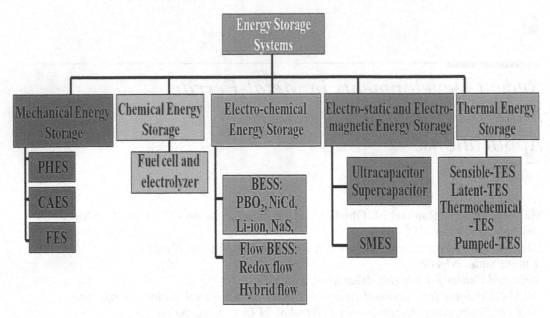

FIGURE 8.1   Forms of energy systems.

the energy is derived originally from the sun. Wind and hydro systems come indirectly from the sun while thermal and photovoltaic energy come directly from the sun. However, other mechanisms such as geothermal and tidal can also be harnessed. Electricity, fuel, and heat can easily be obtained using renewable energy technologies which translate the renewable energy resources to energy form that is usable (Obodo et al. 2019a). Thus, efficient energy storage techniques are required for transformation of the generated energy into diverse forms. The poor energy storage facilities have lead to loss of enormous amount of energy. Thus, researchers are working tirelessly to prevent this energy loss by developing and implementing alternative energy storage mechanisms. Fuel cells, capacitors, solar cells, batteries, and supercapacitors are some of the energy storage devices. Conventional capacitors and supercapacitors are mostly used as energy storage devices because of their long life span, short charging time, elevated power densities, and their eco-friendliness but they have fair energy density (Theerthagiri et al. 2017; Thiagaran et al. 2018). For competent energy storeroom, supercapacitor has been regarded as the utmost choice, but its little energy density and elevated cost of fabrication are the key test in its application (Aneke and Wang, 2016; Obodo et al., 2020a).

8.1.1 Forms of Energy

While maintaining quick charge-discharge rate, high power density, and high cycle stability, researchers have focused on improving the energy density of supercapacitors by fabricating novel electrode materials. The power and energy capacities can be improved drastically by formation of hybrids and miniaturization of the electrochemical capacitors (El-Kady et al. 2015; Muzaffar et al. 2019). With this step, asymmetric/hybrid and nanostructured materials with high prospective are fabricated (Uduh et al. 2014).

Hybrid vehicles, military devices, portable electronics among others are some of the practical applications of supercapacitors having distinguishable characteristics. Commercially produced supercapacitors always have modest energy density challenges, which limit the use of supercapacitor as primary power generating source at the expense of batteries (Gao et al. 2012). Efforts have been made by researchers recently in developing enhanced supercapacitive electrode materials with large operating voltage and specific capacitance (Zhang et al. 2010;; Du et al. 2011; Lee et al. 2012). Use of carbon derivative materials having large surface area such as grapheme (Jeong et al. 2016; Li et al. 2016; Huang et al. 2019a; Ye et al. 2016), fibers (CNFs), or nanotubes (CNTs) (Obodo et al. 2020b), carbon aerogels (Kwon et al. 2015) helps to accumulate charges through the process of reversible ion adsorption at the interface of electrode and electrolyte. The high surface area is used in the fine pores which enhances the capacitance through enlarged ionic diffusion (Ratajczak et al. 2019). The capacitance of supercapacitors can also be enhanced by adding conducting polymers (Kashani et al. 2016; Kurra et al. 2016), metal carbides (Li et al. 2017), and redox active materials (Fic et al. 2018; Kate et al. 2018; Obodo et al. 2019).

Coprecipitation, hydrothermal, thermal decomposition, microwave-assisted, electrode position, and sol gel have been the major techniques employed in the synthesis and deposition of nanomaterials and nanocomposites for supercapacitor applications (Zhang et al. 2018; Obodo 2020b; Obodo et al. 2020b). These materials have produced varying results which depend on the nature of the composites. Thus, the performance of supercapacitive material is determined by the surface area, permeability, porosity, electrical properties, and morphology of the synthesized nanomaterial and also the synthesizing or deposition techniques (Obodo et al. 2020d). Consequently, sulphides, phosphates, ferrites, oxides, etc., of single, double, and triple metals have been fabricated for years.

However, major setbacks are limiting the applications and optimum performance of metal oxide–based electrodes; these include low specific surface area and poor electroconductivity, as metal oxides possess large bandgap compared to their metals, chemical composition, and electrolytes (An et al. 2019). To overcome these constraints, basic strategies including nanostructuring and metal oxide combination with electroconductive resources like polymers and carbon compound (graphene, carbon nanotubes/fibre, activated carbon, aerogels, etc.) have been proposed as a highly effective approach.

This chapter discusses different forms of supercapacitors and current advances and difficulties in nanostructured metal ferrite-based electrode materials for supercapacitor energy storage.

Structural stability and charge transport properties of these electrode materials play significant roles in effective performance, exhibiting high electrical conductivity and good stability.

8.2 Electrochemical Energy Storage Systems

Solar and wind energy can be utilized in meeting global energy demand by generating electricity. Techniques used in storage of electricity have been achieved through these sustainable energy schemes. Due to the sporadic supply of these energy systems, it is pertinent to develop more steady facilities for energy storage which will be in use when the energy supply source is not available within a certain time. These energy storage systems should posses useful methods of storage that will contain the electricity generated intermittently. Even though we have great demand for steady supply of energy in form of electricity during daily cycle, a little instability in its efficiency can create large negative impact, which leads to loss of huge amount of money. Supercapacitors and batteries have been identified as the utmost leading energy storage systems today. These energy storage systems find use in laptops, phones, electric vehicles, plug-in hybrids, hybrid electrical vehicles, among others (Geng et al. 2018; Theerthagiri et al. 2018; Turgut 2018).

Batteries have numerous commercial and domestic applications, which made them one of the prominent devices for electricity storage. The electrochemical cells in a battery supply electrical power when the battery is connected to an external circuit. The battery is made up of a negative electrode called anode that acts as the electron source and a positive electrode called cathode. The electrons from the anode flow during electricity supply from the battery to an external device connected to the battery. This occurs during a redox reaction process where high energy reactants are converted to low energy reactants as the battery is connected to an external load. The electrical energy supplied to an external load emanates from free energy difference of the high- and low-energy products. Batteries find applications in phones, electric vehicles, flashlights, and laptops. Batteries can be classified into primary and secondary types.

Primary batteries are also referred to as primary cells, disposable cells, or single-use batteries, since the energy stored in a battery is not replenished after use. This type of battery can be used only once. The electricity in primary batteries is produced from chemical reactions that are irreversible. Alkaline batteries and Daniel cells are typical examples of primary batteries.

Secondary batteries are typified of their ability to be recharged after use unlike the primary batteries. Electricity is allowed to flow through the cells during charging. During this process the original state of the electrodes is restored by passing current in opposite direction. Secondary batteries such as lithium-ion batteries and lead-acid batteries found their applications in vehicles, photovoltaic system. They are also utilized in laptops and phones which are portable electronic systems.

Supercapacitors are capacitors having elevated energy storage competency with lesser limits of voltage. It can be referred to as *ultracapacitor* or *supercaps*. The energy stored in supercapacitors is more than 100 times the energy stored in batteries. Supercapacitors have greater number of charging and discharging cycles compared to batteries. Supercapacitors can be used to accommodate and utilize the overwhelming percentage of generated power that got wasted as a result of inadequate storage facilities. These advanced energy conversion and storage devices have been employed in commercialization of renewable energy system. The astonishing power density of electrochemical supercapacitors distinguishes them from other devices used for energy storage. The supercapacitors are flexible to work with, are adaptable to temperature variations, have elongated life cycles with high specific capacitance and energy density. Supercapacitors have rapid power delivery, as they are easily charged. The physical charge storage nature of supercapacitors makes their power density to be higher than that of fuel cells and batteries. However, they have relatively low energy density.

The mode of ion migration from electrolytes to surface of the electrode and the mechanism for energy storage determine the class of a supercapacitor. As a result of these, supercapacitors can be classified into hybrid supercapacitors, pseudocapacitors, and electrochemical double-layer capacitors (EDLC).

Just like ordinary capacitors, double-layer capacitors are made of two electrodes, an electrolyte, and a separator. Figure 8.2 (Kate et al. 2018) shows the schematic of EDLC. It contains a mixture of negative and positive ions in a polar solvent. This mixture is known as *electrolyte* and a separator is used to separate the two electrodes. Two electrically charged layers builds up at the boundary between the two electrodes and electrolytes. This is because an electrostatic force pulls the opposing charged ions towards an oppositely charged electrode. These charges are either negative or positive in polarity. The dielectric solvent molecules help to separate the ions from the electrolytic solution. Thus, positively charged ions from the electrolytic solution migrate or get attracted to the negative electrodes while the negatively charged ions get attracted strongly to the positive electrodes. The solvent molecule exerts an opposition force on the ions when they approach the electrodes. With this situation, charge transfer does not actually occur from the electrode to the electrolyte and vice versa.

On the other hand, the supercapacitor exhibits faradic type of energy storage where electrons are transferred between electrolyte and electrode in pseudocapacitor. The process involved in transfer of charges here are intercalation, redox reaction, or electrosorption. In pseudocapacitor, high specific energy density and specific capacitance are obtained when compared to the one in EDLC due to the

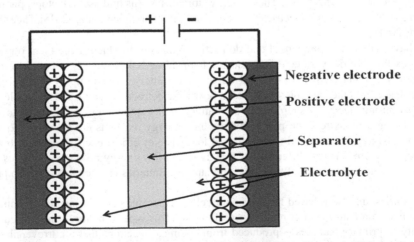

FIGURE 8.2 A typical EDLC supercapacitor. Reproduced with permission from Kate et al. 2018, *Journal of Alloys and Compounds.*

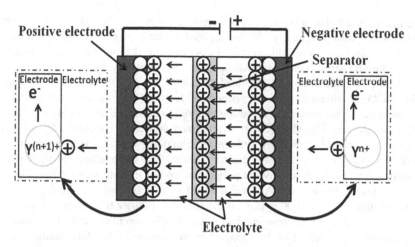

FIGURE 8.3 Diagram of a pseudocapacitor. Metal ion is denoted by Y. Reproduced with permission from Kate et al. 2018, *Journal of Alloys and Compounds.*

various faradic features (Durai et al. 2019). Figure 8.3 shows the diagram of a pseudocarpacitor (Kate et al. 2018).

A hybrid capacitor has intermediate properties of a supercapacitor and EDLC since it combines both Faradic and non-Faradic mechanism in energy storage process. It also has elevated energy and power density than EDLC while the operating voltage is also relatively high.

8.3 Metal Ferrite for Supercapacitor Applications

Metallic nanostructured materials have been prepared for many applications, and this has been the hub for intensive research by researchers because of their technological and domestic applications. Reduction of metallic ions has been the principle upon which metallic nanostructures are formed. A surfactant or an additive is usually added during the reaction. The nature and performance of metallic nanostructured materials are mostly dependent on the morphology, size, and method of preparation.

Metal ferrites (MFs) have been identified as having great importance as supercapacitive electrode materials because of their exceptional features. They have low electronegativity, moderate electrical conductivity, exceptional crystal structures, towering redox actions and outstanding specific capacitance (Zhang et al. 2010; Lee et al. 2012; Ye et al. 2016). The substitution of oxygen in metal oxides by ferrites during synthesis enhances the electrochemical properties of MFs. The sizes, surface morphologies, and structures determine the performances and structures of metal ferrites (Li et al. 2016). The utilization of good metal ferrite nanostructure electrode materials enhances the electrochemical properties of the materials. This is why the MF materials exhibit exceptional power and energy densities when compared to other materials. Many researchers have also worked on the use of single, double, and triple metal ferrites (Jeong et al. 2016; Aydin 2019; Obodo et al. 2020e). The potential in the electrochemical properties of mixed metal ferrites has been noticed through their tremendous electrochemical capacitances. In this chapter, we report the recent progress and setbacks in the study of metal ferrites for supercapacitor applications. We therefore lay emphasis on metal ferrites of manganese, copper, tin, cobalt, and nickel, the challenges and benefits for future use as energy storage devices.

8.4 Manganese Ferrite

Manganese ferrite is a promising material for energy storage facilities due to its extraordinary electrochemical features. During faradic reaction, they exhibit magnificent ionic transport routine on the surface of the electrode. Bashir et al. (2016) incorporated nanosheets of graphene in the fabrication of copper-doped manganese ferrite as supercapacitor electrode material. They observed a synergetic effect on the nanocomposite $Mn_{1-x}Cu_xFe_2O_4$/rGO as opposed to the individual components, and when the electrode material was properly examined, the nanocomposite demonstrated high capacitive performance, moderate faradic performance, and excellent electric double-layer capacitance. Thus, the doping of manganese ferrite with metal copper enhanced the conductivity of the material. The nanocomposite $Mn_{1-x}Cu_xFe_2O_4$/rGO revealed excellent electrochemical features from the electrochemical impendance and cyclic voltammetry. This makes the electrode material to be considered superior for applications in supercapacitor. Singh and Chandra (2019) synthesized low-cost nanorparticles of copper-doped manganese ferrite nanoparticles using hydrothermal techniques which ensured moderate conductivity and great surface area of the material for energy storage. The electrochemical features revealed an exceptional influence when the mixed hybrid polyaniline combines with nanoferrites. The symmetrical supercapacitor at current density of 1 A/g had 478.797 mAh/g highest specific capacitance and high 78% capacitance retention after 5000 cycles on the stability. Fourier transform infrared spectroscopy, field-emission scanning electron microscope, and transition electron microscope were used to study the structural and morphology analysis of the prepared nanocomposites. The fabrication of the electrodes was made easier without usage of binder during the process. Guo et al. (2017) used galvanostatic charge-discharge, electrochemical impedance, and cyclic voltammetry to study the electrochemical properties of $MnFe_2O_4$-based electrode material for supercapacitors energy storage system. They used 230–950 nm range size of $MnFe_2O_4$ colloidal nanocrystal assemblies (CNAs) during the in situ self-assembly of main $MnFe_2O_4$ nanoparticles with varying sizes. The electrochemical activities revealed that the supercapacitive nature of the $MnFe_2O_4$ is relatively attributed to the structures of the assembled nanoparticle sizes. They observed that 420 nm-sized assembled particles of 16 nm nanoparticles at current density of 0.01 A/g demonstrated the utmost capacitance of 88.4 F/g. however, when nanoparticles of 43 nm and 21 nm were used to form CNAs of 950 and 230 nm at the current density of 0.01 A/g, the capacitance increased to 20.02 F/g and 55.8 F/g, respectively. Among the samples, the 420 nm $MnFe_2O_4$ CNAs retained 59.4% of their capacitance during electrochemical stability test. However, when the current density was raised from 0.01 A/g to 2 A/g they observed capacitance retention of 69.2% after 2000 cycles under the current density of 0.2 A/g. An oxidative solution of NaOH having 1M concentration was used to chemically synthesis facile and simple nanoparticles of manganese ferrite ($MnFe_2O_4$) using coprecipitation technique. In their work, Vignesh et al. (2018) used scanning electron microscopy, FTIR spectroscopy, and powder X-ray diffraction to characterize the obtained nanoparticles. They obtained 20–50 nm spherical-shaped particles in the morphology study. They used different electrolytes such as KOH, Li_3PO_4, and $LiNO_3$ of variable molar concentrations to investigate the nanoparticle manganese ferrite's electrochemical properties with the use of electrochemical impedance, galvanostatic charge-discharge, and cyclic voltammetry. Maximum specific capacitances of 173 Fg^{-1}, 430 Fg^{-1}, and 31 Fg^{-1} were obtained using 3.5M KOH, 1M Li_3PO_4, and 1M $LiNO_3$ electrolytes, respectively, in three-electrode systems. The electrolyte medium containing 3.5M KOH had the best performing rate and it recorded 60% retention using 10 Ag^{-1} as a result of the materials' synergistic remark and superior electronic conductivity, and lofty accessible surface of $MnFe_2O_4$ nanoparticles. Additionally, utmost power density, energy density, and specific capacitance of 1207 Wkg^{-1}, 12.6 $Whkg^{-1}$, and 245 Fg^{-1} were attained in that order when $MnFe_2O_4$ material was used as symmetric electrodes. In addition, after 10,000 cycles about 105% specific capacitance was retained using 1.5 A g^{-1} current density and after all the 10,000 cycles about 98% efficiency was maintained. Wang et al. (2014a) investigated the electrochemical abilities of assembled symmetric supercapacitors of $MnFe_2O_4$ with electrochemical impedance spectroscopy, cyclic ability, galvanostatic charge-discharge, and cyclic voltammetry. They formed clusters of $MnFe_2O_4$ colloidal nanocrystals with manageable or adjustable electrolytes. Using 2M electrolytes of Na_2SO_4, LiOH, NaOH, and KOH, 47.4 Fg^{-1}, 74.2 Fg^{-1}, 93.9 Fg^{-1}, and 97.4 Fg^{-1} CNCs-based electrode specific capacitances were obtained, respectively, at 0.1 Ag^{-1} current density. Changing the concentration of the KOH electrolytes from 0.5M to 6M automatically improved the electrode's capacitance from 56.9 Fg^{-1} to 152.5 Fg^{-1}. Among all the electrolytes, the utmost stability result was obtained from the CNC-based $MnFe_2O_4$ supercapacitor in 6M aqueous KOH electrolyte. Singh and Chandra (2020) fabricated supercapacitor electrode materials using a $MnFe_2O_4$@PANI as composite. Hydrothermal process was used to synthesize the doped manganese ferrites and that was followed with aniline polymerization using oxidative in situ chemicals to form composites of the nanoflower. The synergistic property of the $MnFe_2O_4$ nanoscale was enhanced by addition of polyaniline that favors the transition route of ions and charges. This addition will also provide large surface area for active electrochemical process. Pseusocapacitor's charge transfer at the ionic sites of Fe and Mn was improved in the $MnFe_2O_4$ where the proton exclusion/inclusion outside/within the crystal lattice was oxidized by the polyaniline salt. The charge transport channels are provided by feathery state of the PANI while the storage of the charges as a result of the alteration in the active sites is as a result of the manganese ferrites. Using 1 A/g current density, 623 F/g elevated specific capacitance was obtained in this study. Generally, they obtained 179 Wh/kg lofty energy density and 982 Wh/kg utmost power density, and in 10,000 cycles the specific capacitance retention obtained was almost 95%. Furthermore, Elkholy et al. (2017) synthesized nanoparticles of $MnCoFeO_4$ using coprecipitation method, and the samples were characterized using N_2 adsorption/desorption techniques, Fourier transform infrared (FT-IR), high-resolution transmission electron microscopy (HR-TEM), X-ray diffraction (XRD) spectroscopy, and energy-dispersive spectroscopy (EDS). The supercapacitor fabricated with $MnCoFeO_4$ nanocomposites at a scan rate of 1.5 A/g revealed 675 F/g highest specific capacitance. The sample demonstrated 337.50 W/kg and 18.85 Wh/kg power and energy densities, respectively. The resultant stability of the composite showed that approximately 93% of its capacitance was retained after 1000 cycles. Kafshgari, Ghorbani, and Azizi (2018) synthesized nanocomposites of manganese ferrites using hydrothermal, sol gel, and coprecipitation methods. In their study, they determined the magnetic features, morphology, size, and structure of the nanostructured materials using the vibration sample magnetometer (VSM), Fourier transform infrared spectroscopy (FTIR), and X-ray diffraction (XRD) spectroscopy. The average sized 16, 45, and 36 nm crystals were obtained using X-ray diffraction study for the hydrothermal, sol gel and coprecipitation techniques employed, accordingly. Furthermore, the FESEM study revealed that the samples have three different morphologies of rectilinear structure, multiwalled hollow sheets, and spherical shapes. The VSM showed 9.52 emu/g, 41.89 emu/g, and 38.76 emu/g of data magnetization saturation for the samples synthesized using sol gel, hydrothermal, and coprecipitation techniques, respectively. The results from this chapter confirm that various synthesis techniques can be used to synthesize nanoparticles of manganese ferrites so as to obtain different sizes of particles with different magnetic features. Ishaq et al. (2019) presented a facile electrode material of ternary metal oxide comprising of polypyrrole conducting polymer, iron oxide, manganese, and graphene oxide. In order to obtain high areal and gravimetric capacitance, they took the first step of forming binary composite by mixing manganese ferrite with graphene by the technique of coprecipitation. They used NaOH as the reducing agent in the composite. The nanocrystalline composites of $rGO/MnFe_2O_4$ revealed 147 mVs^{-1} gravimetric capacitance and 232 $mFcm^{-2}$ areal capacitance using 5 mVs^{-1} rate of scan. On the second state, polypyrrole conducting polymer was incorporated into ther$GO/MnFe_2O_4$ binary composite to form a ternary composite of $rGO/MnFe_2O_4/Ppy$ electrode. As expected, the $rGO/MnFe_2O_4/Ppy$ nanocomposite exhibited better properties, as it revealed enhanced areal capacitance and gravimetric capacitance of 395 $mFcm^{-2}$ and 232 Fg^{-1} in that order as a result of the synergistic properties of the ternary composite. In an attempt to evaluate the electrochemical features of ternary metal ferrites, Xiong et al. (2014a) designed a two-step approach of the nanostructure. In this study, binary manganese ferrite/graphene (MG) was prepared using hydrothermal techniques during the first step and nanostructure ternary manganese ferrite/graphene/polyaniline (MGP) nanostructure was also synthesized during the second stage by incorporating conducting polymer of polyaniline into the binary composite. The designed nanocomposite showed elevated retention capacity of 75.8% and 76.4% after 5000 cycles at 5 A/g and 2 A/g, respectively, with 454.8 F/g specific capacitance at 0.2 A/g. This demonstrated high-quality capacitance of the composites when compared to the individual components of reduced graphene oxide, manganese ferrite, and polyaniline and also the combined binary composites. This outstanding supercapacitance of the ternary metal ferrite is a characteristic of the nanostructured composite having great synergistic properties. The surface morphology of $MnFe_2O_4$ as studied by Kafshgari, Ghorbani, and Azizi using FESEM revealed three varied particles of coprecipitated nano samples having spherical shape, hydrothermally synthesized samples having hollow nanosheets, and sol gel synthesized samples having rectilinear shape with clusters of large agglomerates as a result of sturdy

TABLE 8.1

Electrochemical Properties of $MnFeO_4$-Based Supercapacitor Electrodes

Electrodes	Methods	Capacitance (F/g)	Current Density (A/g)	Electrolytes	References
$Mn_{1-x}Cu_xFe_2O_4/$ rGO	coprecipitation	300		1M Na_2SO_4	(Bashir et al. 2016)
$Mn_{1-x}Cu_xFe_2O_4$	Hydrothermal	478.797 mAh/g	1	2M KOH	(Singh and Chandra 2019)
$MnFe_2O_4$ CNAs	Hydrothermal	20.02, 55.8	0.01 – 2	1M NaOH	(Guo et al. 2017)
$MnFe_2O_4$	Coprecipitation	173, 430, and 31	10	3.5M KOH, 1M Li_3PO_4, 1M $LiNO_3$	(Vignesh et al. 2018)
$MnFe_2O_4$	Hydrothermal	47.4, 74.2, 93.9, and 97.4	0.1	0.5M to 6M of Na_2SO_4, LiOH, NaOH, and KOH	(Wang et al. 2014)
$MnFe_2O_4$@PANI	Hydrothermal	982 Wh/kg	1	1M H_2SO_4	(Singh and Chandra 2020)
$MnCoFeO_4$	Coprecipitation	675 and 670	1 and 1.5	6M KOH	(Elkholy et al. 2017)
rGO/$MnFe_2O_4$/ Ppy	Coprecipitation	147 and 232	1, 2, 4 and 6	1M H_2SO_4	(Ishaq et al. 2019)
$MnFe_2O_4$/ rGO/ PANI	Hydrothermal	454.8, 425.2, 400.7, 365.6, 354.7, and 344.7	0.2, 0.5, 1, 2, 3, and 5	1M KOH	(Xiong et al. 2014a)

relationship within the magnetic particles and the production of voluminous gases within a short time at the time of combustion process (Wang et al. 2015; Pandav et al. 2016). On the other hand, the elemental distributions of the prepared samples revealed the presence of Mn, Fe, and O, which confirms the deposition of $MnFe_2O_4$ samples (Table 8.1).

Nonaka et al. 2020 synthesized manganese ferrite decorated with crumpled graphene structures in a single-step approach. They used aerosol facilitated compression method making the crumpled ball-shaped nanocomposite to incorporate large quantity of manganese ferrite. The obtained nanoparticles were employed for electrochemical supercapacitor and electrochemical hydrogen peroxide sensor application. In order to ascertain the electrochemical features of the composites, the pure $MnFe_2O_4$ was prepared first and later the rGO:$MnFe_2O_4$ was synthesized as a composite material. The synergistic feature of the rGO:$MnFe_2O_4$ was confirmed through the outstanding electrochemical features it produced.

The crumpled samples were characterized for the morphology and structural features by spraying the precursor on a heated region. The compression of the sample resulted in a crumpled ball-like structure as shown in Figure 8.4(a) having 390±86 nm mean diameter (Luo et al. 2011; Dou et al. 2016). Furthermore, Figure 8.4(b–d) revealed the morphology of the various combination ratios of the crumpled graphene (CG) and manganese ferrites. Figure 8.4(b) shows the SEM micrograph of the CG:Mn:Fe composite with 1:1/2:1/2 ratio. The morphology looks like the SEM micrograph for the pure CG. It was observed that the surface layers were completely covered when more metal oxides are combined with the CG at high percentage combination of the metals. This reduces the visibility of the vertices and ridges as observed for the pure CG. The SEM morphology revealed observable voids in Figure 8.4(d) for the 1:5:5 ratio of CG:Mn:Fe combination, and this signifies high percentage of transition metal oxides. The method employed here produced smaller nanoparticles when compared to conventional coprecipitation and hydrothermal synthesis techniques that produce higher diameters of the $MnFe_2O_4$ nanoparticles (Souza et al. 2016). EDS analysis in Figure 8.4(e and f) showed the distribution or mapping of the elements deposited for the 1:5:5 composite, which marked iron and manganese uniform distribution at the surface of the substrate. Other combination ratio had the same behavior. However, it is observed that there is intense decoration of iron and manganese as the composition of the metals increases in ratios.

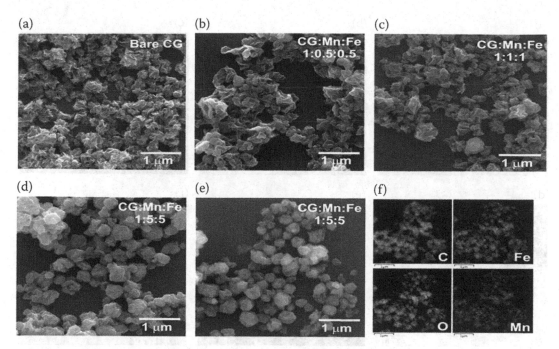

FIGURE 8.4 SEM images of (a) bare crumpled graphene and of the composites CG:Mn:Fe (b) 1:1/2:1/2, (c) 1:1:1, and (d) 1:5:5. SEM images of the composite (e) 1:5:5 and (f) EDS mapping images of the selected area of carbon, iron, oxygen, and manganese. Reproduced with permission from Nonaka et al. 2020, *ACS Appl. Nano Mater.*

HRTEM was used to give thorough investigation of the sample's morphology. There exists an agglomeration of the ball-like crumpled graphene in Figure 8.5(a), which has different vertices and ridges (Luo et al. 2011). Figure 8.5(b) revealed a different morphology for the CG incorporated with some amounts of metals in the ratio of 1:1:1 for CG:Mn:Fe while Figure 8.5(c) showed some diffracted spots as a result of the increasing manganese ferrite nanocomposite. This morphology was confirmed in Figure 8.5(d) with the TEM micrograph having composition of CG:Mn:Fe in the ratio of 1:5:5 indicating huge quantity of the metal composite. In addition to that, 0.263 and 0.256 nm interplanar gaps were observed with manganese ferrite plane corresponding to (311) as seen in Figure 8.5(e and f).

The XRD spectra of the samples are shown in Figure 8.6(a). The figure revealed broad peaks for the graphite at (004) and (002) hkl planes corresponding to 2θ value of 43.3° and 24.6° in that order (Salvatierra et al. 2013; Guerrero-Contreras and Caballero-Briones 2015). Furthermore, the face centered cubic–structured manganese ferrite having 10–319 PDF number revealed reflection planes of (400), (311), (220), and (111) for 2θ value of 42.9°, 35.3°, 29.1°, and 17.6° peaks, respectively (Ravindran Madhura et al. 2017). The broad diffraction peaks observed in the XRD pattern agree with the formation of manganese ferrite with small nanoparticles as confirmed by the HRTEM micrograph. There was no noticeable peak for the 1:1/2:1/2 which shows the masking of the metal ferrite by the CG. Further increase in the ratio of the manganese ferrite is indicated with the peaks for the 1:1:1 and 1:5:5, as (222) hkl planes were seen for the manganese component with 1-1061 PDF number at 2θ value of 33.4° peak (Trusovas et al. 2016). The Raman spectroscopy is used to analyze the Raman scattering effect on the assembled samples as shown in Figure 8.6(b). The stacking of graphene layers lead to the formation of G and D bands which emanated from second-order Raman scattering procedure, and all sp^2 networks exhibit this feature with characteristic disorders (Pimenta et al. 2007). The quantity of this defect or disorder in the graphene is denoted with I_D/I_G ratio where we noticed from the Raman curve that as the manganese ferrite increases the I_D/I_G ratio (Lee et al. 2017).

The CV plots for the CG and CG/MnFe$_2$O$_4$ nanocomposites are shown in Figure 8.7(a) for 50th cycle. Using spray pyrolysis for an inert environment at temperature of 400°C was insufficient to delete all functional groups on the synthesis of the CG for redox-pair within 1.19/−0.54 V. The 1:1/2:1/2 had

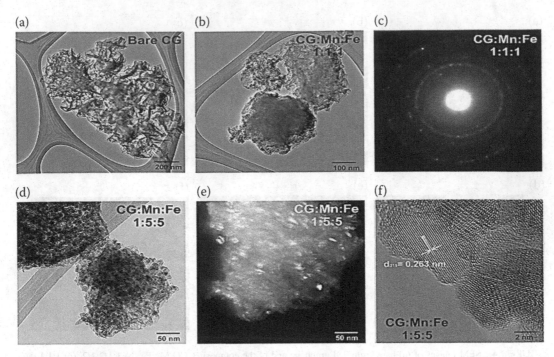

FIGURE 8.5 HRTEM images of (a) bare CG, (b) CG:Mn:Fe 1:1:1, and (d) CG:Mn:Fe 1:5:5. (d) SAED pattern acquired for CG:Mn:Fe 1:1:1. (e) Dark-field mode HRTEM image of CG:Mn:Fe 1:5:5. (f) HRTEM image of the composite CG:Mn:Fe 1:5:5. Reproduced with permission from Nonaka et al. 2020, *ACS Appl. Nano Mater.*

almost the same attribute as the pure CG due to the domineering effect of the CG. However, increasing the ratio of the metal components showed anodic peaks at −0.42 V and−0.44 V. the slight shift in the voltage for high metal component ratio might be attributed to the conductive nature of the metal composite (Chen et al. 2011). On the other hand, the voltammogram electrochemical stability after 50 cycles is shown in Figure 8.7(b) which confirmed that the 1:1/2:1/2 ratio composite showed more stability than other samples (Nagar et al. 2018). The 1:5:5 sample at initial 10 cycles illustrated abrupt reduced electrochemical stability, attaining 61% current density after 50 cycles. However, 1:1:1 composite recorded 87% current density relative to the 1:5:5 composite. Surprisingly, the 1:1/2:1/2 composite revealed outstanding 228% current density after 50 cycles due to high content of the carbon derivative, which increased the surface area of that sample unlike others that have high content of the metal ferrite composite.

As electrochemically effective material, the electrochemical features of the CG/MnFe$_2$O$_4$ composites were evaluated for charge/discharge using 0.05M concentration of KCl (Bashir et al. 2017; Vignesh et al. 2018). At 0.5 A/g current density, the charge/discharge values of the CG/MnFe$_2$O$_4$ composites as well as specific capacitance, power density, and energy density with the help of equations 8.1, 8.2, and 8.3.

$$C_{sp} = \frac{iX\Delta t}{\Delta VXm} \tag{8.1}$$

$$E = \frac{1}{2}C_{sp}V^2 \tag{8.2}$$

$$P = \frac{E}{\Delta t} \tag{8.3}$$

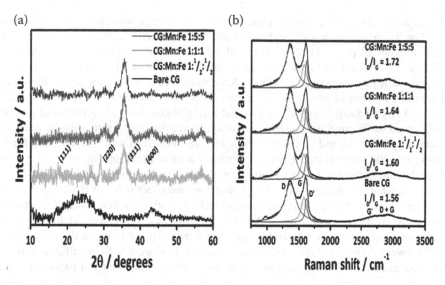

FIGURE 8.6 (a) XRD of bare CG and composites (b) Raman spectra of all materials along with ID/IG ratios. Reproduced with permission from Nonaka et al. 2020, *ACS Appl. Nano Mater.*

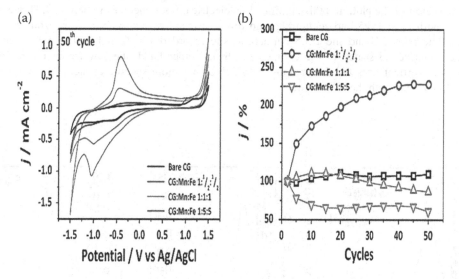

FIGURE 8.7 (a) Cyclic voltammograms (50th cycle) in KCl 0.05 mol L^{-1} at 50 mV s^{-1}. (b) Electrochemical stability of the samples after 50 cyclic voltammograms. Reproduced with permission from Nonaka et al. 2020, *ACS Appl. Nano Mater.*

where the specific capacitance is denoted with C_{sp}, the applied current is i. The change in voltage is denoted with Δv while Δt represents the discharge time. P an E are the power and energy density, respectively, while m is the mass of the deposited material. In addition to the above equations, equation 8.4 represents the pseudocapacitive feature emanating from the nanoparticles of manganese ferrite.

$$EV^2 + (1 - \delta)Fe^{3+} + K^+(aq) + e^- \rightarrow \delta Mn^{2+} + (1 - \delta)Fe^{2+} + K^+(lattice) \quad (8.4)$$

The application of manganese ferrite nanoparticles as energy storage materials is shown in Figure 8.8. The 1:5:5 nanocomposite exhibited the utmost specific capacitance value of 195 F/g, while the composites of pure CG, 1:1/2:1/2, and 1:1:1 ratio combination showed specific capacitance of 24 F/g, 53 F/g, and 54 F/g at 0.5 current density in that order. The high current in the CV plot buttressed the higher capacitance of the 1:5:5 composite, which is a result of high quantity of the metal ferrite (Vignesh et al. 2018). It is important to understand that increasing the current density from 0.5 A/g to 5 A/g kept the charge/discharge plots unchanged, as they had similar shapes as can be observed in Figure 8.8(b). On the contrary, increasing the current density resulted to low retention rate. Figure 8.8 (a and b) did not show the presence of plateau since CG and manganese ferrite exhibit a distinctive electrochemical feature because of the wrapping of some nanoparticles by the CG and reduces the electrolytic reaction with some of the composite.

Using charge/discharge stability check, it was observed that at 5000 cycles the 1:5:5 nanocomposite shown in Figure 8.8(c) exhibited a sharp decrease of stability within the first 100 cycles after which the decrease was slow using power density of 2.0 A/g. This attribute is a typical characteristic of nanocomposite comprising graphene oxide and manganese ferrite. The stability of the composite became almost equal between 4000 and 5000 cycles. The authors also opined that the stability of this composite is highly dependent on the nature of the electrolyte used instead of the material nanocomposite (Wang et al. 2014).

In order to evaluate the impedance ability of the nanocomposite, the Electrochemical Impedance Spectrometry (EIS) was studied. Figure 8.9 (b) showed the EIS plot for the pure CG and the composite of the CG:Mn:Fe 1:5:5. The pure CG instead of showing a semicircle revealed an arc at high frequency region of the plot, and this signifies low electrical resistance of the pure CG. However, the composite with ratio 1:5:5 had semicircle, and this is a result of relatively high resistance than the pure CG. The pure CG and the 1:5:5 nanocomposite recorded 176 Ω and 379 Ω, which confirms the higher relative resistance of the composite. On the other hand, the low frequency region had characteristic vertical lines for the composite, which shows more specific capacitance than the pure CG and this agrees with the CV results.

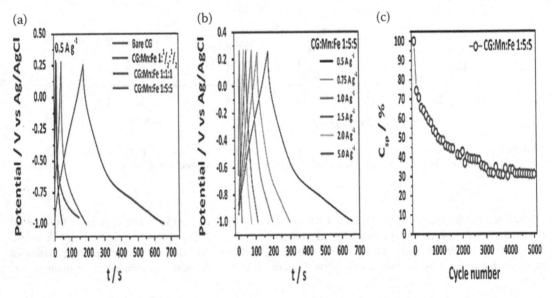

FIGURE 8.8 Charge/discharge curves (3rd cycle) (a) at 0.5 A g−1 for all samples and (b) at different current densities for the composite 1:5:5. (c) cycle stability of the composite 1:5:5 along 5000 galvanostatic charge/discharge cycles at 2 A g−1. Reproduced with permission from Nonaka et al. 2020, *ACS Appl. Nano Mater.*

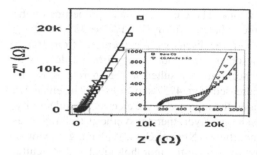

FIGURE 8.9 EIS measurement for bare CG and CG:Mn:Fe 1:5:5. Reproduced with permission from Nonaka et al. 2020, *ACS Appl. Nano Mater.*

8.5 Cobalt Ferrite

Nanoparticle cobalt ferrites belong to the type with $CoFe_2O_4$ chemical formula where Co is a metal with divalent ion (Co^{+2}) Obodo et al. (2020f). The $CoFe_2O_4$ is a very resourceful material because of its useful properties in the fields of magnetism and electricity (Deng et al. 2012; Kumbhar et al. 2012). Their simple fabrication method, elevated theoretical specific capacitance, many oxidation-reduction states, and outstanding chemical stability made $CoFe_2O_4$ to stand out among other materials used for supercapacitor application (Zhang et al. 2015; Obodo et al 2019b; Alshehri et al. 2017; Kennaz et al. 2018). The inverse spinel $CoFe_2O_4$ is renowned for its ions occupying a tetrahedral position whereas the ferrite ions (Fe^{+3}) have their sites at tetrahedral and octahedral positions, thereby necessitating improved redox reaction and great ability in storage of charges (Das and Verma 2019a). Even with the recorded theoretical and experimental findings, it has been observed that the individual $CoFe_2O_4$ has reduced electrical conductivity that limits the transport ability of the electron and hence the performance of the material electrochemically. In order to enhance energy applications using advanced research areas, Ayman et al. (2020) developed an extremely efficient elevated power device from material electrodes. Due to the functional group interaction emanating from the re-stacking of MXene layers, the performance ability in terms of supercapacitance has been limited as an electrode material. However, forming a composite between MXene and $CoFe_2O_4$ for battery-like hybrid supercapacitor tends to ameliorate this problem. The $CoFe_2O_4$ nanoparticles act between the MXene layers as interlayer spacer. This leads to the formation of material with commendable electrochemical properties with the $CoFe_2O_4$/MXene composite. From the work, 594 F/g, 1046.25 F/g, and 1268.75 F/g utmost specific capacitance were obtained from $CoFe_2O_4$, MXene, and nanocomposite of $CoFe_2O_4$/MXene in that order using 1 A/g current density. The material characterized as synthesized revealed magnificent 97% retention of its capacitance for 5000 cycles. Liu et al. (2016) synthesized porous nanomesh arrays of $CoFe_2O_4$ using hydrothermal technique and deposited the material on nickel form substrate. The pores and large surface area helped to expose the materials' active sites for chemical reactions as the supercapacitive and electrocatalysis properties of the material were examined without a binder. For the electrochemical properties, the deposited $CoFe_2O_4$NM-As/Ni electrode at 1 A/g and 2 A/g current density had 1426 F/g and 1024 F/g specific capacitance, respectively, with 92.6% retention of the capacitance after 3000 cycles. The electrochemical performance of cobalt ferrite synthesized using coprecipitation techniques and deposited on stainless steel to form a thin film was studied by Kumbhar et al. (2012). They characterized the films to know the FT-IR, surface morphological and structural features, and the XRD reveals $CoFe_2O_4$ having single phase. The SEM morphology revealed nanoflake formation. Cyclic voltammetry was used to study the electrochemical properties of the $CoFe_2O_4$ electrode; 366 F/g utmost specific capacitance was obtained at 5 mV/s scan rate. They also obtained 1.1 ohms equal series resistance using AC impedance method. Furthermore, Vargas et al. (2020) reported the formation of supercapacitors that have tunable qualities using cobalt ferrite. The sample was synthesized using coprecipitation method where the 0, 25, and 33 mol% concentration were used to obtain approximated 10.2, 12.3, and 8.8 nm sizes of the average particles using SEM analysis. The cobalt ferrite showed cubic phase structure in the X-ray diffraction study. Two flexible electrodes made of graphene are used for the fabrication of flexible supercapacitors, where one of the electrode is coated with cobalt ferrite having different concentrations. The electrochemical properties of the materials showed that at 1 A/g, the specific capacitance changes from 77.8 F/g to 337.1 F/g for different concentrations of the electrode material. However, the discharge time increases from 1650 s to 4957 s for 0 to 33 increase in concentration. On the other hand, fabricating different devices of cobalt ferrite nanomaterial with 33 mo% of cobalt concentration resulted in the increment of the capacitance from 337.1 F/g to 835.7 F/g at 1 A/g also increased from 4957 s to 21969 s. The enhanced surface area of the cobalt ferrite nanoparticles necessitated the increase in the electrical conductivity, specific capacitance, and improvement in the time of discharge. Hydrothermal synthesis technique has been employed by Haratet al. (2016) to synthesize nanoparticles (NPs) of $CoFe_2O_4$ having maximum depositing circumstances. XRD and HRTEM were used to analyze the morphological and structural features of the nanoparticles and the nanoplatelet-shaped material showed 17 nm average-sized nanoparticles with spinel structure having single phase. Cyclic voltammetry was used to determine the electrochemical features of the $CoFe_2O_4$ in electrolyte of 6M KOH where the active material is made of the synthesized samples of different loading weight deposited on a nickel foam substrate. The lowest loaded weight of the cobalt ferrite nanoparticles revealed the maximum specific capacitance of 315 F/g. Coprecipitation technique was employed by Rani et al. (2018) to synthesize clean and replaceable cobalt zinc ferrites $Co_xZn_{0.04-x}Fe_2O_4$ (x = 0, 0.01, 0.02). They characterized the samples comprehensively to ascertain the physical properties of the nanoparticles. Irregular cluster of particles were formed with spherical grains of particles ranging between 30 nm and 150 nm as revealed by the SEM morphology. Zinc ferrite cubic phase was revealed by Raman modes of phonon vibration. At variable scan rate of 377 Fg^{-1} enhanced specific capacitance was recorded during the electrochemical analysis. The electrochemical, optical, and structural features of the composite were dominated with the influence of cobalt substitution. Furthermore, ternary cobalt ferrite comprising of cobalt ferrite (C), graphene (G), and polyaniline (P) was fabricated by Xiong Huang, and Wang (2014) using two-step processes by hydrothermal technique. Firstly, the cobalt ferrite was combined with graphene to form the first composite using hydrothermal method. The second step involves incorporating polyaniline into the graphene/copper ferrite composite to form a larger composite. The composite exhibited maximum specific capacitance for the ensuing ternary metal ferrite. At scan rate of 1 mV/s and current density of 0.1 A/g the ternary metal ferrite revealed 1133.3 F/g and 767.7 F/g specific capacitance when two electrode systems were used, respectively. It is evident that the individual components or binary composites of polyaniline, graphene, $CoFe_2O_4$, graphene/polyaniline, $CoFe_2O_4$/polyaniline, and $CoFe_2O_4$/graphene had lesser attributes in terms of capacitance when compared to the ternary composite. Over 5000 cycles, the nanocomposite had over 96% retention of its capacitance. Tung et al. (2020) presented a supercapacitor hybrid electrode of graphene–cobaltferrite (GP-CoFe) by employing direct writing technique. Cobalt ferrite and sodium dodecyl sulphate aqueous colloid with graphene was used to prepare the depositing ink at the first stage. The second stage involves printing of the ink on substrate graphite paper via express ink printing. Microwave was used to treat the electrodes at different time intervals. The as-printed samples were characterized for morphology, optical, electrical, and electrochemical properties. The sample of $GP-CoFe_2O_4$ treated with microwave for 60 mins revealed the best electrochemical feature by showing 304 F/g maximum specific capacitance at 0.83 A/g current density with highest capacitive retention of 94.68% after 16,000 cycles. Dong et al. (2017) on the other hand, assembled very thin layers of cobalt ferrite successfully on the exterior of reduced graphene oxide(rGO) by regulating ratio of the volume of water and ethanol, which was later treated with calcination. They used 30:10 DI water and ethanol volume ratio to obtain the ultrathin $CoFe_2O_4NSs@rGO$ nanocomposite. As a result, an entire rGO layer enveloped by nanoporous films was formed. The samples were evaluated for supercapacitor and lithium-ion batteries (LIBs) hybrid features which resulted in improved electrochemical attributes. TheLIBs' anode electrode material at 400 mA/g exhibited 200 charge-discharge cycles and 835.6 mAh/g discharge ability reversibly. Additionally, the excellent electrochemical ability of the $CoFe_2O_4NSs@rGO$ was confirmed when after 3000 cycles, the electrode material used for supercapacitor revealed 1120 F/g specific capacitance. Excellent charge storage and liberation of active sites were hydrothermally synthesized with oleic acid of varying concentrations and later annealed at 400C for electrochemical capacitance by Lalwani et al. (2018) The hierarchically prepared cobalt ferrite ($CoFe_2O_4$) nanorods with pores elevated the surface area, thereby boosting the sites for high capacitance. The nanorods of the $CoFe_2O_4$ exhibited 460.5 F/g specific capacitance at current density of 1.0 A/g and after 5000 cycles of stability test, it retains 95.8% of its capacitance. The pseudocapacitative attributes of this sample was facilitated by the diffusion of OH ions from the electrolytic solution. In addition, when these electrodes are used asymmetrically by fabricating the negative electrode with graphene nanoribbons and the positive electrode with cobalt ferrite nanorods, an extraordinary energy density of 33.5 Wh/kg at 727.8 W/kg was obtained from the cell, which is almost two times of 16.5 Wh/kg at 288 W/kg, which is the value obtained from the symmetric configuration. The effect of the electrochemical ability and morphology of cobalt ferrite on PVA, CTAB, PVP, and SDS as template agents was studied by Gao et al. (2020), as they hydrothermally synthesized the cobalt ferrite ($CoFe_2O_4$) and used nickel foam as the growing substrate. X-ray photoelectron spectroscopy, scanning electron microscopy, N_2 adsorption-desorption, and X-ray diffraction were used to study the physicochemical features of the nanocomposite $CoFe_2O_4$. The $CoFe_2O_4$ nanocomposite incorporated with PVA, SDS, PVP and CTAB at 1 A/g revealed 828 F/g, 1100 F/g, and 1148 F/g maximum specific capacitance. Additionally, the asymmetrically prepared samples with negative electrode made of activated carbon and negative electrode made of CFO-A showed 55.42 Wh/kg energy density at 769.7 W/kg power density. It also recorded 56% retention ability at 769.7 W/kg power density using electrolyte of 6M KOH. Gao, Xiang, and Cao (2017) fabricated nanosheets of electrode materials for supercapacitor application using mesoporous cobalt ferrite synthesized with hydrothermal technique and deposited on nickel foam substrate before annealing. Using the fabricated $CoFe_2O_4$ nanosheets as electrode material exhibited 503 F/g specific capacitance at 2 A/g current density. Specific capacitance of 395 F/g was obtained when the current density was reduced to 20 A/g and 78.5% of its capacitance was retained, which indicates that the material is good for supercapacitor. On the other hand, when the negative electrode and positive electrode are made of activated carbon and $CoFe_2O_4$nanosheets, respectively, the asymmetric supercapacitor at 1.2 A/g current density conveyed 73.12 F/g specific capacitance. It also exhibited 22.85 Wh/kg elevated energy density and 98% retention ability of its capacitance after 5000 cycles within 1.5 V potential window. Alshehri et al. (2017) used polymerization of egg white albumin to synthesize cobalt ferrite doped with nitrogen, and they obtained 10 nm average size of the nanoparticles. The sample was later annealed at 700°C for three hours under helium gas flow. Electron microscope, X-ray photoelectron spectroscopy, and powder X-ray diffraction were used to study the morphological and structural features of the nanocomposite. The cobalt ferrite nanocomposite at scan rate of 5 mV/s using 5M KOH revealed improved specific capacitance of approximately 474 Fg^{-1}which is almost five times the specific capacitance of the clean ZFe_2O_4 nanoparticles (Z = Ni, Co, Cu, etc). The composite exhibited a whopping 116 Wh/kg energy density, which is approximately five times the energy density of the individual composites. The pure $CoFe_2O_4$ nanoparticles lost more energy than the $CoFe_2O_4$/C when measured with galvanostatic charge-discharge. However, the discharge time was found to be higher with $CoFe_2O_4$/C than the pure $CoFe_2O_4$ nanoparticles (Table 8.2).

Furthermore, Heydari et al. (2018) used precipitation method to synthesize nanocomposites of carbon/$CoFe_2O_4$ and studied the physical and chemical features of the material using vibrating sample magnetometry, FTIR, TEM, X-ray diffraction, and SEM. They also used galvanic charge/discharge and cycle voltammogram to evaluate the electrochemical features of the nanocomposites. The nanocomposite's XRD spectrum is revealed in Figure 8.10(a) and it shows crystalline lattice plane of (511), (400), and (311) which correspond to the theta values of 57°, 43°, and 37° for the $CoFe_2O_4$ lattice (Chen et al. 2010). The peaks of the XRD pattern are not sharp because of the presence of carbon in the $CoFe_2O_4$ composite. At reflections of (112) and (101), the composite exhibited two theta values of 83° and 44° which correspond to JCPDS card number of 01-075-1621. On the other hand, Figure 8.10(b) revealed the EDX spectrum of the nanocomposite which designates around 0.5 ratios for the Co/Fe composite, and this agrees with the formula of the nanosynthesized $CoFe_2O_4$. The composite carbon, iron, oxygen, and cobalt are well represented with appropriate ratios in the nanocomposite.

Uniform morphology that are not much varied having 209 ± 16 nm sizes were observed with FESEM analysis as shown in Figure 8.11(a). The surface was made highly permissible to electrolytes due to the presence of cracks and porous surfaces during the electrochemical process. The elevated atomic mass of

TABLE 8.2

Electrochemical Properties of $CoFe_2O_4$-Based Supercapacitor Electrodes

Electrodes	Methods	Capacitance (F/g)	Current Density (A/g)	Electrolytes	References
$CoFe_2O_4$/MXene	Coprecipitation	594, 1046.25, and 1268.75	1	0.1M KOH	(Ayman et al. 2020)
$CoFe_2O_4$NM-As/Ni	Hydrothermal	1426 and 1024	1 and 20	0.1M KOH	(Liu et al. 2016)
$CoFe_2O_4$	Coprecipitation	366	1	1M NaOH	(Kumbhar et al. 2012)
$CoFe_2O_4$	Coprecipitation	77.8–835.7	1	1M NaOH	(Vargas et al. 2020)
$CoFe_2O_4$	Hydrothermal	316	1	6M KOH	(Kennaz et al. 2016)
$CoZnFe_2O_4$	Coprecipitation	377	0.2–0.6	1M NaOH	(Rani et al. 2018)
$CoFe_2O_4$/GO/PANI	Hydrothermal	392.3–1133.3	0.1	1M KOH	(Xiong et al. 2014a)
$CoFe_2O_4$/GO	Thermal decomposition	304–1120	0.83	6M KOH	(Tung et al. 2020)
$CoFe_2O_4$NSs@rGO	Coprecipitation	2238.3, 1723.8, 1606.3, 1577.5, 1321.3, and 1225	1, 5, 7, 10, 15, and 20	1M LiPF	(Dong et al. 2017)
$CoFe_2O_4$	Hydrothermal	460.5	1	3M KOH	(Lalwani et al. 2018)
$CoFe_2O_4$	Hydrothermal	828, 1100, and 1148	1–10	6M KOH	(Gao et al. 2020)
$CoFe_2O_4$/AC	Hydrothermal	73.12– 503	1.2–20	3M KOH	(Gao et al. 2017)
NCFC-NCs and $CoFe_2O_4$	Polymeric route	94–474	5 mV/s	5M KOH	(Alshehri et al. 2017)
C/$CoFe_2O_4$	Coprecipitation	36.56, 53.1, 75.0, and 106.5	22, 6.4, 1.4, and 0.64	5M KOH	(Heydari, Kheirmand, and Heli 2018)

the composing particles arranged around the carbon material exhibited non-uniform TEM micrograph. The term *micrograph* reveals about 35.6 ± 7.8 nm average sizes which signify the existence of high surface area in the presence of the carbon constituent. TEM and FESEM structures are confirmed by Figure 8.11(b and c) showing 3.264 nm mesopore diameter sizes.

In order to unravel the functional group and associated bond in the synthesized C/$CoFe_2O_4$ nano-composite, Figure 8.12(a) shows the FTIR spectrum of the synthesized nanocomposite. The spectrum revealed O–H stretching vibration for the watered crystalline ferrite corresponding to 3417 cm^{-1} broad peak. The presence of carbon in the nanocomposite prompted the existence of C–C bond with 1126 and 1621 cm^{-1} peaks. The Co–OH vibrational bond denotes 430 cm^{-1} peak while stretching vibrational mode for the metal–oxygen interaction was shown with 656 cm^{-1} (Andrade et al. 2014). Furthermore, the magnetic hysteresis loop for the sample is shown in Figure 8.12(b), which shows negligible loop as a result of zero remnant magnetization and coercive field. The material is said to have superparamagnetic attributes. In this work, 191.35 G and 9.73 emu/g coercivity and saturation magnetization were obtained in that order whereas 0.7 emu/g was obtained from the residual magnetization.

KOH of 5M concentration was used in galvanostatic charge-discharge curve and cyclic voltammetry to study the electrochemical properties of the C/$CoFe_2O_4$ nanocomposite in a system of three electrodes. Figure 8.13 represents the recorded cyclic voltammogram of the sample at a 50 mV/s rate of potential sweep. An inset of acetylene black electrode and carbon synthesized was also used for comparison, and it showed a very low charging current whereas the original nanocomposite revealed significant capacitive current. The cobalt ferrite nanocomposite also exhibited relatively high current density unlike the inlet counterpart. Equation (8.5) shows the formula for calculating the gravimetric capacitance of the composite (VijayaSankar and KalaiSelvan 2016):

$$C_g = \int i \frac{dV}{2m} v. \; \Delta V \tag{8.5}$$

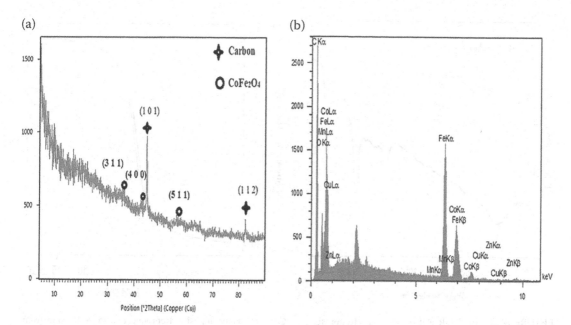

FIGURE 8.10 (a) XRD and (b) EDX spectra of C/CoFe$_2$O$_4$ nanocomposite. Reproduced with permission from Heydari et al. 2019, *International Journal of Green Energy*.

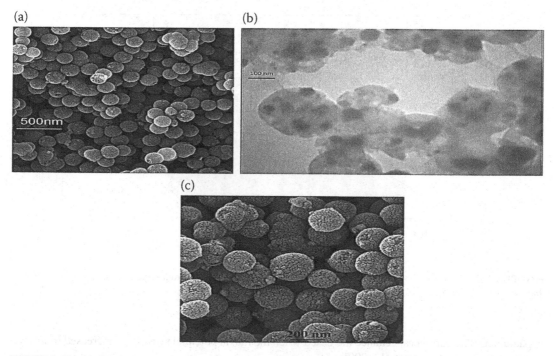

FIGURE 8.11 (a) SEM, (b) TEM, and (c) FESEM micrograph of C/CoFe$_2$O$_4$ nanocomposite. Reproduced with permission from Heydari et al. 2019, *International Journal of Green Energy*.

where the quantity of charge stored in the cyclic voltammogram, m, is denoted by $\int i dV(AV)$, the mass of the electrode deposited is represented by m (g), the difference in potential is referred to as ΔV (V), and the rate of potential sweep is given as v (V/s). The nanocomposite exhibited 6.2 F/g gravimetric

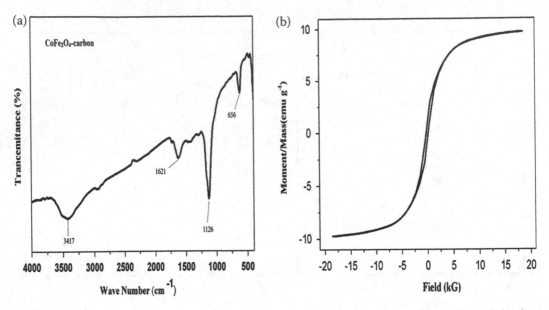

FIGURE 8.12 (a) FTIR and (b) magnetic hysteresis of $C/CoFe_2O_4$ nanocomposite. Reproduced with permission from Heydari et al. 2019, *International Journal of Green Energy*.

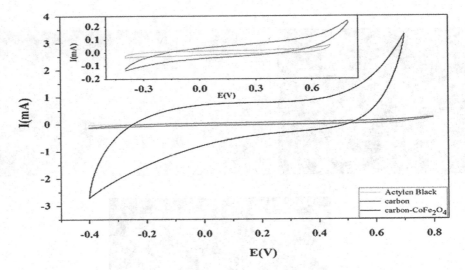

FIGURE 8.13 CV of $C/CoFe_2O_4$ using 5M KOH. Reproduced with permission from Heydari et al. 2019, *International Journal of Green Energy*.

capacitance. The current capacitance increased when the rate in potential sweep was increased following equation (8.3) (Ardizzone et al. 1990):

$$C(v) = C(\infty) + constant / \sqrt{v} \qquad (8.6)$$

where the capacitance rate of potential sweep of infinite and v are denoted with $C(\infty)$ and $C(v)$, accordingly. Here, 10.9 F/g was obtained for the $C(\infty)$, and this accrued at the external surface of the material. This equation can be rearranged to obtain equation (8.7).

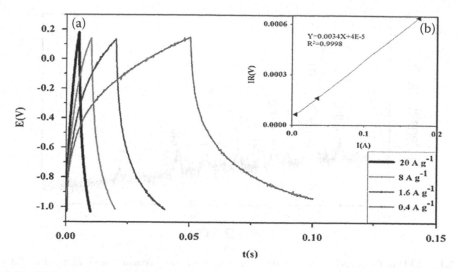

FIGURE 8.14 Galvanostatic charge-discharge plots of the C/CoFe$_2$O$_4$. Reproduced with permission from Heydari et al. 2019, *International Journal of Green Energy.*

$$C(v) = C(0) + constant\ X\sqrt{v} \tag{8.7}$$

where the capacitance at zero rate of potential sweep is denoted by C(0), and it equals the utmost summation of outer and inner capacitance of the composite (VijayaSankar and KalaiSelvan 2016). Respective 67.4 F/g and 78.3 F/g capacitance were obtained for the inner and C(0) using equation (8.7). Furthermore, Figure 8.14 was used to analyze the two electrode system to prove the performing ability of the charge storage using charge-discharge study. It shows the existence of semi-triangular curve for the currents while equation (8.8) was used to estimate the gravimetric capacitance of the sample (Heli and Yadegari 2014):

$$C_{sg} = \frac{i\ X\Delta t}{\Delta V\ X\ m} \tag{8.8}$$

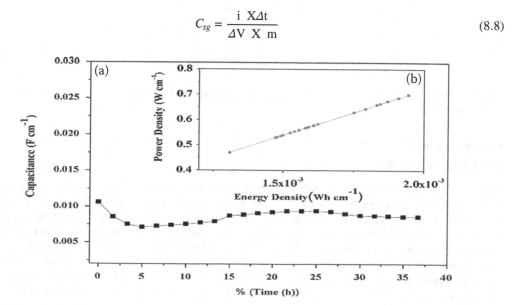

FIGURE 8.15 Cyclic stability of the C/CoFe$_2$O$_4$. Reproduced with permission from Heydari et al. 2019, *International Journal of Green Energy.*

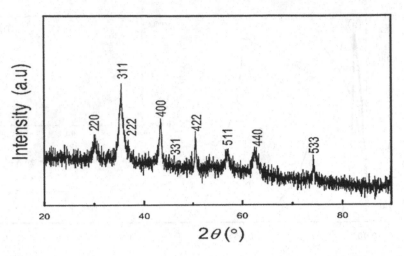

FIGURE 8.16 XRD of C/CoFe2O4. GN nanocomposite. Reproduce with permission from Zhang et al. 2015, *J. ACS Appl. Mater. Interfaces.*

where the respective discharge time and current density are Δt and I. At different current densities of 22, 6.4, 1.4 and 0.64 respective 36.56 A/g, 53.1 A/g, 75.0 A/g, and 106.5 A/g were obtained. This shows that as the decrease in current density leads to corresponding increase in specific capacitance with the same semi-angular shape retained which indicate excellent electrochemical process. During charge-discharge cycles, insignificant loss in energy was confirmed by the reversal low resistant 0.75 Ω edge current for different current densities as shown in Figure 8.15. At 1 mA/cm current density and beyond 6000 charge/discharge cycles, the two-electrode system showed outstanding 81% retention of its capacitance as shown in Figure 8.16. Equations (8.9) and (8.10) were used to estimate the respective energy density and power density (Li et al. 2014):

$$\text{Energy density} = C\frac{V^2}{2} \tag{8.9}$$

$$\text{Power density} = \frac{E}{\Delta t} \tag{8.10}$$

The C/CoFe$_2$O$_4$ nanocomposite revealed appreciable power and energy densities of 7.057 Wh/kg and 31.224 W/kg.

8.6 Copper Ferrite

Presently, copper ferrite is one composite among others that has been used as spinel compounds for water organic pollutant degrading catalyst and electrochemical capacitors (Zhou et al. 2018; Gao et al. 2019; Huang et al. 2019b; Huang et al. 2019; Lia et al. 2019). The strength of metal ferrites' crystal structure and their corresponding three-dimensional diffusion route play a vital role in their supercapacitive prowess (Zhu et al. 2013a; Silva et al. 2019; Zhang et al. 2019a). The combination of copper and iron produces a synergistic property that makes the redox reaction to be enhanced in order to obtain elevated supercapacitive features. This CuFe$_2$O$_4$ material has one of the best spinel structures among the binary transition metals. The CuFe$_2$O$_4$ has a structure composed of bi-cations which are Cu^{2+} and Fe^{3+} that occupy octahedral and tetrahedral positions, respectively (Nilmoung et al. 2016). Many researchers have reported the importance of these nanocomposites in the area of batteries, supercapacitors, catalytic dye degradation immobilized lipases, among others. Jin et al.

(2011) fabricated hollow-shaped carbon-coated $CuFe_2O_4$ ball that exhibited 550 mAhg^{-1} elevated specific capacity by employing the method of polymer template hydrothermal growth. In terms of supercapacitive ability, Zhang et al. (2015) used simple solvothermal technique to fabricate copper ferrite graphene nanosheets ($CuFe_2O_4$-GN) and evaluated the high supercapacitive performance of the electrodes. The result revealed that at 1 A/g the material yielded 576.6 F/g electrochemical capacitance. The result showed that using materials with high specific capacitance can assist in enhancing the performance of supercapacitors and also reduce the environmental problems associated with fossil fuels. Fu et al. (2012) used one-step hydrothermal path to fabricate hetero-architecture $CuFe_2O_4$ having multiple functional features. The samples when irradiated with visible light of the electromagnetic spectrum exhibited elevated photocatalytic activity. It also has excellent magnetic cycling and wonderful electrochemical properties when used as the anode of lithium-ion batteries. The graphene sheets that are decorated and exfoliated with nanoflakes of the hexagonal $CuFe_2O_4$ were studied using TEM. They also observed that when graphene and $CuFe_2O_4$ are combined under the irradiation of visible light, an inert $CuFe_2O_4$ of very high active photocatalytic catalyst for methylene blue degradation was formed. The heteroarchitecture of $CuFe_2O_4$-graphene revealed outstanding magnetic features, since it is magnetically recyclable in a suspension system. Using 25%wt of graphene into heteroarchitecture $CuFe_2O_4$-graphene as a lithium-ion battery anode material revealed elevated 1165 mAh g^{-1} specific reversible capacity with moderate capability rate and cycling stability. The special attributes of the heteroarchitecture nanocomposite $CuFe_2O_4$-graphene shows superior electrochemical performance and photocatalytic activity which offers a notable synergistic effect between the graphene sheets and the $CuFe_2O_4$ nanoflakes. Other metals have also been combined with cobalt to produce effective material for energy storage. In that light, Bhujun et al. (2016) used sol gel technique to fabricate a new nanomaterial of nickel copper ferrite doped with aluminium for supercapacitor application. They employed the services of XRD spectroscopy, energy dispersive spectroscopy, and SEM to study the crystallinity, chemical composition, and morphology of the materials. On the other hand, they used galvanostatic charge/discharge and cyclic voltammetry to investigate the electrochemical features of the deposited materials. Using 1 A/g current density, $Al_{0.2}Ni_{0.4}Cu_{0.4}Fe_2O_4$ material exhibited 412.5 F/g specific capacitance with 57.3 Wh/Kg energy density (sol gel). A fascinating synergistic ternary composite material made of polyaniline-acetylene black-copper ferrite was studied by Das and Verma (2019a, b). They used FTIR, SEM, EDX, and XRD to study the chemical bond, physical morphology, elemental composition, and crystal structure of the composite, respectively. KOH with 1M was used during the electrochemical study. The composite revealed 192.64 F/g specific capacitance at 0.5 A/g. On the other hand fabricating a symmetric hybrid of the supercapacitor yielded specific power and specific energy of 3165.25 Wh/kg and 26.757 Wh/kg in that order Table 8.3.

TABLE 8.3

Electrochemical Properties of $CuFe_2O_4$-Based Supercapacitor Electrodes

Electrodes	Methods	Capacitance (F/g)	Current Density (A/g)	Electrolytes	References
$CuFe_2O_4$/PANI	Sol gel/combustion	192.64	0.5	0.1M KOH	(Das and Vermab)
$CuFe_2O_4$	Hydrothermal	1165			(Fu et al. 2012)
$Al_{0.2}Ni_{0.4}Cu_{0.4}Fe_2O_4$	Coprecipitation	412.5	1	5–100 mV/s	(Bhujun et al. 2016)
$CuFe_2O_4$-GN	Solvothermal	576.6	1	3M KOH	(Zhang et al. 2015)
$CuFe_2O_4$ fibre	Hydrothermal	28	0.5	1M KOH	(Zhao et al. 2012)
$CuFe_2O_4$ film	Coprecipitation	5.7	0.3 μA/cm^2	1M NaOH	(Ham et al. 2009)
$CuFe_2O_4$ nanosphere	Hydrothermal	334	0.6	1M KOH	(Zhu et al. 2013b)
$CuFe_2O_4$ nanowires/CNT	Thermal decomposition	267	0.83	1M KCl	(Giri et al. 2012)

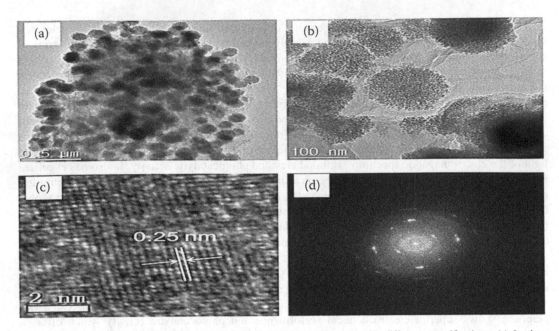

FIGURE 8.17 (a and b): TEM micrographs of CuFe2O4-GN composites with different magnifications, (c) Lattice-resolved HRTEM micrograph of CuFe2O4-GN composite and (d), selected area electron diffraction pattern of CuFe2O4 in the composite reproduce with permission from Zhang et al. 2015, *J. ACS Appl. Mater. Interfaces.*

During the electrochemical study of $CuFe_2O_4$-GN by Zhang et al. (2015) the synthesis was done using solvothermal method. They characterized the samples for the structural, magnetic, electrochemical, and morphological features using X-ray diffractometer, physical property measurement system, electrochemical workstation, and scanning electron microscopy in that order. The XRD spectra for the $CuFe_2O_4$-GN nanocomposite is shown in Figure 8.17. The diffraction peaks of the $CuFe_2O_4$ nanocomposite is represented with 01-077-0010 JCPDS card number which shows the peaks of CuO and CuO_2. The cubic structured $CuFe_2O_4$ had hkl index of (533), (440), (511), (422), (331), (400), (222), (311), and (220) which correspond to 2θ values of 74.2°, 62.6°, 57.3°, 50.8°, 46.4°, 43.4°, 37.0°, 35.5°, and 30.0° respectively. The effect of the GN was observed from the broad peak shown at 2θ value of 23° which corresponds to reflection index of (002), and this indicates the reduction of GO during the synthesis (Nethravathi and Rajamathi 2008; Si and Samulski 2008).

The TEM micrograph of the synthesized nanocomposite of $CuFe_2O_4$-GN is shown in Figure 8.18(a–d). Figure 8.18(a and b) shows $CuFe_2O_4$ nanoparticles that are narrowly arranged with highly dense surfaces clouded with GN. They also have averaged- sized particles of within 100 nm with sphere-like structure. In other to elucidate the effects of the GN on the nanoparticles, HRTEM was employed as shown in Figure 8.18(c) and the nanocrystal revealed lattice resolved HRTEM $CuFe_2O_4$ micrograph. The 0.25 nm spacing of the lattice fringe agreed with the (311) hkl deflection plane for the nanocrystal structure (Fu et al. 2012). Furthermore, Figure 8.18(d) revealed the diffraction rings of the nanoparticles using the selected area electron diffraction (SAED) pattern in Figure 8.18(d) showed diffraction rings of the prepared $CuFe_2O_4$ nanoparticle on GN. The cubic spinel nanostructure was found to be responsible for the rings.

In order to study the electronic structure and chemical composition of the $CuFe_2O_4$-GN nanocomposite, XPS was employed, and it is shown in Figure 8.19(a–d). Figure 8.19(a) revealed O, Cu, C, and Fe as the elemental composition. Furthermore, Figure 8.19(b) shows the presence of C which emanates as a result of C–C bond with 284.5 eV which is associated with the GN. Weak peaks observed in Figure 8.19(b) occurred due to the oxygenation of carbons. Reduction of epoxy groups was prominent among the groups of GN containing oxygen. The uses of ethylene glycol during the solvothermal process as a reducing agent proved that GN can be obtained by reducing GO in the

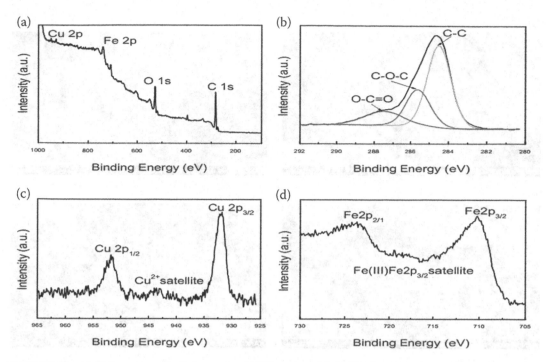

FIGURE 8.18 (a) Wide scan and (b-d) deconvoluted XPS spectra of CuFe2O4-GN. Reproduce with permission from Zhang et al. 2015, *J. ACS Appl. Mater. Interfaces.*

FIGURE 8.19 SEM micrograph of the nanocomposite CuFe2O4-GN wirh (a) 0.5, (b) 1.0, (c), 2.0 and (d) 4.0 g/L GO concentration. Reproduce with permission from Zhang et al. 2015, *J. ACS Appl. Mater. Interfaces.*

(a)   (b)

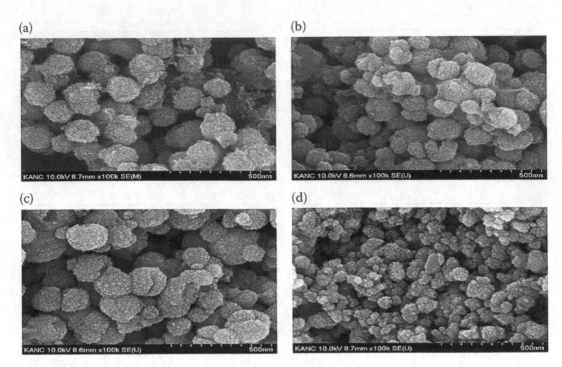

(c)   (d)

FIGURE 8.20 SEM micrograph of CuFe2O4-GN capped with PVP with (a) 0, (b) 50, (c) 100 and (d) 150 g/L concentrations. Reproduce with permission from Zhang et al. 2015, *J. ACS Appl. Mater.*

presence of minute oxygen residual groups (Dreyer et al. 2011; Zhao et al. 2014). The Fe and Cu ions in the nanocomposite have Cu $2p$ and Fe $2p$ binding energy in the high-resolution XPS spectra as shown in Figure 8.19(c and d) which are in consonance with other research articles (Zhu et al. 2013). The presence of copper ions was observed with the binding energy of $Cu_2p^{1/2}$ and $Cu_2p^{3/2}$ at 951.8 eV and 931.9 eV, having 943.5 eV optimal peak of the Cu^{2+} (Fu et al. 2012). The $2p$ spectra of Fe have 724.2 eV and 710.2 eV binding energies corresponding to $2p^{1/2}$ and $2p^{3/2}$. The presence of Fe^{3+} cations shows satellite peaks at around 718.4 eV binding energy and this confirms the nanocomposite of $CuFe_2O_4$.

When different amounts of GO were added to the $CuFe_2O_4$ nanocomposite, the morphological effect of the GO addition was studied as shown in Figure 8.20(a–d). The $CuFe_2O_4$ nanocomposite had uniform sizes of approximately 90 nm when 0.5 g/L GO was introduced to the composite. The low concentration of GO could not mask the appearance of GN in the composite. Addition of 1 g/LGO and 2 g/LGO to the solution as shown in Figure 8.20(b and c) exhibited nanorod structures of the composite as a result of the GO while the nanoparticles of $CuFe_2O_4$ were not observed when the concentration of GO was increased to 4.0 as shown in Figure 8.20(d).

PVP was added to the $CuFe_2O_4$-GN nanocomposite at different concentrations to ascertain its effect on the morphology. The SEM micrograph of the samples with different concentrations of PVP as a stabilizing agent are shown in Figure 8.21(a–d). Figure 8.21(a) shows the SEM micrograph for the sample with zero (0) or no PVP capping agent. Addition of 50 g/L (Figure 8.21(b)) and 100 g/L (Figure 8.21(c)) concentration of PVP altered the uniform sizes of the images which is around 150 nm. Increasing the quantity of the PVP decreases the size of the nanocomposite $CuFe_2O_4$-GN as shown in Figure 8.21(d) and this affects the size-control effect of the nanocomposite (Xia et al. 2009).

The electrochemical feature of the deposited nanocomposites of $CuFe_2O_4$ and $CuFe_2O_4$-GN is shown with CV curve in Figure 8.22. The CV was carried out at a scan rate of 50 mV/s from -0.2 V to 0.7 V. The $CuFe_2O_4$ and $CuFe_2O_4$-GN samples revealed dual redox peaks. This shows that the capacitances of

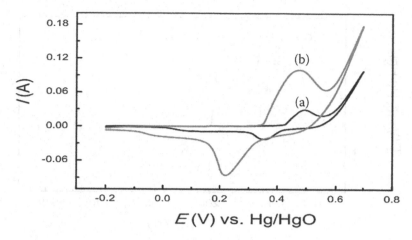

FIGURE 8.21 CV curves of the (a) CuFe2O4-GN and (b) CuFe2O4 at 50 mV/s scan rate using 3 M KOH solution. Reproduce with permission from Zhang et al. 2015, *J. ACS Appl. Mater. Interfaces.*

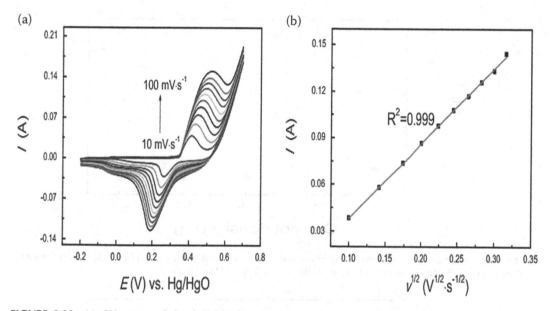

FIGURE 8.22 (a) CV curves of the CuFe2O4-GN at different scan rates using 3 M KOH solution and (b) the i and v1/2 plot. Reproduce with permission from Zhang et al. 2015, *J. ACS Appl. Mater. Interfaces.*

the $CuFe_2O_4$ and $CuFe_2O_4$-GN start from pseudocapacitance. The elevated peak current of $CuFe_2O_4$-GN when compared to $CuFe_2O_4$ denotes relative higher specific capacitance. The $CuFe_2O_4$-GN also exhibited negative shift in redox potential peaks as a result of the extraordinary conductivity nature of the GN in the $CuFe_2O_4$-GN composite. Using 3M concentration of KOH, the CV of the $CuFe_2O_4$-GN was studied at scan rates of 10–100 mV/s, and this gives the relationship between the supercapacitive performance and the scan rate of the $CuFe_2O_4$-GN.

In Figure 8.23(a), the redox process at different rates revealed that at different scan rates, the current peaks increase which indicates moderate composite capability (Chen et al. 2009). On the other hand, Figure 8.22(b) revealed a linear relationship of the peak current to the square root of the scan rate and this confirms diffusion-controlled pattern at the electrode surface during the electrochemical process of the $CuFe_2O_4$-GN.

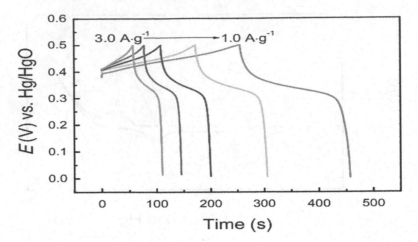

FIGURE 8.23   Charge-discharge curves of CuFe$_2$O$_4$-GN electrode at different current density. Reproduce with permission from Zhang et al. 2015, *J. ACS Appl. Mater. Interfaces*

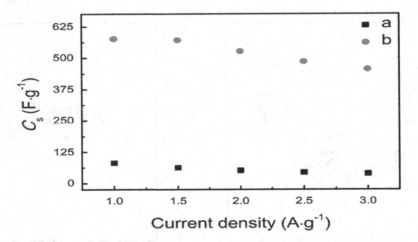

FIGURE 8.24   The relationship between the specific capacitance and current density of (a) CuFe$_2$O$_4$ and (b) CuFe$_2$O$_4$-GN. Reproduced with permission from Zhang et al. 2015, *J. ACS Appl. Mater. Interfaces*.

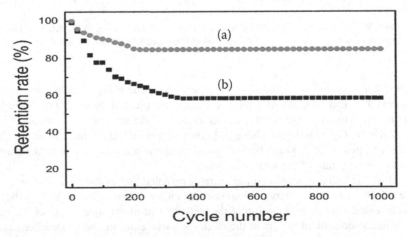

FIGURE 8.25   Stability check on (a) CuFe$_2$O$_4$-GN and (b) CuFe$_2$O$_4$ at 1.0 A/g constant current density. Reproduce with permission from Zhang et al. 2015, *J. ACS Appl. Mater. Interfaces*.

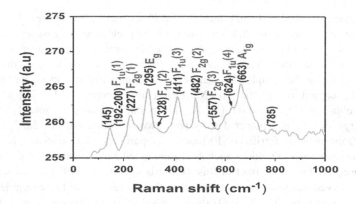

FIGURE 8.26 Raman spectrum of NiFe$_2$O$_4$. Reproduced with permission from Mordina et al. 2020, *Journal of Energy Storage.*

Furthermore, Figure 8.24 shows the specific capacitance of CuFe$_2$O$_4$-GN at 1.0 A/g, 1.5 A/g, 2.0 A/g, 2.5 A/g, and 3.0 A/g varying ranges of increasing discharge current density. The result from the CV analysis shows that CuFe$_2$O$_4$-GN has larger charge capacity.

Figure 8.25 shows the relationship between current density and specific capacitance and the plot revealed that at 1 A/g, the utmost specific capacitance obtained from the CuFe$_2$O$_4$-GN is 576.6 F/g. High specific capacitance can also be obtained from the samples when the discharge current density is reduced and this leads to improved capacitance. Increasing the current density of CuFe$_2$O$_4$-GN from 1.0 A/g to 3.0 A/g retains its capacitance at 79.1% for the capacitance of 576.6 A/g to 456.0 F/g, while the capacitance of CuFe$_2$O$_4$ changes from 81.5 to 39.3 F·g^{-1} denoting 48.2% loss in its capacitive retention. In a nutshell, the result showed that the capacitance of CuFe$_2$O$_4$ was enhanced when GN was incorporated into the composite (Stoller et al. 2008).

The CuFe$_2$O$_4$-GN had 11 times energy density than that of CuFe$_2$O$_4$ and this is shown in the result of the CuFe$_2$O$_4$-GN having 15.8 Wh/kg at energy density at 15.8 Wh/kg power density. The charge-discharge stability check of the CuFe$_2$O$_4$ and CuFe$_2$O$_4$-GN composites for 1000 cycles using 1 A/g is shown in Figure 8.26. the figure shows that after 300 cycles 58% of the CuFe$_2$O$_4$ capacitance was retained while 85% of the CuFe$_2$O$_4$-GN composite capacitance was retained. This confirms that the nanocomposite of CuFe$_2$O$_4$-GN is more stable than that of CuFe$_2$O$_4$.

8.7 Nickel Ferrite

One of the most recognized faradic electrode material is NiFe$_2$O$_4$ and this is due to its naturally accommodating feature in the environment, low cost, and its abundance in earth's crust. Different nanorod and nanoparticles of NiFe$_2$O$_4$ including their various hybrids such as poly(3,4-ethylene-dioxythiophene)-NiFe$_2$O$_4$, PANI-NiFe$_2$O$_4$, and graphene-NiFe$_2$O$_4$ have been used to form composites of NiFe$_2$O$_4$ for applications in supercapacitor (Sen and De 2010; Senthilkumar et al. 2013; Wang et al. 2013). Binder-free electrode has been used to deposit NiFe$_2$O$_4$ nanoparticles on carbon cloth (Yu et al. 2014). Sol gel, polyol-mediated, combustion, and other methods have been used to synthesize NiFe$_2$O$_4$, and their electrochemical performances have been compared (Anwar et al. 2011). Stainless steel, conducting glass, and Ni-foam have been used as depositing substrates for NiFe$_2$O$_4$ nanoparticles, and these substrates play a vital role in determining the overall electrochemical performances of the nanoparticles (Yu et al. 2014).

Sharif, Yazdani, and Rahimi (2020) used hydrothermal technique to synthesize nanoparticles of Ni$_{1-x}$Mn$_x$Fe$_2$O$_4$ and the material was used for supercapacitor application. In their study, the Mn element was used to substitute Mn in a NiFe$_2$O$_4$ composite and the effect of this substitution on the structural, electrochemical, and electronic features was studied using appropriate equipment. The conversion of NiFe$_2$O$_4$ inverse structure to MnFe$_2$O$_4$ was confirmed with the help of Raman spectroscopy while the single-phase structure was observed using the X-ray diffractometry. The spherical shape of the nano-particles obtained was studied using field-emission scanning electron microscopy and it revealed 20 –30 nm sized particles. The study also showed that as the Mn content increases there is a decrease in the optical bandgaps of the samples. The outstanding performance of the samples during the electro-chemical study shows that the samples have attractive cycling stability and as the content of Mn increases the specific capacitance of the $Ni_{1-x}Mn_xFe_2O_4$ also increases. 1221 F/g maximum specific capacitance was obtained for the $MnFe_2O_4$. At this point, the sample showed 88.16 Wh/kg and 473.96 W/kg energy density and power density respectively. Heidari et al. (2015) fabricated super-capacitor with graphene/nickel ferrite(G-NF)-based nanoparticles using one-step facile solvothermal method with varying synthesis conditions such as change in temperature and time of synthesis. HRTEM and XRD were used to study the effect of this synthesis conditions on the structural features of the samples, and this showed that there is enhancement of the crystallinity of these samples as temperature and time of deposition increase. The $NiFe_2O_4$ showed good phase and high reduction of graphene oxide to graphene as these conditions increase. Correspondingly, the highest specific capacitance for the $NiFe_2O_4$ during the electrochemical study at 10 h and 180°C solvothermal process at 5 A/g and 1 A/g current densities were 196 F/g and 312 F/g. for the stability test, the G-NF nanomaterial at 10 A/g current density revealed 105 F/g capacity after 1500 cycles, and this shows extraordinary feature as the material is expected to have long cycle during its electrochemical performance. Bhojane et al. (2017) reported low-cost facile synthesis nanostructure of $NiFe_2O_4$ deposited using chemical bath technique. They obtained porous $NiFe_2O_4$ structure with ammonia-assisted template, which offered superior electrochemical performance to pure $NiFe_2O_4$. The samples were characterized using XRD, HRTEM, FESEM for the morphology and structural properties of the deposited materials. The monitoring of the structural, morphological, and the electrochemical attributes of the composites were performed with the role of ammonia. At a scan rate of 2 mV/s and utmost 541 F/g specific capacitance and this value is higher than the one obtained in $NiFe_2O_4$ without incorporation of carbon derivatives. Furthermore, monodispersed nanospheres of $NiFe_2O_4$ were synthesized by Ghasemia, Kheirmand, and Hel (2019) and the FTIR, XRD, FESEM, and the electrochemical studies were carried out to ascertain the elemental structure and bonds, structure, morphology and electrochemical features of the prepared nanospheres so as to understand their usability as electrode material for supercapacitor application. An elevated 122 F/g specific capacitance at 8.0 A/g current density was obtained from the $NiFe_2O_4$ nanocomposite. The sample also revealed an outstanding specific energy of 16.9 Wh/kg. furthermore, at current density of 4 A/g, maximum specific capacity of 137.2 F/g was obtained from the nanospheres and this result did not decrease even at a higher cycling test which shows its strong stability. The core-shell nanocomposite of $NiFe_2O_4$/PPy with elevated supercapacitive ability was reported by Scindia, Kamble, and Kher (2019) when they employed inexpensive and simple in situ chemical oxidation technique in a medium con-taining aqueous surfactant sodiumdodecyl sulphate (SDS) during the preparation of the electrode ma-terial. They also characterized the prepared electrode material for morphological, thermal, electrical, electrochemical, and structural attributes. Electrochemical impedance spectroscopy, charge-discharge, and cyclic voltammetry were used to characterize the electrochemical features of the nanocomposites. Electrolyte containing 0.1M H_2SO_4 was used during the electrochemical test of the samples, and with different concentrations of the electrolyte solution, the various stability and specific capacitances of the electrode materials were critically analyzed. An excellent 721.66 F/g specific capacitance was obtained from the electrode material. Furthermore, 99.08%, 6.18 kW kg^{-1}, and 51.95 Wh kg^{-1} were revealed to be the coulomb efficiency (η%), specific power, and specific energy in that order, and the result after 1000 cycles of charging and discharging shows that the material is very stable for supercapacitor ap-plication as energy storage device. Nanocomposite of nickel ferrite/graphene was fabricated and its electrochemical features were studied by Soam et al. (2019) for application in supercapacitor. The team of researchers studied the morphological and structural attributes using SEM, TEM, and XRD. Electrochemical impedance spectroscopy, charge-discharge, and cyclic voltammetry were used to measure the electrochemical features of the nanocomposites. The nanosheets of graphene were used to moderate or alter the morphology, as well as improve the electrochemical attributes of the nano-composites. This study showed that 207 F/g specific capacitance was obtained for the graphene in-corporated nanocomposite, and this result is quadruple of the one obtained from the as-deposited

$NiFe_2O_4$. After 1000 cycles, 95% of the capacitance of the nanocomposite was retained during the stability test in an electrolyte containing 1M Na_2SO_4. The graphene nanosheets minimized the resistance of charge transfer as can be observed from the results obtained from the electrochemical impedance spectroscopy of the nanocomposite material. Venkatachalam and Jayavel (2015) used citric acid fuel to synthesize new $NiFe_2O_4$ nanocrystals via combustion method. XRD was used to study the phase and structure of the nanocomposite material, and the XRD study showed that the $NiFe_2O_4$ nanocrystal has a cubic phase. Electrochemical impedance spectroscopy, chronopotentiometry, and cyclic voltammetry were used to study the electrochemical features of the $NiFe_2O_4$ electrode material. The highest specific capacitance obtained from the electrode material is 454 F/g and it showed attributes of pseudocapacitor. The stability test revealed that after 1000 cycles, greater percentage of the capacitance was retained. Co-electrospinning technique was employed in formation of nanocomposite $NiFe_2O_4@CoFe_2O_4$ core–shell nanofibres by Wang et al. (2020) and the electrochemical features were enhanced by the synergistic effect of the combined nanomaterials formed where the nanocomposite revealed 480 F/g specific capacitance at 1 A/g current density. Addition of graphene into nickel ferrite and polyaniline to form nanocomposite of nitrogen-doped graphene/nickel ferrite/polyaniline (NGNP) doped with nitrogen has been studied by Wang et al. (2014). Two-step facile routes were used during the synthesis and different analytical methods were used to evaluate the electrochemical attributes of the nanocomposites formed. The NGNP sample revealed 667.0 F/g specific capacitance at 0.1 A/g current density and 645.0 F/g specific capacitance at 1 mV/s, respectively, using two- and three-electrode system. Furthermore, 110.8 W kg^{-1} and 92.7 Wh kg^{-1} power density and energy density were, respectively, obtained from the two-electrode system and about 10% and 5% of the capacitance were lost, respectively, at 10,000 and 5000 cycles using 5 A/g current density during the stability test (Table 8.4).

Nanoparticles of $NiFe_2O_4$ were synthesized using measurable and simplistic methods by Mordina et al. (2020) and the electrochemical features were evaluated using galvanostatic charge-discharge and cyclic voltammetry equipment. Activated carbon was incorporated into the best electrode material for

TABLE 8.4

Electrochemical Properties of $NiFeO_4$-Based Supercapacitor Electrodes

Electrodes	Methods	Capacitance (F/g)	Current Density (A/g)	Electrolytes	References
$Ni_{1-x}Mn_xFe_2O_4$	Hydrothermal	1221	0.5	3M KOH	(Sharif et al. 2020)
$G/NiFe_2O_4$	Hydrothermal	196 and 312	5 and 1	1M Na_2SO_4	(Heidari et al. 2015)
$NiFe_2O_4$	Chemical bath	541 and 342	0.833 A/g and 2 mV/s	1M KOH	(Bhojane et al. 2017)
$NiFe_2O_4$	Coprecipitation	137.2 and 122	4 and 8	0.1M Na_2SO_4	(Ghasemia et al. 2019)
$NiFe_2O_4/PPy$	Sol gel	624.17, 507.21, and 721.66	1	1M Na_2SO_4	(Scindia et al. 2019)
$NiFe_2O_4$/grapheme	Coprecipitation	60 and 207	5 mV/s	1M Na_2SO_4	(Soam et al. 2019)
$NiFe_2O_4$	Combustion	454	–	1M Na_2SO_4	(Venkatachalam and Jayavel 2015)
$NiFe_2O_4@CoFe_2O_4$	Co-electrospinning	480	1	3M KOH	(Wang et al. 2020)
$G/NiFe_2O_4/PPy$	Hydrothermal	667.0 and 645.0	0.1 A/g and 1 mV/s	1M KOH	(Wang et al. 2014b)
$CnFe_2O_4/CNT$	Hydrothermal	545	1	2M KOH	(Kumar et al. 2018)
$Li_2O/NiFe_2O_4/SiO_2$	Sol gel	32.8	0.5	1M NaOH	(Balamurugan et al. 2018)
$NiFe_2O_4$nanocubes/rGO	Hydrothermal	488	1	(1 g PVA + 1 g KNO_3) gel	(Zhang et al. 2019b)
$NiFe_2O_4$	Coprecipitation	398	1	6M KOH	(Mordina et al. 2020)

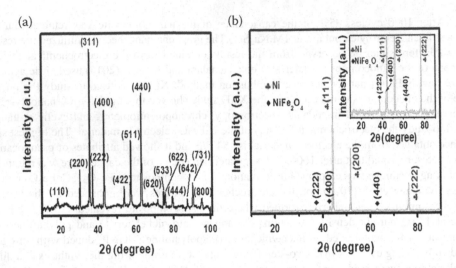

FIGURE 8.27 XRD pattern of NiFe$_2$O$_4$ (a) before cycling and (b) after cycling on Ni-foam. Reproduced with permission from Mordina et al. 2020, *Journal of Energy Storage.*

the fabrication of supercapacitor, and the electrochemical characteristics were also calculated using appropriate methods.

The structural feature of the synthesized NiFe$_2$O$_4$ nanoparticles was characterized using XRD as shown in Figure 8.27(a). The pattern revealed XRD peaks corresponding to hkl lattice orientation of (800), (731), (642), (444), (622), (533), (620), (440), (511), (422), (400), (222), (311), (220), and (111) for 2θ values of 95.38°, 90.49°, 87.51°,79.52°, 75.53°, 74.67°, 71.58°, 63.06°, 57.45°, 53.90°, 43.34°, 37.22°, 35.7°, 30.35°, and 18.35°, respectively. The peaks of diffraction agree with JCPDS card no. 74-1913 which is attributed to the nanocomposite of NiFe$_2$O$_4$. Scherer formula was used to calculate the average crystal size of the NiFe$_2$O$_4$ at (311) diffraction plane. This result revealed an approximate 27.4 nm crystal size. In order to confirm that the NiFe$_2$O$_4$ has a face-centered cubic crystal structure, the diffraction plane at (311) revealed 0.5788 nm^3, 5.37 g/cm^3, and 0.8334 nm volume, density, and lattice parameter, respectively (Kooti and Sedeh 2013).

The structural stability of the synthesized NiFe$_2$O$_4$ on Ni-foam after charge-discharge was shown with the XRD pattern in Figure 8.27(b). To avoid contamination, this analysis was carried out on Ni-foam directly and the samples revealed peaks of XRD corresponding to diffraction planes of (440), (400), and (222) at 2θ values of 62.70°, 43.01°, and 37.08° in that order. At reduced values of 2θ, it was

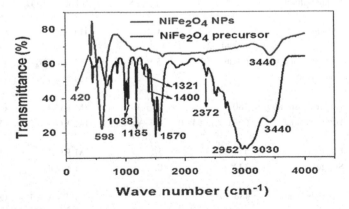

FIGURE 8.28 FTIR spectra of NiFe$_2$O$_4$ precursor and NiFe2O4 nanoparticles. Reproduced with permission from Mordina et al. 2020, *Journal of Energy Storage.*

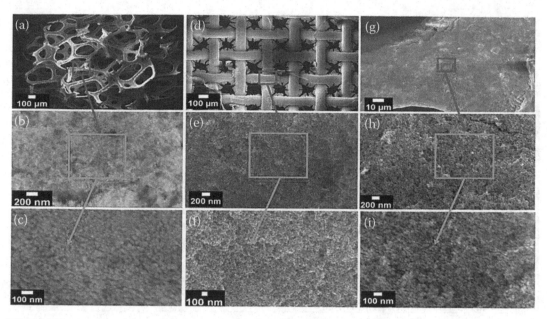

FIGURE 8.29 FESEM micrograph of the $NiFe_2O_4$ deposited on (a, b and c) Ni-foam, (d, e and f) stainless steel mash and (I, j and k).stainless steel foil. Reproduced with permission from Mordina et al. 2020, *Journal of Energy Storage*.

observed that the nanoparticles exhibited small shift in the XRD peaks due to the effect of the cycling. Peaks at lower values of 2θ were found to disappear because of the cycling and also due to the increase in the peaks of the Ni intensity. The dominance of Ni after the cycling presents strong peaks of XRD corresponding to (222), (200), and (111) for 2θ values of 76.32°, 51.78°, and 44.50°, respectively, crystal orientation. The strong XRD peaks at 44.50°, 51.78°, and 76.32° can be ascribed to the diffraction at (111), (200), and (222) crystal planes of Ni, respectively. The $NiFe_2O_4$ nanocomposite revealed 0.5910 nm^3, 5.267 g/cm^3, and 0.8392 nm volume, density, and lattice parameter, respectively, after charge-discharge cycling.

Figure 8.28 gives additional information on the crystal structure of the $NiFe_2O_4$ where the tetrahedral and octahedral sites show the presence of half Fe^{3+} ions and both half Fe^{3+} ions and Ni^{2+} ions, respectively (Dixit et al. 2011). Raman active modes and unitary infrared active mode ($4F_{1u}$) are attributed to five major peaks corresponding to Raman active modes and one infrared active mode (Lazarevic et al. 2013). The oxygen atom with symmetric stretching possesses the most prominent peak at 663 cm^{-1} and it has tetrahedral site for Fe–O with A1g mode. The Fe–O bond of oxygen having asymmetric bending is assigned with peak at 295 cm^{-1}. Combined translation motion of four Fe–O bonds, bending vibration and asymmetric stretching in the tetrahedral and octahedral sites show peaks of $F_2g(3)$, $F_2g(2)$, and $F_2g(1)$, at 557 cm^{-1}, 482 cm^{-1}, and 227 cm^{-1}. The non-equivalent atoms existing in the octahedral site responsible for the peaks at 785 cm^{-1} and 145 cm^{-1} (Lazarevic et al. 2013).

The precursors and nanoparticles obtained from the $NiFe_2O_4$ nanocomposites were subjected to FTIR spectroscopy so as to know the functional groups, bonds, and type of vibrations present in the composite as revealed in Figure 8.29. The figure revealed symmetric and asymmetric C–H stretching bands at 2850 cm^{-1} and 2952 cm^{-1} wavenumbers, which are attributed to the molecules of propyl amine. NH_2 scissoring and C–N stretching have their respective peaks at 1570 cm^{-1} and 1321 cm^{-1} while stretching vibrational mode of N–H has corresponding bands within 3030–3440 cm^{-1} region (Mohapatra et al. 2013; Kooti et al. 2014). The bending vibration of water with H–O–H band and the CH_2 scissoring are attributed to the absorption peaks at 2372 cm^{-1} and 2372 cm^{-1}, respectively (Ahmed et al. 2010). The presence of SO_2 from the molecule of dodecyl sulphate with asymmetric and symmetric vibrations revealed high peaks of absorption, respectively, at 1185 cm^{-1} and 1038 cm^{-1} (Viana et al. 2012). The octahedral and tetrahedral sites from Fe–O (Fe^{3+} bond) presents strong stretching vibration at 598 cm^{-1} absorption band while the octahedral sites from Fe–O (Fe^{2+} bond)

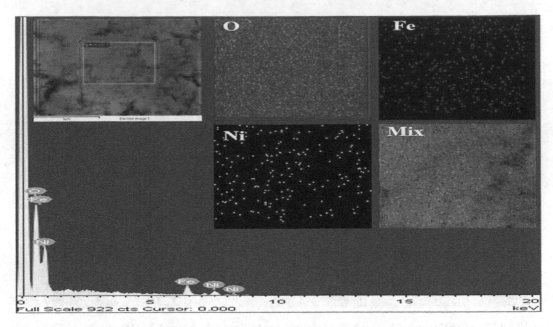

FIGURE 8.30 EDX and elemental distribution of NiFe$_2$O$_4$ nanoparticles. Reproduced with permission from Mordina et al. 2020, *Journal of Energy Storage*.

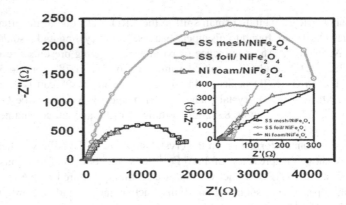

FIGURE 8.31 The EIS spectra of the NiFe$_2$O$_4$ deposited on Ni-foam, stainless steel mesh and stainless steel foil. Reproduced with permission from Mordina et al. 2020, *Journal of Energy Storage*.

having strong stretching vibration also depicts strong peak of absorption at 420 cm^{-1}. The O–H hydroxyl group with stretching vibration has its peak at 3440 cm^{-1} as a result of the presence of water molecules.

The surface morphology of the samples were studied using FESEM and Figure 8.30(a, b, and c), (d, e, and f), and (I, j, and k) shows the FESEM micrograph of theNiFe$_2$O$_4$ deposited on Ni-foam, stainless steel mesh, and stainless steel foil, respectively, with increasing magnifications. An efficient transport pathway is provided by the 3D and 2D skeletal porous nature of Ni-foam and stainless steel mesh thereby making the active material to allow the electrolytes penetrate through them for effective electrochemical study while there is restriction of the electrolyte on the stainless steel foil. Three-dimensional hierarchical choral-like microstructures are shown in Figure 8.30(b, c, e, f, h, and i), as they are formed during the electrode deposition. There is evidence of uniform growth of particles with size range of 20–30 nm on the substrates as can be seen at FESEM high magnification. However, the particle sizes differ from one substrate to another when carefully examined, and this is attributed to variable heat

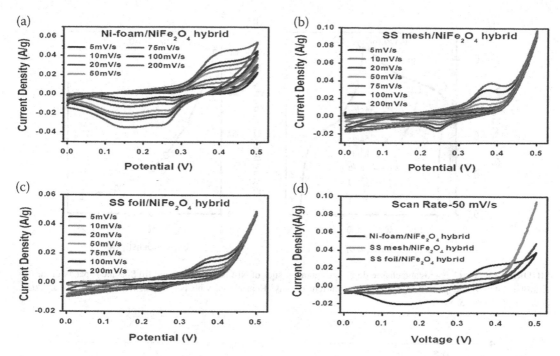

FIGURE 8.32 CV of (a) Ni-foam/NiFe$_2$O$_4$, (b) stainless steel mesh/NiFe$_2$O$_4$, (c) stainless steel foil/NiFe$_2$O$_4$, and (d) the CV of the three samples. Reproduced with permission from Mordina et al. 2020, *Journal of Energy Storage.*

conduction properties and nature of the substrate's porosity for the different substrates. This makes the precursors to be subjected to variable heat flow rate during the annealing of the materials.

Figure 8.31 shows the elemental distribution of the deposited nanoparticles of NiFe$_2$O$_4$. The elemental distribution and EDX revealed the presence of O, Fe, and Ni as the major components of the nanocomposite, and these elements were uniformly distributed all over the substrate. The mapping also confirms that the elemental concentration of Ni is less than that of Fe while the concentration of Fe is less than O as depicted in NiFe$_2$O$_4$ molecular formula.

KOH electrolyte with 6M molar concentration was used to investigate the electrochemical features of the NiFe$_2$O$_4$ nanoparticles using CV. The CVs were carried out at scan rate of 5–200 mVs^{-1} with 0–0.5 V (vs. Ag/AgCl) voltage range. Figure 8.32(a, b, and c) shows the respective CV curves of the deposited NiFe$_2$O$_4$ nanoparticles on (a) Ni-foams (b) stainless steel mesh, and (c) stainless steel foil. It is confirmed that the samples revealed quasi-reversible faradaic reaction since the peak-to-peak separation is over 59 mV, which is a sign of high reduction and oxidation peaks. The peak current is observed to increase in all the curves as the scan rate increases. Because of the quasi-reversible redox reaction as a result of electrode material polarization and internal resistance, there are shifts in the potential of oxidation peaks towards positive direction as scan rate increases (Yan et al. 2012; Athika et al. 2019). The presence of Ni during the electrochemical reaction leads to the formation of double reduction (Athika et al. 2019). At high scan rate of 200 mVs^{-1}, there was significant increase in the electronic and ionic transport (Mai et al. 2011). The comparative CV plots of the three samples at 50 mVs^{-1} scan rate are shown in Figure 8.32(d). The figure shows that the Ni-foam/NiFe$_2$O$_4$ had higher current and higher specific capacitance when compared to stainless steel foil/NiFe$_2$O$_4$ and stainless steel mesh/NiFe$_2$O$_4$. This might be as a result of lower series resistance and 3D porous structure of the Ni-foam/NiFe$_2$O$_4$. On the other hand, the stainless steel foil/NiFe$_2$O$_4$ had the least current value due to the characteristic non-porous nature while stainless steel mesh/NiFe$_2$O$_4$ had intermediate current value as a result of relative porosity of the hybrid, which resulted in intermediate low series resistance (Chen et al. 2014).

(a)   (b)

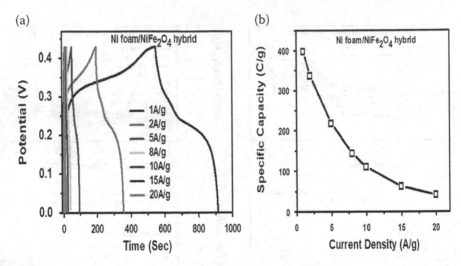

FIGURE 8.33   (a) Galvanostatic charge-discharge characteristics of Ni-foam/NiFe$_2$O$_4$ and (b) Specific capacity of the Ni-foam/NiFe$_2$O$_4$ hybrid plotted as a function of current density. Reproduced with permission from Mordina et al. 2020, *Journal of Energy Storage*.

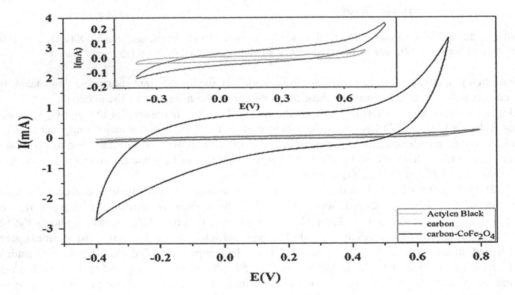

FIGURE 8.34   CV of C/CoFe$_2$O$_4$ using 5 M KOH. Reproduced with permission from Heydari et al. 2019, *International Journal of Green Energy*.

The maximum specific capacitance obtained from the stainless steel foil/NiFe$_2$O$_4$, stainless steel mesh/NiFe$_2$O$_4$, and Ni-foam/NiFe$_2$O$_4$ at scan rate of 5 mVs^{-1} are 126, 166, and 461, respectively. This result shows that the Ni-foam/NiFe$_2$O$_4$ has virtually 2.78 and 3.65 times better specific capacitance than the stainless steel foil/NiFe$_2$O$_4$ and stainless steel mesh/NiFe$_2$O$_4$ and is highly preferred for supercapacitor application.

The EIS spectra of the NiFe$_2$O$_4$ deposited on Ni-foam, stainless steel mesh, and stainless steel foil electrodes are shown in Figure 8.33. The resistance of the solution (Rs) from the EIS curve shows 22.41, 1.37, and 1.143 for the respective stainless steel foil/NiFe$_2$O$_4$, stainless steel mesh/NiFe$_2$O$_4$, and Ni-foam/NiFe$_2$O$_4$ that is magnified. The EIS did not show any semicircle for any of the samples deposited on different substrates. The Ni-foam/NiFe$_2$O$_4$ had the least Warburg impedance followed by stainless steel mesh/$NiFe_2O_4$ and then stainless steel foil/$NiFe_2O_4$ and this shows that the stainless steel mesh/$NiFe_2O_4$ has strong resistance to the diffusion of electrolyte through the foil and thus least electrochemical features among the three substrates.

The Ni-foam/$NiFe_2O_4$ revealed the galvanostatic charge-discharge characteristics for 1–20 A/g ranges of current density as shown in Figure 8.34(a). the figure shows that at current densities of 1 A/g, 2 A/g, 5 A/g, 8 A/g, 10 A/g, 15 A/g, and 20 A/g, the specific capacitances of 398 C/g, 336 C/g, 218 C/g, 144 C/g, 110 C/g, 62 C/g, and 42 C/g were obtained, respectively. Figure 8.34(b) shows the specific capacity of the Ni-foam/$NiFe_2O_4$ hybrid plotted as a function of current density and the results obtained are much higher than the one obtained from literature for some composites (Vijayakumar et al. 2013; Xie et al. 2013).

8.8 Conclusion

In this chapter, metal ferrites of Mn Co, Cu, and Ni have been extensively reviewed and the discussion highlighted the great significance of their applications in supercapacitors. Metal oxides can be synthesized using different methods, including sol gel, hydrothermal, and coprecipitation. Hydrothermal, sol gel, coprecipitation, co-electrospinning, chemical bath, etc., have been used to synthesize metal ferrites and the roles of the synthesis methods were also underlined. On the other hand, various characterization approaches such as XRD, EDX, SEM, CV, FESEM, TEM, HRTEM, EIS, FTIR, Raman spectroscopy were used to study the features of the ferrite materials. The electrochemical properties of the conventional metal ferrite materials have been observed to be poor. However, incorporating other materials like carbon derivatives, conducting polymers, as well as formation of composites with other materials have improved the electrochemical properties of the metal ferrites. These materials helped to make the greater area of the ferrite materials available for reaction during the electrochemical process. Furthermore, the electrolytes penetrated the electrode materials as the porosity of the nanoferrites were enhanced by incorporation of these materials. The results obtained from these articles show that metal ferrites of Mn Co, Cu, and Ni are capable of being effective for supercapacitor applications.

Acknowledgements

RMO and IA humbly acknowledge NCP for their PhD fellowship (NCP-CAAD/PhD-132/EPD) award and COMSATS for a travel grant for the fellowship.

RMO also acknowledge PPSMB Enugu State, Nigeria, for study leave permission granted.

FIE (90407830) affectionately acknowledge UNISA for VRSP Fellowship award and also graciously acknowledge the grant by TETFUND under contract number TETF/DESS/UNN/NSUKKA/STI/VOL.I/B4.33. We thank Engr. Meek Okwuosa for the generous sponsorship of April 2014, July 2016, and July 2018 conferences/workshops on applications of nanotechnology to energy, health and environment, and for providing some research facilities.

REFERENCES

Ahmed, M.A., S.F. Mansour and S.I. El-Dek. 2010. "Investigation of the Physico-Chemical Properties of Nanometric NiLa Ferrite/PST Matrix". *Solid State Ionics* 181: 1149–1155.

Alshehri, S.M., J. Ahmed, A.N. Alhabarah, T. Ahamad and T. Ahmad. 2017. "Nitrogen-Doped Cobalt Ferrite/Carbon Nanocomposites for Supercapacitor Applications". *ChemElectroChem* 4: 2952–2958.

An, C., Y. Zhang, H. Guo and Y. Wang. 2019. "Metal Oxide-Based Supercapacitors: Progress and Prospectives". *Nanoscale Advances* 1(12): 4644–4658.

Andrade, P.L., V.A.J. Silva, J.C. Maciel, M.M. Santillan, N.O. Moreno, L. De Los Santos Valladares, et al. 2014. "Preparation and Characterization of Cobalt Ferrite Nanoparticles Coated with Fucan and Oleic Acid". *Hyperfine Interactions* 224 (1): 211–219.

Aneke, M. and M. Wang. 2016. "Energy Storage Technologies and Real Life Applications: A State of the Art Review". *Applied Energy* 179: 350–377.

Anwar, S., K.S. Muthu, V. Ganesh, and N. Lakshminarasimhan. 2011. "A Comparative Study of Electrochemical Capacitive Behavior of NiFe2O$_4$ Synthesized by Different Routes". *Journal of Electrochemical Society* 158: A976–A981.

Ardizzone, S., G. Fregonara, and S. Trasatti. 1990. " Inner and Outer Active Surface of RuO$_2$ Electrodes". *Electrochemical Acta* 35 (1): 263–267.

Athika, M., A. Prasath, A.S. Sharma, V.S. Devi, E. Duraisamy, and P. Elumalai. 2019. "Ni/NiFe2O$_4$@carbon Nanocomposite Involving Synergistic Effect for High-Energy Density and High-Power Density Supercapattery". *Materials Research Express* 6: 1–12 .

Aydin, M. 2019. "Renewable and Non-renewable Electricity Consumption-Economic Growth Nexus: Evidence from OECD Countries". *Renewable Energy* 136: 599–606.

Ayman, I., A. Rasheed, S. Ajmal, A. 1 Rehman, A. Ali, I. Shakir, and M.F. Warsi. 2020. "CoFe$_2$O$_4$ Nanoparticles-Decorated 2D MXene: A Novel Hybrid Material for Supercapacitors Applications". *Energy & Fuels*. DOI: 10.1021/acs.energyfuels.0c00959.

Balamurugan, S., M.D. Devi, I. Prakash, and S. Devaraj. 2018. "Lithium Ion Doped NiFe2O$_4$/SiO$_2$ Nanocomposite Aerogel for Advanced Energy Storage Devices". *Applied Surface Science* 449: 542–550.

Bashir, B., W. Shaheen, M. Asghar, M.F. Warsi, M.A. Khan, S. Haider, I. Shakir, and M. Shahid. 2016. "Copper Doped Manganese Ferrites Nanoparticles Anchored on Graphene Nano-Sheets for High Performance Energy Storage Applications". *Journal of Alloys and Compounds*. DOI:10.1016/j.jallcom.2016.10.183.

Bashir, B., W. Shaheen, M. Asghar, M.F. Warsi, M.A. Khan, S. Haider, I. Shakir, and M. Shahid. 2017. "Copper Doped Manganese Ferrites Nanoparticles Anchored on Graphene Nano-Sheets for High Performance Energy Storage Applications". *Journal of Alloys and Compounds* 695: 881–887.

Bhojane, P., A. Sharma, M. Pusty, Y. Kumar, S. Sen, and P. Shirage. 2017. "Synthesis of Ammonia-Assisted Porous Nickel Ferrite (NiFe2O$_4$) Nanostructures as an Electrode Material for Supercapacitors". *Journal of Nanoscience and Nanotechnology* 17: 1387–1392.

Bhujun, B., S.A. Shanmugam, and M.T.T. Tan. 2016. "Evaluation of Aluminium Doped Spinel Ferrite Electrodes for Supercapacitors". *Ceramics International* 42: 6457–6466.

Chen, I., C. Wang, and C. Chen. 2010. "Fabrication and Characterization of Magnetic Cobalt Ferrite/ Polyacrylonitrile and Cobalt Ferrite/Carbon Nanofibers by Electrospinning". *Carbon* 48 3: 604–611.

Chen, L., Y. Tang, K. Wang, C. Liu, and S. Luo. 2011. "Direct Electrodeposition of Reduced Graphene Oxide on Glassy Carbon Electrode and Its Electrochemical Application". *Electrochemistry Communication* 13(2): 133–137.

Chen, T., H. Peng, M. Durstock, and L. Dai. 2014. "High-Performance Transparent and Stretch-Able All-Solid Supercapacitors Based on Highly Aligned Carbon Nanotube Sheets". *Sci. Rep.* 4: 1–7.

Chen, Z., Y. Qin, D. Weng, Q. Xiao, Y. Peng, X. Wang, H. Li, F. Wei, and Y. Lu. 2009. "Design and Synthesis of Hierarchical Nanowire Composites for Electrochemical Energy Storage". *Advanced Functional Materials*. 19: 3420–3426.

Das, A.K., S. Sahoo, P. Arunachalam, S. Zhang, and J. Shim. 2016. "Facile Synthesis of Fe$_3$O$_4$ Nanorod Decorated Reduced Graphene Oxide (rGO) for Supercapacitor Application". *RSC Advances* 6: 107057–107064.

Das, T. and B. Verma. 2019a. "High performance ternary polyaniline-acetylene black-cobalt ferrite hybrid system for supercapacitor electrodes". *Synthetic Metals* 251: 65–74.

Das, T. and B. Verma. 2019b. "Synthesis of Polymer Composite Based on Polyaniline-Acetylene Black-Copper Ferrite for Supercapacitor Electrodes". *Polymer*. doi:10.1016/j.polymer.2019.01.058.

Deng, D.H., H. Pang, J.M. Du, J.W. Deng, S.J. Li, J. Chen, and J.S. Zhang. 2012. "Fabrication of Cobalt Ferrite Nanostructures and Comparison of Their Electrochemical Properties". *Cryst. Research and Technology*. 1032–103847.

Dixit, G., J.P. Singh, R.C. Srivastava, and H.M. Agrawal. 2011. "Study of 200 MeV Ag15+ Ion Induced Amorphisation in Nickel Ferrite Thin Films". *Nuclear Instruments and Methods in Physics Research* 269: 133–139.

Dong, B., M. Li, C. Xiao, D. Ding, G. Gao, and S. Ding. 2017. "Tunable Growth of Perpendicular Cobalt Ferrite Nanosheets on Reduced Graphene Oxide for Energy Storage". *Nanotechnology* 28: 055401.

Dou, X., A.R. Koltonow, X. He, H.D. Jang, Q. Wang, Y.-W. Chung, and J. Huang. 2016. "Self-Dispersed Crumpled Graphene Balls in Oil for Friction and Wear Reduction". *Proceedings of the National Academy of Sciences of the United States of America* 113 (6): 1528–1533.

Dreyer, D.R., S. Murali, Y. Zhu, R.S. Ruoff, and C.W. Bielawski. 2011. "Reduction of Graphite Oxide Using Alcohols". *Journal of Materials Chemistry* 21: 3443–3447.

Du, F., D. Yu, L. Dai, S. Gangul, V. Varshney, and K. Roy. 2011. "Preparation of Tunable 3D Pillared Carbon Nanotube–Graphene Networks for High-Performance Capacitance". *Chemistry of Materials* 23 (21): 4810–4816.

Durai, G., P. Kuppusami, T. Maiyalagan, J. Theerthagiri, P. Kumar, and H. Kim. 2019. "Influence of Chromium Content on Microstructural and Electrochemical Supercapacitive Properties of Vanadium Nitride Thin Films Developed by Reactive Magnetron Co-sputtering Process". *Ceramics International* 45: 12643–12653.

Duraisamy, N., N. Arshid, K. Kandiah, J. Iqbal, P. Arunachalam, G. Dhanaraj, K. Ramesh, and S. Ramesh. 2019. "Development of Asymmetric Device Using $Co_3(PO_4)_2$ as a Positive Electrode for Energy Storage Application". *Journal of Material Science: Materials in Electronics* 30: 7435–7446.

El-Kady, M.F., M. Ihns, M. Li, J.Y. Hwang, M.F. Mousavi, L. Chaney, and R.B. Kaner. 2015. "Engineering Three-Dimensional Hybrid Supercapacitors and Microsupercapacitors for High-Performance Integrated Energy Storage". *Proceedings of the National Academy of Sciences* 112 (14): 4233–4238.

Elkholy, A.E., F. El-Taib Heakal, and N.K. Allam. 2017. "Nanostructured spinel manganese cobalt ferrite for high-performance supercapacitors". *Journal of Royal Society of Chemistry* 7: 51888–51895.

Fethi, A. 2019. "Renewable and Non-renewable Categories of Energy Consumption and Trade: Do the Development Degree and the Industrialization Degree Matter". *Energy* 173: 374–383.

Fic, K., A. Platek, J. Piwek, and E. Frackowiak. 2018. "Sustainable Materials for Electrochemical Capacitors". *Materials Today* 21(4): 437–454.

Fu, Y., Q. Chen, M. He, Y. Wan, X. Sun, H. Xia, and X. Wang. 2012. "Copper Ferrite-Graphene Hybrid: A Multifunctional Heteroarchitecture for Photocatalysis and Energy". *Industrial & Engineering Chemistry Research* 51 (36): 11700–11709.

Gao, H., F. Xiao, C.B. Ching, and H. Duan. 2012. "High-Performance Asymmetric Supercapacitor Based on Graphene Hydrogel and Nanostructured MnO_2". *Applied Materials and Interfaces*, 4: 2801–2810.

Gao, H., J. Xiang, and Y. Cao. 2017. "Hierarchically Porous $CoFe_2O_4$ Nanosheets Supported on Ni Foam with Excellent Electrochemical Properties for Asymmetric Supercapacitors". *Applied Surface Science* 413: 351–359.

Gao, H., X. Wang, G. Wang, C. Hao, C. Huang, and C. Jiang. 2019. "Facile Construction of a $MgCo2O_4$@ NiMoO4/NF Core-Shell Nanocomposite for High-Performance Asymmetric Supercapacitors". *Journal of Materials Chemistry* 7: 13267–13278.

Gao, L., E. Han, Y. He, C. Du, J. Liu, and X. Yang. 2020. "Effect of Different Templating Agents on Cobalt Ferrite $(CoFe_2O_4)$ Nanomaterials for High-Performance Supercapacitor". *Nature.* 10.1007/s11581-020-03482-z.

Geng, P., S. Zheng, H. Tang, R. Zhu, L. Zhang, S. Cao, H. Xue, and H. Pang. 2018. "Transition Metal Sulfides Based on Graphene for Electrochemical Energy Storage". *Advanced Energy Materials* 8: 1703259.

Ghasemia, A., M. Kheirmand, and H. Hel. 2019. "Synthesis of Novel NiFe2O4 Nanospheres for High Performance Pseudocapacitor Applications". *Russian Journal of Electrochemistry* 55 (3): 206–214.

Giri, S., D. Ghosh, A.P. Kharitonov, and C.K. Das. 2012. "Study of Copper Ferrite Nanowire Formation in Presence of Carbon Nanotubes and Influence of Fluorination on High Performance Supercapacitor Energy Storage Application". *Functional Materials Letters* 5: 1250046.

Guerrero-Contreras, J. and F. Caballero-Briones. 2015. "Graphene Oxide Powders with Different Oxidation Degree, Prepared by Synthesis Variations of the Hummers Method". *Materials Chemistry and Physics* 153: 209–220.

Guo, P., Z. Li, S. Liu, J. Xue, G. Wu, H. Li, and X.S. Zhao. 2017. "Electrochemical Properties of Colloidal Nanocrystal Assemblies of Manganese Ferrite as the Electrode Materials for Supercapacitors". *Journal of Material Science* 52: 5359–5365.

Ham, D., J. Chang, S.H. Pathan, W.Y. Kim, R.S. Mane, B.N. Pawar, O. Joo, H. Chung, M. Yoon, and S. Han. 2009. "Electrochemical Capacitive Properties of Spray-Pyrolyzed Copper-Ferrite Thin Films". *Current Applied Physics* 9: S98–S100.

Heidari, E.K., A. Ataie, M.H. Sohi, and J.K. Kim. 2015. "Effect of Processing Parameters on the Electrochemical Performance of Graphene/ Nickel Ferrite (G-NF) Nanocomposite". *Journal of Ultrafine Grained and Nanostructured Materials* 48 (1): 27–35.

Heli, H. and H. Yadegari. 2014. "Poly(Ortho-Aminophenol)/Graphene Nanocomposite as an Efficient Supercapacitor Electrode". *Journal of Electron Chemistry* 713: 103–111.

Heydari, N., M. Kheirmand, and H. Heli. 2018. "A Nanocomposite of $CoFe_2O_4$-Carbon Microspheres for Electrochemical Energy Storage Applications". *International Journal of Green Energy*, 10.1080/15435 075.2019.1580198.

Huang, C., Y. Ding, C. Hao, S. Zhou, X. Wang, H. Gao, L. Zhu, and J. Wu. 2019a. "PVP-Assisted Growth of Ni-Co Oxide on N-Doped Reduced Graphene Oxide with Enhanced Pseudocapacitive Behavior". *Journal of Chemical Engineering* 378: 122202.

Huang, C., C. Hao, Z. Ye, S. Zhou, X. Wang, L. Zhu, and J. Wu. 2019b. "In Situ Growth of ZIF-8-Derived Ternary ZnO/ZnCo2O4/NiO for High Performance Asymmetric Supercapacitors". *Journal of Nanoscale* 11: 10114–10128.

Ishaq, S., M. Moussa, F. Kanwal, M. Ehsan, M. saleem, T. Van, and D. Losic. 2019. "Facile Synthesis of Ternary Graphene Nanocomposites with Doped Metal Oxide and Conductive Polymers as Electrode Materials for High Performance Supercapacitors". *Scientific Reports* 9: 597.

Jeong, G.H., S. Baek, S. Lee, and S.W. Kim. 2016. "Metal Oxide/Graphene Composites for Supercapacitive Electrode Materials". *Chemistry – an Asian Journal* 11 (7): 949–964.

Jin, L., Y. Qiu, H. Deng, W. Li, H. Li, and S. Yang. 2011. "Hollow $CuFe_2O_4$ Spheres Encapsulated in Carbon Shells as an Anode Material for Rechargeable Lithium-Ion Batteries". *Electrochimica Acta* 56: 9127–9132.

Kafshgari, L.A., M. Ghorbani, and A. Azizi. 2018. " Synthesis and Characterization of Manganese Ferrite Nanostructure by Co-precipitation, Sol-gel, and Hydrothermal Methods". *Particulate Science and Technology*. 10.1080/02726351.2018.1461154.

Kashani, H., L. Chen, Y. Ito, J. Han, A. Hirata, and M. Chen. 2016. "Bicontinuous Nanotubular Graphene–Polypyrrole Hybrid for High Performance Flexible Supercapacitors". *Nano Energy* 19: 391–400.

Kate, R.S., S.A. Khalate, and R. Deokate. 2018. "Overview of Nanostructured Metal Oxides and Pure Nickel Oxide (NiO) Electrodes for Supercapacitors: A review". *Journal of Alloys and Compounds* 734: 89–111.

Kennaz, H., A. Harat, O. Guellati, D.Y. Momodu, F. Barzegar, J.K. Dangbegnon, N. Manyala, and M. Guerioune. 2018. "Synthesis and Electrochemical Investigation of Spinel Cobalt Ferrite Magnetic Nanoparticles for Supercapacitor Application". *Journal of Solid State Electrochemistry* 22: 835–847.

Kennaz, H., A. Harat, O. Guellati, N. Manyala, and M. Guerioune. 2016. "Synthesis of Cobalt Ferrite Nanoparticles by Hydrothermal Method for Supercapacitors Application". In *Third International Conference on Energy, Materials, Applied Energetics and Pollution*, Constantine, Algeria.

Kooti, M. and A.N. Sedeh. 2013. "Synthesis and Characterization of NiFe2O4 Magnetic Nano-Particles by Combustion Method". *Journal of Material Science and Technology* 29: 34–38.

Kooti, M., P. Kharazi, and H. Motamedi. 2014. "Preparation and Antibacterial Activity of Three-Component NiFe2O4@PANI@Ag Nanocomposite". *Journal of Material Science and Technology* 30: 656–660.

Kumar, N., A. Kumar, G.M. Huang, W.W. Wu, and T.Y. Tseng. 2018. "Facile Synthesis of Mesoporous NiFe2O4/CNTs Nanocomposite Cathode Material for High Performance Asymmetric Pseudocapacitors". *Applied Surface Science* 433: 1100–1112.

Kumbhar, V.S., A.D. Jagadale, N.M. Shinde, and C.D. Lokhande. 2012. "Chemical Synthesis of Spinel Cobalt Ferrite ($CoFe_2O_4$) Nano-Flakes for Supercapacitor Application". *Applied Surface Science* 259: 39–43.

Kurra, N., Q. Jiang, A. Syed, C. Xia, and H.N. Alshareef. 2016. " Micro-pseudocapacitors with Electroactive Polymer Electrodes: Toward AC-Line Filtering Applications". *ACS Applied Materials & Interfaces*, 8 (20): 12748–12755.

Kwon, S.H., E. Lee, B.S. Kim, S.G. Kim, B.J. Lee, M.S. Kim, and J.C. Jung. 2015. "Preparation of Activated Carbon Aerogel and Its Application to Electrode Material for Electric Double Layer Capacitor in Organic Electrolyte: Effect of Activation Temperature". *Korean Journal of Chemical Engineering* 32 (2): 248–254.

Lalwani, S., R.B. Marichi, M. Mishra, G. Gupta, G. Singh, and R.K. Sharma. 2018. "Edge Enriched Cobalt Ferrite Nanorods for Symmetric/Asymmetric Supercapacitive Charge Storage". *Electrochimica Acta* 283: 708–717.

Lazarevic, Z.Z., C. Jovalekic, A. Recnik, V.N. Ivanovski, A. Milutinovic, M. Romcevic, M.B. Pavlovic, B. Cekic, and N.Z. Romcevic. 2013. "Preparation and Characterization of Spinel Nickel Ferrite Obtained by the Soft Mechanochemically Assisted Synthesis". *Material Research Bulletin* 48: 404–415.

Lee, C., E.H. Jo, S.K. Kim, J.H. Choi, H. Chang, and H.D. Jang. 2017. "Electrochemical Performance of Crumpled Graphene Loaded with Magnetite and Hematite Nanoparticles for Supercapacitors". *Carbon* 115: 331–337.

Lee, S.W., B.M. Gallant, Y. Lee, N. Yoshida, D.Y. Kim, Y. Yamada, S. Noda, A. Yamada, and S. Shao-Horn. 2012. "Self-Standing Positive Electrodes of Oxidized Few-Walled Carbon Nanotubes for Light-Weight and High-Power Lithium Batteries". *Energy and Environmental Science*, 5: 5437–5444.

Li, J., X. Huang, L. Cui, N. Chen, and L. Qu. 2016. "Preparation and Supercapacitor Performance of Assembled Graphene Fiber and Foam". *Progress in Natural Science: Materials International* 26 (3): 212–220.

Li, J., X. Yuan, C. Lin, Y. Yang, L. Xu, X. Du, and J. Sun. 2017. "Achieving High Pseudocapacitance of 2D Titanium Carbide (MXene) by Cation Intercalation and Surface Modification". *Advanced Energy Materials* 7 (15): 602–725.

Li, L., Y. Zhang, F. Shi, Y. Zhang, J. Zhang, C. Gu, X. Wang, and J. Tu. 2014. "Spinel Manganese-Nickel-Cobalt Ternary Oxide Nanowire Array for High-Performance Electrochemical Capacitor Applications". *Applied Materials & Interfaces* 6 (20): 18040–18047.

Lia, J., J. Yan, and G. Yao. 2019. "Improving the Degradation of Atrazine in the Three-Dimensional (3D) Electrochemical Process Using $CuFe_2O_4$ as both Particle Electrode and Catalyst for Persulfate Activation". *Journal of Chemical Engineering*. 361: 1317–1332.

Liu, L., H. Zhang, Y. Mu, Y. Bai, and Y. Wang. 2016. "Binary Cobalt Ferrite Nanomesh Arrays as the Advanced Binder-Free Electrode for Applications in Oxygen Evolution Reaction and Supercapacitors". *Journal of Power Sources* 327 (6): 599–609.

Luo, J., H.D. Jang, T. Sun, L. Xiao, Z. He, A.P. Katsoulidis, M.G. Kanatzidis, J.M. Gibson, and J. Huang. 2011. "Compression and Aggregation-Resistant Particles of Crumpled Soft Sheets". *ACS Nano* 11 (5): 8943–8949.

Mai, L.Q., F. Yang, Y.L. Zhao, X. Xu, L. Xu, and Y.Z. Luo. 2011. "Hierarchical $MnMoO_4$/ $CoMoO_4$ Heterostructured Nanowires with Enhanced Supercapacitor Performance". *Nature Communications* 2: 1–5.

Mohapatra, J., A. Mitra, D. Bahadur, and M. Aslam. 2013. "Surface Controlled Synthesis of MFe2O4 (M = Mn, Fe, Co,Ni and Zn) Nanoparticles and Their Magnetic Characteristics". *CrystEngComm* 15: 524–532.

Mordina, B., R. Kumar, N.S. Neeraj, A.K. Srivastava, D.K. Setua, and A. Sharma. 2020. "Binder Free High Performance Hybrid Supercapacitor Device Based on Nickel Ferrite Nanoparticles". *Journal of Energy Storage*, 31, 101677.

Muzaffar, A., M.B. Ahamed, K. Deshmukh, and J. Thirumalai. 2019. "A Review on Recent Advances in Hybrid Supercapacitors: Design, Fabrication and Applications". *Renewable and Sustainable Energy Reviews* 101: 123–145.

Nagar, B., D.P. Dubal, L. Pires, A. Merkoçi, and P. Gómez-Romero. 2018. "Design and Fabrication of Printed Paper-Based Hybrid Micro-Supercapacitor by using Graphene and Redox-Active Electrolyte". *ChemSusChem* 11 (11): 1849–1856.

Nethravathi, C. and M. Rajamathi. 2008. "Chemically Modified Graphene Sheets Produced by The Solvothermal Reduction of Colloidal Dispersions of Graphite Oxide". *Carbon* 46: 1994–1998.

Nilmoung, S., T. Sinprachim, I. Kotutha, P. Kidkhunthod, R. Yimnirun, S. Rujirawat, and S. Maensiri. 2016. "Electrospun Carbon/$CuFe_2O_4$ composite Nanofibers with Improved Electro-Chemical Energy Storage Performance". *Journal of Alloys and Compounds* 688: 1131–1140.

Nonaka, L.H., T.S.D. Almeida, C.B. Aquino, S.H. Domingues, R.V. Salvatierra, and V.H.R. Souza. 2020. "Crumpled Graphene Decorated with Manganese Ferrite Nanoparticles for Hydrogen Peroxide Sensing and Electrochemical Supercapacitors". *ACS Appl. Nano Mater* 3: 4859–4869.

Obodo, R.M., A. Ahmad, G.H. Jain, I. Ahmad, M. Maaza, and F.I. Ezema, 2020a. "8.0 MeV Copper Ion (Cu^{++}) Irradiation-Induced Effects on Structural, Electrical, Optical and Electrochemical Properties of Co_3O_4-NiO-ZnO/GO Nanowires". *Materials Science for Energy Technologies* 3: 193–200.

Obodo, R.M., A.C. Nwanya, A.B.C. Ekwealor, I. Ahmad, T. Zhao, M. Maaza, and F.I. Ezema. 2019a. "Influence of pH and Annealing on the Optical and Electrochemical Properties of Cobalt (III) Oxide (Co_3O_4) Thin Films". *Surfaces and Interfaces* 16: 114–119.

Obodo, R.M., A.C. Nwanya, C. Iroegbu, A.B.C. Ekwealor, I. Ahmad, M. Maaza, and F.I. Ezema. 2020b. "Effects of Swift Copper (Cu^{2+}) Ion Irradiation on Structural, Optical and Electrochemical Properties of Co_3O_4-CuO-MnO_2/GO Nanocomposites Powder". *Advanced Powder Technology*.

Obodo, R.M., A.C. Nwanya, C. Iroegbu, I. Ahmad, A.B.C. Ekwealor, R.U. Osuji, M. Maaza, and F.I. Ezema. 2020c. "Transformation of GO to rGO due to 8.0 MeV Carbon (C++) Ions Irradiation and Characteristics Performance on MnO_2–NiO–ZnO@GO Electrode". *International Journal of Energy Research*: 1–12.

Obodo, R.M., A.C. Nwanya, M. Arshad, C. Iroegbu, I. Ahmad, R. Osuji, M. Maaza, and F.I. Ezema. 2020d. "Conjugated NiO-ZnO/GO Nanocomposite Powder for Applications in Supercapacitor Electrodes Material". *International Journal of Energy Research* 44: 3192–3202.

Obodo, R.M., A.C. Nwanya, Tabassum Hassina, Mesfin Kebede, Ishaq Ahmad, M. Maaza and Fabian I. Ezema. 2019b. Transition Metal Oxide-Based Nanomaterials for High Energy and Power Density Supercapacitor. In *Electrochemical Devices for Energy Storage Applications*. United Kingdom: Taylor & Francis Group, CRC Press, 131–150.

Obodo, R.M., E.O. Onah, H.E. Nsude, A. Agbogu, A.C. Nwanya, I. Ahmad, T. Zhao, P.M. Ejikeme, M. Maaza, F.I. Ezema. 2020e. Performance Evaluation of Graphene Oxide Based Co_3O_4@GO, MnO_2@GO and Co_3O_4/MnO_2@GO Electrodes for Supercapacitors "Electroanalysis" 32: 1–10.

Obodo, R.M., N.M. Shinde, U.K. Chime, S. Ezugwu, A.C. Nwanya, I. Ahmad, M. Maaza, P.M. Ejikeme, and F.I. Ezema. 2020f. "Recent Advances in Metal Oxide/Hydroxide on Three-Dimensional Nickel Foam Substrate for High Performance Pseudocapacitive Electrodes". *Current Opinion in Electrochemistry* 21: 242–249.

Pandav, R.S., R.P. Patil, S.S. Chavan, J.S. Mulla, and R.P. Hankare. 2016. "Magneto-Structural Studies of Sol-Gel Synthesized Nanocrystalline Manganese Substituted Nickel Ferrites". *Journal of Magnetism and Magnetic Materials* 417: 407–412.

Pimenta, M., G. Dresselhaus, M.S. Dresselhaus, L. Cancado, A. Jorio, and R. Saito. 2007. "Studying Disorder in Graphite-Based Systems by Raman spectroscopy". *Physical Chemistry* 9 (11): 1276–1290.

Rani, B.J., G. Ravi, R. Yuvakkumar, V. Ganesh, S. Ravichandran, M. Thambidurai, A.P. Rajalakshmi, and A. Sakunthala. 2018. "Pure and Cobalt-Substituted Zinc-Ferrite Magnetic Ceramics for Supercapacitor Applications". *Applied Physics A* 124: 511.

Ratajczak, P., M.E. Suss, F. Kaasik, and F. Béguin. 2019. " Carbon Electrodes for Capacitive Technologies". *Energy Storage Materials* 16: 126–145.

Ravindran Madhura, T., P. Viswanathan, G. Gnana kumar, and R. Ramaraj. 2017. "Nanosheet-like Manganese Ferrite Grown on Reduced Graphene Oxide for Non-enzymatic Electrochemical Sensing of Hydrogen Peroxide". *Journal of Electroanalytical Chemistry* 792: 15–22.

Salvatierra, R.V., S.H. Domingues, M.M. Oliveira, and A.J.G. Zarbin. 2013. "Tri-layer graphene films produced by mechanochemical exfoliation of graphite". *Carbon* 57 (0): 410–415.

Scindia, S.S., R.B. Kamble, and J.A. Kher. 2019. " Nickel Ferrite/Polypyrrole Core-Shell Composite as an Efficient Electrode Material for High-Performance Supercapacitor". *AIP Advances* 9: 055218.

Sen, P. and A. De. 2010. "Electrochemical performances of poly(3,4-ethylenedioxythiophene) $NiFe_2O_4$ nanocomposite as electrode for supercapacitor". *Electrochimica Acta* 55: 4677–4684.

Senthilkumar, B., K. Vijaya Sankar, C. Sanjeeviraja, and R. Kalai. 2013. "Selvan, Synthesis and Physico-Chemical Property Evaluation of PANI-$NiFe_2O_4$ Nanocomposite as Electrodes for Supercapacitors". *Journal of Alloys and Compounds* 553: 350–357.

Sharif, S., A. Yazdani, and K. Rahimi. 2020. "Incremental Substitution of Ni with Mn in$nife2o4$to Largely Enhance Its Supercapacitance Properties". *Scientific Reports* 10: 10916.

Si, Y. and E.T. Samulski. 2008. "Exfoliated Graphene Separated by Platinum Nanoparticles". *Chemistry of Materials* 20: 6792–6797.

Silva, V.D., L.S. Ferreira, T.A. Simoes, E.S. Medeiros, and D.A. Macedo. 2019. "1.D. Hollow MFe_2O_4, 1D hollow MFe_2O_4(M=Cu, Co, Ni)Fibers by Solution Blow Spinning for Oxygen Evolution Reaction". *Journal of Colloid and Interface Science*. 540: 59–65.

Singh, G. Gita, and S. Chandra. 2019. "Copper Doped Manganese Ferrites PANI for Fabrication of Binder-Free Nanohybrid Symmetrical Supercapacitors". *Journal of the Electrochemical Society* 166(6): A1154–A1159.

Singh, G. Gita, and S. Chandra. 2020. "Nano-Flowered Manganese Doped Ferrite@P ANI Composite as Energy Storage Electrode Material for Supercapacitors". *Journal of Electroanalytical Chemistry* 2020: 114491.

Soam, A., R. Kumar, P.K. Sahoo, C. Mahender, B. Kumar, N. Arya, M. Singh, S. Parida, and R.O. Dusane. 2019. "Synthesis of Nickel Ferrite Nanoparticles Supported on Graphene Nanosheets as Composite Electrodes for High Performance Supercapacitor". *Energy Technology & Environmental Science* 4: 9952–9958.

Souza, V.H.R., S. Husmann, E.G.C. Neiva, F.S. Lisboa, L.C. Lopes, R.V. Salvatierra, and A.J.G. Zarbin. 2016. "Flexible, Transparent and Thin Films of Carbon Nanomaterials as Electrodes for Electrochemical Applications". *Electrochimica Acta* 197: 200–209.

Stoller, M.D., S. Park, Y. Zhu, J. An, and R.S. Ruoff. 2008. "Graphene-Based Ultracapacitors". *Nano Letters* 8:, 3498–3502.

Theerthagiri, J., G. Durai, K. Karuppasamy, P. Arunachalam, V. Elakkiya, P. Kuppusami, T. Maiyalagan, and H.S. Kim. 2018. "Recent Advances in 2-D Nanostructured Metal Nitrides, Carbides and Phosphides Electrodes for Electrochemical Super-Capacitors: A Brief Review". *Journal of Industrial Engineering and Chemistry* 69: 12–27.

Theerthagiri, J., R. Senthil, B. Senthilkumar, A. Polu, J. Madhavan, and M. Ashokkumar. 2017. "Recent Advances in MoS2nanostructured Materials for Energy and Environmental Applications – a Review". *Journal of Solid State Chemistry.* 252: 43–71.

Theerthagiri, J.J., A.P. Murthy, V. Elakkiya, S. Chandrasekaran, P. Nithyadharseni, Z. Khan, R.A. Senthil, R. Shanker, M. Raghavender, P. Kuppusami, J. Madhavan, and M. Ashokkumar. 2018. "Recent Development on Carbon Based Heterostructures for Their Applications in Energy and Environment: A Review". *Journal of Industrial Engineering and Chemistry* 64: 16–59.

Theerthagiri, J.J., Z. Karuppasamy, G. Durai, A. Sarwar Rana, P. Arunachalam, K. Sangeetha, P. Kuppusami, and H.S. Kim. 2018. "Recent Advances in Metal Chalcogenides (MX; X = S, Se) Nanostructures for Electrochemical Supercapacitor Applications: A Brief Review Nanomaterials". *Nanomaterials*, 8: 256.

Thiagaran, K., J. Theerthagiri, R.A. Senthil, P. Arunachalam, J. Madhavan, and M.A. Ghanem. 2018. "Synthesis of $Ni_3V_2O_8$@graphene Oxide Nanocomposite as an Efficient Electrode Material for Supercapacitor Applications". *Journal of Solid State Electrochemistry* 22: 527–536.

Trusovas, R., G. Račiukaitis, G. Niaura, J. Barkauskas, G. Valušis, and R. Pauliukaite. 2016. "Recent Advances in Laser Utilization in the Chemical Modification of Graphene Oxide and Its Applications". *Advanced Optical Materials* 4 (1): 37–65.

Tung, D.T. L.T.T. Tam, H.T. Dung, N.T. Dung, N.T. Dung, T. Hoang, T.D. Lam, et al. 2020. "Direct Ink Writing of Graphene–Cobalt Ferrite Hybrid Nanomaterial for Supercapacitor Electrodes". *Journal of Electronic Materials* 49: 4671–4679.

Turgut, G.M. 2018. "Review of Electrical Energy Storage Technologies, Materials and Systems: Challenges and Prospects for Large-Scale Grid Storage". *Energy and Environmental Science.* 11: 2696–2767.

Uduh, U.C., R.M. Obodo, S. Esaenwi, C.I. Amaechi, P.U. Asogwa, R.U. Osuji, F.I. Ezema. 2014. "Sol-Gel Synthesis, Optical and Structural Characterization of Zirconium Oxysulfide (ZrOS) nanopowder". *Journal of Sol-Gel Science and Technology* 71: 79–85.

Vargas, S.M., A.I. Enriquez, H.F.H. Zuñiga, A. Encinas, and J. Oliva. 2020. "Enhancing the Capacitance and Tailoring the Discharge Times of Flexible Graphene Supercapacitors with Cobalt Ferrite Nanoparticles". *Synthetic Metals* 264: 116384.

Venkatachalam, V. and R. Jayavel. 2015. "Novel Synthesis of Ni-Ferrite (NiFe2O4) Electrode Material for Supercapacitor Applications". *American Institute of Physics* 140016. 10.1063/1.4918225.

Viana, R.B., A.B.F. da Silva, and A.S. Pimentel. 2012. "Infrared Spectroscopy of Anionic, Cationic, and Zwitterionic Surfactants". *Advances in Physical Chemistry* 903272: 1–14.

Vignesh, V., K. Subramani, M. Sathish, and R. Navamathavan. 2018. "Electrochemical Investigation of Manganese Ferrites Preparedviaa Facile Synthesis Route for Supercapacitor Applications". *Colloids and Surfaces* 538: 668–677.

Vijayakumar, S., S. Nagamuthu, and G. Muralidharan. 2013. "Supercapacitor Studies on NiO Nanoflakes Synthesized through a Microwave Route". *ACS Applied Materials and Interfaces* 5: 2188–2196.

VijayaSankar K. and R. KalaiSelvan. 2016. "Fabrication of Flexible Fiber Supercapacitor Using Covalently Grafted CoFe2O4/Reduced Graphene Oxide/Polyaniline and Its Electrochemical Performances". *Electrochimica Acta* 213: 469–481.

Wang, Q., H. Gao, X. Qin, J. Dai, and W. Li. 2020. "Fabrication of NiFe2O4@CoFe2O4core-Shell Nanofibers for High-Performance Supercapacitors". *Materials Research Express* 7: 015020.

Wang, R., Q. Li, L. Cheng, H. Li, B. Wanga, X.S. Zhao, and P. Guo. 2014a. "Electrochemical Properties of Manganese Ferrite-Based Supercapacitors in Aqueous Electrolyte: The Effect of Ionic Radius". *Colloids and Surfaces A: Physicochem. Eng. Aspects* 457: 94–99.

Wang, W., Q. Hao, W. Lei, X. Xia, and X. Wang. 2014b. "Ternary Nitrogen-Doped Graphene/Nickel Ferrite/Polyaniline Nanocomposites for High-Performance Supercapacitors". *Journal of Power Sources* 269: 250–259.

Wang, W., Z. Ding, M. Cai, H. Jian, Z. Zeng, F. Li, and J.P. Liu. 2015. "Synthesis and High-Efficiency Methylene Blue Adsorption of Magnetic PAA/MnFe2O4 Nanocomposites". *Applied Surface Science* 346: 348–353.

Wang, Z., X. Zhang, Y. Li, Z. Liu, and Z. Hao. 2013. "Synthesis of Graphene-NiFe2O4 Nano-composites and Their Electrochemical Capacitive Behaviour". *Journal of Materials Chemistry*: 6393–6399.

Xia, Y., Y. Xiong, B. Lim, and S. Skrabalak. 2009. "E. Shape-Controlled Synthesis of Metal Nanocrystals: Simple Chemistry Meets Complex Physics". *Angewandte Chemie International Edition International Edition*. 48: 60–103.

Xie, X., C. Zhang, M.B. Wu, Y. Tao, W. Lvac, and Q.H. Yang. 2013. "Porous MnO_2 for Use in a High Performance Supercapacitor: Replication of a 3D Grapheme Network as a Reactive Template". *ChemComm* 49: 11092–11094.

Xiong, P., C. Hu, Y. Fan, W. Zhang, J. Zhu, and X. Wang. 2014a. "Ternary Manganese Ferrite/Graphene/Polyaniline Nanostructure with Enhanced Electrochemical Capacitance Performance". *Journal of Power Sources* 266: 384–392.

Xiong, P., J. Huang, and X. Wang. 2014b. "Design and Synthesis of Ternary Cobalt Ferrite/Graphene/Polyaniline Hierarchical Nanocomposites for High-Performance Supercapacitors". *Journal of Power Sources* 245: 937–946.

Yan, J., Z. Fan, W. Sun, G. Ning, T. Wei, Q. Zhang, R. Zhang, L. Zhi, and F. Wei. 2012. "Advanced Asymmetric Supercapacitors Based on $Ni(OH)_2$/Graphene and Porous Graphene Electrodes with High Energy Density". *Adv. Funct. Mater.* 22: 2632–2641.

Ye, S., L. Zhu, I.J. Kim, S.H. Yang, and W.C. Oh. 2016. "Characterization of Expanded Graphene Nanosheet as Additional Material and Improved Performances for Electric Double Layer Capacitors". *Journal of Industrial and Engineering Chemistry* 43: 53–60.

Yu, Z.Y., L.F. Chen, and S.H. Yu. 2014. "Growth of NiFe2O4 nanoparticles on carbon cloth for high performance flexible supercapacitors". *Journal of Material Chemistry* 2: 10889–10894.

Zhang, L.L., R. Zhoua, and X.S. Zhao. 2010. "Graphene-Based Materials as Supercapacitor Electrodes". *Materials Chemistry* 20 (29): 5983–5992.

Zhang, Q.Z., D. Zhang, Z.C. Miao, X.L. Zhang, and S.L. Chou. 2018. "Research Progress in MnO_2–Carbon Based Supercapacitor Electrode Materials". *Small* 14 (24): 1702883.

Zhang, W., B. Quan, C. Lee, S. Park, X. Li, E. Choi, G. Diao, and Y. Piao. 2015a. "One Step Facile Solvothermal Synthesis of Copper Ferrite-Graphene Composite as a High-Performance Supercapacitor Material". *Journal of ACS Applied Materials and Interfaces* 7: 2404–2414.

Zhang, W., Y. Fu, W. Liu, L. Lim, X. Wang, and A. Yu. 2019a. "A General Approach for Fabricating 3D MFe2O4 (M = Mn, Ni, Cu, Co)/Graphitic Carbon Nitride Covalently Functionalized Nitrogen-Doped Graphene Nanocomposites as Advanced Anodes for Lithium-Ion Batteries". *Journal of Nano Energy* 57: 48–56.

Zhang, X., M. Zhu, T. Ouyang, Y. Chen, J. Yan, K. Zhu, K. Ye, G. Wang, K. Cheng, and D. Cao. 2019b. "NiFe2O4 nanocubes Anchored on Reduced Graphene Oxide Cryogel to Achieve a 1.8 Vflexible Solid-State Symmetric Supercapacitor". *Chemical Engineering Journal* 360: 171–179.

Zhang, Z., W. Li, R. Zou, W. Kang, Y. San Chui, M.F. Yuen, C.S. Lee, and W. Zhang. 2015b. "Layer-Stacked Cobalt Ferrite ($CoFe_2O_4$) Mesoporous Platelets for High-Performance Lithium Ion Battery Anodes". *Journal of Material Chemistry* 3: 6990–6997.

Zhao, J., Y. Cheng, X. Yan, D. Sun, F. Zhu, and Q. Xue. 2012. "Magnetic and Electrochemical Properties of $CuFe_2O_4$ Hollow Fibers Fabricated by Simple Electrospinningand Direct Annealing". *CrystEngComm* 14: 5879–5885.

Zhao, Y., G. He, W. Dai, and H. Chen. 2014. "High Catalytic Activity in the Phenol Hydroxylation of Magnetically Separable $CuFe_2O_4$–Reduced Graphene Oxide". *Industrial Engineering and Chemisty Research* 53: 12566–12574.

Zhou, S., C. Hao, J. Wang, X. Wang, and H. Gao. 2018. "Metal-Organic Framework Templated Synthesis of Porous $NiCo_2O_4$/$ZnCo_2O_4$/Co_3O_4 hollow Polyhedral Nanocages and Their Enhanced Pseudocapacitive Properties". *Journal of Chemical Engineering* 351: 74–84.

Zhu, H., C. Hou, Y. Li, G. Zhao, X. Liu, K. Hou, and Y. Li. 2013a. "One-Pot Solvothermal Synthesis of Highly Water-Dispersible Size-Tunable Functionalized Magnetite Nanocrystal Clusters for Lipase Immobilization". *Journal of Chemistry* 8: 1447–1454.

Zhu, M., D. Meng, C. Wang, and G. Diao. 2013b. "Facile Fabrication of Hierarchically Porous $CuFe_2O_4$ Nanospheres with Enhanced Capacitance Property". *ACS Applied Materials and Interfaces* 5: 6030–6037.

9

Advances in Nickel-Derived Metal-Organic Framework-Based Electrodes for High-Performance Supercapacitor

Blessing N. Ezealigo[1], Ebubechukwu N. Dim[2], and Fabian I. Ezema[3]
[1]*Dipartimento di Ingegneria Meccanica, Chimica e dei Materiali,*
 Università degli Studi di Cagliari, Cagliari, Italy
[2]*Department of Science Laboratory Technology, University of Nigeria, Nsukka*
[3]*Department of Physics and Astronomy, University of Nigeria, Nsukka, Nigeria*

9.1 Introduction

Metal-organic frameworks (MOFs) are porous crystalline materials formed by metal ions linked with strong coordinating organic bonds. The divalent or polyvalent organic bonds form 3D structures with metals which leads to exceptional porosity. MOFs have a large surface area (up to $7000\,\mathrm{m^2\,g^{-1}}$) (Farha et al. 2012; 15016–21), ultrahigh porosity (high adsorption), multidimensional nature (1D, 2D, 3D), high permeability, and flexible structural properties (Yang et al. 2014a, 16640–44; Wu and Hsu 2015, 1055–62; Zhao et al. 2016, 35–62). The latter can be further modified through post-synthetic alteration. They can be used for gas capture and storage, drug delivery, optics, and also multi-functional applications (Wu and Hsu 2015, 1055–62; Salunkhe et al. 2017, 5293–308). One of the drawbacks of these materials is that they are unstable and non-conductive or possess low conductivity, thereby restricting their applications, especially in energy storage (Qu et al. 2016, 66–73). To solve this problem, MOFs are functionalized with conductive materials such as metals, carbon, graphene, and polymers which do not necessarily affect their porosity and stability (Wang et al. 2019b, 2063–71). Nickel functionalized MOF (Figure 9.1) has a 2D layered structure, that enables easy penetration of electrolytes, and conjugated pie (π) bond for efficient electronic conduction (Jiao et al. 2016, 13344–51). The unlimited combination of metal ions and organic linkers has resulted in over 20,000 MOFs (Salunkhe et al. 2017, 5293–308). Furthermore, MOFs have an excellent ability to tune their properties at the molecular level.

MOFs can be obtained in three basic ways: Pristine synthesis, destruction of existing MOFs to produce metal oxides, and pyrolyzing MOFs to get their carbon derivatives. From the literature, the latter MOFs show lower capacitance; however, the capacitance of the former (i.e. pristine and metal oxide–derived MOFs) are still low compared to RuO_2 (Wang et al. 2016, 361–81). Furthermore, MOFs can serve as templates or precursors for obtaining composites such as MOF-derived sulphides, phosphides, complex oxides, etc.

MOFs bridge the gap/limitations of conventional supercapacitor materials. Carbon-based materials have long cycle life but low capacitance; on the other hand, metal oxide–based materials have high capacitance but low cycle life while conducting polymers suffer from degradation (Choi et al. 2014, 7451–57). The interaction of the metal and organic components in MOFs strongly influences their

DOI: 10.1201/9781003145585-9

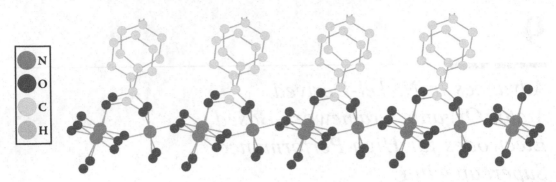

FIGURE 9.1 Schematic of Ni-MOFs (Jiao et al. 2016, 13344–51). Reproduced with permission. Copyright 2014, Royal Society of Chemistry.

supercapacitor properties leading to high capacitance and long cycle life. More interestingly, tuning the organic linkers can lead to an increase in the rate of charge transfer and the redox character creating electron pathways (Yang et al. 2014a, 16640–44). There are various reports in the literature where the different organic linkers for the synthesis of Ni-MOF supercapacitors and their derivates gave better properties (Yang et al. 2014a, 16640–44; Xu et al. 2016, 595–602; Yang et al. 2020a, 1803–06). Also, the layered structure provides favourable facets for electron diffusion which results in high capacitance.

The attribute of the organic linker is that they form 2D/3D structures with metals and give rise to unique pore distribution (Zhao et al. 2016, 35–62). The wide range of choices for metal ions and organic linkers could lead to novel properties (Zain et al. 2018, 665–74).

The outstanding properties (such as excellent electrical conductivity, low cost, and high redox reactions) of conventional transition metal oxide supercapacitors have limited practical application due to the inability to vary their microstructure. Also, the continuous faradaic reaction destroys their structure, phase composition, short cycle life, and finally degrades the capacitor (Salunkhe et al. 2017, 5293–308).

Depending on the preparation route, starting precursor, and device assembly, MOFs can display electric double-layer capacitance (EDLC) (Wang et al. 2016, 361–81; Sheberla et al. 2017, 220–24), pseudocapacitive behaviour (Kang et al. 2014, 957–61; Ma et al. 2019, 9543–47), or a battery-like character (Zhang et al. 2020a, 154069; Zhang et al. 2020b, 23–31). To be able to differentiate between these three electrode types, the cyclic voltammetry and the galvanostatic charge-discharge curves are useful (Figures 9.2 and 9.3). In summary, the capacitance of EDLC is independent of current and voltage and their GCD curve is linear, while for battery (faradaic) the distance between the redox peaks of the CV curves is greater than 0.1–0.2 V and their GCD curve shows a plateau-like shape. Pseudocapacitance falls in between these two properties. Although carbon-based supercapacitors have excellent cycle life, they are limited by their low capacitance. Also, polymers have shown to be promising in energy storage, but they undergo degradation over time. On the other hand, metal oxides show high capacitance but short cycle life. Therefore MOF-based supercapacitors with long cycling life and high specific capacitance arising from the synergistic effect of the metal oxide and the organic components have huge potential in energy storage and conversion (Choi et al. 2014, 7451–57). Supercapacitors are currently being used in the emergency doors of airbus, memory backup, regenerative brake systems, and rapid energy recovery systems (Wang et al. 2020b, 213093).

Amongst transition metal–based MOFs, Ni-MOF is widely studied as a result of their relatively superior performance, high specific capacity, and better stability in alkaline electrolyte.

The poor conductivity of MOFs generally discouraged research on energy storage. However, after the pioneering work of Diaz and coworkers on cobalt-based MOF supercapacitor (Co8-MOF-5) where they reported a capacitance of 3.27 F g^{-1} (Díaz et al. 2012, 126–28), followed by the work of Lee et al. (2012, 163–65) on Co-MOF supercapacitor with 206.76 F g^{-1}, The energy storage potential of MOFs became interesting to be further exploited. In 2016, Sheberla and colleagues reported high conductivity of 5000 S m^{-1} in pristine Ni$_3$(2,3,6,7,10,11-hexaiminotriphenylene)$_2$ [Ni$_3$(HITP)$_2$] even

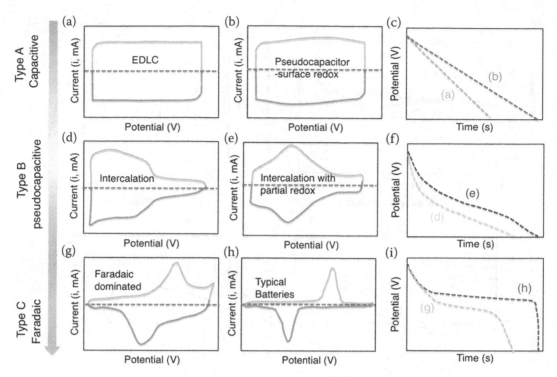

FIGURE 9.2 Schematic cyclic voltammetry curves with their corresponding galvanostatic discharge curves of various kinds of electrode materials (Gogotsi and Penner 2018, 2081–83). Reproduced with permission Copyright 2018, American Chemical Society.

exceeding that of activated carbon and porous graphite. The symmetric dense capacitor possesses capacitance 118 F cm^{-3}(111 F g^{-1} @ 50 mA g^{-1}) (Sheberla et al. 2017, 220–24). After their study, a great improvement has been recorded in the capacitance of pristine and composite Ni-based MOFs. In Table 9.1 the advantages and disadvantages of MOF supercapacitor are presented.

The partial substitution of Ni^{2+}(0. 69 Å) atoms with other elements of similar atomic radius such as Co^{2+}(0.64 Å) atoms create vacancies that lead to the formation of a large volume of holes in the composite material, hence higher electrical conductivity. On the other hand, when Ni is partly substituted by atoms with larger size (e.g. Zn^{2+}(0.74 Å)), it results in a larger layer structure which allows for increased electrolyte ion diffusion and easy intercalation/de-intercalation (Jiao et al. 2017, 1094–1102).

9.2 Methods of Synthesizing MOF-Based Supercapacitor

In preparing MOF supercapacitors, first, the powders are processed and then deposited on conductive substrates. The MOF powders can be obtained by direct synthesis or template method. The processing techniques are discussed in detail in the following subsections.

9.2.1 Powder Preparation

9.2.1.1 Direct Powder Synthesis

Generally, MOFs are prepared using the solution processing route. A hydrothermal or solvothermal method is the popular method of producing these powders.

For Ni-based MOFs, a stock solution containing the precursors, organic linker, and appropriate solvents are mixed.

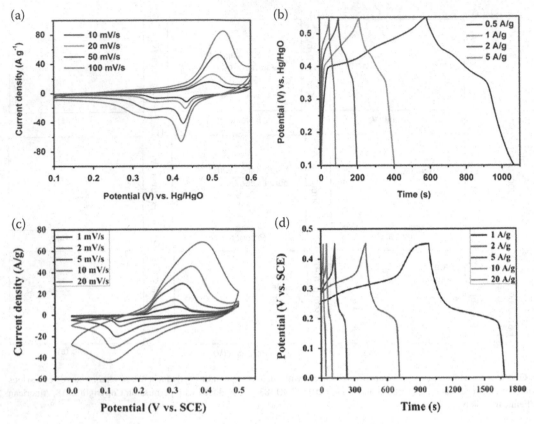

FIGURE 9.3 Schematic cyclic voltammetry curves with their corresponding galvanostatic discharge curves of Ni-MOF-based (a, b) pseudocapacitive (Ma et al. 2019, 9543–47). Reproduced with permission. Copyright 2019, American Chemical Society. (c, d) battery-like electrode materials (Chen et al. 2018b, 114–23). Reproduced with permission copyright 2018, Elsevier.

TABLE 9.1

Advantages and Disadvantages of MOF Supercapacitor

Advantages	Disadvantages
Adjustable pore size and pore distribution	Low chemical stability and cycling performance
Multifunctional capability (sensing and energy storage)	Poor electrochemical conductivity
Controllable morphology	Poor ion insertion arising from steric hindrance

Ni-precursor could be any of the following: nickel chlorate hexahydrate ($NiCl_2 \cdot 6H_2O$), nickel nitrate hexahydrate ($Ni(NO_3)_2 \cdot 6H_2O$), nickel acetate tetrahydrate ($Ni(OAc)_2 \cdot 4H_2O$).

Organic linker: p-benzenedicarboxylic acid (PTA), 1,4-benzenedicarboxylate acid (H_2BDC), DABCO, ADC, H_3btc, 2,3,6,7,10,11-hexahydroxytriphenylene ($H_{12}C_{18}O_6$, HHTP)

Solvent: N,N-dimethylformamide (DMF), deionized water, alcohol

Complexing agent: NaOH aqueous, HOC_6H_4COONa

First, the organic pellet is dissolved in a solvent at ambient temperature then Ni-precursor solution is gradually added to it. The solutions are thoroughly mixed using a magnetic stirrer or ultra-sonication. Afterward, the mixture is transferred into a Teflon-lined stainless-steel autoclave and the temperature is raised to 100–200°C for 8–48 h. The resulting product is cooled, filtered, or centrifuged and washed repeatedly with DMF, alcohol, or deionized water and dried in air at 60–70°C

(a) (b)

FIGURE 9.4 Morphology of (a) pristine Ni-MOF (b) 65% Ni–35% Co MOF (Qu et al. 2017, 1263–69). Reproduced with permission. Copyright 2017, American Chemical Society.

for 12–24 h in the air to obtain the pristine Ni-MOF powders (Yang et al. 2014a, 16640–44; Wu and Hsu 2015, 1055–62).

For doped or composite Ni-MOFs, the precursor of the required dopant is added into the solution before the isothermal step in an oven. Also, a slightly higher temperature may be required for mixing composite materials (Hong et al. 2019, 62–71).

A microwave-assisted reactor can also be utilized to reduce the processing temperature (Choi et al. 2014, 7451–57). Combinatory techniques have also been adopted to prepare composite Ni-based MOFs where hydrothermal and electrospinning were utilized (Zhao et al. 2019, 1824–30).

To produce layered double-hydroxide composites, after the hydrothermal step the resulting product is treated with KOH at room temperature for about 5 h, washed in deionized water, and dried (Cao et al. 2017, 154–60). Figures 9.4 and 9.5 show the morphology, X-ray diffraction pattern, cyclic voltammetry, and galvanostatic charge-discharge curves of Ni-based MOFs.

9.2.1.2 Powder Synthesis Using MOF-Template

First, the MOF-5 template is prepared by mixing triethylamine (TEA), zinc nitrate hexahydrate ($Zn(NO_3)_2$, $6H_2O$), and terephthalic acid (BDC, $C_6H_4(COOH)_2$ in DMF at room temperature with few drops of hydrogen peroxide added to the solution. The solution is filtered and dried to get the MOF powder. Afterward, nickel nitrate hexahydrate ($Ni(NO_3)_2$, $6H_2O$) is dissolved in deionized water and some quantity of the MOF-5 is added to the solution. The mixture is stirred for about 12 h at ambient temperature, filtered, and dried (Wang et al. 2010, 3017–24; Peng et al. 2014, 3213–18).

9.2.2 Device Assembly

The prepared MOFs are deposited on conductive materials to form either a asymmetric device or single electrode system.

9.2.2.1 Deposition

The obtained MOFs can be deposited on conductive substrates such as stainless steel, carbon fibre paper (CFP), carbon cloth by electrophoretic deposition. The substrate is first cleaned with alcohol before deposition. The electrode on which the Ni-MOF is to be deposited is used as the working electrode (negative electrode) (Wu and Hsu 2015, 1055–62).

Alternatively, the substrate to be used as an electrode can be placed in the vial containing the precursor solution during the hydrothermal synthesis. After the reaction, the electrode is then washed and dried in an oven (Chu et al. 2020b, 5669–77).

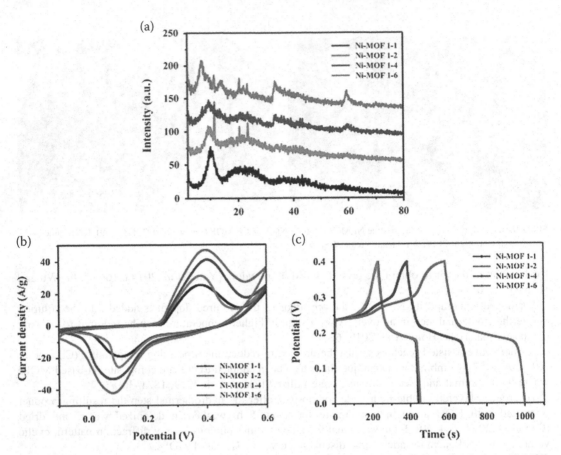

FIGURE 9.5 (a) X-ray pattern (b) cyclic voltammetry curves (c) galvanostatic charge-discharge curve of pristine Ni-MOF at different Ni and MOF ratios (Du et al. 2018, 57–68). Reproduced with permission. Copyright 2018, Elsevier.

For active carbon (AC) electrode commonly used in the asymmetric device assembly, a paste containing 80% AC, 15% acetylene black, and 5% Nafion mixed in ethanol can be pressed on the substrate and dried at 60°C (Deng et al. 2017; Chu et al. 2020b, 5669–77).

An alkaline electrolyte, specifically potassium hydroxide (KOH), is suitable for Ni-based MOF supercapacitors. The gel derivative can be obtained by mixing polyvinyl alcohol (PVA) and KOH in deionized water, then heated at about 90°C. Cellulose paper or any other insulator can be used as a separator (Li et al. 2019a, 37675–84).

Asymmetric device configuration is popularly adopted for Ni-based MOFs. It is important to note that asymmetric capacitors are a special case of hybrid supercapacitors (HSCs), while asymmetric capacitor (ASC) consists of two electrodes of a different type, weight, and thickness. HSC consists of a battery-like electrode as the positive electrode and a carbon material as the negative electrode. On the other hand, symmetric capacitors have two electrodes of the same type/material, weight, and thickness.

9.3 Advances and Optimizations in Ni-Based MOF Supercapacitor

After the pioneering work on Ni-MOF in 2014, tremendous progress has been made in improving the specific capacitance, energy density, and power density of these materials.

MOFs can be applied in three ways: pristine, as precursors, and sacrificial templates: pyrolysis to obtain metal oxides, and their carbon equivalent, and composites (Yang et al. 2014a, 16640–44; Zain et al. 2018, 665–74). Resulting MOFs from these three processes have been implemented in synthesizing Ni-based supercapacitors.

9.3.1 Pristine Ni-Based MOFs

This class of Ni-MOFs contains no additives, they are purely in their natural state. Usually, they possess pseudocapacitive behaviour seen in the reversible redox reaction (Xu et al. 2016, 595–602).

The highly porous nature of MOFs is vital for energy storage application, but on the other hand, their poor electric conductivity and stability have limited their application; hence alternative ways to improve their conductivity and the overall storage capacity are the subject of modern research.

Pristine nanocrystals of Ni-MOF (nNi-MOF-74) film fabricated in a coin-type supercapacitor with polypropylene separator displayed an areal capacitance of 0.415 mF cm^{-2} with 4000 life cycles (Choi et al. 2014, 7451–57). Hydrothermally prepared Ni-MOF with a very high specific capacitance of 1127 Fg^{-1} at 0.5 A g^{-1} using p-benezedicarboxylic acid (PTA) as the organic linker was first reported by Yang et al. (2014a, 16640–44). Afterward, Ni-MOF synthesized by simple solvothermal method had 1457.7 F g^{-1}at 1 A g^{-1} (Zhou et al. 2016, 28904–16); then 1D Ni-based MOF nanorods with salicylic acid as organic linker had a capacitance of 1698 F g^{-1} at 1 A g^{-1} (Xu et al. 2016, 595–602). Strategically apart from the capacitance, the energy density has also been augmented as reported in Table 9.2.

9.3.2 Derived Ni-Based MOFs/Composites

They result from chemical or thermal treatment of MOF precursors of the sacrificial template (Figure 9.6). In addition, the carbonization of MOFs at optimized conditions produces composites that retain the porosity and structure of the parent material. The derived composites can then be further functionalized with nitrogen, sulphur, phosphorus for better performance (Salunkhe et al. 2017, 5293–308).

9.3.2.1 Metal Oxide/Hydroxide

Some advantages of metal oxide derived from MOF include high porosity for mass transport, large surface area, active site exposure, and adjustable composition for better performances (Wang et al. 2020b, 213093). Excellent cyclability (16,000 cycles) was recorded by Du et al. for Nickel-based pillared supercapacitor (Ni-DMOF-ADC) using 1,4-diazabicyclo [2.2.2] octane (DABCO) as pillar

TABLE 9.2

Pristine Ni-MOFs with Their Supercapacitor Properties

S/N	MOF	Specific Capacitance (F g^{-1})	Organic Linker/Asymmetric Capacitor Specific Capacitance (F g^{-1})	Energy Density (Wh kg^{-1})/ Power Density (W kg^{-1})	Cycling Performance	Reference
1	Ni-MOF	1127@0.5 A g^{-1}	p-Benzenedicarboxylic acid/-	19.17/1750	3000	Yang et al. 2014a, 16640–44
2	Ni-MOF	1457.7@1 A g^{-1}	1,4 Benzenedicarboxylic acid/-	–/–	3000	Zhou et al. 2016, 28904–16
3	Ni-hydroxide MOF	1698@1 A g^{-1}	Salicylic acid / 166@1 Ag^{-1}	–/–	1000	Xu et al. 2016, 595–602
4	Ni-MOF	1518.8@1 A g^{-1}	2,5-Thiophenedicarboxylate acid (H$_2$TDA)/244 mF cm^{-2}@0.5 mA cm^{-2}	5.08 x 10^{-2} mW cm^{-2}/4.172 mW cm^{-2}	10,000	Yang et al. 2020a, 1803–06
5	Ni-MOF	1057	Trimesic acid/87@0.5 A g^{-1}	21.05/6030	2500	Du et al. 2018, 57–68
6	Ni-MOF	1668@2 A g^{-1}	H$_3$BTC (1,3,5-benzenetricarboxylic/161@0.2 A g^{-1}	57.29/160	5000	Zhang et al. 2018a, 17747–53
7	Ni-Hsal-CSN	1232.16@2.04 A g^{-1}	Salicylic acid/ 6.32 F cm^{-3} at 7 mA cm^{-2}	2.39 mWh cm^{-3}/ 1.82 W cm^{-3}	15,000	Zhang et al. 2018b, 230–39
8	Ni-MOF	977.04	Trimesic acid (H$_3$BTC)/ 189.23@0.5 A g^{-1}	55.26/362.5	5000	Li et al. 2019b, 1902463
9	Ni$_3$(btc)MOF	726	Trimesic acid (H$_3$BTC)/-	16.5/2078	5000	Kang et al. 2014, 957–61

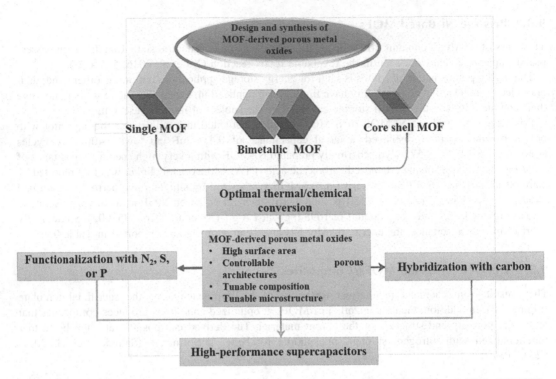

FIGURE 9.6 Schematic of deriving metal oxide and its composites from MOFs (Salunkhe et al. 2017, 5293–308). Adapted with permission. Copyright 2017, American Chemical Society.

ligands and 9,10-anthracenedicarboxylic acid (ADC) as ligands (Du et al. 2018, 57–68). The electrode cycle stability was correlated with Ni-DMOF structure stability (Qu et al. 2016, 66–73). One strategy to facilitate electron transport and diffusion is to use a layered structure: Ni-based double-layer hydroxide (Ni-LDH) specific capacitance of 2148 Fg^{-1} at 1 Ag^{-1} (Zhou et al. 2020, 3281–88). Although it is difficult to integrate different metal ions into one structure, however, ions with similar size, metallic species, charge density, and the same processing conditions are essential for mixed-metal MOF synthesis (Rajak et al. 2020, 11792–818).

9.3.2.2 Mixed Metal (Bimetallic/Ternary) MOFs

Mixed metal oxide MOF nanocomposites can be produced by adding a second layer on an already existing MOF to form a core-shell structure or incorporating two or more metals in a framework. This provides additional structural stability, sufficient electrode/electrolyte contact (shortens the O–H pathway), improves cyclability, enhances ion storage, and increases electrochemical performance due to large charge transfer between different metal ions, and possesses various redox sites (Wang et al. 2020b, 213093). Ni-MOFs incorporating carbon composites have been reported to display enhanced capacitance retention resulting from high dispersity of Ni particles, ease of electron transport with large macropores, and conductivity of carbon facilitates electron transport (Wu and Hsu 2015, 1055–1062; Jiao et al. 2017, 1094–1102).

9.3.3 Hybrid Ni-MOF Supercapacitors

Hybrid supercapacitors (HSCs) are the combination of a capacitor material (e.g. carbon) and a battery-like material (e.g. transition metal oxides, hydroxides, sulphides, and phosphides) as positive and negative electrodes, respectively, in a device. They are adopted to improve the energy density of supercapacitors and the power density of batteries by taking advantage of the synergistic contribution of these two energy storage devices (Jiao et al. 2016, 13344–51; Wang et al. 2020b, 213093) while compensating for the weakness of each system. HSC is influenced by the capacity and working voltage. Interestingly Ni-based oxides due to their excellent redox character, high specific capacity, and alkaline compatibility have made them a promising battery-type electrode. Nickel-based active material combined with carbon provided enormous benefits outlined as follows:

- It acts as a reducing agent
- Enhances dispersity of the metal ion enabling more active site availability/visibility
- High porosity for more redox mass transport (Yang et al. 2019, 905–14).

Majority of the reported Ni-MOF supercapacitors show a battery-like character (from their cyclic voltammetry (CV) curve, hence promoting the capacitance behaviour of the hybrid supercapacitor (Wang et al. 2020b, 213093). To enhance the energy storage capacity, redox materials which have high electrochemical reversibility (e.g. $Fe(CN)_6^{4-}/Fe(CN)_6^{3-}$) can be added to the electrolyte to act as an electron relay for better energy and power density. Table 9.3 shows the capacitance properties of derived and composite MOFs

*BDC = 1,4-benzenedicarboxylic acid; *NF = nickel foam; *CFP = carbon fibre paper
*LDH = layered double hydroxide; *NC = nano carbon; *MDH = metal double-layer hydroxide; *HIH = hollow in hollow (which involves the distribution of Ni-Co-LDH on hollow carbon spheres).

9.4 Challenges

MOFs are usually obtained in powder form basically because they tend to aggregate in solution. The high aggregation tendency of MOFs drastically affects their capacitance. Also, they require binders and conductive materials to enable them to attach properly to the electrode. In return, this affects the surface area, restricts electrolyte penetration, and electrode flexibility which generally decreases their performance (Li et al. 2020, 136139; Zhong et al. 2020, 265–74). A solution to this problem is the use of flexible and binder-free electrodes (e.g. carbon cloth).

Some of the reported hybrid supercapacitor devices display low power density, high cost, and short cyclability (Jiao et al. 2016, 13344–51).

Although the conductivity of MOFs has been improved by the addition of conductive polymers or carbon-based materials, yet they still suffer energy loss due to poor contact; they aggregate away from the conductive particles leading to relatively low conductivity, and hole blocking that affects ion diffusion (Zhang et al. 2019, 24279–87). Though HSC is a very promising energy storage avenue, the choice of suitable electrode materials for device assembly poses a challenge, as some materials display good electrode properties in a standalone setup but when incorporated in a device, their cycle performance and capacitance deteriorate severely.

In MOF-derived metal oxides and their carbon composites, lack of pretreatment in nitrogen gas could destroy the parent morphology. Also, thermal treatment at low temperatures leads to unwanted carbon bonds that increase the resistance of the material. On the other hand, at higher temperatures, the carbon becomes more conductive and the metal oxide is reduced to metal nanoparticles (Salunkhe et al. 2017, 5293–308).

Transition MOFs display poor conductivity at higher charge-discharge rate arising from lack of bulk electron conduction, short cycle life; their crystalline nature limits the electrolyte ion diffusion into the pores compared to their amorphous counterpart (Salunkhe et al. 2017, 5293–308; Xia et al. 2015, 1837–66).

The volume expansion of MOFs during the charge-discharge cycle strongly limits their performance and cycle stability. It is difficult to control the pore size and homogeneity at the same time because of the variety of factors involved during preparation that influence the morphology and the structure of the final products (Wu et al. 2018, 697–716).

TABLE 9.3

Derived/Composite Ni-MOFs with Their Supercapacitor Properties

S/N	MOF	Specific Capacitance (F g⁻¹)	Sacrificial Template/ Asymmetric Capacitance (F g⁻¹)	Energy Density (Wh kg⁻¹)/ Power Density (W kg⁻¹)	Cycles	Reference
MOF-derived metal oxide / hydroxide						
1	Ni-Co oxide/Graphene	2870.8@1 A g⁻¹	ZIF-67/160.6@1 A g⁻¹	50.2/750	5000	Jayakumar et al. 2017, 1603102
2	Co_3O_4@NiO	1877.93@2 A g⁻¹	ZIF-67/164.8@1 A g⁻¹	54.99/780	10,000	Hou et al. 2020, 136053
3	NiO	1863@0.5 A g⁻¹	Ni-MOF/108@0.5 A g⁻¹	38.4/400	5000	Meng et al. 2020, 145077
4	NiCo LDH	1386@1 A g⁻¹	-/94.3@0.5 A g⁻¹	118/108	10,000	Chen et al. 2020, 154794
5	NiCo-MOF	1945.83 @0.5 A g⁻¹	—	—	–	Ma et al. 2019, 9543–47
Composite						
1	Zn doped Ni-MOF	1620	—	27.56/1750		Yang et al. 2014b, 19005–10
2	Ni-MOF (PPy)	715	—	40.1/1500.6		Wang et al. 2020a, 12129–34
3	CC@$NiCo_2O_4$	1055.3@2.5 mA cm⁻²	Ni-Co LDH/89.7@5 mA cm⁻²	31.9/2900	20,000	Guan et al. 2017b, 1602391
4	Ni-Co-Se	1668@1 A g⁻¹	ZIF-67/108.2@1 A g⁻¹	38.5/800.8	5000	Qu et al. 2020, 2007–12
5	Ni-MOF@GO	2192.4@1 A g⁻¹	—	-/-	3000	Zhou et al. 2016, 28904–16
6	Ni-MOF/CNT	1765@0.5 A g⁻¹	-/103@0.5 A g⁻¹	36.6/480	5000	Wen et al. 2015, 13874–83
7	Ni-CAT/NiCo-LDH/NF	3200 mF cm⁻² (2133 F g⁻¹)@1mA cm⁻²	-/435 mF cm⁻²@1 mA cm⁻²	93µWhcm⁻³/18.3 mWcm⁻²	1000	Li et al. 2018, 6202–05
8	Ni-MOF/PANI/NF	36264 mF cm⁻²@2mA cm⁻²	113.6@1 A g⁻¹	45.6/850	10,000	Cheng et al. 2019,4119–23
9	NiCo-MOF	1202.1@1 A g⁻¹	-/158.1 @0.5 A g⁻¹	49.4/562.5	5000	Wang et al. 2019b, 2063–71
10	NiCo-MOF	1220.2@1 A g⁻¹	-/76.67 @1 A g⁻¹	30.9/1132.8	3000	Chen et al. 2018a, 5639–45
11	Co/Ni-MOF@SNC	1970@1 Ag⁻¹	-/156.7@1 A g⁻¹	55.7/800	3000	Tong et al. 2017, 9873–81
12	Co/Ni phosphate	1616@1 A g⁻¹	ZIF-67/116.67@1 A g⁻¹	33.29/150	10,000	Xiao et al. 2019, 1086–92
13	$NiCo_2SO_4$	1382@1 A g⁻¹	ZIF-67//113@1 A g⁻¹	35.3/750	10,000	Cai et al. 2020, 144501
14	CoNi-MOF/CFP	2033@1 A g⁻¹	—	26.8/1450	10,000	Chu et al. 2020b, 5669–77

No.	Material				Cycles	Reference
15	Co-Ni-MOF	$1300@1\ A\ g^{-1}$		$25.92/375$	3000	Wang et al. 2019a, 1158–65
16	Ni-Co-oxide	$1900@2\ A\ g^{-1}$		$52.6/1604$	20,000	Guan et al. 2017a, 1605902
17	Ni-MOF/rGO	$1154.4@1\ A\ g^{-1}$		$20/820$	3000	Kim et al. 2020, 2750–54
18	Ni-Co@Co-MOF/NF	$2697@1\ A\ g^{-1}$	$-/152.9@1\ A\ g^{-1}$	$61.4/853$	10,000	Jiang et al. 2020, 136077
19	Ni-Co-S	$1377.5@1\ A\ g^{-1}$	Ni/Co-MOF/$103.9@1\ A\ g^{-1}$	$24.8/1066.42$	3000	Chen et al. 2018a, 5639–45
20	Ni-Co-Se	$1668@1\ A\ g^{-1}$	ZIF-67/$108.2@1\ A\ g^{-1}$	$38.5/802.8$	5000	Qu et al. 2020, 2007–12
21	$NiCoMoS_x$	$2595@1\ A\ g^{-1}$	$-/136@1\ A\ g^{-1}$	$48.2/807.2$	10,000	Yang et al. 2020b, 154118
22	NC/Ni-Ni_3S_4/CNTs	$1489.9@1\ A\ g^{-1}$	$-/127.5@1\ A\ g^{-1}$	$398/749.8$	15,000	Yang et al. 2020c, 2406–12

Battery-type MOFs

No.	Material				Cycles	Reference
1	Ni-MDH	$875\ C\ g^{-1}\ @1\ A\ g^{-1}$	$-/215@2\ A\ g^{-1}$	$81/1900$	10,000	Qu et al. 2017, 1263–69
2	Zn-doped Ni-MOF	$237.4\ mAh\ g^{-1}\ @1\ A\ g^{-1}$	–	–	4000	Chen et al. 2018b, 114–23
3.	CoCuNi-BDC/NF	$321.3\ mA\ h^{-1}@1\ A\ g^{-1}$	–	–	1000	Zain et al. 2018, 665–74
4	NiCo-LDH@rGO	$750\ C\ g^{-1}\ @0.5\ A\ g^{-1}$	$-/234.88\ C\ g^{-1}@1\ A\ g^{-1}$	$50/780$	3000	Chu et al. 2020a, 130–38
5	Ni-CoMOF	$833\ C\ g^{-1}\ @0.5\ A\ g^{-1}$	$-/172.7@0.5\ A\ g^{-1}$	$77.7/450$	5000	Ye et al. 2019, 4998–5008
6	Ni-Co MDH/GO	$904.3\ C\ g^{-1}@1\ A\ g^{-1}$	ZIF-67/$142.1@1\ A\ g^{-1}$	$50.3/853.3$	5000	Yu et al. 2017, 16865–72
7	HIH-LDH	$156.4\ mA\ h^{-1}@1\ A\ g^{-1}$	ZIF-67/$135.8@1\ A\ g^{-1}$	$34.5/55000$	10,000	Lee et al. 2019, 17637–47
8	$Co_3O_4@CoN_{12}S_4$	$244.4\ mA\ h^{-1}@1\ A\ g^{-1}$	Co-MOF/ $58.3\ mA\ h\ g^{-1}$ @t $5\ mA\ cm^{-2}$	$55.6/884$	10,000	Han et al. 2020, 1428–36

Hybrid supercapcitor

No.	Material				Cycles	Reference
1	$MnNi_2O_4$	$2848\ @1A\ g^{-1}$	$401.7@1\ A\ g^{-1}$	$142.8/800$	5000	Lan et al. 2020, 153546
2	Co/Ni-MOF/ CNT-COOH	$236.1\ mAh\ g^{-1}$ ($849.96\ C\ g^{-1}$)$@1\ A\ g^{-1}$	$-/211.7@1\ A\ g^{-1}$	$61.8/725$	5000	Jiao et al. 2017, 1094–1102
3.	Ni-MOF/ CNT-COOH	$123.4\ mAh\ g^{-1}@1\ A\ g^{-1}$	$96.7\ mAh\ g^{-1}\ @1\ A\ g^{-1}$	$55.8/7000$	3000	Jiao et al. 2016, 13344–51

Finally, the current preparation techniques are not scalable for commercial purposes. Besides, the current state-of-the-art is insufficient for the latter application. Finally, if the above challenges are carefully tackled, transition metal–based MOFs hold great promise for energy storage.

9.5 The Future of MOF-Based Energy Supercapacitor

- The energy storage capability of MOFs can be enhanced by adopting layered structures that enable more electron transport and electrolyte diffusion, tuning the linker structure, and the addition of redox substances in the electrolytes (Yu et al. 2012, 402–07; Wang et al. 2020b, 213093).
- More work is required in optimizing the processing conditions of MOFs especially in those derived from metal oxides and their nanocomposites such that high specific capacitance, energy, and power density can be obtained.
- Matching the pore size of MOFs with electrolytes of specific ion size has the potential to increase their energy storage performance and may also improve potential windows which are currently limited to about 1.2V, though few studies have exceeded this threshold with potential up to 1.6 V (Chu et al. 2020a, 130–38).
- The adoption of flexible electrode materials for MOFs can increase the versatility of their application and make the device assembly simple.
- Modern and innovative methods of synthesis, which are scalable, economic, and facile are important for future progress on MOF-based energy storage devices.
- MOF-based or derived energy storage materials have interesting potential to replace traditional transition metal oxide ones, but certain factors such as cost, appropriate metal, and inorganic components, optimizations, etc., still need to be considered.

REFERENCES

Cai, P., T. Liu, L. Zhang, B. Cheng, and J. Yu. 2020. "ZIF-67 Derived Nickel Cobalt Sulfide Hollow Cages for High-Performance Supercapacitors." *Applied Surface Science* 504: 144501. 10.1016/j.apsusc.2019.144501.

Cao, F., M. Gan, L. Ma, X. Li, F. Yan, M. Ye, Y. Zhai, and Y. Zhou. 2017. "Hierarchical Sheet-like Ni–Co Layered Double Hydroxide Derived from a MOF Template for High-Performance Supercapacitors." *Synthetic Metals* 234: 154–160. 10.1016/j.synthmet.2017.11.001.

Chen, C., M.-K. Wu, K. Tao, J.-J. Zhou, Y.-L. Li, X. Han, and Han, L. 2018a. "Formation of Bimetallic Metal–Organic Framework Nanosheets and Their Derived Porous Nickel–Cobalt Sulfides for Supercapacitors." *Dalton Transactions* 47(16): 5639–5645. 10.1039/c8dt00464a.

Chen, Y., D. Ni, X. Yang, C. Liu, J. Yin, and K. Cai. 2018b. "Microwave-Assisted Synthesis of Honeycomblike Hierarchical Spherical Zn-Doped Ni-MOF as a High-Performance Battery-Type Supercapacitor Electrode Material." *Electrochimica Acta* 278: 114–123. 10.1016/j.electacta.2018.05.024.

Chen, S., L. Zhao, W. Wei, Y. Li, and L. Mi. 2020. "A Novel Strategy to Synthesize NiCo Layered Double Hydroxide Nanotube from Metal Organic Framework Composite for High-Performance Supercapacitor." *Journal of Alloys and Compounds* 831, 154794. 10.1016/j.jallcom.2020.154794.

Cheng, Q., K. Tao, X. Han, Y. Yang, Z. Yang, Q. Ma, and L. Han. 2019. "Ultrathin Ni-MOF nanosheet arrays grown on polyaniline decorated Ni foam as an advanced electrode for asymmetric supercapacitors with high energy density." *Dalton Transactions*, 48(13): 4119–4123. 10.1039/c9dt00386j.

Choi, K.M., H.M. Jeong, J.H. Park, Y.-B. Zhang, J.K. Kang, and O.M. Yaghi,. 2014. "Supercapacitors of Nanocrystalline Metal–Organic Frameworks." *ACS Nano* 8(7): 7451–7457. 10.1021/nn5027092.

Chu, D., F. Li, X. Song, H. Ma, L. Tan, H. Pang, X. Wang, D. Guo, and B. Xiao. 2020a. A "Novel Dual-Tasking Hollow Cube $NiFe_2O_4$-NiCo-LDH@rGO Hierarchical Material for High Preformance Supercapacitor and Glucose Sensor." *Journal of Colloid and Interface Science* 568: 130–138. 10.1016/j.jcis.2020.02.012.

Chu, X., Meng, F., Deng, T., Lu, Y., Bondarchuk, O., Sui, M., Feng, M., Li, H., & Zhang, W. 2020b. "Mechanistic insight into bimetallic CoNi-MOF arrays with enhanced performance for supercapacitors." *Nanoscale* 12(9): 5669–5677. 10.1039/c9nr10473a.

Deng, T., Y. Lu, W. Zhang, M. Sui, X. Shi, D. Wang, and W. Zheng. 2017. "Inverted Design for High-Performance Supercapacitor Via Co(OH)2 -Derived Highly Oriented MOF Electrodes." *Advanced Energy Materials* 8(7): 1702294. 10.1002/aenm.201702294.

Díaz, R., M.G. Orcajo, J.A. Botas, G. Calleja, and Palma, J. 2012. "Co8-MOF-5 as Electrode for Supercapacitors." *Materials Letters* 68: 126–128. 10.1016/j.matlet.2011.10.046.

Du, P., Y. Dong, C. Liu, W. Wei, D. Liu, and Liu, P. 2018. "Fabrication of Hierarchical Porous Nickel Based Metal-Organic Framework (Ni-MOF) Constructed with Nanosheets as Novel Pseudo-Capacitive Material for Asymmetric Supercapacitor." *Journal of Colloid and Interface Science* 518: 57–68. 10.1016/j.jcis.2018.02.010.

Farha, O.K., I. Eryazici, N.C. Jeong, B.G. Hauser, C.E. Wilmer, A.A. Sarjeant, R.Q. Snurr, S.T. Nguyen, A.Ö. Yazaydın, and Hupp, J.T. 2012. "Metal–Organic Framework Materials with Ultrahigh Surface Areas: Is the Sky the Limit?" *Journal of the American Chemical Society* 134(36): 15016–15021. 10.1021/ja3055639.

Gogotsi, Y., & R.M. Penner. 2018. "Energy Storage in Nanomaterials – Capacitive, Pseudocapacitive, or Battery-Like?" *ACS Nano* 12(3): 2081–2083. 10.1021/acsnano.8b01914.

Guan, B.Y., A. Kushima, L. Yu, S. Li, J. Li, and X.W.D. Lou. 2017a. "Coordination Polymers Derived General Synthesis of Multishelled Mixed Metal-Oxide Particles for Hybrid Supercapacitors." *Advanced Materials* 29(17): 1605902. 10.1002/adma.201605902.

Guan, C., X. Liu, W. Ren, X. Li, C. Cheng, and J. Wang. 2017b. "Rational Design of Metal-Organic Framework Derived Hollow $NiCo_2O_4$ Arrays for Flexible Supercapacitor and Electrocatalysis." *Advanced Energy Materials*, 7(12), 1602391. 10.1002/aenm.201602391.

Han, D., J. Wei, Y. Zhao, Y. Shen, Y. Pan, Y. Wei, and L. Mao. 2020. "Metal–Organic Framework Derived Petal-like Co_3O_4@CoNi2S4 Hybrid on Carbon Cloth with Enhanced Performance for Supercapacitors." *Inorganic Chemistry Frontiers* 7(6): 1428–1436. 10.1039/c9qi01681c.

Hong, J., S.-J. Park, and S. Kim. 2019. "Synthesis and Electrochemical Characterization of Nanostructured Ni-Co-MOF/Graphene Oxide Composites as Capacitor Electrodes." *Electrochimica Acta* 311: 62–71. 10.1016/j.electacta.2019.04.121.

Hou, S., Y. Lian, Y. Bai, Q. Zhou, C. Ban, Z. Wang, J. Zhao, and H. Zhang. 2020. "Hollow Dodecahedral Co_3S4@NiO Derived from ZIF-67 for Supercapacitor." *Electrochimica Acta* 341: 136053. 10.1016/j.electacta.2020.136053.

Jayakumar, A., R.P. Antony, R. Wang, and J.-M. Lee. 2017. "MOF-Derived Hollow Cage NixCo3−xO4 and Their Synergy with Graphene for Outstanding Supercapacitors." *Small* 13(11): 1603102. 10.1002/smll.201603102.

Jiang, J., Y. Sun, Y. Chen, Q. Zhou, H. Rong, X. Hu, H. Chen, L. Zhu, and S. Han. 2020. "Design and Fabrication of Metal-Organic Frameworks Nanosheet Arrays Constructed by Interconnected Nanohoneycomb-like Nickel-Cobalt Oxide for High Energy Density Asymmetric Supercapacitors." *Electrochimica Acta* 342: 136077. 10.1016/j.electacta.2020.136077.

Jiao, Y., J. Pei, D. Chen, C. Yan, Y. Hu, Q. Zhang, and G. Chen. 2017. "Mixed-Metallic MOF Based Electrode Materials for High Performance Hybrid Supercapacitors." *Journal of Materials Chemistry A* 5(3): 1094–1102. 10.1039/C6TA09805C.

Jiao, Y., J. Pei, C. Yan, D. Chen, Y. Hu, and G. Chen. 2016. "Layered Nickel Metal–Organic Framework for High Performance Alkaline Battery-Supercapacitor Hybrid Devices." *Journal of Materials Chemistry A* 4(34): 13344–13351. 10.1039/c6ta05384j.

Kang, L., S.-X. Sun, L.-B. Kong, J.-W. Lang, and Y.-C. Luo. 2014. "Investigating Metal-Organic Framework as a New Pseudo-Capacitive Material for Supercapacitors." *Chinese Chemical Letters* 25(6): 957–961. 10.1016/j.cclet.2014.05.032.

Kim, J., S.-J. Park, S. Chung, and S. Kim. 2020. "Preparation and Capacitance of Ni Metal Organic Framework/Reduced Graphene Oxide Composites for Supercapacitors as Nanoarchitectonics." *Journal of Nanoscience and Nanotechnology* 20(5): 2750–2754. 10.1166/jnn.2020.17469.

Lan, M., X. Wang, R. Zhao, M. Dong, L. Fang, and L. Wang. 2020. "Metal-Organic Framework-Derived Porous MnNi2O4 Microflower as an Advanced Electrode Material for High-Performance Supercapacitors." *Journal of Alloys and Compounds* 821: 153546. 10.1016/j.jallcom.2019.153546.

Lee, D.Y., S.J. Yoon, N.K. Shrestha, S.-H. Lee, H. Ahn, and S.-H. Han. 2012. "Unusual Energy Storage and Charge Retention in Co-Based Metal–Organic-Frameworks." *Microporous and Mesoporous Materials*, 153: 163–165. 10.1016/j.micromeso.2011.12.040.

Lee, G., W. Na, J. Kim, S. Lee, and J. Jang. 2019. "Improved Electrochemical Performances of MOF-Derived Ni–Co Layered Double Hydroxide Complexes Using Distinctive Hollow-In-Hollow Structures." *Journal of Materials Chemistry A* 7(29): 17637–17647. 10.1039/c9ta05138d.

Li, G., H. Cai, X. Li, J. Zhang, D. Zhang, Y. Yang, and J. Xiong. 2019a. "Construction of Hierarchical $NiCo_2O_4$@Ni-MOF Hybrid Arrays on Carbon Cloth as Superior Battery-Type Electrodes for Flexible Solid-State Hybrid Supercapacitors." *ACS Applied Materials & Interfaces* 11(41): 37675–37684. 10.1021/acsami.9b11994.

Li, J., W. Cao, N. Zhou, F. Xu, N. Chen, Y. Liu, and G. Du. 2020. "Hierarchically Nanostructured Ni(OH) 2–MnO2@C Ternary Composites Derived from Ni-MOFs Grown on Nickel Foam as High-Performance Integrated Electrodes for Hybrid Supercapacitors." *Electrochimica Acta* 343: 136139. 10.1016/j.electacta.2020.136139.

Li, Y., Y. Xu, Y. Liu, and H. Pang. 2019b. "Exposing {001} Crystal Plane on Hexagonal Ni-MOF with Surface-Grown Cross-Linked Mesh-Structures for Electrochemical Energy Storage." *Small* 15(36): 1902463. 10.1002/smll.201902463.

Li, Y.-L., J.-J. Zhou, M.-K. Wu, C. Chen, K. Tao, F.-Y. Yi, and L. Han. 2018. "Hierarchical Two-Dimensional Conductive Metal–Organic Framework/Layered Double Hydroxide Nanoarray for a High-Performance Supercapacitor." *Inorganic Chemistry* 57(11): 6202–6205. https://doi.org/10.1021/acs.inorgchem.8b00493

Ma, H.-M., J.-W. Yi, S. Li, C. Jiang, J.-H. Wei, Y.-P. Wu, J. Zhao, and Li, D.-S. 2019. "Stable Bimetal-MOF Ultrathin Nanosheets for Pseudocapacitors with Enhanced Performance." *Inorganic Chemistry* 58(15): 9543–9547. 10.1021/acs.inorgchem.9b00937.

Meng, X.-X., J.-Y. Li, B.-L. Yang, and Z.-X. Li. 2020. "MOF-Derived NiO Nanoparticles Prilled by Controllable Explosion of Perchlorate Ion: Excellent Performances and Practical Applications in Supercapacitors." *Applied Surface Science* 507: 145077. 10.1016/j.apsusc.2019.145077.

Peng, M.M., U.J. Jeon, M. Ganesh, A. Aziz, R. Vinodh, M. Palanichamy, and H.T. Jang. 2014. "Oxidation of Ethylbenzene Using Nickel Oxide Supported Metal Organic Framework Catalyst." *Bulletin of the Korean Chemical Society* 35(11): 3213–3218. 10.5012/bkcs.2014.35.11.3213.

Qu, C., Y. Jiao, B. Zhao, D. Chen, R. Zou, K.S. Walton, and M. Liu. 2016. "Nickel-Based Pillared MOFs for High-Performance Supercapacitors: Design, Synthesis and Stability Study." *Nano Energy* 26: 66–73. 10.1016/j.nanoen.2016.04.003.

Qu, C., B. Zhao, Y. Jiao, D. Chen, S. Dai, B.M. deglee, Y. Chen, K.S. Walton, R. Zou, and M. Liu. 2017. "Functionalized Bimetallic Hydroxides Derived from Metal–Organic Frameworks for High-Performance Hybrid Supercapacitor with Exceptional Cycling Stability." *ACS Energy Letters* 2(6): 1263–1269. 10.1021/acsenergylett.7b00265.

Qu, G., X. Zhang, G. Xiang, Y. Wei, J. Yin, Z. Wang, X. Zhang, and X. Xu. 2020. "ZIF-67 Derived Hollow Ni-Co-Se Nano-Polyhedrons for Flexible Hybrid Supercapacitors with Remarkable Electrochemical Performances." *Chinese Chemical Letters* 31(7): 2007–2012. 10.1016/j.cclet.2020.01.040.

Rajak, R., R. Kumar, S.N. Ansari, M. Saraf, and S.M. Mobin. 2020. "Recent Highlights and Future Prospects on Mixed-Metal MOFs as Emerging Supercapacitor Candidates." *Dalton Transactions*, 49(34), 11792–11818. 10.1039/D0DT01676D.

Salunkhe, R.R., Y.V. Kaneti, and Y. Yamauchi. 2017. "Metal–Organic Framework-Derived Nanoporous Metal Oxides toward Supercapacitor Applications: Progress and Prospects." *ACS Nano*, 11(6), 5293–5308. 10.1021/acsnano.7b02796.

Sheberla, D., J.C. Bachman, J.S. Elias, C.-J. Sun, Y. Shao-Horn, and M. Dincă. 2017. "Conductive MOF Electrodes for Stable Supercapacitors with High Areal Capacitance." *Nature Materials* 16(2): 220–224. 10.1038/nmat4766.

Tong, M., S. Liu, X. Zhang, T. Wu, H. Zhang, G. Wang, Y. Zhang, X. Zhu, and H. Zhao, 2017. "Two-Dimensional CoNi nanoparticles@S,N-Doped Carbon Composites Derived from S,N-Containing Co/Ni MOFs for High Performance Supercapacitors." *Journal of Materials Chemistry A* 5(20): 9873–9881. 10.1039/C7TA01008G.

Wang, B., W. Li, Z. Liu, Y. Duan, B. Zhao, Y. Wang, and J. Liu. 2020a. "Incorporating Ni-MOF Structure with Polypyrrole: Enhanced Capacitive Behavior as Electrode Material for Supercapacitor." *RSC Advances* 10(21): 12129–12134. 10.1039/C9RA10467D.

Wang, D.-G., Z. Liang, S. Gao, C. Qu, and R. Zou. 2020b. "Metal-Organic Framework-Based Materials for Hybrid Supercapacitor Application." *Coordination Chemistry Reviews* 404: 213093. 10.1016/j.ccr. 2019.213093.

Wang, H., Q. Gao, and J. Hu. 2010. "Asymmetric Capacitor Based on Superior Porous Ni–Zn–Co oxide/hydroxide and Carbon Electrodes." *Journal of Power Sources* 195(9): 3017–3024. 10.1016/ j.jpowsour.2009.11.059.

Wang, J., Q. Zhong, Y. Xiong, D. Cheng, Y. Zeng, and Y. Bu. 2019a. "Fabrication of 3D Co-Doped Ni-Based MOF Hierarchical Micro-Flowers as a High-Performance Electrode Material for Supercapacitors." *Applied Surface Science* 483: 1158–1165. 10.1016/j.apsusc.2019.03.340.

Wang, L., Y. Han, X. Feng, J. Zhou, P. Qi, and B. Wang. 2016. "Metal–organic frameworks for energy storage: Batteries and supercapacitors." *Coordination Chemistry Reviews* 307: 361–381. 10.1016/j.ccr. 2015.09.002.

Wang, Yanzhong, Y. Liu, H. Wang, W. Liu, Y. Li, J. Zhang, H. Hou, and J. Yang. 2019b. "Ultrathin NiCo-MOF Nanosheets for High-Performance Supercapacitor Electrodes." *ACS Applied Energy Materials* 2(3): 2063–2071. 10.1021/acsaem.8b02128.

Wen, P., P. Gong, J. Sun, J. Wang, and S. Yang. 2015. "Design and Synthesis of Ni-MOF/CNT Composites and rGO/Carbon Nitride Composites for an Asymmetric Supercapacitor with High Energy and Power Density." *Journal of Materials Chemistry A* 3(26): 13874–13883. 10.1039/c5ta02461g.

Wu, M.-S., and W.-H. Hsu. 2015. "Nickel Nanoparticles Embedded in Partially Graphitic Porous Carbon Fabricated by Direct Carbonization of Nickel-Organic Framework for High-Performance Supercapacitors." *Journal of Power Sources* 274: 1055–1062. 10.1016/j.jpowsour.2014.10.133.

Wu, S., J. Liu, H. Wang, and H. Yan. 2018. "A Review of Performance Optimization of MOF-Derived Metal Oxide as Electrode Materials for Supercapacitors." *International Journal of Energy Research* 43(2): 697–716. 10.1002/er.4232.

Xia, W., A. Mahmood, R. Zou, and Q. Xu. 2015. "Metal–Organic Frameworks and Their Derived Nanostructures for Electrochemical Energy Storage and Conversion." *Energy & Environmental Science* 8(7): 1837–1866. 10.1039/c5ee00762c.

Xiao, Z., Y. Bao, Z. Li, X. Huai, M. Wang, P. Liu, and L. Wang. 2019. "Construction of Hollow Cobalt–Nickel Phosphate Nanocages through a Controllable Etching Strategy for High Supercapacitor Performances." *ACS Applied Energy Materials* 2(2): 1086–1092. 10.1021/acsaem.8b01627.

Xu, J., C. Yang, Y. Xue, C. Wang, J. Cao, and Z. Chen. 2016. "Facile Synthesis of Novel Metal-Organic Nickel Hydroxide Nanorods for High Performance Supercapacitor." *Electrochimica Acta* 211: 595–602. 10.1016/j.electacta.2016.06.090.

Yang, C., X. Li, L. Yu, X. Liu, J. Yang, and M. Wei. 2020a. "A New Promising Ni-MOF Superstructure for High-Performance Supercapacitors." *Chemical Communications* 56(12): 1803–1806. 10.1039/ C9CC09302H.

Yang, H.-X., D.-L. Zhao, W.-J. Meng, M. Zhao, Y.-J. Duan, X.-Y. Han, and X.-M. Tian. 2019. "Nickel Nanoparticles Incorporated into N-Doped Porous Carbon Derived from N-Containing Nickel-MOF for High-Performance Supercapacitors." *Journal of Alloys and Compounds* 782: 905–914. 10.1016/ j.jallcom.2018.12.259.

Yang, J., P. Xiong, C. Zheng, H. Qiu, and M. Wei. 2014a. "Metal–Organic Frameworks: A New Promising Class of Materials for a High Performance Supercapacitor Electrode." *J. Mater. Chem. A* 2(39): 16640–16644. 10.1039/c4ta04140b.

Yang, J., C. Zheng, P. Xiong, Y. Li, and M. Wei. 2014b. "Zn-Doped Ni-MOF Material with a High Supercapacitive Performance." *J. Mater. Chem. A* 2(44): 19005–19010. 10.1039/c4ta04346d.

Yang, W., H. Guo, L. Yue, Q. Li, M. Xu, L. Zhang, T. Fan, and W. Yang. 2020b. "Metal-Organic Frameworks Derived MMoSx (M = Ni, Co and Ni/Co) Composites as Electrode Materials for Supercapacitor." *Journal of Alloys and Compounds* 834: 154118. 10.1016/j.jallcom.2020.154118.

Yang, Yan, M.-L. Li, J.-N. Lin, M.-Y. Zou, S.-T. Gu, X.-J. Hong, L.-P. Si, and Y.-P. Cai. 2020c. "MOF-derived Ni3S4 Encapsulated in 3D Conductive Network for High-Performance Supercapacitor." *Inorganic Chemistry* 59(4): 2406–2412. 10.1021/acs.inorgchem.9b03263.

Ye, C., Q. Qin, J. Liu, W. Mao, J. Yan, Y. Wang, J. Cui, Q. Zhang, L. Yang, and Y. Wu. 2019. "Coordination Derived Stable Ni–Co MOFs for Foldable All-Solid-State Supercapacitors with High Specific Energy." *Journal of Materials Chemistry A* 7(9): 4998–5008. 10.1039/C8TA11948A.

Yu, D., L. Ge, X. Wei, B. Wu, J. Ran, H. Wang, and T. Xu. 2017. "A General Route to the Synthesis of Layer-by-Layer Structured Metal Organic Framework/Graphene Oxide Hybrid Films for High-Performance Supercapacitor Electrodes." *Journal of Materials Chemistry A* 5(32): 16865–16872. 10.1039/c7ta04074a.

Yu, H., J. Wu, L. Fan, Y. Lin, K. Xu, Z. Tang, C. Cheng, et al. 2012. "A Novel Redox-Mediated Gel Polymer Electrolyte for High-Performance Supercapacitor." *Journal of Power Sources* 198: 402–407. 10.1016/j.jpowsour.2011.09.110.

Zain, N.K.M., B.L. Vijayan, I.I. Misnon, S. Das, C. Karuppiah, C.-C. Yang, M.M. Yusoff, and R. Jose. 2018. "Direct Growth of Triple Cation Metal–Organic Framework on a Metal Substrate for Electrochemical Energy Storage." *Industrial & Engineering Chemistry Research* 58(2): 665–674.10.1021/acs.iecr.8b03898.

Zhang, C., Q. Zhang, K. Zhang, Z. Xiao, Y. Yang, and L. Wang. 2018a. "Facile Synthesis of a Two-Dimensional Layered Ni-MOF Electrode Material for High Performance Supercapacitors." *RSC Advances* 8(32): 17747–17753. 10.1039/c8ra01002a.

Zhang, F., J. Ma, and H. Yao. 2019. "Ultrathin Ni-MOF Nanosheet Coated NiCo$_2$O$_4$ Nanowire Arrays as a High-Performance Binder-Free Electrode for Flexible Hybrid Supercapacitors." *Ceramics International* 45(18, Part A): 24279–24287. 10.1016/j.ceramint.2019.08.140.

Zhang, F., G. Zhang, H. Yao, Z. Gao, X. Chen, and Y. Yang. 2018b. "Scalable In-Situ Growth of Self-Assembled Coordination Supramolecular Network Arrays: A Novel High-Performance Energy Storage Material." *Chemical Engineering Journal* 338: 230–239. 10.1016/j.cej.2018.01.003.

Zhang, Xiaolong, J. Wang, X. Ji, Y. Sui, F. Wei, J. Qi, Q. Meng, Y. Ren, and Y. He. 2020a. "Nickel/Cobalt Bimetallic Metal-Organic Frameworks Ultrathin Nanosheets with Enhanced Performance for Supercapacitors." *Journal of Alloys and Compounds* 825: 154069. 10.1016/j.jallcom.2020.154069.

Zhang, Xiuli, L. Zhang, G. Xu, A. Zhao, S. Zhang, and T. Zhao. 2020b. "Template Synthesis of Structure-Controlled 3D Hollow Nickel-Cobalt Phosphides Microcubes for High-Performance Supercapacitors." *Journal of Colloid and Interface Science* 561: 23–31. 10.1016/j.jcis.2019.11.112.

Zhao, S., H. Wu, Y. Li, Q. Li, J. Zhou, X. Yu, H. Chen, K. Tao, and L. Han. 2019. "Core–Shell Assembly of Carbon Nanofibers and a 2D Conductive Metal–Organic Framework as a Flexible Free-Standing Membrane for High-Performance Supercapacitors." *Inorganic Chemistry Frontiers* 6(7): 1824–1830. 10.1039/c9qi00390h.

Zhao, Yang, Z. Song, X. Li, Q. Sun, N. Cheng, S. Lawes, and X. Sun. 2016. "Metal Organic Frameworks for Energy Storage and Conversion." *Energy Storage Materials* 2: 35–62. 10.1016/j.ensm.2015.11.005.

Zhong, Y., X. Cao, L. Ying, L. Cui, C. Barrow, W. Yang, and J. Liu. 2020. "Homogeneous Nickel Metal-Organic Framework Microspheres on Reduced Graphene Oxide as Novel Electrode Material for Supercapacitors with Outstanding Performance." *Journal of Colloid and Interface Science* 561: 265–274. 10.1016/j.jcis.2019.10.023.

Zhou, M., F. Xu, S. Zhang, B. Liu, Y. Yang, K. Chen, and J. Qi. 2020. "Low Consumption Design of Hollow NiCo-LDH Nanoflakes Derived from MOFs for High-Capacity Electrode Materials." *Journal of Materials Science: Materials in Electronics* 31(4): 3281–3288. 10.1007/s10854-020-02876-z.

Zhou, Yingjie, Z. Mao, W. Wang, Z. Yang, and X. Liu. 2016. "In-Situ Fabrication of Graphene Oxide Hybrid Ni-Based Metal–Organic Framework (Ni–MOFs@GO) with Ultrahigh Capacitance as Electrochemical Pseudocapacitor Materials." *ACS Applied Materials & Interfaces* 8(42): 28904–28916. 10.1021/acsami.6b10640

10

The Place of Biomass-Based Electrode Materials in Next-Generation Energy Conversion and Storage

Chioma E. Njoku[1], Innocent S. Ike[2,3], Adeolu A. Adediran[4], and Cynthia C. Nwaeju[5]

[1]Department of Materials and Metallurgical Engineering, Federal University of Technology Owerri, Nigeria

[2]Department of Chemical Engineering, Federal University of Technology Owerri, Nigeria

[3]African Centre of Excellence in Future Energies and Electrochemical Systems (ACE-FUELS), Federal University of Technology, Owerri, Nigeria

[4]Department of Mechanical Engineering, Landmark University, Omu-Aran, Kwara State, Nigeria

[5]Department of Mechanical Engineering, Nigerian Maritime University, Okerenkoko, Warri, Nigeria

10.1 Introduction

Due to the rapid increase in population and present consumption rate, the fossil fuels from the found reserves would not last more than 220 years (Tarascon et al. 2010). This implies that within two centuries, most of the reserves might have been exhausted. As these fossil fuels are consumed, there is increased concentration of atmospheric carbon released. This in turn leads to rise in global warming, acid rain, and associated problems, which indicates the dire need for renewable sources of energy. However, there has been rapid development of technologies for sustainable and clean energy which include wind energy, solar energy, hydro energy, and energy from biomass. Nevertheless, the technologies of energy storage and conversion are dominating, especially in applications to fuel cells, batteries, and supercapacitors (Wang, Lin, and Shen 2016). The specific energy and specific power for each of them are displayed in the Ragone plot (Figure 10.1). So many factors affect the efficient performance of the devices used for storing and converting energy, which include the type of electrode materials used, methods of preparation, inherent properties, availability of resources, inexpensiveness, and eco-friendliness.

The three most commonly used electrode materials for energy storage and conversion are metal compounds, conducting polymers, and carbon materials. Metal compounds are naturally abundant and possess multielectron redox ability; however, they have reduced conductivity and easily self-aggregate (Li and Gong 2020). Oftentimes the metal compounds used are mainly metal nitrides and metal oxides. Metal nitrides have been utilized, especially, in electrode materials for electrochemical capacitors and lithium-ion batteries. Some of the metal nitrides used include Mo_2N, TiN, MoN, FeN, NbN, TaN (as supercapacitors and lithium-ion anode materials), and $LiMON_2$ as lithium-ion cathode material (Elder et al. 1992). These metal nitrides possess high electrical conductivities, increased chemical and thermal

DOI: 10.1201/9781003145585-10

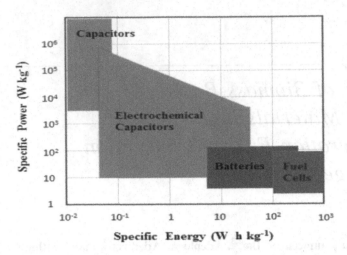

FIGURE 10.1 Ragone plot displaying the specific power in relation to specific energy for different energy storage and conversion systems (Ho et al., 2014). Reprinted with permission copyright 2014, World Scientific publishing company, Pt Ltd.

stabilities, increased specific capacities; however, they are poorly stable and have complex synthesis routes (Zerr et al. 1999). Metal oxides on the other hand can be easily approached from the many available reserves, possess high theoretical specific capacitance, and enormous surface area. Although they are limited in use because they are toxic, they have slow rate of ionic diffusion in the bulk phase, reduced electrical conductivity, and irrepressible volume expansion (Yu et al. 2011). Conducting polymers are also used, for they are relatively inexpensive, and have increased energy density for electrode material applications. Some of the commonly used conducting polymers include polythiophene, polyphenyl, polyphenylacetylene, polypyrrole, and polyaniline (PANI), with polyaniline being the most studied and utilized amongst other conducting polymers. This is because PANI can be easily synthesized, has increased theoretical conductivity (3407 F/g), is an inexpensive monomer, and more stable than other conducting polymers (Bhadra et al. 2009). Nevertheless, the PANI electrodes exhibit reduced cyclic life and mechanical stability (Snook, Kao, and Best 2011). The severe and constant exploitation of the metal compounds and conducting polymers could cause serious geopolitical and environmental issues (Vaalma et al. 2018).

Then, amid the exploited electrode materials (viz., metal compounds, conducting polymers, carbon materials) utilized for energy storage and conversion, carbon materials are most prevalently used. This is because carbon materials are versatile and can exist in various forms (fibres, nanosheets, powders, nanotubes, nanospheres). They possess improved electronic conductivity, amendable morphology/microstructure, and exceptional stability (Jiang, Sheng, and Fan 2018), but have reduced specific capacitance due to non-occurrence of faradaic reactions (Snook, Kao, and Best 2011). The carbon materials exist in several allotropes, which include, the crystalline nature (diamond and graphite), polycrystalline nature (carbon fibres from polyacrylonitrile), disordered and amorphous nature (carbon black and activated carbon), then recently, carbon nanomaterials (graphene, fullerenes, and carbon nanofibres/tubes (Hirsch 2010). Since more than two decades, the use of graphene, fullerenes, and carbon nanotubes has gained tremendous attention in their utilization as electrode materials for energy storage and conversion. They display superb physicochemical characteristics namely high specific surface area, dense electroactive sites, manageable porosity, increased specific surface area and outstanding chemical stability (Wen, Li, and Cheng 2016), excellent functionalization patterns and structures. But the processes involved in the synthesis of these carbon nanomaterials are complex, with toxic and hazardous reagents often being used (Deng, Li, and Wang 2016). Nonetheless, the production of carbon nanomaterials is greatly dependent on precursors from fossil fuels (such as pitch, methane, ethylene, acetylene), and the process routes are severe and energy intensive (such as plasma/laser ablation, chemical vapour deposition, and electric-arc discharge (Dai 2002). Chemical vapour deposition makes use of the deposition method whereby grapheme or carbon nanofibres/tubes are produced at elevated temperatures (about 800°C) from fossil-based gases (such as acetylene, methane, hydrogen gas), as non-renewable carbon resource (Zhao, Zhang, and Essene 2015). On the other hand, in the electric-arc discharge method, carbon nanotubes of high quality are produced at very high temperatures from the use of graphite or coal as the carbon resource. Most awfully is the use of Hummer's technique in grapheme production; strong and toxic oxidant (KMnO$_4$) and acidic reagent (H$_2$SO$_4$) are used, leading to harsh environmental impacts (Hummers and Offeman 1958). In the production of activated carbon, coal or some other non-renewable resources are utilized through several synthetic methods of activation, which include steam and carbon (IV) oxide activation method or the use of chemicals (H$_3$PO$_4$, KOH, and ZnCl$_2$) (W. J. Liu, Jiang, and Yu 2019). All these point to the significant need for clean, renewable, and sustainable carbon materials that have less expensive methods of preparation. Biomass has been anticipated as the renewable resource in the production of functional carbon materials, as they are sustainable, abundant, relatively less expensive, and not harmful, among others.

Biomass can be obtained from plants and animal wastes/residues. Biomass renewable resources are usually naturally hierarchically structured, which makes it possible to derive carbon materials that have preferred properties for applications in fuel cells, batteries, and supercapacitors (Gao et al. 2017a). The very abundant and recyclable biomass resources can be transformed to renewable carbon materials by different less expensive and non-toxic methods of activation, such as physical activation, hydrothermal carbonization, chemical activation method, one-stage pyrolysis, template method, among others (Wang and Kaskel 2012; J. Pang et al. 2017). In the process of pyrolysis of biomass, defects which include edges, vacancies, and heteroatoms can be made, and the doping of sulphur or nitrogen contained in some biomass by the heteroatoms can lead to increased electric conductivity and more active site creation (Jiang et al. 2018). Titirici et al. (2015) reported the successful conversion of malleable, renewable cotton textile into activated carbon textile that possessed the textile's design and the porous (spongy) configuration of the cotton fibre, with exceptional conductivity and elasticity. The carbon obtained from biomass has many distinct desirable characteristics, for they possess outstanding electric conductivity, simple modifiable surface chemistry, high specific surface area, and clear pore size distribution (Deng et al. 2016). Due to the large abundance of biomass recyclable resources from animal and agricultural wastes/residues, the efficient, less expensive, eco-friendly, carbon-rich electrode materials can be obtained for applications in fuel cells, batteries, and supercapacitors. In this chapter, biomass resources would be discussed, with their synthesis and the applications of extracted carbon in electrochemical systems for energy storage and conversion; specifically, fuel cells, batteries, and ultracapacitors.

10.2 Biomass and Its Carbon Derivations

10.2.1 Biomass Reserve

Biomass is the only other naturally abundant, energy-rich carbon resource that is large enough to replace fossil fuels. Biomass can be said to be the organic material that is non-fossil but contains core chemical energy (Balat and Ayar 2005). Biomass includes virgin wood (forestry residues, tree surgery residues, fuel wood), energy crops (grasses and non-woody energy, agricultural energy crops, hydroponics), dry agricultural residues, wet agricultural residues, food waste, industrial waste, and coproducts (woody waste and non-woody waste) (Ladanai and Vinterbäck 2009). Various energy demands have been met by the use of biomass, such as fuelling of vehicles, electricity generation, provision of process heat for industrial facilities, and heating of homes (Sürmen 2003). The energy from biomass can be said to be an aspect of solar energy, in that in the presence of sunlight, there is a combination of photosynthesis with carbon dioxide in the atmosphere and water to produce biomass and release oxygen. This can be expressed using the equation (10.1) (Schuck 2006):

$$CO_2 + 2H_2O \xrightarrow{Sunlight} ([CH_2O] + H_2O) + O_2 \qquad (10.1)$$

where [CH_2O] denotes the biomass (as carbohydrates).

Several techniques and technologies have been employed in the recovery of the energy stored in the biomass (Schuck 2006). Apart from the direct usage of biomass by burning, and the indirect usage of biomass in the production of liquefied or gaseous fuels (for example biodiesel and bioethanol), it can also be utilized in the derivation of several essential chemicals (Sheldon 2014). The compositional makeup of the biomass has a major effect on the properties of the final derived carbon from the biomass. The key constituents of most biomass resources are cellulose, hemicellulose, and lignin (Reis et al. 2020), especially the lignocellulosic biomass (Figure 10.2). The cellulose, hemicellulose, and lignin undergo decomposition at varying periods of time and specific temperature ranges in the course of the thermal behavior of the biomass raw material. These variations in turn lead to provision of derived biomass carbons with different properties, and also cause the production of electrode materials with varying electrochemical behaviors (Reis et al. 2020).

Several biomass-derived carbon materials have been obtained, which include highly permeable carbons and nano carbons with outstanding conductivity and with very high porosity; they are being applied in catalysis, energy storage, and environmental remediation/purification (Titirici et al. 2012; Shi et al. 2014). Activated carbons have many unique features that drive their regular usage, and they include high thermal and chemical stability, high porosity, high surface area, and packing density. Many authors have reported on the use of several biomass resources as electrode materials for energy storage and conversion applications:, pine wood (Huggins et al. 2016), yellow birch wood (Khudzari et al. 2019), rape pollen grass (Li et al. 2017), yeast cells (Xia et al. 2018), coconut shells (Chen et al. 2017), silk (Yun et al. 2013), hemp stems (Yang et al. 2017), aloe vera (Karnan et al. 2016), and saw dust (Liu et al. 2014), among others.

10.2.2 Methods of Carbon Derivation from Biomass

In order to obtain efficient, clean, and sustainable energy from biomass, the method of synthesis/ derivation of the carbon from biomass is very essential. Some methods have been strategically employed in the derivation of carbon from biomass which empirically controls the physical structure and surface chemistry of the large molecules of the resulted carbon materials. Energy can be obtained from biomass following two major procedures, the thermochemical process that includes combustion, pyrolysis, and gasification, and biochemical process which includes anaerobic digestion and fermentation (Gao et al. 2017a). Pyrolysis, physical activation, chemical activation, and hydrothermal carbonization established on different experimental conditions and conversion mechanisms have been described as synthetic methods carbon material could be obtained from renewable biomass materials as shown in Figure 10.3; and the control of the physical structure and surface chemistry that results in carbon materials is given by these methods (Deng et al. 2016; Gao et al. 2017a; Jiang et al. 2018; Thomas et al. 2019).

10.2.2.1 Pyrolysis

Pyrolysis involves the thermal decomposition (degradation) of biomass resources (energy crops, agricultural residues, woody wastes, animal wastes, algae, aquatic plants) without oxygen or air at increased temperature (Schuck 2006). Many stages are involved in pyrolysis and the process is complex; and the by-products obtained include biochar, bio-oil, and bio gas (Zaman et al. 2017). The biochar formed during biomass pyrolysis usually contains greater carbon content as compared to the substrate for volatiles, moisture and non-carbon heteroatoms are eliminated (Januszewicz et al. 2020). The thermal degradations of the constituents (hemicellulose, cellulose, and lignin) contained in the biomass result in the production of the pyrolysis by-products. During the pyrolysis process, moisture removal (dehydration) takes place at temperatures below 200°C; then, the hemicellulose, which is (almost 15 wt.% to 30 wt.% of biomass) a heterogeneously branched polysaccharide, that consists mainly of sugars, which include mannose, xylose, glucose, arabinose, galaturonic and methylglucoronic acids (McKendry 2002) decomposes within the temperature range of 200°C to 260°C (Gírio et al. 2010).

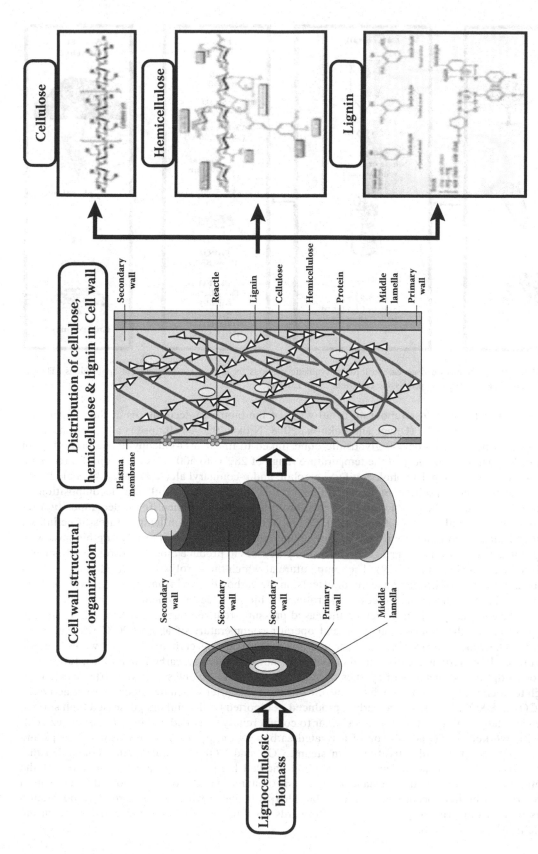

FIGURE 10.2 Pictorial illustration of the lignocellulosic framework (Menon and Rao 2012). Reprinted with permission copyright 2012, Elsevier.

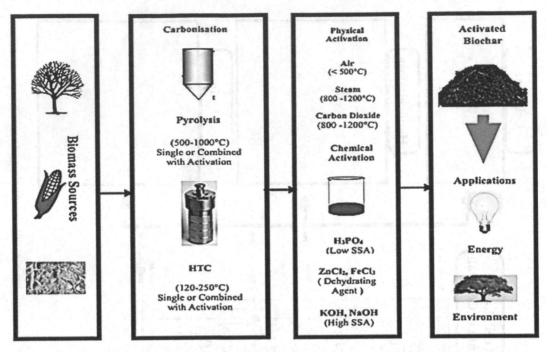

FIGURE 10.3   Biomass-derived carbon through synthetic methods for energy and environment applications (Thomas et al., 2019). Reprinted with permission copyright 2019, Elsevier.

Thereafter the cellulose (about 40 to 50 wt.% of the biomass) decomposes at the temperature range of 240°C to 350°C (Dhepe and Fukuoka 2008; Njoku 2020). Cellulose is a glucose polymer (homopolymer) having only β-glycosidic bond linked to linear β-D-glucopyranose units. Lignin being the last to decompose at the temperature range of 280°C to 500 °C is made up of three main monomers, viz., sinapyl alcohol, coniferyl alcohol, and p-coumaryl alcohol (Zakzeski et al. 2010). The thermal decomposition of hemicellulose yields biogas and bio-oil, the decomposition of cellulose results in the production of biogas, bio-oil, and biochar, then lignin decomposition results in bio-oil and biochar (Zaman et al. 2017). With decomposition temperatures of cellulose, hemicellulose, and lignin about 500°C, low temperatures can be used in pyrolyzing biomass with appropriate activation and surface modification processes to produce biogas, bio-oil, and biomass-derived carbon (Liu et al. 2019). The temperature at which the pyrolysis is carried out, the dwell time, and the rate of heating have major effects on the carbon derived from the biomass (Lua, Lau, and Guo 2006). During the process of pyrolysis of biomass, certain conditions are to be met to obtain larger char yield; they include increased pressure, reduced rate of gas flow, slow heating, larger sizes of the biomass particles, and operating temperature of about 500°C (Zaman et al. 2017). The biochar from biomass can be converted to varying carbon materials which include porous carbon, activated carbon, and also nanocarbons (graphenes, carbon nanotubes, fullerenes) through optimized methods of synthesis (Marriott et al. 2014). Contescu et al. (2018) carried out high temperature pyrolysis on white pine and chestnut oak, after which the biochars were activated in CO_2 at 800°C. The activated carbon produced as reported by the authors possessed high surface area and large volume of mesopores similar to commercially activated carbons. Januszewicz et al. (2020) worked on the production of activated carbon by the pyrolysis of waste wood strips and straw pellets, afterward activation with steam, CO_2, and KOH was done. They concluded that the activated carbon produced by the method showed higher surface area irrespective of the method of activation used. The standard pyrolysis of starch-based biomass followed by activation gave rise to higher specific surface area, but sharply reduced when hydrothermal carbonization was carried out prior to pyrolysis and afterwards no form of activation was carried out on the biochar (Li et al. 2008).

10.2.2.2 Activation

The biochar from pyrolysis does not usually possess high specific surface area and highly porous structure suitable for application in electrochemical energy storage (Zhang and Zhao 2009). Thus, the biochar is often routed through the process of activation (physical or chemical activation). Physical activation consists of a two-stage process whereby the thermal carbonization of the biomass (carbon precursor) is carried out at temperatures below 860°C (pyrolysis), after which activation is done on the resultant biochar at increased temperatures with CO_2, steam, air, or their blends (gasification) (El-Hendawy, Samra, and Girgis 2001). However, chemical activation consists of a one-stage process whereby the carbon (precursor) is mixed with chemical activators and simultaneously carbonized at temperature ranges of 300°C to 900°C (Jiang et al. 2018). Physical activation, a non-homogeneous complex process, entails the transportation of gaseous components to the surface of the sample, disperses into the pores, comes up to the pore surface, reacts with the carbon constituents in the gas, and releases reaction products, then diffuses to the atmosphere (Mattson 1971). In physical activation process, volatiles are burnt up in the first stage of pyrolysis, then during the second stage of gasification, closed pores are opened up as tar-like pyrolysis residue inside the pores and are burnt off (Marta Sevilla and Mokaya 2014). Waste tea biomass was used to prepare activated carbon through physical activation with steam by Zhou, Luo, and Zhao (2018). From the results, the specific surface area of 995 m^2/g was reached as the temperature was increased, and also micropores and mesopores were developed at increased temperature. There is more storage of electrolyte ions in mesopores as compared to micropores, which leads to increased specific surface area with enough storage sites that give room for high energy density, improved capacitance, and power density, and also enhance rate capability, suitable for improved electrochemical performance (Jiang et al. 2018). The hierarchical porous structure of carbon materials derived from biomass resources – micropores (less than 2 nm), mesopores (between 2 to 50 nm), and macropores (more than 50 nm) – helps in the improvement of the diffusion kinetics, and increase in the performance of the electrode materials used in electrochemical storage and conversion applications (Dutta, Bhaumik, and Wu 2014). Precisely, the macropores can be compared to the reservoir for ion shielding that would reduce the pathway for ionic diffusion to the inner surfaces of the carbon materials derived from biomass. Then the mesopores and micropores would provide increased surface area for interaction between the electrolyte and the electrode material, and offer easy pathway for electrolyte ion transfer (Sun et al. 2019). Physical activation is an environmental friendly process; although longer times are required, high temperatures are needed for conversion to take place, and also poorly porous carbons are derived from the process (Jiang et al. 2018).

Chemical activation is generally preferred to physical activation because of high carbon yield, manageable pore size and proper pore development, higher specific surface area, reduced operating temperature, and a one-stage process (Sevilla and Mokaya 2014). However, the use of hazardous chemicals poses environmental challenges. Through chemical activation, porous carbon materials with high specific surface area (more than 2000 m^2/g) and huge pore volume can be achieved. Most of the reagents (activators) used in chemical activation process include KOH, H_3BO_3, $ZnCl_3$, NaOH, H_2SO_4, Na_2CO_3, $KHCO_3$, K_2CO_3, HNO_3, and $FeCl_3$. Among these, KOH is mostly used due to its reduced activation temperature that leads to more carbon yields, distinct distribution of micropore size, and extremely high specific surface area (about 3000 m^2/g) (Zou and Group 2001). When using KOH as the activator, the alkali to biochar ratio varies from 1:1 to 5:1, depending on how concentrated the soaking solution is (Babel and Jurewicz 2004). To discard the extra KOH, the activated carbon is suspended in 0.1M HCl solution, then rinsed with H_2O to achieve a pH of 7. The samples are then dried in the oven ready for electrochemical tests and other required characterizations (Enock et al. 2017). The activation mechanism is quite complex but there are three acceptable steps involved: (a) formed potassium compounds (KOH, K_2CO_3, and K_2O) at calcination react with carbon to form a porous carbon framework; (b) hot steam formed by the mixture of H_2O and CO_2 at high temperature enhances the porous structure by gasification; (c) as-produced metallic potassium gets well intercalated into the carbon framework, thus expanding the carbon lattice, after which the carbon structure becomes more porous after the removal of the metallic potassium and other potassium compounds (Wang and Kaskel 2012). In the chemical activation of bamboo biochar according to Kim et al. (2006), the ratio of KOH to bamboo biochar was increased from 1 to 4, and this led to increase in the BET surface area from 1010 to 1400 m^2/g, micropore percentage increased from 67.8% to 88.7%, and also the total volume of pores increased from 0.123 cm^3/g to 0.708 cm^3/g. With an aqueous electrolytic solution of 30 wt.% of H_2SO_4, the activated carbon with increased mesopore fraction showed improved capacitance in the region of high current density discharge of 300–800 mA/cm^2. The concept was explained in the sense that the larger the mesopores, the more the ionic transfer and diffusion; and the smaller the micropores, the lesser the ionic transfer and diffusion. Ghosh et al. (2019) prepared hard carbons and activated carbons from corn cob, potato starch, and banana stems for utilization in electrochemical energy storage. The banana stems were activated with KOH and H_3PO_4, the corn cob was pyrolyzed, and the potato starch was carbonized. From the results, KOH-activated banana stems (KHC) gave the best results followed by the corn cob–derived hard carbon (CHC), having specific capacitances of 479.23 F/g at 1 mV/s and 309.81 F/g at 2 mV/s, respectively, surface areas of 567.36 m^2/g and 215.42 m^2/g, respectively, pore diameters of 1.205 nm and 1.199 nm, respectively. The coulombic efficiency for KHC was 108.61% for 2 A/g current density, and that of CHC was 93.9%. Chen et al. (2013) produced activated carbon using cotton stalk and HPO as the activator. The results showed a large specific surface area of 1481 cm/g, a pore volume of 0.0377 cm/g, and specific capacitance of 114 F/g at 0.5 A/g due to the way the resulting structure was.

Chemically activated carbon can be enhanced as electrode materials in electrochemical energy applications by treatment with certain agents. To improve the electrochemical performance of activated carbon produced from custard apple tree, fig tree, and olive tree with KOH, the different wood biomasses were treated with melamine, ammonium carbonate, nitric acid, and ammonium persulphate (Elmouwahidi et al. 2019). From the results, the retention capacity for the as-prepared treated activated carbons for all the samples were in the range of 97–100% after 10,000 cycles. Also, among the different treatments used, melamine and ammonium carbonate showed increase in the surface area and the pore volume for all the tested biomass samples. The surface areas for custard apple tree, fig tree, and olive tree were 1706 m^2 g^{-1}, 1669 m^2 g^{-1}, and 1314 m^2 g^{-1}, respectively. The maximum capacitance obtained from samples treated with ammonium persulphate was 312 F g^{-1} at 125 mA g^{-1}, followed by treatments with melamine, ammonium carbonate, and nitric acid having highest capacitances of 283 F g^{-1}, 280 F g^{-1}, and 259 F g^{-1} at 125 mA g^{-1}, respectively. The authors informed that the improved electrochemical performance was because of outstanding and sufficient textural characteristics obtained from the treatments.

10.2.2.3 Hydrothermal Carbonization

Hydrothermal carbonization equally known as *wet torrefaction* or *hydrothermal pretreatment* was used in the production of hydrochars from biomass, and was first reported by Bergius (1913) for it was used to transform cellulose to coal-like materials. Hydrothermal carbonization is a thermochemical process in which the high-moisture content biomass is immersed in water and heated at temperatures of 180–350°C under pressure (2–6 MPa), for a period of 5–240 minutes (Reza et al. 2014; Heidari et al. 2019). The starting biomass materials for hydrothermal carbonization of biomass include ordinary plant resources (lignocelluloses, agricultural wastes/residue, energy crops, and wood), sludge, sewage, animal manure, community solid wastes with different configurations, and carbohydrates such as starch, sugars, glucose, cellulose, hemicellulose, and other products obtained from glucose dehydration like hydroxymethylfurfural and 2-furaldehyde (furfural) (Hu et al. 2010; Heidari et al. 2019). The main by-product of hydrothermal carbonization is the solid hydrochar; the other products include liquid (process water) and gas (majorly CO_2) (Fiori et al. 2014).

The mechanisms of reactions that take place in hydrothermal carbonization are complex, for they are not straightforward and do not occur in series or in order (Sevilla and Fuertes 2009). Various methods are used in the preparation of hydrochar from hydrothermal carbonization. Titirici, Thomas, and Antonietti (2007) suggested a method having two varying paths. In the first path, the cellulose undergoes hydrolysis to become glucose, and dehydration to form furfural (hydroxymethyl). The furfural undergoes polymerization/aromatization to become a carbonized structure. In the second path, the cellulose is decarboxylated, and then further condensed to produce an aromatic structure; this route is

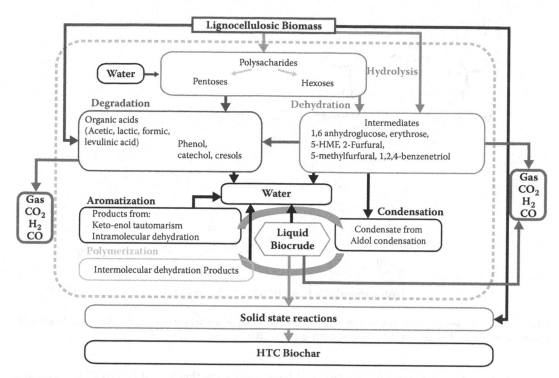

FIGURE 10.4 Reaction pathways for hydrothermal carbonization of lignocellulosic biomass (Reza et al., 2014). Reprinted with permission copyright 2014, Degruyter.

similar to the simple pyrolysis method. Five process reactions are generally accepted to occur during hydrothermal carbonization of biomass materials, viz., hydrolysis, dehydration, decarboxylation, polymerization, and aromatization (Funke and Ziegler 2010). In the case of lignocellulosic biomass, the hydrothermal carbonization can occur via two pathways: the breakdown of the intermediate products via hydrolysis and polymerization, and the pyrolysis-type of decomposition which is dependent on the severity of the HTC conditions (Figure 10.4).

The first approach is in the dissolution of parts of the main constituents (hemicellulose, cellulose, and lignin) at the subcritical region of water. Then the other pathway is the pyrolysis of the undissolved part of the biomass. When the needed activation energy to dissolve the crystalline lignin and crystalline cellulose is not provided by the hydrothermal carbonization, the pyrolysis-type carbonization occurs. Unlike the conventional pyrolysis process, this takes place in a different medium with increased pressure in a homogeneous phase. Two different types of hydrochar are produced from the entire process: (a) hydrochar from condensation reactions (polyfuranic char, phenolic char, and polyaromatic char), and (b) hydrochar from pyrolysis-type process (cellulose and lignin char). Although if a homogeneous phase was maintained during the HTC process, molecular readjustment and dehydration can cause automatic integration, and cohesive HTC products can be formed at increased temperatures and prolonged reaction times (Wang et al. 2018). The hydrolysis of hemicellulose takes place at a much-reduced activation energy than the other reactions. The ester and ester bonds in the hemicellulose, cellulose, and lignin are broken down at varying temperatures (less than 180°C, more than 200°C, and more than 220°C respectively), during hydrolysis reaction (Brunner 2009). Then dehydration and decarboxylation would take place. More water is produced during the dehydration process. During decarboxylation, the degradation of carboxyl and carbonyl groups occur causing the release of CO and CO_2, and the formation of hydroxymethyl furfural (HMF). At that point, condensation polymerization takes place and two highly reactive molecules (furfurals) combine to form a larger molecule (polyfuran polymer) with the release of water. Furthermore, stable aromatic polymer structures which are the basis

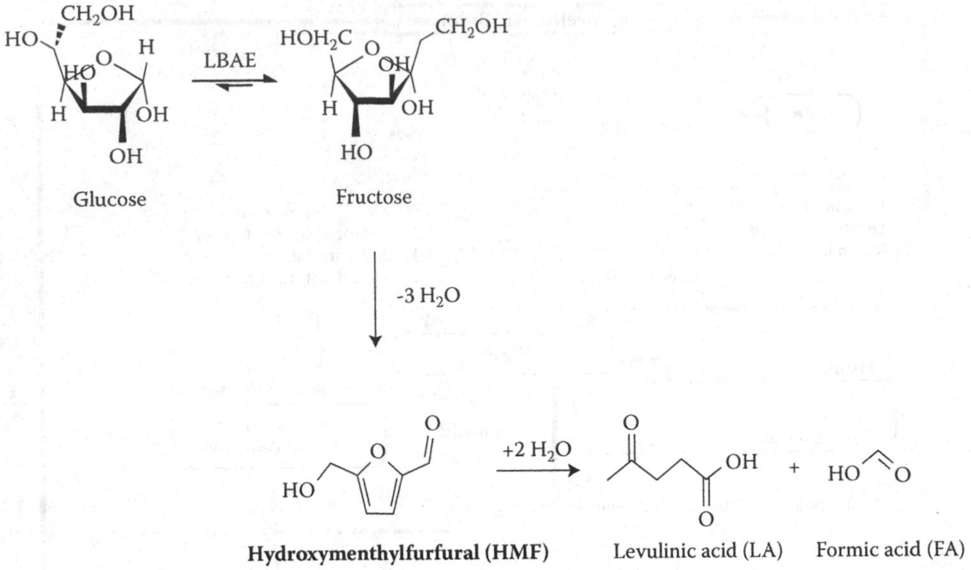

FIGURE 10.5   Representation of glucose isomerization to fructose and further dehydration to form hydroxymethylfurfural (Titirici et al. 2015). Reprinted with permission copyright 2015, RSC.

for the formation of the hydrochar are produced by aromatization (Funke and Ziegler 2010). The hydrochar produced has diverse oxygen-containing functional groups on its surface (Sharma et al. 2020).

In the case of hydrothermal carbonization of carbohydrates, three main stages are needed: dehydration, polymerization, and carbonization/aromatization (Titirici, Funke, and Kruse 2015). For glucose, through the Lobry de Bruyn-Alberda van Ekenstein (LBAE), isomerizes to give fructose. Then the fructose undergoes dehydration to give hydroxymethyl furfural (HMF, which is the major reactive intermediate) (Figure 10.5), and other products, which include aldehydes and organic acids (glyceraldehyde, levulinic acid, formic acid, and also phenols (Kabyemela et al. 1999)). Through aldol reactions, the produced products get assimilated into the carbon framework and become part of the HTC process, thus reducing the entire process pH and encouraging autocatalytic effect in the HTC (Hu et al. 2010). If pentoses are used, the formation of the hydrochar follows through the major reactive furfural intermediate. The HTC process following the HMF or furfural intermediate is dependent on whether the major sugar resource is a 6- or 5-carbon ring monomer (Titirici, Funke, and Kruse 2015). In the case of the pentoses, the products from the furfural ring possess more aromatic character and also reduced fraction of carbonyl and carboxyl groups. Different carbon structures and morphologies are obtained for the pentoses and hexoses during HTC. The HTC from the pentose-based carbon consists of single carbon spheres, while that for the hexose-based carbon is made up of interlocked spherical carbon particles due to high HMF solubility in water (Titirici, Antonietti, and Baccile 2008). Thereafter, the intermediates pass through subsequent polymerization and condensation reactions, with the combination of reverse intermolecular and aldol condensation reactions (Funke and Ziegler 2010; Kumar 2013) (Figure 10.5).

The products obtained by concurrent condensation, polymerization, and aromatization reactions are known as biocrude, which after continuous polymerization and aromatization forms the solid cross-linked hydrochar (Reza et al. 2014). The produced hydrochar has to undergo hydrothermal treatments so as to reduce the contained oxygen and hydrogen constituents for a highly valued carbon-rich constituent to be produced (Funke and Ziegler 2010). Depending on the type of biomass used, hydrochar can be suitable for many applications, which include soil amelioration, waste water pollution remediation, bioenergy production, and carbon sequestration (Saidur et al. 2011). High-powered characterization techniques should be employed in order to determine the structure of hydrothermal carbonization–derived carbon materials. Apart from the carbon precursors, other process parameters that control the hydrothermal carbonization process and the mechanisms of the formation of the hydrochar include temperature, residence time, pH of feedwater, heating rate, and substrate concentration (Wang et al. 2018).

Hydrothermal carbonization has some advantages over other methods of biomass conversion which include the following: pre-drying process is not needed because wet biomasses are used, relatively low reaction temperatures are used, high conversion efficiency, and shorter retention times. Due to the mild temperature conditions employed in hydrothermal carbonization processes, the carbons produced are quite different from the activated carbons. They have more functionalities, thus they are hydrophilic and can easily disperse in water; they come out as microspheres (spheres) with tunable sizes and also very different structures (Kubo et al. 2010). However, the reduced surface area and porosity pose limitations to the use of the hydrochar in many applications. Some techniques have been put in place to overcome the limitations through functionalization processes of the hydrothermal carbons.

10.2.2.4 Functionalization of Hydrothermal Carbons

The functionalization of hydrothermal carbons take place at elevated temperatures in the range of 300–900°C, and this leads to the synthesis of highly functional carbons which include hierarchically ordered carbon materials, activated carbon materials, carbon nanotubes, and graphitic carbon materials (Hu et al. 2010). The functionalization is a post treatment that modifies the chemical and physical property of hydrothermal carbons as shown in Figure 10.6.

To this effect, several methods have been employed to enhance the desired properties, ranging from combination of HTC process and activation methods, use of catalysts, templating methods, among others. Başakçılardan Kabakcı and Baran (2019) reported on the hydrothermal carbonization and

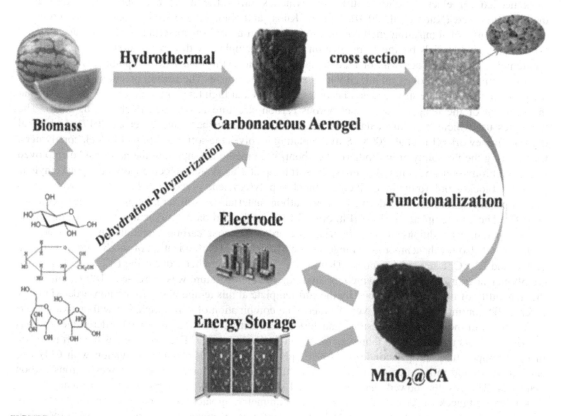

FIGURE 10.6 Hydrothermal carbonization of water melona nd further functionalization (Ren et al. 2014). Reprinted with permission copyright 2014, ACS.

activation of several lignocellulosic biomass materials (wood saw dust, olive pomace, walnut shell, apricot seed, tea stalk, and hazel nut husk). The hydrochars from the biowastes were activated with KOH/hydrochar ratio of 2. The authors reported that all the hydrochars from the plant residues showed increased heating values and increased carbon yield with olive pomace hydrochar having the highest carbon content of 61.1% and heating value of 25.56 MJ/Kg, but reduced in the case of the activated carbon. Also, that the hydrothermal carbonization of the samples prior to activation resulted in positive effects on the surface areas (ranging from 308.9 m^2 g^{-1} for wood sawdust to 666.7 m^2 g^{-1} for tea stalk) and microporous structure (ranging from 0.25 cm^3 g^{-1} for olive pomace to 0.73 cm^3 g^{-1} for wood sawdust). It is also a pointer that the fuel features of the hydrochar relies on the kind of biomass, and the porous structure of the activated hydrochars are seriously affected. It has been reported that the use of catalysts for hydrothermal carbonization and subsequent activation of the hydrochar would lead to increased degree of carbonization, introduction of hetero-atoms in the carbon lattice, and also modification of the physical structure of the hydrochar (Titirici et al. 2007). The hydrothermal carbonization of Salacca peel biomass was done in the presence of citric acid catalyst and later activated using KOH activator by Susanti et al. (2019) to be used as electrode for electrochemical energy storage applications. The authors reported that there was great improvement in the specific surface area of the samples in the presence of the catalyst (1720.210-1635.440 m^2 g^{-1}), and also the structure of the activated hydrocarbon in the presence of the catalyst had a coral reef–like structure with more swelling and more developed porosity. Catalyzed hydrothermal carbonization was used in the synthesis of raw grains in the presence of $[Fe(NH_4)_2 (SO_4)]$ (Cui, Antonietti, and Yu 2006). The obtained carbons were made up of porous carbon, single carbon nanofibres, and carbon microspheres.

Templating method is one of the most effective means of functionalizing hydrothermal carbons to achieve the desired purpose. Templating methods are able to introduce pores of about 1–200 nm pore sizes needed for electrolytic/ionic diffusion, which seems difficult when simple carbon derivative methods are used (Kubo et al. 2010). Gilbert, Knox, and Kaur (1982) were about the first group to attempt the use of templating methods in producing porous carbon materials. They produced glassy porous carbon materials by the impregnation of silver template with a resin mixture of phenol formaldehyde. Their approach was of hard templating (nanocasting). In hard templating, a carbon precursor (commonly phenolics) is mixed with a sacrificial template. The carbon precursor infiltrates into the pores in the template structure and becomes carbonized at high temperatures (above 700°C), then at the removal of the template, a distinct porous carbon structure is obtained (Kubo et al. 2010). The templates used include zeolite (Stadie et al. 2017), silica (Sanchez-Sanchez et al. 2017), and metal organic frameworks (Liu et al. 2008). Soft templating involves the soft assembling of block copolymers. Considering the flexibility of hydrothermal carbonization method, templating methods can be employed in carbon biomass derivation. Oftentimes, the soft templates used are triblock copolymers poly(ethylene glycol) – block – poly(propylene glycol) – block – poly(ethylene glycol) (P123 or F127). Kubo et al. (2011) produced carbohydrate-derived porous carbon materials by soft templating at low temperatures (130°C). The soft template used was Pluronic® F127 and the carbon precursor was fructose. Due to the low temperature conditions of the synthesis, the ordered porous carbon materials obtained were not uniform and also for the temperature range for the hydrothermal carbonization of saccharides is between 150°C and 220°C. Later on the authors (Kubo et al. 2013) did another work using F127 and dual-block copolymer-latex as templates, in which the hydrothermal temperature was increase to 180°C, but due to the instability of the micelles formed by the soft template at this temperature, rather non-ordered highly layered 3D continuous porosities were formed. The combination of soft templating with activation has shown to yield more improved results. Sanchez-Sanchez et al. (2018) synthesized ordered mesoporous carbon materials using mimosa tannin as the precursor and Pluronic® F127 solution as the soft template in mild temperature and pH conditions (30°C, pH 3), after which physical activation with CO_2 was carried out to add more microporosity to one of the obtained hierarchical ordered mesoporous carbon materials. The results showed that the activated ordered mesoporous carbons showed greater gravimetric capacitances of 286 F g^{-1} by cyclic voltammetry, greater capacitance retention of 50% at 12 A g^{-1}, greater energy density of 5.41 W kg^{-1}, and greater power density of 5.03 KWkg by galvanostatic charge-discharge than the ordered mesoporous carbons that were not activated and the former are very suitable for electrochemical energy applications.

In order to overcome the unstable nature of micelles formed when soft templates (P123 or F127) are used at high temperatures in hydrothermal carbonization of saccharides, both soft and hard templating methods can be employed. Xiao et al. (2017) synthesized ordered mesoporous hydrothermal carbon materials from saccharide carbon precursor (which include, glucose, D-maltose, D-fructose, hydrochloride, D-glucosamine, sucrose, and β-cylodextrin) by the combination of the soft (triblock copolymer) and hard template (tetraethyl orthosilicate (TEOS)). During the production process, the triblock copolymer released micelles which were shielded by the silica layer produced by the hydrolysis of TEOS and successive condensation of orthosilicic acid. In the course of the process, the carbon precursors cleaved to the silica surface, and also reacted with themselves by various reactions (dehydration, cycloaddition, and polymerization). After which calcination (at 600°C) and hydrofluoride etching leads to the production of ordered mesoporous hydrothermal carbon materials. The results from TEM and SAXS (Small Angle X-ray Scattering) characterizations confirmed that the ordered mesoporous hydrothermal carbon materials possessed ordered 2D hexagonal structure, and also BET surface area of 330–620 m^2/g. Matching the polarity of silica templates with those of carbon precursors can aid in the proper infiltration of the pores of hard templates. (Titirici et al. (2007) synthesized various mesoporous hydrothermal carbons by carrying out hydrothermal carbonization technique on furfural carbon precursor using nanostructured silica templates. To ensure that the carbonization products from the HTC process infiltrated into the pores of the silica beads template, the polarity of the template surface was matched with the polarity of the furfural precursor, and this resulted in hydrothermal carbons with varying morphologies: hollow carbon spheres, macroporous carbon casts, mesoporous carbon shells, and mesoporous carbon microspheres. It was evident that the nature of the template also affected the resulting carbonaceous material, for mesoporous silica template resulted to mesoporous carbon shells, whilst non-porous templates led to hollow carbon spheres with vigorous coating of the carbon. The authors recommended the produced carbon materials for use as electrode materials. Several templates have been used in hard templating (nanocasting) (silica beads (Yang et al. 2005), silica monoliths (Nakanishi, Takahashi, and Soga 1992), silica nanoparticles (Li and Jaroniec 2003), alumina membrane (Kubo et al. 2010), tellurium nanowires (Qian et al. 2006)); in the production of hierarchical mesoporous carbon spheres, carbon nanotubes, nanofibres, hexagonal hollow ordered carbon materials, among others, although, the templates are relatively expensive, and toxic reagents such as HF and H_3PO_4 are used. However, Sanchez-Sanchez et al. (2017) carried out a single-step hard templating method to avoid the use of toxic reagents during the carbonization process. The authors (Sanchez-Sanchez et al. 2017) produced hierarchically porous, oxygen-doped, ordered mesoporous carbons (OMCs) from plant-derived polyphenols (phloroglucinol (PhC), garlic acid (GaC), catechin (Cat C), and mimosa tannin (Tan C)), through nanocasting method (hard templating). Silica template was used and the infiltration was done by a single-step process in which there was an addition of the solution of the bioprecursors and the removal of the solvent under vacuum conditions (Figure 10.7). Toxic solvents were not used during the infiltration and also there were no extended polymerization-stabilization times. The obtained OMCs showed high rate capabilities, with good cycling stabilities up to 5000 cycles, maximum energy densities within 15.8 W kg^{-1} and 8 W kg^{-1} to 22.1 KW kg^{-1}, respectively, and the current density ranging from 0.1 A g^{-1} to 12 A g^{-1} from galvanostatic charge-discharge. Amongst the polyphenols used, garlic acid–derived electrode showed the utmost electrical conductivity and the fastest response to frequency.

10.3 Applications of Biomass-Based Electrode Materials

10.3.1 Applications in Fuel Cells

A fuel cell can be said to be a device that converts chemical energy to electricity through redox reactions (Deng et al. 2016). Generally, in a fuel cell, hydrogen or alcohol oxidation reaction takes place at the anode, while oxygen reduction reaction takes place at the cathode, and within the electrolyte, current is passed in between the electrodes by ions. The electrolyte may either be a liquid, molten salt, polymer, or ceramic; and oftentimes, the fuel is hydrogen gas, methanol, or methane (Sekhon et al. 2012; Pollet,

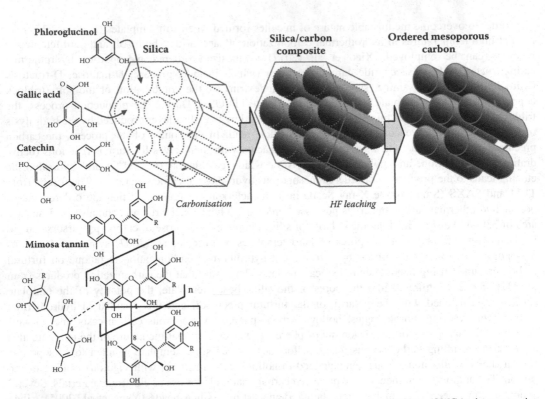

FIGURE 10.7 Representation procedure of the production of ordered mesoporous carbons (OMCs) via nanocasting method using tannin-related carbon precursors (through a single-step process) (Sanchez-Sanchez et al., 2017). Reprinted with permission copyright 2017, Elsevier.

Staffell, and Shang 2012). Based on the temperature of operation, fuel cells are classified as low-temperature (less than 200°C) and high-temperature (more than 450°C) fuel cells. Then based on the type of electrolyte used, fuel cells can be classified as microbial fuel cell (MFC), alkaline fuel cell (AFC), polymer electrolyte membrane fuel cell (PEMFC), phosphoric acid fuel cell (PAFC), direct methanol fuel cell (DMFC), solid oxide fuel cell (SOFC), and molten carbonate fuel cell (MCFC) (Kaur, Verma, and Sekhon 2019). The main issue in fuel cell technology is to explore catalysts that will carry out alcohol electro-oxidation and oxygen reduction reactions. Noble metals and their oxides (which include Pt, RuO_2, Pd, and IrO_2) are often used in catalyst synthesis, but their expensiveness and non-abundance pose hindrances to their usage. Thus the need to manufacture carbon-based materials with high specific capacitance, high porous and tunable structures, and functionalized surfaces for use as electrochemical catalysts becomes paramount (Zhu et al. 2016). In order to achieve the desired porous structure, heteroatom doping, production of composites with metal and metal oxides, and other measures have been considered.

10.3.1.1 Electrocatalytic Alcohol Oxidation and Oxygen Reduction Reaction

Oxidation of hydrogen or methanol fuels takes place at the anodes, while free air is reduced at the cathode. Oxygen reduction reaction (ORR) is the key reaction that occurs at the cathode of fuel cells (Dai et al. 2015). The ORR progresses through a more effective four-electron pathway or a less effective two-electron pathway. The four-electron pathway is much preferred, for O_2 directly combines with protons and electrons to produce the end product, H_2O, whilst in the two-electrode pathway, H_2O_2 is produced as an intermediate, which results in active site degradation and cell voltage reduction (Kaur, Verma, and Sekhon 2019). Pt/C alloys are the conventional and very effective ORR catalysts, but they have associated limitations such as scarcity, high cost, low tolerances to carbon monoxide, and methanol crossover (Gong et al. 2009). However, carbon-based catalysts have shown to be better replacements to the Pt-based catalysts. Biomass-derived carbons that are abundantly available, sustainable, relatively less expensive, possessing inherent interconnected and porous structures have attracted much attention as suitable carbon-based electrochemical catalysts for the replacement of Pt-based catalysts. The heteroatom doping of biomass-based carbons was acceptable to significantly improve the performances of carbon catalysts. Nitrogen doping of biomass-derived carbons has showed much more improvement on the carbon catalyst than the other heteroatoms (Deng et al. 2016; Yan et al. 2017a). During the nitrogen doping of biomass-derived carbons, four different functional groups are formed; pyridinic, pyrrolic, graphitic-quaternary, and oxidized nitrogen functional groups (Bao et al. 2013). It has been accepted that pyridinic-nitrogen and graphitic-nitrogen functional groups enhance the oxygen reduction reactions (Guo et al. 2016). The enhanced ORR activity obtained in nitrogen-doped biomass-derived carbons can be due to some reasons: (i) the carbon structures are highly porous containing micro-, meso-, and macro-pores which enhances ionic transport; (ii) natural biomass rich in nitrogen can be used to achieve high nitrogen doping levels, thus providing more ORR active sites; (iii) due to the presence of the micro- and/or meso-pores contained in the biomass-derived carbons, large specific surface area is created for exposure to more ORR active sites (Wang et al. 2018; Deng et al. 2017). Wang et al. (2018) developed metal-free nitrogen-doped hierarchical porous carbons (PPC-NaZnFe) using pomelo peel biowaste as the carbon precursor. The pomelo peels were first treated by hypersaline environments (containing saturated NaCl, $ZnCl_2$, and $FeCl_3$), after which carbonization, activation, and ammonia treatments were carried out. High specific surface area of up to 3092 m^2 g^{-1} and high pore volume of 2.05 cm^3 g^{-1} were obtained. The produced PPC-NaZnFe catalyst showed excellent ORR activity (more than other reported metal-free N-doped biomass-derived carbon), maximum durability, and outstanding methanol tolerance, as compared to the costly Pt catalyst. The authors reported that the outstanding electrochemical performance of the produced PPC-NaZnFe catalyst was due to the effect of synergy from enlarged total volume of pores, increased content on pyridinic nitrogen doping, and reduced stacking defects of carbon graphite layers, increased by the hypersaline effects on the pomelo peels.

The biomass-derived nitrogen-doped carbon can be used as catalyst supports in fuel cells. This is because more active sites are introduced on the surface by the nitrogen-doping of the biomass-derived carbon, which act as anchor sites and thus result in enhanced bonding between the nanoparticles and the carbon, and also improved uniformity in the dispersion of the nanoparticles (Zhao et al. 2014). Zhao et al. (2014) produced Pt dendrites having average particle size of 20 nm when soybean-derived nitrogen-doped carbon was used as the catalyst support. The produced Pt dendrites showed more improved electro-oxidation of methanol when compared to the conventional Pt/C catalysts. Biomass chitosan and carbon black were used as precursors in the production of nitrogen-doped carbon (NC), and the Pt was deposited on the nitrogen-doped carbon to form Pt/NC catalyst (Zhao et al. 2014). The produced catalyst was compared to the commercial Pt/C, and it showed greater anode peak current, improved tolerance to adsorbed carbon monoxide, and maximum electrocatalytic stability. Apart from the modification of Pt structure when nitrogen is incorporated in biomass-derived nitrogen-doped carbons, OH absorber is formed on the catalyst surface at reduced potential as water is activated, and thus the onset potential is lowered and the stability increased (Yan et al. 2017b). Co- or tri-doping of heteroatoms in biomass-derived carbon can also increase the electrochemical performance (especially ORR activity) of fuel cells, for most biomass-derived carbon contain enough of the heteroatoms, viz. O, P, S, and N (Fang et al. 2016). Also, the addition of non-precious transition metals (for example, Fe and Co) in the nitrogen-doped biomass-derived carbon can enhance cathodic oxygen reduction reaction in both alkaline and acidic media (Yang et al. 2015). Other ways of improving electrocatalytic oxidation of alcohols include the use of carbides in forming the catalyst (Pt/graphitic carbide, to enhance desorption of CO) (Wang et al. 2015), and replacement of Pt with Pd-based N-doped carbon catalysts (Shen et al. 2011).

On the other hand, undoped carbons are said to exhibit reduced ORR catalytic activities, but there are exceptions to that (Jin et al. 2015; Jia et al. 2016). Hao et al. (2018) reported the preparation of a series of defective carbon catalysts through facile and scalable "N-doping-removal" process with seaweed biomass sodium aliphate (SA) as the carbon precursor. The optimized defective porous carbon catalysts (at 900°C pyrolysis) (D-PC-1(900)) produced rich ORR-active defects, large surface area of

1377 m^2 g^{-1}, rich hierarchical porosity, and good conductivity. The produced D-PC-1(900) showed very good ORR activity and selectivity in 0.1M KOH that can be compared to conventional Pt/C catalyst, and also exhibited about the highest significant ORR activity in 0.5 MH$_2$SO$_4$ for defective carbons and can be compared to nitrogen-doped carbons. Above all, the D-PC-1(900) showed much improved methanol tolerance and stability in both acidic and alkaline media than the commercial Pt/C catalysts, demonstrating superb candidate for non-precious metal ORR catalyst for fuel cell applications. Despite the excellent impact on the catalytic performance of biomass-derived carbon materials, they greatly improve the environment.

10.3.2 Applications in Li Batteries

Lithium-sulphur battery possessing ultrahigh specific capacity of 1650 mAh g^{-1} and energy density as high as 2600 Wh kg^{-1} is obviously very promising for next-generation high-energy batteries, particularly for use in fields that demand higher energy delivery like electric vehicles (Gao et al. 2017b). More so, sulphur being the main component in lithium-sulphur battery is freely available, cheaper, and environmentally friendly (Gao et al. 2017b). Several efforts have been channelled to design and fabricate hierarchically porous carbons/sulphur composites based on biomass. These carbon/sulphur composites have the capacity to control "shuttle effects" for improved stability, while larger pores facilitate ion transport for enhanced rate performance in lithium-sulphur battery applications (Gao et al. 2017b). All types of biomass precursors like pig bone, fish scales, shrimp shell, litchi shells, olive stones, cotton, silk cocoon, bamboo, wheat straw, mango stone, pomelo peels, banana peels, gelatin, cassava, bark of plane trees, starch have been utilized in preparation of hierarchical porous carbons by well-patterned carbonization processes(Wei et al. 2011; Li et al. 2014; Qin et al. 2014; Wang et al. 2014; Zhang et al. 2014, 2016; Cheng et al. 2015; Gu et al. 2015; Liu, et al. 2015; Moreno et al. 2014; Qu et al. 2015; Qua et al. 2016; Xu et al. 2016; Yang et al. 2016; Zhang et al. 2014). All of these biomass-based carbons are employed as conductive host of sulphur for lithium-sulphur battery with enhanced electrochemical performances.

Most importantly, majority of biomass materials like leaves, barks, and shells often have complex arrangements and intriguing microstructures inheritable via well-patterned synthesis methods. For example, complex carbon structure with interconnected micropore-macropore has been obtained from natural bark to achieve carbon/sulphur composite for lithium-sulphur battery. The carbon microsheets and macropores were conserved during the process of activation, and then become physical reservoirs for sulphur and electrolyte storage. Also, the interconnected architectures created an integrated three-dimensional conductive network, which enhances fast transfer and migration of electrons and ions, respectively. Moreover, the micrometre-scale walls of veins can equally allow volumetric change during electrochemical reaction. All of these components were integrated to ensure the conductivity and stability of the integrated carbon/sulphur cathode (Xu et al. 2016).

Most biomass materials have been employed in the production of porous carbon with modified surface chemistry due to their low cost, natural abundance, high accessibility, as well as their rich protein content, that obviously serves as nitrogen and oxygen resource to achieve heteroatom doping/co-doping (Gao et al. 2017b).

10.3.3 Applications in Supercapacitors

Supercapacitors (SCs) or electrochemical capacitors are energy storage devices that give high power density and increased cycle efficiency, but have reduced energy density of about 4–10 Wh kg^{-1} (Zhang et al. 2017). The reduced energy density poses limitations in their utilization as energy storage devices. Supercapacitors are made up of electrode materials, electrolytes, and separators. To increase the power density and the energy density, the specific capacitance and operating voltage need to be increased, while the equivalent series resistance needs to be reduced. The choice of a suitable electrolyte improves the voltage, whereas a suitably chosen electrode material enhances the specific capacitance (Conway 1999). The need to provide electrode materials with high power and high energy densities is of high importance, for there would be increased pathways for electron movement and also shortened distances between the electrode and the electrolyte (Wang et al. 2017). According to the mechanism of energy storage, supercapacitors can be classified into two: (a) electrochemical double- layer capacitors (EDLC) and (b) pseudocapacitors. The EDLC operate by adsorbing/storing of electrolyte ions on the electrode surface without any redox reaction, while the pseudocapacitors store energy by the reversible redox reactions that occur at the electrode/electrolyte interface, motivated by faradaic mechanism (Reis et al. 2020). Both the EDLC and pseudocapacitative mechanisms depend on the structure and behaviour of the electrode material, such as porosity, surface functionalities, and surface chemistry and how conductive the material is, and carbon-based materials have been utilized due to their inherent properties (Frackowiak and Béguin 2002). Both activated and non-activated carbons have been used as electrode materials in the production of supercapacitors. Electrodes that are highly porous with large surface areas, excellent pore size distribution and pore development for enhanced conductivity, and improved physicochemical stability are needed and utilized in order to achieve optimal interaction between electrode and electrolyte, and also enhance electrolytic adsorption (Zhang et al. 2016). The biomass-derived carbon materials with their highly porous and interconnected structure fittingly meet the need. Moreover the biomass-derived carbons are environmentally friendly, sustainable, and relatively inexpensive (Enock et al. 2017). The biomass-derived carbons for use as supercapacitors should be characterized by the following attributes: (a) large specific capacitance for increased power and energy densities; (b) reduced equivalent series resistance so as to reduce drop in voltage and power loss; (c) large surface area and excellent pore size distribution to enable increased conductivity and improved rate performance; (d) good cyclability and long lifespan for increased electrode stability during repeated charge-discharge cycles and varying temperatures (Liu et al. 2018; Zhang et al. 2016). To achieve the desired properties in the biomass-derived carbon, many factors are put into consideration – the choice of the precursor, method of derivation, process conditions, post-production treatments, and functionalizations, among others.

The porous structure, surface functionalities, and heteroatomic doping of biomass-derived carbons play key roles in the high performance of electrodes used in supercapacitors. High specific surface area (SSA) which are mainly macropores are associated with biomass-derived carbons produced through activation processes, and these macropores hinder ionic movement (Li et al. 2014). The geometry and dimension of the pores enhance the performance of the supercapacitors (Xia et al. 2008). Thus, there is need to tune the pore structure and also increase the degree of graphitization for improved performance (Zheng et al. 2015). Sun et al. (2013) used coconut shell as the carbon biomass precursor in synthesizing graphene-like porous nanosheets with $ZnCl_2$ as the activator and $FeCl_3$ as the graphitic catalyst by concurrent activation and graphitization. The synthesized graphene-like porous nanosheets showed great specific surface area of 1874 m^2 g^{-1} with pore volume of 1.21 cm^3 g^{-1}. The high degree of graphitization resulted in good electrical conductivity, with specific capacitance of 268 F g^{-1} at 1 A g^{-1} in KOH electrolyte without the use of additives. The nanosheets produced exhibited energy density of 54.7 KW kg^{-1} at a power density of 10 KW kg^{-1}. Hierarchically 2D porous carbon nanosheets were synthesized, using glucose precursor by Zheng et al. (2014). Activation and melt templating were done using potassium species, and ordered nanosheets were obtained having specific capacitance of 257 F g^{-1} at 0.5 A g^{-1}, specific surface area of 2600 m^2 g^{-1}. Due to the interconnected arrangement of the hierarchically two-dimensional (2D) porous structure, there was enhanced ionic transport which led to ultrahigh rate capability. Low sulphonated alkali lignin was used as precursor in preparing activated carbon fibres with NaOH and KOH activators (Hu et al. 2014). The produced activated carbon fibres showed large-sized nanographites and also good electrical conductivity. The KOH-obtained activated carbon demonstrated higher specific capacitance of 344 F/g at 10 mV/s in comparison with the NaOH-activated carbon fiber due to more microporous size distribution and reduced pore size.

Heteroatom doping is considered as one of the most active ways of effectively wetting and introducing pseudocapacitance in carbon materials. Many types of heteroatom doping have been carried out which include N, P, B, F, Cl, Si, S-doping, among others (Liu et al. 2017; Paraknowitsch and Thomas 2013). Heteroatom doping of biomass-derived carbon can be by self-doping, mixing of the biomass precursor with heteroatom doping precursor (Hulicova-Jurcakova et al. 2009), and also by post-treatment (Hulicova-Jurcakova et al. 2009). Sun et al. (2020) synthesized self-nitrogen-doped porous carbon from heteroatom-rich quinoa biomass using KOH as the activating agent. The nitrogen-doped porous carbon produced showed high specific surface area of 2597 m^2/g and high specific capacitance of

330 F/g in 6M KOH at a density of 1 A/g, and the rate capability was good. The authors suggested the application of the obtained quinoa-derived nitrogen-doped porous carbon in high-capacitance super-capacitors. Co-doping of heteroatoms in biomass-derived carbon can effectively enhance the storage sites and diffusion mechanisms of produced electrodes than sole heteroatom doping (Gao et al. 2017a). Nitrogen-doping is the most often used type of doping, for it rapidly increases electrical conductivity and the diffusion processes of the carbon material. Also co-doping with other heteroatoms such as O, S, B, or P might improve the pseudocapacitance, thus enlarging the distance between layers and filling up storage sites (Pang et al. 2015; Qin et al. 2018). B and N co-doped porous carbon materials were synthesized from dandelion fluff by direct pyrolysis by Zhao et al. (2017). The co-doping of B (4.6 at%) and N (2.2 at%) into the carbon lattice resulted in added double-layer capacitance and more pseudo-capacitance for total storage site's improvement. The produced carbon exhibited high volumetric energy density of 12.15 Wh//L at 699.84 W/L.

Researchers have moved on to the production of binder-free electrodes for supercapacitors. Due to the related challenges in their usage which include increase in overall cost and weight, reduction in specific capacitance, power and energy densities of supercapacitors, the binders (such as poly(vi-nyldene fluoride) (PVDF), poly(tetrafluoroethylene) (PTFE), acetylene black and carbon black) are being avoided (Cheng et al. 2018). This is to achieve very fast mass to charge transport and release enough active sites, and still maintain high volumetric performance (Li et al. 2014). Many approaches have been used in the production of binder-free carbon materials; they include gelation, vacuum filtration, and electrospinning, among others (Wang et al. 2019). Sahu et al. (2015) produced a binder-free heavily doped nitrogen grapheme from silk cocoon film through pyrolysis at 400°C in argon-filled atmosphere. The specific capacitance obtained was 220.5 F/g at 0.8 A/g, and the supercapacitor achieved energy density and power density of 9.8 Wh/kg and 9.9 KW/kg in aqueous solution of 1M H_2SO_4, respectively.

10.3.4 Advantages of Biomass-Based Electrode Materials over Other Sources

The biomass-derived carbons for electrode materials are relatively inexpensive compared to other sources of electrode materials. They are plentiful, sustainable, have excellent electrical conductivities, good specific surface areas, superb electrochemical stabilities, and modifiable dimensions (Yang et al. 2019). Using biomass-derived carbon as electrode material production can reduce the rapid upsurge in the percentage of carbon dioxide emitted to the atmosphere (Pandey 2019). Biomass-based electrode materials can be synthesized via one or two stages of process synthesis. Also they can be easily coated on various substrates using techniques (Bhat, Supriya, and Hegde 2020). It is possible to produce bendable (flexible) electrodes in form of foams and fibres from biomass-derived carbons (Lu and Zhao 2017). Flexible energy devices that can be worn can also be produced.

Biomass-derived electrode materials possess naturally given highly porous structure that is important for power-related applications and also the active surfaces are rich in functional groups that would enhance increased specific surface capacity and, thus give greater number of storage sites for charge (Bi et al. 2019; Li et al. 2020). Biomass-derived electrodes show improvement in energy density of su-percapacitors, and also in reaction blocking of intermediates formed in lithium-sulphur batteries. Furthermore, biomass-derived carbon materials offer scaffolds used in the deposition of active materials for supercapacitor applications, and also host sulphur in applications involving lithium-sulphur batteries for effective sulphur utilization (Gao et al. 2017a). Bearing in mind the superb benefits of electro-negative heteroatom doping (especially nitrogen) in increasing the performance of electrochemical energy storage electrode materials, the use of naturally occurring heteroatom containing biomass-based electrodes can aid in the elimination of post-treatment high temperature challenges, heat treatment of costly synthetic chemicals that contain heteroatoms, as well as electrochemical processes. Moreover, heteroatom-doped biomass-derived carbons are richly available, benign, and friendly to the environment (Jiang Deng et al 2016; Yu et al. 2019). Biomass-derived carbons can display varied specific morphologies and also possess tunable specific morphologies, which enhance the excellent performance of the produced electrode materials (Jiang et al. 2018).

Various biomass-derived carbons can be made into electrodes free of binders, and this process eliminates the use of toxic reagents such as polymers based on fluorine (Conder et al. 2020). Apart from this, the binders have some associated demerits: (a) there may be presence of pores in the binders leading to reduction in accessing the electrolyte (Ruiz et al. 2007),(b) the binders do not conduct electricity, thus the electrical conductivity of the electrode is reduced (Weng and Teng 2001), (c) when binders are used, extra processing steps are needed, and also costs are increased (Kalyani and Anitha 2013).

10.3.5 The Place of Biomass-Based Electrode Materials in Next-Generation Energy Conversion and Storage

The development of an alternative biowaste-based electrochemical energy electrode material helps in disposal of waste, converting waste to a useful product, and gives cheaper route for sustainable electrochemical energy storage technology (Ike et al. 2015; Ike and Iyuke 2016; Ike et al. 2019a, 2019b).

There has recently been a growth in synthesis of biowaste-based materials for sustainable development (Hill 2017; Maharjan et al. 2017; Azwar et al. 2018; Benedetti, Patuzzi, and Baratieri 2018; Guardia et al. 2018; Zhang et al. 2019; Veerakumar et al. 2020). Many biowaste sources presented in the literature for synthesis of activated carbon for energy conversion and storage devices are animal, mineral, plant, vegetables, etc. (Misnon et al. 2015; Gong et al. 2016; Zhang et al. 2016; Na et al. 2018; Parveen, Al-Jaafari, and Han 2019; Zhang et al. 2019).

In this view, renewable biomass and their derivatives are obviously potential alternatives capable of replacing conventional non-sustainable electrode materials, due to their inherent features and merits like environment-friendliness, natural abundance, different architecture, intrinsic mechanical strength, flexibility, and their flexibility to form hybrids with other functional materials. Several researches have been channelled for the development of sustainable and high-performance biomass-based electrode materials and their derivatives for energy conversion and storage devices (Titirici et al. 2014; Abioye and Ani 2015) with excellent capacity. Renewable biomass-based carbon materials have shown great potential in enhancing the electrode performance for energy conversion and storage devices. In fact, biomass-based materials have been employed as separators, binders, and electrodes for energy conversion and storage devices (Lixue Zhang et al. 2015).

Most commercial nanostructured carbon electrodes are derived from fossil fuel precursors, which are not renewable and are quite expensive, prepared under unfriendly and complex conditions like high temperature and high vacuum that are energy consuming. Judging from energy and environmental contexts, the development of facile, reproducible, and environmentally benign techniques for carbon electrode materials production is inevitable. Several researchers have studied and reported the production of renewable biomass-based carbon materials or their by-products. Those biomass-based electrode materials often demonstrated promising electrochemical properties, to a greater extent compared to activated carbons, carbon nanotubes, carbon fibres, and graphene. Table 10.1 presents the performance parameter values for biowaste-based electrode materials.

Different biowaste employed in the production of activated carbon for electrochemical energy conversion and storage are presented in Table 10.2

Many biomass resources, such as all types of wood and plant tissues, agricultural wastes, even industrial wastes and municipal wastes have been employed as starting materials to synthesize carbon electrodes for energy conversion and storage, and have recently drawn growing attention because of their abundant availability, environmental friendliness, and low cost (; Zhi et al. 2014; Abioye and Ani 2015; Divyashree and Hegde 2015).

10.4 Conclusion and Future Outlooks

The global quest for renewable energy sources has drawn great interest for many years now. Development and optimization of new materials for energy conversion and energy is important to achieve clean energy via renewable sources. Biowaste-based materials have steadfastly permeated into

TABLE 10.1

Performance Parameter Values for Biowaste-Based Electrode Materials (Mensah-Darkwa et al. 2019)

Biowaste	Process	Material Form	Electrolyte	Electrode Configuration	BET Surface Area (m^2/g)	Measurement Protocol	Specific Capacitance (F/g)
Bamboo	Carbonization and KOH activation	Activated biomass carbon	3M KOH	3 electrodes	2221	0.5 A/g	293
Bamboo	KOH activation	Activated carbon	6M KOH	3 electrodes	3000	5 A/g	300
Corncob residue	Steam activation without precarbonization	Porous carbon	6M KOH	3 electrodes	1210	1 A/g	314
Coconut kernel pulp (milk-free)	KOH activation	Activated carbon	1M Na_2SO_4	2 electrodes	1200	10mV/s	173
Corn stalk core	KOH activation	Activated carbon	–	3 electrodes	2350	1A/g	140
Corn syrup (high fructose)	Self-physical	Activated carbon	KOH	2 electrodes	1473	0.2 A/g	168
Endothelium corneum *Gigeriae galli*	Carbonized	Nitrogen-doped porous carbon	6M KOH	3 electrodes	2150	1 A/g	198
Fish gill	Carbonization and thermal activation	Activated carbon	6M KOH	3 electrodes	2082	2 A/g	334
Gelatin	Hydrothermal	Porous carbon nanosheets	6M KOH	3 electrodes	1620	50 A/g	183
Leaves (fallen)	Activations of (KOH and K_2CO_3)	Porous active carbon	6M KOH	2 electrodes	1076	0.3 A/g	242
Starch (porous)	Carbonization and KOH activation	Porous carbon microspheres	6M KOH	2 electrodes	3251	0.05 A/g	304
Sugar cane bagasse	Chemical activation with $ZnCl_2$	Activated carbon	1M Na_2SO_4	2 electrodes	1452	50 A/g	300
Sugar cane bagasse	Calcium chloride ($CaCl_2$) activation	Nitrogen-rich porous carbons	6M KOH	2 electrodes	806	30 A/g	213
Waste tea leaves	Carbonization and KOH activation	Activated carbons	2M KOH	3 electrodes	2841	1 A/g	330

TABLE 10.2

Summary of Main Performance Parameters for Biowaste-Based Activated Carbon for Electrochemical Capacitors (Mensah-Darkwa et al. 2019)

Biowaste	Energy Density (Wh/kg)	Power Density (W/kg)	Cycles	Percentage Retention (%)
Bamboo	3.3	2250	3000	91
Bamboo biochar	–	–	150	95
Banana peel	40.7	8400	1000	88.7
Banana peel	–	–	5000	~100
Banana peel waste	0.75	31	–	–
Bradyrhizobium japonicum with soybean leaf separator	–	–	8000	91
Celtuce leaves	–	–	2000	92.6
Coconut shells	38.5	–	>3000	93
Coconut shells	69	–	2000	85
Coconut shell	–	–	3000	97.2
Coffee beans	10–20	6000	>10,000	–
Coffee bean	15	75	10,000	82
Coffee ground	34	21500	–	–
Corncob residue	5.3–15	2276 - 2827	100,000	82
Cotton (natural)	–	–	20,000	97
Dead neem leaves (*Azadirachta indica*)	55	569	–	–
Eucalyptus tree leaves	–	–	15,000	97.7
Fibres from oil palm empty fruit bunches	4.297	173	–	–
Garlic peel			100	95–98
Garlic skin	14.65	310.67	5000	94
Gelatin	7.43	263.5	5000	92
Human hair	29	2243	>20,000	~100
Indian cake rusk	47.1	22,644	6000	95
Lemon peel	6.61	425.26	3000	92
Lignocellulosic waste, fruit stones	13	3410	20,000	99
Oil palm kernel shell	–	–	1000	95–97
Orange peel	23.3	2334.3	–	–
Paulownia flower (PF)	44.5~22.2	247~3781	1000	93
Pea skin	19.6	254,000	5000	75
Peanut shell and rice husk	19.3	1007	–	–
Pistachio nutshell	–	–	4000	~100
Pistachio nutshell	10–39	52000–286000	–	–
Potato starch	–	–	900	86
Rape flower stems	–	–	1000	96
Raw rice brans	70	1223	10,000	~97
Rice husk	–	–	10,000	97–99
Rice husk	5.11	–	10,000	90
Sago bark	5	400	1700	94
Shells of broad beans	–	–	3000	90
Soybean residue	12	2000	5000–10,000	90
Soybean Root	100.5	63,000	10,000	98
Spent coffee grounds	–	–	~2000	98
Sugarcane bagasse	5	35,000	1000	90

(Continued)

TABLE 10.2 (Continued)

Summary of Main Performance Parameters for Biowaste-Based Activated Carbon for Electrochemical Capacitors (Mensah-Darkwa et al. 2019)

Biowaste	Energy Density (Wh/kg)	Power Density (W/kg)	Cycles	Percentage Retention (%)
Sugar cane bagasse	5.9	10,000	5000	83
Sugar industry spent wash waste	–	41,4000	1000	~100
Sunflower seed shell	4.8	24,000	–	–
Waste tea leaves	–	–	2000	92
Wood sawdust	5.7–7.8	250–5000	10,000	94.2

the field of energy conversion and storage technology. In this chapter, various biowaste-based activated carbons for energy conversion and storage devices were presented, and the different performance metrics and their storage mechanisms were discussed.

Specifically, specific capacitance, energy/power densities, and cyclic stability were presented along with their requirements for an application. The development of an alternative biowaste-based energy conversion and storage electrode material assists in (1) disposal of waste; conversion of waste to useful products and (2) creates cheaper route for sustainable energy conversion and storage devices technology.

Due to the cheap and ready availability of the precursor materials in the form of biomass, as well as relatively low-energy HTC technique, it is most likely that bio-based electrode materials will in the long run replace their fossil predecessors.

The key to the advancement of electrochemical energy conversion and storage devices is materials design. The envisaged breakthroughs rely on the development of novel electrode materials with regulated morphologies, improved properties, and optimized functionalities.

It was identified that porous carbons having great surface areas, porosity, and polished surface chemistry are important for enhancing electrochemical features more. Different carbon materials, like activated carbons, porous carbons, carbon fibres, and nanostructured carbons (graphene, carbon nano-tubes (CNTs), and fullerene), have been extensively employed in fabrication of electrochemical energy storage devices. Unfortunately, most of these carbon materials are often derived from non-renewable resources under harsh environments and at high cost.

In the context of non-renewability, high cost, and unfriendly environment for production of most traditional carbon materials, naturally abundant biomass resources with microstructures have peculiar potential to change traditional energy storage design concepts, leading to great avenues for fabricating novel next-generation electrochemical energy conversion and storage devices.

The features of biomass-derived carbon materials like specific surface area, pore size distribution, porosity, morphology, and uniform surface chemistry have been tuned and directed by various activation processes using various activation agents. The biomass starting materials often have improved porous architecture from the pristine traditional materials. One of the best examples is the activated carbon fibres from cotton. For instance, the activated cotton fibres are versatile and can be a flexible backbone for fabricating wearable energy storage devices. Biomass-based carbons are successful in enhancing electrochemical capacitors' energy density and in stopping the dissolution of reaction intermediates in lithium-sulphur batteries. Obviously, employing biomasses is the right direction towards making renewable carbon nanomaterials for next-generation energy storage devices. Future researches ought to be directed to fundamental examination and devices design and optimization with aim to attain high degree of sustainability. This demands the cross-cutting collaboration among materials scientists, environmental scientists, chemists, physicists, economists, and social scientists. It is our belief that this chapter provides inspiration to the development of new renewable carbon materials for next-generation energy storage devices.

Acknowledgements

The financial assistance of the African Centre of Excellence in Future Energies and Electrochemical Systems (ACE-FUELS), Federal University of Technology, Owerri, Nigeria, and the World Bank towards this research is hereby acknowledged. Opinions expressed and conclusions arrived at are those of the authors and are not necessarily to be attributed to the African Centre of Excellence in Future Energies and Electrochemical Systems (ACE-FUELS), Federal University of Technology, Owerri, Nigeria, and the World Bank.

REFERENCES

Divyashree, A. and Gurumurthy Hegde. 2015. "Activated Carbon Nanospheres Derived from Bio-Waste Materials for Supercapacitor Applications – a Review." *RSC Advances* 5 (107): 88339–88352. 10.1039/C5RA19392C.

Abioye, Adekunle Moshood, and Farid Nasir Ani. 2015. "Recent Development in the Production of Activated Carbon Electrodes from Agricultural Waste Biomass for Supercapacitors: A Review." *Renewable and Sustainable Energy Reviews* 52 (C): 1282–1293.

Azwar, Elfina, Wan Adibah Wan Mahari, Joon Huang Chuah, Dai-Viet N. Vo, Nyuk Ling Ma, Wei Haur Lam, and Su Shiung Lam. 2018. "Transformation of Biomass into Carbon Nanofiber for Supercapacitor Application – a Review." *International Journal of Hydrogen Energy* 43 (45): 20811–20821. 10.1016/j.ijhydene.2018.09.111.

Babel, K., and K. Jurewicz. 2004. "KOH Activated Carbon Fabrics as Supercapacitor Material." *Journal of Physics and Chemistry of Solids* 65:275–280. Pergamon. 10.1016/j.jpcs.2003.08.023.

Balat, Mustafa, and Günhan Ayar. 2005. "Biomass Energy in the World, Use of Biomass and Potential Trends." *Energy Sources* 27 (10): 931–940. 10.1080/00908310490449045.

Bao, Xiaoguang, Xiaowa Nie, Dieter Von Deak, Elizabeth J. Biddinger, Wenjia Luo, Aravind Asthagiri, Umit S. Ozkan, and Christopher M. Hadad. 2013. "A First-Principles Study of the Role of Quaternary-N Doping on the Oxygen Reduction Reaction Activity and Selectivity of Graphene Edge Sites." *Topics in Catalysis* 56: 1623–1633. Springer. https://doi.org/10.1007/s11244-013-0097-z.

Başakçılardan Kabakcı, Sibel, and Sümeyra Seniha Baran. 2019. "Hydrothermal Carbonization of Various Lignocellulosics: Fuel Characteristics of Hydrochars and Surface Characteristics of Activated Hydrochars." *Waste Management* 100: 259–268. 10.1016/j.wasman.2019.09.021.

Benedetti, Vittoria, Francesco Patuzzi, and Marco Baratieri. 2018. "Characterization of Char from Biomass Gasification and Its Similarities with Activated Carbon in Adsorption Applications." *Applied Energy*, Transformative Innovations for a Sustainable Future – Part III, 227 (October): 92–99. 10.1016/j.apenergy.2017.08.076.

Bhadra, Sambhu, Dipak Khastgir, Nikhil K. Singha, and Joong Hee Lee. 2009. "Progress in Preparation, Processing and Applications of Polyaniline." *Progress in Polymer Science (Oxford)*. 10.1016/j.progpolymsci.2009.04.003.

Bhat, Vinay S., Supriya S., and Gurumurthy Hegde. 2020. "Review – Biomass Derived Carbon Materials for Electrochemical Sensors." *Journal of The Electrochemical Society* 167 (3): 037526. 10.1149/2.0262003jes.

Bi, Zhihong, Qingqiang Kong, Yufang Cao, Guohua Sun, Fangyuan Su, Xianxian Wei, Xiaoming Li, Aziz Ahmad, Lijing Xie, and Cheng Meng Chen. 2019. "Biomass-Derived Porous Carbon Materials with Different Dimensions for Supercapacitor Electrodes: A Review." *Journal of Materials Chemistry A* 7 (27): 16028–16045. 10.1039/c9ta04436a.

Brunner, G. 2009. "Near Critical and Supercritical Water. Part I. Hydrolytic and Hydrothermal Processes." *Journal of Supercritical Fluids*. 10.1016/j.supflu.2008.09.002.

Chen, Mingde, Xueya Kang, Tuerdi Wumaier, Junqing Dou, Bo Gao, Ying Han, Guoqing Xu, Zhiqiang Liu, and Lu Zhang. 2013. "Preparation of Activated Carbon from Cotton Stalk and Its Application in Supercapacitor." *Journal of Solid State Electrochemistry* 17 (4): 1005–1012. 10.1007/s10008-012-1946-6.

Chen, Zhao-Hui, Xue-Li Du, Jian-Bo He, Fang Li, Yan Wang, Yu-Lin Li, Bing Li, and Sen Xin. 2017. "Porous Coconut Shell Carbon Offering High Retention and Deep Lithiation of Sulfur for

Lithium–Sulfur Batteries." *ACS Applied Materials & Interfaces* 9 (39): 33855–33862. 10.1021/acsami.7b09310.

Cheng, Chunfeng, Shuijian He, Chunmei Zhang, Cheng Du, and Wei Chen. 2018. "High-Performance Supercapacitor Fabricated from 3D Free-Standing Hierarchical Carbon Foam-Supported Two Dimensional Porous Thin Carbon Nanosheets." *Electrochimica Acta* 290 (November): 98–108. 10.1016/j.electacta.2018.08.081.

Cheng, Youmin, Shaomin Ji, Xijun Xu, and Jun Liu. 2015. "Wheat Straw Carbon Matrix Wrapped Sulfur Composites as a Superior Cathode for Li–S Batteries." *RSC Advances* 5 (121): 100089–100096. 10.103 9/C5RA21416E.

Conder, Joanna, Krzysztof Fic, and Camelia Matei Ghimbeu. 2020. Supercapacitors (electrochemical capacitors). Mejdi Jeguirim; Lionel Limousy. Char and Carbon Materials Derived from Biomass. Production, Characterization and Applications, Elsevier, pp. 383-427.

Contescu, Cristian, Shiba Adhikari, Nidia Gallego, Neal Evans, and Bryan Biss. 2018. "Activated Carbons Derived from High-Temperature Pyrolysis of Lignocellulosic Biomass." *C* 4 (3): 51. 10.3390/c4030051.

Conway, B.E. 1999. "Introduction and Historical Perspective." In *Electrochemical Supercapacitors*, 1–9. Boston, MA: Springer US. 10.1007/978-1-4757-3058-6_1.

Cui, Xianjin, Markus Antonietti, and Shu Hong Yu. 2006. "Structural Effects of Iron Oxide Nanoparticles and Iron Ions on the Hydrothermal Carbonization of Starch and Rice Carbohydrates." *Small* 2 (6): 756–759. 10.1002/smll.200600047.

Dai, Hongjie. 2002. "Carbon Nanotubes: Synthesis, Integration, and Properties." *Accounts of Chemical Research* 35 (12): 1035–1044. 10.1021/ar0101640.

Dai, Liming, Yuhua Xue, Liangti Qu, Hyun Jung Choi, and Jong Beom Baek. 2015. "Metal-Free Catalysts for Oxygen Reduction Reaction." *Chemical Reviews*. American Chemical Society. 10.1021/cr5003563.

Deng, Jiang, Mingming Li, and Yong Wang. 2016. "Biomass-Derived Carbon: Synthesis and Applications in Energy Storage and Conversion." *Green Chemistry* 18: 4824–4854. 10.1039/c6gc01172a.

Deng, Jinxing, Tingmei Wang, Jinshan Guo, and Peng Liu. 2017. "Electrochemical Capacity Fading of Polyaniline Electrode in Supercapacitor: An XPS Analysis." *Progress in Natural Science: Materials International* 27 (2): 257–260. 10.1016/j.pnsc.2017.02.007.

Dhepe, Paresh L., and Atsushi Fukuoka. 2008. "Cellulose Conversion under Heterogeneous Catalysis." *ChemSusChem*. 10.1002/cssc.200800129.

Dutta, Saikat, Asim Bhaumik, and Kevin C.W. Wu. 2014. "Hierarchically Porous Carbon Derived from Polymers and Biomass: Effect of Interconnected Pores on Energy Applications." *Energy and Environmental Science*. Royal Society of Chemistry. 10.1039/c4ee01075b.

Elder, S.H., Linda H. Doerrer, F.J. DiSalvo, J.B. Parise, D. Guyomard, and J.M. Tarascon. 1992. "LiMoN2: The First Metallic Layered Nitride." *Chemistry of Materials* 4 (4): 928–937. 10.1021/cm00022a033.

El-Hendawy, Abdel Nasser A., S.E. Samra, and B.S. Girgis. 2001. "Adsorption Characteristics of Activated Carbons Obtained from Corncobs." *Colloids and Surfaces A: Physicochemical and Engineering Aspects* 180 (3): 209–221. 10.1016/S0927-7757(00)00682-8.

Elmouwahidi, Abdelhakim, Esther Bailón-García, Luis A. Romero-Cano, Ana I. Zárate-Guzmán, Agustín F. Pérez-Cadenas, and Francisco Carrasco-Marín. 2019. "Influence of Surface Chemistry on the Electrochemical Performance of Biomass-Derived Carbon Electrodes for Its Use as Supercapacitors." *Materials* 12 (15). 10.3390/ma12152458.

Enock, Talam Kibona, Cecil K. King'ondu, Alexander Pogrebnoi, and Yusufu Abeid Chande Jande. 2017. "Status of Biomass Derived Carbon Materials for Supercapacitor Application." *International Journal of Electrochemistry* 2017: 1–14. 10.1155/2017/6453420.

Fang, Yajun, Hongjuan Wang, Hao Yu, and Feng Peng. 2016. "From Chicken Feather to Nitrogen and Sulfur Co-Doped Large Surface Bio-Carbon Flocs: An Efficient Electrocatalyst for Oxygen Reduction Reaction." *Electrochimica Acta* 213 (September): 273–282. 10.1016/j.electacta.2016.07.121.

Fiori, Luca, Daniele Basso, Daniele Castello, and Marco Baratieri. 2014. "Hydrothermal Carbonization of Biomass: Design of a Batch Reactor and Preliminary Experimental Results." *Chemical Engineering Transactions* 37. 10.3303/CET1437010.

Frackowiak, Elzbieta, and François Béguin. 2002. "Electrochemical Storage of Energy in Carbon Nanotubes and Nanostructured Carbons." *Carbon* 40 (10): 1775–1787. 10.1016/S0008-6223(02)00045-3.

Funke, Axel, and Felix Ziegler. 2010. "Hydrothermal Carbonization of Biomass: A Summary and Discussion of Chemical Mechanisms for Process Engineering." *Biofuels, Bioproducts and Biorefining*. John Wiley & Sons, Ltd. 10.1002/bbb.198.

Gao, Zan, Yunya Zhang, Ningning Song, and Xiaodong Li. 2017a. "Biomass-Derived Renewable Carbon Materials for Electrochemical Energy Storage." *Materials Research Letters* 5 (2): 69–88. 10.1080/21 663831.2016.1250834.

Gao, Zan, Yunya Zhang, Ningning Song, and Xiaodong Li. 2017b. "Biomass-Derived Renewable Carbon Materials for Electrochemical Energy Storage." *Materials Research Letters* 5 (2): 69–88. 10.1080/ 21663831.2016.1250834.

Ghosh, Sourav, Ravichandran Santhosh, Sofia Jeniffer, Vimala Raghavan, George Jacob, Katchala Nanaji, Pratap Kollu, Soon Kwan Jeong, and Andrews Nirmala Grace. 2019. "Natural Biomass Derived Hard Carbon and Activated Carbons as Electrochemical Supercapacitor Electrodes." *Scientific Reports* 9 (1): 1–15. 10.1038/s41598-019-52006-x.

Gilbert, M.T., J.H. Knox, and B. Kaur. 1982. "Porous Glassy Carbon, a New Columns Packing Material for Gas Chromatography and High-Performance Liquid Chromatography." *Chromatographia* 16 (1): 138–146. 10.1007/BF02258884.

Gírio, F.M., C. Fonseca, F. Carvalheiro, L.C. Duarte, S. Marques, and R. Bogel-Łukasik. 2010. "Hemicelluloses for Fuel Ethanol: A Review." *Bioresource Technology*. https://doi.org/10.1016/ j.biortech.2010.01.088.

Gong, Chengcheng, Xinzhu Wang, Danhua Ma, Huifeng Chen, Shanshan Zhang, and Zhixin Liao. 2016. "Microporous Carbon from a Biological Waste-Stiff Silkworm for Capacitive Energy Storage." *Electrochimica Acta* 220 (December): 331–339. 10.1016/j.electacta.2016.10.120.

Gong, Kuanping, Feng Du, Zhenhai Xia, Michael Durstock, and Liming Dai. 2009. "Nitrogen-Doped Carbon Nanotube Arrays with High Electrocatalytic Activity for Oxygen Reduction." *Science* 323 (5915): 760–764. 10.1126/science.1168049.

Gu, Xingxing, Yazhou Wang, Chao Lai, Jingxia Qiu, Sheng Li, Yanglong Hou, Wayde Martens, Nasir Mahmood, and Shanqing Zhang. 2015. "Microporous Bamboo Biochar for Lithium-Sulfur Batteries." *Nano Research* 8 (1): 129–139. 10.1007/s12274-014-0601-1.

Guardia, Laura, Loreto Suárez, Nausika Querejeta, Covadonga Pevida, and Teresa A. Centeno. 2018. "Winery Wastes as Precursors of Sustainable Porous Carbons for Environmental Applications." *Journal of Cleaner Production* 193 (August): 614–624. 10.1016/j.jclepro.2018.05.085.

Guo, Donghui, Riku Shibuya, Chisato Akiba, Shunsuke Saji, Takahiro Kondo, and Junji Nakamura. 2016. "Active Sites of Nitrogen-Doped Carbon Materials for Oxygen Reduction Reaction Clarified Using Model Catalysts." *Science* 351 (6271): 361–365. 10.1126/science.aad0832.

Hao, Yajuan, Xu Zhang, Qifeng Yang, Kai Chen, Jun Guo, Dongying Zhou, Lai Feng, and Zdeněk Slanina. 2018. "Highly Porous Defective Carbons Derived from Seaweed Biomass as Efficient Electrocatalysts for Oxygen Reduction in Both Alkaline and Acidic Media." *Carbon* 137: 93–103. 10.1016/j.carbon. 2018.05.007.

Heidari, Mohammad, Animesh Dutta, Bishnu Acharya, and Shohel Mahmud. 2019. "A Review of the Current Knowledge and Challenges of Hydrothermal Carbonization for Biomass Conversion." *Journal of the Energy Institute* 92 (6): 1779–1799. 10.1016/j.joei.2018.12.003.

Hill, Josephine M. 2017. "Sustainable and/or Waste Sources for Catalysts: Porous Carbon Development and Gasification." *Catalysis Today* 285 (May): 204–210. 10.1016/j.cattod.2016.12.033.

Hirsch, Andreas. 2010. "The Era of Carbon Allotropes." *Nature Materials* 9: 868–871.

Ho, M. Y., Khiew, P. S., Isa, D.,Tan, T. K., Chiu, W. S., & Chia, C. H. (2014). A review of metal oxide composite electrode materials for electrochemical capacitors. In Nano (Vol. 9, Issue 6). World Scientific Publishing Co. Pte Ltd. https://doi.org/10.1142/S1793292014300023

Hu, Bo, Kan Wang, Liheng Wu, Shu Hong Yu, Markus Antonietti, and Maria Magdalena Titirici. 2010. "Engineering Carbon Materials from the Hydrothermal Carbonization Process of Biomass." *Advanced Materials* 22 (7): 813–828. 10.1002/adma.200902812.

Hu, Sixiao, Sanliang Zhang, Ning Pan, and You-Lo Hsieh. 2014. "Publication Date High Energy Density Supercapacitors from Lignin Derived Submicron Activated Carbon Fibers in Aqueous Electrolytes." *Journal Journal of Power Sources* 270. 10.1016/j.jpowsour.2014.07.063.

Huggins, Tyler, Albert Latorre, Justin Biffinger, and Zhiyong Ren. 2016. "Biochar Based Microbial Fuel Cell for Enhanced Wastewater Treatment and Nutrient Recovery." *Sustainability* 8 (2): 169. 10.3390/su8020169.

Hulicova-Jurcakova, Denisa, Mykola Seredych, Gao Qing Lu, and Teresa J. Bandosz. 2009. "Combined Effect of Nitrogen- and Oxygen-Containing Functional Groups of Microporous Activated Carbon on Its Electrochemical Performance in Supercapacitors." *Advanced Functional Materials* 19 (3): 438–447. 10.1002/adfm.200801236.

Hummers, William S., and Richard E. Offeman. 1958. "Preparation of Graphitic Oxide." *Journal of the American Chemical Society* 80 (6): 1339. 10.1021/ja01539a017.

Ike, Innocent S., and Sunny Iyuke. 2016. "Mathematical Modelling and Simulation of Supercapacitors." In *Nanomaterials in Advanced Batteries and Supercapacitors*, edited byKenneth I. Ozoemena and Shaowei Chen, 515–562. Nanostructure Science and Technology. Springer International Publishing. 10.1007/978-3-319-26082-2_15.

Ike, Innocent S., Iakovos Sigalas, Sunny Iyuke, and Kenneth I. Ozoemena. 2015. "An Overview of Mathematical Modeling of Electrochemical Supercapacitors/Ultracapacitors." *Journal of Power Sources* 273 (January): 264–277. 10.1016/j.jpowsour.2014.09.071.

Ike, Innocent S., Iakovos J. Sigalas, Sunny E. Iyuke, Iakovos J. Sigalas, and Egwu E. Kalu. 2019a. "The Contributions of Electrolytes in Achieving the Performance Index of Next-Generation Electrochemical Capacitors (ECs)." In *Electrochemical Devices for Energy Storage Applications*. CRC Press. December 11, 2019. 10.1201/9780367855116-11.

Ike, Innocent S., Iakovos J. Sigalas, Sunny E. Iyuke, Iakovos J. Sigalas, and Egwu E. Kalu. 2019b. "The Role of Modelling and Simulation in the Achievement of Next-Generation Electrochemical Capacitors." *Electrochemical Devices for Energy Storage Applications*. CRC Press. December 11, 2019. 10.1201/9780367855116-8.

Januszewicz, Katarzyna, Paweł Kazimierski, Maciej Klein, Dariusz Kardaś, and Justyna Łuczak. 2020. "Activated Carbon Produced by Pyrolysis of Waste Wood and Straw for Potential Wastewater Adsorption." *Materials* 13 (9): 2047. 10.3390/MA13092047.

Jia, Yi, Longzhou Zhang, Aijun Du, Guoping Gao, Jun Chen, Xuecheng Yan, Christopher L. Brown, and Xiangdong Yao. 2016. "Defect Graphene as a Trifunctional Catalyst for Electrochemical Reactions." *Advanced Materials* 28 (43): 9532–9538. 10.1002/adma.201602912.

Jiang, Lili, Lizhi Sheng, and Zhuangjun Fan. 2018. "Biomass-Derived Carbon Materials with Structural Diversities and Their Applications in Energy Storage." *Science China Materials* 61 (2): 133–158. 10.1007/s40843-017-9169-4.

Jin, Huile, Huihui Huang, Yuhua He, Xin Feng, Shun Wang, Liming Dai, and Jichang Wang. 2015. "Graphene Quantum Dots Supported by Graphene Nanoribbons with Ultrahigh Electrocatalytic Performance for Oxygen Reduction." *Journal of the American Chemical Society* 137 (24): 7588–7591. 10.1021/jacs.5b03799.

Kabyemela, Bernard M., Tadafumi Adschiri, Roberto M. Malaluan, and Kunio Arai. 1999. "Glucose and Fructose Decomposition in Subcritical and Supercritical Water: Detailed Reaction Pathway, Mechanisms, and Kinetics." *Industrial and Engineering Chemistry Research* 38 (8): 2888–2895. 10.1021/ie9806390.

Kalyani, P., and A. Anitha. 2013. "Biomass Carbon & Its Prospects in Electrochemical Energy Systems." *International Journal of Hydrogen Energy* 38 (10): 4034–4045. 10.1016/j.ijhydene.2013.01.048.

Karnan, M., K. Subramani, N. Sudhan, N. Ilayaraja, and M. Sathish. 2016. "Aloe Vera Derived Activated High-Surface-Area Carbon for Flexible and High-Energy Supercapacitors." *ACS Applied Materials and Interfaces* 8 (51): 35191–35202. 10.1021/acsami.6b10704.

Kaur, Prabhsharan, Gaurav Verma, and S.S. Sekhon. 2019. "Biomass Derived Hierarchical Porous Carbon Materials as Oxygen Reduction Reaction Electrocatalysts in Fuel Cells." *Progress in Materials Science* 102: 1–71. 10.1016/j.pmatsci.2018.12.002.

Kim, Yong Jung, Byoung Ju Lee, Hiroaki Suezaki, Teruaki Chino, Yusuke Abe, Takashi Yanagiura, Ki Chul Park, and Morinobu Endo. 2006. "Preparation and Characterization of Bamboo-Based Activated Carbons as Electrode Materials for Electric Double Layer Capacitors." *Carbon* 44 (8): 1592–1595. 10.1016/J.CARBON.2006.02.011.

Kubo, Shiori, Rezan Demir-Cakan, Li Zhao, Robin J. White, and Maria Magdalena Titirici. 2010. "Porous Carbohydrate-Based Materials via Hard Templating." *ChemSusChem* 3 (2): 188–194. 10.1002/cssc.200900126.

Kubo, Shiori, Robin J. White, Klaus Tauer, and Maria Magdalena Titirici. 2013. "Flexible Coral-like Carbon Nanoarchitectures via a Dual Block Copolymer-Latex Templating Approach." *Chemistry of Materials* 25 (23): 4781–4790. 10.1021/cm4029676.

Kubo, Shiori, Robin J. White, Noriko Yoshizawa, Markus Antonietti, and Maria Magdalena Titirici. 2011. "Ordered Carbohydrate-Derived Porous Carbons." *Chemistry of Materials* 23 (22): 4882–4885. 10. 1021/cm2020077.

Kumar, Sandeep. 2013. "Sub- and Supercritical Water Technology for Biofuels." In *Advanced Biofuels and Bioproducts*:147–183. Springer New York. 10.1007/978-1-4614-3348-4_11.

Ladanai, Svetlana, and Johan Vinterbäck. 2009. "Global Potential of Sustainable Biomass for Energy SLU." *Institutionen för energi och teknik*, Swedish University of Agricultural Sciences, Department of Energy and Technology, Report 013, ISSN 1654-9406, Uppsala 2009.

Li, Jie, Furong Qin, Liyuan Zhang, Kai Zhang, Qiang Li, Yanqing Lai, Zhian Zhang, and Jing Fang. 2014. "Mesoporous Carbon from Biomass: One-Pot Synthesis and Application for Li–S Batteries." *Journal of Materials Chemistry A* 2 (34): 13916–13922. 10.1039/C4TA02154A.

Li, Q.Y., Wang, H.Q., Dai, Q.F., Yang, J.H., Zhong, Y.L. 2008. "Novel Activated Carbons as Electrode Materials for Electrochemical Capacitors from a Series of Starch." *Solid State Ionics* 179 (7): 269–273.

Li, Ruizi, Yanping Zhou, Wenbin Li, Jixin Zhu, and Wei Huang. 2020. "Structure Engineering in Biomass-Derived Carbon Materials for Electrochemical Energy Storage." *Research* 2020: 1–27. /10.34133/2020/8685436.

Li, Wenbin, Jianfeng Huang, Liangliang Feng, Liyun Cao, Yijie Ren, Ruizi Li, Zhanwei Xu, Jiayin Li, and Chunyan Yao. 2017. "Controlled Synthesis of Macroscopic Three-Dimensional Hollow Reticulate Hard Carbon as Long-Life Anode Materials for Na-Ion Batteries." *Journal of Alloys and Compounds* 716: 210–219. 10.1016/j.jallcom.2017.05.062.

Li, Zhenghui, Dingcai Wu, Yeru Liang, Ruowen Fu, and Krzysztof Matyjaszewski. 2014. "Synthesis of Well-Defined Microporous Carbons by Molecular-Scale Templating with Polyhedral Oligomeric Silsesquioxane Moieties." *Journal of the American Chemical Society* 136 (13): 4805–4808. 10.1021/ja412192v.

Li, Zhihua, and Liangjun Gong. 2020. "Research Progress on Applications of Polyaniline (PANI) for Electrochemical Energy Storage and Conversion." *Materials* 1313 (3): 548. 10.3390/MA13030548.

Li, Zuojiang, and Mietek Jaroniec. 2003. "Synthesis and Adsorption Properties of Colloid-Imprinted Carbons with Surface and Volume Mesoporosity." *Chemistry of Materials* 15 (6): 1327–1333. 10.1021/cm020617a.

Liu, Bo, Hiroshi Shioyama, Tomoki Akita, and Qiang Xu. 2008. "Metal-Organic Framework as a Template for Porous Carbon Synthesis." *Journal of the American Chemical Society* 130 (16): 5390–5391. 10. 1021/ja7106146.

Liu, M., Y. Chen, K. Chen, N. Zhang, X. Zhao, F. Zhao, Z. Dou, X. He, and L. Wang. 2015. "Biomass-Derived Activated Carbon for Rechargeable Lithium-Sulfur Batteries: BioResources." 2015. https://bioresources.cnr.ncsu.edu/.

Liu, Wu Jun, Hong Jiang, and Han Qing Yu. 2019. "Emerging Applications of Biochar-Based Materials for Energy Storage and Conversion." *Energy and Environmental Science* 12 (6): 1751–1779. 10.1039/c9ee00206e.

Liu, Wu Jun, Ke Tian, Yan Rong He, Hong Jiang, and Han Qing Yu. 2014. "High-Yield Harvest of Nanofibers/Mesoporous Carbon Composite by Pyrolysis of Waste Biomass and Its Application for High Durability Electrochemical Energy Storage." *Environmental Science and Technology* 48 (23): 13951–13959. 10.1021/es504184c.

Liu, Yang, Jiareng Chen, Bin Cui, Pengfei Yin, and Chao Zhang. 2018. "Design and Preparation of Biomass-Derived Carbon Materials for Supercapacitors: A Review." *C* 4 (4): 53. 10.3390/c4040053.

Liu, Yuxi, Zechuan Xiao, Yongchang Liu, and Li Zhen Fan. 2017. "Biowaste-Derived 3D Honeycomb-like Porous Carbon with Binary-Heteroatom Doping for High-Performance Flexible Solid-State Supercapacitors." *Journal of Materials Chemistry A* 6 (1): 160–166. 10.1039/c7ta09055b.

Liu, Zaichun, Xinhai Yuan, Shuaishuai Zhang, Jing Wang, Qinghong Huang, Nengfei Yu, Yusong Zhu, et al. 2019. "Three-Dimensional Ordered Porous Electrode Materials for Electrochemical Energy Storage." *NPG Asia Materials* 11 (1). 10.1038/s41427-019-0112-3.

Lu, Hao, and X.S. Zhao. 2017. "Biomass-Derived Carbon Electrode Materials for Supercapacitors." *Sustainable Energy and Fuels* 1 (6): 1265–1281. 10.1039/C7SE00099E.

Lua, Aik Chong, Fong Yow Lau, and Jia Guo. 2006. "Influence of Pyrolysis Conditions on Pore Development of Oil-Palm-Shell Activated Carbons." *Journal of Analytical and Applied Pyrolysis* 76 (1–2): 96–102. 10.1016/j.jaap.2005.08.001.

Maharjan, Makhan, Arjun Bhattarai, Mani Ulaganathan, Nyunt Wai, Moe Ohnmar Oo, Jing-Yuan Wang, and Tuti Mariana Lim. 2017. "High Surface Area Bio-Waste Based Carbon as a Superior Electrode for Vanadium Redox Flow Battery." *Journal of Power Sources* 362 (September): 50–56. 10.1016/j.jpowsour.2017.07.020.

Marriott, A.S., A.J. Hunt, E. Bergström, K. Wilson, V.L. Budarin, J. Thomas-Oates, J.H. Clark, and R. Brydson. 2014. "Investigating the Structure of Biomass-Derived Non-Graphitizing Mesoporous Carbons by Electron Energy Loss Spectroscopy in the Transmission Electron Microscope and X-Ray Photoelectron Spectroscopy." *Carbon* 67: 514–524. 10.1016/j.carbon.2013.10.024.

Mattson, James. 1971. *Activated Carbon: Surface Chemistry and Adsorption from Solution*. New York: M. Dekker.

McKendry, Peter. 2002. "Energy Production from Biomass (Part 1): Overview of Biomass." *Bioresource Technology* 83 (1): 37–46. 10.1016/S0960-8524(01)00118-3.

Md Khudzari, Jauharah, Yvan Gariépy, Jiby Kurian, Boris Tartakovsky, and G. S.Vijaya Raghavan. 2019. "Effects of Biochar Anodes in Rice Plant Microbial Fuel Cells on the Production of Bioelectricity, Biomass, and Methane." *Biochemical Engineering Journal* 141 (January): 190–199. 10.1016/j.bej.2018. 10.012.

Mensah-Darkwa, Kwadwo, Camila Zequine, Pawan K. Kahol, and Ram K. Gupta. 2019. "Supercapacitor Energy Storage Device Using Biowastes: A Sustainable Approach to Green Energy." *Sustainability* 11 (2): 414. 10.3390/su11020414.

Menon, V., & Rao, M. (2012). Trends in bioconversion of lignocellulose: Biofuels, platform chemicals & biorefinery concept. In Progress in Energy and Combustion Science (Vol. 38, Issue 4, pp. 522–550). Elsevier Ltd. https://doi.org/10.1016/j.pecs.2012.02.002

Misnon, Izan Izwan, Nurul Khairiyyah Mohd Zain, Radhiyah Abd Aziz, Baiju Vidyadharan, and Rajan Jose. 2015. "Electrochemical Properties of Carbon from Oil Palm Kernel Shell for High Performance Supercapacitors." *Electrochimica Acta* 174 (August): 78–86. 10.1016/j.electacta.2015.05.163.

Moreno, Noelia, Alvaro Caballero, Lourdes Hernán, and Julián Morales. 2014. "Lithium–Sulfur Batteries with Activated Carbons Derived from Olive Stones." *Carbon* 70 (April): 241–248. 10.1016/j.carbon. 2014.01.002.

Na, Ruiqi, Xinyu Wang, Nan Lu, Guanze Huo, Haibo Lin, and Guibin Wang. 2018. "Novel Egg White Gel Polymer Electrolyte and a Green Solid-State Supercapacitor Derived from the Egg and Rice Waste." *Electrochimica Acta* 274 (June): 316–325. 10.1016/j.electacta.2018.04.127.

Nakanishi, Kazuki, Ryoji Takahashi, and Naohiro Soga. 1992. "Dual-Porosity Silica Gels by Polymer-Incorporated Sol-Gel Process." *Journal of Non-Crystalline Solids* 147–148 (C): 291–295. 10.1016/ S0022-3093(05)80632-5.

Njoku, Chioma E., Joseph A. Omotoyinbo, Kenneth K. Alaneme, and Michael O. Daramola. 2020. "Physical and Abrasive Wear Behaviour of Urena Lobata Fiber-Reinforced Polymer Composites." *Journal of Reinforced Plastics and Composites*. 10.1177/0731684420960210.

Pandey, Garima. 2019. "Biomass Based Bio-Electro Fuel Cells Based on Carbon Electrodes: An Alternative Source of Renewable Energy." *SN Applied Sciences* 1 (5): 1–10. 10.1007/s42452-019-0409-4.

Pang, Jie, Wenfeng Zhang, Jinliang Zhang, Gaoping Cao, Minfang Han, and Yusheng Yang. 2017. "Facile and Sustainable Synthesis of Sodium Lignosulfonate Derived Hierarchical Porous Carbons for Supercapacitors with High Volumetric Energy Densities." *Green Chemistry* 19 (16): 3916–3926. 10.1039/c7gc01434a.

Pang, Quan, Juntao Tang, He Huang, Xiao Liang, Connor Hart, Kam C. Tam, and Linda F. Nazar. 2015. "A Nitrogen and Sulfur Dual-Doped Carbon Derived from Polyrhodanine@Cellulose for Advanced Lithium-Sulfur Batteries." *Advanced Materials* 27 (39): 6021–6028. 10.1002/adma. 201502467.

Paraknowitsch, Jens Peter, and Arne Thomas. 2013. "Doping Carbons beyond Nitrogen: An Overview of Advanced Heteroatom Doped Carbons with Boron, Sulphur and Phosphorus for Energy Applications." *Energy and Environmental Science*. The Royal Society of Chemistry. 10.1039/c3ee41444b.

Parveen, Nazish, A. Ibrahim Al-Jaafari, and Jeong In Han. 2019. "Robust Cyclic Stability and High-Rate Asymmetric Supercapacitor Based on Orange Peel-Derived Nitrogen-Doped Porous Carbon and

Intercrossed Interlinked Urchin-like NiCo2O4@3DNF Framework." *Electrochimica Acta* 293 (January): 84–96. 10.1016/j.electacta.2018.08.157.

Pollet, Bruno G, Iain Staffell, and Jin Lei Shang. 2012. "Electrochimica Acta Current Status of Hybrid, Battery and Fuel Cell Electric Vehicles: From Electrochemistry to Market Prospects." *Electrochimica Acta* 84: 235–249. 10.1016/j.electacta.2012.03.172.

Qian, Hai Sheng, Shu Hong Yu, Lin Bao Luo, Jun Yan Gong, Lin Feng Fei, and Xian Ming Liu. 2006. "Synthesis of Uniform Te@carbon-Rich Composite Nanocables with Photoluminescence Properties and Carbonaceous Nanofibers by the Hydrothermal Carbonization of Glucose." *Chemistry of Materials* 18 (8): 2102–2108. 10.1021/cm052848y.

Qin, Decai, Zhanying Liu, Yanzhang Zhao, Guiyin Xu, Fang Zhang, and Xiaogang Zhang. 2018. "A Sustainable Route from Corn Stalks to N, P-Dual Doping Carbon Sheets toward High Performance Sodium-Ion Batteries Anode." *Carbon* 130 (April): 664–671. 10.1016/j.carbon.2018.01.007.

Qin, Furong, Kai Zhang, Jing Fang, Yanqing Lai, Qiang Li, Zhian Zhang, and Jie Li. 2014. "High Performance Lithium Sulfur Batteries with a Cassava-Derived Carbon Sheet as a Polysulfides Inhibitor." *New Journal of Chemistry* 38 (9): 4549–4554. 10.1039/C4NJ00701H.

Qu, Yaohui, Zhian Zhang, Xiahui Zhang, Guodong Ren, Yanqing Lai, Yexiang Liu, and Jie Li. 2015. "Highly Ordered Nitrogen-Rich Mesoporous Carbon Derived from Biomass Waste for High-Performance Lithium–Sulfur Batteries." *Carbon* 84 (April): 399–408. 10.1016/j.carbon.2014.12.001.

Qua, J., Lv Siyuan, Peng Xiyue, Tian Shuo, Wang Jia, and Gao Feng. 2016. "Nitrogen-Doped Porous 'Green Carbon' Derived from Shrimp Shell: Combined Effects of Pore Sizes and Nitrogen Doping on the Performance of Lithium Sulfur Battery." *Journal of Alloys and Compounds* 671 (June): 17–23. 10.1016/j.jallcom.2016.02.064.

Reis, Glaydson Simões Dos, Sylvia H. Larsson, Helinando Pequeno de Oliveira, Mikael Thyrel, and Eder Claudio Lima. 2020. "Sustainable Biomass Activated Carbons as Electrodes for Battery and Supercapacitors – a Mini-Review." *Nanomaterials* 10 (7): 1–22. 10.3390/nano10071398.

Ren, Y., Xu, Q., Zhang, J., Yang, H., Wang, B., Yang, D., Hu, J., & Liu, Z. (2014). Functionalization of biomass carbonaceous aerogels: Selective preparation of MnO$_2$@CA composites for supercapacitors. *ACS Applied Materials and Interfaces*, 6 (12): 9689–9697. https://doi.org/10.1021/am502035g.

Reza, M. Toufiq, Janet Andert, Benjamin Wirth, Daniela Busch, Judith Pielert, Joan G. Lynam, and Jan Mumme. 2014. "Hydrothermal Carbonization of Biomass for Energy and Crop Production." *Applied Bioenergy* 1 (1): 11–29. 10.2478/apbi-2014-0001.

Ruiz, V., C. Blanco, R. Santamaría, J.M. Ramos-Fernández, M. Martínez-Escandell, A. Sepúlveda-Escribano, and F. Rodríguez-Reinoso. 2007. "An Activated Carbon Monolith as an Electrode Material for Supercapacitors." *Journal of Applied Electrochemistry* 37: 717–740.

Sahu, Vikrant, Sonia Grover, Brindan Tulachan, Meenakshi Sharma, Gaurav Srivastava, Manas Roy, Manav Saxena, et al. 2015. "Heavily Nitrogen Doped, Graphene Supercapacitor from Silk Cocoon." *Electrochimica Acta* 160 (April): 244–253. 10.1016/j.electacta.2015.02.019.

Saidur, R., E.A. Abdelaziz, A. Demirbas, M.S. Hossain, and S. Mekhilef. 2011. "A Review on Biomass as a Fuel for Boilers." *Renewable and Sustainable Energy Reviews*. Pergamon. 10.1016/j.rser.2011.02.015.

Sanchez-Sanchez, A., Maria Teresa Izquierdo, Jaafar Ghanbaja, Ghouti Medjahdi, Sandrine Mathieu, Alain Celzard, and Vanessa Fierro. 2017. "Excellent Electrochemical Performances of Nanocast Ordered Mesoporous Carbons Based on Tannin-Related Polyphenols as Supercapacitor Electrodes." *Journal of Power Sources* 344: 15–24. 10.1016/j.jpowsour.2017.01.099.

Sanchez-Sanchez, Angela, Maria Teresa Izquierdo, Ghouti Medjahdi, Jaafar Ghanbaja, Alain Celzard, and Vanessa Fierro. 2018. "Ordered Mesoporous Carbons Obtained by Soft-Templating of Tannin in Mild Conditions." *Microporous and Mesoporous Materials* 270 (February): 127–139. 10.1016/j.micromeso.2018.05.017.

Schuck, S. 2006. "Biomass as an Energy Source." *International Journal of Environmental Studies* 63 (6): 823–836. 10.1080/00207230601047222.

Sekhon, S.S., D.P. Kaur, J.S. Park, and K. Yamada. 2012. "Ion Transport Properties of Ionic Liquid Based Gel Electrolytes." *Electrochimica Acta* 60 (January): 366–374. 10.1016/j.electacta.2011.11.072.

Sevilla, M., and A.B. Fuertes. 2009. "The Production of Carbon Materials by Hydrothermal Carbonization of Cellulose." *Carbon* 47 (9): 2281–2289. 10.1016/j.carbon.2009.04.026.

Sevilla, Marta, and Robert Mokaya. 2014. "Energy Storage Applications of Activated Carbons: Supercapacitors and Hydrogen Storage." *Energy and Environmental Science* 7:1250–1280. Royal Society of Chemistry. 10.1039/c3ee43525c.

Sharma, Ronit, Karishma Jasrotia, Nicy Singh, Priyanka Ghosh, Shubhangi Srivastava, Neeta Raj Sharma, Joginder Singh, Ramesh Kanwar, and Ajay Kumar. 2020. "A Comprehensive Review on Hydrothermal Carbonization of Biomass and Its Applications." *Chemistry Africa* 3 (1): 1–19. 10.1007/s42250-019-00098-3.

Sheldon, Roger A. 2014. "Green and Sustainable Manufacture of Chemicals from Biomass: State of the Art." *Green Chemistry*. Royal Society of Chemistry. 10.1039/c3gc41935e.

Shen, Pei Kang, Zaoxue Yan, Hui Meng, Mingmei Wu, Guofeng Cui, Ruihong Wang, Lei Wang, Keying Si, and Honggang Fu. 2011. "Synthesis of Pd on Porous Hollow Carbon Spheres as an Electrocatalyst for Alcohol Electrooxidation." *RSC Advances* 1 (2): 191–198. 10.1039/c1ra00234a.

Shi, Kaiqi, Jiefeng Yan, Edward Lester, and Tao Wu. 2014. "Catalyst-Free Synthesis of Multiwalled Carbon Nanotubes via Microwave-Induced Processing of Biomass." *Industrial and Engineering Chemistry Research* 53 (39): 15012–15019. 10.1021/ie503076n.

Snook, Graeme A., Pon Kao, and Adam S. Best. 2011. "Conducting-Polymer-Based Supercapacitor Devices and Electrodes." *Journal of Power Sources*. Elsevier. 10.1016/j.jpowsour.2010.06.084.

Stadie, Nicholas P., Shutao Wang, Kostiantyn V. Kravchyk, and Maksym V. Kovalenko. 2017. "Zeolite-Templated Carbon as an Ordered Microporous Electrode for Aluminum Batteries." *ACS Nano* 11 (2): 1911–1919. 10.1021/acsnano.6b07995.

Sun, Fei, Kunfang Wang, Lijie Wang, Tong Pei, Jihui Gao, Guangbo Zhao, and Yunfeng Lu. 2019. "Hierarchical Porous Carbon Sheets with Compressed Framework and Optimized Pore Configuration for High-Rate and Long-Term Sodium and Lithium Ions Storage." *Carbon* 155 (December): 166–175. 10.1016/j.carbon.2019.08.051.

Sun, Li, Chungui Tian, Meitong Li, Xiangying Meng, Lei Wang, Ruihong Wang, Jie Yin, and Honggang Fu. 2013. "From Coconut Shell to Porous Graphene-like Nanosheets for High-Power Supercapacitors." *Journal of Materials Chemistry A* 1 (21): 6462–6470. 10.1039/c3ta10897j.

Sun, Yao, Jianjun Xue, Shengyang Dong, Yadi Zhang, Yufeng An, Bing Ding, Tengfei Zhang, Hui Dou, and Xiaogang Zhang. 2020. "Biomass-Derived Porous Carbon Electrodes for High-Performance Supercapacitors." *Journal of Materials Science* 55 (12): 5166–5176. 10.1007/s10853-019-04343-5.

Susanti, R. F., Arie, A. A., Kristianto, H. et al. (2019). Activated carbon from citric acid catalyzed hydrothermal carbonization and chemical activation of salacca peel as potential electrode for lithium ion capacitor's cathode. *Ionics*, 25, 3915–3925. https://doi.org/10.1007/s11581-019-02904-x

Sürmen, Yusuf. 2003. "The Necessity of Biomass Energy for the Turkish Economy." *Energy Sources* 25 (2): 83–92. 10.1080/00908310390142145.

Tarascon, Jean Marie, Nadir Recham, Michel Armand, Jean Noël Chotard, Prabeer Barpanda, Wesley Walker, and Loic Dupont. 2010. "Hunting for Better Li-Based Electrode Materials via Low Temperature Inorganic Synthesis." *Chemistry of Materials*. 10.1021/cm9030478.

Thomas, Paul, Chin Wei Lai, and Mohd Rafie Bin Johan. 2019. "Recent Developments in Biomass-Derived Carbon as a Potential Sustainable Material for Super-Capacitor-Based Energy Storage and Environmental Applications." *Journal of Analytical and Applied Pyrolysis*. Elsevier B.V. 10.1016/j.jaap.2019.03.021.

Titirici, Maria Magdalena, Markus Antonietti, and Niki Baccile. 2008. "Hydrothermal Carbon from Biomass: A Comparison of the Local Structure from Poly- to Monosaccharides and Pentoses/Hexoses." *Green Chemistry* 10 (11): 1204–1212. 10.1039/b807009a.

Titirici, Maria Magdalena, Axel Funke, and Andrea Kruse. 2015. "Hydrothermal Carbonization of Biomass." *Recent Advances in Thermochemical Conversion of Biomass*, 325–352. 10.1016/B978-0-444-63289-0.00012-0.

Titirici, Maria Magdalena, Arne Thomas, and Markus Antonietti. 2007. "Replication and Coating of Silica Templates by Hydrothermal Carbonization." *Advanced Functional Materials* 17 (6): 1010–1018. 10.1002/adfm.200600501.

Titirici, Maria Magdalena, Robin J. White, Nicolas Brun, Vitaliy L. Budarin, Dang Sheng Su, Francisco Del Monte, James H. Clark, and Mark J. MacLachlan. 2015. "Sustainable Carbon Materials." *Chemical Society Reviews*. Royal Society of Chemistry. 10.1039/c4cs00232f.

Titirici, Maria Magdalena, Robin J. White, Camillo Falco, and Marta Sevilla. 2012. "Black Perspectives for a Green Future: Hydrothermal Carbons for Environment Protection and Energy Storage." *Energy and Environmental Science* 5: 6796–6822. 10.1039/c2ee21166a.

Titirici, Maria-Magdalena, Robin J. White, Nicolas Brun, Vitaliy L. Budarin, Dang Sheng Su, Francisco del Monte, James H. Clark, and Mark J. MacLachlan. 2014. "Sustainable Carbon Materials." *Chemical Society Reviews* 44 (1): 250–290. 10.1039/C4CS00232F.

Vaalma, Christoph, Daniel Buchholz, Marcel Weil, and Stefano Passerini. 2018. "A Cost and Resource Analysis of Sodium-Ion Batteries." *Nature Reviews Materials*. Nature Publishing Group. 10.1038/natrevmats.2018.13.

Veerakumar, Pitchaimani, Thandavarayan Maiyalagan, Balasubramaniam Gnana Sundara Raj, Kuppuswamy Guruprasad, Zhongqing Jiang, and King-Chuen Lin. 2020. "Paper Flower-Derived Porous Carbons with High-Capacitance by Chemical and Physical Activation for Sustainable Applications." *Arabian Journal of Chemistry* 13 (1): 2995–3007. 10.1016/j.arabjc.2018.08.009.

Wang, Hongqiang, Zhixin Chen, Hua Kun Liu, and Zaiping Guo. 2014. "A Facile Synthesis Approach to Micro–Macroporous Carbon from Cotton and Its Application in the Lithium–Sulfur Battery." *RSC Advances* 4 (110): 65074–65080. /10.1039/C4RA12260G.

Wang, Huanhuan, Jianyi Lin, and Ze Xiang Shen. 2016. "Polyaniline (PANi) Based Electrode Materials for Energy Storage and Conversion." *Journal of Science: Advanced Materials and Devices* 1 (3): 225–255. 10.1016/j.jsamd.2016.08.001.

Wang, Jiacheng, and Stefan Kaskel. 2012. "KOH Activation of Carbon-Based Materials for Energy Storage." *Journal of Materials Chemistry*. The Royal Society of Chemistry. 10.1039/c2jm34066f.

Wang, Jie, Ping Nie, Bing Ding, Shengyang Dong, Xiaodong Hao, Hui Dou, and Xiaogang Zhang. 2017. "Biomass Derived Carbon for Energy Storage Devices." *Journal of Materials Chemistry A*. Royal Society of Chemistry. 10.1039/c6ta08742f.

Wang, Nan, Tuanfeng Li, Ye Song, Jingjun Liu, and Feng Wang. 2018. "Metal-Free Nitrogen-Doped Porous Carbons Derived from Pomelo Peel Treated by Hypersaline Environments for Oxygen Reduction Reaction." *Carbon* 130: 692–700. 10.1016/j.carbon.2018.01.068.

Wang, Tengfei, Yunbo Zhai, Yun Zhu, Caiting Li, and Guangming Zeng. 2018. "A Review of the Hydrothermal Carbonization of Biomass Waste for Hydrochar Formation: Process Conditions, Fundamentals, and Physicochemical Properties." *Renewable and Sustainable Energy Reviews*. 10.1016/j.rser.2018.03.071.

Wang, Yan Jie, Nana Zhao, Baizeng Fang, Hui Li, Xiaotao T. Bi, and Haijiang Wang. 2015. "Carbon-Supported Pt-Based Alloy Electrocatalysts for the Oxygen Reduction Reaction in Polymer Electrolyte Membrane Fuel Cells: Particle Size, Shape, and Composition Manipulation and Their Impact to Activity." *Chemical Reviews*. American Chemical Society. 10.1021/cr500519c.

Wang, Yulin, Qingli Qu, Shuting Gao, Guosheng Tang, Kunming Liu, Shuijian He, and Chaobo Huang. 2019. "Biomass Derived Carbon as Binder-Free Electrode Materials for Supercapacitors." *Carbon* 155: 706–726. 10.1016/j.carbon.2019.09.018.

Wei, Shaochen, Hao Zhang, Yaqin Huang, Weikun Wang, Yuzhen Xia, and Zhongbao Yu. 2011. "Pig Bone Derived Hierarchical Porous Carbon and Its Enhanced Cycling Performance of Lithium–Sulfur Batteries." *Energy & Environmental Science* 4 (3): 736–740. 10.1039/C0EE00505C.

Wen, Lei, Feng Li, and Hui-Ming Cheng. 2016. "Carbon Nanotubes and Graphene for Flexible Electrochemical Energy Storage: From Materials to Devices." *Advanced Materials* 28 (22): 4306–4337. 10.1002/ADMA.201504225.

Weng, To-Chi, and Hsisheng Teng. 2001. "Characterization of High Porosity Carbon Electrodes Derived from Mesophase Pitch for Electric Double-Layer Capacitors." *Journal of The Electrochemical Society* 148 (4): A368. 10.1149/1.1357171.

Xia, Kaisheng, Qiuming Gao, Jinhua Jiang, and Juan Hu. 2008. "Hierarchical Porous Carbons with Controlled Micropores and Mesopores for Supercapacitor Electrode Materials." *Carbon* 46 (13): 1718–1726. 10.1016/j.carbon.2008.07.018.

Xia, Yang, Haoyue Zhong, Ruyi Fang, Chu Liang, Zhen Xiao, Hui Huang, Yongping Gan, Jun Zhang, Xinyong Tao, and Wenkui Zhang. 2018. "Biomass Derived Ni(OH)2@porous Carbon/Sulfur Composites Synthesized by a Novel Sulfur Impregnation Strategy Based on Supercritical CO_2

Technology for Advanced Li-S Batteries." *Journal of Power Sources* 378 (February): 73–80. 10.1016/j.jpowsour.2017.12.025.

Xiao, Pei Wen, Li Zhao, Zhu Yin Sui, Meng Ying Xu, and Bao Hang Han. 2017. "Direct Synthesis of Ordered Mesoporous Hydrothermal Carbon Materials via a Modified Soft-Templating Method." *Microporous and Mesoporous Materials* 253: 215–222. 10.1016/j.micromeso.2017.07.001.

Xu, Jiaqi, Kuan Zhou, Fang Chen, Wei Chen, Xiangfeng Wei, Xue-Wei Liu, and Jiehua Liu. 2016. "Natural Integrated Carbon Architecture for Rechargeable Lithium–Sulfur Batteries." *ACS Sustainable Chemistry & Engineering* 4 (3): 666–670. 10.1021/acssuschemeng.5b01258.

Yan, Litao, Jiuling Yu, Jessica Houston, Nancy Flores, and Hongmei Luo. 2017a. "Biomass Derived Porous Nitrogen Doped Carbon for Electrochemical Devices." *Green Energy and Environment* 2 (2): 84–99. 10.1016/j.gee.2017.03.002.

Yan, Litao, Jiuling Yu, Jessica Houston, Nancy Flores, and Hongmei Luo. 2017b. "Biomass Derived Porous Nitrogen Doped Carbon for Electrochemical Devices." *Green Energy and Environment*. KeAi Publishing Communications Ltd. 10.1016/j.gee.2017.03.002.

Yang, Chia Min, Claudia Weidenthaler, Bernd Spliethoff, Mamatha Mayanna, and Ferdi Schüth. 2005. "Facile Template Synthesis of Ordered Mesoporous Carbon with Polypyrrole as Carbon Precursor." *Chemistry of Materials* 17 (2): 355–358. 10.1021/cm049164v.

Yang, Hui, Shewen Ye, Jiaming Zhou, and Tongxiang Liang. 2019. "Biomass-Derived Porous Carbon Materials for Supercapacitor." *Frontiers in Chemistry* 7 (April): 1–17. 10.3389/fchem.2019.00274.

Yang, Kai, Qiuming Gao, Yanli Tan, Weiqian Tian, Weiwei Qian, Lihua Zhu, and Chunxiao Yang. 2016. "Biomass-Derived Porous Carbon with Micropores and Small Mesopores for High-Performance Lithium–Sulfur Batteries." *Chemistry: A European Journal* 22 (10): 3239–3244. 10.1002/chem.2015 04672.

Yang, Minho, Dong Seok Kim, Seok Bok Hong, Jae Wook Sim, Jinsoo Kim, Seung Soo Kim, and Bong Gill Choi. 2017. "MnO_2 Nanowire/Biomass-Derived Carbon from Hemp Stem for High-Performance Supercapacitors." *Langmuir* 33 (21): 5140–5147. https://doi.org/10.1021/acs. langmuir.7b00589.

Yang, Wenxiu, Xiangjian Liu, Xiaoyu Yue, Jianbo Jia, and Shaojun Guo. 2015. "Bamboo-like Carbon Nanotube/Fe3C Nanoparticle Hybrids and Their Highly Efficient Catalysis for Oxygen Reduction." *Journal of the American Chemical Society* 137 (4): 1436–1439. 10.1021/ja5129132.

Yu, Fang, Shizhen Li, Wanru Chen, Tao Wu, and Chuang Peng. 2019. "Biomass-Derived Materials for Electrochemical Energy Storage and Conversion: Overview and Perspectives." *Energy & Environmental Materials* 2 (1): 55–67. 10.1002/eem2.12030.

Yu, Guihua, Liangbing Hu, Nian Liu, Huiliang Wang, Michael Vosgueritchian, Yuan Yang, Yi Cui, and Zhenan Bao. 2011. "Enhancing the Supercapacitor Performance of Graphene/MnO_2 Nanostructured Electrodes by Conductive Wrapping." 10.1021/NL2026635.

Yun, Young Soo, Se Youn Cho, Jinyong Shim, Byung Hoon Kim, Sung Jin Chang, Seung Jae Baek, Yun Suk Huh, et al. 2013. "Microporous Carbon Nanoplates from Regenerated Silk Proteins for Supercapacitors." *Advanced Materials* 25 (14): 1993–1998. 10.1002/adma.201204692.

Zakzeski, Joseph, Pieter C.A. Bruijnincx, Anna L. Jongerius, and Bert M. Weckhuysen. 2010. "The Catalytic Valorization of Lignin for the Production of Renewable Chemicals." *Chemical Reviews* 110 (6): 3552–3599. 10.1021/cr900354u.

Zaman, Chowdhury Zaira, Kaushik Pal, Wageeh A. Yehye, Suresh Sagadevan, Syed Tawab Shah, Ganiyu Abimbola Adebisi, Emy Marliana, Rahman Faijur Rafique, and Rafie Bin Johan. 2017. "Pyrolysis: A Sustainable Way to Generate Energy from Waste." In *Pyrolysis*. InTech. 10.5772/intechopen.69036.

Zerr, Andreas, Gerhard Miehe, George Serghiou, Marcus Schwarz, Edwin Kroke, Ralf Riedel, Hartmut Fueß, Peter Kroll, and Reinhard Boehler. 1999. "Synthesis of Cubic Silicon Nitride." *Nature* 400 (6742): 340–342. 10.1038/22493.

Zhang, Bin, Min Xiao, Shuanjin Wang, Dongmei Han, Shuqin Song, Guohua Chen, and Yuezhong Meng. 2014. "Novel Hierarchically Porous Carbon Materials Obtained from Natural Biopolymer as Host Matrixes for Lithium–Sulfur Battery Applications." *ACS Appl. Mater. Interfaces* 6 (15): 13174–13182. 10.1021/am503069j.

Zhang, Li, and X.S. Zhao. 2009. "Carbon-Based Materials as Supercapacitor Electrodes." *Chemical Society Reviews* 38 (9): 2520–2531. 10.1039/b813846j.

Zhang, Lixue, Zhihong Liu, Guanglei Cui, and Liquan Chen. 2015. "Biomass-Derived Materials for Electrochemical Energy Storages." *Progress in Polymer Science*, 43 (April): 136–164. 10.1016/j.progpolymsci.2014.09.003.

Zhang, Songtao, Mingbo Zheng, Zixia Lin, Nianwu Li, Yijie Liu, Bin Zhao, Huan Pang, Jieming Cao, Ping He, and Yi Shi. 2014. "Activated Carbon with Ultrahigh Specific Surface Area Synthesized from Natural Plant Material for Lithium-Sulfur Batteries." *Journal of Materials Chemistry A* 2 (38): 15889–15896. 10.1039/C4TA03503H.

Zhang, Songtao, Mingbo Zheng, Zixia Lin, Rui Zang, Qingli Huang, Huaiguo Xue, Jieming Cao, and Huan Pang. 2016. "Mango Stone-Derived Activated Carbon with High Sulfur Loading as a Cathode Material for Lithium–Sulfur Batteries." *RSC Advances* 6 (46): 39918–39925. 10.1039/C6RA05560E.

Zhang, Xiyue, Haozhe Zhang, Ziqi Lin, Minghao Yu, Xihong Lu, and Yexiang Tong. 2016. "Recent Advances and Challenges of Stretchable Supercapacitors Based on Carbon Materials 基于碳材料的可伸缩型超级电容器的研究进展." *Science China Materials* 59 (6): 475–494. 10.1007/s40843-016-5061-1.

Zhang, Ying, Xiaolan Song, Yue Xu, Haijing Shen, Xiaodong Kong, and Hongmei Xu. 2019. "Utilization of Wheat Bran for Producing Activated Carbon with High Specific Surface Area via NaOH Activation Using Industrial Furnace." *Journal of Cleaner Production* 210 (February): 366–375. 10.1016/j.jclepro.2018.11.041.

Zhang, Yunqiang, Xuan Liu, Shulan Wang, Li Li, and Shixue Dou. 2017. "Bio-Nanotechnology in High-Performance Supercapacitors." *Advanced Energy Materials* 7 (21): 1700592. https://doi.org/10.1002/aenm.201700592.

Zhang, Yunya, Zan Gao, Ningning Song, and Xiaodong Li. 2016. "High-Performance Supercapacitors and Batteries Derived from Activated Banana-Peel with Porous Structures." *Electrochimica Acta* 222 (December): 1257–1266. 10.1016/j.electacta.2016.11.099.

Zhao, Donggao, Youxue Zhang, and Eric J Essene. 2015. "Electron Probe Microanalysis and Microscopy: Principles and Applications in Characterization of Mineral Inclusions in Chromite from Diamond Deposit." 10.1016/j.oregeorev.2014.09.020.

Zhao, Jing, Yiju Li, Guiling Wang, Tong Wei, Zheng Liu, Kui Cheng, Ke Ye, Kai Zhu, Dianxue Cao, and Zhuangjun Fan. 2017. "Enabling High-Volumetric-Energy-Density Supercapacitors: Designing Open, Low-Tortuosity Heteroatom-Doped Porous Carbon-Tube Bundle Electrodes." *Journal of Materials Chemistry A* 5 (44): 23085–23093. 10.1039/c7ta07010a.

Zhao, Xiao, Jianbing Zhu, Liang Liang, Chenyang Li, Changpeng Liu, Jianhui Liao, and Wei Xing. 2014. "Biomass-Derived N-Doped Carbon and Its Application in Electrocatalysis." *Applied Catalysis B: Environmental* 154–155 (July): 177–182. 10.1016/j.apcatb.2014.02.027.

Zheng, Xiaoyu, Jiayan Luo, Wei Lv, Da Wei Wang, and Quan Hong Yang. 2015. "Two-Dimensional Porous Carbon: Synthesis and Ion-Transport Properties." *Advanced Materials* 27 (36): 5388–5395. 10.1002/adma.201501452.

Zheng, Xiaoyu, Wei Lv, Ying Tao, Jiaojing Shao, Chen Zhang, Donghai Liu, Jiayan Luo, Da Wei Wang, and Quan Hong Yang. 2014. "Oriented and Interlinked Porous Carbon Nanosheets with an Extraordinary Capacitive Performance." *Chemistry of Materials* 26 (23): 6896–6903. 10.1021/cm503845q.

Zhi, Mingjia, Feng Yang, Fanke Meng, Minqi Li, Ayyakkannu Manivannan, and Nianqiang Wu. 2014. "Effects of Pore Structure on Performance of An Activated-Carbon Supercapacitor Electrode Recycled from Scrap Waste Tires." *ACS Sustainable Chemistry & Engineering* 2 (7): 1592–1598. 10.1021/sc500336h.

Zhou, Jiazhen, Anran Luo, and Youcai Zhao. 2018. "Preparation and Characterisation of Activated Carbon from Waste Tea by Physical Activation Using Steam." *Journal of the Air and Waste Management Association* 68 (12): 1269–1277. 10.1080/10962247.2018.1460282.

Zhu, Chengzhou, He Li, Shaofang Fu, Dan Du, and Yuehe Lin. 2016. "Highly Efficient Nonprecious Metal Catalysts towards Oxygen Reduction Reaction Based on Three-Dimensional Porous Carbon Nanostructures." *Chemical Society Reviews*. Royal Society of Chemistry. 10.1039/c5cs00670h.

Zou, Yong, and Bu-Xing Han Group. 2001. "Preparation of Activated Carbons from Chinese Coal and Hydrolysis Lignin." *Adsorption Science & Technology* 19 (1): 59–72.

11

Synthesis and Electrochemical Properties of Graphene

Sylvester M. Mbam[1], **Raphael M. Obodo**[1,2,3], **Assumpta C. Nwanya**[1,4,5], **A. B. C. Ekwealor**[1], **Ishaq Ahmad**[2,3], **and Fabian I. Ezema**[1,4,5]

[1]*Department of Physics and Astronomy, University of Nigeria, Nsukka, Enugu State, Nigeria*
[2]*National Center for Physics, Islamabad, Pakistan*
[3]*NPU-NCP Joint International Research Center on Advanced Nanomaterials and Defects Engineering, Northwestern Polytechnical University, Xi'an, China*
[4]*Nanosciences African Network (NANOAFNET) iThemba LABS-National Research Foundation, Somerset West, Western Cape Province, South Africa*
[5]*UNESCO-UNISA Africa Chair in Nanosciences/Nanotechnology, College of Graduate Studies, University of South Africa (UNISA), Muckleneuk Ridge, Pretoria, South Africa*

11.1 Introduction

Recently, energy storage systems have attracted tremendous research interests which can be linked to the increasing technological innovations, greater demands for portable and reliable power sources, coupled with the need to curb the over-dependency on fossil fuels that pose harm to the environment. Batteries and supercapacitors are among the most deeply researched options for efficient energy storage and utilization. The major drawbacks in batteries such as low power density, low cycle life, and short range of working temperature have however opened an overwhelming insight into the development of supercapacitors (Obodo et al. 2019).

Supercapacitors, therefore, link the gap between the ordinary capacitors, which have lower energy densities, and batteries that have lower power densities (Obodo et al. 2020). Supercapacitors have found efficient applications in systems requiring a charging-discharging interval in the range of few seconds and minutes, whereas conventional capacitors have been mostly utilized in systems (like power converters) where the cycle interval is in the range of milliseconds to seconds, while batteries with the cycle interval in the range of hours have been more suitable in such applications as Photovoltaic systems (Aneke and Wang 2016). Different applications/innovations requiring supercapacitors have been reported by various authors; this is ascribed to its fascinating properties such as higher power density, swift electrical response, wide range of operating temperatures, and extended cycle life, contrary to batteries (Berrueta et al. 2019; Obodo et al. 2019).

Enhancing both the energy and power densities of supercapacitors has recently involved various researches aimed at increasing the electrochemical activities of the electrodes' active materials to achieve optimal electrode–electrolyte interactions. There are three basic mechanisms of charge storage in energy storage devices, which include electrochemical double-layer capacitance (EDLC), pseudo-capacitance, and faradaic mechanism. EDLC technique involves an accumulation of the electrolyte ions

DOI: 10.1201/9781003145585-11

on the electrode–electrolyte boundary. The density of the charges on this interface is dependent on the voltage applied as the charges are electrostatically trapped at the electrode's edges (Parida et al. 2017). The EDLC technique is associated with electrodes made of carbon sources (such as graphene, carbon aerogels, carbon nanotubes, carbon black, activated carbon) (Jian et al. 2016). The pseudocapacitance mechanism emanates from a quick and reversible ion migration through the electrode-electrolyte layer. This charge storage mechanism can be ascribed to redox-active metal oxides, conducting polymers, sulphides, and nitrides (Broussea et al. 2015). Faradaic mechanism originates from the redox activities of metal ions inside the crystalline configuration of the electrode; this is because of the intercalation and de-intercalation of the positive metal ions which in turn triggers a redox reaction (Uduh et al. 2014). This mechanism is associated with battery-like electrodes. EDL and pseudocapacitive mechanism being the basic storage technique in supercapacitors can offer a remarkable specific capacitance and energy density. However, pseudocapacitative electrodes face two major challenges – less power density and fragile cyclability – caused by the irreversibility and reaction dynamics of the electroactive material (Wu et al. 2012).

Supercapacitors utilizing a hybrid mechanism (EDLC and pseudocapacitance) have been reported by several researchers to have improved electrochemical performance when compared to a single mechanism. Precisely, the composites of metal oxides and graphene have recently attracted greater attention over other hybrid options. This is caused by an optimal synergistic effect between metal oxides and graphene, which consequently enhances charge retention, strong cyclability, and the overall electrochemical performance (Muzaffar et al. 2019). Besides, graphene is known for such fascinating properties as abundant surface area, very thin and flexible, exceptional electrical conductivity (with zero energy gap), chemical and thermal stability, coupled with its oxygen-rich functional groups, and wide range of operating window. Thus, the ever-increasing interest in graphene-based energy storage systems has been aimed at utilizing these excellent intrinsic properties (Obodo et al. 2020). Chemical exfoliation of graphite has been reported to be among the most facile synthesis methods for large-scale production of graphene oxide (GO) and reduced graphene oxide (rGO) (Hummers and Offeman 1958).

We hereby present a review on the synthesis and electrochemical applications of graphene, this is in essence of providing clearer insight and prospects for an optimal and innovative utilization of the tremendous performance of graphene especially in the energy storage field.

11.2 Nanostructures of Carbon

Carbon is among the most abundant element on earth, it readily exists in a form of diamond, graphite, and amorphous carbon (Tsai and Tu 2010; Janani et al. 2015). Nanostructures of carbon appear in different dimensions as shown in Figure 11.1; such as zero dimension observed in fullerenes (Veerappan et al. 2011), one-dimensional configuration as seen in carbon nanotubes (Harris 1999), two-dimensional structure as seen in graphene (Obodo et al. 2020), and three-dimensional graphite structure (Fugallo et al. 2014). The two-dimensional graphene thin sheet is a building block for all the graphitic/carbon nanostructures; it becomes fullerene when cut and bent into a spherical structure thus exhibiting a zero-dimensional shape, rolling up a graphene sheet turns it into the one-dimensional structure known as carbon nanotube (CNT). Depending on the angle through which the roll-up occurs, the resulting CNT can appear in such a shape as the armchair, chiral, and zig-zag (Popov 2004). Also, the metallic or semiconducting properties of the CNT are determined by the diameter and the spiral arrangement of the parent graphite ring. When various CNTs are joined, it results in molecular wires that can be suitably applied in sensors (Dai et al. 2002), transistors consisting of a single molecule (Dai et al. 1996), tips of a scanning electron microscope (Rueckes et al. 2000), electrochemical and gas storage (Nwanya et al. 2020), and electron field flat panel displays (Smith et al. 2005). These remarkable applications are ascribed to the excellent electron mobility, high mechanical strength, and optical and magnetic behaviours of the molecular wires (Chen et al. 2008; Eda and Chhowalla 2009).

However, graphene constitutes sp^2 hybrid atoms of carbon that are closely arranged in a single layer to form a flat honeycomb lattice with only two dimensions (Gilje et al. 2007). Due to this unique

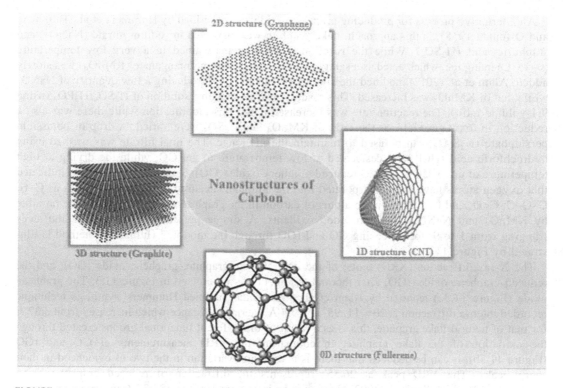

FIGURE 11.1 The various forms of carbon nanostructures indicating; fullerene (zero-dimensional structure); Carbon nanotube (1-dimensional structure); graphene (2-dimensional structure); and the 3-dimesional graphite network (multiple-stacked graphene sheets).

structure, graphene possesses charge carriers that move like massless Dirac fermions and consequently demonstrate a quantum hall effect coupled with ambipolar electric field phenomenon (Novoselov et al. 2004; Novoselov et al. 2005; Zhang et al. 2005). Graphene also contains a wide range of π interfaces, thus making it exhibit highly remarkable electrical, thermal, and magnetic properties that afford it a huge suitability for various applications in different fields such as electronics, sensors, energy storage, photovoltaics, photocatalysis, biotechnology, military, etc.

11.3 Graphene Layer, Graphene Oxide (GO), and Reduced Graphene Oxide (rGO) Synthesis

A pure and stand-alone graphene layer was first demonstrated by Geim et al. (Obodo et al. 2020). Their experiment involved using a normal adhesive tape to extract a single layer of graphite out of a bulk graphite as contained in normal pencils. Using this mechanical exfoliation of graphite they were able to obtain a first known material to possess one atom in thickness (being two-dimensional in nature). Subsequent experiments to synthesize pure and large-scale graphene had broadly involved either the bottom-up techniques (chemical vapour deposition (CVD), unzipping of carbon nanotubes, epitaxial growth on crystalline silicon substrates) or the top-down approach (chemical reduction of exfoliated graphene oxide, and liquid-phase exfoliation of graphite). Oxidizing of natural flake graphite powder is the oldest and most utilized route to producing graphene oxide (GO), as flake graphite is a readily available natural mineral that can be made pure through the removal of its heteroatomic impurity. This technique was first demonstrated by Brodie (1859) where he obtained graphene oxide from a mixture of potassium chlorate in a solution of graphite and nitric acid. Staudenmaier (Staudenmaier 1898) amended this procedure by using a combination of sulphuric and nitric acids together and gradually introducing chlorate into the reaction.

An alternative process for producing graphene oxide was described by Hummers et al. (Hummers and Offeman 1958); in this approach, flake graphite was dissolved in sodium nitrate ($NaNO_3$) and sulphuric acid (H_2SO_4). While the reaction was being maintained in a very low temperature (0–15°C) using ice, which acted as a safety measure, potassium permanganate ($KMnO_4$) was slowly added. Alam et al. (2017) modified the Hummers' method by introducing a low quantity of $NaNO_3$ while that of $KMnO_4$ was increased. This reaction was done with a solution of H_2SO_4/H_3PO_4 (using 9:1 volume ratio). The reaction rate was increased under this modification while there was also a reduction in toxic gas yield, as the ratio of $KMnO_4$ and H_2SO_4 were varied. A drop of potassium persulphate ($K_2S_2O_8$) can be used to maintain the pH range. The final filtrate was washed using hydrochloric acid (HCl), and desiccated at low temperature to get GO, while the drying at high temperature of up to 350°C yielded reduced graphene oxide (rGO). Spectroscopic analysis indicated that oxygen atoms have been incorporated into the graphite sheath to produce such bonds as C–H, C–O–C, C=O, and COOH; with the atoms of carbon in the graphite sheath this was made possible by $KMnO_4$ and $NaNO_3$ which are good oxidants. A simple presentation illustrating the steps (ranging from 1 to 9) for producing GO and rGO through the modified Hummers' method is illustrated by Figure 11.2.

The X-ray diffraction (XRD) plots; of the natural flake graphite, graphene oxide (GO), and the reduced graphene oxide (rGO) after thermal treatment are represented in Figure 11.3. The graphene oxide (Figure 11.3c) reported by Alam et al. through the modified Hummers' synthesis technique recorded intense diffraction at $2\theta = 11.95°$ and 7.4 Å interlayer distance which increased from 3.37 Å for that of natural flake graphite, this is as a result of the different functional groups created through the oxidation of the flake graphite. In comparing the XRD measurements of GO and rGO (Figure 11.3b), it can be deduced that there is a structural variation in the two as evidenced in their sharp peaks. The XRD peak at 26.55° represents the diffraction of reduced graphene oxide (Figure 11.3d) which lies in the (002) plane with an interlayer distance of 3.37 Å in the bearing of the c-axis. Thus, the high thermal mortification of graphene oxide at 350°C caused the disappearance of the diffraction peak (located at $2\theta = 11.95°$) which signifies the complete exfoliation of the manifold sheaths of graphene oxide under the thermal treatment. This technique was also detailed fully by (Guo et al. 2009; Ju et al. 2010; Chen et al. 2013).

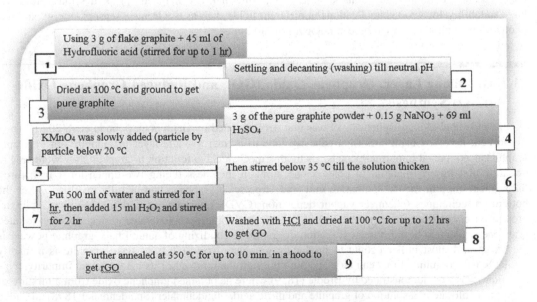

FIGURE 11.2 Simple presentation of the steps (1 to 9) for producing graphene oxide (GO), and reduced graphene oxide (rGO) through the modified Hummers' method.

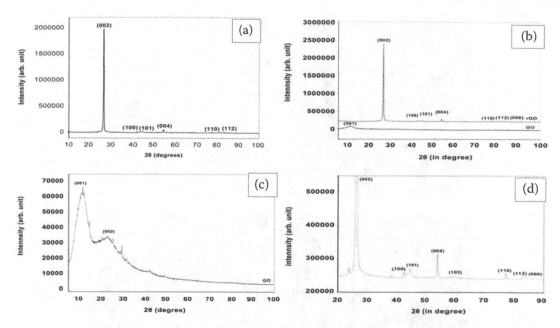

FIGURE 11.3 XRD plots of; (a) Natural flake graphite; (b) compared plots of GO and rGO; (c) graphene oxide (GO), (d) reduced graphene oxide (rGO). Ref. Alam et al. (2017). Reprinted with permission from Ref. Alam et al. (2017), copyright 2017, Scientific Research Publishing Inc.

11.4 Electrochemical Applications of Graphene and Reduced Graphene Oxide

The intrinsic features of graphene have afforded it an interesting material for various applications ranging from electronics (Meng et al. 2013) to energy systems (energy harvesting and storage) (Wang, Zhi, and Müllen 2008; Zhang et al. 2010) with more suitability for energy storage. This is basically due to its vast surface area amounting up to 2670 m^2g^{-1} which can as well be modified, chemically stable, and mechanically tough, coupled with significant thermal and electrical conductivity (Booth et al. 2008; Lee et al. 2008Xia et al. 2009) computed the quantum capacitance of a distinct layer-graphene and recorded a value of 21 $\mu F\ cm^{-2}$ as the theoretical specific capacitance which has an equivalent specific capacitance of 550 F g^{-1} for the whole surface area. Consequently, pure graphene still retains a lower capacitive performance practically. This can be caused by clustering of its particles during the synthesis and fabrication of the electrode. One of the major ways of enhancing the electrochemical behaviour of graphene-based electrode materials reported by several authors had involved making it a conductive support/matrix for optimal redox activities of metal oxides, polymers, nitrides, and hydroxides (Wei et al. 2012; Ke and Wang 2016; Kumar et al. 2018). Graphene-based electrodes (being carbonaceous) possess a double-layer capacitive behaviour induced by the accumulated charges over the electrode–electrolyte edges caused by desorption and adsorption of electrolyte ions over that interface. Combining this effect with redox-active metal oxides can create an optimal synergistic effect with the ability to boost the electrochemical performance of the graphene-based electrodes. This is basically due to an even scattering of the nanoparticles/ions of the metal oxides/hydroxides unto the conducting graphene layers. The metal oxides, hydroxides, nitrides, and conducting polymers in turn create a remarkable pseudocapacitance effect into the electrode's storage mechanism and thereby retain higher capacitance and stability (Wu et al. 2012; Ma et al. 2015; Zhao et al. 2017).

The availability of oxygen-rich functional groups on the surface and boundaries of graphene oxide and reduced graphene oxide has made it very advantageous over other closely related carbon derivatives such as carbon nanotubes and graphite. Consequently, the morphology, particle size, and dispersal of the metal oxide nanostructures can easily be affected by these functional groups. This implies that using an appropriate ratio of the metal oxide solution/ions and the graphene content can produce the desired

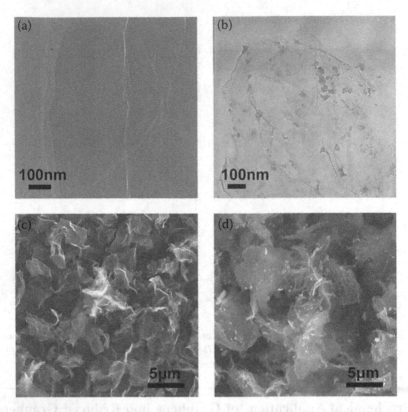

FIGURE 11.4 Transmission Electron Microscope images (a, b), and Scanning Electron Microscope images (c, d) of graphene sheets (a, c), and a RuO₂/graphene sheet composite having 38.3 wt% ruthenium. Ref. Wu et al. (2010). Reprinted with permission from Ref. Wu et al. (2010), copyright 2010, John Wiley & Sons, Inc.

composite with optimal electrochemical performance. Metal oxides interact with graphene oxide in two mechanisms, which involve reactive adsorption of the metal oxides' particles on the graphene's functional groups (like HO–C=O and –OH); this adsorption thus fills the centres having carboxyl/hydroxyl groups at oxygen-defect spots of the metal oxides, the second mechanism is the van der Waals interactions of the pristine area of graphene and metal oxides (Kamat 2010; Lightcap, Kosel, and Kamat 2010),. Wu et al. compared the anchoring rate of ruthenium oxide (RuO₂) on oxygen-rich graphene produced by chemical exfoliation of graphite oxide and oxygen-free graphene layer that contain no oxygen groups produced through a chemical vapour deposition. It was observed through the Transmission Electron Microscope (TEM) and Scanning Electron Microscope (SEM) images that the surface of the GO anchored a significant and highly distributed amount of RuO₂ nanoparticles (having a size less than 5 nm). While that of the CVD-made graphene film anchored a large and poorly distributed RuO₂ nanoparticles (in the range of 100 nm) (Wu et al. 2010) as presented in Figure 11.4. These studies thus confirmed the basic role of oxygen-rich functional groups in the formation and spreading of the metal oxides' nanostructures on the graphene conductive support.

11.4.1 Graphene-Based Electrode Materials for Supercapacitors

In a quest to boost both the specific capacitance and power density of supercapacitors, graphene has over the years been a deeply researched electro-active material for supercapacitor electrodes. This is aimed at utilizing its remarkable qualities such as huge surface area, high conductivity, remarkable thermal and chemical stability, oxygen-rich functional groups, and a large range of operating voltage (Lv et al. 2009; Wang et al. 2009). The electrochemical behaviour of graphene-based electrodes for utilization in supercapacitors was first studied by Stoller et al. and Vivekchand et al. Their studies

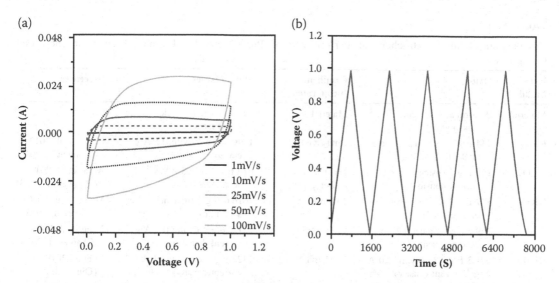

FIGURE 11.5 Electrochemical performance of graphene material (hydrazine-reduced graphene oxide). (a) Cyclic voltammogram curves at different scan rates; (b) the galvanostatic charging-discharging curve at 100 mA g^{-1} current density. Ref. Wang et al. (2009). Reprinted with permission from Ref. Wang et al. (2009), copyright 2009, American Chemical Society.

recorded a value of 75 F g^{-1} specific capacitance and about 31.9 Wh kg^{-1} energy density using an ionic liquid electrolyte (Stoller et al. 2008), and while using organic and aqueous electrolytes the electrode offered corresponding values of 99 F g^{-1} and 135 F g^{-1} specific capacitance (Vivekchand et al. 2008). Wang et al. (2009) reported an electrochemical behaviour of reduced graphene oxide which recorded a specific capacitance value of 205 F g^{-1}, and an energy density of 28.5 Wh kg^{-1} using KOH electrolyte. The electrochemical performance of the fabricated electrode is represented by Figure 11.5 (11.5a denotes the cyclic voltammogram curve having a rectangular-like curl – the EDLC behaviour of carbon materials, while 11.5b is the charge-discharge curve at 100 mA g^{-1} current density). The electrode also attained a charge retention of up to 90% after 1200 cycles. Using an exfoliation technique at mild temperature (about 200°C), Lv et al. produced a graphene active material for supercapacitor electrode and achieved a significant specific capacitance value of 260 F g^{-1} over a scan rate of 10 mV s^{-1} using an aqueous electrolyte (Lv et al. 2009). A reduced graphene oxide obtained through microwave heating of graphene oxide using propylene carbonate solvent yielded a specific capacitance value of 120 F g^{-1} and 191 F g^{-1} in tetraethylammonium tetraflfluoroborate (TEABF$_4$) and KOH electrolytes, correspondingly (Zhu et al. 2010). However, calculating the inherent capacitance of a single graphene nanosheet can be a bit challenging due to the solvent-eroded layers. Thus, the graphene's interfacial capacitance is a function of the total of its layers being deduced from the individual surface area (Wang et al. 2009).

Conventional capacitors have been an effective device for smoothening alternate current (AC) signals; however, they are not yet fit for microelectronics applications as they are colossal. Several graphene-based nanostructures (like graphene-carbon nanotube sheet (Obodo et al. 2020), reduced graphene oxide (Sheng et al. 2012), graphene quantum dots (Liu et al. 2013)) can be utilized in designing a more reliable AC filter that operates within seconds. Miller et al. (2010) had previously reported a graphene nanosheet grown on nickel substrate for fabricating an AC filtering electrochemical double-layer (EDL) capacitor. The electrode effectively filtered up to 120 Hz current under a resistor-capacitor operating interval of 0.2 ms which is quite higher than that of conventional capacitors.

Recently, hybrid supercapacitors comprising of pseudocapacitive and EDL charge storage mechanisms have been widely reported by various researchers. The hybrid electrode materials made of graphene have been confirmed to possess improved electrochemical activities when compared to the single components. For instance, the specific capacitance value of RuO$_2$ was improved from

TABLE 11.1

Comparison of the Electrochemical Performance of Pure Metal Oxides and Their Graphene-Based Composites

Pure Metal	Oxides	Graphene-Incoporated	Metal Oxides	References
Material	Performance (specific capacitance)	Material	Performance (specific capacitance)	
RuO_2	234 F g^{-1} (with a scan rate of 0.005 V s^{-1})	RuO_2/rGO	1099.6 F g^{-1} (with 0.5 A g^{-1} current density)	(Dubal et al. 2013), (Wang et al. 2016)
MnO_2	140 F g^{-1} (using 0.1 A g^{-1} current density)	MnO_2/Defect-free graphene	679 F g^{-1} (in 25 mV s^{-1} scan rate)	(Attias et al. 2017), (Hong et al. 2018)
NiO	337 F g^{-1} (using a scan rate of 2 mV s^{-1})	NiO/rGO	1016.6 F g^{-1} (in 1 mV s^{-1} scan rate)	(Meher et al. 2010), (Cao et al. 2015)
CuO	269.6 F g^{-1} (under 0.25 A g^{-1} current density)	CuO-GNS	331.9 F g^{-1} (using 0.6 A g^{-1} current density)	(Yang et al. 2018), (Zhao et al. 2013)
Bi_2O_3	691.3 F g^{-1} (applying 2.0 A g^{-1} current density)	Bi_2O_3/rGO	1423 F g^{-1} (applying 1 A g^{-1} current density)	(Li et al. 2019), (Qiu et al. 2018)
V_2O_5	298 F g^{-1} (in 1 A g^{-1} current density)	V_2O_5/rGO	537 F g^{-1} (using 1 A g^{-1} current density)	(Hou et al. 2018), (Li et al. 2013)
Co_3O_4	476 F g^{-1} (using 0.5 A g^{-1} current density)	Co_3O_4/rGO	531 F g^{-1} (under 0.1 A g^{-1} current density)	(Deori et al. 2013), (Hout et al. 2017)
CeO_2	154.875 F g^{-1} (using 0.5 A g^{-1} current density)	CeO_2/Graphene nanocomposite	208 F g^{-1} (with 1 A g^{-1} current density)	(Wang et al. 2018), (Wang et al. 2011)

234 F g^{-1} (Dubal et al. 2013) to 441 F g^{-1} when incorporated with graphene (Thangappan et al. 2018), and 1099.6 F g^{-1} when incorporated with rGO (Wang et al. 2016). Attias et al. (2017) recorded a chemically prepared manganese oxide (MnO_2) for application in supercapacitor electrode with a specific capacitance value of 140 F g^{-1} under a current density of 0.1 A g^{-1}. The capacitance was significantly improved to a high value of 679 F g^{-1} at 25 mV s^{-1} scan rate when incorporated with graphene oxide (Hong et al. 2018). Similarly, the specific capacitance of nickel oxide (NiO) drastically increased from 337 F g^{-1} (Meher et al. 2010) to a high value of 1016.6 F g^{-1} when incorporated with rGO (Cao et al. 2015). Table 11.1 illustrates the comparison of the electrochemical performance of pure metal oxides and their graphene-based composites, while Figure 11.6 indicates the plots of the specific capacitance offered by pure metal oxides and graphene-based metal oxide supercapacitor electrode materials.

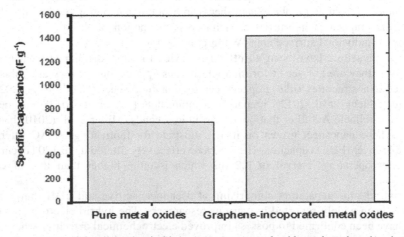

FIGURE 11.6 Comparison of the electrochemical behaviour of pure metal oxides and graphene-based metal oxides.

11.4.2 Graphene-Based Battery Electrodes

The most commercially utilized anode material has been graphite; however it still has a low theoretical capacity (about 372 mAh g^{-1} which restricts it from applications requiring much energy capacity (Winter et al. 1998). Graphene has some remarkable properties such as the high specific surface area and flexibility that can be utilized in place of graphite or carbon nanotubes for charge storage purposes (Obodo et al. 2020). In comparison to graphite, which hinders ion diffusion due to its bulk nature, the huge surface-to-volume ratio coupled with the fine pores in graphene can easily enable swift ion diffusion and transport which signifies long stability and high charge retention. Also, the less impurity and more oxygen-rich functional groups in graphene are among its major benefit over carbon nanotubes and graphite as energy storage materials. The specific capacity of up to 540 mAh g^{-1} was previously recorded for graphene nanosheet by Yoo et al. This is a higher value than that of graphite. The authors also identified that the lithium-ion (Li$^+$) storage capacity of the graphene nanosheet could be boosted by inserting fullerene or carbon nanotube macromolecules into the graphene layers; the increased interlayer spacing, therefore, provides enough sites for retaining the Li$^+$ (Yoo et al. 2008). Wang et al. prepared a graphene nanosheet having a flower-like morphology which yielded an improved capacity and cyclic stability for Li+ storage. Their report also confirmed that the lithium ions can be stored on both the edges, bonding sites, and adsorbed on both surfaces of the graphene nanosheet (Wang et al. 2009). Boron/nitrogen-doped graphene has been reported as a tool for achieving an increased charge retention capacity and stability in lithium-ion batteries. The report by Wu et al. specified that the improved electrochemical performance of the doped graphene is ascribed to such remarkable properties as rough surface morphology, fairly stacked layers, heteroatomic defects, the two-dimensional structure of graphene, synergistic electrode–electrolyte interaction, coupled with a more electrical and thermal conductivity (Wu et al. 2011).

The swift electron and ion diffusion associated with graphene is a breakthrough technique in developing a battery that can charge and discharge in a few seconds. Flexible and transparent energy storage materials have also been made possible using graphene (El-Kady et al. 2012; Wang and Shi 2015; Obodo et al. 2020). Li et al. fabricated a lithium-ion battery made of hybrid electrodes (formed by stacking LiFePO$_4$ cathode and Li$_4$Ti$_5$O$_{12}$ anode materials into an elastic graphene foam) without adding any binder, it was observed that the battery was able to charge completely within 18 s due to the high conductivity and morphology of the electrodes (Li et al. 2012). Similarly, Lin et al. reported an aluminium-ion battery that was able to get fully charged within one minute due to its intercalation into a 3-dimensional graphene foam (Lin et al. 2015). Apart from replacing the graphite in the battery anodes with highly efficient graphene materials such as fine-pore graphene films (Liu et al. 2012), flexible graphene paper (Mukherjee et al. 2012), and solvated gaphene frameworks (Xu et al. 2015), which have recorded an outstanding electrochemical performance far higher than the conventional graphite anode and more applicability in systems requiring longer power usage such as portable electronics and electric vehicles, graphene can also be utilized as a conductive and mechanical support for metal oxide ions, as well as enhancing its surface area with improved electrochemical properties. Zhou et al. (2011) reported a composite anode made of Fe$_2$O$_3$ wrapped with reduced graphene oxide, which offered a specific capacity that is many times greater than that of a single Fe$_2$O$_3$. Similarly, a hybrid cathode comprising of LiFePO$_4$ and graphene has also been reported to retain up to twice the charges as that of pure LiFePO$_4$, coupled with an improved cycle life (Hu et al. 2013).

11.4.3 Innovative Features Associated with Graphene Electroactive Material

In comparison with the conventional electrode materials, graphene-based energy storage electrodes are inherently associated with optimal performance (which includes increased charging and discharging rate, higher energy and power density, increased cycle life, and increased charge retention/capacitance). The ever increasing interest in graphene-based electrode systems can be ascribed to the following intrinsic properties:

1. High surface area
2. Chemically and thermally stable

3. A highly thin layer (enabling an efficient diffusion of ions)

4. Excellent electrical conductivity

5. Mechanically flexible (enabling flexible electrodes)

6. Large surface-to-volume ratio (enabling optimum ion adsorption)

7. Oxygen-rich functional groups (making it hydrophilic in aqueous electrolytes and enabling heteroatomic binding)

8. Wide operating potential window (enabling an improved energy density)

9. Cost-effective synthesis option (through chemical exfoliation of graphite)

Intriguingly, these fascinating properties of graphene have facilitated various outstanding innovations into the energy storage devices; these include flexible and transparent energy storage systems (Facchetti and Marks 2010; Shao et al. 2015) , fast-charging and durable batteries (Mukherjee et al. 2012; Xu et al. 2015), stretchable and wearable energy storage systems (Chen et al. 2013; Kou et al. 2014), super-cpacitors having a similar energy density as batteries (El-Kady et al. 2015), very light batteries for microelectronics (Ye et al. 2014), supercapcitors with efficient AC signal-filtering ability (Wu et al. 2015), efficient electrodes without any binder/additives (Obodo et al. 2020).

11.5 Conclusion

We hereby presented a review of the intrinsic properties of graphene. Graphene being a one-atom-thick, flexible, and transparent material was first produced through a mechanical exfoliation of graphite. Having a zero-energy gap, it acts as a semi-metal, and thus is highly conductive. We deduced that the chemical exfoliation of graphite yields graphene oxide, while a thermal mortification of GO precisely at 350°C results in reduced graphene oxide. The Hummers' method is the most utilized synthesis technique for chemically exfoliating graphite into GO; this is because of its facile and cost-effective nature. Defect-free and large-quantity graphene oxide is also feasible using the popular Hummers' technique. We also reviewed and reported the electrochemical properties of graphene-based energy storage systems (supercapacitors and batteries). It can be concluded that the excellent electrochemical performance associated with graphene-based energy storage electrodes can be ascribed to such fascinating core properties as huge surface area, superb electrical conductivity, thin and flexible structure, chemical and thermal stability, coupled with its oxygen-rich functional groups, and wide range of operating potential. It was also recorded that incorporating graphene into redox-active (pseudocapacitive) materials such as metal oxides boosts its electrochemical applications and performance, due to the synergistic/hybrid effect (EDL- and pseudo-capacitance) between the graphene material and metal oxides. We also reported some key innovations (like flexible and transparent energy storage systems, fast-charging and durable batteries, stretchable and wearable energy storage systems, supercapacitors having a similar energy density as batteries, very light batteries for microelectronics, supercapacitors with efficient AC signal-filtering ability, efficient electrodes without any binder/additives) which were made possible in the energy storage devices because of the intrinsic and electrochemical properties of graphene.

Acknowledgements

RMO and IA humbly acknowledge NCP for their PhD fellowship (NCP-CAAD/PhD-132/EPD) award and COMSATS for a travel grant for the fellowship.

RMO also acknowledge PPSMB Enugu State, Nigeria, for study leave permission granted.

FIE (90407830) affectionately acknowledge UNISA for VRSP Fellowship award and also graciously acknowledge the grant by TETFUND under contract number TETF/DESS/UNN/NSUKKA/STI/VOL.I/B4.33. We thank Engr. Emeka Okwuosa for the generous sponsorship of April 2014, July 2016, and July 2018 conferences/workshops on applications of nanotechnology to energy, health and environment, and for providing some research facilities.

REFERENCES

Alam, S. N., N. Sharma, and L. Kumar. 2017. "Synthesis of Graphene Oxide (GO) by Modified Hummers Method and Its Thermal Reduction to Obtain Reduced Graphene Oxide (rGO)." *Graphene* 6, no. 1: 1–18.

Aneke, M., and M. Wang. 2016. "Energy Storage Technologies and Real Life Applications: A State of the Art Review." *Appl. Energy* 179: 350–377.

Attias, R D. Sharon, A. Borenstein, D. Malka, O. Hana, S. Luski, and D. Aurbach. 2017. "Asymmetric Supercapacitors Using Chemically Prepared MnO2 as Positive Electrode." *Materials J Electrochem Soc* 164: A2231–A2237.

Berrueta, A., A. Ursua, I. San Martin, A. Eftekhari, and P. Sanchis. 2019. "Supercapacitors: Electrical Characteristics, Modeling, Applications, and Future Trends." *Ieee Access* 7: 50869–50896.

Booth, T. J., P. Blake, R. R. Nair, D. Jiang, E. W. Hill, U. Bangert, A. Bleloch, et al. 2008. "Macroscopic Graphene Membranes and Their Extraordinary Stiffness." *Nano Letters* 8: 2442–2446.

Brodie, B. C. 1859. "On the Atomic Weight of Graphite." *Philosophical Transactions of the Royal Society of London* 149: 249–259.

Broussea, T., D. Bélangerc, and J. W. Long. 2015. "To Be or Not to Be Pseudocapacitive?" *J. Electrochem Soc.* 162, no. 5: A5185–A5189.

Cao, P., L. Wang, Y. Xu, Y. Fu, and X. Ma. 2015. "Facile Hydrothermal Synthesis of Mesoporous Nickel Oxide/Reduced Graphene Oxide Composites for High Performance Electrochemical Supercapacitor." *Electrochim Acta* 157: 359–368.

Chen, T., Y. Xue, A. K. Roy, and L. Dai. 2013. "Transparent and Stretchable High-Performance Supercapacitors Based on Wrinkled Graphene Electrodes." *ACS Nano* 8: 1039–1046.

Chen, J., B. Yao, C. Li, and G. Shi. 2013. "An Improved Hummers Method for Eco-Friendly Synthesis of Graphene Oxide." *Carbon* 64: 225–229.

Chen, H., M. B. Müller, K. J. Gilmore, G. G. Wallace, and D. Li. 2008. "Mechanically Strong, Electrically Conductive, and Biocompatible Graphene Paper." *Advanced Materials* 20, no. 18: 3557–3561.

Dai, L., P. Soundarrajan, and T. Kim. 2002. "Sensors and Sensor Arrays Based on Conjugated Polymers and Carbon Nanotubes." *Pure and Applied Chemistry* 74, no. 9: 1753–1772.

Dai, H., J. H. Hafner, A. G. Rinzler, D. T. Colbert, and R. E. Smalley. 1996. "Nanotubes as Nanoprobes in Scanning Probe Microscopy." *Nature* 384, no. 6605: 147–150.

Deori, K., S. K. Ujjain, R. K. Sharma, and S. Deka. 2013. "Morphology Controlled Synthesis of Nanoporous Co_3O_4 Nanostructures and Their Charge Storage Characteristics in Supercapacitors." *ACS Appl Mater Interfaces* 5, no. 21: 10665–10672.

Dubal, D. P., G. S. Gund, R. Holze, H. S. Jadhav, C. D. Lokhande, and C. J. Park. 2013. "Solution-Based Binder-free Synthetic Approach of RuO2 Thin Films for All Solid State Supercapacitors." *Electrochim Acta* 103: 103–109.

Eda, G., and M. Chhowalla. 2009. "Graphene-Based Composite Thin Films for Electronics." *Nano letters* 9, no. 2: 814–818.

El-Kady, M. F., V. Strong, S. Dubin, and R. B. Kaner. 2012. "Laser Scribing of High-performance and Flexible Graphene-based Electrochemical Capacitors." *Science* 335: 1326–1330.

El-Kady, M. F., Melanie Ihns, Mengping Li, Jee Youn Hwang, Mir F. Mousavi, Lindsay Chaney, Andrew T. Lech, and Richard B. Kaner. 2015. "Engineering Three-dimensional Hybrid Supercapacitors and Microsupercapacitors for High-performance Integrated Energy Storage." *Proc. Natl Acad. Sci. USA* 112: 4233–4238.

Facchetti A., and T., Marks. (Eds.), 2010. *Transparent electronics: from synthesis to applications.* John Wiley & Sons.

Fugallo, G., A. Cepellotti, L. Paulatto, M. Lazzeri, N. Marzari, and F. Mauri. 2014. "Thermal Conductivity of Graphene and Graphite: Collective Excitations and Mean Free Paths." *Nano letters* 14, no. 11: 6109–6114.

Gilje, S., S. Han, M. Wang, K. L. Wang, and R. B. Kaner. 2007. "A Chemical Route to Graphene for Device Applications." *Nano letters* 7, no. 11: 3394–3398.

Guo, H. L., X. F. Wang, Q. Y. Qian, F. Wang, and X. H. Xia. 2009. "A Green Approach to the Synthesis of Graphene Nanosheets." *ACS Nano* 3: 2653–2659.

Harris, P. J. F. 1999. *Carbon Nanotubes and Related Structures: New Materials for the Twenty First Century.* Cambridge: Cambridge University Press.

Hong, S. B., J. M. Jeong, H. G. Kang, D. Seo, Y. Cha, H. Jeon, G. Y. Lee, et al. "Fast and Scalable Hydrodynamic Synthesis of MnO2/Defect-Free Graphene Nanocomposites with High Rate Capability and Long Cycle Life." *ACS Appl Mater Interfaces* 10, no. 41: 35250–35259.

Hou, Z. Q., Z. G. Yang, Gao., and Y. P. 2018. "Synthesis of Vanadium Oxides Nanosheets as Anode Material for Asymmetric Supercapacitor." *Chem Pap* 72: 2849–2857.

Hout, S. I. E., C. Chen, T. Liang, L. Yang, and J. Zhang. 2017. "Cetyltrimethylammonium Bromide Assisted Hydrothermal Synthesis of Cobalt Oxide Nanowires Anchored on Graphene as an Efficient Electrode Material for Supercapacitor Applications." *Mater Chem Phys* 198: 99–106.

Hu, L. H., F. Y. Wu, C. T. Lin, A. N. Khlobystov, and L. J. Li. 2013. "Graphene-Modified LiFePO$_4$ Cathode for Lithium Ion Battery beyond Theoretical Capacity." *Nat. Commun.* 4, no. 1687.

Hummers, W., and R. Offeman. 1958. "Preparation of Graphitic Oxide." *Journal of The American Chemical Society* 80: 13–39.

Janani, M., P. Srikrishnarka, S. V. Nair, and A. S. Nair. 2015. "An In-depth Review on the Role of Carbon Nanostructures in Dye-sensitized Solar Cells." *Journal of Materials Chemistry A* 3, no. 35: 17914–17938.

Jian, X., S. Liu, Y. Gao, W. Tian, Z. Jiang, X. Xiao, ,Hui Tang, and L. Yin. 2016. "Carbon-Based Electrode Materials for Supercapacitor: Progress, Challenges and Prospective Solutions." *J. Electr. Eng.* 4: 75–87.

Ju, H., S. H. Choi, and S. H. Huh. 2010. "X-Ray Diffraction Patterns of Thermally-Reduced Graphenes." *Journal of the Korean Physical Society* 57: 1649–1652.

Kamat, P. V. 2010. "Graphene—A Physical Chemistry Perspective."*J. Phys. Chem. Lett.*1: 587–588.

Ke, Q., and J. Wang. 2016. "Graphene-Based Materials for Supercapacitor Electrodes – A Review." *J Materiomics* 2: 37–54.

Kou, L., Tieqi Huang, Bingna Zheng, Yi, Han, et al. 2014. "Coaxial Wet-Spun Yarn Supercapacitors for High-Energy Density and Safe Wearable Electronics." *Nat. Commun* 5, no. 3754, 1–10.

Kumar, K. S., N. Choudhary, Y. Jung, and J. Thomas. 2018. "Recent Advances in Two-Dimensional Nanomaterials for Supercapacitor Electrode Applications." *ACS Energy Letters* 3: 482–495.

Lee, C., X. D. Wei, J. W. Kysar, and J. Hone. 2008. "Measurement of the Elastic Properties and Intrinsic Strength of Monolayer Graphene." *Science* 321: 385–387.

Li, N., Z. Chen, W. Ren, F. Li, and H. M. Cheng. 2012. "Flexible Graphene-based Lithium Ion Batteries with Ultrafast Charge and Discharge Rates." *Proc. Natl Acad. Sci. USA* 109: 17360–17365.

Li, M., G. Sun, P. Yin, C. Ruan, and K. Ai. 2013. "Controlling the Formation of Rodlike V2O5 Nanocrystals on Reduced Graphene Oxide for High-Performance Supercapacitors." *ACS Appl Mater Interfaces* 5, no. 21: 11462–11470.

Li, J., S. Huang, J. Gu, Q. Wu, D. Chen, and C. Zhou. 2019. "Facile Synthesis Of Well-dispersed Bi2O3 Nanoparticles and rGO as Negative Electrode for Supercapacitor." *J Nanopart Res* 21, no. 56: 1–8.

Lightcap, I. V., T. H. Kosel, and P. V. Kamat. 2010. "Anchoring Semiconductor and Metal Nanoparticles on a Two-dimensional Catalyst Mat. Storing and Shuttling Electrons with Reduced Graphene Oxide." *Nano Letters* 10, no. 2: 577–583.

Lin, M. C., Ming Gong, Bingan Lu, Yingpeng Wu, Di-Yan Wang, Mingyun Guan, Michael Angell, 2015. "An Ultrafast Rechargeable Aluminium-Ion Battery." *Nature* 520: 324–328.

Liu, F., S. Song, D. Xue, and H. Zhang. 2012. "Folded Structured Graphene Paper for High Performance Electrode Materials." *Adv. Mater* 24: 1089–1094.

Liu, W. W., Y. Q. Feng, X. B. Yan, J. T. Chen, and Q. J. Xue. 2013. "Superior Micro-Supercapacitors Based on Graphene Quantum Dots." *Adv. Funct. Mater* 23: 4111–4122.

Lv, W., D. M. Tang, Y. B. He, C. H. You, Zhi-Qiang Shi, Xue-Cheng Chen, Cheng-Meng Chen, Peng-Xiang Hou, Chang Liu, and Quan-Hong Yang. "Low-Temperature Exfoliated Graphenes: Vacuum-promoted Exfoliation and Electrochemical Energy Storage." *ACS Nano* 3, no. 11: 3730–3736.

Ma, Y., H. Chang, M. Zhang, and Y. Chen. 2015. "Graphene-Based Materials for Lithium-Ion Hybrid Supercapacitors." *Adv. Mater* 27: 5296–5308.

Meher, S. K., P. Justin, and G. R. Rao. 2010. "Pine-cone morphology and pseudocapacitive behavior of nanoporous nickel oxide." *Electrochim Acta* 55, no. 28: 8388–8396.

Meng, Y., Y. Zhao, C. Hu, H. Cheng, Y. Hu, and Z. a. Zhang. 2013. "Graphene Core-sheath Microfibres for All-Solid-State, Stretchable Fibriform Supercapacitors and Wearable Electronic Textiles." *Adv Mater* 25: 2326–2331.

Miller, J. R., R. A. Outlaw, and B. C. Holloway. 2010. "Graphene Double-layer Capacitor With AC Line-Filtering Performance." *Science* 329, no. 5999: 1637–1639.

Mukherjee, R., A. V. Thomas, A. Krishnamurthy, and N. Koratkar. 2012. "Photothermally Reduced Graphene as High-power Anodes for Lithium-ion Batteries." *ACS Nano* 6: 7867–7878.

Muzaffar, A., M. B. Ahamed, K. Deshmukh, and J. Thirumalai. 2019. "A Review on Recent Advances in Hybrid Supercapacitors: Design, Fabrication and Applications." *Renew. Sustain. Energy Rev.* 101: 123–145.

Novoselov, K. S., A. K. Geim, S. V. Morozov, D. Jiang, et al. 2005. "Two-Dimensional Gas of Massless Dirac Fermions in Graphene." *Nature* 438, no. 7065: 197–200.

Novoselov, K. S., A. K. Geim, S. V. Morozov, D. Jiang, Y. Zhang, S. V. Dubonos, and A. A.Firsov. 2004. "Electric Field Effect in Atomically Thin Carbon Films." *Science* 306, no. 5696: 666–669.

Nwanya, A. C., M. M. Ndipingwi, C. O. Ikpo, R. M. Obodo, S. C. Nwanya, S. Botha, F. I. Ezema, E. I. Iwuoha, and M. Maaza. 2020. "Zea Mays Lea Silk Extract Mediated Synthesis of Nickel Oxidenanoparticles as Positive Electrode Material for Asymmetric Supercabattery." *Journal of Alloys and Compounds* 822: 153581.

Obodo, R. M., A. Ahmad, G. H. Jain, I. Ahmad, M. Maaza, and F. I. Ezema. 2020. "8.0 MeV Copper Ion (Cu^{++}) Irradiation-Induced Effects on Structural, Electrical, Optical and Electrochemical Properties of Co$_3$O$_4$-NiO-ZnO/GO Nanowires." *Materials Science for Energy Technologies* 3: 193–200.

Obodo, R. M., Assumpta C. Nwanya, Tabassum Hassina, Mesfin Kebede, Ishaq Ahmad, M. Maaza, and Fabian I. Ezema. 2019. "Transition Metal Oxide-based Nanomaterials for High Energy and Power Density Supercapacitor." In Electrochemical Devices for Energy Storage Applications, edited by Mesfin A., Kebede, and Fabian I. Ezema. United Kingdom: Taylor & Francis Group, CRC Press, 131–150.

Obodo, R. M., A. C. Nwanya, A. B. C. Ekwealor, I. Ahmad, T. Zhao, M. Maaza, and F. I. Ezema. 2019. "Influence of pH and Annealing on the Optical and Electrochemical Properties of Cobalt (III) Oxide (Co$_3$O$_4$) Thin Films." *Surfaces and Interfaces* 16: 114–119.

Obodo, R. M., A. C. Nwanya, C. Iroegbu, I. Ahmad, Azubike B. C. Ekwealor, Rose U. Osuji, Malik Maaza, and Fabian I. Ezema. 2020. "Transformation of GO to rGO due to 8.0 MeV Carbon (C++) Ions Irradiation and Characteristics Performance on MnO$_2$–NiO–ZnO@GO Electrode." *Int J Energy Res.* 44: 6792 – 6803.

Obodo, R. M., A. C. Nwanya, M. Arshad, C. Iroegbu, I. Ahmad, R. U. Osuji, M. Maaza, and F. I. Ezema. 2020. "Conjugated NiO-ZnO/GO Nanocomposite Powder for Applications in Supercapacitor Electrodes Material." *Int J Energy Res* 2020, no. 44: 3192–3202.

Obodo, R. M., A. C. Nwanya, C. Iroegbu, B. A. Ezekoye, A. B. C. Ekwealor, I. Ahmad, M. Maaza, and F. I. Ezema. 2020. "Effects of Swift Copper (Cu^{2+}) Ion Irradiation on Structural, Optical And Electrochemical Properties of Co$_3$O$_4$-CuO-MnO$_2$/GO Nanocomposites Powder." *Advanced Powder Technology* 31: 1728 – 1735.

Obodo, R. M., M. Asjad, A. C. Nwanya, I. Ahmad,Tingkai Zhao, A. B. Ekwealor, Paul M. Ejikeme, Maalik Maaza, and Fabian I . Ezema. 2020. "Evaluation of 8.0 MeV Carbon (C^{2+}) Irradiation Effects on Hydrothermally Synthesized Co$_3$O$_4$-CuO-ZnO@GO Electrodes for Supercapacitor Applications." *Electroanalysis* 32: 1–12,

Obodo, R. M., N. M. Shinde, U. K. Chime, S. Ezugwu, A. C. Nwanya, I. Ahmad, M. Maaza, P. M. Ejikeme, and F. I. Ezema. 2020. "Recent Advances in Metal Oxide/hydroxide on Three-dimensional Nickel Foam Substrate for High Performance Pseudocapacitive Electrodes." *Current Opinion in Electrochemistry* 21: 242–249.

Obodo, R. M., E. O. Onah, H. E. Nsude, A. Agbogu, A. C. Nwanya, I. Ahmad, T. Zhao, P. M. Ejikeme, M. Maaza, F. I. Ezema. 2020. "Performance Evaluation of Graphene Oxide Based Co$_3$O$_4$@GO, MnO$_2$@GO and Co$_3$O$_4$/MnO$_2$@GO Electrodes for Supercapacitors." *Electroanalysis* 32: 1–10.

Parida, K., V. Bhavanasi, V. Kumar, J. Wang, and P. S. Lee. 2017. "Fast Charging Self-powered Electric Double Layer Capacitor." *J. Power* 342: 70–78.

Popov, V. N. 2004. "Carbon Nanotubes: Properties and Application." *Material science and Engineering R* 43: 61–102.

Qiu, Y., H. Fan, X. Chang, H. Dang, Q. Luo, and Z. Cheng. 2018. "Novel Ultrathin Bi2O3 Nanowires for Supercapacitor Electrode Materials with High Performance." *Appl Surf Sci.* 434: 16–20.

Rueckes, T., K. Kim, E. Joselevich, G. Y. Tseng, C. L. Cheung, and C. M. Lieber. 2000. "Carbon Nanotube-Based Nonvolatile Random Access Memory for Molecular Computing." *Science* 289, no. 5476: 94–97.

Shao, Y.Lisa J. Wang, Qinghong Zhang, Yaogang Li, Hongzhi Wang, Mir F. Mousaviae, and Richard B. Kaner. 2015. "Graphene-Based Materials for Flexible Supercapacitors." *Chem. Soc. Rev.* 44: 3639–3665.

Sheng, K., Y. Sun, C. Li, W. Yuan, and G. Shi. 2012. "Ultrahigh-Rate Supercapacitors Based on Eletrochemically Reduced Graphene Oxide for AC Line-Filtering." *Sci. Rep.* 2, no. 1: 1–5.

Smith, R. C. , D. C. Cox, and S. R. P. Silva. 2005. "Electron Field Emission from a Single Carbon Nanotube: Effects of Anode Location." *Applied Physics Letters* 87, no. 10: 103–112.

Staudenmaier, L. 1898. "Process for the Preparation of Graphitic Acid." *Reports of the German Chemical Society* 31, no. 2: 1481–1487.

Stoller, M. D., S. Park, Y. Zhu, J. An, and R. S. Ruoff. 2008. "Graphene-Based Ultracapacitors." *Nano letters* 8, no. 10: 3498–3502.

Thangappan, R., M. Arivanandhan, R. Dhinesh Kumar, and R. Jayavel. 2018. "Facile Synthesis of RuO_2 Nanoparticles Anchored on Graphene Nanosheets for High Performance Composite Electrode for Supercapacitor Applications." *J Phys Chem Solids* 121: 339–349.

Tsai, J. L., and J. F. Tu. 2010. "Characterizing Mechanical Properties of Graphite Using Molecular Dynamics Simulation." *Materials & Design* 31, no. 1: 194–199.

Uduh, U. C., R. M. Obodo, S. Esaenwi, C. I. Amaechi, P. U. Asogwa, R. U. Osuji, F. I. Ezema 2014. "Sol-Gel Synthesis, Optical and Structural Characterization of Zirconium Oxysulfide (ZrOS) Nanopowder." *J Sol-Gel Sci Technol* 71: 79 – 85.

Veerappan, G., K. Bojan, and S. W. Rhee. 2011. "Sub-micrometer-sized Graphite as a Conducting and Catalytic Counter Electrode for Dye-sensitized Solar Cells." *ACS applied materials & interfaces* 3, no. 3: 857–862.

Vivekchand, S. R. C., C. S. Rout, K. S. Subrahmanyam, A. Govindaraj, and C. N. R. Rao. 2008. "Graphene-Based Electrochemical Supercapacitors." *Journal of Chemical Sciences* 120, no. 1: 9–13.

Wang, X., and G. Shi. 2015. "Flexible Graphene Devices Related to Energy Conversion and Storage." *Energy Environ. Sci.* 8: 790–823.

Wang, X., L. J. ZhiKlaus Müllen. 2008. "Transparent, Conductive Graphene Electrodes for Dye-sensitized Solar Cells." *Nano Lett* 8: 323–327.

Wang, G., X. Shen, J. Yao, and J. Park. 2009. "Graphene Nanosheets for Enhanced Lithium Storage in Lithium Ion Batteries." *Carbon* 47, no. 8: 2049–2053.

Wang, D. W., F. Li, Z. S. Wu, W. Ren, and H. M. Cheng. 2009. "Electrochemical Interfacial Capacitance in Multilayer Graphene Sheets: Dependence on Number of Stacking Layers." *Electrochemistry Communications* 11, no. 9: 1729–1732.

Wang, Y., C. X. Guo, J. Liu, T. Chen, H. Yanga, and C. M. Li. 2011. "CeO_2 Nanoparticles/Graphene Nanocomposite-Based High Performance Supercapacitor." *Dalton Trans* 40: 6388–6391.

Wang, P., H. Liu, Y. Xu, Y. Chen, J. Yang, and Q. Tan. 2016. "Supported Ultrafine Ruthenium Oxides with Specific Capacitance up to 1099 F g–1 for a Supercapacitor." *Electrochim Acta* 194: 211–218.

Wang, H., M. Liang, D. D. Zhang X, W. Shi, Y. Song, and Z. Sun. 2018. "Novel CeO_2 Nanorod Framework Prepared by Dealloying for Supercapacitors Applications." *Ionics* 24: 2063–2072.

Wang, Y., Z. Shi, Y. Huang, Y. Ma, C. Wang, M. Chen, and Y. Chen. 2009. "Supercapacitor Devices Based on Graphene Materials." *The Journal of Physical Chemistry C* 113, no. 30: 13103–13107.

Wei, D., L. Grande, V. Chundi, R. White, C. Bower, P. Andrew, and T. Ryhänen. 2012. "Graphene from Electrochemical Exfoliation and Its Direct Applications in Enhanced Energy Storage Devices." *Chemical Communications* 48(9): 1239–1241.

Winter, M., J. O. Besenhard, M. E. Spahr, and P. Novak. 1998. "Insertion Electrode Materials for Rechargeable Lithium Batteries." *Advanced materials* 10, no. 10: 725–763.

Wu, Z. S., W. Ren, L. Xu, F. Li, and H. M. Cheng. 2011. "Doped Graphene Sheets as Anode Materials with Superhigh Rate and Large Capacity for Lithium Ion Batteries." *ACS Nano* 5: 5463–5471.

Wu, Z. S., G. Zhou, L. C. Yin, W. Ren, and F. Li. 2012. "H. and M. Cheng' Graphene/Metal Oxide Composite Electrode Materials for Energy Storage." *N. Energy* 1: 107–31.

Wu, Z. S., Z. Liu, K. Parvez, X. Feng, and K. Müllen. 2015. "Ultrathin Printable Graphene Supercapacitors with AC Line-Filtering Performance." *Adv. Mate* 27: 3669–3675.

Wu, Z. S., G. Zhou, L. C. Yin, W. Ren, F. Li, and H. M. Cheng. 2012. "Graphene/Metal Oxide Composite Electrode Materials for Energy Storage." *Nano Energy* 1, no. 1: 107–131.

Wu, Z. S., D. W. Wang, W. Ren, J. Zhao, G. Zhou, F. Li, and H. M. Cheng. 2010. "Anchoring Hydrous RuO2 on Graphene Sheets for High-Performance Electrochemical Capacitors." *Advanced Functional Materials* 20, no. 20: 3595–3602.

Xia, J. L., F. Chen, J. H. Li, and N. J. Tao. 2009. "Measurement of the Quantum Capacitance of Graphene." *Nat Nanotechnol* 4: 505–509.

Xu, Y., Zhaoyang Lin, Xing Zhong, Ben Papandrea, Yu Huang, and Xiangfeng Duan. 2015. "Solvated Graphene Frameworks as High-performance Anodes for Lithium-ion Batteries." *Angew. Chem. Int. Ed. Engl* 127: 5435–5440.

Yang, F., X. Zhang, Y. Yang, S. Hao, and L. Cui. 2018. "Characteristics and Supercapacitive Performance of Nanoporous Bamboo Leaf-like CuO." *Chem Phys Lett* 691: 366–372.

Ye, M., Zelin Dong, Chuangang Hu, Huhu Cheng, Nan Chen, and Liangti Qu. 2014. "Uniquely Arranged Graphene-on-Graphene Structure as a Binder-Free Anode for High-Performance Lithium-Ion Batteries." *Small* 10: 5035–5041.

Yoo, E., J. Kim, E. Hosono, H. S. Zhou, T. Kudo, and I. Honma. 2008. "Large Reversible Li Storage of Graphene Nanosheet Families for Use in Rechargeable Lithium Ion Batteries." *Nano Letters* 8, no. 8: 2277–2282.

Zhang, L. L., R. Zhou, and X. S. Zhao. 2010. "Graphene-Based Materials as Supercapacitor Electrode." *J Mater Chem.* 20: 5983–5992.

Zhang, Y., Y. W. Tan, H. L. Stormer, and P. Kim. 2005. "Experimental Observation of the Quantum Hall Effect and Berry's Phase in Graphene." *Nature* 438, no. 7065: 201–204.

Zhao, B., D. Chen, X. Xiong, B. Song, R. Hu, Q. Zhang, and B. H. Rainwater, et al. 2017. "A High-energy, Long Cycle-life Hybrid Supercapacitor Based on Graphene Composite Electrodes." *Energy Storage Materials* 7: 32–39.

Zhao, B., P. Liu, H. Zhuang, Z. Jiao, T. Fang, W. Xu, B. Lub, and Y. Jiang. 2013. "Hierarchical Self-assembly of Microscale Leaf-like CuO on Graphene Sheets For High-performance Electrochemical Capacitors." *J Mater Chem A* 1, no. 2: 367–373.

Zhou, W., Jixin Zhu, Chuanwei Cheng, Jinping Liu 2011. "A General Strategy toward Graphene Metal Oxide Core–Shell Nanostructures for High-performance Lithium Storage." *Energy Environ. Sci.* 4: 4954–4961.

Zhu, Y., S. Murali, M. D. Stoller, A. Velamakanni, R. D. Piner, and R. S. Ruoff. 2010. "Microwave Assisted Exfoliation and Reduction of Graphite Oxide for Ultracapacitors." *Carbon* 48, no. 7: 2118–2122.

Zhu, Y., M. D. Stoller, W. Cai, A. Velamakanni, R. D. Piner, D. Chen, and R. S. Ruoff. 2010. "Exfoliation of Graphite Oxide in Propylene Carbonate and Thermal Reduction of the Resulting Graphene Oxide Platelets." *ACS nano* 4, no. 2: 1227–1233.

12

Dual Performance of Fuel Cells as Efficient Energy Harvesting and Storage Systems

Agnes Chinecherem Nkele[1], Fabian I. Ezema[1,2,3], and Mesfin A. Kebede[4,5]

[1]Department of Physics & Astronomy, University of Nigeria, Nsukka, Enugu, Nigeria
[2]Nanosciences African Network (NANOAFNET), iThemba LABS-National Research Foundation, Somerset West, Western Cape Province, South Africa
[3]UNESCO-UNISA Africa Chair in Nanosciences/Nanotechnology, College of Graduate Studies, University of South Africa (UNISA), Muckleneuk Ridge, Pretoria, South Africa
[4]Energy Centre, Council for Scientific & Industrial Research, Pretoria, South Africa
[5]Department of Physics, Sefako Makgatho Health Science University, South Africa

12.1 Introduction

The world has been faced with energy and pollution challenges that need to be sorted out (Kordesch and Simader 1996). Industrialization and the fast-growing economy have led to the discharge of industrial and organic wastes into water bodies ("Fuel Cells" 2020). Fuel cells are systems that generate electricity from the chemical energy stored in hydrogen or other fuels. Hydrogen as gas has the highest energy content of any fuel. Fuel cells can convert hydrogen (obtained from water splitting) to electricity that would be useful in powering trucks and cars (https://Secure.gravatar.com/Avatar/5863d93f56930946502b44f6bc5b9ee5?s=60 et al. 2019). Here, oxygen and fuel are continuously supplied for the sustenance of its chemical reactions and continuous generation of electricity. Fuel cells do not involve burning, have no moving parts, and do not need to be recharged as they have similar operations to batteries. They have energy efficiencies ranging from 40% to 60% and the highest theoretical efficiency of 83% at low power densities ("Fuel Cell" 2020). Although losses could be incurred from transport, production, and storage of fuels, fuel cells are wise paths toward providing a nation with electric power. Stacks of fuel cells are very reliable and efficient as there are no moving parts to constitute pollution or noise. Fuel cells are vastly applied in power production, heat generation, stationary materials, stand-alone power plants, parks, automobiles, spacecraft, boats, emergency power devices, industrial lift trucks, communication centres, transport media, etc. (Kordesch and Simader 1996; "Fuel Cell" 2020).

Fuel cells generate electricity from external fuel supplies. Catalysts and electrodes used in fuel cells undergo degradation after activation and are left on standby (Kordesch et al. 2000). Platinum can be utilized as a catalyst component to improve the activity of fuel cells. Increased high potential encountered under open-circuit conditions is responsible for the oxidation of carbon. Electrolytes serve as reactors that prevent reagents from mixing, while the electrodes are the reaction catalysts (https://Secure.gravatar.com/Avatar/5863d93f56930946502b44f6bc5b9ee5?s=60 et al. 2019). Solid electrolytes are useful in fuel cells for improving cell performance (Pasciak et al. 2001). Electrolytes with ion-exchange membranes have recorded more durability and increased efficiency, thereby giving rise

DOI: 10.1201/9781003145585-12

to more experimental procedures, transport models, and electrode assemblies. The type of electrolyte used and the time difference for start-up determine the kind of fuel cell that would be produced ("Fuel Cell" 2020). Fuel cells make an efficient output of about 80% after capturing the released heat in a cogeneration manner.

12.2 Working Principle of Fuel Cells

Fuel cells are made up of anode/negative electrode and cathode/positive electrode wrapped in an electrolyte as shown in Figure 12.1. The anode (usually made of platinum) splits the fuel into ions and electrons; cathode releases waste chemicals upon conversion of ions; and the electrolyte consists of phosphoric acid, potassium hydroxide, or salt carbonates. These three components form an interface where chemical reactions occur and result in fuel consumption, current creation, and water production. Fuel and air are respectively fed into the anode and cathode where a catalyst at the anode separates the molecules into protons and electrons. The protons are transported towards the cathode through the electrolyte, while the electrons create electricity flow as they migrate through an external circuit. The electrolyte gives room for ionic motion at both sides of the fuel cell. At the anode, fuel is oxidized by a catalyst as oxidation reactions triggered by the fuel produce electrons and ions. The ions get transported to the cathode through the electrolyte from the anode. Another catalyst at the cathode triggers a reaction between ions, oxygen, and electrons to form water and other by-products like heat, nitrogen(IV) oxide. Losses from ohmic loss, activation energy, and mass transport lead to the reduction of voltage with an increase in current.

12.3 Advantages and Disadvantages of Fuel Cells

Every energy system has advantages and disadvantages associated with it as outlined in Table 12.1 ("Fuel Cell" 2020).

12.4 Classifications of Fuel Cells

Although there are different classes of fuel cells, they operate on the same principle. The classifications of fuel cells discussed here are alkaline fuel cells, proton exchange membrane fuel cell, direct methanol fuel cell, microbial fuel cell, polymer electrolyte fuel cells, photocatalytic fuel cell, solid acid fuel cell, phosphoric acid fuel cell, and molten carbonate fuel cell. Other kinds of a fuel cell include metal hydride fuel cell, electrogalvanic fuel cell, direct formic acid fuel cell, upflow microbial fuel cell, regenerative

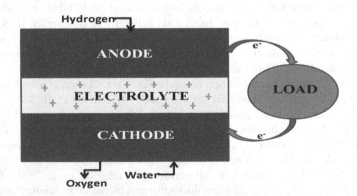

FIGURE 12.1 Schematic diagram of a fuel cell.

TABLE 12.1

Some Advantages and Disadvantages of Fuel Cells

Advantages	Disadvantages
Useful in energy generation	Very expensive to manufacture
Highly efficient in converting power	
Keeps generating electricity as long as there is a fuel supply	Less durable
Generates power with different kinds of fuels	Size and weight barriers to commercialization of fuel cells
Vast areas of applications	
No poisonous emissions	
Less noise emitted during operating activities	Fuel cells require temperature regulation
No need for recharging	
Has minimal transmission losses for a disturbed system	The cost of storing fuel is expensive
Quiet mode of operation	
Easy mode of transporting gases to the power stations	Transportation losses could be incurred
Low running costs	
Able to co-generate	Fuel cell technology is not widely available
Low maintenance is required	
High power density	

fuel cell, direct borohydride fuel cell, redox fuel cell, tubular solid oxide fuel cell, planar solid oxide fuel cell, enzymatic biofuel cells, magnesium air fuel cells, etc.

12.4.1 Alkaline Fuel Cells (AFCs)

In the early days, AFCs used liquid electrolytes before the introduction of hydrogen-oxygen fuel cells. ELENCO group in 1996 developed alkaline fuel cells with circulating electrolyte until after a year when the ZEVCO group took over (Kordesch et al. 2000). The fuel cell city car in Kordesch was constructed with a carbon(IV) oxide air absorber and no power converter. The lifespan increases as the electrolyte circulates. AFCs suffer limitations like immobile electrolytes, accumulated carbonate residues, deteriorating separators, and crystallization of stored gases if not properly maintained. Ammonia is an efficient fuel used in AFCs because of its hydrogen content and strong odour that allows for leakage detection (Kordesch et al. 2000). This detection is possible because the ammonia which is insensitive to the electrolyte is circulated around the heating chamber without any need for an oxidizer or converter.

12.4.2 Proton Exchange Membrane Fuel Cells (PEMFCs)

PEMFCs that are optimized at high temperature and low humid conditions are more catalytically stable with fast electrode kinetics (Carollo et al. 2006). This optimization allows newly formed membranes to move protons without assistance from water molecules. PEMFCs comprise of catalyst, bipolar plates, membrane, electrodes, and current collectors. The different stages from the design, manufacturing, assembling, and testing of proton exchange membrane fuel cells are represented in Figure 12.2 (Chen et al. 2019). Dissociation of hydrogen into electrons and protons occurs at the anode. The reaction between the protons and oxidants leads to the formation of proton membranes. The protons migrate to the cathode through the formed membrane while the electrons move through an external circuit for reaction with oxygen. The polymer membrane used is highly conductive to the proton, encloses the electrolyte, and separates the electrodes. Nafion is an acid electrolyte that is highly stable and conductive, except at very high temperatures where it becomes very resistant with low adhesive power. Zirconium phosphate/Nafion composites, titanium/Nafion composite organic

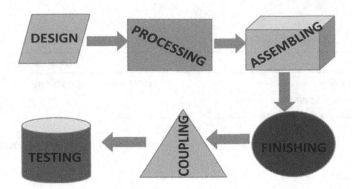

FIGURE 12.2 Different stages from the design to testing of PEMFCs.

metal oxides, and acid-doped polybenzimidazoles have protons that are highly conductive even in very low humid environments (Carollo et al. 2006). Zirconium phosphate/ionic liquids can also be adopted to improve proton conductivity (Mohammed et al. 2019). Incorporating membranes with ionic liquids make it highly stable, non-flammable, and very conductive. The improved conductivity is due to the intercalation of the ionic liquids, exfoliation of hydrogen sites, and increased water uptake (Mohammed et al. 2019). Doping with acid, use of inorganic fillers, organic metallic fillers, and fabrication of porous films soaked in acids increase the conductivity of protons at very low temperatures (Carollo et al. 2006). Acid fillers are used to reduce the losses associated with proton conduction (Cozzi et al. 2014). Decay in fuel cells could occur due to gas insufficiency during cell operations. This gas insufficiency exists because of uneven gas distribution, water mismanagement, hydrogen insufficiency, quick shutdown and loading occurrences, reflux of air, high permeability of hydrogen, and presence of gaseous impurities (Chen et al. 2019). The performance of PEMFCs gets degraded due to idle load changing, the reduced surface area of the catalyst, and high power situations. PEMFCs are also used for stationary applications.

12.4.3 Direct Methanol Fuel Cells (DMFCs)

DMFCs have high fuel densities, are easily transported, and relatively cheap because of methanol (Cozzi et al. 2014). Methanol may be obtained from coal, natural gas, and natural gas. The efficiencies of DMFCs are reduced because methanol lowers the oxidation kinetics, efficiency, and power density of fuel cells. However, the cell's performance can be increased by operating at higher temperatures such that the rate of reaction can be increased, the permeability level of methanol decreased, and the anode tolerance level increased. Adding metal oxides of silica, zirconia, and titania enhances the performance and stability of the cell (Cozzi et al. 2014). Reformed methanol fuel cells have a polymer membrane, power output between 5 to 100 kilo Watts, and output efficiency above 50% ("Fuel Cell" 2020).

12.4.4 Microbial Fuel Cells (MFCs)

Microbial fuel cells are bio-electrochemical devices that generate an electric current through the activities of microorganisms (Mathuriya and Yakhmi 2016). MFCs engage wastewater as the fuel for electricity creation. Here, the electrodes are separated by a cationic membrane, and the chemical energy obtained from organic materials is converted to electrical energy via several metabolisms. Electrons and protons are generated in the anode upon oxidation of fuel by microorganisms. The electrons get migrated to an external circuit while the protons migrate to the cathode where the protons and electrons react together to produce water. MFCs have improved performance activities and high efficiencies even at ambient temperatures, and aid the conversion of chemical to electrical energy (Mathuriya and Yakhmi 2016). Figure 12.3 shows the schematic of a microbial fuel cell with a proton exchange membrane (PEM). The MFC has a proton exchange membrane that separates the cathode and anode chambers that are connected to an external circuit for current to flow. Microbial fuel cells are used to generate

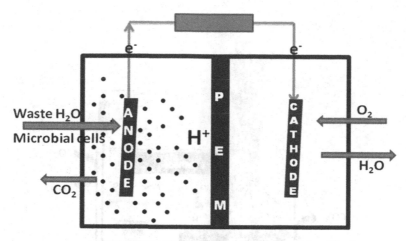

FIGURE 12.3 Schematic of a two-chamber microbial fuel cell.

bioelectricity from bacteria, waste materials, sludge, and other microorganisms. MFCs are also useful in nitrification, softening water, decolourizing dyes, bioremediation, generation of biofuels, treating wastewater, and different pollutants, as they do not involve much energy usage or sophisticated systems (Mathuriya and Yakhmi 2016). The limitation posed by electron transport at the anode can be resolved by using a photoanode of high performance. Several research studies have shown that MFCs are also used for stationary applications. The low power output of microbial fuel cells can be enhanced by adopting functional materials like graphene in designing the anode. Graphene is an efficient anode in MFCs because it creates a wide area of surface for reaction, is highly conductive, increases the rate of moving electrons, and enhances power density (Zhu et al. 2019).

Bioenergy produced from biological cathode MFCs has helped alleviate renewable energy problems. This bioenergy is obtainable from biological systems like bacteria, after oxidation of the organic materials at the anode and oxygen reduction at the cathode (Song, Zhu, and Li 2019). The bacteria serve as a biocatalyst with a high density of electrons. Biocathodes use metal ions, inorganic salts, oxygen, carbon(IV) oxide, contaminants, or transition metals as electron acceptors. The different MFC configurations are usually based on different compositions/materials that the cathode is made of like liquid, air, or soluble materials; while the anode contains the microorganisms (Song, Zhu, and Li 2019). During aerobic reactions, iron and manganese can be used to mediate the transfer of oxygen for efficient electron transport. Biocathodes are relatively cheap with large areas of surfaces, minimal pollution, improved power output, and increased potential at the cathode. Biocathodes are applied in microbial fuel cells and in treating wastewater. Microbial fuel cells can also use roots of plants inserted into the substrate and used in fuelling bacteria at the anode for bioenergy production while the released oxygen becomes the biocathode as illustrated in Figure 12.4 (Kabutey et al. 2019). Plant MFCs can be designed into single or double chambers and have no harmful effects on foods produced in farms or indoors. The choice of plants is dependent on the species of plants available and the location under study (Kabutey et al. 2019).

12.4.5 Polymer Electrolyte Fuel Cells (PEFCs)

PEFCs have high power densities for energy production. Some polymer electrolytes like sulphonated polyetheretherketones and Nafion retain less water at increased temperatures, show reduced proton conduction, and are also less stable to heat (Charradi et al. 2019). Some fillers like particles of zirconia, titania, silica can be mixed with the electrolyte to improve its retention of water and stability to heat. Other fillers like materials made from clay can also be explored to enhance proton conductivity due to the water present in the clay particles (Charradi et al. 2019). Anion exchange membranes like silica-based or graphene-based composite membranes can be incorporated into PEFCs for their cost-effectiveness and high efficiency (Vijayakumar and Nam 2019).

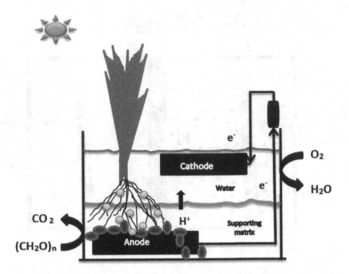

FIGURE 12.4   Diagram of a plant microbial fuel cell (Kabutey et al. 2019). Reproduced with permission. Copyright Elsevier 2019.

12.4.6 Photocatalytic Fuel Cells (PFCs)

Photocatalytic fuel cells have been developed to solve energy scarcity and minimize pollution effects by properly utilizing the energy and carbon obtained from organic compounds. These fuel cells are obtained after the decomposition of organic pollutants and the conversion of chemical energy into electricity (Li et al. 2019). Photocatalytic FC devices consist of electrocatalyst carrier/cathode, photocatalyst carrier/photoanode, substrates, reactor, electrolyte, etc. PFCs have an easy method of construction and operation, as their functions are dependent on the type and properties of electrode used. From Figure 12.5, photodecomposition occurs when the light illuminates the anode, and synchronously produces hydrogen at the cathode after electrons are transferred via an external circuit. PFCs can create energy while degrading organic pollutants. They also have simple development methods, high efficient output, and a practically obtainable system. PFCs are useful in treating wastewater (Li et al. 2019).

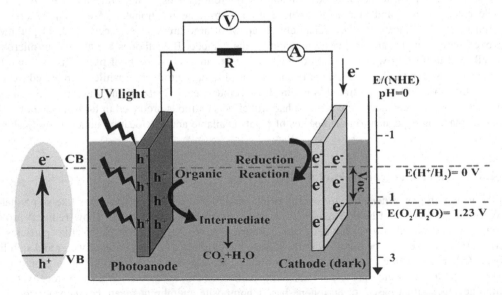

FIGURE 12.5   Diagram showing the working principle of a photocatalytic fuel cell (Li et al. 2019). Reproduced with permission. Copyright Elsevier 2019.

12.4.7 Solid Acid Fuel Cells (SAFCs)

Solid acid fuel cells use a solid acid as the electrolyte. Solid acids have well-arranged molecules at low temperatures which get disordered at higher temperatures ("Fuel Cell" 2020). Cesium dihydrogen phosphate is currently used as the cell electrolyte for energy production.

12.4.8 Phosphoric Acid Fuel Cells (PAFCs)

Phosphoric acid fuel cells work effectively between 150°C and 200°C and use phosphoric acid as an electrolyte because it allows positive ions to be easily moved from the anode through the electrolyte to the cathode without conduction ("Fuel Cell" 2020). The time needed for constructing and installing PAFCs is small. The efficiency of phosphoric acid FCs does not depend on the load, but on the energy production rate of the fuel used. Platinum is incorporated as a catalyst to boost the rate of producing hydrogen ion at the anode. However, the acidic electrolyte causes the cell components to corrode easily.

12.4.9 Molten Carbonate Fuel Cells (MCFCs)

MFCFs convert fossil fuel into gases rich in hydrogen and emit carbon dioxide gases. They operate at high temperatures, withstand impurities, are highly efficient, and use lithium potassium carbonate salt as the electrolyte. This salt aids the motion of charges in the cell and changes to liquid at high temperatures. Carbonate ions combine with the hydrogen component of the gas to yield carbon dioxide, electrons, and water. The electrons generate electricity after passing through an external circuit, before returning to the cathode. The electrolyte is refilled upon reaction between the recycled oxygen/carbon dioxide and electrons. The high temperature at which the cells operate slows down the time for start-up and degrades the components and reduces their lifespan.

12.5 Dual Functions of Fuel Cells

Fuel cells can harvest energy, convert hydrogen into electrical power, and store the harvested energy through diverse hydrogen storage systems.

12.5.1 Fuel Cells as Energy Harvesters

Fuel cells can harvest energy from different materials for energy production and electricity generation. Harvesting of energy is essential, as it allows energy to be produced from external materials like solar energy, kinetic sources, thermal energy, and microbial fuel cells, using reduced-power electronic materials (Osorio de la Rosa et al. 2019). Energy harvesters require light, and available, portable, and cheap sources of energy that can be harnessed and stored (Vullers et al. 2009). Such harvesters should be attached to wireless sensors that would receive and transmit useful signals for processing at a base station. Every energy harvester has an operating point at which the harvested electrical energy is at the highest level and regulated by a controller. The controller prevents the discharge of power when low energy is harvested but releases power when the available energy is sufficient (Vullers et al. 2009). Microbial fuel cells are small-sized energy harvesting devices that generate power from bacteria-converted biomass. Mink et al. fabricated MFCs to harvest energy with the use of human saliva as fuel, graphene as anode for generating current, and a cathode (Mink et al. 2014). A conductivity of about 2.5 mS/cm, improved conductivity, higher electron mobility, and increased current density were achieved. The cathode made of air eliminates the use of external chemicals as oxygen is obtained from air. The saliva-powered fuel yielded increased current density and more power. The obtained results make it useful to be applied in microchips, bioelectronics, and diagnostic equipment (Mink et al. 2014). Osorio et al. harvested energy using plant microbial fuel cell (PMFC) to develop self-powered devices (Osorio de la Rosa et al. 2019). Although PMFCs suffer from fluctuations in power supply and charging rates, it is a clean means of energy generation. Power density, maximum power point, and current density values of 3.5 mW/cm^2, 0.71 V, and 5 mA/cm^2 were, respectively, obtained from the process and can be applied in electrical appliances and biosensors (Osorio de la Rosa et al. 2019).

In practical terms, devices for capturing, storing, and boosting energy like power management systems (made of diodes, boosters, capacitors, inductors, charge pumps, and potentiometer) and maximum power point tracking (MPPT) should be incorporated into harvesting systems (Wang, Park, and Ren 2015). MPPTs directly harness energy at the maximum power (without external hindrances) for optimal energy to be produced. Charge pumps and boost converters can also be attached with microbial fuel cells to increase the cell harvesting rate. Harvesting of energy takes place in the charging and discharging phases (Wang, Park, and Ren 2015). During charging, the controller drains energy from the microbial fuel cell for storage in the inductor. The energy is then moved from the controller and stored in capacitors for future use. Implantable glucose alongside abiotic catalysts like activated carbon and noble metals can be incorporated into fuel cells to supply energy for medical implantation systems (Kerzenmacher et al. 2008). The glucose gets oxidized into carbon dioxide and water, releases electrons, and is useful in power generators. This approach doesn't require an external recharging process and works based on the electrochemical reaction between glucose and oxygen. Such cells were very stable with high power densities and increased performance (Kerzenmacher et al. 2008). Energy harvested can be converted into power after rectifying and converting the energy by power management circuits. Transducers can also harvest energy from resonant vibration and motion, which results in a potential change across the capacitor, flow of current through the external circuit, and electricity generation (Vullers et al. 2009). The type of transducer used would determine if the voltage would be low (as in electromagnetic transducers) or high (as in electrostatic transducers).

12.5.2 Fuel Cells as Energy Storage Systems

Fuel cells are ideal materials to solve the problem of renewable energy storage that the research industry seeks. Energy storage devices based on fuel cells activate individual units that collectively store the harvested energy. Fuel cells can convert fuel into energy through electrochemical processes (don 2020). In a bid to store energy, batteries can be replaced with energy storage devices that have large energy densities in fuel cells. Most kinds of fuel cells have hydrogen as their medium of storing energy that has been generated through water electrolysis. Although fuel cells do not store energy like batteries, their energy storage can be activated by combining with electrolyzers to form a regenerative fuel cell (RFC) (Vullers et al. 2009). RFC converts electrical energy into a form of fuel that can be stored and made available when needed as seen in Figure 12.6. RFC's effective structure lies in it to be able to distinguish between their energy storage and power conversion roles. For regenerative fuel cells, hydrogen is transported to the fuel cell to operate the load during fuel necessities while the pressure in the storage tank (properly kept safe externally) is indicative of the quantity of hydrogen left (Smith 2000). The hydrogen stored does not degrade irrespective of temperature or number of times it gets discharged. After hydrogen is discharged from the tank, the tank can be refilled from its source via electrolysis of water. Energy storage devices can also be buffered for continual working operations during energy-scarce periods (Vullers et al. 2009).

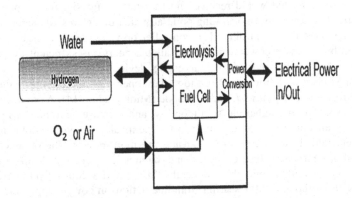

FIGURE 12.6 A system of regenerative fuel cell (Smith 2000). Reproduced with permission. Copyright Elsevier 2000.

Dong, Chen, Zhang, and Cui fabricated transition metal nitrides for storing energy in fuel cells (Dong et al. 2013). Better electrochemically performing cells with high electron kinetics were obtained. Lukic et al. developed energy storage systems to develop large sources of power for automotive use (Lukic et al. 2008). Soloveichik studied the use of regenerative fuel cells in producing fuel and generating power (Soloveichik 2014). The active material was the anode while the organic hydride served as the storage medium for the hydrogen. Barbir, Molter, and Dalton obtained high efficiency of 34%, reduced cell voltage, and high specific energy of 555 Wh/kg from less massive regenerative fuel cell (Barbir, Molter, and Dalton 2005). Fuel cells are useful sources of power production in commercial and residential environments; emergency power systems; and in vehicles like automobiles, boats, forklifts, etc. ("Fuel Cell" 2020). Despite being reliable with minimal noise and toxic emissions, fuel cells suffer some limitations. They are less durable and very costly, hence affordable only to the elites.

12.6 Conclusion

A fuel cell is an electrochemical energy conversion device that converts oxygen and hydrogen into heat, electricity, and water. This chapter has discussed basic information, working principle, different classifications, advantages, and disadvantages of fuel cells. Fuel cells are made up of anode/negative electrode and cathode/positive electrode wrapped in an electrolyte. These three components form an interface where chemical reactions occur. Fuel cells do not involve burning, have no moving parts, and do not need to be recharged as they have similar operations like batteries. The same principle applies to all the classes of fuel cells. Fuel cells can harvest energy from different materials for energy production and electricity generation. Although fuel cells do not store energy like batteries, their energy storage can be activated by combining with electrolyzers to form a regenerative fuel cell (RFC). RFC converts electrical energy into a form of fuel that can be stored and made available when needed. Fuel cells are vastly applied in power production, heat generation, stationary materials, stand-alone power plants, parks, automobiles, spacecraft, etc. Future perspectives should delve into diverse means of energy storage and further commercialization of fuel cells to make it cheaper and more affordable.

REFERENCES

Barbir, Frano, Trent Molter, and Luke Dalton. 2005. "Efficiency and Weight Trade-off Analysis of Regenerative Fuel Cells as Energy Storage for Aerospace Applications." *International Journal of Hydrogen Energy* 30, no. 4. Elsevier: 351–357.

Carollo, A., E. Quartarone, C. Tomasi, P. Mustarelli, F. Belotti, A. Magistris, F. Maestroni, M. Parachini, L. Garlaschelli, and P. P. Righetti. 2006. "Developments of New Proton Conducting Membranes Based on Different Polybenzimidazole Structures for Fuel Cells Applications." *Journal of Power Sources* 160, no. 1. Elsevier: 175–180.

Charradi, Khaled, Zakarya Ahmed, Pilar Aranda, and Radhouane Chtourou. 2019. "Silica/Montmorillonite Nanoarchitectures and Layered Double Hydroxide-SPEEK Based Composite Membranes for Fuel Cells Applications." *Applied Clay Science* 174. Elsevier: 77–85.

Chen, Huicui, Xin Zhao, Tong Zhang, and Pucheng Pei. 2019. "The Reactant Starvation of the Proton Exchange Membrane Fuel Cells for Vehicular Applications: A Review." *Energy Conversion and Management* 182. Elsevier: 282–298.

Cozzi, Dafne, Catia de Bonis, Alessandra D'Epifanio, Barbara Mecheri, Ana C. Tavares, and Silvia Licoccia. 2014. "Organically Functionalized Titanium Oxide/Nafion Composite Proton Exchange Membranes for Fuel Cells Applications." *Journal of Power Sources* 248. Elsevier: 1127–1132.

Dong, Shanmu, Xiao Chen, Xiaoying Zhang, and Guanglei Cui. 2013. "Nanostructured Transition Metal Nitrides for Energy Storage and Fuel Cells." *Coordination Chemistry Reviews* 257, no. 13–14). Elsevier: 1946–1956.

"Fuel Cell." 2020a. *Wikipedia*. Accessed April 12 2020. https://en.wikipedia.org/w/index.php?title=Fuel_cell&oldid=950529654.

"Fuel Cells." 2020b. *Energy.Gov*. Accessed October 23 2020. https://www.energy.gov/eere/fuelcells/fuel-cells.

"Fuel Cells – Alternate Energy Storage." *Energy Matters*. Accessed October 27 2020. https://www.energymatters.com.au/components/fuel-cells/.

Kabutey, Felix Tetteh, Qingliang Zhao, Liangliang Wei, Jing Ding, Philip Antwi, Frank Koblah Quashie, and Weiye Wang. 2019. "An Overview of Plant Microbial Fuel Cells (PMFCs): Configurations and Applications." *Renewable and Sustainable Energy Reviews* 110. Elsevier: 402–414.

Kerzenmacher, S., J. Ducrée, R. Zengerle, and F. von Stetten. 2008. "Energy Harvesting by Implantable Abiotically Catalyzed Glucose Fuel Cells." *Journal of Power Sources* 182, no. 1: 1–17. doi:10.1016/j.jpowsour.2008.03.031.

Kordesch, Karl, Viktor Hacker, Josef Gsellmann, Martin Cifrain, Gottfried Faleschini, Peter Enzinger, Robert Fankhauser, Markus Ortner, Michael Muhr, and Robert R. Aronson. 2000. "Alkaline Fuel Cells Applications." *Journal of Power Sources* 86, no. 1–2. Elsevier: 162–165.

Kordesch, Karl, and Günter Simader. 1996. *Fuel Cells and Their Applications*. Vol. 117. Germany: VCh Weinheim.

Li, Mohua, Yanbiao Liu, Liming Dong, Chensi Shen, Fang Li, Manhong Huang, Chunyan Ma, Bo Yang, Xiaoqiang An, and Wolfgang Sand. 2019. "Recent Advances on Photocatalytic Fuel Cell for Environmental Applications: The Marriage of Photocatalysis and Fuel Cells." *Science of The Total Environment* 668, no. June: 966–978. doi:10.1016/j.scitotenv.2019.03.071.

Lukic, Srdjan M., Jian Cao, Ramesh C. Bansal, Fernando Rodriguez, and Ali Emadi. 2008. "Energy Storage Systems for Automotive Applications." *IEEE Transactions on Industrial Electronics* 55, no. 6. IEEE: 2258–2267.

Mathuriya, Abhilasha Singh, and J. V. Yakhmi. 2016. "Microbial Fuel Cells–Applications for Generation of Electrical Power and Beyond." *Critical Reviews in Microbiology* 42, no. 1. Taylor & Francis: 127–143.

Mink, Justine E., Ramy M. Qaisi, Bruce E. Logan, and Muhammad M. Hussain. 2014. "Energy Harvesting from Organic Liquids in Micro-Sized Microbial Fuel Cells." *NPG Asia Materials* 6, no. 3. Nature Publishing Group. doi:10.1038/am.2014.1.

Mohammed, Hanin, Amani Al-Othman, Paul Nancarrow, Yehya Elsayed, and Muhammad Tawalbeh. 2019. "Enhanced Proton Conduction in Zirconium Phosphate/Ionic Liquids Materials for High-Temperature Fuel Cells." *International Journal of Hydrogen Energy* 46. no. 6. Elsevier: 4857–4869.

Osorio de la Rosa, Edith, Javier Vázquez Castillo, Mario Carmona Campos, Gliserio Romeli Barbosa Pool, Guillermo Becerra Nuñez, Alejandro Castillo Atoche, and Jaime Ortegón Aguilar. 2019. "Plant Microbial Fuel Cells–Based Energy Harvester System for Self-Powered IoT Applications." *Sensors* 19, no. 6. Multidisciplinary Digital Publishing Institute: 1378. doi:10.3390/s19061378.

Pasciak, G., K. Prociow, W. Mielcarek, B. Gornicka, and B. Mazurek. 2001. "Solid Electrolytes for Gas Sensors and Fuel Cells Applications." *Journal of the European Ceramic Society* 21, no. 10–11. Elsevier: 1867–1870.

Smith, W. 2000. "The Role of Fuel Cells in Energy Storage." *Journal of Power Sources* 86, no. 1: 74–83. doi:10.1016/S0378-7753(99)00485-1.

Soloveichik, Grigorii L. 2014. "Regenerative Fuel Cells for Energy Storage." *Proceedings of the IEEE* 102, no. 6. IEEE: 964–975.

Song, Hai-Liang, Ying Zhu, and Jie Li. 2019. "Electron Transfer Mechanisms, Characteristics and Applications of Biological Cathode Microbial Fuel Cells–A Mini Review." *Arabian Journal of Chemistry* 12, no. 8. Elsevier: 2236–2243.

Vijayakumar, Vijayalekshmi, and Sang Yong Nam. 2019. "Recent Advancements in Applications of Alkaline Anion Exchange Membranes for Polymer Electrolyte Fuel Cells." *Journal of Industrial and Engineering Chemistry* 70. Elsevier: 70–86.

Vullers, R. J. M., R. van Schaijk, I. Doms, C. Van Hoof, and R. Mertens. 2009. "Micropower Energy Harvesting." *Solid-State Electronics*, Papers Selected from the 38th European Solid-State Device Research Conference – ESSDERC'08, 53, no. 7: 684–693. doi:10.1016/j.sse.2008.12.011.

Wang, Heming, Jae-Do Park, and Zhiyong Jason Ren. 2015. "Practical Energy Harvesting for Microbial Fuel Cells: A Review." *Environmental Science & Technology* 49, no. 6. ACS Publications: 3267–3277.

Zhu, Weihuang, Haoxiang Gao, Fei Zheng, Tinglin Huang, Fengchang Wu, and Huan Wang. 2019. "Electrodeposition of Graphene by Cyclic Voltammetry on Nickel Electrodes for Microbial Fuel Cells Applications." *International Journal of Energy Research* 43, no. 7. Wiley Online Library: 2795–2805.

13

The Potential Role of Electrocatalysts in Electrofuel Generation and Fuel Cell Application

Xolile Fuku[1], Andile Mkhohlakali[1,2], Nqobile Xaba[1,2], and Mmalewane Modibedi[1]
[1]Energy Center, Smart Places, Council for Scientific and Industrial Research (CSIR), Pretoria, South Africa
[2]Chemistry Department, University of the Western Cape Bellville, Cape Town, South Africa

13.1 Introduction and Background

The ever-increasing greenhouse gases and ecological threats arising from the role of fossil fuel as energy sources have become issues of serious concern. Ever since the second industrial revolution fossil fuels have boosted economic growth. However, their unforeseeable impact on the environment and humankind has led to global warming. Moreover, the high CO_2 content which almost exceeds (a record-breaking) 410 ppm (from 280 ppm, in the 1800s) in the atmosphere, which is a consequence of fossil fuel combustion and other industrial activities (Anmin Liu et al. 2020) makes the global warming a pressing problem. Another strategy to mitigate CO_2 from the atmosphere is associated with CO_2-capture by pumping it into mid-ocean depth. However, excess storage of CO_2 is unpredicted and there is potential leakage and CO_2 eventually returns to the atmosphere (Anmin Liu et al. 2020). In this view, the carbon capture and utilization (CCU) has emerged as a potential solution to alleviate the carbon footprint. Photolectrocatalysis and photocatalysis strategies also have served to lessen greenhouse gases by mitigating the excess CO_2 and converting it into valuable fuels and useful chemicals (Yang et al. 2016; Qiao et al. 2019). The re-usable value-added chemicals and fuels are classified as formate C_1: (C:HCOO⁻/HCOOH), (syngas: CO, CH_4), MeOH, C_2:(EtOH, oxalic acid), and they offer an attractive "carbon-neutral" energy. In this regard, many researchers have focused on the electrochemical reduction of carbon dioxide (E-CO_2R) (Qiao et al. 2019). The main challenge for CO_2RR is low product selectivity and faradaic efficiency. In addition, the new technology approach for carbon dioxide is co-electrolysis. Co-electrolysis could be defined as electrolysis of water and CO_2 at the same time to produce valuable fuels and chemicals. The efficient hydrogen molecule produced is used with syngas for the production of hydrocarbons and other fuels; these fuels are also known as *electrofuels*. Another alternative that can electrolyze CO_2 and hydrogen to produce the power-to-liquid are the high-temperature solid oxide electrolysis cells (SOECs). SOECs are electrochemical devices that can electrochemically transform water, carbon dioxide, or both into hydrogen, carbon monoxide, and synfuels. These technologies are considered as clean energy, alternative approaches due to high technology maturity and low cost. SOECs can be combined with other chemical synthesis techniques to recycle CO_2 and H_2O into liquid fuels such as gasoline, methanol, and other natural gases (Summary 2020).

DOI: 10.1201/9781003145585-13

13.2 Electrofuels and Pathways: Power-to-x

An *electrofuel* is referred to as process rather than the fuel itself, which includes hydrogen production through water splitting (electrolysis) and the reaction of the hydrogen with CO_2 (CO_2R) (Brynolf et al. 2017). Electrofuels (carbon-based fuels) are produced from carbon dioxide and water decomposition using electricity as the primary energy source (Brynolf et al. 2017). Electrochemical water splitting (electrolysis) is the core technology of power-to-x, where x can represent the syngas, synthetic fuel, and hydrogen. The product forms the energy carrier such as synfuels (methane) and liquid fuels (gasoline, diesel). Electrofuels have been classified as power-to-liquid (Pt-L), power-to-gas (Pt-G), biomass-to-liquid (Bt-L), power-to-synthetic fuels (Pt-x, x as value-added chemicals) (Schemme et al. 2017). There are several CO_2 sources including combustion, air, and seawater (Chen et al. 2018). The electrofuel process is considered to have high potential to alleviate the CO_2 and efficiently produce valuable fuels such as liquid fuels and gaseous fuels. In view and simplicity, hydrogen produced from renewable energy sources (solar, wind-powered electrochemical water splitting) play a significant role as a reductant for hydrogenation of CO_2 (Natalia and Martin 2017). Figure 13.1 displays the general synthetic production pathways using electrofuel process. The process is an analogue of synthetic fuel production using SOECs, see Figure 13.2.

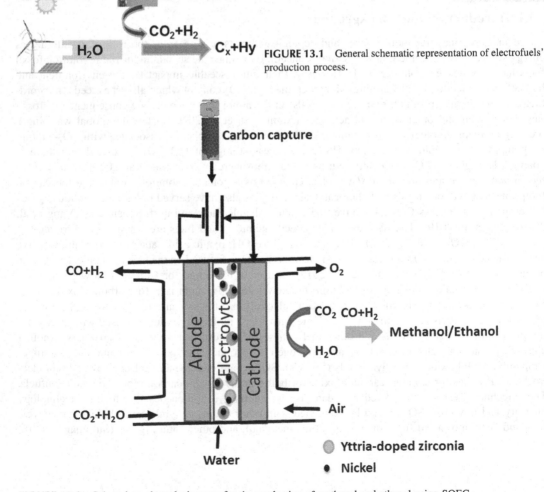

FIGURE 13.1   General schematic representation of electrofuels' production process.

FIGURE 13.2   Selected co-electrolysis route for the production of methanol and ethanol using SOEC.

13.2.1 Power-to-Hydrogen (H_2): H_2-Based Synthetic Fuel

The concept of power-to-hydrogen is that excess generated electricity is used to produce hydrogen molecules via water electrolysis (Andika et al. 2018). There has been a motivation to promote clean energy (electricity) and hydrogen economy. Hydrogen can provide environmental friendly energy to transport while being a versatile energy carrier to be utilized across many sectors. Although the hydrogen molecule is found in abundance on the earth in the form of compounds such as H_2O (water), its production and separation requires a large amount of overvoltage. Due to these challenges associated with hydrogen (H_2) such as local H_2 infrastructure, storage and transportation are required. As a result H_2 consumption should be onsite, and H_2 is used as the raw material or fuel.

13.2.2 Power to Liquid Fuels (Methanol and Ethanol): C_1-C_2-Based Synthetic Fuels Using Solid Oxide Electrolysis Cell

Co-electrolysis or *syntrolysis* has emerged as an attractive and efficient approach in electrofuel process through converting the CO_2 into liquid fuels through the sufficient supply of hydrogen from renewable water splitting (Andika et al. 2018). The process (electrofuel) can sufficiently convert CO_2 to liquid fuels using a solid oxide electrolysis cell (SOEC). SOEC is an electrochemical device that generates adequate amount of hydrogen from various sources. Its working principle can be considered as the reverse operation of the solid oxide fuel cell (SOFC) (Pandiyan et al. 2019). SOEC is used by Andika et al. for power-to-methanol (PtM) and other synthetic fuels through the electrochemical conversion of CO_2 (Andika et al. 2018). Fouil et al. reported the production of ethanol from recycling CO_2 using SOEC technology.

A single SOEC is comprised of three porous layers, which represent the ionic oxide electrolyte, mostly Yttria-doped zirconia, sandwiched between the anode and the cathode, as displayed in Figure 13.2. As aforementioned, SOECs's working principle is reverse to that of SOFCs. In SOECs, water steam reduces at the anode to form hydrogen molecule (H_2) and oxygen ion (O_2^{2-}) which in turn propagates through the solid electrolyte to the cathode to form oxygen molecule (O_2) which form water steam at the anode and steam by releasing the electrons.

Currently, many technologies are under discussion for the implementation of power-to-synfuels (i.e. liquid fuels), especially alcohols and other small organic fuels are of great interest for use as fuel in transport sector (Schemme et al. 2019). Liquid fuels may address the issue of hydrogen infrastructure, fossil fuel consumption, and also have a high potential for application in fuel cell devices (direct liquid fuel cells). Liquid organic fuels include alcohols, hydrocarbons, and inorganic liquids. This leads to the emerging of direct liquid fuel cells (DLFCs) which use MeOH, EtOH, ethylene glycol, glycerol, and formic acid as fuel (Ong et al. 2017). DLFC directly converts the chemical energy of liquid organic molecules into electrical energy (Soloveichik 2014). DLFCs are classified according to the type of liquid fuel used, for example, methanol in direct methanol fuel cells (DMFCs), ethanol in direct ethanol fuel cells (DEFCs), glycerol in direct glycerol fuel cells (DGFCs), and so on (Ong et al. 2017). Among DLFCs, direct alcohol fuel cells (DAFCs) have attracted most interest in liquid fuel cells, because ethanol (EtOH), as sustainable energy carrier, is regarded as a renewable biofuel that possesses several advantages including ease of storage of liquid fuel, allowing storage in conventional tanks. EtOH is a liquid hydrocarbon just like gasoline; it can be spread widely via the existing infrastructure unlike pure compressed or liquid hydrogen (Modibedi et al. 2011). These factors of EtOH make DAFCs more attractive in economic viability and environmental perspective. DAFCs also address the issue of global warming crisis, because they do not emit greenhouse gases (GHGs) (Badwal et al. 2015). EtOH has attracted great interests in direct liquid fuel cell (DEFCs) technology due to the significant advantages of high theoretical energy density (8.3 kWh g^{-1}) (Huang et al. 2009), safer as compared to MeOH (6.1 kWh g^{-1}), formic acid (2.104 kWh g^{-1}) (Zhang et al. 2019; Jana et al. 2016), gaseous hydrogen fuels (3.8 kWh kg^{-1}) (Modibedi et al. 2011) (Ong et al. 2017; Chen et al. 2016), environment friendly in contrast to the traditional fuel gasoline (Wang et al. 2015; Friedl and Stimming 2013; Soloveichik 2014; Yu et al. 2010). As a clean energy source, DEFC has been studied extensively over the past decades due to low pollution, liquid

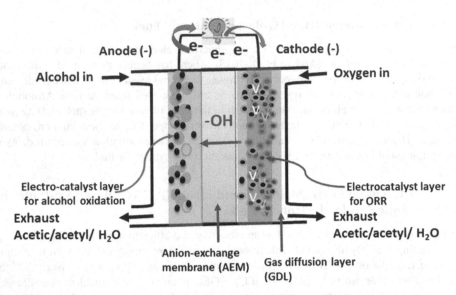

FIGURE 13.3 Schematic representation of anion membrane (AEM) direct alcohol fuel cell (DAFCs) ; the basic working principle for AEM-DAFC.

ethanol (as fuel), and high energy density (Zhang et al. 2019). In addition, EtOH can be produced in large quantities from biomass (Badwal et al. 2015). DEFCs (Figure 13.3) convert chemical energy stored in ethanol into electricity. DEFCs can be classified depending on the membrane used such as proton membrane DEFCs and anion membrane DEFCs. It can also be subdivided into two sub-categories, based on the electrolyte used, i.e. acidic-type DEFCs and alkaline-type DEFCs (anion membrane electrolyte fuel cell (AEMFCs)). The main challenge of DEFCs' development and commercialization is to develop a highly efficient anode material electrocatalyst for the complete EOR to 12e (Cai et al. 2014; Huang et al. 2008; Wang et al. 2015; Cai et al. 2013), according to the reaction illustrated in quations (13.1) and (13.2).

$$C_2H_5OH + 3H_2O \rightarrow CO_2 + 12H^+ + 12e^- \quad \text{(Complete oxidation)} \tag{13.1}$$

$$2C_2H_5OH + H_2O \rightarrow C_2H_4O_2 + C_2H_4O + 6H^+ + 6e^- \quad \text{(Partial oxidation)} \tag{13.2}$$

The incomplete oxidation of EtOH produces acetic acid and acetaldehyde with 2e- and 4e-, respectively, instead of the required 12e- (Courtois et al. 2014; Huang et al. 2008). Pt-based catalysts are known to suffer from the poisoning by EOR intermediates, such as carbonaceous (CO_{ads}) species (Grozovski et al. 2013). It is therefore necessary to find the new and efficient electrocatalyst that could cleave the C–C bond of ethanol completely and relatively enhance the overall kinetic for EOR (Jin et al. 2019) and produce high power output (Anna Zalineeva et al. 2014). The recent developments of alkaline-type DEFCs (anions exchange membrane fuel cell ((AEM)-FCs)) have grabbed wide attention due to their flexibility to the wide choice of non-platinum metals such as Pd and Ag, to name a few (Moraes et al. 2016; Jiang et al. 2014). Among platinum group metals (PGMs), the Pt and Pd are primarily active electrocatalysts for electro-oxidation of ethanol (Mkhohlakali et al. 2019). However, the high cost associated with Pt limits its large-scale application and Pt-based DEFCs' commercialization (Sequeira et al. 2017; Wang et al. 2011). Several attempts have been done to reduce the Pt loading and to replace it with the cheaper metal while maintaining its performance and efficiency. Alloying the Pd with the second metal or adding the third metal has been the common approach to improve electrocatalyst activity (Modibedi et al. 2014; Jin et al. 2019). The incorporation of promoter species such as Bi, Te, Tl, and Sn on Pd and Pt enhance the activity and stability towards EOR (Walock et al. 2013; Ong et al. 2017; Cai et al. 2013) (Table 13.1).

For the sole purpose of the results, the AEM-DAFC system will be discussed and studied further.

TABLE 13.1

Theoretical Energy Densities of Liquid Fuels

Energy Source	Energy (Wh L^{-1})
H_2	380
Methanol	4820
Diesel	9444
Ethanol	8030

13.3 Nanomaterials and Nanotechnology

The nanotechnology concept was first described by an American physicist and Nobel Laureate Richard Feynman in 1959 at the annual symposium (meeting) of American Physical Society at Calfornia Institute of Technology (Park et al. 2013; Balasooriya et al. 2017). In his talk (lecture) titled "There is a plenty of room at the bottom", he stated "The principles of physics, as far as I can see, do not speak to against the possibility of manipulating things into atoms by atoms", in which he introduced a concept of manoeuvring matter at atomic level (Balasooriya et al. 2017; Hulla et al. 2015; Kalia and Nepovimova 2020). The Tokyo University professor Novio Tanogushi in 1974 (Khan et al. 2017) referred nanotechnology as the studying, creation, manipulating, measure of useful materials, devices, and systems at billionth meter (10^{-9} m) regime, the atomic and molecular level. Nanotechnology and nanomaterials have contributed towards the carbon economy, which entails carbon dioxide reduction and fuels cell technologies, through manipulating the activity of the electrocatalyst (Xu et al. 2020). Nanomaterials as the driving force of nanotechnology hold the futuristic technological developments, next technology generation such as the fourth industrial revolution (4IR) (Rai and Rai 2015; Lee et al. 2018; Liao et al. 2016; Dey and Jain 2004; Zhang et al. 2017). The nanometres concept was first proposed by Richards Zsigmondy, the 1925 Nobel Laureate in Chemistry, as he measured gold colloids using a microscope (Hulla et al. 2015), and the observed properties involving magnetism, optical, heat, and fusion enable them for many applications due to manipulation of the metal bulk material into the <100 nm size (Choy 2003; Carrera-Carritos et al. 2018; Qiao and Li 2010). Many nanoscale objects have existed in the past centuries in nature due to biological processes such as protein units assembly, macromolecules, quasi-inorganic systems (shells and bones), and photosynthesis; also, nanoclay minerals such as vermiculite, montmorillonite, kaolinit,e to name a few, exist in nature (natural nanoparticles) (Griffin et al. 2018; Islam et al. 2012; Zangari 2015). Researcher's efforts have also been devoted to the utilization of nanostructured materials in healthcare (biotechnology), water purification, energy storage (battery and capacitors), optoelectronics, medicine, corrosion inhibitors due to their unique physicochemical properties that significantly differ from the bulk (large) material (Grozovski et al. 2013; Pitkethly 2004; Petrii 2015; Thomas et al. 2014; Liao et al. 2016; Dey and Jain 2004; Zhang et al. 2017). Generally nanostructure materials can be prepared via the main manufacturing approaches, namely "top-down" and "bottom-up" (Ozin et al. 2009) (Ozin et al. 2009; Whitesides and Grzybowski 2002) as illustrated in Figure 13.4, aiming the near-atomic/molecular level (Ozin et al. 2009). The size of the nanomaterials is of great interest, due to their huge influences in fine-tuning the electrocatalytic performance due to an enhanced electrochemical surface area and electric properties compared to the bulk material counterparts (Qiao and Li 2010). The aforementioned fabrication methods follow varied mechanisms, such as self-assembly of atoms and molecules (Thiruvengadathan et al. 2013) (nucleation and growth process) and chemical and mechanical etching/lithographic design for bottom-up and top-down, respectively (Biswas et al. 2012). The bottom-up surpasses the "top-down" due to the expenses associated with top-down and pin-hole (surface defects). A number of fabrication techniques include chemical reduction and vapour phase deposition: chemical vapour deposition (CVD) (Messier 1988), physical vapour (PVD), metal organic chemical vapour deposition (MOCVD), molecular beam epitaxy (MBE) (Banga et al. 2014; Qiao and Li 2010; Griffin et al. 2018), sol gel

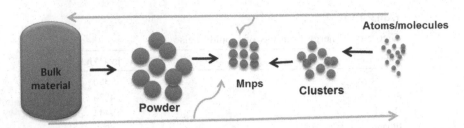

FIGURE 13.4 Schematic sketch illustrating "top down" and "bottom up" fabrication approaches of nanomaterials.

synthesis (Andrabi et al. 2016; Mello et al. 2003; Ingale et al. 2015), and atomic layer deposition (ALD) (Puurunen 2014; Dasgupta et al. 2010; Travis and Adomaitis 2013; Wu et al. 2015). Among other deposition methods, it is a considerable technique due to homogeneity, composition and mild processing conditions (Chekin et al. 2015), microwave polyol (Huang et al. 2008), hydrothermal (Sheridan et al. 2013) electrodeposition, and green synthesis. However, due to lack of control over the size, shape, and structure among other limitations associated with these methods, the research efforts have devoted to electrodeposition. Among other approaches, the green synthesis and E-ALD are versatile methods and they share common attracting features such as cost-effective and room temperature deposition process (Matinise et al. 2020; Kim et al. 2012), simple and cost-effective methods. Below is the schematic (Figure 13.4) which illustrates the fabrication approaches of nanomaterials (Keat et al. 2015; Gates et al. 2005; Lu et al. 2016).

Another different method of producing the nanomaterial is via the green bio-inspired method. The green synthesis prepares the nanoparticle with controlled size and desirable shape through optimizing the pH, temperature, incubation time of plant extract, and concentration of M^{z+}, without utilizing the hazardous chemical (solvents). The conditions and effectiveness of the green synthesis confirm that this method is eco-friendly, pollution-free, simple, and efficient as compared to other chemical reduction methods. The green synthesis has attracted considerable attention for the synthesis of metal oxides. Different catalysts have been prepared for CO_2 electroreduction, fuel cells, and the study shows the interaction of the prepared Ni, Cu, and Pd- based nanocatalyst with the used reducing agents such as punicaligin, penduculagin, etc. Scheme 13.1 clearly articulates the surface coordination of the set processes. The main chemical constituents are the derivatives of major compounds such as penducu-lagin, punicalagin (Scheme 13.1), and one of ellagitannins (ETs).

13.3.1 Preparation of AC and Pd-Based Nanocatalysts

Pyrolysis and activation methods were used to prepare activated carbon from banana peels, and spectroscopic and microscopic analyses were carried out. Loading of AC with Pd was done as reported by S. Bas et al. The NiO-modified AC and NiO-modified PdAC catalysts were prepared by co-impregnation method with the solutions containing palladium chloride and nickel salts. Structure, texture, and morphology were studied (Figure 13.5a,b) Moreover, the morphological structures of the prepared nanopowered NiO NP changed from pyramid-like structures to vertically doubled platelet-like structures, 2 nm in length (500°C). According to the HRSEM photographs, it could be concluded that the bioreduction method had successfully overcome the problem of agglomeration and is suitable in obtaining pure NiO NPs with smaller and highly crystalline sizes. Furthermore, the micrographs elucidate that the high calcination temperatures are an important factor in the formation of pure, highly crystalline, and well-distributed NiO particles. The prepared nanostructured NiO have been used in different application areas such as thin films for solar absorbers, catalysis, photovoltaic and energy storage, as well as in bio-mimicry. In this view, the as-prepared nanomaterial can be applied as a co-catalyst in fuel cell, solar absorbers, and many other applications.

On the other hand, the presence and proof of purely synthesized nanostructured NiO was equally important. EDS spectrum (Figure 13.5) revealed the main constituents of nanostructured NiO NPs which emerged at (0.8–7.8 keV, Ni) and (0.6 keV, O) energies. The energy values of the labelled Ni and

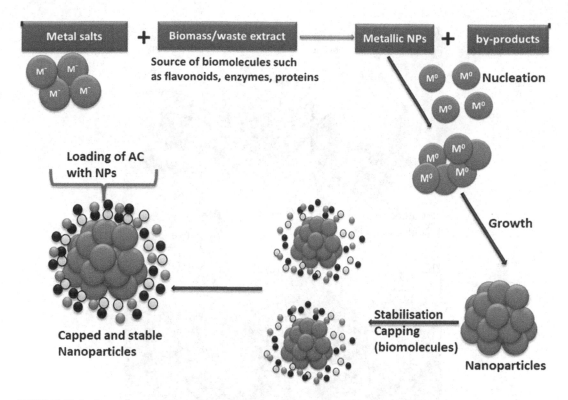

SCHEME 13.1 Representation of material synthesis.

FIGURE 13.5 SEM micrographs of the green prepared nanocatalysts.

O elements in Figure 13.5 are in accord with the reported literature values (Sequeira et al. 2017). The elemental composition of the prepared nanostructured sample provides evidence of the pure bioreduced NiO NPs.

13.3.2 Application of the Green Prepared Nanocatalysts: MEA Fabrication and Cell Performance Tests

Cell performance tests were carried out with an in-house passive fuel cell (DEFC). The alkaline DEFC (Figure 13.6) was made of an MEA, with an active area of 2.25 cm × 2.25 cm, sandwiched between two bipolar plates, which were fixed by two conducting plates. Both plates were made of stainless steel, and a single circuitous flow field, 1.0 mm wide, 0.5 mm deep, and 1.0 mm wide in channel ribs, was formed on each fixture plate. The MEA consisted of an anion exchange membrane and two electrodes. The cathode electrode (carbon cloth) consisted of a FeCo–C and Pd/PANI/C catalyst whilst the anodes were

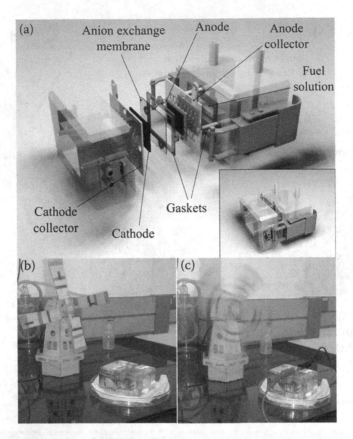

FIGURE 13.6 Passive fuel cell.

composed of Pd/C and Pd/NiO/C. To form the anode and cathode electrode, an anode ink was prepared by mixing 5 wt.% (anode and cathode materials) with 5 wt.% PTFE emulsion as a binder in 2-propanol. Each catalyst ink was well-dispersed by the ultrasonic process and brushed on a nickel foam (Hohsen Corp., Japan) and carbon cloth electrodes. Cell performance tests were conducted at ambient temperatures, in an aqueous solution of 1M ethanol/1M KOH.

Electrochemical Performance of Anode Catalysts, the potentiodynamic and power density curves shown in Figure 13.7a,b were recorded for air-breathing monoplanar passive fuel cells using 1M KOH + 1M ethanol solution. The passive MEA was tested by using both synthesized and commercialized catalyst as anode and cathode, Figure 13.7. Anode electrodes were prepared by coating a nickel foam support with a homogeneous mixture of PTFE, Pd/C, and Pd/NiO/C (Pd loading of 1.5 mg cm^{-2}) and cathode materials on carbon cloth cathode: FeCo/C, Pd/PANI/C. Cell potential scans and constant current experiments were carried out at ambient cell temperatures.

The obtained i-E curves (Figure 13.7a) show that the open circuit potential (OCV) of the fuel cell using the 1:2 Pd/PANI/C catalyst was 0.49 V, while that of the fuel cell using the 32 wt.% FeCo catalyst was 0.43 V. The OCV of the 1:2 Pd/PANI/C fuel cell was 60 mV higher than that of the 3.2 wt.% FeCo fuel cell. Although the OCV of the 1:2 Pd/PANI/C catalyst is higher than that of the 3.2 wt.% FeCo catalyst, it is however not comparable to the OCV reported in literature for Pd-based catalysts (Anmin Liu et al. 2020; Yang et al. 2016). This low OCV value could be attributed to the use of a low Pd metal loading of 1 mg cm^{-2} in the MEAs instead of the mostly used Pd metal loading of 2 mg cm^{-2} (Anmin Liu et al. 2020; Yang et al. 2016). This catalyst behaviour (lower performance) could also be due to limited amount of OH^{-} with respect to ethanol concentration thus reducing membrane conductivity. This was also evident by lower current (6 mA cm^{-2}) and power (1.7 mW cm^{-2}) densities compared to the commercial catalyst (current (10.5 mA cm^{-2}) and power (2 mW cm^{-2}) densities. Figure 13.7b shows the MEA tested by using synthesized Pd-NiO/C and commercial Pd/C as anode catalyst while using

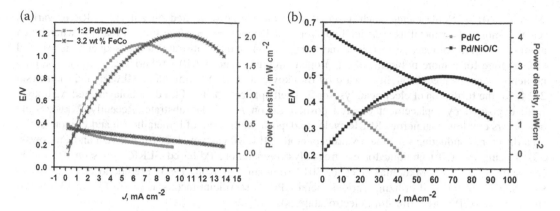

FIGURE 13.7 Polarization curves of the (a) in-house 1:2 Pd/PANI/C (cathode) and commercial 3.2 wt. % FeCo/C (cathode) and (b) in-house anode material Pd/NiO/C with the commercial anode material Pd/C catalyst under passive state at air-breathing room-temperature DEFC fed with 3M EtOH and 3M KOH.

FeCo/C as the cathode. Compared to Pd/C, Pd-NiO/C exhibited fair OC potential (0.67 V), current density (68 mA cm^{-2}), and power density (2.7 mW cm^{-2}). In addition, the current observed was directly proportional to the cell voltage thus revealing ohmic polarization.

We further evaluate the synthesis of Pb-based electrocatalyst via the electrochemical atomic layer deposition (E-ALD) method. Electrochemical deposition of nanostructured materials has attracted extensive attention and been used for fabrication of metallic mirrors and corrosion inhibitors, to name a few. Electrodeposition reactor is composed of an electrolyte containing metal ion as a precursor, working electrode on which the desired deposition takes place, and a counter electrode. Among electrodeposition methods, the E-ALD method was found to be versatile (Johnson et al. 2014), low cost, low temperature and pressure condition during its operation, and it also uses small concentration of precursor solutions (Modibedi et al. 2013; Mathe et al. 2005). E-ALD was found by Stickney in the 1970s using silver (Ag) on single-crystal gold Au (111) and has been identified as an efficient, potential method for fabrication of catalysts because it produces small nanoparticles, well-dispersed, stable, and self-controlled thin-film epitaxial layers. It is an electrochemical analogue of ALD in which underpotential deposition (UPD) is involved (Thambidurai et al. 2010). UPD phenomenon refers to a surface-confined reaction and provides adlayer deposition of metal onto the foreign metal substrate (S) at potentials more positive with respect to the reduction potential Nernst equilibrium potential (Sudha and Sangaranarayanan 2005). UPD refers to the surface-limited protocol that restricts the atomic layer-by-layer deposition in sequence fashion. The electrochemistry (UPD) is unique to each element and it requires a careful investigation before the deposition. The UPD potential region is of paramount importance, as it involves the strong interaction between (M-S) more than weak metal-metal (M-S) interaction as explained by Kolb in 1974 (Kolb et al. 1974; Nishizawa et al. 1997; Goric 2001). The strong M-S promotes and limits the deposition at atomic layer regime, whereas M-M prefers the nucleation and 3D growth deposition using overpotential deposition (OPD) (Jennings and Laibinis 1996). The requirements for underpotential deposition are confined to overpotential that is greater than zero ($\eta > 0$) and the overvoltage (E°) is more positive than potential predicted Nernst equilibrium E>>E M^{2+}/M $_{(bulk)}$, quation (13.3) below is being used to calculate the overpotential (η).

$$\text{For peak potential } E_P = E_{\frac{1}{2}} - 109\frac{RT}{nF} = \frac{28.5}{mV} \text{ at 25} \tag{13.3}$$

In most studies metal UPD is Cu-$_{UPD}$, which occurs positive to Nernst positive potential and has been used for deposition of various noble metals and to determine the surface area, in special cases (Sheridan et al. 2013). E-ALD is an electrochemical method for the formation of a compound, a monolayer at a time (Wang et al. 2015; Friedl and Stimming 2013; Yu et al. 2010; Cai et al. 2014). It involves an exchange of alternated electrodeposition of the atomic layer of metal UPD (Brankovic et al. 2001;

Al et al. 2016). Noble metals such as Pt, Pd, Ag have been deposited on Au via SLRR of various reactive, non-nobler metal as sacrificial adlayer. SLRR involves a surface limited reaction whereby an atomic layer (AL) of reactive sacrificial metal deposited first on a substrate and subsequently is exposed to exchange for a more noble metal at an open-circuit voltage (OCP) (Achari et al. 2017). The phenomenon was invented/ brought forward by Brankovic et al. (J. Y. Kim 2008; Mkwizu and Cukrowski 2014) for the formation of electrocatalyst for fuel cell application. The Pt, Pd monolayers and Ag bilayer were deposited by replacement of predeposited Cu on Au (111) substrate. Recently, Mkwizu and coworkers employed numerous SLRR cycles to deposit multilayers of bimetallic Ru and Pt monolayers on a different conducting substrate (Modibedi et al. 2014; Mkwizu et al. 2010; Mkwizu and Cukrowski 2014) using E-ALD technique for electrocatalyst preparation. Pd-based SLRR cycles on Au and Pt (Hossain et al. 2012; Mkhohlakali et al. 2019; Baldauf and Kolb 1993; Previdello et al. 2017) as substrates were reported utilizing various metal-UPD as sacrificial metals as see in Figure 13.8 below (preparation of Pd-based thin-film electrocatalysts).

Figure 13.8a Illustrates the front view of the computer controlled equipment setup, including the automated electrochemical thin flow cell system that is used to control the laminar flow of the metal precursor. The electrochemical thin flow electroformation system is comprised of the computer controlled variables (connected using RS232 and USB interface) including, speed pump, five-way valves, solution containers, potentiostat, and the software (sequencer 4), all designed for E-ALD cycles. The E-ALD cycles deposition is traced/monitored through the time-potential-current graph (Figure 13.8b).

Figure 13.9 represents the AFM topographic 2D micrographs of (a) TePd, and (b) CuTePd, their corresponding 3D view, and texture profiles. The 2D TePd micrographs exhibit the high conformity and well-dispersed grain distribution; the grain distribution is also clearly seen and corroborates with mapping (insert) and 3D view (Figure 13.9a (iii)). This finding is associated with the layer-by-layer formation of Te on Pd-covered Au. Two-dimensional images of CuTe reveals an even grain distribution,

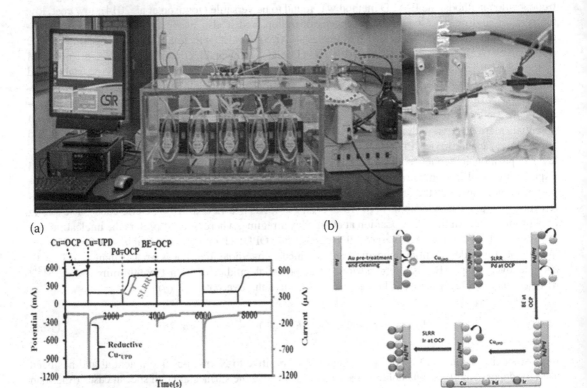

FIGURE 13.8 Representative picture of computer controlled E-ALD setup (a), magnified thin-flow cell (insert) the, typical time-potential-current plot recorded for two E-ALD (SLRR) cycles and the sketch of SLRR (b).

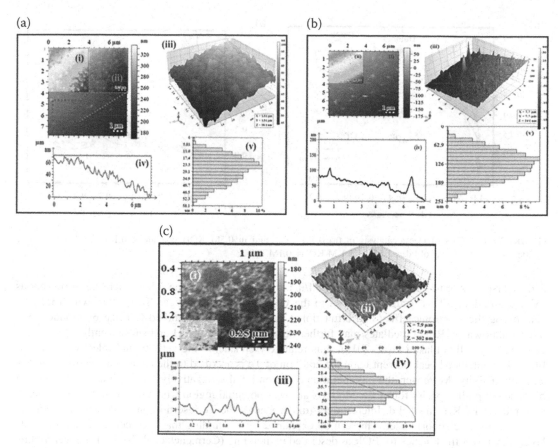

FIGURE 13.9 Representative AFM images of (a) TePd, (b) CuTePd, (c) and CuPd, with the corresponding line scan profile of their size distribution histogram.

which forms cone-like island clearly observed with 3D view. The islands are manifested on a texture profile (Figure 13.9b (iv)). The texture profile results indicate the inhomogeneity and turbulent grain distribution with varied phases, which may signify the formation of a ternary compound. CuTePd exhibit the higher surface roughness (Sa = 51.1 nm, Sq = 60.7 nm) as compared to its analogues TePd (Sa = 48.5 nm; Sq = 57.5 nm). The cross-section (east to west) analysis of the whole TePd and CuTePd AFM micrograph shows that some grains of trimetallic CuTePd developed needle-like islands with height from 50–100 nm ad displayed in Figure 13.9b (iii–iv). This suggests some 3D growth process for trimetallic CuTePd.

Figure 13.10 compares the CVs of CuTePd, CuPd, TePd, and Pd electrocatalysts in KOH (a) and in 0.5M KOH containing 0.1M EtOH solutions. Figure 13.10a exhibits well-pronounced Pd signatures as labelled, a clear double-layer (Helmholtz) charging region, Pd oxidation (PdOx), reduction (PdOr), and hydrogen adsorption/desorption (H_{adz}, H_{des}) (Radmilovi 2016). The double region is due to anchoring potassium ions (K+) and hydroxyl anions (-OH) (Zhang et al. 2015) on the Pd surface as displayed by a sketch (insert) according to the theory of inner/outer Helmoltz. PdOx and PdOr regions are due to the formation of oxide ($^-O,\,^-OH$) layer and dissociation (Yang et al. 2016; Bommersbach et al. 2007). It can be observed that the insertion of Te on Pd and CuPd broadens and increases the charge for the PdOr due to the high Te coverage on Pd surface. This finding is attributed to the strong electron interaction between electrodes and Te, which confirms the insertion of Te in Pd crystal structure. The insertion of Te on Pd disturbs the Pd, d-level according to d-band theory (Radmilovi 2016). In addition Te is an oxygen species, hence the increase in the oxide layer (Cai et al. 2014).

The EOR was performed in Pd-based electrocatalysts as illustrated in Figure 13.10b. The effect of Te on Pd and on CuPd electrocatalyst enhances the forward peak current and improves the ratio of forward

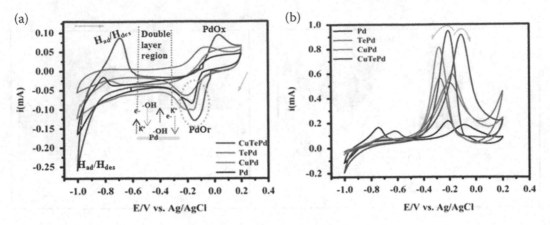

FIGURE 13.10 CVs of electrocatalysts in (a) 0.5M KOH and in 0.5M KOH containing 0.1M EtOH at scan rate of 30mVs⁻¹. (b) CV plots of Pd catalyst in 0.1 M KOH+ 0.1M EtOH.

and reverse peak current (if/ir). The higher if/ir ratio may be due to the Pd-M tolerance to poisonous EOR by-products. This result is ascribed to the OH associated to oxygen Te species, which may be confirming the electronic effect and bifunctional mechanism. The adsorbed CO-like by-products and other unknown EOR intermediates may further be stripped off by –OH and subsequently increase the peak current (Mkhohlakali et al. 2019), thus relatively enhancing the activity and selectivity towards EOR. The enhanced peak current upon the addition of Te is ascribed to the enhanced electron mobility and conductivity. Moreover, the incorporation of Te on Pd also negatively affects (improves) the onset because oxyphilic species forms the M(OH)x at lower potential region which results in the least energy required for EOR using TePd. Interestingly (discrepancy), the Te insertion on CuPd exhibited an unexpected positive (poor) onset. This can be explained by the finding that high concentration of oxide layer may lower the kinetics for EOR as described in literature (Cermenek et al. 2019). The layer finding can be attributed to the two oxophilic (Cu and Te) species on Pd surface which delays the C–C breaking by Pd during EOR, hence the later onset. Further improvement can be done, through optimizing and decreasing the number of Te or Cu cycles during the deposition. The trend of EOR indices are summarized in Table 13.2.

The Pd is generally known as one of the best electrocatalysts for EtOH electro-oxidation under alkaline condition even compared with Pt (Xu and Liu 2007; Zheng et al. 2007; Tao et al. 2006). The EOR involves 12 electrons as in following reaction (equation 13.4):

$$C_2H_5OH + 3H_2O \rightarrow 2CO_2 + 12H^+ + 12e^- \tag{13.4}$$

However, the main product of the EOR on Pd surface is not CO_2 but acetate ions because the Pd has a limited capacity for C–C bond cleavage. Therefore, previous literature had proposed that EtOH oxidation is determined by the extent of CH_3CO_{ads} and OH_{ads} coverage (Liu et al. 2007; Liang et al. 2009) as following mechanism (Liu et al. 2007; Liang et al. 2009; Truong et al. 2009; Zhang et al. 2011):

TABLE 13.2

Summary of Electro-Oxidation Activity of Pd-Based Electrocatalyst Towards Ethanol in Alkaline Media

Thin Films	Forward Peak Current; I_f (mA)	Reverse Peak Current, I_b (mA)	Onset (V)	I_f/I_b Ratio
Pd	0.13	0.15	–0.39	0.938
CuPd	0.587	0.811	–0.45	0.75
TePd	0.527	0.559	–0.535	0.94
CuTePd	0.9089	0.942	–0.432	0.964

$$Pd + OH^- \leftrightarrow Pd - OH_{ads} + e^- \tag{13.5}$$

$$Pd + CH_3CH_2OH \leftrightarrow Pd - (CH_3CH_2OH)_{ads} \tag{13.6}$$

$$Pd - (CH_3CH_2OH)_{ads} + 3OH^- \rightarrow Pd - (CH_3CO)_{ads} + 3H_2O + 3e^- \tag{13.7}$$

$$Pd - (CH_3CO)_{ads} + Pd - 3OH_{ads} \rightarrow Pd - CH_3COOH + Pd \tag{13.8}$$

$$Pd - CH_3COOH + OH^- \rightarrow Pd + CH_3COO^- + H_2O \tag{13.9}$$

It has been suggested that the rate determining step is quation (13.8) (Liu et al. 2007; Truong et al. 2009). Accordingly, the OH_{ads} adsorbed on the electrode is imperative for EOR because the adsorbed EOR intermediate species is stripped by adsorbed oxygen species (hydroxyl (^-OH)) ions and release free active catalytic sites via bifunctional mechanism. In this case, the EOR that occurs on the alloyed Pd (TePd, CuTePd) surface may be explained by aforementioned bifunctional mechanism. The incorporation of the promoter species (Te and Cu) and atoms increases the concentration of oxygen species and aids the activation of H_2O dissociation to form ^-OH on the Pd surface, which may further assist in oxidizing the poisonous CO_{ads} intermediates and relatively enhance electrocatalytic activity.

Even though the proposed study is considered successful, the research can continue for future work. The recommended future research will be to use this suitable electrocatalyst in the fabrication of the Membrane Electrode Assembly (MEA) and testing it in a unit cell under passive and active conditions.

13.4 Application of the Nanomaterials Electrocatalysts for Energy Conversion: Carbon Dioxide Reduction

Electrochemical activity of the Bi/Au electrode was investigated using cyclic voltammetry (CV). A potentiostat (Metrohm, PGSTAT302) equipped with NOVA 2.0 software was used. A standard three-electrode system was used, where Pt rod (Metrohm), Ag/AgCl/ 3M KCl, and Bi/Au connected using Cu tape served as counter, reference, and working electrodes, respectively. CV scans were done in 0.5M $NaHCO_3$ solution (pH 8.62) between 1.3 V and -1.3 V at a scan rate of 50 mV/s. Each sample was scanned in air, N_2, and CO_2 saturated $NaHCO_3$ solution. The by-products were identified by CV scans in methanol, ethanol, formic acid, and ethylene glycol N_2 saturated solutions. Electrodeposited thin films of Bi on Au substrates were evaluated for their ability to electrochemically reduce carbon dioxide into useful fuels. This was done by exposing the prepared Bi on Au (Bi/Au) electrode to CO_2 saturated solution (0.5M $NaHCO_3$, pH 8.62).

The Bi/Au electrode was further studied by scanning the electrode in the presence of N_2 and CO_2 saturated $NaHCO_3$ solution as shown in Figure 13.10, including an insert of the CV for Au electrode. The redox peaks attributed to Au substrate are denoted by A and B, and the Bi/Au catalyst peaks denoted as C and D as explained earlier were found. Further, as we move from high to low potentials, oxidation reduction peaks E and F at -0.76 V and -0.90 V, respectively, for oxygen reduction were observed. These peaks were significantly reduced due to the presence of N_2, CO_2, and the formation of by-products. The hydrogen evolution began at -1.1 V. All peaks were found to slightly shift to more negative potentials when the solution is purged with N_2 but remained in the same peak positions in the presence of CO_2 saturated solutions.

The comparison of the CV scan between CO_2 saturated electrolyte solution and that obtained in urea solution is revealed by Figure. 13.11. The formed products were identified by cyclic voltammetry scan of Bi/Au electrode in solutions containing different possible products (respectively), and were compared to the CV scan of CO_2 saturated $NaHCO_3$ solution. The products were methanol, ethanol, formic acid, ethylene glycol, and urea. From the many possible products, urea was more eminent, Figure 13.11a.

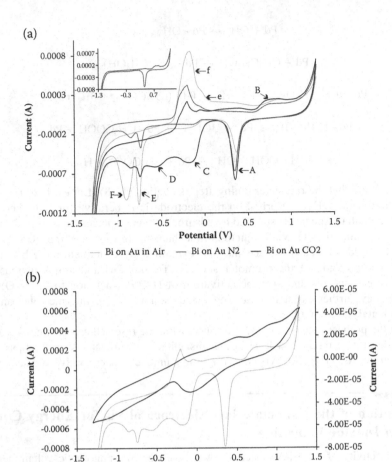

FIGURE 13.11 (a) CV of Bi on Au electrode in 0.5M NaHCO$_3$ at 50 mV/s vs. Ag/AgCl and an insert of the CV of Au electrode, (b) CV of a Bi on Au electrode in 0.5M NaHCO$_3$ and in 0.1M urea at 50 mV/s vs. Ag/AgCl.

The CV scan of Bi on Au electrode in urea was found to have a reduction peak at -0.1 V, which resembled the shape of peak C, but found at more positive potential than peak C, Figure 13.11b. Therefore, urea was identified as the main products formed.

Urea is one of the many possible products formed during reduction of CO$_2$, others include methanol, methane, polymers, and other higher hydrocarbons (Lee et al. 2017). The formation of urea occurs during the simultaneous reduction of CO$_2$ and nitrite ions as shown in equations. 13.10–13.12 (DiMeglio and Rosenthal 2013; Jones et al. 2014; Lee et al. 2017; Shibata et al. 1998). This study was based on the formation of urea, CO, formic acid, and ammonia at gas diffusion electrodes with Au, Cu, and Ag catalysts. They were unable to detect the formation of ammonia precursor (equation 13.12), and concluded that the reaction mechanism of urea formation (DiMeglio and Rosenthal 2013; Lee et al. 2017; Walker et al. 2015) still applied as the possibility.

$$CO_{2\,(g)} + 2H^+ + 2e^- \rightarrow CO + H_2O \tag{13.10}$$

$$N_2^- + 6H^+ + 5e^- \rightarrow NH_2 + H_2O \tag{13.11}$$

$$CO + NH_2 \rightarrow (NH_2)_2CO \tag{13.12}$$

13.5 Conclusion and Recommendations

The feasible method such as atomic layer deposition and green route displayed a reliable synthetic approach to synthesis high active robust Pd thin films electrocatalysts and nanocatalysts toward ethanol electro-oxidation and production of synfuel/chemicals. Furthermore, the AFM and SEM of these results were coherent with structural analysis from EDS, which shows some existence of layer-by-layer formation (epitaxial) relationship. The later results display the desirable electrocatalytic properties, which may be possessed by 2D materials with high access of active sites. On the other hand, the prepared nanocatalyst reveals good power density together with commercial catalysts. Conclusively, electrofuel, particularly synthetic fuel-based alcohols could be a potential energy carriers for direct liquid fuel cells.

13.5.1 Recommendations

Biomass-derived liquid alcohol represents the ultimate fuel in a future sustainable society and can be converted to useful electrical energy in DAFCs. For future purposes, providing more evidence by using capable techniques such as GCMS, NMR, etc., will assist in determining the specific products. In addition, the catalyst encourages new approach for a wide range of materials for smart electronic devices and manufacturing processes, especially in fuel cell (e.g. direct alcohol fuel cells – DAFCs) applications.

Acknowledgements

The authors would like to extend gratitude to CSIR and NRF (PDP) for financial support.

REFERENCES

Achari, I., S. Ambrozik, and N. Dimitrov. 2017. "Electrochemical Atomic Layer Deposition of Pd Ultrathin Films by Surface Limited Redox Replacement of Underpotentially Deposited H in a Single Cell." *J. Phys. Chem. C* 121: 4404–4411.

Al, Z., M. P. Mercer, and N. Vasiljevic. 2016. "Surface Limited Redox Replacement Deposition of Platinum Ultrathin Films on Gold: Thickness and Structure Dependent Activity towards the Carbon Monoxide and Formic Acid Oxidation reactions." *Electrochim. Acta* 210: 520–529.

Andika, R., A. Bayu, D. Nandiyanto, Z. A. Putra, M. R. Bilad, Y. Kim, C. Mun, and M. Lee. 2018. "Co-electrolysis for Power-to-Methanol Applications." *Renew. Sustain. Energy Rev.* 95, no. July: 227–241.

Andrabi, S. M. A. L. I., M. A. Shah, and K. P. Road. 2016. "A Study of Catalyst Preparation Methods for Synthesis of Carbon Nanotubes." 5, no. 1: 1–7.

Anmin Liu, T. M., Mengfan Gao, Xuefeng Ren, Fanning Meng, Yenan Yang, Liguo Gao, and Qiyue Yang. 2020. "Current Progress in Electrocatalytic Carbon Dioxide Reduction of Fuels on Heterogeneous Catalyasts." *J. Mater. Chem. A* 1–56.

Anna Zalineeva, K., Alexey Serov, Monica Padilla, Ulises Martinez and and P. B. A. Stève Baranton, and Christophe Coutanceau. 2014. "Self-Supported PdxBi Catalysts for the Electrooxidation of Glycerol in Alkaline Media." *J. Am. Chem. Soc.* 136: 3937–3945.

Badwal, S. P. S., S. Giddey, A. Kulkarni, J. Goel, and S. Basu. 2015. "Direct Ethanol Fuel Cells for Transport and Stationary Applications – A Comprehensive Review." *Appl. Energy* 145: 80–103.

Balasooriya, E. R., C. D. Jayasinghe, U. A. Jayawardena, R. Weerakkodige, D. Ruwanthika, R. M. De Silva, and P. V. Udagama. 2017. "Honey Mediated Green Synthesis of Nanoparticles: New Era of Safe Nanotechnology." *J. Nanomater.* 2017: 1–10.

Baldauf, M. and D. M. Kolb. 1993. "A Hydrogen Adsorption and Absorption Study with Ultrathin Pd Overlayers on Au (111) and Au (100)." *Electr. Acta* 38, no. 15: 2145–2153.

Banga, D., B. Perdue, and J. Stickney. 2014. "Electrodeposition of a PbTe/CdTe Superlattice by Electrochemical Atomic Layer Deposition (E-ALD)." *J. Electroanal. Chem.* 716: 129–135.

Biswas, A., I. S. Bayer, A. S. Biris, T. Wang, E. Dervishi, and F. Faupel. 2012. "Advances in top-down and bottom-up surface nanofabrication: Techniques, applications & future prospects." *Adv. Colloid Interface Sci.* 170, no. 1–2: 2–27.

Bommersbach, P., M. Mohamedi, and D. Guay. 2007. "Electro-Oxidation of Ethanol at Sputter-Deposited Platinum – Tin Catalysts Electro-oxidation of Ethanol at Sputter-Deposited Platinum – Tin." *J. Electrchemical Soc.* 154, no. January 2014: B876–B882.

Brankovic, S. R., J. X. Wang, and R. R. Adzic. 2001. "Metal Monolayer Deposition by Replacement of Metal Adlayers on Electrode Surfaces." *Surf. Sci.* 474: 73–79.

Brynolf, S., M. Taljegard, M. Grahn, and J. Hansson. 2017. "Electrofuels for the Transport Sector: A Review of Production Costs." *Renew. Sustain. Energy Rev.* 81, no. February: 1–19.

Cai, J., Y. Huang, and Y. Guo. 2013. "Bi-modified Pd / C Catalyst via Irreversible Adsorption and Its Catalytic Activity for Ethanol Oxidation in Alkaline Medium." *Electrochim. Acta* 99: 22–29.

Cai, J., Y. Huang, and Y. Guo. 2014. "PdTe x / C Nanocatalysts with High Catalytic Activity for Ethanol Electro-Oxidation in Alkaline Medium." *Applied Catal. B, Environ.* 150–151: 230–237.

Cai, J., Y. Huang, and Y. Guo. 2014. "PdTex/C Nanocatalysts with High Catalytic Activity for Ethanol Electro-oxidation in Alkaline Medium." *Appl. Catal. B Environ.* 150–151: 230–237.

Carrera-Carritos, F.-R.R. , C. Salazar-Hernandez, I. R. Galindo-Esquidvel. 2018. "Effect of the Reduction Temperature of PdAg Nanoparticles during the Polyol Process in the Ethanol Electrooxidation Reaction." *J. Nanomater.* 2018. 1–9.

Cermenek, B., J. Ranninger, B. Feketeföldi, I. Letofsky-papst, N. Kienzl, B. Bitschnau, and V. Hacker. 2019. "Novel Highly Active Carbon Supported Ternary PdNiBi Nanoparticles as Anode Catalyst for the Alkaline Direct Ethanol Fuel Cell." *Nano Res.* 12, no.1: DOI: 10.1007/s12274-019-2277-z

Chekin, F., S. Bagheri, and S. B. A. Hamid. 2015. "Synthesis and Spectroscopic Characterization of Palladium-doped Titanium Dioxide Catalyst." *Bull. Mater. Sci.* 38, no. 2: 461–465.

Chen, C., J. F. K. Kotyk, and S. W. Sheehan. 2018. "Progress toward Commercial Application of Electrochemical Carbon Dioxide Reduction." *Chem Rev* 4, no. 11: 2571–2586.

Chen, H., Z. Xing, S. Zhu, L. Zhang, Q. Chang, J. Huang, W. Cai, N. Kang, C. Zhong, and M. Shao. 2016. "Palladium Modified Gold Nanoparticles as Electrocatalysts for Ethanol Electrooxidation." *J. Power Sources* 321: 264–269.

Choy, K. L. 2003. "Chemical vapour deposition of coatings." *Prog. Mater. Sci.* 48: 57–170.

Courtois, J., W. Du, E. Wong, X. Teng, and N. A. Deskins. 2014. "Screening Iridium-based Bimetallic Alloys as Catalysts for Direct Ethanol Fuel Cells." *Appl. Catal. A Gen.* 483: 85–96.

Dasgupta, N. P., J. F. MacK, M. C. Langston, A. Bousetta, and F. B. Prinz. 2010. "Design of an Atomic Layer Deposition Reactor for Hydrogen Sulfide Compatibility." *Rev. Sci. Instrum.* 81, no. 4: 1–6.

Dey, B. S., and V. K. Jain. 2004. "Platinum Group Metal Chalcogenides." *Platin. Met. Rev.* 2, no. 5: 16–29.

DiMeglio, J. L., and J., Rosenthal. 2013. "Selective Conversion of CO_2 to CO with High Efficiency Using an Inexpensive Bismuth-Based Electrocatalyst." *J. American Chemical Society* 135: 8798–8801.

Friedl, J., and U. Stimming. 2013. "Model Catalyst Studies on Hydrogen and Ethanol Oxidation for Fuel Cells." *Electrochim. Acta* 101: 41–58.

Gates, B. D., Q. Xu, M. Stewart, D. Ryan, C. G. Willson, and G. M. Whitesides. 2005. "New Approaches to Nanofabrication: Molding, Printing, and Other Techniques." *Chem. Rev.* 105, no. 4: 1171–1196.

Goric, T. 2001. "Reactions of Copper on the Au (111) Surface in the Underpotential Deposition Region from Chloride Solutions." *Langmuir* 17, no. 111: 4347–4351.

Griffin, S., M. I. Masood, M. J. Nasim, M. Sarfraz, A. P. Ebokaiwe, K. Schäfer, C. M. Keck, and C. Jacob. 2018. "Natural Nanoparticles: A Particular Matter Inspired by Nature." *Antioxidants* 7: 1–21.

Grozovski, V., F. J. Vidal-iglesias, E. Herrero, and J. M. Feliu. 2013. "Surface Structure and Anion Effects in the Oxidation of Ethanol on Platinum Nanoparticles." *J. Mater. Chem. A* 1: 7068–7076.

Hossain, M. A., K. D. Cummins, Y. Park, M. P. Soriaga, and J. L. Stickney. 2012. "Layer-by-Layer Deposition of Pd on Pt (111) Electrode: an Electron Spectroscopy – Electrochemistry Study." *Electrocatal* 3: 183–191.

Huang, M., L. Li, and Y. Guo. 2009. "Microwave Heated Polyol Synthesis of Pt 3 Te/C Catalysts." *Electrochim. Acta* 54: 3303–3308.

Huang, M., F. Wang, L. Li, and Y. Guo. 2008. "A Novel Binary Pt 3 Te x / C Nanocatalyst for Ethanol Electro-oxidation." *J. Power* 178: 48–52.

Hulla, J. E.., S. C. Sahu, and A. W. Hayes. 2015. "Nanotechnology: History and Future." *Hum. Exp. Toxicol.* 34, no. 12: 1318–1321.

Ingale, S. V., P. B. Wagh, D. Bandyopadhyay, I. K. Singh, R. Tewari, and S. C. Gupta. 2015. "Synthesis of Nanosized Platinum Based Catalyst Using Sol-gel Process." *IOP Conf. Ser. Mater. Sci. Eng.* 73: 012076.

Islam, K. N., A. B. Z. Zuki, M. E. Ali, M. Zobir, B. Hussein, M. M. Noordin, M. Y. Loqman, H. Wahid, M. A. Hakim, S. Bee, and A. Hamid. 2012. "Facile Synthesis of Calcium Carbonate Nanoparticles from Cockle Shells." *J. Nanomater.* 2012, no. 1: 1–5.

Jana, R., S. Dhiman, and S. C. Peter. 2016. "Facile Solvothermal Synthesis of Highly Active and Direct Ethanol Fuel Cell Applications Facile Solvothermal Synthesis of Highly Active and Robust Pd 1. 87 Cu 0. 11 Sn Electrocatalyst towards Direct Ethanol Fuel Cell Applications." *Mater. Res. Express* 3, no. 8: 1–9.

Jennings, G. K. and P. E. Laibinis. 1996. "Underpotentially Deposited Metal Layers of Silver Provide Enhanced Stability to Self-Assembled Alkanethiol Monolayers on Gold." 7463, no. 11: 6173–6175.

Jiang, R., D. T. Tran, J. P. Mcclure, and D. Chu. 2014. "A Class of (Pd – Ni – P) Electrocatalysts for the Ethanol Oxidation Reaction in Alkaline Media." *ACS Catal.* 4: 2577–2586.

Jin, L., H. Xu, C. Chen, T. Song, C. Wang, Y. Wang, H. Shang, and Y. Du. 2019. "PdCu Coated Te Nanowires as Efficient Catalysts for Electrooxidation of Ethylene Glycol." *J. Colloid Interface Sci.* 540: 265–271.

Johnson, R. W., A. Hultqvist, and S. F. Bent. 2014. "A Brief Review of Atomic Layer Deposition: From Fundamentals to Applications." *Biochem. Pharmacol.* 17, no. 5: 236–246.

Jones, J. P., G., Prakash, and G. A., Olah. 2014. "Electrochemical CO_2 Reduction: Recent Advances and Current Trends." *Israel J. Chemistry* 54: 1451–1466.

Kalia, A. and E. Nepovimova. 2020. "Flower-Based Green Synthesis of Metallic Nanoparticles: Applications beyond Fragrance." *Nanomaterialsanomaterials* 10: 766.

Keat, C. L., A. Aziz, A. M. Eid, and N. A. Elmarzugi. 2015. "Biosynthesis of Nanoparticles and Silver Nanoparticles." *Bioresour. Bioprocess.* 2: 1–11.

Khan, I., K. Saeed, and I. Khan. 2017. "Nanoparticles: Properties, Applications and Toxicities." *Arab. J. Chem.* 12, no. 7: 908–931.

Kim, J. Y., Y. Kim, and J. L. Stickney. 2008. "Cu Nanofilm Formation by Electrochemical Atomic Layer Deposition (ALD) in the Presence of Chloride Ions." *J. Electroanaltical Chem.* 621: 205–213.

Kim, Y., J. L. Stickney, N. Jayaraju, D. Vairavapandian, Y. G. Kim, and D. Banga. 2012. "Electrochemical Atomic Layer Deposition (E- ALD) of Pt Nanofilms Using SLRR Cycles." *J. Electrochem. Soc.* 159, no. February 2015: 616–622.

Kolb, D. M., M. Przasnyski, and H. Gerischer. 1974. "Underpotential Deposition of Metals and Work Function Differences." *J. Electroanal. Chem.* 54, no. 1: 25–38.

Lee, C. W., J. S., Hong, K. D., Yang, K., Jin, J. H., Lee, H.-Y., Ahn, H., Seo, N.-E., Sung, and K. T., Nam. 2017. "Selective Electrochemical Production of Format From Carbon Dioxide with Bismuth-Based Catalysts in an Aqueous Electrolyte." *ACS Catalysi* 2: 931–937.

Lee, M., J. J. Yun, A. Pyka, D. Won, et al. 2018. "How to Respond to the Fourth Industrial Revolution, or the Second Information Technology Revolution? Dynamic New Combinations between Technology, Market, and Society through Open Innovation." *J. Open Innov. Technol., Mark. Complex.* 4, no. 3: 21.

Liang, Z. X., T. S. Zhao, J. B. Xu, and L. D. Zhu. 2009. "Mechanism Study of the Ethanol Oxidation Reaction on Palladium in Alkaline Media." *Electrochim. Acta* 54: 2203–2208.

Liao, J., W. Ding, S. Tao, Y. Nie, W. Li, G. Wu, S. Chen, and L. Li. 2016. "Carbon Supported IrM (M = Fe, Ni, Co) Alloy Nanoparticles for the Catalysis of Hydrogen Oxidation in Acidic and Alkaline Medium Article (Special Issue on Electrocatalysis Transformation) Carbon Supported IrM (M = Fe, Ni, Co) Alloy Nanoparticles." *Chinese J. Catal.* 37, no. July: 1142–1148.

Liu, J., J. Ye, C. Xu, S. P. Jiang, and Y. Tong. 2007. "Kinetics of Ethanol Electrooxidation at Pd Electrodeposited on Ti." *Electrochem. commun.* 9, no. 9: 2334–2339.

Lu, J., J. W. Elam, and P. C. Stair. 2016. "Atomic Layer Deposition - Sequential Self-limiting Surface Reactions for Advanced Catalyst 'Bottom-up' Synthesis." *Surf. Sci. Rep.* 71, no. 2: 410–472.

Mathe, M. K., S. M. Cox, V. Venkatasamy, U. Happek, and J. L. Stickney. 2005. "Formation of HgSe Thin Films Using Electrochemical Atomic Layer Epitaxy." *J. Electrochem. Soc.* 152: C751.

Matinise, N., N. Mayedwa, K. Kaviyarasu, Z. Y. Nuru, I. G. Madiba, and N. Mongwaketsi. 2020. "Zinc zirconate (ZnZrO3) Nanocomposites Bimetallic Designed by Green Synthesis via Moringa Olefeira Extract for High-performance Electrochemical Applications." *Mater. Today Proc.*, xxxx.

Medina-Ramos, J., J. L., DiMeglio, and J, Rosenthal. 2014. "Efficient Reduction of CO2 to CO with High Current Density Using In Situ or Ex Situ Prepared Bi-Based Materials." *Journal of the American Chemical Society* 136: 8361–8367.

Mello, R. L. S., F. I. Mattos-Costa, H. de las M Villullas, and L. O de S Bilhões. 2003. "Preparation and Electrochemical Characterization of Pt Nanoparticles Dispersed on Niobium Oxide." *Eclética química* 28, no. 2: 69–76.

Messier, R. 1988. "Thin Film Deposition Processes." *MRS Bull* 13, no. 11: 18–21.

Mkhohlakali, A. C., X. Fuku, R. M. Modibedi, L. E. Khotseng, S. C. Ray, and M. K. Mathe. 2019. "Electrosynthesis and Characterization of PdIr Using Electrochemical Atomic Layer Deposition for Ethanol Oxidation in Alkaline Electrolyte." *Appl. Surf. Sci.* 502, no. August 2019: 144–158.

Mkwizu, T. S. and I. Cukrowski. 2014. "Physico – Chemical Modelling of Adlayer Phase Formation via Surface – limited Reactions of Copper in Relation to Sequential Electrodeposition of Multilayered Platinum on Crystalline Gold." *Electr. Acta* 147: 432–441.

Mkwizu, T. S., M. K. Mathe, and I. Cukrowski. 2010. "Electrodeposition of Multilayered Bimetallic Nanoclusters of Ruthenium and Platinum via Surface-limited Redox-replacement Reactions for Electrocatalytic Applications." *Langmuir* 26, no. 1: 570–580.

Modibedi, R. M., T. Masombuka, and M. K. Mathe. 2011. "Carbon Supported Pd-Sn and Pd-Ru-Sn Snanocatalysts for Ethanol Electro-Oxidation in Alkaline Medium." *Int. J. Hydrogen Energy* 36: 4664–4672.

Modibedi, M. R. M., L. E., Motsoeneng, M. K., Khotseng 2014. "Electrodeposition of Pd Based Binary Catalysts on Carbon Paper via Surface Limited Redox-Replacement Reaction for Oxygen Reduction Reaction." *Electr. Acta* 128, no. May: 406–411.

Modibedi, R. M., E. K. Louw, M. K. Mathe, and K. I. Ozoemena. 2013. "The Electrochemical Atomic Layer Deposition of Pt and Pd Nanoparticles on Ni Foam for the Electro Oxidation of Alcohols." *ECS Trans.* 50, no. 21: 9–18.

Modibedi, R. M., M. K. Mathe, R. G. Motsoeneng, L. E. Khotseng, K. I. Ozoemena, and E. K. Louw. 2014. "Electro-Deposition of Pd on Carbon Paper and Ni Foam via Surface Limited Redox-replacement Reaction for Oxygen Reduction Reaction." *Electrochim. Acta* 128, no. May: 406–411.

Moraes, L. P. R., B. R. Matos, C. Radtke, E. I. Santiago, F. C. Fonseca, S. C. Amico, and C. F. Malfatti. 2016. "Synthesis and Performance of Palladium-based Electrocatalysts in Alkaline Direct Ethanol Fuel Cell." *Int. J. Hydrogen Energy* 41, no. 15: 6457–6468.

Natalia, S. and M. Martin. 2017. "Catalytic Hydrogenetation of CO2 to Methane over Supported Pd, Rh and Ni Catalysts." *Catal. Sci. Technol.* 7: 1086–1094.

Nishizawa, M., T. Sunagawa, and H. Yoneyama. 1997. "Underpotential Deposition of Copper on Gold Electrodes through Self-Assembled Monolayers of Propanethiol." 7463, no. c: 5215–5217.

Ong, B. C., S. K. Kamarudin, and S. Basri. 2017. "Direct Liquid Fuel Cells: A review." *Int. J. Hydrogen Energy* 42, no. 15: 10142–10157.

Ozin, G. A., K. Hou, B. V. Lotsch, L. Cademartiri, D. P. Puzzo, F. Scotognella, A. Ghadimi, and J. Thomson. 2009. "Nanofabrication by Self-Assembly." *Mater. Today* 12, no. 5: 12–23.

Pandiyan, A., A. Uthayakumar, S. Cha, S. Babu, and K. Moorthy. 2019. "Solid Oxide Electrolysis Cell: A Clean Energy Strategy for Hydrogen Review of Solid Oxide Electrolysis Cells: A Clean Energy Strategy for Hydrogen Generation." no. January.

Park, H., Y. Park, W. Kim, and W. Choi. 2013. "Surface Modification of TiO2 Photocatalyst for Environmental Applications." *J. Photochem. Photobiol. C Photochem. Rev.* 15: 1–20, Jun.

Petrii, O. A. 2015. "Electrosynthesis of Nanostructures and Nanomaterials." *Russ. Chem. Rev.* 84: 159.

Pitkethly, M. J. 2004. "Nanomaterials – the Driving Force." *Mater. Today* 7, no. 12: 20–29.

Previdello, B. A. F., E. Sibert, M. Maret, and Y. Soldo-olivier. 2017. "Palladium Electrodeposition onto Pt (100): Two-Layer Underpotential Deposition." *Langmuir* 33, no. 100: 2087–2095.

Puurunen, R. L. 2014. "A Short History of Atomic Layer Deposition: Tuomo Suntola's Atomic Layer Epitaxy." *Chem. Vap. Depos.* 20, no. 10–12: 332–344.

Qiao, Yan, and Chang Ming Li. 2010. "Nanostructured Catalysts in Fuel Cells." *J. Mater. Chem.* 21 (6), p. 062001.

Qiao, S., J. Wu, Z. Yin, and R. Amal. 2019. "Surface Strategies for Catalytic CO2 Reduction: from Two-Dimensional Materials to Nanoclusters to Single Atoms." *Chem. Soc. Rev* 48, no. 21: 5310–5349.

Radmilovi, V. R. 2016. "Electrochemical Oxidation of Ethanol on Palladium-Nickel Nanocatalyst in Alkaline Media." *Appl. Catal. B Environ* 189: 110–118.

Rai, S. and A. Rai. 2015. "Review: Nanotechnology - The Secret of Fifth Industrial Revolution and the Future of Next Generation." 7, no. 2: 61–66.

Schemme, S., R. Can, R. Peters, and D. Stolten. 2017. "Power-to-Fuel as a Key to Sustainable Transport Systems – An Analysis of Diesel Fuels Produced from CO2 and Renewable Electricity." *Fuel* 205: 198–221.

Schemme, S., J. Lucian, K. Maximilian, S. Meschede, F. Walman, R. C. Samsun, and R. Peters. 2019. "H2-Based Synthetic Fuels: A Techno-Economic Comparison of Alcohol, Ether and Hydrocarbon Production." *Int. J. Hydrogen Energy.* 45, no. 8: 5395–5414.

Sequeira, C. A. C., D. S. P. Cardoso, M. Martins, and L. Amaral. 2017. "Novel Materials for Fuel Cells Operating on Liquid Fuels." *AIMS Energy* 5, no. May: 458–481.

Sheridan, L. B., D. K. Gebregziabiher, J. L. Stickney, and D. B. Robinson. 2013. "Formation of Palladium Nanofilms Using Electrochemical Atomic Layer Deposition (E-ALD) with Chloride Complexation." *Langmuir* 29, no. 5: 1592–1600.

Sheridan, L. B., Y. Kim, B. R. Perdue, K. Jagannathan, J. L. Stickney, and D. B. Robinson. 2013. "Hydrogen Adsorption, Absorption, and Desorption at Palladium Nano fi lms formed on Au(111) by Electrochemical Atomic Layer Deposition (E-ALD): Studies using Voltammetry and In Situ Scanning Tunneling Microscopy." *J. Phys. Chem. C* 117, no. 111: 15728–15740.

Shibata, M., K., Yoshida, and N., Furuya. 1998 "Electrochemical Synthesis of Urea at Gas-Diffusion Electrodes: Part II. Simultaneous Reduction of Carbon Dioxide and Nitrite Ions at Cu, Ag and Au Catalysts." *J. Electroanalytical Chemistry* 442: 67–72.

Soloveichik, G. L. 2014. "Liquid Fuel Cells." *Beilstein J. Nanotechnol.* 5: 1399–1418.

Sudha, V. and M. V. Sangaranarayanan. 2005. "Underpotential Deposition of Metals – Progress and Prospects in Modelling." *J. Chem. Sci.* 117, no. 3: 207–218.

Summary, R. 2020. "Recent advances in solid oxide cell technology for electrolysis." *Science* 370, no. 6513: eaba6118 eaba6118. https://doi.org/10.1126/science.aba6118

Tao, H., Y. Li, S. Chen, and P. Kang. 2006. "Effect of Support on the Activity of Pd Electrocatalyst for Ethanol Oxidation." *J. Power Sources* 163: 371–375.

Thambidurai, C., D. K. Gebregziabiher, X. Liang, Q. Zhang, V. Ivanova, P. Haumesser, and J. L. Stickney. 2010. "E-ALD of Cu Nanofilms on Ru / Ta Wafers Using Surface Limited Redox Replacement." *J. Electrochem. Soc.* 157: 466–471.

Thiruvengadathan, R., V. Korampally, A. Ghosh, N. Chanda, K. Gangopadhyay, and S. Gangopadhyay. 2013. "Nanomaterial Processing Using Self-assembly-bottom-up Chemical and Biological Approaches." *Reports Prog. Phys.* 76, no. 6: 066501.

Thomas, J., V. S. Saji, and J. Thomas. 2014. "Nanomaterials for Corrosion Control." *Curr. Sci.* 92, no. April: 1–51.

Travis, C., and R. Adomaitis. 2013. "Dynamic Modeling for the Design and Cyclic Operation of an Atomic Layer Deposition (ALD) Reactor." *Processes* 1, no. 2: 128–152.

Truong, S., H. Mung, H. Tien, N. Kristian, S. Wang, S. Hwa, and X. Wang. 2009. "Enhancement Effect of Ag for Pd/C towards the ethanol electro-oxidation in alkaline media." *Appl. Catal. B Environ* 91: 507–515.

Vickers, J. W., Alfonso, D., and Kauffman, D. R. 2017. "Electrochemical Carbon Dioxide Reduction at Nanostructured Gold, Copper, and Alloy Materials." *Energy Technology* 5: 775–795.

Walker, R. J., A., Pougin, F. E., Oropeza, I. J., Villar-Garcia, M. P., Ryan, J., Strunk, and D. J., Payne 2015. "Surface Termination and CO_2 Adsorption onto Bismuth Pyrochlore Oxides." *Chemistry of Materials* 28: 90–96.

Walock, M. J., P. Dziugan, S. Karski, and A. V. Stanishevsky. 2013. "The Structure of Pd – M Supported Catalysts Used in the Hydrogen Transfer Reactions (M = In, Bi and Te)." *Appl. Surf. Sci.* 273, no. 273: 330–342.

Wang, Y., S. Zou, and W. Cai. 2015. "Recent Advances on Electro-Oxidation of Ethanol on Pt- and Pd-Based Catalysts: From Reaction Mechanisms to Catalytic Materials." *Catalysts* 5: 1507–1534.

Wang, Y., S. Zou, and W. Cai. 2015. "Recent Advances on Electro-Oxidation of Ethanol on Pt- and Pd-Based Catalysts: From Reaction Mechanisms to Catalytic Materials." *Catalysts* 5: 1507–1534.

Wang, Y., K. S. Chen, J. Mishler, S. C. Cho, and X. C. Adroher. 2011. "A Review of Polymer Electrolyte Membrane Fuel Cells: Technology, Applications, and Needs on Fundamental Research." *Appl. Energy* 88, no. 4: 981–1007.

Whitesides, G. M. and B. Grzybowski. 2002. "Self-assembly at all Scales." *Sci. (New York, NY)* 295, no. 5564: 2418–2421.

Wu, Y., D. Döhler, M. Barr, E. Oks, M. Wolf, L. Santinacci, and J. Bachmann. 2015. "Atomic Layer Deposition from Dissolved Precursors." *Nano Lett.* 15, no. 10: 6379–6385.

Xu, C. and Y. Liu. 2007. "Ethanol Electrooxidation on Pt/C and Pd/C Catalysts Promoted with Oxide." *J. Power Source* 164: 527–531.

Xu, H., D. Rebollar, H. He, L. Chong, Y. Liu, C. Liu, C. Sun, et al. 2020. "Highly Selective Electrocatalytic CO_2 Reduction to Ethanol Metallic Cluster Dynamically Formed from Atomically Dispersed Copper." *Nat. Energy*, 10.1038/s4.

Yang, N., S. R. Waldvogel, and X. Jiang. 2016. "Electrochemistry of Carbon Dioxide on Carbon Electrodes." *ACS Appl. Mater. Interfaces* 8: 28357–28371.

Yang, T., Y. Ma, Q. Huang, and G. Cao. 2016. "Palladium – Iridium Nanocrystals for Enhancement of Electrocatalytic Activity toward Oxygen Reduction Reaction." *Nano Energy* 19: 257–268.

Yu, E. H., U. Krewer, and K. Scott. 2010. "Principles and Materials Aspects of Direct Alkaline Alcohol." *Energies* 3: 1499–1528.

Zangari, G. 2015. "Electrodeposition of Alloys and Compounds in the Era of Microelectronics and Energy Conversion Technology." *Coatings* 5: 195–218.

Zhang, Z., L. Xin, K. Sun, and W. Li. 2011. "Pd-Ni Electrocatalysts for Efficient Ethanol Oxidation Reaction in Alkaline Electrolyte." *Int. J. Hydrogen Energy* 36, no. 20: 12686–12697.

Zhang, W., A. D. Bas, E. Ghali, and Y. Choi. 2015. "Passive Behavior of Gold in Sulfuric Acid Medium." *Trans. Nonferrous Met. Soc. China* 25, no. 6: 2037–2046.

Zhang, P., D. Su, J. G. Chen, and Z. Chen. 2019. "Enhancing C–C Bond Scission for Efficient Ethanol Oxidation using PtIr Nanocube Electrocatalysts." *ACS Catal.* 9: 7618–7625.

Zhang, Y., Q. Zhou, J. Zhu, Q. Yan, S. X. Dou, and W. Sun. 2017. "Nanostructured Metal Chalcogenides for Energy Storage and Electrocatalysis." *Adv. Funct. Mater.* 27, no. 35: 1–34.

Zheng, H. T., S. Chen, and P. K. Shen. 2007. "Spontaneous Formation of Platinum Particles on Electrodeposited Palladium." *Electrochem. commun.* 9: 1563–1566.

14

Reliability Study of Solar Photovoltaic Systems for Long-Term Use

Zikhona Tshemese[1], Farai Dziike[2], Linda Z. Linganiso[2], and Kittessa Roro[3]

[1]Department of Chemistry, Faculty of Science and Agriculture, University of Zululand, KwaDlangezwa, KwaZulu Natal, South Africa

[2]Research and Postgradute Support Directorate, Durban University of Technology, Durban, South Africa

[3]Smart Plces Cluster: EnergyCentre, Council for Scientific and Industrial Research (CSIR), Meiring Naude Rd, Brummeria, Pretoria

14.1 Introduction

A global gradual increase in human population and the growing economy are the reasons why energy consumption has extraordinarily increased. On the other hand, large amounts of CO_2 emissions give rise to an increasing global temperature which leads to global warming (Dubey and Guruviah 2019, 1419–1445). Mitigation measures must be put in place to reduce CO_2 emissions as a means of preventing further damage to the universe. Moreover, realization that fossil fuels are a limited resource on earth has facilitated a need for searching for alternative renewable energy sources. In this regard, energy can be converted effectively from its infinite sources such as solar, biomass, wind, geothermal, and water power to its forms such as electricity and fuel (Mahmood et al. 2014, 15; Fichtner 2015, 1601–1602; Kazemi et al. 2020, 3769–3775). This energy conversion is popularly known as *energy harvesting* where minute amounts of energy are captured, accumulated, and stored for future use (Kim et al. 1129–1149).

Recently, there has been a rapid progress of renewable resources usage in the energy sector, which has been advanced by several factors, such as the significant decrease in system costs, the invention of committed national strategies, and initiatives for adopting renewable energy (McCrone 2016). Accordingly, new opportunities for centralized and disseminated renewable energy markets are expected to appear globally (Energy Information Administration EIA:International Energy Outlook 2016 With Projections to 2040). The group of seven (G7) the group of twenty (G20) governments have made high-level agreements as an attempt to speed up access to renewable energy and stimulate energy efficiency on the demand side. Likewise, the United Nations General Assembly has adopted clear sustainable development goals which encourage sustainable energy projects (REN21 2009 global status report).

Recent technological development in "smart energy storage" enables a system wherein energy storage devices such as batteries, super capacitors, and fuel cells charge and reserve energy at one point and release electricity at peak times during the times of increased load demand. The smart storage technology has found a wide range of applications including at micro scale in miniaturized surveillance cameras, codeless audio devices, and microchip detectors in security and medical applications. The macro levels include power banks such as industrial-scale batteries. These technologies include chemical and electrical storage, mechanical storage, thermal storage, fuel stored energy, and biomass stored energy and batteries. Competing technologies are innovations in one or more of the following energy

DOI: 10.1201/9781003145585-14

storage technologies, including lithium-ion batteries, Compressed Air and Liquid Fuel Energy Storages (CAES; LFES), Vanadium/Zinc Flow Batteries (VFB; ZFB), super capacitors, and photovoltaics (PV). The US Department of Energy and industry research show that PV modules have high reliability due to a substantial multi-year effort which is not always true of PV systems (Maish et al. 1997).

The long-term strength of the photovoltaic industry needs that the PV systems work as anticipated. An impression of high maintenance costs, poor system reliability, and low availability is usually assumed when sketchy reports of inverter, battery, and other constituent failures are presented. Many systems are currently sold simply based on initial cost, but end users of these systems need to consider the full system's life cycle cost which includes primary cost, cost of non-availability and operation, and maintenance cost. In most cases, the latter two expenses are not well known. Knowledge of operation and maintenance costs can be helpful to manufacturers, suppliers, as well as customers even for the system that operates reliably. This knowledge will diminish technology risk to the customer while improving likelihood of commitment to PV projects (Maish et al. 1997).

14.2 PV Technology Description

A photovoltaic system is a power system designed to provide usable solar power through the usage of photovoltaics. This technology consists of an arrangement of several components which include solar panels to absorb and convert sunlight to electricity, a solar inverter to convert the output from direct current to alternating current and other apparatus. Photovoltaic systems can be categorized by a variety of aspects such as grid-connected, stand-alone rooftop, and ground-mounted, and distributed and centralized systems. Some parts of the world use solar power systems which are connected to the electrical grid whereas some use off-grid systems. PV system capacities vary from a few kilowatts to hundreds of megawatts (Figure 14.1) (Residential, Commercial and Utility Scale Photovoltaic (PV)

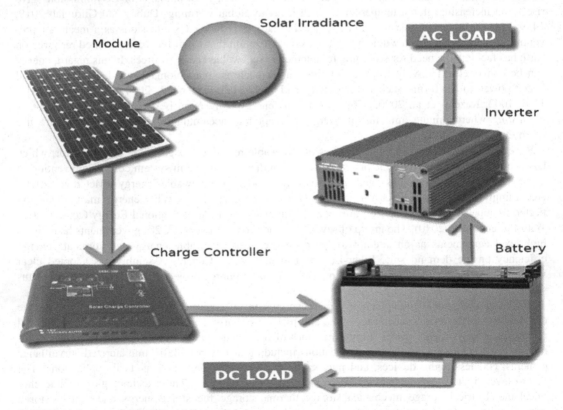

FIGURE 14.1 Layout of an off-grid photovoltaic system.

System Prices in the United States 2012; Global market outlook for Photovoltaics 2014). A major component of a PV system and module is a cell and its constituent materials. Many PV cell technologies exist in the market with different efficiencies; the common ones are monocrystalline silicon (c-Si), polycrystalline silicon (x-Si), amorphous silicon (a-Si), cadmium telluride (CdTe), copper selenium indium (CSI), and copper indium gallium selenium (CIGS) (Ndiaye et al. 2013, 140–151).

In the past two decades, photovoltaic systems have become a point of focus of scientific research due to their potential as a source of renewable energy. In order to contribute to the growth of the photovoltaic industry in South Africa and abroad, it is imperative to precisely forecast power delivery of PV systems over a period. This is affected by two major sources, i.e. the efficiency with which sunlight is converted into power and how this relationship changes over time (Jordan and Kurtz 2012). Measuring the degradation rate of a photovoltaic system is an essential element of both the business aspect and the research area of photovoltaics. Quantification of degradation of a photovoltaic system is vital, since a higher degradation rate is directly linked to reduced power production level. Equally important is to have a full understanding with respect to the degradation mechanisms as they might ultimately lead to failure (Meeker and Escobar 1998).

Degradation of a photovoltaic system is usually evidenced at individual levels, i.e. cell, module, array and the overall system with different factors apparent at each stage. These factors include precipitation, humidity, temperature, solar irradiation, dust, and snow. There can be additional other factors such as module mismatches, shading, corrosion, contact stability, cracked cell, and light-induced degradation at each level of a PV system giving rise to different kinds of degradation mechanisms (Wohlgemuth et al. 2011; Jordan et al. 2012, 24–28;). A great reduction of durability of a PV system is often manifested due to the significant stress over a lifetime imposed by the aforementioned factors (Phinikarides et al. 2014, 143–152). It is important to deal with degradation mechanisms separately as different types affect the PV systems differently.

14.3 Different Technologies Used in PV Systems

Photovoltaic cells made from crystalline (both monocrystalline – c-Si and polycrystalline – x-Si) silicon dominate the photovoltaic research and market presumably due to the superior stability and performance of these semiconductor materials. These two features affect the fiscal feasibility of PV installations (Quintana et al. 2002; Khan and Kim 2019, 4047). It is however also evident that other semiconductor materials such as amorphous silicon (a-Si), cadmium telluride (CdTe), copper indium selenium (CIS), copper indium gallium diselenide (CIGS), to name but a few, also find use in the PV system composition.

14.3.1 Crystalline Silicon

Photovoltaic modules have been made from different forms of silicon namely monocrystalline, polycrystalline and amorphous silicon with their electric power outputs being monitored. A typical crystalline silicon semiconductor material (Figure 14.2) is prepared by a silicon purification process, ingot production, wafer slicing, etching, and doping, which forms a PNP junction.

The PNP junction traps photons resulting in the discharge of electrons within the junction barrier thus creating a flow of current. Crystalline silicon cell is very stable in chemical structure and has high durability (Figure 14.3). This cell has an advantage of great efficiency compared to other products, but the challenge is its high price due to the pure silicon material it contains. Research shows that solar cells made from crystalline silicon do not easily degrade (Kirmani and Kalimullah 2017, 212–219).

Ordinary c-Si modules that are available in the market have a guarantee of 25 years in operation with 80% provision of its initial power after this period (Khan and Kim 2019, 4047).

Therefore, there is a level of degradation that cannot be escaped. Degradation differs from cell to cell, module to module, which explains why it is helpful to deal with a population of degrading modules than the entire PV system. Even though performance loss is related with the module package, some degradation can be traced back to the semiconductor itself (Balaska et al. 2017, 53–60; Limmanee et al. 2016, 12–17; Silvestre et al. 2016, 599-607). One of the factors that diminish the

(a)   (b)

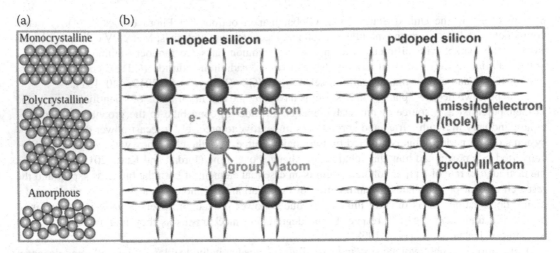

FIGURE 14.2 (a) Packing in pure crystalline silicon (b) The n-doped and p-doped silicon wafers.

FIGURE 14.3 Crystalline silicon sheets.

efficacy of a PV module is the potential-induced degradation (PID). Kirmani et al. give insights on this factor; the conditions that make PV systems prone to it, how the degradation process builds up, and lastly the effects of PID and prevention measures. The conclusions drawn from their work included that the efficiency of the panel is dependent on both pollution and particulate material accumulated on the surface (Kirmani and Kalimullah 2017, 212–219).

Hacke and others worked out systematic experiments to study the relationship between two parameters measured in situ and at the stress temperature (Ts) over the course of PID progression in conventional *p*-type crystalline silicon PV modules. Khan and coworkers, on the other hand, investigated the effects of high temperature and humidity on crystalline silicon modules. In their work, crystalline silicon PV modules were set up on the concrete slab on the contraction of roadways. The outcome of their study clearly showed that the concrete back layer plays a key role in the protection of the PV modules from the destructive effects of high temperature and humidity (Khan and Kim 2019, 4047). Luo et al. (2019) studied the long-term stability of crystalline silicon modules and their focus was on the performance degradation rates of the modules based on outdoor measurements at given time intervals. In the same study they also checked degradation modes and root causes of performance loss. The results showed that the products met the manufacturers' long-term service contract and were appropriate for applications in the tropical environment (Luo et al. 2019).

14.3.2 Cadmium Telluride (CdTe)

Cadmium telluride is another most popular PV technology based on the use of a thin film of CdTe for the absorption and conversion of sunlight into electricity. CdTe has become the second-best used

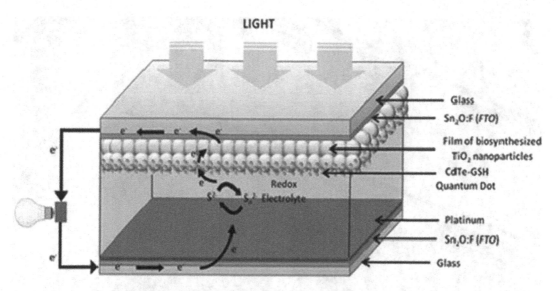

FIGURE 14.4 CdTe PV cell and its mechanism of photo energy generation.

material worldwide, following silicon. Notably, solar panels made from this material became the first in thin film technology to compete with the price of crystalline silicon PV in multi-kilowatt systems (Bosio et al., 2020). History and research of CdTe can be traced back to the 1950s due to the material being virtually perfectly matched to the photons' distribution in the solar spectrum with regards to the ideal conversion to electricity (Fthenakis et al., 2020). Advantages of CdTe include its ease of manufacturing, low cost, being a good match with sunlight, and the fact that Cd is abundant, and conversely the drawbacks of this material are lower efficiency levels as well as the scarcity of tellurium supply which makes it expensive (Ghosh et al., 2020, Maani et al., 2020).

Silvestre and coworkers checked the performance stabilization of CdTe PV modules using bias and light. In their study they used three different stabilization techniques, i.e. temperature-dependent, post-light-exposure, and bias-only stabilization to examine how each method affects the modules' state. One of their findings was that the performance of the CdTe modules, stabilized using light exposure, depends on the temperature of module during light exposure (Silvestre et al. 2016). Another study by Rawat et al. was based on the long-term performance analysis of CdTe PV module in real operating conditions. The long-term analysis of the study showed a degradation of 2.29% monthly average value of the initial exposure (Rawat et al. 2018, 23210–23217).

14.3.3 Copper Indium Selenide (CIS)

$CuInSe_2$ has been known as the efficient PV material for more than two decades, but its production preliminary line levels have only started recently. The slow development in this technology can be accounted to the complexity of the $CuInSe_2$ solar cell and the whole process. At the beginning, the solar cells were made from pure $CuInSe_2$ single crystals with a n-type window sheet deposited by the vaporization of CdS. Outdoor surveys of CIS-based modules have been done, and the outcomes showed that these modules are as stable as the crystalline silicon modules in principle. The properties of the two technologies have been proven to be comparable with respect to effects of temperature and light intensity (Dimmler et al. 2002, 149–157).

14.3.4 Copper Indium Gallium Diselenide (CIGS)

Thin-film technologies (CIGS in particular) have a great potential of reducing the cost of clean energy production through solar PV. CIGS thin-film technology is manufactured directly in module form as a result of monolithic integration technique. The five-alloy component CIGS PV is a result of increasing

FIGURE 14.5   CIGS modules as facade elements on the ZSW building in Stuttgart, Germany.

the small band gap of the CIGS through addition of gallium and/ or sulphur. This alloy system opened many opportunities for device optimization, for example, the incorporation of sulphur at the surface of the absorber helps to increase the open circuit voltage. On the other hand, the addition of gallium absorber layer increases the process tolerance as well as the open circuit voltage. Studies showed that CIGS PV stands as a promising technology based on its performance at the cell as well as module level. However, this technology is still not mature enough with regards to module manufacturing (Shafarman 2017).

One of the great achievements made through the CIGS was to get an efficiency of 22.6%, which is the highest of any thin-film technology; this efficiency exceeds even that of multicrystalline silicon (21.9%). The achievements have not only qualified the CIGS technology as an alternative option to lower the production costs, but it is also a high efficiency competitor in the PV market. Another possible factor that may help with reduction of production costs is the standardization of equipment and experience gained from the rise of CIGS-based PV, as this possibility has not yet been accessed. Research shows that the development with regards to CIGS technology has not yet reached its peak; rather it is expected to continue. This PV technology has promising prospects making it appear as a staid contender in the PV market future (Powalla et al. 2017, 445–451). A photograph of this kind of PV is shown in Figure 14.5.

14.4 Performance Analysis of PV Modules

According to the 2018 PV status report, there has been a 2% increase of new investments in renewable power capacity in 2017 compared to 2016. Also, a total of new installed renewable power capacity increase from 138.5 GW in 2016 to 157 GW in 2017 has been recorded and solar power accounts for about 62% of this capacity (Frankfurt School - UNEP Centre/BNEF 2018). It is worth noting that solar power has attracted major share of new investments in renewable energies and accounted for 58% which was followed by wind power for eight consecutive years. PV production increased by three orders of magnitude from 46 MW to over 100 GW in 2017 since 1990. By 2017, statistically documented growing installations accounted for 408 GW in the whole world. As much as there had been a steady increase in the solar power generation, between 2004 and 2008 a temporary shortage in silicon

TABLE 14.1

Statistical Estimates of PV Growth for a 25-year Period

Year	PV Power (GW)
1992–1997	0–~0.6
1998–2002	~0.6–1.2
2003–2007	1.2–10
2008–2012	10–100
2013–2017	100–401

feedstock was witnessed. This led to several new technologies entering the market (Jäger-Waldau 2018) (Table 14.1).

Investment analysts and industry projections show that solar energy will continue to develop at high rates in the forthcoming years. Different PV industry associations and alliances have established scenarios for the future projection growth of the PV systems. One great initiative towards future market development of renewable power (particularly solar PV) was to form a union of nations and states which are committed to moving the world away from burning of coal and rather into cleaner generation. According to the country members, 55 GW of coal burned generation capacity needs to be phased out at least by 2030, and this coal generated electricity should be replaced by solar PV power (approximately 250 GW). Besides, there is an overall hiking of energy prices as well as the need to stabilize the climate. These factors continue to keep solar power systems in high demand. Long-term forecast shows that PV growth rates will constantly be high even when the economic conditions vary locally (Jäger-Waldau 2018) (Table 14.2).

A continuous change from multi MW industry which aims to be a mass-producing industry (GW production) is realized in the PV industry. This expected development is associated with the increasing industry partnership, presenting a high risk as well as opportunity at the same time. It is up to the big solar cell companies, if they could use their cost advantages to offer products with low prices so that customers could buy more solar systems, since the PV market is expected to grow at an accelerated rate. On the contrary, competitiveness of small and medium companies will be influenced by this development. Small and medium companies can overcome the negative influence weighed by the development through offering technologically advanced cost-effective solar cell concepts and by focusing in niche markets with high value added in their products (Jäger-Waldau 2011).

There has been a continuous decrease in the cost of production for renewable energy technologies due to sharp learning curves. Renewable energies and photovoltaics are still observed as being costly in the market than conventional energy sources since external energy charges, supports in conventional energies, and price instability risks are mostly not yet taken into consideration. However, electricity

TABLE 14.2

Evolution Scenarios of the Worldwide Cumulative Solar Electrical Capacities Until 2040

Year	2017 (GW)	2020 (GW)	2025 (GW)	2030 (GW)	2040 (GW)
Actual installations	**408**				
Estimated growth: Greenpeace (reference scenario)		332	413	494	635
Greenpeace (advance scenario)		844	2000	3725	6678
LUT 100% RES Power Sector		1168	3513	6980	13805
BNEF NEO 2018		759	1353	2144	4527
IEA New Policy Scenario 2016		481	715	949	1405
IEA 450ppm Scenario 2016		517	814	1278	2108
IEA New Policy Scenario 2018		665	1109	1589	2540
IEA Sustainable Development Scenario 2018		750	1472	2346	4240

production from photovoltaic solar systems has previously demonstrated that it can be less expensive compared to peak prices in the electricity interchange in a wide range of countries plus they are closing in on suburban consumer prices. Lastly, renewable energy sources are the only kind of energy to offer a decrease in prices instead of an increase in future as opposed to traditional energy sources (Jäger-Waldau 2011).

14.5 Degradation Analysis of PV Modules

Photovoltaic industry need PV systems that are able to perform for approximately 25 years or even longer in the field and reliable means to determine that practicability. These features are needed for product development as well as warranty considerations. On the other hand, reliability checks are important relative to risk to financial stakeholders and general economic viability and key safety concerns. Some of the reliability concerns related to the PV technologies include quick connector reliability, improper insulation leading to loss of grounding, corrosion leading to a loss of grounding, delamination, bypass diode failure, glass fracture, moisture ingress, and inverter reliability. Other reliability issues can be found specifically under individual technologies (Zielnik 2009).

PV technology is continuously growing throughout the world due to its potential as a diffused solution to residential installations, but there are issues pertaining to their dependability. PV system design usually lacks accurate dependability analysis, and this is caused by the shortage of related ready-to-use tools as well as specific methodological approaches. Garro and Barrara studied the reliability analysis of residential PV system. In their study, they presented a methodological approach for the reliability analysis of these systems which was centered on the application of a series-parallel system reliability analysis model. The presented model was used in residential and grid-connected PV systems; however, it can be modified and extended for use as a guide to many other systems such as commercial, industrial, and residential, off-grid, on-grid, and hybrid (Garro and Barrara 2011, 9) (Figure 14.6).

Singh and Fernandez did the reliability evaluation on a solar PV system under two cases: with and without battery storage. In their study, the development of a model for PV system was done considering variable performance of solar resources and the outages caused by hardware failure of panel. Their methodology was applied to a remote PV system taking the solar resource and load profile of a case study area situated at a certain district in India. The analysis was carried out using the Monte Carlo through the evaluation of loss of load probability, and the results are discussed in their article (Singh and Fernandez 2015). The highest reported efficiencies for small PV cell performance which are tested in laboratory environment is about 32% while commercial PV modules have a maximum efficiency that is

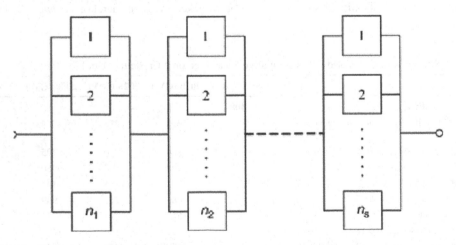

FIGURE 14.6   The series-parallel system diagram.

TABLE 14.3

Confirmed Single-Junction Terrestrial Cell Efficiencies Measured Under the Global AM1.5 Spectrum (1000 W/m^2) at 25°

PV Cell	Efficiency (%)
Si (crystalline)	26.7
GaAs (thin film cell)	29.1
InP (crystalline cell)	24.2
CIGS (Cd-free cell)	23.35
CdTe	21.0
CZTSSe	11.3
Organic	13.45

close to 22%. In operation of PV installations, the reported efficiencies decline more to a range of 13% to 17% due to various factors (Reinders et al. 2018) (Table 14.3).

14.6 Failure Mode and Effect Analysis (FMEA) for PV Systems

Failure Mode and Effect Analysis (FMEA) is described as a bottom-up approach of preventive quality assurance (Wirth et al. 1996, 19–29). The technique permits to identify and analyse the entire systems' faults, evaluate their significance in the reliability of the system, focus on maintenance practices and their effect on system reliability though it does not allow the evaluation of the reliability function of a complex system. Also, FMEA allows to deal with uncertainty which are due to inaccuracy associated with the complexity of the systems as well as the ambiguity of human judgement (Villarini et al. 2017, 1–12). FMEA has been used in different fields and industries, but Hanley and coworkers outline its precise uses, i.e. (i) to make provisions available to subsidize these sorts of improvements; (ii) the individuation of components most exposed to failure, in order to apply an appropriate succession of improvement; (iii) to improve the procedures regarding protective equipment and the monitoring of warning systems; (iv) the provision of the correct specifications from suppliers; and (v) to gain a knowledge of the components in need of major quality control (Henley and Kumamoto 1996).

Cristaldi and others did a root cause analysis and risk evaluation of PV balance of system failures, where they applied two approaches: the failure mode effects criticality analysis and the Markov process (Cristaldi, Khalil and Soulatiantork 2017, 113–120). FMEA as they outlined looked at the root causes of failures and presented prioritization numbers to highlight critical components for a balance of system. On the other hand, Markov process is a reliability approach that aims to predict the possibility of success and failure for a balance of system. The novelty of their proposed methodologies was based from analyzing the roots of failure sources of balance of system constituents and estimating the probability of failure for those constituents in order to improve the early improvement of a balance of system, boost maintenance management, and fulfil the demanding reliability by electric utilities (Cristaldi et al. 113–120).

A study of premature failure of PV systems using FMEA and integrated ISM approaches was done in Malaysia. It was concluded from the study that poor cooling system have the maximum risk priority number. The least depending element for premature failure to happen was found to be poor workmanship, thus it requires most attention. Poor monitoring and maintenance were deemed as highest driving forces of premature failure. Additional focus should be given to the premature failure throughout the planning for operation and maintenance because of its severity and impact (Han et al. 2018). Basu also did FMEA of a rooftop PV system and recommended that the same methodology could be used to other PV systems. This FMEA report can be used to advance the systems' reliability. The proposed methodology functions as a comprehensive, straightforward, and versatile technique for smooth set-up of a rooftop PV system (Basu 2015).

14.7 Conclusions and Future Projections

Photovoltaic technology has been proven to be the solution to the ongoing environmental crisis brought about by the burning of coal for the production of energy and many other forms of energy productions. It is worth noting that energy produced from the PV technologies does not compare to its counter competitors yet; however, there has been a steady increase in this production since its inception. Energy production from PV technologies is expected to be between 494 GW and 6980 GW by 2030 and 635 GW and 13,805 GW by 2040 globally (Jäger-Waldau 2018). In this regard, the world's energy demand will slowly be covered, which will in turn mean minimizing the negative effects of global warming to the environment. The general escalating energy prices and the necessity to stabilize the climate will keep the demand for solar-power systems relatively high.

Currently there are no reliability test practices that can predict with 100% certainty how a module will perform in specific climatic regions for 25–30 years. Consequently, development of an improved methodology which can closely simulate real conditions in which the system performs is necessary. The global available protocols that are used by the industry for qualification tests in certifying PV modules are mostly not capable of predicting long-term performance. An evolution in the application of accelerated lifetime testing to predict the PV modules and systems performances has occurred because of this. Therefore, a study that will envisage, understand, and quantify short- and long-term effects of climatic conditions on PV modules and/or systems is crucial.

REFERENCES

Balaska, A., A. Tahri, F. Tahri, and A. B. Stambouli. 2017. "Performance Assessment of Five Different Photovoltaic Module Technologies under Outdoor Conditions in Algeria." *Renewable Energy*, 107: 53–60.

Basu, J. B. 2015. "Failure Modes and Effects Analysis (FMEA) of a Rooftop PV System." *International Journal of Scientific Engineering and Research (IJSER)* 3: 9.

Cristaldi, L., M. Khalil, and P. Soulatiantork. 2017. "A Root Cause Analysis and a Risk Evaluation of PV Balance of System Failures."*Journal of the International Measurement Confederation (IMEKO)*, 6, no. 4: 113–120.

Dimmler, B., M. Powalla, and H. W. Schock. 2002. "CIS-Based Thin-Film Photovoltaic Modules: Potential and Prospects." *Prog. Photovolt: Res. Appl.* 10: 149–157.

REN21. 2009. *Renewables Global Status Report: 2009 Update*. Paris: REN21 Secretariat.

Dubey, R., and V. Guruviah. 2019. "Review of Carbon-Based Electrode Materials for Supercapacitor Energy Storage." *Ionics* 25: 1419–1445.

Fichtner, M. 2015. "Materials for Sustainable Energy Production, Storage, and Conversion." *Beilstein Journal of Nanotechnology* 6: 1601–1602.

Frankfurt School - UNEP Centre/BNEF. 2018. Global Trends in Renewable Energy Investment (2018).

Garro, A., and F. Barrara. 2011. "Reliability Analysis of Residential Photovoltaic Systems." *RE & PQJ* 1: 9.

Ghafoor, A., and A., Munir 2015. "Design and Economics Analysis of an Off-Grid PV System for Household Electrification." *Renewable and Sustainable Energy Reviews* 42: 496–502.

Global market outlook for Photovoltaics. 2014. European Photovoltaic Industry Association.

Green, A. M., E. D. Dunlop, J. Hohl-Ebinger, M. Yoshita, N. Kopidakis, and A. W. Y. Ho-Baillie. 2020. "Solar Cell Efficiency Tables (Version 55)." *Prog Photovolt Res Appl* 28: 3–15.

Han, T. D., M. Rosman, M. Razif, and S. A. Sulaiman. 2018. "Study on Premature Failure of PV Systems in Malaysia using FMEA and Integrated ISM Approaches." *MATEC Web of Conferences* 225, 04004.

Henley, E., and H. Kumamoto. 1996. *Risk Assessment and Management for Engineers and Scientists*. New York: IEE Press.

Honsberg, C. B., and S. G. Bowden. 2019. "Photovoltaics Education Website." www.pveducation.org/pvcdrom/introduction/solar-energy.

Jäger-Waldau, A. 2011. "Photovoltaics: Status and Perspectives until 2020." *Green*, November.

Jäger-Waldau, A. 2018. PV Status Report. *EUR 29463 EN*. Luxembourg: Publications Office of the European Union.

Jordan, D. C., and Kurtz, S. R. 2012. "Photovoltaic Degradation Rates: An Analytical Review." *Progress in Photovoltaics: Research and Applications.*

Jordan, D. C. Wohlgemuth, J. H., and Kurtz, S. R. 2012. "Technology and Climate Trends in PV Module Degradation." Presented at the 27th European Photovoltaic Solar Energy Conference and Exhibition Frankfurt, Germany. September 24–28.

Energy Information Administration (EIA): International Energy Outlook. 2016. "With Projections to 2040." *In Statistical report U.S. energy information administration office of energy analysis U.S.* Washington, DC: Department of Energy.

Kazemi, M., A. Kianifar, and H. Niazmand. 2020. "Nanoparticle Loading Effect on the Performance of the Paraffin Thermal Energy Storage Material for Building Applications." *Journal of Thermal Analysis and Calorimetry* 139: 3769–3775.

Khan, F., and J. H. Kim. 2019. Performance Degradation Analysis of c-Si PVModules Mounted on a Concrete Slab under Hot-Humid Conditions Using Electroluminescence Scanning Technique for Potential Utilization in Future Solar Roadways. *Materials* 12: 4047.

Kim, H. S., K. Joo-Hyong, and J. Kim. 2011. "A Review of Piezoelectric Energy Harvesting Based on Vibration." *International Journal of Precision Engineering and Manufacturing* 12, no. 6: 1129–1141.

Kirmani, S., and M. Kalimullah. 2017. "Degradation Analysis of a Rooftop SolarPhotovoltaic System: A Case Study." *Smart Grid and Renewable Energy* 8: 212–219.

Limmanee, A., N. Udomdachanut, S. Songtrai, S. Kaewniyompanit, Y., Sato, M. Nakaishi, S. Kittisontirak, K. Sriprapha, and Y. Sakamoto. 2016. "Field Performance andDegradation Rates of Different Types of Photovoltaic Modules: A Case Study in Thailand." *Renewable Energy* 89: 12–17.

Luo, W., Y. Khoo, S. P. Hecke, D. Jordan, L. Zhao, S. Ramakrishna, A. G., Aberle, and T. Reindl. 2019. "Analysis of the Long-Term Performance Degradation of Crystalline Silicon Photovoltaic Modules in Tropical Climates." IEEE Journal of Photovoltaics 9, no. 1:266–271.

Mahmood, N., C. Zhang, H. Yin, and Y. Hou 2014. "Graphene-Based Nanocomposites for Energy Storage and Conversion in Lithium Batteries, Supercapacitors and Fuel Cells." *J. Mater. Chem. A* 2: 15.

Maish, A. B., C. Atcitty, S. Hester, D. Greenberg, D. Osborn, D. Collier, and M., Brine. 1997. "Photovoltaic System Reliability." Proceedings of the 26th IEEE Photovoltaic Specialists Conference, Anaheim CA.

McCrone, A. 2016. "Frankfurt School-UNEP Centre and Bloomberg New Energy Finance Report." https://www.actu-environnement.com/media/pdf/news-26477-rapport-pnue-enr.pdf

Meeker, W. Q., and Escobar, L. A. 1998. *Statistical Methods for Reliability Data.* New York: John Wiley & Sons.

Ndiaye, A., A. Charki, A. Kobi, C. M. F. Ke´be´, P. A. Ndiaye, and V. Sambou. 2013. "Degradations of Silicon Photovoltaic Modules: A Literature Review. *Solar Energy* 96: 140–151.

Órdenes-Aenishanslins, N. A. L. A. Saona, V. M. Durán-Toro, J. P. Monrás, D. M. Bravo, and J. M. Pérez-Donoso. 2014. "Use of Titanium Dioxide Nanoparticles Biosynthesized by Bacillus Mycoides in Quantum Dot Sensitized Solar Cells." *Microbial Cell Factories (2014)* 13: 90.

Phinikarides, A., N., Kindyni, G., Makrides, and G. E., Georghiou 2014. Review of Photovoltaic Degradation Rate Methodologies." *Renewable and Sustainable Energy Reviews* 40, 143–152.

Powalla, M., S., Paetel, D., Hariskos, R., Wuerz, F., Kessler, P., Lechner, W., Wischmann, and T. M., Friedlmeier 2017. "Advances in Cost-Efficient Thin-Film Photovoltaics Based on Cu(In,Ga)Se$_2$." *Engineering* 3: 445–451.

Quintana, M. A., King, D. L., McMahon, T. J., and Osterwald, C. R. (2002). "Commonly observed degradation in field-aged photovoltaic modules." Conference Record of the Twenty-Ninth IEEE Photovoltaic Specialists Conference, 1436–1439.

Rau, U., and H. W. Schock. 1999. "Electronic Properties of Cu(In,Ga)Se$_2$ Heterojunction Solar Cells–Recentachievements, Current Understanding, and Future Challenges." *Appl. Phys. A* 69: 131–147.

Rawat, R., S. C. Kaushik, O. S. Sastry, B. Bora, and Y. K. Singh. 2018. "Long-Term Performance Analysis of CdTe PV module in Real Operating Condition." *Materials Today: Proceedings* 5 23210–23217.

Reehal, H. S. *Thin Film Crysalline Silicon Solar Cells.* UK: South Bank University.

Reinders, A., D. Moser, W. V. Sark, G. Oreski, N. P. A. Scognamiglio, and J. Leloux. 2018. *"Introducing 'PEARL-PV': Performance and Reliability of Photovoltaic Systems: Evaluations of Large-Scale Monitoring Data,"* 2018 IEEE 7th World Conference on Photovoltaic Energy Conversion (WCPEC) (A Joint Conference of 45th IEEE PVSC, 28th PVSEC & 34th EU PVSEC), 2018, 0762-0766

REN21. 2016. Renewables 2016 Global Status Report (Paris: REN21 Secretariat). Residential, Commercial and Utility Scale Photovoltaic (PV) System Prices in the United States. Technical Report NREL/TP-6A20-53347 February 2012.

Shafarman, W. 2017. "CIGS Module Design and Manufacturing." In *Photovoltaic Solar Energy: From Fundamentals to Applications*, edited by Angèle Reinders, Pierre Verlinden, Wilfried van Sark, and Alexandre Freundlich. 1st Edition. https://doi.org/10.1002/9781118927496.ch20

Silvestre, S. S. Kichou, L. Guglielminotti, G. Nofuentes, and Alonso-Abella M. 2016. "Degradation Analysis of Thin Film Photovoltaic Modules under Outdoor Long Term Exposure in Spanish Continental Climate Conditions." *Solar Energy*, 139, 2016, 599-607.

Singh, S. S., and Fernandez, E. 2015. *Reliability Evaluation of a Solar Photovoltaic System with and without Battery Storage*. 2015 Annual IEEE India Conference (INDICON), 2015, 1-6.

Villarini, M., V. Cesarotti, L. Alfonsi, and V. Introna. 2017. "Optimization of Photovoltaic Maintenance Plan by Means of a Fmea Approach Based on Real Data." *Energy Conversion and Management* 152: 1–12.

Wagner, S., J. L. Shay, and P. Migliorato. 1974. "CuInSe$_2$/CdS Heterojunction Photovoltaic Detectors." In *Bell Telephone Laboratories*. New Jersey: Murray Hill, 07974.

Wirth, R., B. Berthold, A. Krämer, and G. Peter. 1996. "Knowledge-Based Support of System Analysis for the Analysis of Failure Modes and Effects." *Eng Appl Artif Intell* 9: 219–229.

Wohlgemuth, J. H. et al. 2011. "Using Accelerated Testing to Predict Module Reliability." 37th Photovoltaic Specialists Conference, Seattle, Washington.

Yun-Chia, L., C. Yi-Ching. 2007. "Redundancy Allocation of Series-Parallel Systems Using a Variable Neighborhood Search Algorithm." *Reliability Engineering and System Safety* 92: 323–331.

Zielnik, A. P.v. 2009. "Durability and Reliability Issues." *Atlas Material Testing*. https://www.renewableenergyworld.com/solar/pv-durability-and-reliability-issues/#gref

15

Physical Methods to Fabricate TiO$_2$ QDs for Optoelectronics Applications

Cyril Oluchukwu Ugwuoke[1,2], Sabastine Ezugwu[3], S. L. Mammah[4], A.B.C. Ekwealor[1], Mutsumi Suguyima[5], and Fabian I. Ezema[1,6,7]

[1]*Department of Physics and Astronomy, University of Nigeria, Nsukka, Enugu State, Nigeria*
[2]*Science and Engineering Unit, Nigerian Young Researchers Academy, 430231 Onitsha, Anambra State, Nigeria*
[3]*Department of Physics and Astronomy, The University of Western Ontario, London, Ontario, Canada*
[4]*Department of Science Laboratory Technology, School of Applied Sciences, Rivers State Polytechnic Bori, Bori, Nigeria*
[5]*Faculty of Science and Technology, Department of Electrical Engineering, Tokyo University of Science, 20641 Yamazaki, Noda, Japan*
[6]*Nanosciences African Network (NANOAFNET), iThemba LABS-National Research Foundation, Somerset West, Western Cape Province, South Africa*
[7]*UNESCO-UNISA Africa Chair in Nanosciences/Nanotechnology, College of Graduate Studies, University of South Africa (UNISA), Muckleneuk Ridge, Pretoria, South Africa*

15.1 Introduction

Quantum dots (QDs) are semiconductor nanoparticles with a few nano meters in size having electrons or holes tightly confined in all three spatial dimensions (Gangasani 2016, Brazis 2017). The confinement is actualized by fabricating a small size of semiconductors with atoms per particle ranging from hundreds to thousands. Apart from the confinement in the three dimensions, there are other semiconductor quantum dots (SQDs) confinement such as quantum wire and quantum well. In quantum wires, electrons or holes are confined in two spatial dimensions and the third dimension has free propagation. The confinement of electrons or holes in quantum well occur only in one dimension and the other two dimensions have free propagation (Harrison and Valavanis 2016).

There are three major categories of SQDs: (a) the III–V SQDs are made of elements from group III on the periodic table (e.g. In, Ga, Al, B) and group V (Bi, Sb, As, P, N). In this group, GaAs is the best-known example of a semiconductor quantum dot. (b) II–Vi SQDs: these are made of elements from group II, i.e. transition metals (e.g. Cd, Zn) and group Vi (Te, Se, S, O). Here, zinc oxide (ZnO), cadmium telluride (CdTe), and cadmium selenide (CdSe) are the best-known examples of semiconductor quantum dot materials. (c) Si QD: they are made of the element of silicon, which is the standard material of the semiconductor and chip industry. Beside the above-named categories, there are other semiconductor quantum dot materials, which are composed of oxides of transition metals such as TiO$_2$ and the subject of this chapter.

The titanium(IV) dioxide (TiO$_2$) is one of the most promising abundant elements on earth. The titanium compound that is most widely used is TiO$_2$, which is due to its stability (including acids), low toxicity, low cost, easiness for structural modification (such as doping, generation of nanostructures, etc.) (Štengl, Martin, and Kormunda 2017). Titanium(IV) dioxide at atmospheric pressure exhibits three polymorphs: anatase, rutile, and brookite (Nyankson et al. 2013). Anatase occurs as a black solid, although the purest form is usually colourless or white. The bulk form of rutile TiO_2 occurs as a stable phase but anatase and brookite are metastable, whereas nanocrystalline brookite and anatase are stable because of their smaller surface energy (Hanaor et al. 2019).

At room temperature, TiO_2 has the properties of an n-type semiconductor due to the oxygen vacancies, and at elevated temperature a defect occurs in Ti^+ which creates vacancies. Therefore, it behaves like a p-type semiconductor although at such elevated temperatures. Nowotny et al. (2010) have demonstrated the p-type property of bulk TiO_2 at room temperature. Studies that are more recent have revealed that p-type TiO_2 semiconductor materials are achieved either by doping with elements or moving from bulk TiO_2 to nano-restructured crystals such as nanowire, nanorods, nanotubes, etc., and this bridges the gap between bulk and quantum dots of TiO_2. Wu, Shih, and Wu (2005) suggest that during the chemical synthesis of such nanocrystals, additional heat treatment is required to enhance its crystalline structure. One of the major properties of TiO_2 in bulk form is the band gap energy (Eg). This is the energy generated during the excitations of an electron from the valence band to the conduction band when it absorbs a photon. When bulk TiO_2 absorbs photon greater than Eg there is a generation of a hole in the valence band as a result of excitation of an electron. When the exciton (the force bounding electron-hole pair) is so small that electrons and holes are confined, this leads to the formation of TiO_2 QD. TiO_2 QDs have perfect tuning and are achieved by the confinement of electrons and holes. The tuning emits fluorescent light throughout the visible spectrum, and it enhances its application in optoelectronic devices. The extent of the emission of light in the TiO_2 QD is based on the overlapping of electron and hole wave function, while the bulk TiO_2 has no confinement of exciton, it moves freely and can easily separate. This is the major shortcoming of bulk TiO_2. The small size of TiO_2 QD enhances the electronic and optical properties, which arises from the observable random off and on blinking demonstrated by TiO_2 QD. In addition, TiO_2 QD has a major application in photovoltaic devices because of its ability to have carrier multiplication when it absorbs excess photons (Toyoda and Shen 2012).

TiO_2 quantum dot has been synthesized using several techniques such as solvothermal/hydrothermal method (Wategaonkar et al. 2020), atomic layer deposition (ALD) (Rodrigues et al. 2019), electrodeposition (Endrődi et al. 2018), chemical vapour deposition (CVD) (Ding et al. 2000), sol gel method (Nateq and Ceccato 2019), sputtering technique (Hassanien and Akl 2019), microwave technique (Bai et al. 2014), spin coating (Huang et al. 2020), thermal (or vacuum) evaporation (Pulker, Paesold, and Ritter 1976; Oboudi et al. 2014a). The sol gel technique involves hydrolysis of precursor in a basic or acidic solvent followed by polycondensation and calcination. One of the major limitations of the sol gel technique is that precipitate formed is usually an amorphous substance, and it has to be calcined at a very high temperature to yield crystalline TiO_2 QD. This process causes an increase in grain size, stimulates the phase change, and reduces the surface area of the formed TiO_2 QD. ALD is a low-temperature technique used for synthesizing TiO_2 QD. It is made of four major steps: (a) the titanium (Ti) precursor is exposed in the chamber, (b) cleaning and removal of the by-product, (c) introduction of oxygen as a reactant precursor and this reacts with titanium surface, (d) cleaning of reactant and removal of by-product from the chamber. The major shortcoming of ALD is that the deposition is a fraction of monolayers per cycle and hence it takes time. In addition, the confinement is only enhanced by the size of the reaction chamber. On the other hand, chemical vapour deposition (CVD) tends to be a promising technique because it yield very pure and high quality films of TiO_2 QD. This technique involves the exposure of heated substrate in the presence of volatile precursors. The precursor diffuses and deposits on the surface of heated substrate forming a desirable film. Despite having the advantage of producing quality film, there is difficulty in controlling stoichiometry when a multiple source precursor is used, and this is partly due to different vaporization rates of the precursors. This is the major limitation of CVD technique. In the case of solvothermal technique, the deposition process is very hard to control and the stainless steel autoclave is very expensive. Similarly, electrodeposition technique is very expensive as well and takes time because multiple coatings are required.

In spite that TiO$_2$ QD has been deposited using different methods, there are still several shortcomings. Thermal evaporation technique is an outstanding technique used in depositing TiO$_2$ QD. Compared to other techniques, the thermal evaporation is a low-cost deposition method used frequently to achieve fast deposition of high purity films in an (ultra) high vacuum chamber. In such a system, the thin-film thickness can be observed in situ using the thickness monitor. The deposition of TiO$_2$ thin film using thermal evaporation was first recorded in 1978 by Morris (1978). However, from our earlier reported literature, Pulker, Paesold, and Ritter (1976) demonstrated the reactive deposition of TiO$_2$ films using high vacuum evaporation plants. They were able to study the refractive indices of TiO$_2$ films obtained from different source materials, e.g. Ti, TiO, Ti$_2$O$_3$, Ti$_3$O$_5$, and TiO$_2$. Similarly, Oboudi et al. (2014a) were able to deposit TiO$_2$ thin film using the same method and studied the TiO$_2$ film transmittance using the UV-VIS technique. The TiO$_2$ described here are quantum dots, which present unique properties compared to the thin films. The deposition was achieved by passing high voltage on a resistive wire using powdered Ti or TiO$_2$ as source material under a low-pressure vacuum of the order of 10^{-5} or 10^{-6} torrs, to prevent the reaction between the vapour and atmosphere in the chamber.

In this chapter, we discuss the thermal evaporation deposition of TiO$_2$ QD using a glass substrate, which can be applied in optoelectronic devices such as solar cells, memory devices, sensing devices, transistors, etc. The optoelectronic property of deposited TiO$_2$ QD thin film can be adjusted by varying the shape and size of TiO$_2$ QD formed. Lager diameter gives longer wavelength and vice versa.

15.2 Device Fabrication

15.2.1 Solar Cell

This is a device used to harness light energy and convert it to electrical energy. Different solar cells or photovoltaic devices (e.g. organic, inorganic, and perovskite) have been fabricated using different materials, which enhance the efficiency of solar energy conversion whilst having a reduction in cost and enabling the use of semiconductor nanomaterials (Mather and Wilson 2017). Over the years, Si has been the major material used in the fabrication of photovoltaic cells, but recently attention has been shifted towards the use of quantum dot material because it maintains its transparency and flexibility. Researchers have shown that SQD oxides with band gap up to 3 eV have salient features which aid their use in the fabrication of solar cell and other electronic devices. TiO$_2$ QD has this property with additional features such as good stability, eco-friendly, and abundance (Rathee, Arya, and Kumar 2011).

15.2.1.1 Organic Solar Cell (OSC)

This is one of the promising solar cells that is made of different layers of organic (polymers) and semiconductor materials. Each is formed by deposition of their thin film (Krebs 2009), where metal oxides or semiconductor nanoparticles, polymers, e.g. poly (3-hexylthiophene or P3HT) act as an absorbing layer (Hosel, Angmo, and Krebs 2013). It works with the principle of bulk heterojunction, and there is a blend of semiconductor materials (p-type and n-type) which are sandwiched between the two electrodes, i.e. anode and cathode (Greiner, Chai, and Lu n.d.). The p-type donates the electron and the n-type accepts the electron (Kuznetsov, Akkuratov, and Troshin 2019). Most organic materials do not have electrons occupying the valence band except at a portion called the highest occupied molecular orbital (HOMO) and the lowest unoccupied molecular orbital (LUMO). LUMO acts as the conduction band and the difference between HOMO and LUMO is called the band gap (Landerer et al. 2019). In the fabrication of OSC, the choice of material used is determined by the fabrication technique. H. Hoppea and N Sariciftci (2004) reported that if thermal evaporation is used, polymers with good stability are required. Figure 15.1 shows the steps in the fabrication of OSC. Firstly, ZnO that is used as a buffer layer is deposited on the ITO substrate using the sol gel technique. Organic semiconductor materials, i.e. [6,6]-phenyl-C$_{61}$butyric acid methyl ester (PCBM) and poly (3 hexalthiopene) P3HT are used as

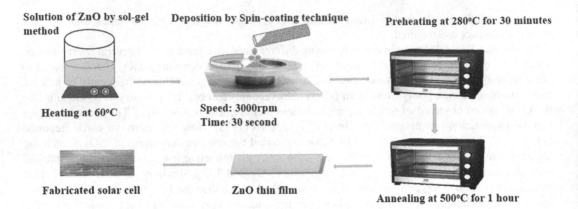

Solution of ZnO by sol-gel method

Deposition by Spin-coating technique

Preheating at 280°C for 30 minutes

Heating at 60°C

Speed: 3000rpm
Time: 30 second

Fabricated solar cell

ZnO thin film

Annealing at 500°C for 1 hour

FIGURE 15.1 Schematic diagram of OSC fabrication. Reproduced with permission from Sahdan et al. 2015, copyright Elsevier. Adapted from Ref. Sahdan et al. 2015.

electron acceptors and donors, respectively. The fabrication of organic cell heterojunction is done by mixing P3HT and PCBM in 1, 2-diclorobenzene ($C_6H_4Cl_2$) and stirred at room temperature. Thereafter, the blend of P3HT:PCBM is deposited on the ZnO buffer layer using spin coating technique followed by annealing and the deposition of anode electrode (Au) using DC sputter.

In recent years, there is a quest to provide materials that have tunable electronic property, are cost-effective and eco-friendly, that aid the fabrication of OSC (Erik Perzon et al. 2005; Dennler, Sariciftci, and Brabec 2015). Some transition metallic oxide films such as TiO_2 QD have such unique properties. It raises the interest in the use for charge generation or recombination and electrode material in OSC (Meyer et al. 2012). The ability to tune the electronic property and the nature of wide band gap energy of TiO_2 QD improves its transparency, which aids its application as an electrode in OSC. During fabrication of OSC, TiO_2 QD is used to cover the entire surface of the OSC and still permit the active layer to receive light incident on it. This is because organic semiconductor material has a low conductivity, and it requires a very small gap between the electrode and charge carriers. Gershon (2011) reported that the major shortcoming of OSC is electrical shunting or short circuiting; this menace can be curbed using TiO_2 QD. OSC is the combination of p-type and n-type organic semiconductor materials, and if they are in direct contact with the two electrodes it creates shunting. This reduces the performance of OSC devices, but TiO_2 QD serves as material for charge blocking.

15.2.1.2 Inorganic Solar Cell

This is like the organic solar cell but here inorganic semiconductor material is used as a light absorber. It has p-type as the donor layer and n-type as acceptor material forming a p-n junction. Inorganic solar cells or PV technology is generally grouped into three generations: (a) first generation: c-Si is the dominant absorber (b) second generation: Si thin film is the major absorber (c) third generation: the tandem solar cells comprising multiple layers of light absorbing material such as amorphous Si (Donne et al. 2013).

Ever since the use of semiconductor quantum dot (SQD) material in the fabrication of inorganic solar cell, Si QD has been an outstanding material when compared with other materials because it has very good efficiency of solar power conversion (Miles, Zoppi, and Forbes 2007), but due to the fact that it is toxic and difficult to produce it is seldomly used. A lot of researchers have looked for an alternative in developing inorganic solar cells using a material with unique properties such as good efficiency and eco-friendliness. TiO_2 QD is one of the best SQD materials that possess such properties. Hayre et al. (2006) reported the fabrication of $CuInS_2$ (CIS) solar cell using nanostructured n-type TiO_2 as an electron transport layer. Nanostructured TiO_2-based solar cell was observed to have good efficiency, but it was affected by charge recombination at the TiO_2 interface and this effected its commercialization. When the thickness of the nanostructured TiO_2 increases, it dramatically has better power conversion efficiency (Nanu and Schoonman 2004; Bai et al. 2014). The charge recombination at the interface limits the use of TiO_2 QD in the fabrication of inorganic solar cells until early 2020 when Professor Ziqi Sun and his team from Queensland University of Technology, Australia, developed a prototype using TiO_2 QD (Sun 2020; Emiliano Bellni 2020). They were able to reduce the interface in the band of TiO_2 and the device yielded maximum power output. It was observed that TiO_2 QD produces power efficiency up to 24%, which is very close to the reported theoretical value of 33 % efficiency of Si QD.

15.2.1.3 Perovskite Solar Cell

The innovation of the perovskite solar cell has been outstanding in photovoltaic technology. Perovskite material was first utilized as sensitizers in dye-sensitized solar cells in 2009 with the efficiency of solar power conversion as 3.8% (Raj 2018) but as at 2019 H. Baig et al. (2019) reported the efficiency to be 21.6% and a whooping efficiency of 25.2% in early 2020 (Gao et al. 2020). Perovskite materials are formed from compounds of calcium titanate ($CaTiO_3$). Perovskite, e.g. $CH_3NH_3PbI_3$ or $MAPbI_3$, is described by molecular structure ABX_3 where X_3 is halogen anion, B is the inorganic cation, and A is the organic cation (Lu et al. 2018). Perovskite solar cell is a blend of inorganic and organic semiconductor materials which are sandwiched in between the compact layers (hole transport layer (HTL) and the electron transport layer (ETL)). Figure 15.2 shows the processes of synthesizing perovskite solar cells using $CH_3NH_3PbI_3$ as perovskite. Poly(3,4-ethylenedioxythiophene)–poly(styrene sulphonate) (PEDOT–PSS) was spin-coated on a heated FTO substrate followed by annealing. The perovskite material was deposited and annealed as well. The annealing effect results in the formation of the natural dark brown colour of the perovskite layer. However, PCBM was deposited directly on the dark brown perovskite layer. Finally, CdS electron transport layer and metal electrode are deposited sequentially using thermal evaporation.

In perovskite solar cells, ETL is usually composed of metal oxides such as TiO_2, Al_2O_3, SnO_2, and ZnO. The efficiency of perovskite solar cell is improved by the thinness of compact layer (Hu et al. 2020), and this aids the use of TiO_2 QD which has few micrometres of thickness. The thinness nature of TiO_2 QD film aids its transparency, which is essential so that light could pass through and fetched up by the absorbing layer. In addition, TiO_2 QD ETL enhances the efficiency of perovskite solar cells because it decreases the resistance during current transport by extracting electrons and repelling holes.

15.2.2 Memory Devices

In recent years, there has been an increase in interest in the use of TiO_2 QD in the fabrication of electronic devices such as memory units, which has helped to increase the number of chips in conventional flash drives, some built-in memory arrays for microcontrollers, dynamic random access memory (DRAM), and resistive random access memory (ReRAM) (Dimitrakis, Normand, and Bonafos

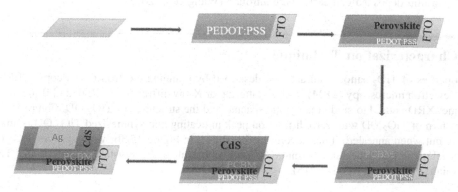

FIGURE 15.2 Step-by-step illustration of perovskite solar cell fabrication. Adapted from Ref. Qasuria, Alam, and Qureshi 2018.

2013). DRAM is a form of memory device that is mainly used in i-pad and most modern computers because of its low cost and effectiveness in memory retention (S. K. Kim, Kim, and Jeong 2012). Its storage mechanism is based on a transistor or capacitor, which can either be charged or discharged. Because of the charge leakage in the capacitor, it is always advisable to either refresh or give it a new electronic charge. DRAM is fabricated by forming the capacitor as well as its dielectric and vertical transistor followed by forming the gate electrode or gate of the transistor (Schlosser and Hofmann 2002). TiO_2 QD is used as capacitor dielectric and this drastically reduces the effect of charge leakage. Similarly, data are stored in ReRAM using charged atoms which create changes in resistive switching (Zou et al. 2016). Two electrodes, i.e metal-oxide-metal (e.g. Al-TiO_2-Al), is adopted in ReRAM. The ionic switching effect in TiO_2 QD is described by a filament which is composed of the migration of oxygen vacancies. Hence, oxygen ion combines with the anode to produce O_2 (Gale 2014). More so, the wide band gap of TiO_2 QD also enhances the fabrication of transparent ReRAM.

15.2.3 Transistor Devices

The special type of field-effect transistor is the metal oxide thin films transistor (TFT). It is an outstanding device used in the fabrication of electronic devices because of their unique properties such as high mobility of electrons, optical transparency, and easiness to fabricate (Pal et al. 2009). TFT is made of three components (a) semiconductor thin film (b) the layers of insulating material (c) the electrodes which comprise of source, drain, and gates. TFT is fabricated by the deposition of metal oxide thin films, dielectric, and the metal contacts, which serve as electrodes on a glass or plastic or silicon substrate (Sharma et al. 2019).

Metal oxides have shown some relevant performance in the fabrication of TFTs, but some of them require very high operating voltage because of their low dielectric property. The introduction of TiO_2 QD eliminates the effect of voltage consumption because it behaves like an insulator with high dielectric constant property up to 100 (Okyay et al. 2017).

15.2.4 Gas Sensor

This is a device used to detect the presence of toxic gasses such as exhaust fumes in the atmosphere. There are different types of gas sensors, and they are determined by the material used during fabrication, e.g. organic-based (Yan et al. 2015), carbon-based (Chen et al. 2013), metal oxide semiconductor-based (Xu et al. 2015), gas sensors. Among these, metal oxide semiconductor-based gas sensor has been outstanding, having cost-effectiveness, easy to fabricate, and high efficiency of gas sensing power. The use of TiO_2 QD in gas sensor provides better sensor efficiency and stability when operating in a high temperature environment (Iftimie et al. 2009). The operation of the gas sensor is based on the heterojunction structure (Kim, Kim, and Yong 2012). TiO_2 QD-based gas sensor is best fabricated using atomic layer deposition (ALD), where all the deposition is done in the reactor autoclave and the reactant residues are cleaned using inert gas purge (Galstyan 2017). ALD is considered the best technique because the film deposited can easily be controlled (Wang et al. 2017).

15.3 Characterization Technique

The properties of TiO_2 nanocrystal are best described by scanning electron microscopy (SEM), transmission electron microscopy (TEM), X-ray scattering or X-ray diffraction (XRD), and Bragg diffraction technique. XRD is used to study the atomic spacing and the structures of TiO_2 QD. Figure 15.3a is the XRD pattern of TiO_2 QD with zero diffraction peak indicating that synthesized TiO_2 QD is amorphous in nature but when annealed, it has several peaks (shown in Figure 15.3b) which proves TiO_2 QD as a nanocrystal material. Scherrer's formula is used to calculate the crystallite size of the TiO_2 QD (Komaraiah, Poloju, and Reddy 2015).

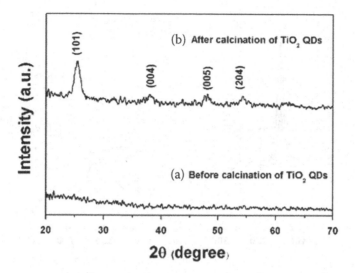

FIGURE 15.3 XRD pattern of TiO₂ QDs (a) amorphous TiO₂ QD (b) calcined TiO₂ QDs. (Reprinted with permission from Ref. Gnanasekaran 2015. Copyright© 2015, Elsevier.)

$$D = \frac{0.9\lambda}{\beta Cos\theta} \tag{15.1}$$

where β = the full width, D = the crystallite size, λ = the wavelength of the incident X-ray, and θ = the Bragg angle. Figure 15.4 shows the XRD pattern of annealed TiO₂ QD polymorphs. Figure 15.4a is the X-ray diffraction pattern of anatase with maximum diffraction peak at 101 planes when $2\theta = 25.28°$ having the crystallite size as 0.352 nm. The peaks between 101 and 200 planes show the preferred orientation of annealed TiO₂ QD. The peaks proved that annealing improves the properties of TiO₂ QD (Venkatachalam et al. 2011). The anatase XRD pattern strongly overlaps the diffraction pattern of brookite (shown in

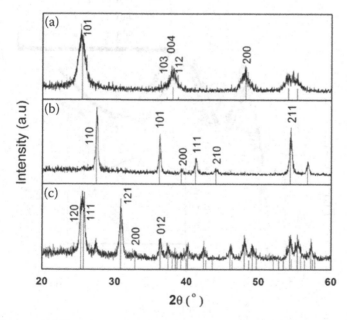

FIGURE 15.4 XRD pattern of TiO₂ polymorph (a) anatase (b) rutile (c) brookite. (Reprinted with permission from Ref. Reyes-Coronado et al. 2008. Copyright© 2008, IOP Publishing Ltd.)

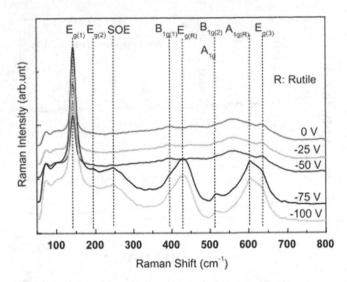

FIGURE 15.5 Raman spectra of TiO$_2$. (Reprinted with permission from Ref. Nezar et al. 2016). Copyright© 2016, Elsevier.)

Figure 15.4c) at 120 planes when $2\theta = 25.34°$ and 111 plane when $2\theta = 25.69°$. The crystallite size is 0.351 nm (Chem et al. 2012,Di et al. 2008). Figure 15.4(b) shows the XRD peak of rutile at 110 planes.

On the other hand, the phase, stress or strain, chemical structure, molecular interactions of TiO$_2$ QDs can also be studied using Raman spectroscopy. This spectrum acts as a chemical fingerprint which a material possesses for ease of identification. Choi, Jung, and Kim (2005) suggest that Raman spectroscopy is one of the best methods to study the particle size of nanocrystals, due to its ability to detect variation in spectra when there is a decrease in particle size. There are several peaks in Raman spectroscopy, and each is equivalent to molecular bond vibration. Figure 15.5 shows the Raman peaks of

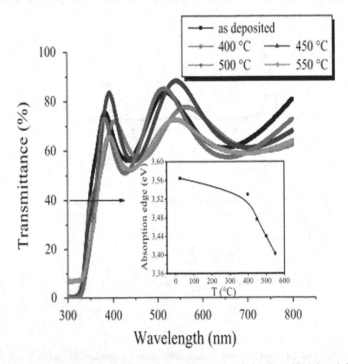

FIGURE 15.6 TiO$_2$ QD UV-vis transmittance spectra. Reproduced with permission from Hanini et al. 2013. Copyright IOSR.

TiO$_2$ QD (anatase, rutile, and brookite) structure and the polymorphs were studied within 100–900 cm^{-1} spectra rang. Anatase has six different Raman peaks Eg$_{(3)}$, B$_{1g(2)}$, A$_{1g}$, B$_{1g(1)}$, Eg$_{(2)}$, Eg$_{(1)}$, and this predicts the tetragonal structure of anatase TiO$_2$ QD. Nezar et al. (2016) proved that those peaks occurred at near frequencies 143 cm^{-1}, 195 cm^{-1}, 396 cm^{-1}, 513 cm^{-1}, 517 cm^{-1}, and 639 cm^{-1}, respectively. The variation in the Raman peaks situated at SOE, E$_{g(R)}$ and A$_{1g(R)}$ is for rutile, which suggests the tetragonal structure of rutile TiO$_2$ QD. The optical property of TiO$_2$ QD is determined using UV-vis technique. In the UV-vis technique, the transmittance and reflectance of TiO$_2$ QD are dependent on wavelength. Figure 15.6 shows before and after annealing of TiO$_2$ QD UV-vis transmittance spectra. It indicates that the transmittance of TiO$_2$ QD decreases with an increasing annealing temperature and the full transparency is achieved in the visible region but suddenly falls between 300 nm and 400 nm in UV region. The transmittance absorption edge inserted in Figure 15.6 shows that the annealing temperature increases and shifts the absorption edge curve in the direction of low energy, indicating a reduction in band gap energy.

15.4 Structural, Optical, and Electrical Properties of TiO$_2$ QDs

The structure of TiO$_2$ polymorphs is composed of distorted TiO$_6$ octahedral but in a separate form as shown in Figure 15.7. Rutile with space group D_{4h}^{14} has a tetragonal structure and each of the two opposite edges of the octahedral structure is shared forming a linear chain towards the direction [001], while the TiO$_6$ chains are joined with each other by connecting their edges. The tetragonal structure of anatase with space group D_{4h}^{19} does not have their corners shared but the four edges of the octahedral structure are interconnected, which leads to the formation of zigzag chains. The brookite has an orthorhombic structure with an atomic space group D_{2h}^{15}. The three edges, as well as corners of the octahedral, are shared. The prevailing structural character is the chain of edge sharing (Li, Ishigaki, and Sun 2007; Zhang et al. 2015). The lattice parameter of anatase are a = b = 3.785 A°, c = 9.514 A° (Fan et al. 2016), rutile a = b = 4.593 A°, c = 2.959 A° and brookite a = 5.143 A°, b = 5.456 A°, c = 9.182 A° (Chem et al. 2012). The polymorphs of TiO$_2$ have different band gaps and this is due to the structural difference. Anatase has high band gap energy compared with other polymorphs, which is due to large Ti-Ti structural distance as well as short Ti-O structural distance (Rahimi, Pax, and Gray 2016). Each of the band structure of the polymorphs has direct and indirect transitions. The direct band gap is achieved when the transition of the conduction band is directly above the valence band and when the transition of the conduction band is offset in momentum space to the valence band gives the indirect band gap. The TiO$_2$ QD optical band gap energies are calculated using Tauc and Davis-Mott relation (Tauc, Grigorovici, and Vancu 1966)

$$(\alpha h\upsilon)^n = K(h\upsilon - E_z) \tag{15.2}$$

(a)   (b)   (c)   (d)

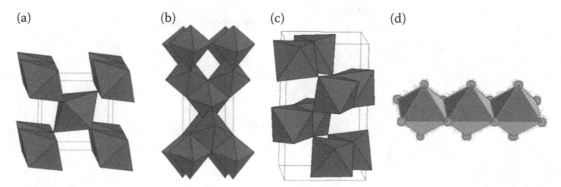

FIGURE 15.7 The schematic diagram showing the sequence of octahedral order in (a) rutile, (b) anatase, (c) brookite (d) the rutile [001] direction of octahedral filament. Reproduced with permission from Yang 2012/ Copyright University of Pittsburgh.

where K = the independent energy constant, α = absorption coefficient, $h\upsilon$ is the incident energy of a photon, E_g = the optical band gap of the semiconductor, and n = the nature of transmission. Direct band gap materials have the value of n = 2 while indirect band gap has n = ½. The direct band gap is determined by the plot of $(\alpha h\upsilon)^2$ against $h\upsilon$. The tangent of the plot is used to estimate the direct band gap value of TiO$_2$ QD. The direct band gap for anatase is 3.2 eV, 3.0 eV for rutile, and brookite is from 3.0 eV to 3.2 eV (Rodr 2008, Shibata et al. 2004). The tangent of $(\alpha h\upsilon)^{1/2}$ and $h\upsilon$ plot gives the indirect band gap (Zhao et al. 2007). Li, Ishigaki, and Sun (2007) reported the indirect band gap as 2.90 eV, 2.81 eV, and 2.85 eV for anatase, rutile, and brookite, respectively.

When the annealing temperature is less than 500°C, the properties of TiO$_2$ QD remain unchanged (Tang et al. 2012) but an increase in annealing temperature leads to the reduction of transmittance due to the rough surface of annealed TiO$_2$ QD. Tahir et al. (2017) report that the refractive index of TiO$_2$ QD increases with the annealing temperature. It was noted that the refractive index of TiO$_2$ QD changes from 2.4 to 2.5 when annealing temperature increases, and this is very close to the refractive index of bulk TiO$_2$. The effect of annealing temperature affect the forbidden band gap of TiO$_2$ QD, and it results in the formation of sub-band (Bakardjieva and Murafa 2009). The sub-band decreases the band gap energy and facilitates the use of TiO$_2$ QD in fabrication of photovoltaic devices. Chiad et al. (2014) demonstrate the temperature annealing effect on band energy. They observed 3.25 eV band gap energy at an annealing temperature of 200°C and 2.95 eV at 250°C. Similarly, due to the quantum confinement, the absorption energy is always shifted to the higher energy and it increases with an increase in annealing temperature due to the surface roughness (Zhao et al. 2007). However, the thickness of TiO$_2$ QD film affects transparency and resistivity. Khan et al. (2017) demonstrate the thickness effect on resistivity using multilayered TiO$_2$ QD. They observed that the resistivity of a single layer to be 2.3 × 10^7 (Ω-m) and it decreases to 1.14 × 10^7 (Ω-m) for four layers. TiO$_2$ QD absorbs UV radiations and transmits visible light, which helps to improve the life span of optoelectronic devices. The transparency of TiO$_2$ QD film depends on the average transmittance which falls within the visible light and as the film thickness increases, the transparency reduces. Increase in an annealing temperature results in the decrease in transmittance due to formation of rough surface.

15.5 Mechanism of TiO$_2$ QD Formation

The process of TiO$_2$ QD formation using the thermal evaporation technique has been the most attractive technique because it produces high-quality uniform film with a very good deposition speed. During film deposition, a high vacuum ranging from 10^{-5} to 10^{-9} torr is needed. The high vacuum assists the formation of purity film. Figure 15.8 shows the mechanism of achieving high vacuum in the vacuum chamber.

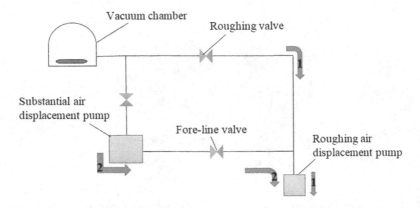

FIGURE 15.8 The processes of achieving a high vacuum in a vacuum coating system ("Thin Film Coating System").

The formation of TiO_2 QD can be achieved using different source materials, e.g. Ti (Oboudi et al. 2014b), TiO, Ti_2O_3, Ti_3O_5, and TiO_2 (Pulker, Paesold, and Ritter 1976). In the formation of TiO_2 QD using TiO_2 powder as source material, the substrate is hung in a vacuum chamber using the substrate holder and TiO_2 powder is loaded on a crucible and then heated very gently by resistive heating method, i.e. passing a large current through a tungsten filament. This heats up the TiO_2 powder until a high vapour pressure is attained; the atoms of powdered TiO_2 melt and begin to evaporate. The evaporated atom moves across the vacuum chamber and condenses directly on the surface of the substrate where it forms thin film coatings. Similarly, TiO_2 QD can also be formed using the reactive evaporation method. In the reactive evaporation technique, titanium metal (Ti) and oxygen gas are used as the source material. Ti is placed on the crucible and heated until it forms vapour. Oxygen gas is then introduced in the chamber, which reacts with Ti vapor on the surface of the substrate forming TiO_2 QD. Usually, the deposited TiO_2 QD is in amorphous form, and it must be annealed to achieve the crystalline form. The annealing temperature ranges from $50^{\circ}C$ to $500^{\circ}C$. Chiad et al. (2014) demonstrate the reactive vacuum deposition using powdered Ti as the source material. After achieving a high vacuum of 10^{-5} mbar, Ti powder was heated until it formed vapour. TiO_2 QD was formed on the surface of the glass substrate by thermal oxidation at 500 °C for 45 min.

15.6 Challenges and Possible Enhancement of TiO_2 QD-Based Device

From the discussion above, TiO_2 QD has unique properties that make it suitable as a material used in the fabrication of optoelectronic devices such as PV cells, transistors, sensors, and memory devices. However, there are still some limitations affecting the full realization of high-performance TiO_2 QD-based optoelectronic devices. For instance, the nature of TiO_2 QD wide band gap makes it incompatible in full utilization of the solar spectrum. TiO_2 QD needs ultraviolet (UV) radiation for activation and sunlight has just 4–5% of UVA, indicating that TiO_2 QD has a low output when it absorbs natural light (Removal 2019). Shockley–Queisser proposed that to achieve high efficiency in PV cells, the p-n junction should be composed of materials which have maximum band gap energy to be 1.4eV (Ru 2016) while TiO_2 QD has the band gap energies fall between 3.0 eV and 3.2 eV. However, this effect can be curbed by doping with either transition metal, noble metal, or non-metals. Duan et al. (2012) reported the Sn-doped TiO_2 as photoanode, and they observed the efficiency of the PV cell increasing from 8.31% to 12.1%. There are other reported doping materials such as Al (Lou et al. 2016), CuO (Rokhmat, Wibowo, and Abdullah 2017), samarium (Liu, Hou, and Qu 2017), etc., but the high cost of these materials make them unattractive. Therefore doping with non-metal such as nitrogen (Barkul et al. 2016), sulphur (Seo et al. 2016), boron (Shi et al. 2019), (Ho, Lin, and Wang 2015), and carbon (Park, Kim, and Bard 2006; Park et al. 2015) present a better feature. They narrow the band gap of TiO_2 QD by displacing oxygen in the lattice. On the other hand, another drawback of TiO_2 QD is the charge recombination at the interface, and this reduces photovoltage in PV cells thereby limiting its application in solar cell fabrication. Inorganic insulating materials like Al_2O_3, In_2S_3, and Nb_2O_5 are used as a buffer layer, which reduces the charge recombination at the interface (Kopidakis, Neale, and Frank 2006). The buffer layer is sandwiched in between the region of n-type and p-type layers.

15.7 Feature Scope

The technology breakthrough on TiO_2 QD has done great work in enhancing the fabrication of high-quality optoelectronic devices with salient features such as low cost, high efficiency, eco-friendly, etc. TiO_2 QD has been applied in solar cells as an electron transport layer, electrodes, and charge blocking layers. Nano-TiO_2 is an n-type semiconductor, but it has been on record that undoped bulk and nano-TiO_2 material can show the p-type properties of semiconductors (Anitha, Banerjee, and Joo 2015; Wang et al. 2015; Lin et al. 2019). There is a need for researchers to explore more on the p-type nano-TiO_2 semiconductor material and show if it can be used as an absorbing layer on solar cells and also fabricating a solar cell having p-n junction with both p-type and n-type nano-TiO_2 material. This will pave the way for the fabrication of very low-cost solar cell devices.

15.8 Conclusion

TiO_2 QD has been an outstanding SQD due to its unique properties, which makes it applicable in the fabrication of optoelectronic devices such as solar cell, transistor, memory, gas sensing, etc. In the solar cell, the thinness and transparency of TiO_2 QD aid its application as ETL, charge blocking layer, and photoelectrodes. From the discussion above, it has been seen that the wide band gap nature of TiO_2 QD has affected its UVA absorbance, but it is usually engineered to achieve the required band gap. Similarly, an increase in annealing temperature also reduces the band gap and a reduction in band gap energy improves the efficiency of solar power conversion.

REFERENCES

Anitha, V. C., Arghya Narayan Banerjee, and Sang Woo Joo. 2015. "Recent Developments in TiO_2 as N- and p-Type Transparent Semiconductors: Synthesis, Modification, Properties, and Energy-Related Applications." *Journal of Materials Science*. https://doi.org/10.1007/s10853-015-9303-7.

Bai, Yu, Filippo De Angelis, Juan Bisquert, and Peng Wang. 2014. "Titanium Dioxide Nanomaterials for Photovoltaic." *Applications American Chemical Society* 114, no. 19: 10095–10130. https://doi.org/10.1021/cr400606.

Baig, Hasan, Hiroyuki Kanda, and Abdullah M Asiri. 2019. "Sustainable Energy & Fuels Increasing e Ffi Ciency of Perovskite Solar Cells Using Low Concentrating Photovoltaic Systems." *Sustainable Energy and Fuels*. https://doi.org/10.1039/c9se00550a.

Bakardjieva, Snejana, and Nataliya Murafa. 2009. "Preparation and Photocatalytic Activity of Rare Earth Doped TiO_2 Nanoparticles Václav Stengl." *Materials Chemistry and Physics* 114: 217–226. https://doi.org/10.1016/j.matchemphys.2008.09.025.

Barkul, R. P., V. B. Koli, V. B. Shewale, M. K. Patil, and S. D. Delekar. 2016. "Visible Active Nanocrystalline N-Doped Anatase TiO_2 Particles for Photocatalytic Mineralization Studies." *Materials Chemistry and Physics*, 1–10. https://doi.org/10.1016/j.matchemphys.2016.01.035.

Brazis, Paul W. 2017. "Quantum Dots and Their Potential Impact on Lighting and Display Applications."

Chem, J Mater, Yulong Liao, Wenxiu Que, Qiaoying Jia, Yucheng He, and Peng Zhong. 2012. "Controllable Synthesis of Brookite/Anatase/Rutile TiO_2 Nanocomposites and Single-Crystalline Rutile Nanorods Array." *Journal of Materials Chemistry*, 7937–7944. https://doi.org/10.1039/c2jm16628c.

Chen, Zhuo, Ahmad Umar, Yao Wang, Tong Tian, Ying Shang, Yuzun Fan, Qi Qi, Dongmei Xu, and Lei Jiang. 2013. "Supramolecular Fabrication of Multilevel Graphene- Based Gas Sensors with High NO2 Sensibility." *Royal Society of Chemistry*. https://doi.org/10.1039/C5NR01770J.

Chiad, Sami Salman, Saad Farhan Oboudi, Nabeel A Bakr, and Nadir Fadhil Habubi. 2014. "Electronic Transitions and Dispersion Parameters of Annealed TiO_2 Films Prepared by Vacuum Evaporation Technique." *Materials Focus* 3: 23–27. https://doi.org/10.1166/mat.2014.1134.

Choi, Hyun Chul, Young Mee Jung, and Seung Bin Kim. 2005. "Size Effects in the Raman Spectra of TiO_2 Nanoparticles" 37: 33–38. https://doi.org/10.1016/j.vibspec.2004.05.006.

Dennler, Gilles, Niyazi Serdar Sariciftci, and Christoph J. Brabec. 2015. "Conjugated Polymer-Based Organic Solar." *Chemical Reviews*. https://doi.org/10.1021/cr050149z.

Di, Agatino, Giovanni Cufalo, Maurizio Addamo, Marianna Bellardita, Renzo Campostrini, Marco Ischia, Riccardo Ceccato, and Leonardo Palmisano. 2008. "Photocatalytic Activity of Nanocrystalline TiO_2 (Brookite, Rutile and Brookite-Based) Powders Prepared by Thermohydrolysis of TiCl4 in Aqueous Chloride Solutions" 317: 366–376. https://doi.org/10.1016/j.colsurfa.2007.11.005.

Dimitrakis, P., P. Normand, and C. Bonafos. 2013. "Quantum Dots for Memory Applications." *Physica Status Solidi* 1504: 1490–1504. https://doi.org/10.1002/pssa.201300029.

Ding, Zhe, Xijun Hu, Gao Q. Lu, Po-lock Yue, Paul F. Greenfield, Clear Water Bay, and Hong Kong. 2000.

"Novel Silica Gel Supported TiO$_2$ Photocatalyst Synthesized by CVD." *American Chemical Society*, 6216–6222. https://doi.org/10.1021/la000119.

Donne, A Le, A. Scaccabarozzi, S. Tombolato, S. Marchionna, P. Garattini, B. Vodopivec, M. Acciarri, and S. Binetti. 2013. "State of the Art and Perspectives of Inorganic Photovoltaics." *ISRN Renewable Energy*. https://doi.org/10.1155/2013/830731.

Duan, Yandong, Nianqing Fu, Qiuping Liu, Yanyan Fang, Xiaowen Zhou, Jingbo Zhang, and Yuan Lin. 2012. "Sn-Doped TiO$_2$ Photoanode for Dye-Sensitized Solar Cells." *Journal of Physical Chemistry* 116: 8–13.

Emiliano Bellni. 2020. "A Titanium Solar Cell with 24% Efficiency." PV Magazine International. Australia. Accessed May 2020. https://www.pvmagazineinternational.com/2020/02/13/a-titanium-solar-cell-with-24-efficiency/.

Endrődi, Balázs, Egon Kecsenovity, Krishnan Rajeshwar, and Csaba Janáky. 2018. "One-Step Electrodeposition of Nanocrystalline TiO$_2$ Films with Enhanced Photoelectrochemical Performance and Charge Storage." *Applied Energy Materials*. https://doi.org/10.1021/acsaem.7b00289.

Erik Perzon, Xiangjun Wang, Fengling Zhang, Wendimagegn Mammo, Juan Luis Degado, Pilar de la Cruz, Olle Inganas, Fernado Langa, and Mats R. Andersson. 2005. "Design, Synthesis and Properties of Low Band Gap Polyfluorenes for Photovoltaic Devices." *Elsevier Ltd* 154: 53–56. https://doi.org/10.1016/j.synthmet.2005.07.011.

Fan, Zhenghua, Fanming Meng, Miao Zhang, Zhenyu Wu, Zhaoqi Sun, and Aixia Li. 2016. "Solvothermal Synthesis of Hierarchical TiO$_2$ Nanostructures with Tunable Morphology and Enhanced Photocatalytic Activity." *Applied Surface Science* 360: 298–305. https://doi.org/10.1016/j.apsusc.2015.11.021.

Gale, Ella. 2014. "TiO$_2$-Based Memristors and ReRAM: Materials, Mechanisms and Models (a Review)." *Semiconductor Science and Technology* 29: 1–29. https://doi.org/10.1088/0268-1242/29/10/104004.

Galstyan, Vardan. 2017. "Porous TiO$_2$ -Based Gas Sensors for Cyber Chemical." *Sensors* 17. https://doi.org/10.3390/s17122947.

Gangasani, Srilaxmi. 2016. "A Study on Quantum Dot and Its Applications." *International Journal of Innovative Research in Science, Engineering and Technology* 5: 8128–8133. https://doi.org/10.15680/IJIRSET.2016.0505119.

Gao, Xiao-xin, Wen Luo, Yi Zhang, Ruiyuan Hu, Bao Zhang, and Andreas Züttel. 2020. "Stable and High-Efficiency Methylammonium-Free Perovskite Solar Cells." *Advanced Functional Materials*, 1–9. https://doi.org/10.1002/adma.201905502.

Gershon, T. 2011. "Metal Oxide Applications in Organic-Based Photovoltaics." *Materials Science and Technology*. https://doi.org/10.1179/026708311X13081465539809.

Gnanasekaran, Lalitha. 2015. "Synthesis and Characterization of TiO$_2$ Quantum Dots for Photocatalytic Application." *Journal of Saudi Chemical Society*. https://doi.org/10.1016/j.jscs.2015.05.002.

Greiner, Mark T., Lilly Chai, and Zeneng-Hong Lu. n.d. "Organic Photovoltaics: Transition Metal Oxides Increase Organic Solar-Cell Power Conversion." Accessed May 16, 2020. https://www.laserfocusworld.com/detectors-imaging/article/16549532/organic-photovoltaics-transition-metal-oxides-increase-organic-solarcell-power-conversion.

Hanaor, Dorian, Charles Sorrell, Dorian Hanaor, Charles Sorrell, Dorian A H Hanaor, and Charles C Sorrell. 2019. "Review of the Anatase to Rutile Phase Transformation." *Jounal of Material Sciences* 46: 855–874. https://doi.org/10.1007/s10853-010-5113-0.

Hanini, F., A. Bouabellou, Y. Bouachiba, F. Kermiche, A. Taabouche, M. Hemissi, and D. Lakhdari. 2013. "Structural, Optical and Electrical Properties of TiO$_2$ Thin Films Synthesized by Sol – Gel Technique." *IOSR Journal of Engineering* 3, no. 11: 21–28. https://doi.org/10.9790/3021-031112128.

Harrison, Paul, and Alex Valavanis. 2016. "Simple Models of Quantum Wires and Dot." *Theoritical and Computational Physics of Semiconductor Nanostructures*, 249–278. https://doi.org/10.1002/9781118923337.ch8.

Hassanien, Ahmed Saeed, and Alaa A Akl. 2019. "Optical Characterizations and Refractive Index Dispersion Parameters of Annealed TiO$_2$ Thin Films Synthesized by RF-Sputtering Technique at Different Flow Rates of the Reactive Oxygen Gas." *Journal of Physics of Condensed Matter*. https://doi.org/10.1016/j.physb.2019.411718.

Hayre, By Ryan O, Marian Nanu, Joop Schoonman, Albert Goossens, Qing Wang, and Michael Grätzel. 2006. "The Influence of TiO$_2$ Particle Size in TiO$_2$/CuInS 2 Nanocomposite Solar Cells." *Advanced Functional Materials* 16: 1566–1576. https://doi.org/10.1002/adfm.200500647.

Ho, Ching-yuan, J. K. Lin, and Hong-wen Wang. 2015. "Characteristics of Boron Decorated TiO_2 Nanoparticles for Dye-Sensitized Solar Cell Photoanode." *International Journal of Photoenergy*, 2015: 1–9. https://doi.org/10.1155/2015/689702.

Hoppe, Harald, and Niyazi Serdar Sariciftci. 2004. "Organic Solar Cells: An Overview." *Materials Research Society* 19: 1924–1945. https://doi.org/10.1557/JMR.2004.0252.

Hosel, M., D. Angmo, and F. C. Krebs. 2013. "Organic Solar Cells (OSCs)." In *Hand Book of Organic Materials for Optical and (Opto)Electronic Devices*, 473–507. https://doi.org/10.1533/9780857098764.3.473.

Hu, Zhelu, Jose Miguel Garcia-martin, Yajuan Li, Laurent Billot, Baoquan Sun, Antonio García-martín, María Ujué González, Lionel Aigouy, and Zhuoying Chen. 2020. "TiO_2 Nanocolumn Arrays for More Efficient and Stable Perovskite Solar Cells." *Applied Materials & Interfaces.* https://doi.org/10.1021/acsami.9b21628.

Huang, Pao-hsun, Chien-wu Huang, Chih-chieh Kang, and Chia-hsun Hsu. 2020. "The Investigation for Coating Method of Titanium Dioxide Layer in Perovskite Solar Cells." *Crystals.* https://doi.org/10.3390/cryst10030236.

Iftimie, Nicoleta, D. Luca, Felicia Lacomi, Mihaela Girtan, and Diana Mardare. 2009. "Gas Sensing Materials Based on TiO_2 Thin Films Gas Sensing Materials Based on TiO_2 Thin Films," no. March 2015. https://doi.org/10.1116/1.3021050.

Khan, M. I., K. A. Bhatti, Rabia Qindeel, Hayat Saeed Althobaiti, and Norah Alonizan. 2017. "Structural, Electrical and Optical Properties of Multilayer TiO_2 Thin Films Deposited by Sol-Gel Spin Coating." *Results in Physics.* https://doi.org/10.1016/j.rinp.2017.03.023.

Kim, Jaehyun, Wooseok Kim, and Kijung Yong. 2012. "CuO/ZnO Heterostructured Nanorods: Photochemical Synthesis and the Mechanism of H 2 S Gas Sensing." *Journal of Physical Chemistry* 116: 15682–15691. https://doi.org//10.1021/jp302129j.

Kim, Seong Keun, Kyung Min Kim, and Doo Seok Jeong. 2012. "Titanium Dioxide Thin Films for Next-Generation Memory Devices." *Materials Research Society* 28: 313–325. https://doi.org/10.1557/jmr.2012.231.

Komaraiah, Durgam, Madhukar Poloju, and M V Ramana Reddy. 2015. "Structural and Optical Properties of Nano Structured TiO_2 Thin Films Prepared by Sol-Gel Spin Coating Technique." *International Journal of Research* 3: 176–182.

Kopidakis, Nikos, Nathan R Neale, and Arthur J Frank. 2006. "Effect of an Adsorbent on Recombination and Band-Edge Movement in Dye-Sensitized TiO_2 Solar Cells: Evidence for Surface Passivation." *Journal of Physical Chemistry* 110: 12485–12489.

Krebs, Frederik C. 2009. "Solar Energy Materials & Solar Cells Fabrication and Processing of Polymer Solar Cells: A Review of Printing and Coating Techniques." *Solar Energy Materials & Solar Cells* 93: 394–412. https://doi.org/10.1016/j.solmat.2008.10.004.

Kuznetsov, Iliya E, Alexander V Akkuratov, and Pavel A Troshin. 2019. "Polymer Nanocomposites for Solar Cells: Research Trends and Perspectives." In *Nanomaterials for Solar Cell Applications*, 557–600. Elsevier Inc. https://doi.org/10.1016/B978-0-12-813337-8.00015-1.

Landerer, Dominik, Christian Sprau, Bernd Ebenhoch, and Alexander Colsmann. 2019. "Solar Cells: Stability and PerformanceNew Directions for Organic Thin-Film Solar Cells: Stability and Performance." In *Advanced Micro- and Nanomaterials for Photovoltaics*, 195–244. Elsevier Inc. https://doi.org/10.1016/B978-0-12-814501-2.00009-8.

Li, Ji-guang, Takamasa Ishigaki, and Xudong Sun. 2007. "Anatase, Brookite, and Rutile Nanocrystals via Redox Reactions under Mild Hydrothermal Conditions: Phase-Selective Synthesis and Physicochemical Properties." *Journal of Physical Chemistry*, 4969–4976. https://doi.org/10.1021/jp0673258.

Lin, Chia-Hua, Ching-Han Liao, Wei-Hao Chen, Chia-Yuen Chou, and Cheng-Yi Liu Department. 2019. "Fabrication of P-Type TiO_2 and Transparent $p-TiO_2/n-ITO$ p-n Junctions." *AIP Advances* 9. https://doi.org/10.1063/1.5092782.

Liu, Meihua, Yuchen Hou, and Xiaofei Qu. 2017. "Enhanced Power Conversion Efficiency of Dye-Sensitized Solar Cells with Samarium Doped TiO_2 Photoanodes." https://doi.org/10.1557/jmr.2017.357.

Lou, Yanyan, Zhuyi Wang, Dongdong Li, and Liyi Shi. 2016. "Enhancement of Power Conversion Efficiency of Dye Sensitized Solar Cells by Modifying Mesoporous TiO_2 Photoanode with Al-Doped TiO_2 Layer." *Journal of Photochemistry & Photobiology, A: Chemistry.* https://doi.org/10.1016/j.jphotochem.2016.01.002.

Lu, Nianduan, Jiawei Wang, Di Geng, Ling Li, and Ming Liu. 2018. "Understanding the Transport Mechanism of Organic-Inorganic Perovskite Solar Cells: The Effect of Exciton or Free-Charge on Diffusion Length Nianduan." *Organic Electronics*. https://doi.org/10.1016/j.orgel.2018.12.007.

Mather, Robert R, and John I B Wilson. 2017. "Fabrication of Photovoltaic Textiles." *Coatings* 7: 1–21. https://doi.org/10.3390/coatings7050063.

Meyer, Jens, Sami Hamwi, Michael Kröger, Wolfgang Kowalsky, Thomas Riedl, and Antoine Kahn. 2012. "Transition Metal Oxides for Organic Electronics: Energetics, Device Physics and Applications." *Advanced Matererials* 40: 5408–5427. https://doi.org/10.1002/adma.201201630.

Miles, Robert W, Guillaume Zoppi, and Ian Forbes. 2007. "Inorganic Photovoltaic Cells." *Materials Today* 10: 20–27. https://doi.org/10.1016/S1369-7021(07)70275-4.

Morris, Henry B. 1978. "Method of Depositing Titanium Dioxide (Rutile) as a Gate Dielectric for MIS Device Fabrication." *US Patent*.

Nanu, By Marian, and Joop Schoonman. 2004. "Inorganic Nanocomposites of N- and p-Type Semiconductors: A New Type of Three-Dimensional Solar Cell." *Advanced Matererials* 16: 453–456. https://doi.org/10.1002/adma.200306194.

Nateq, Mohammad Hossein, and Riccardo Ceccato. 2019. "Sol-Gel Synthesis of TiO₂ Nanocrystalline Particles with Enhanced Surface Area through the Reverse Micelle Approach." *Advances in Materials Science and Engineering*. https://doi.org/10.1155/2019/1567824.

Nezar, Sawsen, Nadia Saoula, Samira Sali, Mohammed Faiz, Mogtaba Mekki, Nadia Aïcha Laoufi, and Nouar Tabet. 2016. "Properties of TiO₂ Thin Films Deposited by Rf Reactive Magnetron Sputtering on Biased Substrates." *Applied Surface Science*. https://doi.org/10.1016/j.apsusc.2016.08.125.

Nowotny, M. K., P. Bogdanoff, T. Dittrich, S. Fiechter, A. Fujishima, and H. Tributsch. 2010. "Observations of P-Type Semiconductivity in Titanium Dioxide at Room Temperature." *Materials Letters* 64: 928–930. https://doi.org/10.1016/j.matlet.2010.01.061.

Nyankson, E., J. Asare, E. Annan, E. R. Rwenyagila, D. S. Konadu, and A. Yaya. 2013. "Nanostructured TiO₂ and Their Energy Applications: A Review." *Journal of Engineering and Applied Sciences* 8: 871–886.

Oboudi, Saad Farhan, Nadir Fadhil Habubi, Ali Hussein Niíma, and Sami Salmann Chiad. 2014a. "Optical Study of Titanium Dioxide Thin Films Prepared by Vacuum Evaporation Technique." *Nano Science and Nano Technology* 8: 320–327. https://doi.org/10.1063/1.2382456.

Oboudi, Saad Farhan, Nadir Fadhil Habubi, Ali Hussein Niíma, and Sami Salmann Chiad. 2014b. "Optical Study of Titanium Dioxide Thin Films Prepared by Vacuum Evaporation Technique." *Nano Science and Nano Technology* 8: 320–327. https://doi.org/10.1063/1.2382456.

Okyay, Ali K, Feyza B Oruç, Furkan Çimen, and Levent E Aygün. 2017. "TiO₂ Thin Film Transistor by Atomic Layer Deposition" 8626: 1–7. https://doi.org/10.1117/12.2005528.

Pal, Bhola N, Bal Mukund Dhar, Kevin C See, and Howard E Katz. 2009. "Solution-Deposited Sodium Beta-Alumina Gate Dielectrics for Low-Voltage and Transparent Field-Effect Transistors." *Nature Materials* 8: 898–903. https://doi.org/10.1038/nmat2560.

Park, Jong Hyeok, Sungwook Kim, and Allen J Bard. 2006. "Novel Carbon-Doped TiO₂ Nanotube Arrays with High Aspect Ratios for Efficient Solar Water Splitting." *Nono Letters* 6: 24–28. https://doi.org/10.1021/nl051807y.

Park, Su Kyung, Jin Seong Jeong, Tae Kwan Yun, and Jae Young Bae. 2015. "Preparation of Carbon-Doped TiO₂ and Its Application as a Photoelectrodes in Dye-Sensitized Solar Cells." *Journal of Nanoscience and Nanotechnology* 15, no. 2: 1529–1532. https://doi.org/10.1166/jnn.2015.9338.

Pulker, H. K., G. Paesold, and E. Ritter. 1976. "Refractive Indices of TiO₂ Films Produced by Reactive Evaporation of Various Titanium-Oxygen Phases." *Applied Optics* 15: 2986–2991. https://doi.org/10.1364/AO.15.002986.

Qasuria, Tahseen Amin Khan, Shahid Alam, and Nabeel Anwar Qureshi. 2018. "Fabrication of Inverted Perovskite Solar Cell with Cadmium Sulphide As Electron Transport Layer." *International Journal Of Applied And Fundamental Research*, no. 3. http://www.science-sd.com/475-25393.

Rahimi, Nazanin, Randolph A Pax, and Evan Maca Gray. 2016. "Review of Functional Titanium Oxides. I: TiO₂ and Its Modifications." *Progress in Solid State Chemistry*, 1–56. https://doi.org/10.1016/j.progsolidstchem.2016.07.002.

Raj, Vidya. 2018. "Heterojunction Perovskite Solar." In *Perovskite Photovoltaics*, 323–340. Elsevier Inc. https://doi.org/10.1016/B978-0-12-812915-9.00010-1.

Rathee, Davinder, Sandeep K Arya, and Mukesh Kumar. 2011. "Analysis of TiO_2 for Microelectronic Applications: Effect of Deposition Methods on Their Electrical Properties." *Front Optoelectronics China* 4: 349–358. https://doi.org/10.1007/s12200-011-0188-z.

Removal, Contaminants. 2019. "N – TiO_2 Photocatalysts: A Review of Their Characteristics and Capacity for Emerging Contaminants Removal." *Water* 11: 373–408. https://doi.org/10.3390/w11020373.

Reyes-Coronado, R. D., G. Rodr´ıguez-Gattorno, M. E. Espinosa-Pesqueira, C. Cab, R. de Coss, and G. Oskam. 2008. "Phase-Pure TiO_2 Nanoparticles: Anatase, Brookite and Rutile." *Nanotechnology*. doi.org/10.@@1088/0957-4484/19/14/145605

Rodr, G. 2008. "Phase-Pure TiO_2 Nanoparticles: Anatase, Brookite and Rutile." *Nanotechnology* 19. https://doi.org/10.1088/0957-4484/19/14/145605

Rodrigues, Bruno V. M., Vanessa M. Dias, Mariana A. Fraga, S. Argemiro, Silva Sobrinho, Anderson O. Lobo, Homero S. Maciel, and Rodrigo S. Pessoa. 2019. "Atomic Layer Deposition of TiO_2 Thin Films on Electrospun Poly (Butylene Adipate-Co-Terephthalate) Fibers: Freestanding TiO_2 Nanostructures via Polymer Carbonization." *Materials Today* 14: 656–662. https://doi.org/10.1016/j.matpr.2019.02.003

Rokhmat, Mamat, Edy Wibowo, and Mikrajuddin Abdullah. 2017. "Performance Improvement of TiO_2/CuO Solar Cell by Growing Copper Particle Using Fix Current Electroplating Method." *Procedia Engineering* 170: 72–77. https://doi.org/10.1016/j.proeng.2017.03.014.

Ru, Sven. 2016. "Tabulated Values of the Shockley–Queisser Limit for Single Junction Solar Cells." *Solar Energy* 130: 139–147. https://doi.org/10.1016/j.solener.2016.02.015.

Sahdan, M. Z., M. F. Malek, M. S. Alias, S. A. Kamaruddin, and C. A. Norhidayah. 2015. "Fabrication of Inverted Bulk Heterojunction Organic Solar Cells Based on Conjugated P3HT:PCBM Using Various Thicknesses of ZnO Buffer Layer." *International Journal for Light and Electron Optics* 126, no. 6: 645–648. https://doi.org/10.1016/j.ijleo.2015.01.017.

Schlosser, Till, and Franz Hofmann. 2002. "Dream Cell Configuration and Fabrication Method." *United States Patent* 1.

Seo, Hyunwoong, Sang-hun Nam, Naho Itagaki, Kazunori Koga, Masaharu Shiratani, and Jin-hyo Boo. 2016. "Effect of Sulfur Doped TiO_2 on Photovoltaic Properties of Dye-Sensitized Solar Cells." *Electron. Mater. Lett* 12, no. 4: 530–536. https://doi.org/10.1007/s13391-016-4018-8.

Sharma, Anand, Nitesh K Chourasia, Nila Pal, Sajal Biring, and Bhola N Pal. 2019. "Role of Electron Donation of TiO_2 Gate Interface for Developing Film Transistor Using Ion-Conducting Gate Dielectric." Research-article. *The Journal of Physical Chemistry* 123: 20278–20286. https://doi.org/10.1021/acs.jpcc.9b04045.

Shi, Xiaoqiang, Yong Ding, Shijie Zhou, Bing Zhang, Molang Cai, and Jianxi Yao. 2019. "Enhanced Interfacial Binding and Electron Extraction Using Boron-Doped TiO_2 for Highly Efficient Hysteresis-Free Perovskite Solar Cells." *Advanced Science* 1901213: 2–11. https://doi.org/10.1002/advs.201901213.

Shibata, Tatsuo, Hiroshi Irie, Masahiro Ohmori, Akira Nakajima, Toshiya Watanabe, Kazuhito Hashimoto, and Ceramics Science. 2004. "Comparison of Photochemical Properties of Brookite and Anatase TiO_2 Films." *Physical Chemistry Chemcal Physics*, 1359–1362. https://doi.org/10.1039/b315777f.

Štengl, Václav, S. Martin, and Martin Kormunda. 2017. "Fast and Straightforward Synthesis of Luminescent Titanium (IV) Dioxide Quantum Dots." *Journal of Nanomaterials*. https://doi.org/10.1155/2017/3089091.

Sun, Ziqi. 2020 "New Titanium Solar Cell Is More Powerful and Less Toxic." Accessed May 19, 2020. https://www.qut.edu.au/institute-for-future-enviroments/about/news?id=158631.

Tahir, Muhammad Bilal, Khalid Nadeem, M. Hafeez, and Shamsa Firdous. 2017. "Review of Morphological, Optical and Structural Characteristics of TiO_2 Thin Film Prepared by Sol Gel Spin-Coating Technique." *Indian Journal of Pure & Applied Physics* 55: 716–721.

Tang, H., K. Prasad, R. Sanjinès, P. E. Schmid, and F. Lévy. 2012. "Electrical and Optical Properties of TiO_2 Anatase Thin Films." *Journal Applied Physics* 2042. https://doi.org/10.1063/1.356306.

Tauc, J., R. Grigorovici, and A. Vancu. 1966. "Optical Properties and Electronic Structure of Amorphous Germanium." *Physica Status Solidi* 627: 627–637.

"Thin Film Coating System." Accessed June 2, 2020. https://shinmaywa.co.jp/vac/english/vacuum/vacuum_2.html.

Toyoda, Taro, and Qing Shen. 2012. "Quantum-Dot-Sensitized Solar Cells: Effect of Nanostructured TiO_2

Morphologies on Photovoltaic Properties." *Journal of Physical Chemistry Letters* 3: 1885–1893. https://doi.org/10.1021/jz3004602.

Venkatachalam, Thangamuthu, Sakthivel Kris, Narayanasamy Ramaswamy, and Padmanabhan Rupa. 2011. "Structural and Optical Properties of TiO₂ Thin Films." *Applied Physics: Materials Science & Processing* 1391: 764–766. https://doi.org/10.1063/1.3643673.

Wang, Songbo, Lun Pan, Jia-jia Song, Wenbo Mi, Ji-jun Zou, Li Wang, and Xiangwen Zhang. 2015. "Titanium-Defected Undoped Anatase TiO₂ with p – Type Conductivity, Room-Temperature Ferromagnetism, and Remarkable Photocatalytic Performance." *Journal of the American Chemistry Society*, 2975–2983. https://doi.org/10.1021/ja512047k.

Wang, Songling, and Michael H K Leung. 2014. "Microwave Synthesis of Monodisperse TiO₂ Quantum Dots and Enhanced Visible-Light Photocatalytic Properties" 70: 142–146. https://doi.org/10.7763/IPCBEE.

Wang, Yuan, Tao Wu, Yun Zhou, Chuanmin Meng, Wenjun Zhu, and Lixin Liu. 2017. "TiO₂-Based Nanoheterostructures for Promoting Gas Sensitivity Performance: Designs, Developments, and Prospects." *Sensors* 17: 1–35. https://doi.org/10.3390/s17091971.

Wategaonkar, S. B., R. P. Pawar, V. G. Parale, D. P. Nade, and B. M. Sargar. 2020. "Synthesis of Rutile TiO₂ Nanostructures by Single Step Hydrothermal Route and Its Characterization." *Materials Today* 23: 444–451. https://doi.org/10.1016/j.matpr.2020.02.065.

Wu, Jyh-Ming, Han C. Shih, and Wen-Ti Wu. 2005. "Electron Field Emission from Single Crystalline TiO₂ Nanowires Prepared by Thermal Evaporation." *Chemical Physics Letters* 413: 490–494. https://doi.org/10.1016/j.cplett.2005.07.113.

Xu, Shuang, Jun Gao, Linlin Wang, Kan Kan, Yu Xie, Peikang Shen, Li Li, and Keying Shi. 2015. "Role of the Heterojunctions in In2O3-Composited SnO2 Nanorods Sensor and Their Remarkable Gas-Sensing Performance for NOx at Room Temperature." *Nanoscale* 7: 14643–14651. https://doi.org/10.1039/b000000x.

Yan, Ye, Cynthia Wladyka, Junichi Fujii, and Shanthini Sockanathan. 2015. "Prdx4 Is a Compartment-Specific H2O2 Sensor That Regulates Neurogenesis by Controlling Surface Expression of GDE2." *Nature Communications*, 1–12. https://doi.org/10.1038/ncomms8006.

Yang, Mengjin. 2012. "Band Gap Engineering and Carrier Transport in TiO₂ for Solar Energy Harvesting." Unpublished manuscript, University of Pittsburgh.

Zhang, Yanyan, Zhelong Jiang, Jianying Huang, Linda Y. Lim, Wenlong Li, Jiyang Deng, Dangguo Gong, Yuxin Tang, Yuekun Lai, and Zhong Chen. 2015. "Titanate and Titania Nanostructured Materials for Environmental and Energy Applications: A Review." *RSC Advances* 5: 79479–79510. https://doi.org/10.1039/C5RA11298B.

Zhao, Yin, Chunzhong Li, Xiuhong Liu, Feng Gu, Haibo Jiang, Wei Shao, Ling Zhang, and Ying He. 2007. "Synthesis and Optical Properties of TiO₂ Nanoparticles." *Materials Letters* 61: 79–83. https://doi.org/10.1016/j.matlet.2006.04.010.

Zou, Lilan, Wei Hu, Wei Xie, and Dinghua Bao. 2016. "Uniform Resistive Switching Properties of Fully Transparent TiO₂-Based Memory Devices." *Journal of Alloys and Compounds*. https://doi.org/10.1016/j.jallcom.2016.10.009.

16

Chemical Spray Pyrolysis Method to Fabricate CdO Thin Films for TCO Applications

M. Anusuya[1] and V. Saravanan[2]
[1]*Indra Ganesan College of Engineering Professor/Registrar, Trichy, Tamilnadu, India*
[2]*Sri Meenatchi Vidiyal Arts and Science College Professor, Valanadu, Trichy, Tamilnadu, India*

16.1 Introduction

Most optically transparent conducting oxides (TCO) are binary or ternary compounds, containing one or two metallic elements. Their resistivity could be as low as 10^{-4} cm, and their extinction coefficient k in the optical visible range (VIS) could be lower than 10^{-4}, owing to their wide optical band gap (E_g) that could be greater than 3eV. This remarkable combination of conductivity and transparency is usually impossible in intrinsic stoichiometric oxides; however, it is achieved by producing them with a non-stoichiometric composition or by introducing appropriate dopants. Badeker discovered that thin CdO films possess such characteristics (Baedeker 1907). Later, it was recognized that thin films of ZnO, SnO_2, In_2O_3, and their alloys were also TCOs (Haacke 1977). The actual and potential applications of TCO thin films include transparent electrodes for flat panel displays, transparent electrodes for photovoltaic cells, low emissive windows, window defrosters, transparent thin films transistors, light emitting diodes, and semiconductor lasers. TCO thin films are essential part of technologies that require both large area electrical contact and optical access in the visible portion of the light spectrum. The various TCOs include the oxides of Sn, In, Zn, Cd, and their alloys.

CdO is one of the promising transparent conducting oxides from II to VI group of semiconductors having high absorption and emission capacity of radiation in the energy gap. CdO has special features such as high conductivity, high transmission, and low band gap which has made it applicable in photodiodes (Lokhande 2004), phototransistors (Su et al. 1984), photovoltaic cells (Champness et al. 1985), transparent electrodes (Benko and Koffyberg 1986), liquid crystal displays, IR detectors, and anti-reflection coatings (Ocampo et al. 1993). CdO thin films have been obtained by different techniques such as chemical bath deposition (Herrero et al. 2000), SILAR (Mane and Han 2005), sol gel (Cruz et al. 2005), eletrodeposition (Han et al. 2005), dc reactive sputtering (Subramanyam et al. 2001), reactive vacuum evaporation process (Eze 2005), metal organic vapour-phase epitaxy (Perez et al. 2004), and spray pyrolysis (Uplane et al. 2005) techniques. Among these techniques spray pyrolysis technique is simple, amendable to large growth, conformal coverage, non line of sight, and wide range of materials. This chapter studies the effect of molarity on the structural, optical, and surface morphological characteristics of CdO thin films prepared at 230°C by using spray pyrolysis technique.

DOI: 10.1201/9781003145585-16

16.2 Application of TCOs

TCOs are a technologically important class of materials that combine electrical conductivity and optical transparency. TCOs are essential for many photovoltaic and optoelectronic applications. Some of these applications are briefly described in the following sections. Generally TCO materials can be categorized according to their electronic properties. N-type TCO material contains excess free electrons in the structure, while there is a deficiency of electrons in a p-type TCO material. It should be noted that there are already many known n-type TCOs, but the lack of good p-type TCOs limits the presence of active electronic devices which can be fabricated with both n- and p-type TCOs. Delafossite $CuXO_2$ (X = Al, Ga, and In) oxides were a first group of p-type materials which are studied since 1997 and triggered the development of a series of p-type TCOs.

16.3 Experimental Details

The precursor solution used to form CdO thin films was obtained by dissolving the salts of cadmium acetate $[Cd(CH_3COO)_2 2H_2O\ 99.99\%]$ in three different concentrations (0.025M, 0.05M, and 0.1M) in double distilled water. The amount of solution was made together as 50 ml and an optically plane cleaned glass plate (7.5 cm × 2.5 cm) was placed over the hot plate. The aqueous solution was then sprayed on the preheated glass substrate maintained at 230°C±2°C. Compressed dry air at a pressure of 2 kg/cm^2 from an air compressor via an air filter cum regulator was used as the carrier gas, and spray rate of the solution was maintained at 3 ml/min.

The flow rate of the solution is kept minimum due to the fact that at lower flow rates the spray will get sufficient time to react endothermically at the heated substrate surface to give the final film in the near stoichiometric phase of CdO. The distance between the spray nozzle and the substrate is 35 cm. Film is obtained due to endothermic thermal decomposition that takes place at the hot surface of the substrate. The overall reaction process can be expressed as decomposition of cadmium acetate to form cadmium oxide onto the substrate as a strongly adherent film, and they appeared yellowish in colour. The deposited film was subsequently annealed in air at 300°C for 1 hour. Film thickness of CdO was determined by gravimetric weighing method (Shinde et al. 2007). The film thickness was determined to be between 330 nm and 680 nm for the solution molarity of 0.025M, 0.05M, and 0.1M. This was consequently verified by the cross-sectional studies of the film using SEM. Here, the film is mounted vertically to measure the thickness directly (Chen et al. 1995).

The structural study was determined by X-ray diffractometer (Rigaku Model RAD II A) with Cukα radiation (λ = 1.54056 Å). The surface morphologies of the films were determined by using SEM (TESCAN – VEGA 3 SEM). Optical transmittance and band gap energy were measured by UV-VIS single beam spectrophotometer (ELICO-159). The experimental accuracy for absorbance is ±0.005 abs and of wavelength is ± 0.05 nm. The resistivity (ρ) and Hall coefficient (R_H) were measured by a standard four-probe technique (ECOPIA HMS5000 Hall system) and silver contacts were used for all electrical measurements.

16.4 Results and Discussion

16.4.1 XRD and Surface Morphology Studies

The molarity had a significant effect on the X-ray diffraction structural analysis of CdO thin films as shown in Figure 16.1

Observation of film shows smooth surface and well adhered to substrate. The peaks observed in the diffractograms confirm the polycrystalline nature of the CdO film. XRD pattern showed that the films have mainly preferential orientation with the cubic face-centered structure along the c-axis, (111) perpendicular to the substrate plane. No other impurities peaks are observed. Also the intense peak oriented along (111) lattice plane indicates that the growth of the grains is parallel to the substrate.

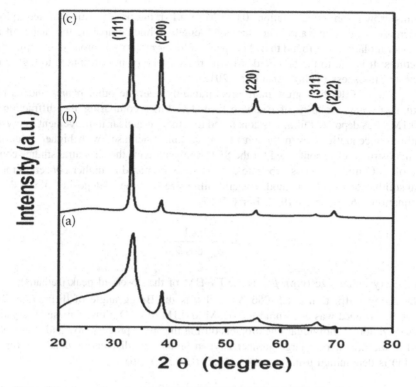

FIGURE 16.1 X-ray diffraction pattern for (a) 0.025M, (b) 0.05M, (c) 0.1M CdO thin films.

It was found to be polycrystalline in nature with five peaks at $2\theta = 33.08°$, $38.57°$, $55.67°$, $66.57°$, and $69.58°$ identified for the film deposited in 0.025M. The corresponding peaks are good in agreement with JCPDS data (65–2908) which are attributed to (111), (200), (220), (311), and (222) planes. However, the peak position of the preferential (111) orientation is close to the JCPDS (#65–2908) file data for the films grown from 0.025M to 0.1M of precursor solution as in Figure 16.2.

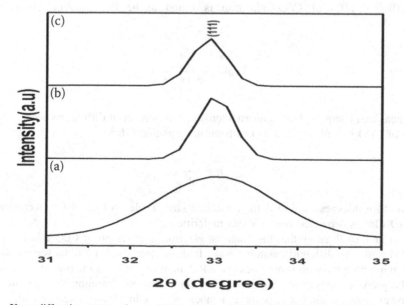

FIGURE 16.2 X-ray diffraction pattern close to (111) plane for 0.025M, (b) 0.05M, (c) 0.1M CdO thin films.

For the films grown from concentrations 0.025 M to 0.1M, the peak position shifted to lower angles indicating increase of "d" value and hence the bond length. While comparing the shapes of the peak of higher and lower molarity, it is found that (111) peak of lower molarity is broad confirming the presence of nanostructures. It is due to the atomic density increase in the planes attributed to higher molarity of the precursor and thickness (Thiagarajan et al. 2012).

The (111) plane of CdO film grew more predominantly than the other planes and its intensity increased with increasing molarity of cadmium acetate. While comparing the diffraction pattern of 0.025M to 0.1M of as deposited film, it is concluded that there is a clear improvement in crystallinity for higher solution concentration. Even though the crystallinity increases with higher solution molarity, presence of nanostructures is confirmed for the SEM analysis. Thus the structural studies confirmed that the spraying of CdO films on glass substrate can be easily prepared at higher concentration than other conventional solid-state reaction method. The crystallite size D of the as-deposited film is determined by Scherrer's equation (16.1) (Bragg 1912; Berry 1967),

$$D = \frac{0.9\lambda}{\beta_{hkl} \cos \theta_{hkl}} \tag{16.1}$$

where D is the crystallite size (nm), β_{hkl} is the FWHM of the observed peak (radians), λ is the wave length of the X-ray diffraction (=1.54056 Å), and θ is the Bragg angle of diffraction. The average crystallite size determined was 6–23 nm for 0.025M to 0.1M of CdO films. This small crystallite size is due to the evaporation of individual fine droplets during the spray process (Ma and Bube 1977). It may be attributed to the presence of large number oxygen faults. The dislocation density δ for preferential reflection (111) is determined using equation (16.2) (Cullity 1956)

$$\delta = \frac{1}{D^2} \tag{16.2}$$

The strain ε is calculated from the equation (16.3) (Cullity 1956)

$$\varepsilon = \frac{\beta \cos \theta}{4} \tag{16.3}$$

Texture Coefficient (TC) of CdO thin film is found to be the equation (16.4) (Barret and Massalski 1980)

$$Tc_{hkl} = \frac{\frac{I_{hkl}}{I_{0hkl}}}{N^{-1} \sum \frac{I_{hkl}}{I_{0hkl}}} \tag{16.4}$$

where I is a measured intensity, I_o is standard intensity, N is number of diffraction peaks. The number of crystallites n of CdO thin film is determined from the equation (16.5),

$$n = \frac{t}{D^3} \tag{16.5}$$

Where t is the film thickness and D is the crystallite size. Table 16.1 gives the microstructural parameters of CdO film as deposited from various molarities.

Figure 16.3a and 16.3b show that the scanning electron micrograph of CdO thin films prepared at 230°C in 0.1M with two different magnifications. It shows that the CdO films were smooth and heterogeneous having porosity with some pinholes. XRD analysis confirms the presence of nanostructures in the thin film prepared with increasing the solution concentration. Scanning electron micrographs also implies in lower magnification it appears as nanoclusters as in Figure 16.3a. As the magnification

TABLE 16.1

Microstructural Parameters of CdO thin Films at (111) Plane

Sample	Lattice Constant (a) (Å)	Interplanar Distance (d) (Å)	Crystallite Size (D) (nm)	Dislocation Density (δ) (10^{15}) Lines/m^a	Strain (ε) (10^{-3}) Lines$^{-2}m^{-4}$	Texture Coefficient (TC)	No. of Crystallites (n) (10^{16}/Unit Area)
0.025M	4.65	2.704	6	28.9	5.6	1.99	162
0.05M	4.695	2.711	19	2.71	1.8	3.25	7
0.10M	4.711	2.718	23	1.87	1.5	2.42	5

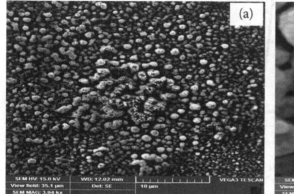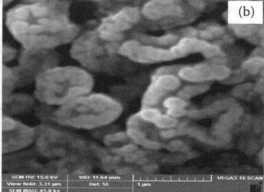

FIGURE 16.3 SEM image of CdO thin film prepared at 0.1M for magnifications (a) 10 μm (b) 1 μm.

increases the clear structure of the nanocluster is visible as shown in Figure 16.3b. And it is found that nanoclusters are interconnected to form a nanorod of CdO.

16.4.2 Optical Studies

The absorption spectra of CdO thin films prepared at 230°C on glass substrate were recorded as a function of wavelength range 200 nm–1100 nm with glass as reference as shown in Figure 16.4. It shows the representatives of optical absorbance which reveals that the absorbance of the film decreases gradually with increase in wavelength.

It is clear from the graph that in the visible region there is no significant change in band edge by increasing the molarity of the solution from 0.025M to 0.1M. This implies that the basic crystal structure is not changed (Agarwal et al. 2006). It also shows that as the solution concentration of the film increases from 0.025M to 0.1M there is a drastic increase in absorbance. The overall increase in absorbance with increase in solution concentration may be associated with the increase in film thickness. This is because in the thicker the films more atoms are present in the film so more states will be available for the photons to be absorbed (Nadeem and Ahmed 2000). Samples of CdO thin films of 0.025M, 0.05M, and 0.1M showed peak absorbance in the visible region at 310 nm with maximum value of 0.42, 0.94, and 1.92, respectively. The film deposited with higher concentration of cadmium acetate showed highly increasing absorption, which then decreases up to 560 nm and attains saturation. So, the higher concentration of precursor solution is more preferable in the preparation of CdO films for better quality on glass substrate. The optical absorption values are in line with XRD and SEM results, the nanocrystallite sizes could be obtained in the films deposited of higher concentration, that produce higher absorption results.

Figure 16.5 shows the optical transmittance pattern of the films in the wavelength region ranging from 200 nm to 1000 nm. The results indicate the transmission over the whole spectral range investigated is lowered with increasing solution concentration. This is due to the higher film absorption associated with

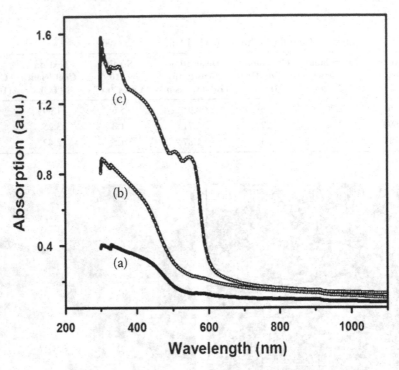

FIGURE 16.4   Absorption vs. Wavelength for (a) 0.025M, (b) 0.05M, (c) 0.1M CdO thin films.

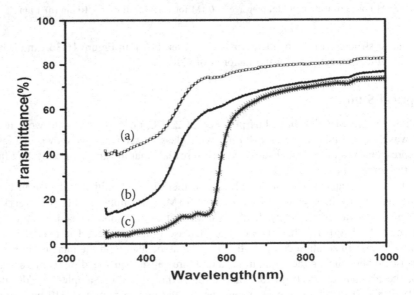

FIGURE 16.5   Transmittance vs. Wavelength for (a) 0.025M, (b) 0.05M, (c) 0.1M CdO thin films.

stress in the film produced by structural defect. It is observed that the increase in transmittance with increase in wavelength is not sharp. Films prepared with concentration of precursor 0.1M show transmission exceeding 70% over the spectral range investigated with a well-defined absorption edge lying in the UV region. Transmission spectra reveal that CdO films exhibit transmittance in the range of 80% in the visible region of 0.025M of solution concentration. Similar results were also reported for CdO films prepared by thermal evaporation (Dantus et al. 2008), spray pyrolysis (Vigil et al. 2001), and vacuum evaporation (Dakhel and Henari 2003). The absorption coefficient α is calculated using

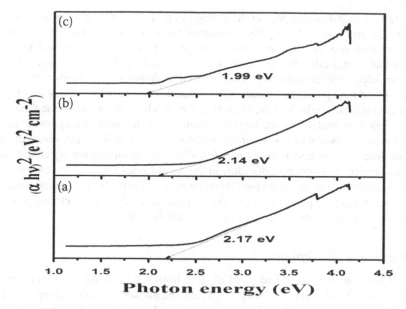

FIGURE 16.6 Band gap energy for (a) 0.025M, (b) 0.05M, (c) 0.1M CdO thin films.

Lambert's law (Mehner et al. 1997), and it is found to be in the order of 10^6 cm^{-1}. The optical direct band gap of the films were calculated by the equation (16.6),

$$\alpha = A(h\upsilon - E_g)^{1/2} \tag{16.6}$$

where A is a constant, E_g is Energy band gap, υ is the frequency of the incident light, h is the Planck's constant. The typical plots of $(\alpha h\upsilon)^2$ versus $h\upsilon$ for CdO thin films deposited on glass substrate is shown in Figure 16.6.

It was observed that increase in concentration of the precursor solution yields slight shrinkage in optical band gap (2.17 eV–1.99 eV). This shrinkage was generally attributed to Moss-Burstein shift (Brustien 1954; Moss 1954). These values are good agreement for previous literature (Lamb and Irvine 2011).

16.4.3 Non-linear Optical Studies

16.4.3.1 Physical Mechanisms of Optical Non-Linearities in Undoped CdO Thin Films

Optical nonlinear response of the undoped CdO thin film partly depends on the laser characteristics, in particular, on the laser pulse duration and on the excitation wavelength, and partly on the material itself. The optical non-linearities usually fall in two main categories: the instantaneous and accumulative non-linear effects. If the non-linear response time is much less than the pulse duration, the non-linearity can be regarded justifiably as responding instantaneously to optical pulses. On the contrary, the accumulative non-linearities may occur in a time scale longer than the pulse duration. Besides, the instantaneous non-linearity (for instance, two-photon absorption and optical Kerr effect) is independent of the laser pulse duration, whereas the accumulative non-linearity depends strongly on the pulse duration. Examples of such accumulative non-linearities include excited-state non-linearity, thermal effect, and free-carrier non-linearity. The simultaneous accumulative non-linearities and inherent non-linear effects lead to the huge difference of the measured non-linear response on a wide range of time scales.

16.4.3.2 Non-linear Refraction

The physical mechanisms of non-linear refraction in the undoped CdO thin films mainly involve thermal contribution, optical electrostriction, population redistribution, and electronic Kerr effect. The thermal heat leads to refractive index changes via the thermal-optic effect. The non-linearity originating from thermal effect will give rise to the negative non-linear refraction. In general, the thermal contribution has a very slow response time (nanosecond or longer). On the picosecond and femtosecond time scales, the thermal contribution to the change of the refractive index can be ignored for it is much smaller than the electronic contribution. *Optical electrostriction* is a phenomenon in which the inhomogeneous optical field produces a force on the molecules or atoms comprising a system resulting in an increase of the refractive index locally. This effect has the characteristic response time of nanosecond order. When the electron undergoes real transition from the ground state of a system to an excited state by absorbing the single photon or two identical photons, electrons will occupy real excited states for a finite period of time. This process is called *population redistribution* and mainly contributes the whole refractive non-linearity of Al-doped CdO films in the picoseconds regime. The electronic Kerr effect arises from a distortion of the electron cloud about atom or molecule by the optical field. This process is very fast, with typical response time of tens of femtoseconds. The electronic Kerr is the main mechanism of the refractive non-linearity in the femtosecond time scale.

16.4.3.3 Non-linear Absorption

In general, non-linear absorption in undoped CdO thin films can be caused by two-photon absorption, three-photon absorption, or saturable absorption. When the excitation photon energy and the band gap of the film fulfil the multiphoton absorption requirement [$(n-1)\, h\nu < Eg < nh\nu$] (here n is an integer. $n=2$ and 3 for two and three-photon absorption, respectively), the material simultaneous absorbs "n" identical photons and promotes an electron from the ground state of a system to a higher lying state by virtual intermediate states. This process is referred to a one-step "n" photon absorption and mainly contributes to the absorptive non-linearity of most undoped CdO films. When the excitation wavelength is close to the resonance absorption band, the transmittance of materials increases with increasing optical intensity. This is the well-known saturable absorption. Accordingly, the material has a negative non-linear absorption coefficient.

The non-linear absorption coefficient β of undoped CdO thin films was determined from Z-scan open aperture measurements.

Figure 16.7 shows the Z-scan results of films corresponding to the normalized transmittance T as a function of distance from the beam focus Z/Z_0. The transmission is symmetric with respect to the

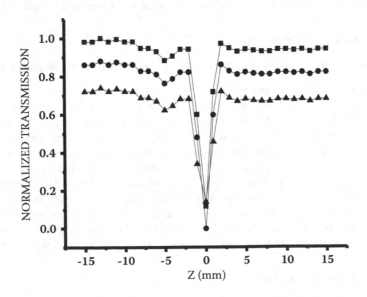

FIGURE 16.7 Open aperture Z-scan response; the dots represent the measurement data; the line corresponds to a fit of equation 16.7.

focus ($Z = 0$), where it has minimum transmission. For a Gaussian beam, the normalized transmittance for open Z-scan (Dantus et al. 2008) is

$$T_{open} = 1 - \frac{\Delta\emptyset}{1 - x^2}$$ (16.7)

where $x = \frac{z}{z_0}$, $z_0 = \frac{\pi\omega_0^2}{\lambda}$, ω_0- is a beam waist, $\Delta\emptyset = \frac{\beta I_0 l}{2}$, and $l = \frac{[1 - \exp(-\alpha_0' d)]}{\alpha_0'}$; $l = \frac{[1 - \exp(-\alpha_0' d)]}{\alpha_0'}$ is a linear absorption coefficient at 632.8 nm, d is the sample thickness, l is the effective thickness of the sample, I_0 the intensity of the laser beam at the focus, and β is the non-linear absorption coefficient. A fit of equation 16.7 to the experimental data is depicted in Figure 16.7 and gives the value of the non-linear optical absorption coefficient β for 0.025M, 0.05M, and 0.1M are 4.82×10^{-3}m/W, 6.96×10^{-3}m/W, and 6.82×10^{-3}m/W, respectively.

16.4.4 Electrical Studies

At room temperature, Hall Effect was measured with the magnetic field applied perpendicular to film surface in the Van der Pauw configuration (Igasaki and Kanma 2001). The film resistivity has been determined by taking the product of resistance and film thickness. The carrier concentration N_d and the Hall mobility μ were calculated from the electrical resistivity ρ and the Hall coefficient R_H by using the following equations (16.8) and (16.9) (Schroder 1990).

$$\mu = \frac{1}{ne\rho}$$ (16.8)

$$N_d = \frac{1}{eR_H}$$ (16.9)

The changes in electrical parameters as a function of various solution concentrations of CdO thin film are presented in Table 16.2 and Figure 16.8.

It was noted that the molarity of solution concentration increases as the resistivity decreases. The minimum resistivity 1.23×10^{-3} Ωcm obtained from the present work is comparable with the value of 1.5×10^{-3} Ωcm reported by Reddy et al. (1998) for reactive evaporation method. The maximum resistivity 16.5×10^{-3} Ωcm obtained from this work is comparable with the value of 20.1×10^{-3} Ωcm reported by Zhao et al. (2002). Thus higher molarity is taken as the most suited molarity for preparing CdO films using chemical spray pyrolysis technique. On the other hand, carrier concentration of as-deposited films also increases with increase in molarity. This result is consistent with the observation from XRD, and SEM studies show that the grain size increases up to 0.1M. Thus, as the molarity increases the crystallinity and grain size improvement are attributed to reduction in the scattering of grain boundary. It is connected with mobility and conductivity to improve the nanocrystalline nature of the as-deposited film. At higher molarity, the mobility value is high and it may be due to low grain boundary scattering.

TABLE 16.2

Measured Electrical Parameters for CdO thin Films at (111) Plane

Sample	Resistivity ρ ($\times 10^{-3}$ Ωcm)	Hall Mobility μ (cm²/Vs)	Carrier Concentration N_d (10^{20} cm⁻³)
0.025M	16.5	11.7	0.322
0.05M	5.71	13.7	7.96
0.1M	1.23	34.2	14.9

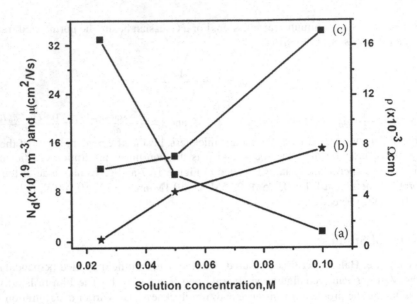

FIGURE 16.8 Electrical parameters of CdO thin films (a) Electrical resistivity (ρ) (b) Carrier concentration (N_d) (c) Hall mobility (μ) with different solution concentrations.

16.5 Conclusion

The highly conducting CdO thin films were obtained by using chemical spray pyrolysis method on glass substrates at 230°C. The XRD studies have confirmed that the films are nanocrystalline in nature, and CdO crystallites have cubic structure of (111) preferable orientation. The structural parameters like crystallite size, dislocation density, number of crystallites per unit area, texture coefficient, and strain were calculated from XRD pattern. The 0.1M of SEM indicated for lower magnification appear as nanoclusters, and at higher magnification it is found that the nanoclusters are interconnected to form a nanorod of CdO. Transmission of CdO thin films is found to be 80% in visible and near infrared region. Energy band gap of thin films is found to be in the range of 2.17–1.99 eV. The carrier concentration of 14.9×10^{19} cm^{-3} and resistivity 1.23×10^{-3} Ωcm are obtained for the 0.1M of CdO thin film.

REFERENCES

Agarwal, D.C., Amit Kumar, S.A. Khan, and D. Kabiraj. 2006. "SHI Induced Modification of ZnO Thin Films." *Nuclear Instruments and Methods in Physics Research* B 244, no. 1: 136–140.

Baedeker, K. 1907. "Electrical Conductivity and Thermoelectric Power of Some Heavy Metal Compounds." *Annuals Physics. (Leipzig)* 22, no. 4: 749–766.

Barret, C., and T.B. Massalski. 1980. *Structure of Metals*. Oxford, U.K: Pergamon. 204.

Benko, F.A., and F.P. Koffyberg. 1986. "Quantum Efficiency and Optical Transitions of CdO Photoanodes." *Solid State Communication* 57, no. 12: 901–903.

Berry, L.G., (Ed.). 1967. ASTM Powder Diffraction File, Sets 1 to 5, Inorganic, Vol.PDIS-5LRB.

Bragg, W.L.. 1912. "The Diffraction of Short Electromagnetic Waves by a Crystal." *Proceedings of the Cambridge Philosophical Society* 17: 43–57.

Brustien, E. 1954. "Anomalous Optical Absorption Limit in InSb." *Physics Review* 93, no. 3: 632–633.

Champness, C.M., K. Ghoneim, and J.K. Chen. 1985. "Optimization of CdO Layer in a Se-CdO Photovoltaic Cell." *Canadian Journal of Physics* 63, no. 6: 767–771.

Chen, O., Y. Quian, Z. Chen, and Y. Zang. 1995. "Fabrication of Ultrafine SnO2 Thin Films by the Hydrothermal Method." Thin Solid Films 264, no. 1: 25–27.

Cruz, J.S., G.T. Delgado, R.C. Perez, S.J. Sandoval, O.J. Sandoval, C.I.Z. Romero, J.M. Marin, and O.Z. Angel. 2005. "Dependence of Electrical and Optical Properties of Sol-gel Prepared Undoped CdO Thin Films on Annealing Temperature." *Thin Solid Films* 493, no. 1–2: 83–87.

Cullity, B.D.1956. *Elements of X-Ray Diffraction, Natural Sciences; Physics; Physical nature of matter.* Massachusetts: Addison-Wesley Publications Company Inc. Reading.

Dakhel, AA., and F.Z. Henari. 2003. "Optical Characterization of Thermally Evaporated Thin CdO Films." *Crystal Research Technology* 38, no. 11: 979–985.

Dantus, C., G.G. Rusu, M. Dobromit, and M. Rusu. 2008. "Preparation and Characterization of CdO Thin Films Obtained by Thermal Oxidation of Evaporated Cd Thin Films." *Applied Surface Science* 255, no. 5: 2665–2670.

Eze, F.C.2005. "Oxygen Partial Pressure Dependence of the Structural Properties of CdO Thin Films Deposited by a Modified Reactive Vacuum Evaporation Process." *Materials Chemistry and Physics* 89, no. 2: 205–210.

Haacke, G. 1977. "Transparent Conducting Coatings." *Annual Review of Materials Science* 7: 73–93.

Han, X., R. Liu, Xu. Zhude, W. Chen, and Y. Zheng. 2005. "Room Temperature Deposition of Nanocrystalline Cadmium Peroxide Thin Film by Electrochemical Route." *Electrochemistry Communications* 7, no. 12: 1195–1198.

Herrero, J., M.T. Gutierrez, C. Gullen, J.M. Dona, M.A. Martinez, A.M. Chaparro, and R. Bayon. 2000. *Thin Solid Films.* 28, no. 33: 361–362.

Igasaki, Y., and H. Kanma. 2001. "Appl. Argon Gas Pressure Dependence of the Properties of Transparent Conducting ZnO:Al Films Deposited on Glass Substrates." *Applied surface science.* 169-170: 508–511.

Lamb, DA, and S.J.C. Irvine. 2011. "A Temperature Dependent Crystal Orientation Transition of CdO Films Deposited by Metal Organic Chemical Vapour Deposition." *Journal of Crystal Growth.* 332, no. 1: 17–20.

Lokhande, B. 2004. "Studies on Cadmium Oxide Sprayed Thin Films Deposited through Non-Aqueous Medium." *Materials Chemistry and Physics* 84, no. 2–3: 238–243.

Ma, Y.Y., and R.H. Bube. 1977. "Properties of CdS Films Prepared by Spray Pyrolysis." *Journal of the Electrochemical Society* 124: 1430–1435.

Manc, R.S., and S.H. Han. 2005. "Growth of Limited Quantum Dot Chains of Cadmium Hydroxide Thin Films by Chemical Route." *Electrochemistry Communications* 7, no. 2: 205–208.

Mehner, A., H. Klümper-Westkamp, F. Hoffmann, and P. Mayr. 1997. "Crystallization and Residual Stress Formation of Sol-Gel-Derived Zirconia Films." *Thin Solid Films* 308-309: 363–368.

Moss, TS1954. "The Interpretation of the Properties of Indium Antimonite." *Proceedings of the Physical Society, London. B* 67, no. 10: 775–782,

Nadeem, M.Y., and Waqas Ahmed. 2000. "Optical Properties of ZnS Thin Films." *Turkey Journal of Physics.* 24, 651–659.

Ocampo, I.M., M. Ferandez, and P.J. Sabastian. 1993. "Transparent Conducting CdO Films Formed by Chemical Bath Deposition." *Semiconductor Science and Technology* 8, no. 1: 750–751.

Perez, J.Z., C. Munuera, C. Ocal, and V.M. Sanjose. 2004. "Structural Analysis of CdO Layers Grown on r-Plane Sapphire by Metal Organic Vapor-Phase Epitaxy." *Journal of Crystal Growth* 271: 223–228.

Ramakrishna Reddy, KT, C. Sravani, and R.W. Miles. 1998. "Characterization of CdO Thin Films Deposited by Activated Reactive Evaporation." *Journal Crystal Growth* 184/185: 1031–1034.

Schroder, DK 1990. *Semiconductor Material and Device Characterization.* New York: Wiley-IEEE Press, 784.

Shinde, V.R., T.R. Gujar, and C.D. Lokhande. 2007. "LPG Sensing Properties of ZnO Films Prepared by Spray Pyrolysis Method: Effect of Molarity of Precursor Solution." *Sensors and Actuators B* 120, no. 2: 551–559.

Su, L.M., N. Grote, and F. Schmitt. 1984. "Diffused Planar InP Bipolar Transistor with a Cadmium Oxide Film Emitter." *Electron Letters* 20, no. 18: 716–717.

Subramanyam, T.K., B.S. Naidu, and S. Uthanna. 2001. "Studies on dc Magnetron Sputtered Cadmium Oxide Films." *Applied Surface Science* 169-170: 529–534.

Thiagarajan, R., M. MahaboobBeevi, M. Anusuya, and T. Ramesh. 2012. "Influence of Reactant Concentration on Nanocrystalline PbS Thin Films Prepared by Chemical Spray Pyrolysis." *Optoelectronics and Advanced Materials – Rapid Communication* 6, no. 1–2: 132–135.

Uplane, M.D., P.N. Kshirsagar, B.J. Lokhande, and C.H. Bhosale. 2005. "Characteristic Analysis of Spray Deposited Cadmium Oxide Thin Films." *Materials Chemisry and Physics* 64, no. 1: 75–78.

Vigil, O., F. Cruz, A. Morales-Acevedo, G. Contreras-Puente, L. Vaillant, and G. Santana. 2001. "Structural and Optical Properties of Annealed CdO Thin Films Prepared by Spray Pyrolysis." *Material Chemistry and Physics* 68, no. 1: 249–252.

Zhao, Z., D.L. Morel, and C.S. Ferekides. 2002. "Electrical and optical properties of tin-doped CdO films deposited by atmospheric metal organic chemical vapor deposition." *Thin Solid Films* 413, no. 1–2: 203–211.

17

Photovoltaic Characteristics and Applications

Mkpamdi N. Eke
Department of Mechanical Engineering, University of Nigeria, Nsukka

17.1 Introduction

Solar energy comes in the form of solar irradiance from the sun. It is appropriate, renewable, and unlimited energy which, using photovoltaic (PV) technology, can be converted directly to electricity. The biggest benefit of solar energy is that it gives no carbon emission through renewable energy. Depending on the environmental and climatic conditions, it is irregular and unpredictable. The demand for producing solar power has experienced fast and enormous growth. Alternative energy has been a top priority through the worldwide revolution for sustainable growth. With its enormous promise and availability, solar power is at the forefront of renewable energy. Three generations of solar cells exist: bulk silicon, thin film, and organic. Bulk silicon P cells are the most common substrate for solar panels and have been invented before decades. Thin-film cells are comparatively recent developments and are currently hitting the market, and finally organic cells are also in the testing process and some difficult ones are still facing some challenging problems until being economically viable (Chen et al. 2015); reliability and longevity need to be addressed. Photovoltaic energy (PV) is the electrical energy produced by the photovoltaic effect directly by the sun's radiation, which was discovered by the French physicist Alexandre-Edmond Becquerelin in 1839. This influence is seen in semiconductor materials that are distinguished by the intermediate electrical conductivity between the conductor and the insulator. Electrons are captured as the incident radiation enters the material in the form of photons, resulting in higher energy content, and as a threshold value called "band gap" is surpassed, their nucleus can severe their nucleus ties and spread through the material. This electron movement provides a potential between terminals, and when an electrical field is applied to the semiconductor, the electrons travel in the direction of the field and generate an electrical current (Zaidi 2018).

The solar cell is built using materials that are semiconducting and differs from system to system. Solar cells mostly contain two kinds of semiconducting materials, p-type and n-type semiconductors, leading to a p-n junction. When sunlight impinges the solar cell with the appropriate wavelength, energy is absorbed by promoting electrons into the conductive band of the semiconductor and leaving the valence band behind a split. The photovoltaic (PV) panel transforms solar power into electricity by using semiconductor materials such as silicon and cadmium telluride, which absorb sunlight.

17.2 Semiconductors

Semiconductors are materials whose electrical properties are intermediate between those of conductors, which give limited resistance to the passage of electric current movement, and insulators, which impede the flow of electricity. They are usually made from silicon. Semiconductors are used in many electrical

DOI: 10.1201/9781003145585-17

circuits since, for example, with regulating current, we can regulate the movement of electrons in that material. For other special properties, they are often used. A solar cell is actually composed of semi-conductors that are susceptible to light radiation. The amount of electrical current produced by semi-conductors that make up the solar cells will be determined by the amount of light energy that reaches the semiconductors. There are of two types of silicon semiconductors used in photovoltaics: n-type (negative) and p-type (positive).

The n-type semiconductors are made of "doped"crystalline silicon with tiny quantities of an impurity, usually phosphorus, such that in the doped material there is a surplus of free electrons. Because electrons have a negative electrical charge, doped silicon is thus known as *n form semiconductor*.

The semiconductors of the p-type are doped with very small amounts of some impurity, usually boron, allowing the material to have a deficit of free electrons. These missing electrons are called "holes". Although the absence of a negatively charged electron can be assumed to be identical to a positively charged atom, the semiconductor silicon-doped p-type is known in this way.

17.3 The P-n Junctions

A p-n junction is formed by joining these dissimilar negative (n) and positive (p) semiconductors. This creates an electric field in the junction region that allows negatively charged particles to go in one direction, and positively charged particles to go in the opposite direction.

Light can consist of a stream of tiny energy particles called *photons*. When light photons pass into a p-n junction of appropriate wavelength, they can transfer their energy to each of the substance's electrons, thereby "promoting" them to a greater degree of energy. Usually, by forming so-called "valence" bonds with neighbouring atoms, these electrons help keep the materials together, and they do not pass in this "valence band". But these electrons have been lifted to an energy level known as the "conduction band" in their current "excited" state and become ready to conduct electric current by flowing through the material. The energy difference between the "valence" band and the conduction band is the energy difference called the *band gap*.

17.4 Materials Used for the Construction of Photovoltaic Cells

For the construction of photovoltaic cells, special materials are used. In general, these are called *semiconductors*. Semiconducting materials must have certain properties to survive the sunlight. The substance derived from it is named after the solar cells or PV cells. Some cells are built to endure sunlight that reaches the surface of the earth, while others are optimized for use in space. In order to take advantage of different absorption and charge separation processes, solar cells can be made from just one layer of light-absorbing material or use several physical combinations. It is possible to classify solar cells in terms of first-, second-, and third-generation cells. The first-generation cells that are also referred to as *conventional*, *traditional*, or *wafer-based cells* are crystalline silicon, the widely common PV technology that includes materials such as monocrystalline silicon, poly-crystalline silicon, ribbon silicon, etc. The most widely used element in the manufacture of PV cells of solar cells is crystalline silicon.

Using silicon as a solar cell material, a range of earliest photovoltaic (PV) devices were made; silicon is still the most common material for solar cells today.

Single-crystal silicon has a homogeneous chemical composition. For effectively moving electrons through the material, this uniformity is optimum. However, silicon has to be "doped" with other elements in order to create an efficient photovoltaic cell.

Multicrystalline silicon is typically known to be less effective in comparison with single-crystal silicon. Multicrystalline devices, on the other hand, are less expensive to manufacture. The casting technique is a popular method of processing multicrystalline silicon on a commercial scale.

Amorphous silicon containing thin-film solar cells, cadmium telluride (CdTe) and copper indium gallium serenaded (CIGS) cells are the second-generation cells and are critical for commercial use in photovoltaic plants or in small stand-alone power systems. Amorphous silicon is silicon without crystalline structure. A thin-film solar cell is produced with it. The cells are formed by vapour depositing silicon on a metal or glass frame in a very thin film. Thin-film technology minimizes the volume of substance that is active in a cell. The majority of models sandwich between two glass panes of active material. Because only one sheet of glass is used for silicon solar panels, thin-film panels despite having a low environmental impact are almost twice as heavy as crystalline silicon panels (Pearce and Lau 2002; Marika 2012)

A variety of thin-film technologies sometimes identified as advanced photovoltaics is used in the third generation of solar cells. Many of them have not been applied commercially and are only used in the research or production process. In order to minimize the amount of light absorbing material required to manufacture solar cells, various thin-film technologies are currently being produced. This could lead to a decrease in cost of manufacturing, but it could also lead to a reduction in the conversion of quality energy. Most use organic ingredients, sometimes both organometallic compounds and inorganic compounds. Despite the fact that their efficiencies were poor and that the stability of the absorber material was always too limited for practical applications, a great deal of research has been expended in these technologies, aiming to meet the target of manufacturing low-cost, high-efficiency solar cells.

17.5 Photovoltaic Panel or Module

In applications such as calculator and satellites, photovoltaic cells (PVCs), more generally referred to as *solar cells*, are found. Photovoltaic cells have been used almost exclusively for the first time and are mostly used for more general applications. In simple terms, photovoltaic cells are devices that convert light energy into electrical energy. Photovoltaic cells are available In several shapes and sizes. Photovoltaic panel or module is a combination of several solar cells. The cell is the unit where the transitions of photon-electron energy is about to take place. The combination of various panels is called an *array*.

17.6 Types of Photovoltaic Panels

Different solar panels are available depending on the material of the semiconductor and manufacturing methods. They can be grouped as per their final shape (www.sitiosolar.com/paneles %20fotovoltaicas.htm#Tiposdepaneles). They include monocrystalline, polycrystalline, amorphous, and tandem panels. It is possible to classify the various PVCs that have been built up to date into four main categories called *generations* (Jayawardena et al. 2013). First generation: based on both monocrystalline and polycrystalline crystalline silicon technology, as well as gallium arsenide (GaAs).

Using Si in the production of PVC has many advantages:

a. It is the second most available element in the earth's crust (Kuhlmann 1963) indicating that the raw material supply will be appropriate in the future and that the raw material costs of production will be minimized.

b. It is a stable and non-toxic chemical element whose characteristics delay pollution processes and the loss of durability that may occur when used as a cell material.

c. Si PVCs are easily compatible with Si-based microelectronics (i.e. integrated circuits, transistors, etc.), well-known and well-developed technologies can be used (Sampaio and González 2017).

17.6.1 Classification based on Materials and Manufacturing Methods

17.6.1.1 Gallium Arsenide

Gallium arsenide or GaAs: gallium and arsenic are two elements. Gallium is rarer than gold and a by-product of other metals, particularly the smelting of aluminium and zinc. On the other hand arsenic is not rare, but is dangerous. Gallium arsenide also has a very high absorptivity and to absorb sunlight just requires a cell of a few microns thick. GaAs-based cells can have several layers of a slightly different structure; allowing for the creation and capturing of electrons and holes more efficiently than silicon cells, which are constrained to doping threshold differences in order to achieve the same result. One of GaAs' key benefits for PVC implementation is that it provides a large variety of possible design choices. GaAs cells are not affected by sun, and are extremely resistant to radiation exposure. This makes it suitable for applications in space and concentrator systems.

Amorphous silicon (a-Si) and microcrystalline silicon (μc-Si) thin films solar cells, cadmium telluride/cadmium sulphide (CdTe/CdS) and copper indium gallium selenide (CIGS) solar cells are used in the second-generation photovoltaic solar cells. The thin-film technologies are planned to reduce the high cost of crystalline silicon by using smaller volume of material and improved efficiency, packed on cheap substrates.

17.6.1.2 Cadmium Telluride

Another well-known polycrystalline thin-film material is cadmium telluride or CdTe. CdTe also has a very high absorption ability and can be produced using low-cost process similar to copper indium diselenide. It is possible to modify the properties of CdTe by adding alloying elements such as mercury and zinc, which will change CdTe's properties.

17.6.1.3 Copper Indium diselenide

Copper indium diselenide, CIS for short, has an exceptionally high absorptive capacity. This indicates that the first micrometer of the substance would absorb 99% of the light illuminated on the CIS. The addition of a small amount of gallium would improve the efficiency of the photovoltaic system. This is usually referred to as the *photovoltaic cell* or *CIGS* or *copper indium gallium diselenide*.

The general benefits of second-generation PV panels are as follows:

  i. Cheaper than the Si-based solar cells
 ii. A drastic reduction in the amount of required materials. Only a one-micron thick layer is required sometimes.
 iii. High coefficient of absorption
 iv. Vacuum and non-vacuum processes may be used
  v. Most of technologies allow direct integration into a higher voltage module (i.e. a-Si), which decreases the number of stages of processing relative to PVCs of the first generation.

They still offer some drawbacks, though:

  i. Less quality
 ii. Deterioration caused by light in the first phases of outdoor use. Higher deterioration in outdoor uses: the ion flow in the glass may be produced by the semiconductor deposited on the glass. This issue will arise in the case of amorphous silicon, even though the problem is the substrate is not glass. Contamination of the atmosphere begins from production process
 iii. The supply of processing of the processing products may not be plentiful in such technologies

The third generation comprises developments focused on newer compounds such as nanocrystalline films, active quantum dots, tandem or stacked multilayers of III–V based inorganic materials such as gallium arsenide/gallium indium phosphide (GaAs/GaInP), solar cells focused organic (polymer), dye-sensitized solar cells, perovskite cells, etc.

17.6.1.4 Perovskite Materials

Hybrid organic-inorganic lead or tin-halide compounds, such as methylammonium lead halide, are typically perovskite structured compounds used in solar cells. This product can be treated by solution, thereby allowing cheap and easy manufacturing. The output of solar cells based on perovskite has been gradually growing and is estimated to be over 20%. The power to consume sunlight through the whole visible spectrum is one of the main benefits of these materials.

17.6.1.5 Organic/polymer Materials

In solar cells, semiconducting polymers such as polyphenylenevinylene (PPV) and small organic small molecules such as phthalocyanines, polyacenes, and squarenes are also used. In the visible and near infrared regions, these strongly conjugated organic molecules have extensive absorption. These materials are deposited either by vacuum deposition or solvent processing as thin films, they are versatile. The performance of these cells is still modest, however just a little over 10%, which is why they have not yet been marketed.

17.6.1.6 Quantum dots

Another type of evolving material used in solar cells is the nanoparticles, some nm in size; other type of new materials used in solar cells are called quantum dots. They are semiconductor materials with a low band gap, such as CdS, CdSe, and PbS. By changing the particle size, their energy band gaps can be tuned over a large spectrum. Many traditional materials such as Cd and Pb used for the manufacture of quantum dots are known to be poisonous, so other alternative materials, such as copper indium selenide, are manufactured.

17.6.1.7 Dye-sensitized Materials

Another type of solar cell material that can absorb a wide variety of the visible areas of sunlight is a tiny molecule dye, such as a ruthenium metal organic dye. A coating of inorganic mesoporous nanoparticles, typically titanium dioxide, expands the space for light absorption of radiation. Using the solution process, solar cells use these materials, making them simple to produce.

The main benefits of third-generation PVCs are:

i. Ideal for large-scale manufacturing.
ii. Mechanical strength.
iii. High efficiency at high temperatures.

Their main task, however, is to reduce the cost of solar electricity delivered per watt.

The PV panels of the fourth generation integrate the low cost/flexibility of polymer thin films with the stability of novel inorganic nanostructures such as metal nanoparticles and metal oxides or organic non-material such as carbon nanotubes, grapheme, and their derivatives.

17.6.2 Classification based on final shape

17.6.2.1 Monocrystalline Panels

Monocrystalline or single-crystal solar PV panels are one of the oldest, most effective, and most powerful ways to generate electricity from solar energy. Here a single silicon crystal is made of a PV module. For individual cells to be produced, the silicon is filtered, heated, and then crystallized into ingots that are then sliced into thin wafers. Monocrystalline PV module as seen in Figure 17.1 is normally black or iridescent

FIGURE 17.1   Monocrystalline panel. (Source: Martin D. Vonka/shutterstock.com)

blue. Without degradation, the life cycle of crystalline silicon cells is more than 25 years, making it suitable for producing commercial solar electricity. It produces an energy conversion efficiency of up to 22%, the highest of all the current maximum made. Crystal silicon is filled with chemicals such as silicon nitride or titanium dioxide to increase performance even more by minimizing reflected light.

- **Advantages of Solar Monocrystalline Panels:** The main advantages of monocrystalline solar PV panels are the following:

  i. Longevity
  ii. The performance is high in the 15–24% range, as they are made from the highest quality of silicon, making them cost- effective in the long run
  iii. Reduced cost for installation
  iv. Effective in volume
  v. Naturally non-hazardous
  vi. Improved heat tolerance
  vii. More strength
  viii. Embodied energy

- **Disadvantages:** The major drawbacks of monocrystalline solar panels are high initial costs and fragility.

17.6.2.2 Polycrystalline Panels

Polycrystalline or multicrystalline panels use solar cells with several facets that are made of silicon crystals. Like monocrystalline solar cells, polycrystalline cells are not standardized in nature. They have a surface with a random pattern of crystal borders, rather than a solid colour with one crystal cell.

Polycrystalline panels are formed by fragments that have been structured as disordered crystals from a silicon bar. They are very identifiable physically, and they have a granulated appearance. Since low-cost silicon is used to manufacture polycrystalline cells, their performance usually is 12–14%, a value

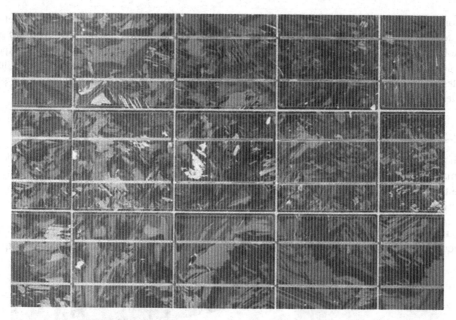

FIGURE 17.2 Polycrystalline solar panel.

marginally lower than monocrystalline cells, but much higher than solar technologies including thin film. The key factor for this reduced performance is the poorer consistency of the material due to grain borders and defects, and the higher impurity content (Figure 17.2).

- **Advantages of Solar Polycrystalline Panels:** Main benefits of multicrystalline panels are as follows:

  i. The development process is quick, cost-effective, and reduces silicon waste compared to single crystal panels.
  ii. The heat response is significantly smaller than that of single-crystal panels.
  iii. The related green house impacts are fewer.
  iv. It's less costly.

- **Disadvantages:** The following are the drawbacks of polycrystalline panels:

  i. Poorer conversion efficiency than monocrystalline panels owing to low-purity silicon use.
  ii. Reduced space-efficiency.
  iii. Contrasted with single-crystal and thin-film panels, less aesthetically appealing.

17.6.2.3 Amorphous Panels

The thickness of these panels is substantial. Silicon with another structure or other semiconductor materials can be used to obtain thinner and flexible panels. These panels make it easier to conform to uneven surfaces in certain situations. They are referred to as *Amorphous PV Solar Panels* or *PV thin-film modules* and they can be categorized according to the material used.

17.6.2.4 Amorphous Silicon Panels

This was created with silicon as well, but in contrast to former cases, in this example the material has no crystal structure. They are commonly used for small electronic instruments (calculators, watches) and small compact panels of this type. Amorphous silicon solar panels achieve just around 7% efficiency leading to substrate deterioration when exposed to sunlight. Amorphous silicon can absorb 40 times as much solar radiation than single-crystal silicon. This is one of the key reasons why the expense of photovoltaic can be minimized by amorphous silicon. Amorphous silicon can be coated on low-cost substrates such as plastics and glass. This makes amorphous silicon perfect for photovoltaic products embedded in buildings (Figure 17.3).

17.6.2.5 Tandem Panels

Tandem panels also exist and combine two different types of semiconductor materials. Just a portion of electromagnetic spectrum of solar radiation is absorbed by each type of material, and a mixture of two or three types of materials may also be used to capture electromagnetic spectra. This panel form can be as effective as 35% (Figure 17.4).

- **Conversion efficiency of PV panel:** This is the most symbolic criterion used for PV technology calculation and assessment. It is expressed as the solar energy input to electrical energy output ratio.

FIGURE 17.3 Amorphous thin-film panel.

FIGURE 17.4 Tandem solar panel.

Performance aggregates all of the system's constituent parameters, such as the short circuit current, the open circuit voltage, and the fill factor, which in turn depend on material's fundamental properties and the manufacturing defects. Besides the defects affecting individual solar cells, the engineering needed to install solar cells into a solar panel will cause more performance deficits.

The **short circuit current** from a solar cell is the highest current that occurs when the voltage across the device is zero. Consequently, short circuit current is the maximum current that can be drawn from the solar cell. The **open circuit voltage** is the highest solar cell voltage available, and this occurs at zero current.

The **fill factor** is the ratio of actual maximum power obtainable to the product of open circuit voltage and short circuit current.

Transmission losses and thermalization losses, with separate relative contributions as a result of the semiconductor band gap, are the main energy losses during the conversion of solar energy to electrical power by a solar cell.

Transmission losses are losses encountered during energy conversion in system from one component to another – solar panels (solar energy to direct current (DC) electrical energy); charge controller; solar batteries (direct current (DC) electrical energy to chemical energy); Inverter (direct current (DC) electrical energy to alternating current (AC) electrical energy); electrical appliances such as fan, bulb, or television.

Thermalization loss is excess energy produced in semiconductor crystal as heat, as high energy photons are excited to a higher energy state outside the band gap in the ultraviolet, visible, and near infrared spectrum.

There are many methods to reduce such losses, such as multijunction cells, solar hot-carrier cells, intermediate band solar cells, or multiple exciton generation. The semiconductor's band gap is the minimum energy necessary to excite an electron trapped in its bound state into a free state where it can engage in conduction. The low energy level of a semiconductor is called the "valence band" (EV), and the "conduction band" (EC) is the level of energy at which an electron can be considered free. The interval between the bound state and the free state, between the valence band and conduction band, is the energy band gap (EG). Good conductors have zero band gap in that they always have free electrons available for conduction when the least electric potential is being applied. Insulators have very high band gaps because to break them loose from the parent atoms their electrons will need very high energy. Semiconductors are in the order of electron volts with intermediate band gap energies. Many customers and people in the solar industry consider the reliability of solar panel the most critical criterion when determining the capacity of a solar panel.

17.7 Factors Influencing Conversion Performance

Not all of the sunlight that hits a PV cell is converted into power. A large part of it is currently destroyed. Many factors play roles in reducing the cell's ability to transform the illumination it receives in the design of solar cell. How to achieve higher efficiencies is to plan with these considerations in mind.

a. **Wavelength:** Light is made of photons with a large variety of wavelengths and energies. The radiation reaching the surface of the earth has wavelengths ranging from ultraviolet to infrared in the visible spectrum. As light reaches the surface of a solar cell, some photons are absorbed, while some flow right through. The energy of each of the absorbed photons has been transformed into heat. The others have the energy required to separate electrons from their atomic bonds to produce charge carriers and electric current.

b. **Recombination:** One way for electric current to flow in a semiconductor is for a "charge carrier", such as a negatively charged electron, to flow across the material. Any such charge carrier is known as a "hole", implying an absence within the material of an electron and serving as a positive charge carrier. When an electron finds a vacuum, they will recombine, thereby cancelling their contributions to the electrical current. Direct recombination reverses the process by which the solar cell produces electricity, where light-induced electrons and holes interact, recombine, and emit a photon. It is one of the basic variables that restrict performance. Indirect recombination is a mechanism by the electrons or holes that induce recombination and release of their energy as heat is attained by an impurity, a flaw in the crystal structure, or an interface.

c. **Temperature:** Generally at low temperatures, solar cells function well. Higher temperatures cause the properties of the semiconductor to change, resulting in a slight current increase but a much higher voltage. Severe temperature changes can also affect semiconductor properties, resulting in shorter operating lifetimes. Since sunlight shines on cells, good thermal control increases both performance and lifespan.

d. **Reflection:** By minimizing the amount of light that reflects off the cell's surface, the performance of a cell can be improved. For instance, over 30% of incident light is absorbed by untreated silicon. Coatings and textured materials that have anti-reflection surfaces help minimize reflection. A heavily impaired cell can look dark blue or black.

17.8 Factors Affecting the Performance of Photovoltaic Panels

It is usually expected that solar panels are built to achieve the most appropriate performance or maximum output. Factors controlling the assessment of optimal performance or optimum yield can be divided into two categories, namely, changeable variables and unchangeable variables. In order to adapt to changing requirements, for implementation, the variables that can be modified have design consistency, while the unchangeable variables must be configured by nature. The multiple changeable and unchangeable variables influence the configuration and design of a solar panel, the construction and function of a solar panel, and play an important part in solar panel output production. Two factors that influence the efficiency of a PV module are solar irradiance and cell temperature. In addition, the quantity of electricity provided by a PV module depends on other aspects such as the functionality of other parts of the whole device, as well as other environmental conditions. In addition to the environment, several other factors influence PV performances, such as the photovoltaic technologies used, solar panel inclination angle, array mismatch, inverter quality, solar panel spacing configuration (shading), soiling of PV panels, etc. Therefore, experience and awareness of the performance of PV module are of great importance for proper product selection and reliable energy consumption calculation under the operating conditions of the execution site.

a. **Solar cell temperature:** Solar cell temperature is one of the key factors responsible for lower PV output module (Jakhrani et al. 2011). This adjusts the capacity of the system and power supply. It depends on the module encapsulating material, the thermal dissipation, the absorption properties, and the modules' operating point as well as the irradiance power. A quantitative analysis of the effect of temperature on characteristics of PV modules, such as voltage, current, and efficiency under real operating conditions is therefore significant. Higher temperatures have an impact on all PV technologies. But some types of modules are more resilient to temperature changes than others. The panel age will be improved by long- term heat treatment, but some materials will not be able to tolerate extremely high temperature, short peak.

b. **Solar Irradiance:** Solar irradiance plays a vital role in improving the efficiency of a PV module. The higher the irradiance, the higher the current and the higher the produced power.

c. **PV technology and cell configuration:** Transmittance is received differently by various PV technologies. This includes monocrystalline panel, polycrystalline panel, tandem panel etc. PV panels' performance varies according to the hardware.

d. **Solar panel inclination angle:** The power incident on a PV panel, in addition to power contained in sunlight, depends on the angle between the module and the sunrays (Pantic et al. 2013). In order to really perform the task of increasing the performance of photoelectric conversion, solar panels need to receive sunlight at the optimum angle. The optimum angle of inclination of the solar panel would also have change considerably in various seasons, geographical positions, and sunlight levels; so that the angle of inclination of the solar panels can be constantly changed due to changes in seasons, latitude and longitude, and sunshine hours. The fixed inclination should be specified as the inclination angle depending on the maximum power generation throughout the year. It order to produce more output, it is the best choice to place the PV panel at the optimum tilted angle. Its energy capacity can be maximized by a strong orientation and tilt angle for PV modules. For the optimum performance of your system in the year, specific PV modules should be aligned to the true south (in the northern hemisphere). In the case that the PV modules are not at all likely to be moved, the ideal tilt angle is adjusted to the latitude of the site for optimal amount direct irradiance.

e. **Array mismatch:** PV modules with different current–voltage characteristics (I–V) have a combined output power lower than the power achieved by summing up the output power given by each of the modules when attached together. Connecting together a "strong" and a "weak" module constrains the performance of the "strong" module to the "weaker" module level. This poses a problem: increasing the strings' length eventually leads to a higher degree of mismatch and thus lack of energy loss. Loss of mismatch is due to an existing mismatch or mismatch in voltage. The current mismatch occurs where a sequence string includes low current modules or when a part of the string is shaded; if the modules' current and voltage are not balanced, a low-output module decides the total array output. There is a voltage imbalance when the cells are reduced. Cells are connected in series or run at the same current or voltage in parallel, resulting in considerable energy loss when the cell power with the lowest peak output is limited to the activity of the cell. So having modules of the same bin class together in strings is essential. A minor mistake can significantly affect the amount of power that comes out.

Mismatch losses in solar photovoltaic arrays can be attributed to many reasons:

i. Responsiveness of the producer in cell characteristics.
ii. Ecological pressures.
iii. Trouble with shadow.

f. **Inverter quality:** After being converted by the solar panels into direct current (DC) electrical power, the electricity from light is passed through the inverter. The inverter's basic purpose is to transform electrical energy from DC to electrical energy from AC. The efficiency of a inverter is a linear factor in the energy yield of the system. As a consequence, for good system performance, high efficiency across the entire power spectrum is important. The efficiency of the inverter depends on the DC voltage, and the output and lifetime of the inverter also depend on operating temperature. Adequate spacing and ventilation help to keep losses low and to reduce the risk of failures for large PV plants. Inverters are typically constantly refrigerated in warm climates.

g. **Climate condition:** The modules of solar panel are exposed for a long time to the natural environment, and conditions such as wind and lightning will affect the solar cell. Sun, wind, temperature, and so on can alter solar panel's (cells) photoelectric conversion efficiency, and certain variables can also harm the solar panel's functions and structures. The meteorological and environmental tracking data must be thoroughly gathered in the design process for the solar power system.

h. **Solar panel spacing design (shading):** When the solar panel is shielded, it impacts the solar system's generating capability. Since PV module's output current is a function of solar irradiance, the PV module's performance will be influenced by a decrease in solar irradiance due to partial or full shading (Wenham 2011). Ideally solar panels can be positioned in such a manner that they never have shadows because a shadow can have a remarkably significant influence on the production on such a small portion of the panel. Partially shaded solar panels create hotspots. Hotspots are localized decrease in efficiency, which results in lower power output and degradation in the affected area of the panel. Therefore, when designing the spacing of solar panel

FIGURE 17.5   A dusty photovoltaic panel.

array, the shading of the building between solar panels and the self-shading between solar panels should be considered.

 i. **Soiling:** This is the accumulation of dust over the PV modules and other minute particles as shown in Figure 17.5.

One of the major loss factors affecting the PV performance is soiling. The soiling loss is the energy loss occurring from mud, snow, dirt, and other particles coating the PV modules. Dust is a thin film coating the surface of the solar array and the particles are less than 10 mm in diameter, although this depends on the area and its environment. The accumulating dust enhances the soiling effect over time. There are currently two types of soiling, namely natural and artificial soiling.

 a. **Natural Soiling:** Dust deposition on solar panels while exposed to the sun is the natural soiling of PV panels. The main source of environmental soiling are dust particles dispersed in the atmosphere because of air pollution, bird droppings, deforestation, construction works, pedestrian and vehicular movements.

 b. **Artificial Soiling:** In order to analyze the influence of soiling on electrical and optical parameters of PV panels, artificial soiling of PV panel is performed. Dust samples will be processed in the laboratory and spread in a constrained setting on the PV panels. Compared to natural soiling, artificial soiling research is a rapid process.

17.9 Ways of Regulating the Variables That Affect the PV Panel's Performance

Sunlight consists of charged particles called photons which derive from the hydrogen atom interaction in the centre of the sun (Padden 2014). They transmit their energy onto the electrons in the cell when these photons damage semiconductive materials of the panels (solar cells). These electrons pass about and are gradually captured and channelled into a stream of electrons producing electricity because they are excited to a higher energy level.

 a. **Temperature:** Excessively high temperatures can greatly impair the performance of solar panels. So, for example, if the panels are in the middle of desert, there output will be diminished. To reduce the high temperature problem, solar panels are placed marginally on roofs of buildings in order to increase air circulation and efficiently cool the panels.

b. **Season:** Any green technology requires an energy source, and a lot of sunlight is required for solar power systems. The same amount of solar radiation is not produced across all sites on earth. Insolation – the amount of solar radiation emitted in a given area – is determined by variables such as local temperature, time of year, and most importantly the local latitude. In the dry season, the sun will be higher in the sky than in the rainy season.

c. **Panels' orientation:** Panels need to be pointed at the sun in order to maximize production. This is accomplished by system designers who orient the panels so that the most annual exposure is permitted in the pitch and azimuth. The slope or relative steepness of the roof is the roof pitch. Since the sun changes in latitude throughout the year, it is important to turn the panels so that they receive the most sunlight on average. This angle is usually equal to latitude of the locality. The local latitude minus the roof pitch (the tilt angle) will be the angle between the panel and the roof. The other angle to be taken into account is angle of the azimuth, usually determined from a cardinal direction such as north or south.

d. **Shade:** shade is the enemy of solar panel because it is the blocking the sunlight needed to generate energy, much like clouds. Trees covering the photovoltaic device must be removed or trimmed to allow sunshine penetrate the panel. The explanation why shade presents such a problem has to do with how internally and to one another solar panels are wired. In a panel, individual cells are wired in series to maximize the module's voltage, but they are just as effective as their weakest connection. The whole array will experience a reduction in power if a small portion of a module is shaded. This is true for the system as a whole, as the panels themselves are connected generally in series. Using power optimizer that separates an individual module, this problem can be solved such that its output does not affect the other modules in an array.

e. **Cleaning of PV panels:** In general, solar panels are self-cleaning, but over time, dust and other substances such as bird droppings can build up and decrease the amount of electricity produced by a module, particularly in dry areas or where panel tilt is minimal. The efficiency of the module is impaired by soiling as a result of the deposition of dew, snow, dirt, dust, bird dropping, etc. (Maghami et al. 2016). To achieve optimum performance, the PV modules should be cleaned regularly. Active and passive techniques for photovoltaic cleaning are available. In order not only to achieve the highest efficiency but also to achieve cleaning effect by rainfall, the PV panels should be held at optimal tilt angles. PV panels would be kept protected by heavy precipitation. The light rain will, however, slightly wash away the dust and the lower side of the PV panels will collect dust. Chemical washing by high-pressure water sprayers, mechanical washing by wipers, the use of electrodynamic screens and hydrophilic or hydrophobic surfaces are other strategies to reduce soil settlement. Active methods of self-cleaning include water cleaning, mechanical cleaning, and electrodynamic coatings. Natural cleaning by rain, super hydrophobic and hydrophilic surfaces on the PV panels are some of the passive self-cleaning process techniques. Advanced strategies under review are robotics for cleaning of PV panels from soiling. Cleaning with robotics is useful because of its feature such as rapid reaction, low power consumption, safe operation, and more efficiency than other cleaning methods (Juzaili et al. 2017).

17.10 Conclusion

Solar energy is plentiful, renewable, and limitless energy and can be transformed directly through photovoltaic (PV) technology into electricity. It offers renewable energy with no greenhouse emissions, drives markets, contributes to technological growth, saves power costs for businesses and households. The photovoltaic (PV) panels transform solar power into electricity by semiconductor materials which absorb sunlight, such as silicon and cadmium telluride.

The solar cell or photovoltaic cell is built using semiconducting materials that differ from system to system. Semiconductors are materials with an intermediate electrical property between conductors, with limited resistance to electric current flow, and insulators, that inhibit electrical movement. A combination of multiple solar cells is a photovoltaic panel or module and a combination of different panels is considered an array. Materials such as monocrystalline silicon, polycrystalline silicon, and ribbon silicon form the commercially prevalent PV technology. Two factors influencing the performance of a PV module are cell temperature solar irradiance. The PV output is influenced by environmental and many other variables such as the photovoltaic technology used, solar pane angle of inclination, array mismatch, inverter quality, solar panel spacing configuration (shading), soiling of PV panels. The effects of these factors impacting PV performance can be mitigated in many ways.

REFERENCES

Chen, F. L., D. J. Yang and H. M. Yin. 2015. Applications. In *Semiconductor Materials for for P hotovoltaic Cells*, edited by M. Parans Paranthaman, Winnie Wong-Ng, and Raghu N. Bhattacharya. New York: Springer Series in Material Science Book series.

Jakhrani, A. Q., A. K. Othman and S. R. Samo. 2011. "Comparison of Solar Photovoltaic Module Temperature Models." *World Applied Sciences Journal* 14: 1–8.

Jayawardena K. D. G. I., L. J. Rozanski, C. A. Mills, M. J. Beliatis, N. A. Nismy and S. R. Silva. 2013. "'Inorganics-in-Organics': Recent Developments and Outlook for 4G Polymer Solar Cells." *Nanoscale* 5: 8411–8427. doi: 10.1039/c3nr02733c.

Juzaili, W., H. Abdul, S. Shaari, and Z. Salam. 2017. "*Performance* Degradation of Photovoltaic Power System: Review on Mitigation Methods." *Renew. Sustain. Energy Rev.* 67: 876–891.

Kuhlmann A. M. 1963. "The Second Most Abundant Element in the Earth's Crust." *JOM* 15: 502–505. doi: 10.1007/BF03378936.

Maghami, M. R., Hashim Hizam, Chandima Gomes, MohdAmran Radzi, Mohammad Ismael Rezadad and Shahrooz Hajighorbani. 2016. "Power Loss due to Soiling on Solar Panel: A Review." *Renewable and Sustainable Energy Reviews*: 1302–1316.

Marika, Edoff. 2012. "Thin Film Solar Cells: Research in an Industrial Perspective." *AMBIO* 41, no. 2: 112–118. doi: 10.1007/s13280-012-0265-6. ISSN 0044-7447. PMC 3357764. PMID 22434436.

Padden, Danor. 2014. *Factors That Affect Solar Panel Efficiency*: Solar Energy.

Pantic, S. L., M. T. Pavlovic and D. D. Milosavljevic. 2013. A Practical Field Study of Performances of Solar Modules at various positions in Serbia. *Thermal Sciencie*, 19, Issue Suplement 2, 5511–5523.

Pearce, J. and A. Lau. 2002. Net Energy Analysis for Sustainable Energy Production from *Silicon* Based cells. *Solar Energy*: 181. doi:10.1115/SED2002-1051. ISBN 978-0-7918-1689-9.

Sampaio, P. G. V. and M. O. A. González. 2017. "Photovoltaic Solar Energy: Conceptual Framework." *Renew. Sustain. Energy Rev* 74: 590–601. doi:

Wenham, S. R. 2011. *Applied Photovoltaics*. New York: Routledge.

www.sitiosolar.com/paneles%20fotovoltaicas.htm#Tiposdepaneles

Zaidi, B. 2018. *Solar Panels and Photovoltaic Materials*. London, UK: InTech Open.

18

Comparative Study of Different Dopants on the Structural and Optical Properties of Chemically Deposited Antimony Sulphide Thin Films

Patrick A. Nwofe and Chinenye Ezeokoye
Department of Industrial Physics, Faculty of Science, Ebonyi State University, Nigeria, Abakaliki, Nigeria

18.1 Introduction

Copper and cobalt are transition metals and thus possess similar chemical characteristics. Application of copper-based devices is common in electronic, optoelectronic, and sustainable energy devices, to mention but a few. Cobalt is mostly applied in fabrication of super alloys, magnetic recording, rechargeable batteries, electronics, catalysts, alloys, medicine, biomedical applications, etc. Antimony sulphide belongs to the chalcogenide family. The need to investigate the potentials of other low-cost inorganic materials that are more environmental friendly compared to those used in the second-generation solar cells is very imminent. The effect of copper impurities on the properties of antimony sulphide thin films for use in photovoltaic solar cell applications has been reported by different research groups (Ornelas-Acosta et al. 2015; Rath et al. 2015). Copper antimony sulphide thin films offer flexibility in device fabrication, as they can be produced using dip coating (Dekhil et al. 2019), spray pyrolysis (Ramos Aquino et al. 2016), chemical bath deposition (Loranca-Ramos et al. 2018), microwave-assisted (Ghanwat et al. 2014), thermal evaporation (Garza et al. 2011; Suriakarthick et al. 2015; Kumar et al. 2020), laser ablation (Shaji et al. 2019; Azizur Rahman et al. 2019), ink-printing (Cho et al. 2017), rapid thermal processing (Vinayakumar et al. 2019), surfactant-mediated solvothermal synthesis (Bincy et al. 2017,2018), solution-processed@technique (Satoshi et al. 2015), and mechanical alloying techniques (Zhang et al. 2016). The use of copper-antimony thin films in various optoelectronic devices including photodetectors (Vinayakumar et al. 2018), photodiodes (Yıldırım et al. 2019; Sivagami et al. 2018), solar cells (Garza et al. 2011; Ramasamy 2014a; Krishnan, Shaji, and Ornelas 2015; Liu et al. 2016; Bincy et al. 2018; Ishaq et al. 2019) has been reported in the literature. The solar conversion efficiency of copper-antimony-based thin-film heterojunction solar cells is still at 4.61% (Ishaq et al. 2019). This implies that there is a need to investigate more on the materials, electrical and device properties, in order to ascertain the optimized conditions needed for maximal device performance. The influence of cobalt impurities on the properties of antimony sulphide thin films is relatively rare in the literature (Cerdán-Pasarán et al. 2019). The present investigation is a baseline study on the properties of chemically deposited cobalt antimony sulphide thin films, and as well extends further understanding of the materials' properties of copper antimony sulphide–based thin films for use as absorber layers in photovoltaic solar cells and related optoelectronic applications.

DOI: 10.1201/9781003145585-18

18.2 Materials and Methods

18.2.1 Materials

The source materials include antimony trichloride powder (SbCl$_3$), sodium thiosulphate pentahydrate (Na$_2$SO$_3$.5H$_2$O), acetone (CH$_3$COCH$_3$), cobalt chloride (CoCl$_2$), copper chloride (CuCl$_2$), distilled water, soda lime glass, synthetic foam, hydrochloric acid, beakers, cellotape, mild detergent solution. All the source materials (chemicals) were of high purity (99.99%) and were all used as supplied by Sigma Aldrich UK through local suppliers.

18.2.2 Method

A suitable quantity of a mild detergent solution was used to clean the soda lime glass. The cleaned glass was subjected to a 5% hydrochloric acid etch for 300 min. The glass slides were further cleaned with double-distilled water, dried in air, and then stored in a vacuum environment. A solution of 750 mg of SbCl$_3$ in 3 ml of CH$_3$COCH$_3$ was prepared in a beaker labelled as beaker A. A solution of 25 ml of IM Na$_2$SO$_3$.5H$_2$O in 72.5 ml of deionized water was prepared in a separate beaker labelled as B. The beaker labelled C contained 7.5 ml of 1M CuCl$_2$. The solutions from beakers B and C were added to the solution in beaker A, and stirred continuously for 15 min. The pH was observed to be in the alkaline range (8.5). The precleaned soda lime glasses which were earlier inserted in the synthetic foam attached to the locally fabricated substrate holder were carefully inserted in the beaker (A), and left for a deposition time of 120 min. Similar method also applies for the cobalt-doped case. For preparation of the undoped layers, solution from beaker B was carefully added in beaker A, and kept under same deposition conditions. The pH was observed to be in the acidic range (5.5). All depositions were done at room temperature conditions.

18.2.3 Growth Mechanism of CuSb$_2$ Thin Films

SbCl$_3$ is soluble in organic solvent; hence it will dissolve in acetone to form a complex as indicated in eqns 18.1–18.2. Na$_2$SO$_3$.5H$_2$O will dissolve easily in water to release S^{2-} as given in eqns 18.3–18.6. CuCl$_2$ also dissociates as given by equation 18.7. In the beaker containing the Cu^{2+}, Sb^{2+}, and S^{2-}, the complexation of the Sb^{2+} ions by acetone and the role of Na$_2$SO$_3$.5H$_2$O in solution with metal ions as indicated in (Grozdanov 1994; Maghraoui-Meherzi et al. 2010; Osuwa and Osuji 2011; Onyia, Nnabuchi, and Chima 2020) ensured the release of the Cu^{2+}, Sb^{2+}, and S^{2-} ions in a controlled manner. The gradual hydrolytic decomposition of the thiosulphate ions, and ionic exchange between the cations and anions in the reaction bath will lead to nucleation and subsequent growth of the CuSbS$_2$ thin films. The formation of the CuSbS$_2$ thin films is intiated when the ionic product (IP) of the cations and anions [Cu^{2+}][Sb^{3+}][S^{2-}]^2 is greater than the solubility product of CuSbS$_2$. The CuSbS$_2$ thin films were expected to form by the complex ion-by-ion reaction mechanism. The schematic diagram for the reaction mechanism for formation of CuSbS$_2$ thin films is illustrated in Figure 18.1. A similar reaction kinetics is expected to occur in reaction bath of the cobalt impurities.

$$2SbCl_3 + CH_3COCH_3 \Leftrightarrow Sb_2(CH_3COCH_3)^{3+} + 3Cl_2 \quad (18.1)$$

$$Sb_2(CH_3COCH_3)^{3+} \rightarrow 2Sb^{3+} + CH_3COCH_3 \quad (18.2)$$

$$Na_2S_2O_3 \rightarrow 2Na^+ + S_2O_3^{2-} \quad (18.3)$$

$$6S_2O_3^{2-} \rightarrow 3S_4O_6^{2-} + 6e^- \quad (18.4)$$

$$3S_2O_3^{2-} + 3H^+ \rightarrow 3HSO_3 + 3S \quad (18.5)$$

FIGURE 18.1 Schematic diagram of the reaction mechanism for formation of CuSbS$_2$ thin films.

$$3S + 6e^- \rightarrow 3S^{2-} \tag{18.6}$$

$$CuCl_2 \rightarrow Cu^{2+} + 2Cl^- \tag{18.7}$$

18.3 Results and Discussion

18.3.1 Structural Analysis

Figure 18.2 shows typical XRD plots for the 0.1M co-doped, 0.1M Cu-doped, and undoped Sb$_2$S$_3$ thin films. The films all exhibit polycrystalline behaviour independent of the growth conditions. The undoped film was indexed with ICDD powder diffraction file: 01-078-1347 that corresponds to the Sb$_2$S$_3$ orthorhombic stibnite phase. The copper-doped layers crystallized in the orthorhombic crystal structure belonging PDF: 00-044-1417. The well-defined diffraction peaks of the CuSbS$_2$ phase are clearly seen at 2θ = 25.84°, 28.64°, 29.84°, 31.42°, 39.10°, and 50.06°, corresponding to the (400), (400), (111), (020), (220), (501), and (801) diffraction planes. It is pertinent to note that due to the novelty of the Co^{2+} doped films it was not possible to index the observed peaks with a particular ICDD powder diffraction file. However, the effect of the cobalt impurities is very significant, as only few peaks are observed in the former compared to the latter. The undoped film exhibited more texturing compared to the doped layers in the respective doping environments. The observation is in agreement with the report of other research groups in CuSbS$_2$ thin films irrespective of the deposition techniques (Lakhdar et al. 2014; Rath et al. 2015; Vinayakumar et al. 2018; Wang et al. 2020).

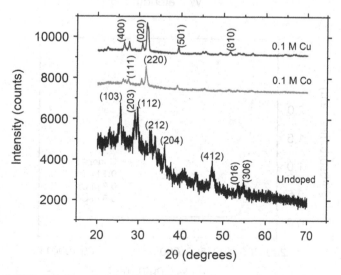

FIGURE 18.2 Typical XRD plots for undoped, 0.1M CuSbS$_2$, and 0.1M CoSbS$_2$ thin films.

18.3.2 Optical Analysis

Figure 18.3a shows the plots for change of absorbance with wavelength within the range of 350–1000 nm for copper treated films while Figure 18.3b gives that of the films grown with cobalt impurities at the respective concentrations. The absorbance values were generally higher at higher energy regions (shorter wavelengths), exhibiting a steep fall down to a critical wavelength and then decreased consistently down the lower energy regions (longer wavelengths) as indicated in Figures 18.3a and 18.3b, respectively. However films treated with copper impurities indicated slightly higher values of absorbance for films doped with 0.1M of copper impurities compared to undoped films, and films for doping concentration >0.1 M. In the cobalt-doped films, the absorbance was optimum for the undoped film and decreased with the varying concentration of the cobalt impurities. The observed differences were attributed to the atomic rearrangement during film formation. Further, the differences in mass densities of the dopants could trigger detectable changes in the films at varying impurity concentrations. Recent report by Onyia and coauthors (2020) observed similar decreasing trend of absorbance with wavelength in chemically deposited $CuSbS_2$ thin films and attributed it to the effect of the different media used in the growth matrix.

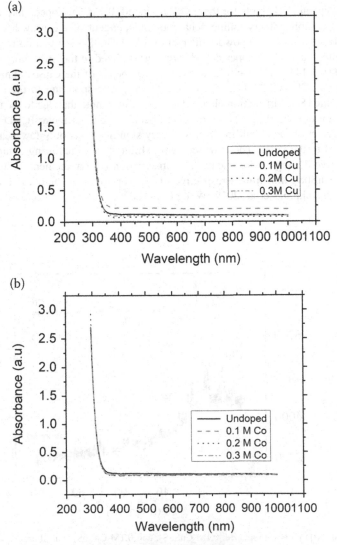

FIGURE 18.3   (a) Absorbance vs. Wavelength of Cu^{2+}-doped films. (b) Absorbance vs. Wavelength of Co^{2+} doped films.

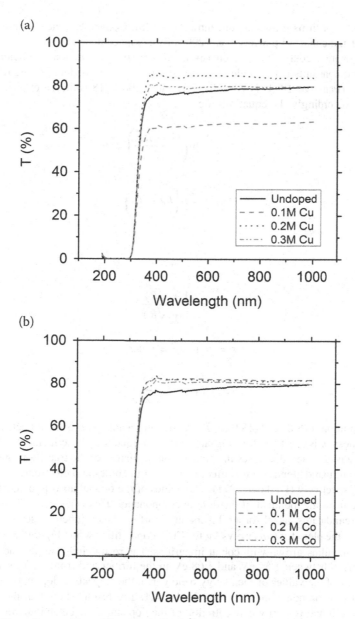

FIGURE 18.4 (a) Transmittance (T) vs. Wavelength for Cu^{2+}-doped films. (b) Transmittance (T) vs. Wavelength for Co^{2+}-doped films.

Figure 18.4a shows the variation of the transmittance with wavelength plots for Cu^{2+}-doped films while Figure 18.4b gives that of the Co^{2+}-doped films. As indicated in Figure 18.4a, the transmittance was least for films grown by inclusion of 0.1M of the copper impurities which exhibited the highest absorbance as indicated earlier. For films grown with copper impurities $\geq 0.2M$, the transmittances were higher than that of the undoped films. The variation in transmittance due to the doping environments is an indication that the films have been tuned to more versatility in device designs including optoelectronic applications and mirrors. Films grown with the cobalt impurities exhibited higher transmittances compared to the undoped films. This behaviour is as expected, since the absorbance of the films was lower than that of the undoped layers (see Figure 18.3b). However, the transmittances of the films grown with the copper impurities were relatively higher compared to those grown with the cobalt impurities especially for films grown at concentration ≥0.2M. Conversely, the transmittance of the film grown by 0.1M of the cobalt impurities was >20% of the copper-doped films.

The optical absorption coefficient (α), energy band gap (Eg), extinction coefficient (k), refractive index (n), dielectric constants (ε), and electrical resistivity (σe) were deduced using relevant equations from the literature (Pankove 1971; Nwofe, Robert, and Agbo 2018; Ikhioya et al. 2020; Nwofe and Sugiyama 2020). Accordingly the equations are

$$\alpha = \frac{1}{t} \ln\left(\frac{(1 - R)^2}{T} \right) \tag{18.8}$$

$$(\alpha h\nu)^y = B\left(h\nu - E_g \right) \tag{18.9}$$

$$k = \frac{\alpha\lambda}{4\pi} \tag{18.10}$$

$$n = \frac{(1 + \sqrt{R})}{(1 - \sqrt{R})} \tag{18.11}$$

$$\varepsilon = (n + ik)^2 = \varepsilon_i + \varepsilon_r \tag{18.12}$$

$$\sigma_e (\Omega. \, cm^{-1}) = \frac{2\pi}{\lambda nc} \tag{18.13}$$

As indicated in equations (18.8) and (18.9), α, T, retain their meanings, R is the reflectance in percentage, t is the film thickness, h is the Planck's constant, ν is the frequency of the incident radiation, Eg is the energy band gap, y is an index that gives information on the nature of the transition, and B is an energy-independent constant that depends on the effective mass of the holes and electrons, and refractive index of the materials, respectively (Pankove 1971). The values of the optical absorption coefficient show α > 10^4 cm^{-1} in all the films independent of the doping components. This is a clear indication that the films will make good candidates as absorber layers in thin-film heterojunction solar cell applications. Figure 18.5a shows the plots for $(\alpha h \, \nu)^2$ vs $h\nu$ for Cu^{2+}-doped films while Figure 18.5b shows the plots for $(\alpha h \, \nu)^2$ vs $h\nu$ for films grown with cobalt impurities. The results indicate that the energy band gap was direct and ranged between 1.55 eV and 1.68 eV in the former and from 1.50 eV to 1.60 eV in the latter. The summary of the values of the energy band gap at the respective doping concentrations in the copper and cobalt environments are shown on Table 18.1. The results show that the energy band gap varied with increase in the impurity concentration of the dopants. This behaviour was attributed to the increase in the crystallite size with corresponding increase in the impurity concentrations. The blue shift (shorter wavelengths) observed as the dopant concentration was increased clearly indicates that the doping treatments improved the quality of the films and enhanced their use in different technological applications. However, the values of the energy band gap were relatively lower in the cobalt-doped films. Rabhi and Kanzari (2011) reported a variation in energy band gap induced by annealing treatments in thermally evaporated CuSbS$_2$ thin films while Birkett et al. (2018) observed a temperature-dependence (direct gap rises from 1.608 eV to 1.694 eV between 300 K and 4.2 K) on energy band gap of CuSbS$_2$ thin films. Other authors (Rodríguez-Lazcano et al. 2005; Rath et al. 2015; Popovici and Duta 2012; Liu et al. 2014; Chen et al. 2015; Suriakarthick et al. 2015; Bincy et al. 2017; Vinayakumar et al. 2018; Bincy et al. 2019) have made similar observation on the variation of energy band gap with different deposition variables.

The extinction coefficient gives information on the variation of the electromagnetic waves as it traverses the films during optical measurements. The extinction coefficient is directly proportional to the

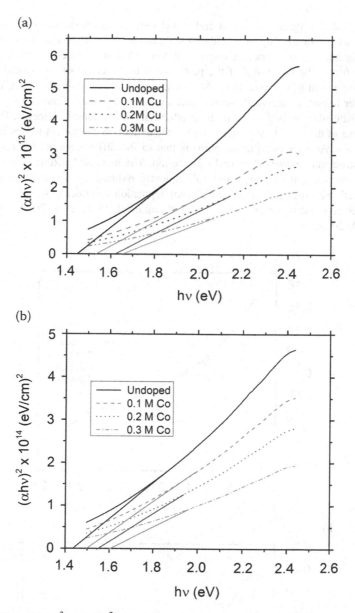

(a)

(b)

FIGURE 18.5 (a) Plots of $(\alpha h\nu)^2$ vs. $h\nu$ Cu^{2+}-doped films. (b) Plots of $(\alpha h\nu)^2$ vs. $h\nu$ Co^{2+}-doped films.

TABLE 18.1

Energy Band Gap of the Sb$_2$S$_3$ thin Films at Different Processing Conditions

Processing Conditions	Energy Bandgap (eV)	
Concentration (M)	Cu^{2+} Ions	Co^{2+} Ions
Undoped	1.43	1.43
0.1	1.55	1.50
0.2	1.62	1.58
0.3	1.68	1.60

product of the optical absorption coefficient and wavelength under consideration, where the inverse of 4π is the constant of proportionality as elucidated in equation 18.10. Figures. 18.6a and 18.6b give the plots for the variation of the extinction coefficient with photon energies for the respective doping treatments on the films. The behaviour of the plots is due to the effect of the optical absorption coefficient in the formula used to compute the values. The values of the extinction coefficient in the films treated with copper impurities are very much reduced compared to that of the cobalt-doped films. This behaviour was attributed to the difference in the conductivity of the doping elements. This will affect the absorbing potentials of the films to significantly alter the extinction coefficient values as observed in the present investigation. Another possible scenario is that of the differences in texturing in the films as indicated in the structural analysis (Figure 18.2) section. The increased texturing in the copper-doped films triggered increased transmittance and subsequently reduced the extinction coefficient values. Variation in extinction coefficient caused by different deposition variables in same/related chalcogenide thin films has been reported by other authors in the literature (Lakhdar et al. 2014; Vinayakumar et al. 2018; Onyia et al. 2020).

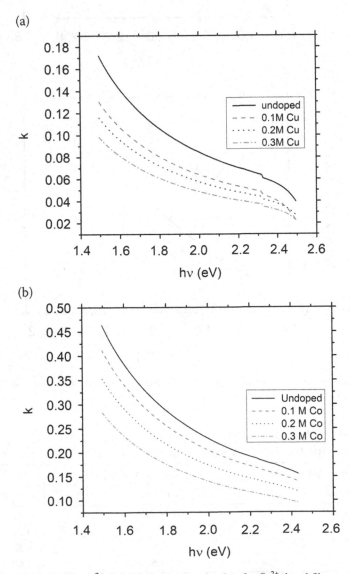

FIGURE 18.6   (a) k vs. $h\nu$ plots for Cu^{2+}-doped films. (b) k vs. $h\nu$ plots for Co^{2+}-doped films.

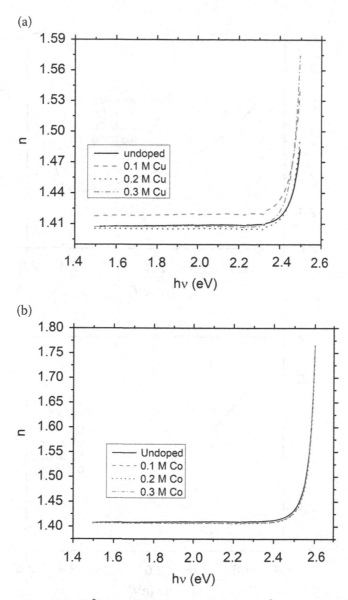

(a)

(b)

FIGURE 18.7 (a) *n* vs. *hv* plots for Cu^{2+}-doped films. (b) *n* vs. *hv* plots for Co^{2+}-doped films.

Figure 18.7a shows the variation of the refractive index (n) with photon energy (hv) plots for Cu^{2+}-doped films, while Figure 18.7b gives that of the Co^{2+}-doped films. The behaviour of the plots is as expected as the refractive indices increase with increasing photon energies, which implies that *n* increases with decreasing wavelength. This behaviour is usually attributed to the effect of free carrier concentration at the longer wavelength regime. Further, the values of the refractive indices were higher in the films doped with cobalt impurities, and decreased uniformly with increase in the concentration of the Co^{2+} impurities. This behaviour was due to the reduced texturing in cobalt-doped films as indicated earlier. Variation of the refractive indices induced by doping treatment, annealing effects, and related phenomena has been reported by other research groups (Lakhdar et al. 2014; Birkett et al. 2018; Onyia et al. 2020).

The dielectric constant ε was evaluated from the optical data using equation 18.12 where *n* and *k* retains their meanings, ε_i and ε_r are the imaginary and real dielectric constants, respectively. Figures 18.8a and 18.8b give the plots for the variation of the dielectric constant with photon energies

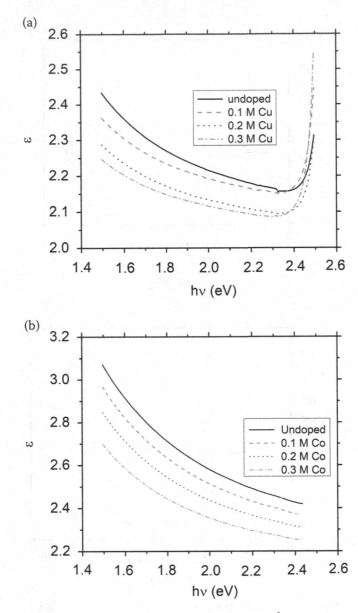

FIGURE 18.8 (a) ε vs. $h\nu$ plots for Cu^{2+}-doped films. (b) ε vs. $h\nu$ plots for Co^{2+}-doped films.

for the different doping environments. The values of the dielectric constant give an insight into the applicability of materials in different device designs. Generally the graphs show that the dielectric constants decreased with increase in the doping concentration in the respective doping environments. However, films grown in the Co-doped environment exhibited more uniform trend with relatively higher values of the dielectric constants. The values of the dielectric constants are within the range suitable for application of the films in devices with low-capacitance requirements.

The electrical resistivities (inverse of conductivity) of the films were evaluated from the optical measurements using relevant equations from the literature as indicated in equation 18.13. In equation 18.13, λ is the wavelength in nanometres, C is the speed of light in vacuum, n retains its meaning, and 2π is a constant. The electrical resistivity is inversely related to the product of the wavelength and the refractive index where $2\pi/C$ is the constant of proportionality. Figure 18.9a shows the plots for the change of the electrical resistivity with photon energy in the Cu^{2+}-doped films, while Figure 18.9b

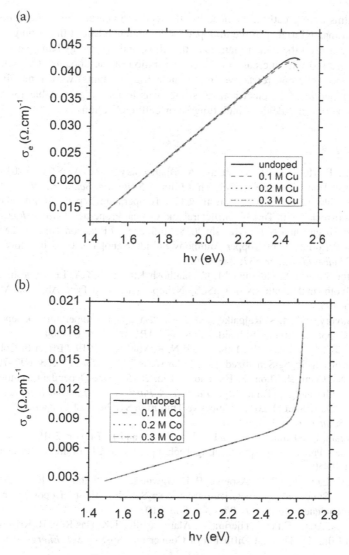

FIGURE 18.9   (a) σ_e vs. $h\nu$ plots for Cu^{2+}-doped films. (b) σ_e vs. $h\nu$ plots for Co^{2+}-doped films.

shows the plots for the change of electrical resistivity with photon energy in the co-doped layers. The results show that the resistivity was of the order of 10^3 Ωcm in both doping environments and increased with decreasing wavelength (increasing photon energies). The resistivities of the films are within the values suitable for applications in semiconductor devices and related optoelectronic applications. The resistivity values obtained in this work are within the range reported by other authors in same/related chalcogenides in the literature (Sze 1985; Yu et al. 2015; van Embden et al. 2102; Ramasamy et al. 2014b; Obodo et al. 2019; Nwanya et al. 2015; Nwanya et al. 2016; Ezema et al. 2007; Nwofe 2015; Islam and Thakur 2020).

18.4 Conclusion

Structural and optical properties of thin films of chemically deposited Co-doped antimony sulphide grown by varying the concentration of the dopants as a pioneering report was investigated, and compared with that of Cu-doped films grown under same deposition conditions. The results show that the properties of the films were greatly influenced by the deposition parameters. The doped films exhibited reduced texturing compared to that of the undoped layers, independent of the doping environments. The doping treatments increased the transmittances of the films, and the energy band gaps values were direct with values that increased with increase in the concentration of the dopants. The values of the optical constants (extinction coefficient, refractive index, dielectric constants) were typically higher for films doped with Co^{2+} impurities. The electrical resistivity was in the range suitable for application of the films' different optoelectronic devices including solar cell applications.

REFERENCES

Azizur Rahman, A., E. Hossain, H. Vaishnav, A. Bhattacharya, and A. Sarma. 2019. "Laser Induced Structural Phase Transitions in Cu_3SbS_4 Thin Films." *Semicond. Sci. Technol.* 34: 105026.

Bincy, J., G. Genifer Silvena, and A. Leo Rajesh. 2017. "Temperature Dependent Solvothermal Synthesis of Cu-Sb-S Nanoparticles with Tunable Structural and Optical Properties." *Mater. Res. Bull.* 95: 267–276.

Bincy, J., G. Genifer Silvena, and A. Leo Rajesh. 2018. "Influence of Reaction Time on the Structural, Optical and Electrical Performance of Copper Antimony Sulfide Nanoparticles Using Solvothermal Method." *Physica B: Condensed Matter* 537: 243–250.

Bincy, J., G. Genifer Silvena, S. Hussain, M. C. Santhosh Kumar, and A. Leo Rajesh. 2019. "Surfactant-Mediated Solvothermal Synthesis of $CuSbS_2$ Nanoparticles as p-Type Absorber Material." *Indian J. Phys.* 93: 185–195.

Birkett, M., C. N. Savory, and M. K. Rajpalke et al. 2018. "Band Gap Temperature-Dependence and Exciton-like State In Copper Antimony Sulphide, $CuSbS_2$." *APL Mater.* 6: 084904.

Cerdán-Pasarán, A., T. López-Luke, X. Mathew, and N. R. Mathews. 2019. "Effect of Cobalt Doping on the Device Properties of Sb_2S_3-Sensitized TiO_2 Solar Cells." *Solar Energy* 183: 697–703.

Chen, K., J. Zhou, W. Chen, P. Zhou, F. He, and Y. Liu. 2015. "Size-Dependent Synthesis of $Cu_{12}Sb_4S_{13}$ Nanocrystals with Bandgap Tunability." *Particle & Particle Systems Characterization* 32: 999–1005.

Cho, A., S. Banu, and K. Kim et al. 2017. "Selective Thin Film Synthesis of Copper-Antimony-Sulfide Using Hybrid Ink." *Solar Energy* 145: 42–51.

Dekhil, S., H. Dahman, F. Ghribi, H. Mortada, N. Yaacoub and L. El Mir. 2019. "Study of $CuSbS_2$ Thin Films Nanofibers Prepared by Spin Coating Technique Using Ultra Pure Water as a Solvent." *Mater. Res. Exp.* 6: 086450.

Ezema, F. I., A. B. C. Ekwealor, P. U. Asogwa, P. E. Ugwuoke, C. Chigbo, and R. U. Osuji. 2007. "Optical Properties and Structural Characterizations of Sb_2S_3 Thin Films Deposited by Chemical Bath Deposition Technique." *Turkish J. Phys.* 31: 205–210.

Garza, C., Shaji. S., A. Arato, E.Perez Tijerina, G. Alan Castillo, T.K. Das Roy, B. Krishnan. 2011. "p-Type $CuSbS_2$ Thin Films by Thermal Diffusion of Copper into Sb_2S_3." *Sol. Energy Mater. Solar Cells* 95: 2001–2005.

Ghanwat, V. B., S. S. Mali, and S. D. KharadeNita B. Pawar, Satish V. Patil, Rahul M. Mane, Pramod S. Patil, Chang Kook , Hong, and Popatrao N. Bhosale. 2014. "Microwave Assisted Synthesis, Characterization and Thermoelectric Properties of Nanocrystalline Copper Antimony Selenide Thin Films." *RSC Advances* 4: 51632–51639.

Grozdanov, I. 1994. "A Simple and Low-Cost Technique for Electroless Deposition of Chalcogenide Thin Films." *Semicond. Sci. Technol.* 9: 1234.

Ikhioya, I. L., A. C. Nkele, S. N. Ezema, M. Maaza, and F. Ezema. 2020. "A Study on the Effects of Barying Concentrations on the Properties of Ytterbium-Doped Cobalt Selenide Thin Films." *Opt. Mater.* 101: 109731.

Ishaq, M., Deng, H., and U. Farooq, Huan Zhang, Xiaokun Yang, Usman Ali Shah, and Haisheng Song. 2019. "Efficient Copper-Doped Antimony Sulfide Thin-Film Solar Cells via Coevaporation Method." *Solar RRL* 3: 1900305.

Islam, M. T., and A. K. Thakur. 2020. "Two Stage Modelling of Solar Photovoltaic Cells Based on Sb_2S_3 Absorber with Three Distinct Buffer Combinations." *Solar Energy* 202: 304–315.

Krishnan, B., S. Shaji, and R. E. Ornelas. 2015. "Progress in Development of Copper Antimony Sulfide Thin Films as an Alternative Material for Solar Energy Harvesting." *J. Mater. Sci.: Mater. Electron.* 26: 4770–4781.

Kumar. B. H., S. Shaji, and M. S. Kumar. 2020. "Effect of Substrate Temperature on Properties of Co-evaporated Copper Antimony Sulfide Thin Films." *Thin Solid Films* 697: 137838.

Lakhdar, M. H., B. Ouni, and M. Amlouk. 2014. "Thickness Effect on the Structural and Optical Constants of Stibnite Thin Films Prepared by Sulfidation Annealing of Antimony Films." *Optik – Int. J. Light Electron Opt.* 125: 2295–2301.

Liu, S., L. Chen, L. Nie, X. Wang, and R. Yuan. 2014. "The Influence of Substrate Temperature on Spray-deposited $CuSbS_2$ Thin Films." *Chalcogenides Letts* 11: 639–644.

Liu, Z., J. Huang, J. Han, T. Hong, J. Zhang, and Z. Liu. 2016. "$CuSbS_2$: A Promising Semiconductor Photo-absorber Material for Quantum Dot Sensitized Solar Cells." Phys. Chem. Chem. Phys. 18: 16615–16620.

Loranca-Ramos, F. F., Diliegros-Godines, C. J., González, R. S., and M. Pal. 2018. "Structural, Optical and Electrical Properties of Copper Antimony Sulfide Thin Films Grown by a Citrate-Assisted Single Chemical Bath Deposition." *Appl. Surf. Sci.* 427: 1099–1106.

Maghraoui-Meherzi, H., T. B. Nasr, N. Kamoun, and M. Dachraoui. 2010. "Structural, Morphology and Optical Properties of Chemically Deposited Sb2S3 Thin Films." *Physica B: Condensed Matter*, 405: 3101–3105.

Nwanya, A. C., P. R. Deshmukh, R. U. Osuji, M. Maaza, C. D. Lokhande, and F. I. Ezema. 2015. "Synthesis, Characterization and Gas-Sensing Properties of SILAR Deposited ZnO-CdO Nano-Composite Thin Film." *Sensors and Actuators B: Chemical* 206: 671–678.

Nwanya, A. C., D. Obi, K. I. Ozoemena, Rose U. Osuji, Chawki Awada, Andreas Ruediger, Malik Maaza, Federico Rosei, and Fabian I. Ezema. 2016. "Facile Synthesis of Nanosheet-like CuO Film and Its Potential Application as a High-performance Pseudocapacitor Electrode." *Electrochimica Acta* 198: 220–230.

Nwofe, P. A. 2015. "Influence of Film Thickness on the Optical Properties of Antimony Sulphide Thin Films Grown by the Solution Growth Technique." *European J. Appl. Eng. Sci. Res.* 4: 1–6

Nwofe, P. A., B. J. Robert, and P. E. Agbo. 2018. "Influence of Processing Parameters on the Structural and Optical Properties of Chemically Deposited Zinc Sulphide Thin Films." *Mater. Res. Exp.* 5: 106405.

Nwofe, P. A., and M. Sugiyama. 2020. "Microstructural, Optical, and Electrical Properties of Chemically Deposited Tin Antimony Sulfide Thin Films for Use in Optoelectronic Devices." *Phys. Status Solidi A*, 1900881.

Obodo, R. M., A. C. Nwanya, and A. B. C. Ekwealor. 2019. "Influence of pH and Annealing on the Optical and Electrochemical Properties of Cobalt (III) Oxide (Co_3O_4) Thin Films." *Surfaces and Interfaces* 16: 114–119.

Onyia, A. I., M. N. Nnabuchi, and A. I. Chima. 2020. "Electrical and Optical Characteristics of Copper Antimony Sulphide Thin Films Fabricated in Chemical Baths of Different Growth Media." *Amer. J. Nanosci.* 5: 1–5.

Ornelas-Acosta, R. E., S. Shaji, D. Avellaneda, G. A. Castillo, T. D. Roy and B. Krishnan. 2015. Thin Films of Copper Antimony Sulfide: A Photovoltaic Absorber Material." *Mater. Res. Bull.* 61: 215–225.

Osuwa, J. C., and R. U. Osuji. 2011. "Analysis of Electrical and Microstructural Properties of Annealed Antimony Sulphide (Sb_2S_3) Thin Films." *Chalcogenide Letts* 8: 51–57.

Pankove, J. I. 1971. *Optical Processes in Semiconductors.* Englewood Cliffs, New Jersey: Prentice-Hall.

Popovici, I., and A. Duta. 2012. "Tailoring the Composition and Properties of Sprayed $CuSbS_2$ Thin Films by Using Polymeric Additives." *Int. J. Photoenergy* 2012: 962649.

Rabhi, A., and M. Kanzari. 2011. "Effect of Air Annealing on $CuSbS_2$ Thin Film Grown by Vacuum Thermal Evaporation." *Chalcogenides Letts* 8: 255–262.

Ramasamy, K., H. Sims, W. H. Butler, and A. Gupta. 2014a. "Mono-, Few-, and Multiple Layers of Copper Antimony Sulfide ($CuSbS_2$): A Ternary Layered Sulfide." *J. Amer. Chem. Soc.* 136: 1587–1598.

Ramasamy, K., H. Sims, W. H. Butler, and A. Gupta. 2014b. "Selective Nanocrystal Synthesis and Calculated Electronic Structure of All Four Phases of Copper–Antimony–Sulfide." *Chemistry of Materials* 26: 2891–2899.

Ramos Aquino, J. A., D. L. Rodriguez Vela, S. Shaji, D. A. Avellaneda, and B. Krishnan. 2016. "Spray Pyrolysed Thin Films of Copper Antimony Sulfide as Photovoltaic Absorber." *Phys. Stat. Solidi (c)* 13: 24–29.

Rath, T., A. J. MacLachlan, M. D. Brown, and S. A. Haque. 2015. "Structural, Optical and Charge Generation

Properties of Chalcostibite and Tetrahedrite Copper Antimony Sulfide Thin Films Prepared from Metal Xanthates." *J. Mater. Chem. A* 3: 24155–24162.

Rodríguez-Lazcano, Y., M. T. S. Nair, and P. K. Nair. 2005. "Photovoltaic p-i-n Structure of Sb_2S_3 and $CuSbS_2$ Absorber Films Obtained via Chemical Bath Deposition." *J. Electrochem. Soc.* 152: G635–G638.

Satoshi, S., K. Horita, M. Yuasa , Tooru Tanaka , Katsuhiko Fujita, Yoichi Ishiwata, Kengo Shimanoe, and Tetsuya Kida. 2015. "Synthesis of Copper–Antimony-Sulfide Nanocrystals for Solution-Processed Solar Cells." *Inorganic Chemistry* 54: 7840–7845.

Shaji, S., V. Vinayakumar, B. Krishnan, and Jacob Johny. 2019. "Copper Antimony Sulfide Nanoparticles by Pulsed Laser Ablation in Liquid and Their Thin Film for Photovoltaic Application." *Appl. Surf. Sci.* 476: 94–106.

Sivagami, A. D., K. Biswas, and A. Sarma. 2018. "Orthorhombic $CuSbS_2$ Nanobricks: Synthesis and Its Photo Responsive Behaviour." *Mater. Sci. Semicond. Proc.* 87: 69–76.

Suriakarthick, R., V. N. Kumar, T. S. Shyju, and R. Gopalakrishnan. 2015. "Effect of Substrate Temperature on Copper Antimony Sulphide Thin Films from Thermal Evaporation." *J. Alloys Compds* 651: 423–433.

Sze, S. M. 1985. *Semiconductor Devices: Physics and Technology*. New York: John Wiley & Sons.

van Embden, J., and Y. Tachibana. 2012. "Synthesis and Characterisation of Famatinite Copper Antimony Sulfide Nanocrystals." *J. Mater. Chem.* 22: 11466–11469.

Vinayakumar, V., S. Shaji, D. Avellaneda, J. A. Aguilar-Martínez, and B. Krishnan. 2018. "Copper Antimony Sulfide Thin Films for Visible to Near Infrared Photodetector Applications." *RSC Advances* 8: 31055–31065.

Vinayakumar, V., S. Shaji, D. Avellaneda, J. A. Aguilar-Martínez, and B. Krishnan. 2019. "Highly Oriented $CuSbS_2$ Thin Films by Rapid Thermal Processing of Pre-annealed Sb_2S_3-Cu Layers for PV Applications." *Mater. Sci. Semicond. Proc.* 91: 81–89.

Wang, W., G. Zhi, J. LiuL. Hao, Lu Yang, Yijie Zhao, and Yingfei Hu. 2020. "Effect of PVP Content on Photocatalytic Properties of $CuSbS_2$ Particles with Chemical Etching." *Journal of Nanoparticle Research* 22: 294.

Yıldırım, M., A. Kocyigit, A. Sarılmaz, and F. Ozel. 2019. "The Effect of the Triangular and Spherical Shaped $CuSbS_2$ Structure on the Electrical Properties of $Au/CuSbS_2/p$-Si Photodiode." *J. Mater. Sci. Mater. Electron.* 30: 332–339.

Yu, L., M. Han, Y. Wan, J. Jia, and G. Yi. 2015. "Synthesis of Stoichiometric $AgSbS_2$ Nanospheres via One-Step Solvothermal Chemical Process." *Mater. Letts.* 161: 447.

Zhang, H., Q. Xu, and G. Tan. 2016. "Physical Preparation and Optical Properties of $CuSbS_2$ Nanocrystals by Mechanical Alloying Process. Electron." *Mater. Letts.* 12: 568–573.

19

Research Progress in Synthesis and Electrochemical Performance of Bismuth Oxide

Sylvester M. Mbam[1], Raphael M. Obodo[1,2,3], Assumpta C. Nwanya[1,4,5], A. B. C. Ekwealor[1], Ishaq Ahmad[2,3], and Fabian I. Ezema[1,4,5]

[1]*Department of Physics and Astronomy, University of Nigeria, Nsukka, Enugu State, Nigeria*
[2]*National Center for Physics, Islamabad, Pakistan*
[3]*NPU-NCP Joint International Research Center on Advanced Nanomaterials and Defects Engineering, Northwestern Polytechnical University, Xi'an, China*
[4]*Nanosciences African Network (NANOAFNET) iThemba LABS-National Research Foundation, Somerset West, Western Cape Province, South Africa*
[5]*UNESCO-UNISA Africa Chair in Nanosciences/Nanotechnology, College of Graduate Studies, University of South Africa (UNISA), Muckleneuk Ridge, Pretoria, South Africa*

19.1 Introduction

The unique properties associated with nanostructured materials have over the years made it a highly beneficial research field, offering wide applications. The basis of these properties is, therefore, mainly due to their shapes, sizes, and length (Gao et al. 2011). This implies that designing the preferred morphologies of these materials is of great importance (Lu et al. 2012).

Bismuth, being a semimetal, has good electronic properties such as high charge-carrier mobility, low carrier density, and high free mean path; it also shows anisotropic properties on the direction of its Fermi level, thus displaying a great magnetoresistance feature (Du et al. 2005); these intrinsic properties originate from the basic spatial arrangement of its atoms (Postel and Duñach 1996). However, bismuth can change to a semiconductor when the thickness of its film ranges between 28 nm (Gribanov et al. 2011) and 55 nm (nanowires) (Konopko et al. 2010). Nanostructured bismuth materials possess excellent intrinsic properties. It has a quantum effect that is highly needed in optical devices, power generators, magnetic sensing, and thermoelectric coolers (Du et al. 2005; Gao et al. 2011) Being less toxic, bismuth films can serve as an electrode in a biosensor to detect nitrogen in organic compounds (Timur and Anik 2007), even heavy metals (Lee et al. 2010), and as suitable electrodes materials in place of mercury which is highly toxic. Bismuth materials have also been reviewed to be a suitable disinfectant agent in cosmetics and pharmaceuticals (Bedoya et al. 2012); it serves as a refrigerant and, due to its high charge retention capacity, is used in automotive lead-acid batteries (Chen et al. 2001).

Bismuth oxide (Bi_2O_3), a semiconductor of the p-type category, is characterized by large dielectric permittivity, large refractive index, highly photoconductive and photoluminescent behaviour, emanating from its large band gap structure which is ascribed by sequencing sheets of bismuth atoms being located in a parallel direction to the (100) plane, as the oxide ions are domiciled in the direction of c-plane (Shu et al. 2016). These intrinsic properties have made this material very suitable for applications in

DOI: 10.1201/9781003145585-19

sensors and catalysis (Hwang et al. 2008; Pandey et al. 2018), including radiation shielding/coatings (Chen et al. 2019; Al-Kuhaili et al. 2020), optoelectronics (Condurache-Bota 2018), fuel cells (Gong et al. 2011), energy harvesting/storage (Mahmoud and Al-Ghamdi 2011; Miller and Bernechea 2018), also for therapeutic applications (Zulkifli et al. 2018). These wide applications are also attributed to the stable nature of this metal oxide (Dubey, Nemade, and Waghuley 2012).

There have been five crystal forms (polymorphs) of bismuth oxide (Bi_2O_3) reported by several authors; the variants are due to the difference in their annealing temperatures. The phases include the alpha, beta, gamma, delta, omega phases (α-, β-, γ-, δ-, ω-Bi_2O_3), respectively. Each phase has varying properties. α-Bi_2O_3 is monoclinic while δ- Bi_2O_3 is a face-centered cubic, and is respectively stable at low and high temperatures. β- Bi_2O_3 is tetragonal, γ- Bi_2O_3 is body-centered cubic, ω- Bi_2O_3 is triclinic; and all are metastable (Qiu et al. 2011; Wu et al. 2011). The wide range applications of Bi_2O_3 materials have attracted an increasing interest/deeper research on its phase transition and characterizations.

Synthesis of Bi_2O_3 is cost-effective and environmental friendly, the methods also involve the general techniques for preparing other metal oxides. In the laboratory, Bi_2O_3 can be synthesized through any of the following methods: chemical methods (which can be sol gel, hydrothermal/solvothermal, coprecipitation), photolysis method, electrodeposition, microwave-assisted synthesis, polyol process, and vacuum impregnation/roasting in a quartz tube, also through Inkjet printing, simple deposition, and conventional solid-state approach. Large-scale production of Bi_2O_3 mostly employs the chemical method/solid-state technique (Kelly and Ojebuoboh 2002; Brezesinski et al. 2010; Jhaa, Pasrichab, and Ravic 2015). To obtain pure Bi_2O_3 with a very low amount of hydroxide (OH) groups, low pH and high annealing temperature are generally required (Bahmani et al. 2012). The specific surface area and other physical properties vary depending on the synthesis method employed. Among the trending applications of Bi_2O_3 materials, the energy storage efficiency of its nanostructures has been reported by many authors. This deep interest can also be linked to its high theoretical and specific capacitance. However, high capacitance, multiple redox activity, and high electrochemical stability, which are common to metal oxides, are among the intrinsic factors for potential electrode materials for charge retention purposes (Obodo et al. 2020). The layered arrangement of the bismuth atoms has a role in offering redox sites during the electrochemical reaction. Thus researchers have found Bi_2O_3 as high-performance energy storage materials mostly in supercapacitors and batteries (as an electrode-active material).

We hereby present a brief review of the basic properties, synthesis approaches, and various electrochemical reports on Bi_2O_3 nanostructures, and its composites. This chapter, in the essence, offers concise information and clarity on the potentials and the ever-increasing interest in the Bi_2O_3 nanostructures.

19.2 Phases and Properties

Different phases of Bi_2O_3 were first analyzed by Narang et al. (1994) and Hardcastle and Wachs (1991). They include α-Bi_2O_3 (monoclinic), β-Bi_2O_3 (tetragonal), γ-Bi_2O_3 (body-centered cubic), δ-Bi_2O_3 (cubic), ω-Bi_2O_3 (triclinic). The polymorphs occur due to temperature variations – α-Bi_2O_3 turns into δ-Bi_2O_3 at about 729°C; when δ-Bi_2O_3 phase cools, it turns to β-Bi_2O_3 at 650°C or the γ-Bi_2O_3 at 639°C. Thus, α-phase is of low temperature while δ-phase is of high temperature which are both stable. The other phases (β-Bi_2O_3 and γ-Bi_2O_3) have high temperatures and thus are metastable, but can be made stable at low temperature by adding impurities. These mixed phases in Bi_2O_3 mainly hinder the proper control of the properties, together with the applications. Precisely, Bi_2O_3 thin films and nanowires reported by Leontie et al. (2002) and Kumari et al. (2007), respectively, confirmed a mixed phase of α- and β-Bi_2O_3. Due to the high technological importance of this material, there has been an increased interest in the synthesis and characterization with the applications involving micro, and stretched structures like the β-Bi_2O_3 nanorods and nanowires. These are mainly obtained by chemical vapour deposition (Moniz et al. 2012), (Shen et al. 2007), and thermal evaporation of a bismuth compound (Liu et al. 2015). Phase nanowires/micro-nanoribbons are also of huge interest; these can, respectively, be obtained through chemical routes (Gou et al. 2009) and vapour transport techniques (Ling et al. 2010). Bi_2O_3 core incorporating a ZnO shell has been synthesized into coaxial nanorods which modified the optical features of the Bi_2O_3 material (Jin et al. 2010). Various optical applications of Bi_2O_3 structures have over the years generated deep research into its luminescence performance (Vila et al. 2013). However, there is still a lack of comprehensive details about the band gap/effects of intrinsic defects on the luminescence radiation. Through optical absorption, α- and β- phases have been reported by Hashimoto et al. (2016), Jalalah et al. (2015), Sun et al. (2012), and Leontie et al. (2002) to possess band gap values of about 2.40–3.0 eV, respectively. Selvapandiyan and Sathiyaraj (2019) also reported direct band gap values of 2.51–3.20 eV and indirect band gap values of about 1.53–1.70 eV, respectively, for the α- and β- phases, with crystallite sizes of about 17 nm to 64 nm, obtained through sol gel technique. An average crystallite size of about 50–80 nm was previously obtained by Jhaa, Pasrichab, and Ravic (2015) using the urea–nitrate process of synthesizing bismuth oxide nanoparticles. Nevertheless, the studies by Gou et al. (2009), through optical absorption of α-Bi_2O_3 nanowires of varying sizes for application in gas sensing, recorded a wide band gap range of up to 3.20 eV. This is, however, in contrast with the values gotten by Xiong et al. (2008) and Dong and Zhu (2003) through a luminescence study of the Bi_2O_3 nanocrystals which recorded a photoluminescence peak of about 2.8 eV and 3.10 eV, respectively; in the α-phase, recombination of band gap can be the cause of this variations. Luminescence peaks occurring at lower energy levels were explained by Bordun et al. (2008), which they attributed to a radiation by Bi ions at varying states. Precise information on the effect of defects and luminescence peak was recorded by Vila et al. (2012), where a cathode luminescence peak of 2.10 eV observed in α-Bi_2O_3 was associated to oxygen defects, while the one at 3.25 eV was related to band gap recombination. In the study of Bi_2O_3, Raman spectroscopy is an effective tool. This identifies the distinct polymorphs and their instinct features/defects (Denisov et al. 1997; Salazar-Pérez et al. 2005; Rubbens et al. 2007; Vila et al. 2012).

Recently, Wang et al. (2019) studied the size-dependent thermodynamic behaviours of α- to δ-Bi_2O_3 phases with their variations, synthesized through hydrothermal techniques. They employed a core-shell model for crystal phase transition and melting of nanospheres to theoretically analyse the proportionality of the phases in varying sizes and their thermodynamic parameters. They recorded a similar result with the experimental analysis, which affirms a reciprocal relationship between the particle diameter with temperature variation and the thermodynamic properties of the transition from the α- to δ-phases. This technique can generally give a piece of relevant information for the synthesis and applications of Bi_2O_3 nanoparticles and other microstructures.

Besides the crystalline phases of Bi_2O_3, there also exist phases that are non-stoichiometric. These include $Bi_2O_{2.33}$ and $Bi_2O_{2.75}$ (Schuisky and Hårsta 1996; Gujar et al. 2005; Fang et al. 2010). However, these two phases have very few reports about them as they mainly exist as an impurity in the Bi_2O_3 thin films (Leontie et al. 2001; Leontie et al. 2002; Choudhary and Waghuley 2018). The production of these phases in large-scale using easy and available techniques is still challenging.

Generally, the sizes and behaviours of nanoparticles vary widely from those of bulk materials; with their sizes ranging about less than 100 nm and thus larger surface area. In many applications involving bismuth oxide nanoparticles, intrinsic properties, such as the band gap, size/morphology, photo-conductive/luminescent properties, with the phase transition are the major considerations depending on the application.

19.3 Synthesis Methods

There have been different synthesis techniques employed by researchers for the production of Bi_2O_3 nanostructures; these include the already available efficient methods for synthesizing other metal oxides. They are categorized into chemical methods (which include sol gel, hydrothermal/solvothermal method, coprecipitation), photolysis, electrodeposition, microwave-assisted synthesis, polyol process, and vacuum impregnation/roasting in a quartz tube. These techniques can synthesize Bi_2O_3 in different forms (such as nanorods, nanotubes/nanofibres, nanowires).

The chemical methods which include sol gel, hydrothermal, solvothermal, coprecipitation methods are the most common routes for synthesizing Bi_2O_3 nanoparticles. These methods can suitably control both the morphology and chemical composition of the nanoparticles, with a reduction in their sintering temperature as well, as it requires a metal ion from a polyfunctional carboxyl acid, like citric acid. Mallahi et al. (2014) prepared Bi_2O_3 nanopowders through sol gel process, starting with an equal molar ratio of a mixture of bismuth nitrate $(Bi(NO_3)_3 \cdot 5H_2O)$ in citric acid. The process generally involves the formation of a uniform solution, gel formation, and drying/calcination of the gel to obtain a raw powder. A surfactant (PEG600) was introduced to the solution to curb agglomeration. This method utilizes a lower sintering temperature when compared to the conventional solid-state technique (Yu et al. 2018); the particles obtained recorded an average size of 20 nm. A monoclinic phase Bi_2O_3 (nanowires) having a diameter of 40 nm was synthesized by Wu et al. (2011) using a hydrothermal approach. This employed a precipitation reaction of $Bi(NO_3)_3 \cdot 5H_2O$ and Na_2SO_4 in distilled water to produce $Bi_2O(OH)SO_4$, followed by a hydrothermal mechanism of $Bi_2O(OH)SO_4$ and NaOH under 120°C for 12 h to produce the monoclinic phase Bi_2O_3 nanowires; this procedure required no addition of any surfactant. To obtain complete Bi_2O_3 nanowires, the effect of Na_2SO_4 in the reaction was examined by conducting a control experiment with no Na_2SO_4; it was therefore recorded that in the absence of Na_2SO_4, Bi_2O_3 with a microrods morphology was obtained as shown through the scanning electron microscope image in Figure 19.1f, while Bi_2O_3 nanowires was obtained in the reaction having Na_2SO_4 (Figure 19.1a). The authors recorded that Na_2SO_4 enhanced the formation of $Bi_2O(OH)SO_4$ which in turn lowers the reaction rate of $Bi(NO_3)_3$ to yield Bi_2O_3 nanowires that have a close morphology to the $Bi_2O(OH)SO_4$ as indicated by SEM and TEM images in Figure 19.1. The hydrothermal method is highly cost-effective and simple, with a huge prospect for industrial/mass production of bismuth oxide (Bi_2O_3) nanostructures. Mozdianfard and Masoodiyeh (2018) optimized the supercritical water hydrothermal technique of bismuth Bi_2O_3 synthesis. This involved estimating the influence of the experimental factors such as temperature, pH, and concentration on the particle size and distribution, using two mathematical tools (Response Surface Methodology and population balance equation) to reduce the number of experiments and validate the variables. They reported a tetragonal Bi_2O_3 nanocrystal with average particle size and

FIGURE 19.1 Field emission scanning electron microscope images of (a) $Bi_2O(OH)SO_4$, as-prepared sample under (b) room temperature, (c) 120°C. High resolution transmission electron microscope images of (d)–(e) sample produced under 120°C, (f) scanning electron microscope image of Bi_2O_3 obtained at 120°C without the addition of Na_2SO_4 (Wu et al. 2011). Reprinted with permission from Ref. Wu et al. (2011), copyright 2011, Elsevier.

distribution that aligned perfectly with the mathematical model analysis. Solvothermal approach to producing Bi_2O_3 nanostructures employs the same steps as the hydrothermal technique, but here a chemical solvent (such as N,N-dimethyl formamide (DMF), toluene, ethyl glycol, ethanol, etc.) can rather be chosen in the place of water, as a solvent for the reactants. Yan et al. (2014) synthesized α- and β-Bi_2O_3 microspheres using a solvothermal approach. The step involved dissolving Bi $(NO_3)_3 \cdot 5H_2O$ in a mixture containing ethanol and glycerol (equal volume ratio); the mixture was hence coupled into an autoclave (Teflon-lined stainless steel) and reacted for some hours at 160°C, then cooled, washed, and dried. Two samples were calcined at 270°C and 350°C to yield the α- and β-Bi_2O_3, respectively. The β-Bi_2O_3 phase recorded a band gap of 2.36 eV, with a better photocatalytic activity while the α-Bi_2O_3 has a band gap of 2.85 eV. Yang et al. employed a precipitation method; in controlling the morphologies of Bi_2O_3 hierarchical nanostructures, this was possible through adjusting a suitable DMF/H_2O ratio. In this case, a uniform mixture of $Bi(NO_3)_3 \cdot 5H_2O$ and DMF was introduced into another mixture that contains KOH with water (H_2O). The mixture was stirred briefly, allowed to coagulate, then filtered, washed, and dried at a low temperature (60°C), hence without further thermal treatment. In this report, the quantity of H_2O affects the extent of the hydrolysis of $Bi(NO_3)_3$, thereby determining the morphology and the basic architecture of the produced Bi_2O_3 nanostructure. An energy gap in the range of 2.70 – 2.88 was recorded for the obtained Bi_2O_3 samples (Yang et al. 2014).

The electrodeposition method for synthesizing Bi_2O_3 nanostructures employs a galvanostatic mechanism using chronopotentiometry to deposit thin films on a conducting substrate (current collector). This process initiates a reduction-oxidation reaction on the electrolyte bath as a certain current passes into it, and thus enabling the transfer of bismuth ions from the electrolyte bath onto the current collector (this forms the working electrode). Gujar et al. (2005) and Fang et al. (2017) produced a bismuth oxide thin film using an electrodeposition process. $Bi(NO_3)_3$ was recorded as the preparatory material, and tartaric acid as a complexing agent. These were mixed as an electrolyte bath and pH maintained alkaline by introducing an aqueous NaOH. A potential window in the range of 0 V and 1 V at a scan rate of 20 mV/s enabled the hydroxide ions of the alkaline electrolyte solution to migrate onto the cathode (copper substrate) which has a greater potential, thereby producing a thin film of Bi_2O_3 from $Bi(OH)_2$. The paper also reported a current of 12 mA/cm^2 as the deposition current. Using a gravimetric weight difference technique, a maximum film thickness of 0.74 nm was also recorded for a 15 minutes deposition time. Bi_2O_3 nanobelts with a rippled morphology were reported by Zheng et al. (2010), employing the electrodeposition technique. The starting material contained a solution of $Bi(NO_3)_3$, Na_2EDTA, and sucrose. Applying a current density of 4 mA/cm^2 within 60 min, Bi_2O_3 nanobelts with a hierarchical rippled morphology were recorded. In this paper, the Bi_2O_3 nanobelts obtained are 250–300 nm in width, 10–30 nm thick, with a length ranging from 1–5 nm as analyzed through a transmission electron microscope (TEM). The work also realized a high specific area and electrochemical performance for the sample having a hierarchical rippled morphology than the nanobelts possessing a smooth surface.

Microwave-assisted synthesis of Bi_2O_3 nanostructures requires the use of microwave radiation on the solution containing $Bi(NO_3)_3$ as a starting material. Sonkusare et al. (2018) employed a microwave-assisted approach to produce a nanoparticle of Bi_2O_3. Their reported procedure involved the mixing of bismuth nitrate pentahydrate $Bi(NO_3)_3.5H_2O$, polyvinylpyrrolidone (PVP), and ethyl glycol, PVP being a surfactant. The homogeneous solution obtained was irradiated in a microwave oven for 10 minutes and 15 minutes, respectively, for separate samples. Afterwards, hydrazine hydrate (N_2H_4) was introduced into the mixture, cooled, and centrifuged. A precipitate having a black colour was recorded after drying under 80°C. Two samples, α-Bi_2O_3 (monoclinic micro flowers) and γ-Bi_2O_3 (cubic microspindles), were reported by the authors. The α-Bi_2O_3 (monoclinic micro flowers) which was irradiated for 10 minutes was reported to possess higher photocatalytic and antibacterial activities than the sample irradiated for 15 minutes. This was ascribed to be the result of the variation in particle size, surface area, and band gap energy of the two samples.

Another technique for synthesizing the nanostructures of Bi_2O_3 is the simple polyol process (Wang et al. 2017), where the solution containing $Bi(NO_3)_3$ was reported to be stirred under a mild temperature of about 60°C for few hours, thereafter washed thoroughly, and air-dried overnight.

In this method, no thermal treatment was reported. Vacuum impregnation/roasting in a quartz tube (Wang et al. 2014) involves the use of a vacuum oven to directly dry off a mixture containing Bi $(NO_3)_3$, followed by heat treatment/roasting at the required temperature in a muffle stove to obtain the respective Bi_2O_3 phase.

19.4 Applications

Bismuth oxide nanoparticles have gained much interest in technological and research applications. This can be attributed to its intrinsic excellent properties. The various fields requiring the applications of the Bi_2O_3 material include chemical, energy storage, sensor, catalysis/ photocatalysis, photovoltaic cells, semiconductor/electronics, and pharmaceutical.

19.4.1 Energy Storage

Over the years, various researches aimed at harnessing the electrochemical properties of Bi_2O_3 nanoparticles have been reported in many literatures. This is targeted at utilizing its high redox activity, long cycling/electrochemical stability, band gap structure/sequential layers of its atoms (Qiu et al. 2006). Bismuth oxide possesses remarkable ionic conductivity, with relatively huge specific capacitance and power density when used as an electrode material. Therefore, it has a great prospect for asymmetric supercapacitors. It has also offered excellent applications in fuel cell and battery technology, having a high theoretical capacity of 690 mA h/g (6280 mA h/cm^3) (Li et al. 2013). Owing to its multiple oxidation states, bismuth oxide has been a very good material for supercapacitor electrode. It has also shown an increased super capacitative behaviour when combined with carbon derivatives such as graphene, carbon nanotube, and activated carbon. Table 19.1 illustrates a summary of the various reports on the applications of Bi_2O_3 nanostructures in energy storage. Furthermore, the non-toxic nature and facile synthesis routes of Bi_2O_3 have also made it a nice substitute over ruthenium oxide (RuO) in applications involving energy storage.

Ultrathin Bi_2O_3 nanowires were produced by Qiu et al. (2018), using the oxidative metal vapour transport technique for a supercapacitor electrode. They obtained a specific capacitance of about 671 F/g using a current density of 2 A/g and 6M KOH electrolyte. The nanowires have a specific surface area of about 7.34 m^2/g deduced through the N_2 adsorption-desorption experiment. The nanowires also have a tetragonal crystal arrangement (metastable β-Bi_2O_3 phase) with the diameters and length that range from 10 nm to tens of micrometres. The outstanding electrochemical behaviour of this electrode can be attributed to the efficient morphology of the active materials (the ultrathin Bi_2O_3 nanowires), and significant electrode-electrolyte interaction which has a low charge transfer resistance (Rct) of 1.23 ohms. Capacitance retention of up to 76% was recorded after 3000 cycles. Figure 19.2 illustrates the electrochemical properties of the ultrathin Bi_2O_3 nanowires. Surface morphology and its effects on the electrochemical properties of the Bi_2O_3 electrode were studied by Zheng et al. The Bi_2O_3 nanobelts having a smooth surface attained a specific capacitance of 61 F/g, while the sample having a rough surface reached a specific capacitance of 250 F/g. This is due to the more available redox site present in the rippled Bi_2O_3 sample than the sample with a smooth surface (Zheng et al. 2010). A non-stoichiometric phase of bismuth oxide ($Bi_2O_{2.33}$) having an orange-like morphology; micropores in the range of 3 mm to 5 mm were synthesized by Huang et al. for utilization in supercapacitor electrodes. The $Bi_2O_{2.33}$ material offered a significant electrochemical performance recording a high specific capacitance of 893 F/g at 0.1 A/g current density. This high performance can be linked to the availability of micropores present in the electrode active material ($Bi_2O_{2.33}$), which allows the influx of electrolyte ions and optimum redox activity (Huang et al. 2013). Ma et al. (2016) synthesized the metastable phase (β-Bi_2O_3) for usage in a supercapacitors' negative electrode. A huge specific capacity value of about 871 C/g was recorded under a potential window of -1.5 V to 1.5 V. This is contributed by the large surface area of the active material, which aided a smooth electrode–electrolyte interaction.

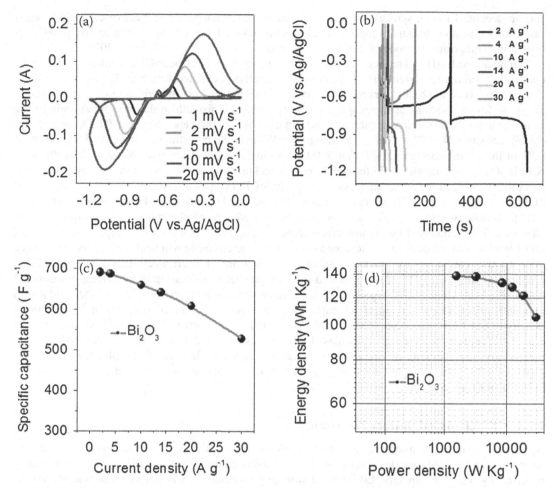

FIGURE 19.2 Electrochemical plots of ultrathin Bi_2O_3 nanowires (a) Cyclic voltammetry curves at various scan rates; (b) Charge-discharge curves at varying current densities; (c) Specific capacitance calculated at various current densities; (d) The Ragone plots of the electrode (Qiu et al. 2018). Reprinted with permission from Ref. Qiu et al. (2018), copyright 2018, Elsevier.

19.4.2 Bi_2O_3-Based Composite Electrodes

Basic factors are limiting the maximum performance of metal oxide electrodes; these are low electroconductivity, low surface area, and their compositions (Obodo et al. 2019). Overcoming these factors require strategies that incorporate conductive and high surface area materials like polymers and carbon compounds (graphene, carbon nanotubes/fibres, activated carbon, and carbon aerogels, etc.) into the metal oxide (Obodo et al. 2020), (Obodo et al. 2020). The mechanism of charge storage in carbon materials requires an adsorption and desorption of ions at the interface between electrolyte and electrode, thereby initiating an electronic double-layer capacitance (Jian et al. 2016). However, metal oxides, nitrides, sulphides, and conducting polymers having pseudocapacitive properties store charges through multiple reduction-oxidation mechanisms (Jiang and Liu 2019), (Augustyn et al. 2014). Electrode materials consisting of composites have offered a highly modified electrochemical performance better than their single components. This is due to the efficient scattering of the metal oxide particles on the large surface area conductive carbon/polymer material, thus increasing the redox sites for improved charge retention and capacitance (Veerakumar et al. 2020).

Wang et al. (2014) modified the capacitance of Bi_2O_3 with activated carbon for application in supercapacitors. The specific capacitance of 332.6 F/g was recorded in 6M KOH electrolyte using 1 A/g current density. Li et al. synthesized the composite of Bi_2O_3 nanowires and active carbon for super-capacitor application, through a facile sol gel process. Using 1 A/g current density in 6M KOH electrolyte the composite electrode offered a specific capacitance of 466 F/g (Li et al. 2015).

Graphene oxide (GO) has been an interesting material in electrochemical studies owing to its outstanding properties (large surface area, high conductivity, coupled with zero band gap) (El-Kady et al. 2016). Hwang (2008) worked on the synthesis of reduced graphene/bismuth oxide composite electrode for applications in supercapacitors. They employed bismuth oxalate precipitation in oxalic acid using the hydrothermal technique; and obtained a specific capacitance of 94 F/g at 0.2 A/g current density and 55 F/g under a scan rate of 5 mV/s over the potential window of 0–1 V, and about 90% of the charges being retained after 3000 cycles in 6M KOH electrolyte. Wang et al. synthesized rGO/Bi_2O_3 composite electrode through a polyol technique in a short reaction time and low temperature. This electrode achieved a specific capacity of 773 C/g using 0.2 A/g density and a potential of 1 V. Also, there was 75% charge retention after 800 cycles in 6M KOH electrolyte (Wang et al. 2017). Gurusamy et al. (2017) worked on graphene oxide/bismuth subcarbonates for a supercapacitor electrode. They employed the hydrothermal technique, and recorded a specific capacitance value of 860 F/g with 100 mg of rGO at a scan rate of 5 mV/s. The electrode withheld 85% of its capacitance after 1000 cycles using a current density of 5 A/g and 6M of KOH electrolyte. Liu et al. (2018) prepared a composite of bismuth oxide and reduced graphene oxide using solvothermal reaction and obtained a high value of 617 F/g as the specific capacitance at a current density of 1 A/g and a scan rate of 5 mV/s. Thirty percent of the capacitance was retained after 1000 cycles at 5 A/g current density in 6M KOH electrolyte. They also found out that as the current density is raised to 10 A/g the rate of the capacitance retention is increased to 79.5%. Sarma et al. (2013) used the electrodeposition method for the synthesis of titania/bismuth oxide composite electrode for applications in supercapacitors. This electrode gave a specific capacitance of 430 mF/cm^2 at a current density of 5 mA/cm^2 in a 1M NaOH electrolyte.

19.4.3 Bi_2O_3-Based Battery Electrodes

Applications of Bi_2O_3 materials as a battery electrode (mostly Li^{2+} storage (Fang et al. 2016), aqueous rechargeable batteries (Zuo et al. 2016), Na-ion battery (Su et al. 2015)) have been widely reported by many researchers. Lithium-ion storage performance of Bi_2O_3 material was at first recorded by Fiordiponti et al. (1978) many years ago. A remarkable capacity of about 660 mA h/g was recorded using a $LiAsF_6$-HF electrolyte. Zuo et al. (2016), explored the performance of Bi_2O_3 electrode material for applications in aqueous rechargeable batteries. The electrode exhibited good electrochemical performance in different aqueous metal ion electrolytes. It also achieved a high specific capacity value of 357 mA/g in 0.72 C current density, with a high rate capability of up to 75,000 mA/g in a mixed Li^{2+} electrolyte – $LiMn_2O_4//Bi_2O_3$ full cell. The capacity retention of about 68% was recorded after 100 cycles under 0.25 A/g current density. Liu et al., through sulphurization of Bi_2O_3 nanosheets, fabricated a Bi_2O_3-Bi_2S_3 heterostructure for application in lithium-ion batteries. In comparison to other bismuth-based materials, this structure offered a remarkable electrochemical behaviour recording a capacity of 433 mAh/g at 600 mA/g current density, with about 84% coulombic efficiency after 100 cycles. The highly porous and large surface area of the electrode's active material, coupled with its heterogeneous phase, was the basic factor for the excellent electrochemical performance (Liu et al. 2015) (Table 19.1).

The chemical methods (sol gel, hydrothermal, solvothermal, chemical precipitation) have been the most popular techniques for synthesizing Bi_2O_3-based electrodes. These methods have also offered an excellent electrochemical performance with relatively high specific capacitances (Figure 19.3). Each segment in Figure 19.3a denotes a single synthesis technique for producing a nanostructure of Bi_2O_3 for electrochemical applications. The specific capacitance values of the Bi_2O_3-based electrodes produced by the respective synthesis methods are represented by each bar in Figure 19.3b. Furthermore, compositing/hybridizing the Bi_2O_3 nanostructure tremendously enhanced its electrochemical performance.

TABLE 19.1

Summary of the Electrochemical Performance of Bi_2O_3-Based Electrodes for Supercapacitor and Battery Applications

Material	Synthesis Method	Specific Capacity	Current Density	Electrolyte	Capacity Retention	Number of Cycles	Ref.
Bi_2O_3 nanowires	Oxidative metal vapour transport	671 F/g	2 A/g	6M KOH	76%	3000	(Qiu et al. 2018)
Rippled Bi_2O_3	Electrodeposition technique	250 F/g	–	1M Na_2SO_4	–	1000	(Zheng et al. 2010)
$Bi_2O_{2.33}$ microspheres	Chemical precipitation	893 F/g	0.1 A/g	6M KOH	96%	1000	(Huang et al. 2013)
β-Bi_2O_3	Hydrothermal method	871 C/g	1 A/g	1M Na_2SO_4	79%	1000	(Ma et al. 2016)
Bi_2O_3-AC composite	Vacuum impregnation and roasting process	332.6 F/g	1 A/g	6M KOH	59 %	1000	(Wang et al. 2014)
Bi_2O_3/active-carbon (AC)	Chemical precipitation	466 F/g	1 A/g	6M KOH	70%	3000	(Li et al. 2015)
rGO/Bi_2O_3 composite	Hydrothermal technique	94 F/g	0.2 A/g	6M KOH	90%	3000	(Ciszewski et al. 2015)
rGO/Bi_2O_3 composite	Polyol rocess	773 C/g	0.2 A/g	6M KOH	75%	800	(Wang et al. 2017)
$Bi_2O_2CO_3$-rGO	Hydrothermal method	860 F/g	–	6M KOH	85%	1000	(Gurusamy et al. 2017)
Bi_2O_3-rGO	Solvothermal method	617 F/g	1 A/g	6M KOH	30%	1000	(Liu et al. 2018)
T-NT/Bi_2O_3 composite	Electrodeposition method	430 mF/cm^{-2}	5 mA/cm^{-2}	1M NaOH	75%	500	(Sarma et al. 2013)
rGO/Bi_2O_3 composite	Hydrothermal method	216 F/g	1 A/g	1M KOH	–	–	(Yang and Lin 2019)
GO/Bi_2O_3 Composite	Solvothermal method	255 F/g	1A/g	6M KOH	65%	1000	(Wang et al. 2010)
Bi_2O_3/GO Composite	Sol gel method	136.76 F/g	0.5 A/g	6M KOH	95%	1000	(Deepi et al. 2018)
Bi_2O_3/HOMC	Simple deposition	232 F/g	5 mV/s scan rate	6M KOH	70%	1000	(Yuan et al. 2009)
Mesoporous carbon@Bi_2O_3	Microwave method	386 F/g	250 mA/g	6M KOH	68%	1000	(Xia et al. 2011)
Bi-Bi_2O_3/CNT (carbon nanotube)	Solvothermal method	850 F/g	1 A/g	6M KOH	72.9 %	1000	(Wu et al. 2020)
AC-Bi_2O_3	Inkjet printing method	0.5127 F cm^{-2}	2.5 mA/cm^2	6M KOH	92.2%	20,000	(Sundriyal and Bhattacharya 2019)
$BiMn_2O_5$- MWCNT	Hydrothermal method	6.0 F cm^{-2}	2 mV/s scan rate	0.5M Na_2SO_4	90%	1000	(Liu and Zhitomirsky 2015)
Bi_2MoO_6	Hydrothermal method	182 F/g	1 A/g	3M KOH	80%	3000	(Yu et al. 2018)
Bi_2O_3-MnO_2	Solid-state chemical synthesis	161 F/g	1 A/g	1M NaOH	95%	10,000	(Singh et al. 2019)

(Continued)

TABLE 19.1 (Continued)

Summary of the Electrochemical Performance of Bi_2O_3-Based Electrodes for Supercapacitor and Battery Applications

Material	Synthesis Method	Specific Capacity	Current Density	Electrolyte	Capacity Retention	Number of Cycles	Ref.
Bi_2O_3@MnO_2	Hydrothermal method	93.1 F/g	1 A/g	1M Na_2SO_4	112%	1000	(Ma et al. 2015)
Bi_2O_3/MnO_2	Hydrothermal method	79.4 F/g	1 A/g	1M Na_2SO_4	95%	1000	(Ng et al. 2018)
Strontium bismuth oxides (SBOs)	Impregnation-calcination method	1228.7 F/g	1 A/g	6M KOH	75.1%	3000	(Han et al. 2018)
Bi_2O_3 thin Films	Electrodeposition	98 F/g	–	1M NaOH	–	1000	(Gujar et al. 2006)
$\alpha - Bi_2O_3$@PPy composite	Hydrothermal method	634 F/g	3 A/g	1M Na_2SO_4	–	50	(Xavier et al. 2019)
Bi_2O_3 nanoplates	Spray-pyrolysis	322.5 F/g	5 mV/s scan rate	1M Na_2SO_4	99.9 %	5000	(Ambare et al. 2018)
Bi_2S_3 nanoparticles	Solvothermal method	550 F/g	0.5 A/g	6M KOH	87%	500	(Miniach and Gryglewicz 2018)
CQD– Bi_2O_3	Hydrothermal method	343 C/g	0.5 A/g	3M KOH	100%	2500	(Prasath et al. 2019)
Bi_2O_3@N-G	Solvothermal method	650 mA h/g	200 mA/g	1M $LiPF_6$	99%	100	(Fang et al. 2016)
Bi@NC composite	Galvanic replacement reaction	285 mA h/g	80 mA/g	1M $LiPF_6$	65%	100	(Zhong et al. 2018)
Bi_2O_3@C film	Hydrothermal method	120 mA h/ g	5 mV/s scan rate	3M NaOH	83%	1,200	(Ba et al. 2019)
Bi@graphene nanocomposite	Hydrothermal synthesis	561 mA h/g	40 mA/g	1M $NaClO_4$	–	–	(Su et al. 2015)
Bi_2O_3@CMK-3@PPy	Nanocasting and chemical polymerization method	321 mA h/g	6 A/g	1M $LiPF_6$	100%	1000	(Fang et al. 2017)
Bi_2O_3@C nanocomposite	Redox reaction and carbon coating technique	392 mA h/g	3200 mA/g	1M $NaClO_4$	74%	100	(Fang et al. 2017)
$LiMn_2O_4$/Bi_2O_3	Hydrothermal method	357 mA/g	0.72 C	Mixed Li^{2+} electrolyte	68%	100	(Zuo et al. 2016)
Bi_2O_3-Bi_2S_3 heterostructure	Sonochemical method	433 mA h/g	600 mA/g	1M $LiPF_6$	84%	100	(Liu et al. 2015)
p- Bi_2O_3/Ni	Polymer-assisted technique	668 mA/g	800 mA/g	1M $LiPF_6$	68%	40	(Li et al. 2013)

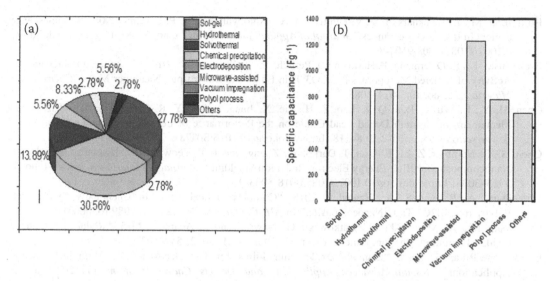

FIGURE 19.3   Chart representation of (a) the synthesis methods, (b) the specific capacitance of each synthesis method.

19.5 Conclusion

We have reviewed the intrinsic properties of bismuth oxide nanostructure; eight most efficient synthesis routes were also analyzed and reported. The physical properties of the bismuth oxide nanoparticles were found to be dependent on the method of synthesis employed. The chemical routes have been the most widely adopted synthesis approach. This is due to its facile synthesis technique, low reaction temperature, and low average particle size of the obtained sample. However, its low reaction temperature is a hindrance to getting a bismuth oxide nanostructure that has low impurity. We can also conclude that the excellent electrochemical properties of bismuth oxide nanostructures are always modified when incorporated with higher electroconductive materials. The specific advantage of bismuth oxide application in batteries is to withstand long cycle life, greater charge retention, and reduced hydrolysis. Because of the reported remarkable properties of bismuth oxide, it shows increasing suitability for a wide range of research advancements and technological innovations.

REFERENCES

Al-Kuhaili, M. F., M. E. Daoud, and M. B. Mekki. 2020. "Spectrally Selective Energy-Saving Coatings Based on Reactively Sputtered Bismuth Oxide Thin Films." *Optical Materials Express* 10, no. 2: 449–463.

Ambare, R. C., P. Shinde, U. T. Nakate, B. J. Lokhande, and R. S. Mane. 2018. "Sprayed Bismuth Oxide Interconnected Nanoplate Supercapacitor Electrode Materials." *Applied Surface Science* 453: 214–219. https://doi.org/10.1016/j.apsusc.2018.05.090

Augustyn, V., P. Simon, and B. Dunn. 2014. "Pseudocapacitive Oxide Materials for High-Rate Electrochemical Energy Storage." *Energy & Environmental Science* 7: 1597–1614.

Ba, D., Y. Li, Y. Sun, Z. Guo, and J. Liu. 2019. "Directly Grown Nanostructured Electrodes for High-power and High-stability Alkaline Nickel/bismuth Batteries." *Science China Materials* 62, no. 4: 487–496. https://doi.org/10.1007/s40843-018-9326-y

Bahmani, A., M. Sellami, and N. Bettahar. 2012. "Synthesis of Bismuth Mixed Oxide by Thermal Decomposition of a Coprecipitate Precursor." *Journal of thermal analysis and calorimetery* 107, no. 3: 955–962. https://doi.org/10.1007/s40843-017-9095-4

Bedoya, H., M. Claudia, C. Pinzón, J. Manuel, O. Alfonso, E. Jose, P. Restrepo, O. F. Elisabeth, and J. Jhon. 2012. "Physical-Chemical Properties of Bismuth and Bismuth Oxides: Synthesis, Characterization and Applications." *DYNA* 79, no. 176: 139–148. http://www.scielo.org.co/scielo.php?script=sci_arttext&pid=S0012-73532012000600017&lng=en&tlng=en.

Bordun, O. M., I. I. Kukharskii, V. V. Dmitruk, V. G. Antonyuk, and V. P. Savchin. 2008. "Luminescence centers in α-Bi 2 O 3 ceramics." *Journal of Applied Spectroscopy* 75, no. 5: 681–684. https://doi.org/1 0.1007/s10812-008-9104-8

Brezesinski, K., R. Ostermann, P. Hartmann, J. Perlich, and T. Brezesinski. 2010. "Exceptional Photocatalytic Activity of Ordered Mesoporous β-Bi2O3 Thin Films and Electrospun Nanofiber Mats." *Chemistry of Materials* 22, no. 10: 3079–3085.

Chen, H. Y., L. Wu, C. Ren, Q. Z. Luo, Z. H. Xie, X. Jiang, … and Y. R. Luo. 2001. "The Effect and Mechanism of Bismuth Doped Lead Oxide on the Performance of Lead-Acid Batteries." *Journal of power sources* 95, no. 1-2: 108–118. https://doi.org/10.1016/S0378-7753(00)00640-6

Chen, S., S. Nambiar, Z. Li, E. Osei, J. Darko, W. Zheng, and J. T. Yeow. 2019. "Bismuth Oxide-Based Nanocomposite for High-Energy Electron Radiation Shielding." *Journal of Materials Science* 54, no. 4: 3023–3034. https://doi.org/10.1007/s10853-018-3063-0

Choudhary, A. R., and S. A. Waghuley. 2018. "Complex Optical Study of Chemically Synthesized Polypyrrole-Bi2O3-TiO2 Nanocomposite." In *AIP Conference Proceedings* 1989 (030005).

Ciszewski, M., A. Mianowski, P. Szatkowski, G. Nawrat, and J. Adamek. 2015. "Reduced Graphene Oxide–Bismuth Oxide Composite as Electrode." *Ionics* 21, no. 2: 557–563.

Condurache-Bota, S. 2018. "Bismuth Oxide Thin Films for Optoelectronic and Humidity Sensing Applications." *Bismuth-Advanced Applications and Defects Characterization*: 171–204. https:// doi.org/10.5772/intechopen.75107

Deepi, A., G. Srikesh, and A. S. Nesaraj. 2018. "Electr@@@ochemical Performance of Bi2O3 Decorated Graphene Nano Composites for Supercapacitor Applications." *Nano-Structures & Nano-Objects* 15: 10–16. https://doi.org/10.1016/j.nanoso.2018.03.003

Denisov, V. N., A. N. Ivlev, A. S. Lipin, B. N. Mavrin, and V. G. Orlov. 1997. "Raman Spectra and Lattice Dynamics of Single-Crystal." *Journal of Physics: Condensed Matter* 9, no. 23: 49–67.

Dong, W., and C. Zhu. 2003. "Optical Properties of Surface-Modified Bi2O3 Nanoparticles." *Journal of Physics and Chemistry of Solids* 64, no. 2: 265–271. https://doi.org/10.1016/S0022-3697(02)00291-3

Du, X., S. Tsai, D. L. Maslov, and A. F. Hebard. 2005. "Metal – Insulator - Like Behavior in Semimetallic Bismuth and Graphite." *Physical Review Letters* 94: 166601–166604.

Dubey, R. G., K. R. Nemade, and S. A. Waghuley. 2012. "Transition Model Analysis of Bismuth Oxide Quantum Dots." *Scientific Reviews and Chemical Communications* 2, no. 3: 436–440.

El-Kady, M. F., Y. Shao, and R. B. Kaner. 2016. "Nature Reviews Materials." *Graphene for batteries, supercapacitors and Beyond* 1, no. 7: 1–14. https://doi.org/10.1038/natrevmats.2016.33

G. Fang, G. Chen, J. Liu, and X. Wang. 2010. "Ultraviolet-Emitting Bi2O2. 33 Nanosheets Prepared by Electrolytic Corrosion of Metal Bi." *The Journal of Physical Chemistry C* 114, no. 2: 864–867.

Fang, W., N. Zhang, L. Fan, and K. Sun. 2016. "Bi2O3 Nanoparticles Encapsulated by Three-dimensional Porous Nitrogen-doped Graphene for High-rate Lithium Ion Batteries." *Journal of Power Sources* 333: 30–36. http://dx.doi.org/10.1016/j.jpowsour.2016.09.155

Fang, W., N. Zhang, L. Fan, and K. Sun. 2017. "Preparation of Polypyrrole-Coated Bi2O3@ CMK-3 Nanocomposite for Electrochemical Lithium Storage." *Electrochimica Acta* 238: 202–209. http:// dx.doi.org/10.1016/j.electacta.2017.04.032

Fang, W., L. Fan, Y. Zhang, Q. Zhang, Y. Yin, N. Zhang, and K. Sun. 2017. "Synthesis of Carbon Coated Bi2o3 Nanocomposite Anode for Sodium-ion Batteries." *Ceramics International* 43, no. 12: 8819–8823. http://dx.doi.org/10.1016/j.ceramint.2017.04.014

Fiordiponti, P., G. Pistoia, and C. Temperoni. 1978. "Behavior of Bi2O3 as a Cathode for Lithium Cells." *Journal of The Electrochemical Society* 125, no. 1: 14. https://doi.org/10.1149/1.2131381

Gao, Z., H. Qin, T. Yan, H. Liu, and J. Wang. 2011. "Structure and Resistivity of Bismuth Nanobelts In Situ Synthesized on Silicon Wafer through an Ethanol-Thermal Method." *Journal of Solid State Chemistry* 184, no. 12: 3257–3261. https://doi.org/ 10.1016/j.jssc.2011.10.010

Gong, Y., W. Ji, L. Zhang, B. Xie, and H. Wang. 2011. "Performance of (La, Sr) MnO3 Cathode Based Solid Oxide Fuel Cells: Effect of Bismuth Oxide Sintering Aid in Silver Paste Cathode Current Collector." *Journal of Power Sources* 196, no. 3: 928–934.

Gou, X., R. Li, G. Wang, Z. Chen, and D. Wexler. 2009. "Room-Temperature Solution Synthesis of Bi2O3 Nanowires for Gas Sensing Application." *Nanotechnology* 20, no. 49: 495501.

Gribanov, E. N., O. I. Markov, and Y. V. Khripunov. 2011. "When Does Bismuth Become a Semimetal?" *Nanotechnologies in Russia* 6: 536–596.

Gujar, T. P., V. R. Shinde, C. D. Lokhande, and S. H. Han. 2006. "Electrosynthesis of Bi2O3 Thin Films and Their Use in Electrochemical Supercapacitors." *Journal of Power Sources* 161, no. 2: 1479–1485. https://doi.org/10.1016/j.jpowsour.2006.05.036

Gujar, T. P., V. R. Shinde, C. D. Lokhande, R. S. Mane, and S. H. Han. 2005. "Bismuth Oxide Thin Films Prepared by Chemical Bath Deposition (CBD) Method: Annealing Effect." *Applied Surface Science* 250, no. 1-4: 161–167.

Gurusamy, L., S. Anandan, and J. J. Wu. 2017. "Synthesis of reduced graphene oxide supported flower-like bismuth subcarbonates microsphere (Bi2O2CO3-RGO) for supercapacitor application." *Electrochimica Acta* 244: 209–221.

Han, Y., L. Li, Y. Liu, X. Li, X. Qi, and L. Song. 2018. "Fabrication of Strontium Bismuth Oxides as Novel Battery-type Electrode Materials for High-performance Supercapacitors." *Journal of Nanomaterials*. https://doi.org/ 10.1155/2018/5078473

Hardcastle, F. D., and I. E. Wachs. 1991. "Molecular Structure of Molybdenum Oxide in Bismuth Molybdates by Raman Spectroscopy." *The Journal of Physical Chemistry* 95, no. 26: 10763–10772. https://doi.org/10.1021/j100179a045

Hashimoto, T., H. Ohta, H. Nasu, and A. Ishihara. 2016. "Preparation and Photocatalytic Activity of Porous Bi2O3 Polymorphisms." *International Journal of Hydrogen Energy* 41, no. 18: 7388–7392. https://doi.org/10.1016/j.ijhydene.2016.03.109

Huang, X., J. Yan, F. Zeng, X. Yuan, W. Zou, and D. Yuan. 2013. "Facile Preparation of Orange-like Bi2O2. 33 Microspheres for High Performance Supercapacitor Application." *Materials Letters* 90: 90–92.

Hwang, G. H., W. K. Han, J. S. Park, and S. G. Kang. 2008. "An Electrochemical Sensor Based on the Reduction of Screen-printed Bismuth Oxide for the Determination of Trace Lead and Cadmium." *Sensors and Actuators B: Chemical* 135, no. 1: 309–316.

Jalalah, M., M. Faisal, H. Bouzid, J. G. Park, S. A. Al-Sayari, and A. A. Ismail. 2015. "Comparative Study on Photocatalytic Performances of Crystalline α-and β-Bi2O3 Nanoparticles under Visible Light." *Journal of Industrial and Engineering Chemistry* 30: 183–189. https://doi.org/10.1016/j.jiec.2015.05.020

Jhaa, R. K., R. Pasrichab, and V. Ravic. 2015. "Synthesis of Bismuth Oxide Nanoparticles Using Bismuth Nitrate and Urea." *Ceramics International* 31): 495–497.

Jian, X., S. Liu, Y. Gao, W. Tian, Z. Jiang, X. Xiao, and L. Yin. 2016. "Carbon-Based Electrode Materials for Supercapacitor: Progress, Challenges and Prospective Solutions." *J. Electr. Eng* 4: 75–87.

Jiang, Y., and J. Liu. 2019. "Definitions of Pseudocapacitive Materials: A Brief Review." *Energy & Environmental Material* 2: 30–37. https://doi.org/10.1002/eem2.12028

Jin, C., H. Kim, K. Baek, H. W. Kim, and C. Lee. 2010. "Structure and Photoluminescence Properties of Bi2O3/ZnO Coaxial Nanorods." *Journal of the Korean Physical Society* 57, no. 61: 1634–1638. https://doi.org/10.3938/jkps.57.1634

Kelly, Z., and F. Ojebuoboh. 2002. "Producing Bismuth Trioxide and Its Applications in Fire Assaying." *JOM* 54, no. 4: 42–45.

Konopko, L., T. Huber, and A. Nikolaeva. 2010. "Quantum Interference and Surface States Effects in Bismuth Nanowires." *Journal of Low Temperature Physics* 158: 523–529.

Kumari, L., J. H. Lin, and Y. R. Ma. 2007. "One-Dimensional Bi2O3 Nanohooks: Synthesis, Characterization and Optical Properties." *Journal of Physics: Condensed Matter* 19, no. 40: 406204.

Lee, G. J., C. K. Kim, M. K. Lee, and C. K. Rhee. 2010. "Effect of Phase Stability Degradation of Bismuth on Sensor Characteristics of Nano-Bismuth Fixed Electrode." *Talanta* 83, no. 2: 682–685. https://doi.org/1 0.1016/j.talanta.2010.10.007

Leontie, L., M. Caraman, M. Delibaş, and G. I. Rusu. 2001. "Optical Properties of Bismuth Trioxide Thin Films." *Materials Research Bulletin* 36, no. 9: 1629–1637.

Leontie, L., M. Caraman, M. Alexe, and C. Harnagea. 2002. "Structural and Optical Characteristics of Bismuth Oxide Thin Films." *Surface Science* 507: 480–485.

Leontie, L., M. Caraman, M. Alexe, and C. Harnagea. 2002. "Structural and Optical Characteristics of Bismuth Oxide Thin Films." *Surface Science* 507: 480–485. https://doi.org/10.1016/S0039-6028(02) 01289-X

Leontie, L., M. Caraman, M. Alexe, and C. Harnagea. 2002. "Structural and Optical Characteristics of Bismuth Oxide Thin Films." *Surface Science* 507: 480–485.

Li, J., Q. Wu, and G. Zan. 2015. "A High-Performance Supercapacitor with Well-Dispersed Bi2O3 Nanospheres and Active-Carbon Electrodes." *European Journal of Inorganic Chemistry* 2015, no. 35: 5751–5756.

Li, Y., M. A. Trujillo, E. Fu, B. Patterson, L. Fei, Y. Xu, and H. Luo. 2013. "Bismuth Oxide: A New Lithium-Ion Battery Anode." *Journal of Materials Chemistry A* 1, no. 39: 12123–12127. https://doi.org/10.1039/c3ta12655b

Ling, B., X. W. Sun, J. L. Zhao, Y. Q. Shen, Z. L. Dong, L. D. Sun, and S. Zhang. 2010. "One-Dimensional Single-Crystalline Bismuth Oxide Micro/Nanoribbons: Morphology-Controlled Synthesis and Luminescent Properties." *Journal of Nanoscience and Nanotechnology* 10, no. 12: 8322–8327. https://doi.org/10.1166/jnn.2010.3051

Liu, Y., and I. Zhitomirsky. 2015. "Electrochemical Supercapacitor Based on Multiferroic BiMn2O5." *Journal of Power Sources* 284: 377–382. https://doi.org/10.1016/j.jpowsour.2015.03.050

Liu, M., C. Y. Nam, and L. Zhang. 2015. "Seedless Growth of Bismuth Nanowire Array via Vacuum Thermal Evaporation." *JoVE (Journal of Visualized Experiments)* 106, no. e53396. https://doi.org/10.3791/53396

Liu, S., Y. Wang, and Z. Ma. 2018. "Bi2O3 with reduced graphene oxide composite as a supercapacitor electrode." *International Journal of Electrochemical Science* 13: 12256–12265. https://doi.org/10.20964/2018.12.10

Liu, T., Y. Zhao, L. Gao, and J. Ni. 2015. "Engineering Bi2O3-Bi2S3 Heterostructure for Superior Lithium Storage." *Scientific Reports* 5: 9307. https://doi.org/10.1038/srep09307

Lu, H., S. Wang, L. Zhao, B. Dong, Z. Xu, and J. Li. 2012. "Surfactant-Assisted Hydrothermal Synthesis of Bi2O3 Nano/Microstructures with Tunable Size." *RSC advances* 2, no. 8): 3374–3378. https://doi.org/10.1039/C2RA01203K

Ma, X. J., W. B. Zhang, L. B. Kong, Y. C. Luo, and L. Kang. 2016. "β-Bi2O3: An Underlying Negative Electrode Material Obeyed Electrode Potential over Electrochemical Energy Storage Device." *Electrochimica Acta* 192: 45–51.

Ma, J., S. Zhu, Q. Shan, S. Liu, Y. Zhang, F. Dong, and H. Liu. 2015. "Facile Synthesis of Flower-Like (BiO)2CO3@ MnO2 and Bi2O3@ MnO2 Nanocomposites for Supercapacitors." *Electrochim. Acta* 168: 97–103. https://doi.org/10.1016/j.electacta.2015.04.018

Mahmoud, W. E., and A. A. Al-Ghamdi. 2011. "Synthesis and Properties of Bismuth Oxide Nanoshell Coated Polyaniline Nanoparticles for Promising Photovoltaic Properties." *Polymers for Advanced Technologies* 22, no. 6: 877–881. https://doi.org/10.1002/pat.1591

Mallahi, M., A. Shokuhfar, M. R. Vaezi, A. Esmaeilirad, and V. Mazinani. 2014. "Synthesis and Characterization of Bismuth Oxide Nanoparticles via Sol-Gel Method." *AJER* 3, no. 4: 162–165.

Miller, N. C., and M. Bernechea. 2018. "Research Update: Bismuth Based Materials for Photovoltaics." *APL Materials* 6, no. 8: 084503. https://doi.org/10.1063/1.5026541

Miniach, E., and G. Gryglewicz. 2018. "Solvent-Controlled Morphology of Bismuth Sulfide for Supercapacitor Applications." *Journal of Materials Science* 53, no. 24: 16511–16523. https://doi.org/10.1007/s10853-018-2785-3

Moniz, S. J., D. Bhachu, C. S. Blackman, A. J. Cross, S. Elouali, D. Pugh, Raul Quesada Cabrera and S. Vallejos. 2012. "A Novel Route to Pt–Bi2O3 Composite Thin Films and Their Application in Photo-Reduction of Water." *Inorganica Chimica Acta* 380: 328–335. https://doi.org/10.1016/j.ica.2011.09.029

Mozdianfard, M. R., F. Masoodiyeh, and J. Karimi-Sabet. 2018. "Supercritical Water Hydrothermal Synthesis of Bi2O3 Nanoparticles: Process Optimization Using Response Surface Methodology Based on Population Balance Equation." *The Journal of Supercritical Fluids* 136: 144–156. https://doi.org/10.1016/j.supflu.2018.02.017

Narang, S. N., N. D. Patel, and V. B. Kartha. 1994. "Infrared and Raman Spectral Studies and Normal Modes of α-Bi2O3." *Journal of Molecular Structure* 327, no. 2-3: 221–235. https://doi.org/10.1016/0022-2860(94)08160-3

Ng, C., H. N. Lim, S. Hayase, Z. Zainal, S. Shafie, and N. Huang. 2018. "Effects of Temperature on Electrochemical Properties of Bismuth Oxide/Manganese Oxide Pseudocapacitor." *Industrial & Engineering Chemistry Research* 57: 2146–2154. https://doi.org/10.1021/acs.iecr.7b04980

Obodo, R. M., A. Ahmad, G. H. Jain, I. Ahmad, M. Maaza, and F. I. Ezema. 2020. "8.0 MeV Copper Ion (Cu++) Irradiation-Induced Effects on Structural, Electrical, Optical and Electrochemical Properties of Co3O4-NiO-ZnO/GO Nanowires." *Materials Science for Energy Technologies* 3: 193–200.

Obodo, R., C. Nnwanya, T. Hassina, M. A. Kebede, A. Ishak, M. Maaza, and F. I. Ezema. 2019. "Transition Metal Oxides-Based Nanomaterials for High Energy and Power Density Supercapacitor."

Obodo, R. M., A. C. Nwanya, C. Iroegbu, I. Ahmad, A. B. C. Ekwealor, R. U. Osuji, M. Maaza, and F. I. Ezema. 2020. "Transformation of GO to rGO due to 8.0 MeV Carbon (C++) Ions Irradiation and Characteristics Performance on MnO2–NiO–ZnO@GO Electrode." *International Journal of Energy Research*, 44 (8): 6792–6803.

Obodo, R. M., A. C. Nwanya, C. Iroegbu, B. A. Ezekoye, A. B. C. Ekwealor, I. Ahmad, F. I. Ezema, Malik Mazza, et al. 2020. "Effects of Swift Copper (Cu2+) Ion Irradiation on Structural, Optical and Electrochemical Properties of Co3O4-CuO-MnO2/GO Nanocomposites Powder." *Advanced Powder Technology* 31 (40): 1728–1735.

Pandey, J., A. Sethi, S. Uma, and R. Nagarajan. 2018. "Catalytic Application of Oxygen Vacancies Induced by Bi3+ Incorporation in ThO2 Samples Obtained by Solution Combustion Synthesis." *ACS Omega* 3, no. 7: 7171–7181.

Postel, M., and E. Duñach. 1996. "Bismuth Derivatives for the Oxidation of Organic Compounds." *Coordination Chemistry Review* 155: 127–144.

Prasath, A., M. Athika, E. Duraisamy, A. Selva Sharma, V. Sankar Devi, and P. Elumalai. 2019. "Carbon Quantum Dot-anchored Bismuth Oxide Composites as Potential Electrode for Lithium-ion Battery and Supercapacitor Applications." *ACS Omega* 4, no. 3: 4943–4954. https://doi.org/10.1021/acsomega. 8b03490

Qiu, Y., D. Liu, J. Yang, and S. Yang. 2006. "Controlled Synthesis of Bismuth Oxide Nanowires by an Oxidative Metal Vapor Transport Deposition Technique." *Advanced Materials* 18, no. 19: 2604–2608.

Qiu, Y., H. Fan, X. Chang, H. Dang, Q. Luo, and Z. Cheng. 2018. "Novel Ultrathin Bi2O3 Nanowires for Supercapacitor Electrode Materials with High Performance." *Applied Surface Science* 434: 16–20. https://doi.org/10.1016/j.apsusc.2017.10.171

Qiu, Y., M. Yang, H. Fan, Y. Zuo, Y. Shao, Y. Xu, and S. Yang. 2011. "Phase-Transitions of α-and β-Bi2O3 Nanowires." *Materials Letters* 65, no. 4: 780–782.

Rubbens, A., M. Drache, P. Roussel, and J. P. Wignacourt. 2007. "Raman Scattering Characterization of Bismuth Based Mixed Oxides with Bi2O3 Related Structures." *Materials Research Bulletin* 42, no. 9: 1683–1690.

Salazar-Pérez, A. J., M. A. Camacho-López, R. A. Morales-Luckie, V. Sánchez-Mendieta, F. Ureña-Núñez, and J. Arenas-Alatorre. 2005. "Structural Evolution of Bi2O3 Prepared by Thermal Oxidation of Bismuth Nano-particles." *Superficies y vacío* 18, no. 3: 4–8.

Sarma, B., A. L. Jurovitzki, Y. R. Smith, S. K. Mohanty, and Misra, M. 2013. "Redox-Induced Enhancement in Interfacial Capacitance of the Titania Nanotube/Bismuth Oxide Composite Electrode." *ACS Applied Materials & Interfaces* 5 (5): 1688–1697.

Schuisky, M., and A. Hårsta. 1996. "Epitaxial Growth of Bi2O2. 33 by Halide Cvd." *Chemical Vapor Deposition* 2, no. 6: 235–238.

Selvapandiyan, M., and K. Sathiyaraj. 2019. "Synthesis, Preparation, Structural, Optical, Morphological and Elemental Analysis of Bismuth Oxides Nanoparticles." *Silicon*: 1–7. https://doi.org/10.1007/s12633-01 9-00327-x

Shen, X. P., S. K. Wu, H. Zhao, and Q. Liu. 2007. "Synthesis of Single-Crystalline Bi2O3 Nanowires by Atmospheric Pressure Chemical Vapor Deposition Approach." *Physica E: Low-dimensional Systems and Nanostructures* 39, no. 1: 133–136. https://doi.org/10.1016/j.physe.2007.02.001

Shu, Y., W. Hu, and Z. e. a. Liu. 2016. "Coexistence of Multiple Metastable Polytypes in Rhombohedral Bismuth." *Scientific Reports* 6 (1): 1–8.

Singh, S., R. K. Sahoo, N. M. Shinde, J. M. Yun, R. S. Mane, and K. H. Kim. 2019. "Synthesis of Bi2O3-MnO2 Nanocomposite Electrode for Wide-Potential Window High Performance Supercapacitor." *Energies* 12, no. 17: 3320. https://doi.org/10.3390/en12173320

Sonkusare, V. N., R. G. Chaudhary, G. S. Bhusari, A. R. Rai, and H. D. Juneja. 2018. "Microwave-Mediated Synthesis, Photocatalytic Degradation and Antibacterial Activity of α-Bi2O3 Microflowers/Novel γ-Bi2O3 Microspindles." *Nano-Structures & Nano-Objects* 13: 121–131. https://doi.org/10.1016/j.nanoso.2018.01.002

Su, D., S. Dou, and G. Wang. 2015. "Bismuth: A New Anode for the Na-Ion Battery." *Nano Energy* 12: 88–95. http://dx.doi.org/10.1016/j.nanoen.2014.12.012

Sun, Y., W. Wang, L. Zhang, and Z. Zhang. 2012. "Design and controllable synthesis of α-/γ-Bi2O3 homojunction with synergetic effect on photocatalytic activity." *Chemical Engineering Journal* 211: 161–167. https://doi.org/10.1016/j.cej.2012.09.084

Sundriyal, P., and S. Bhattacharya. 2019. "Scalable Micro-Fabrication of Flexible, Solid-State, Inexpensive, and High-Performance Planar Micro-supercapacitors through Inkjet Printing." *ACS Applied Energy Materials* 2, no. 3: 1876–1890. https://doi.org/10.1021/acsaem.8b02006

Timur, S., and U. Anik. 2007. "α-Glucosidase Based Bismuth Film Electrode for Inhibitor Detection." *Analytica Chimica Acta* 598, no. 1: 143–146. https://doi.org/10.1016/j.aca.2007.07.019

Veerakumar, P., A. Sangili, S. Manavalan, P. Thanasekaran, and K. C. Lin. 2020. "Research Progress on Porous Carbon Supported Metal/Metal Oxide Nanomaterials for Supercapacitor Electrode Applications." *Industrial & Engineering Chemistry Research* 59, no. 14: 6347–6374. https://doi.org/10.1021/acs.iecr.9b06010

Vila, M., C. Díaz-Guerra, and J. Piqueras. 2012. "Luminescence and Raman Study of α-Bi2O3 Ceramics." *Materials Chemistry and Physics* 133, no. 1: 559–564.

Vila, M., C. Díaz-Guerra, and J. Piqueras. 2013. "α-Bi2O3 Microcrystals and Microrods: Thermal Synthesis, Structural and Luminescence Properties." *Journal of alloys and compounds* 548: 188–193. https://doi.org/10.1016/j.jallcom.2012.08.133

Wang, S. X., C. C. Jin, and W. J. Qian. 2014. "Bi2O3 with Activated Carbon Composite as a Supercapacitor Electrode." *Journal of Alloys and Compounds* 615: 12–17.

Wang, Y., Z. Cui, Y. Xue, R. Zhang, and A. Yan. 2019. "Size-Dependent Thermodynamic Properties of Two Types of Phase Transitions of Nano-Bi2O3 and Their Differences." *The Journal of Physical Chemistry C* 123, no. 31: 19135–19141.

Wang, H. W., Z. A. Hu, Y. Q. Chang, Y. L. Chen, Z. Q. Lei, Z. Y. Zhang, and Y. Y. Yang. 2010. "Facile Solvothermal Synthesis of a Graphene Nanosheet–Bismuth Oxide Composite and Its Electrochemical Characteristics." *Electrochimica Acta* 55, no. 28: 8974–8980. https://doi.org/10.1016/j.electacta.2010.08.048

Wang, J., H. Zhang, M. R. Hunt, A. Charles, J. Tang, O. Bretcanu, and L. Šiller. 2017. "Synthesis and Characterisation of Reduced Graphene Oxide/Bismuth Composite for Electrodes in Electrochemical Energy Storage Devices." *ChemSusChem* 10, no. 2: 363. https://doi.org/10.1002/cssc.201601553

Wu, H., J. Guo, and D. A. Yang. 2020. "Facile Autoreduction Synthesis of Core-Shell Bi-Bi2O3/CNT with 3-Dimensional Neural Network Structure for High-Rate Performance Supercapacitor." *Journal of Materials Science & Technology* 47: 169–176. https://doi.org/10.1016/j.jmst.2020.02.007

Wu, C., L. Shen, Q. Huang, and Y. C. Zhang. 2011. "Hydrothermal Synthesis and Characterization of Bi2O3 Nanowires." *Materials Letters* 65, no. 7: 1134–1136. https://doi.org/10.1016/j.matlet.2011.01.021

Xavier, F. F., C. G. Bruziquesi, W. S. Fagundes, E. Y. Matsubara, J. M. Rosolen, A. C. Silva, and F. A. Amaral. 2019. "New Synthesis Method for a Core-Shell Composite Based on α-Bi2O3@ PPy and Its Electrochemical Behavior as Supercapacitor Electrode." *Journal of the Brazilian Chemical Society* 30, no. 40: 727–735. http://dx.doi.org/10.21577/0103-5053.20180192

Xia, N., D. Yuan, T. Zhou, J. Chen, S. Mo, and Y. Liu. 2011. "Microwave Synthesis and Electrochemical Characterization of Mesoporous Carbon@ Bi2O3 Composites." *Materials Research Bulletin* 46, no. 5: 687–691. https://doi.org/10.1016/j.materresbull.2011.01.022

Xiong, Y., M. Wu, J. Ye, and Q. Chen. 2008. "Synthesis and Luminescence Properties of Hand-like α-Bi2O3 Microcrystals." *Materials Letters* 62, no. 8-9: 1165–1168. https://doi.org/10.1016/j.matlet.2007.08.004

Yan, Y., Z. Zhou, Y. Cheng, L. Qiu, C. Gao, and J. Zhou. 2014. "Template-Free Fabrication of α-and β-Bi2O3 Hollow Spheres and Their Visible Light Photocatalytic Activity for Water Purification." *Journal of Alloys and Compounds* 605: 102–108. http://dx.doi.org/10.1016/j.jallcom.2014.03.111

Yang, W. D., and Y. Lin. 2019. "Preparation of rGO/Bi2O3 Composites by Hydrothermal Synthesis for Supercapacitor Electrode." *J. Electr. Eng.* 70: 101–106. https://doi.org/ 10.2478/jee-2019–0049

Yang, L. L., Q. F. Han, J. Zhao, J. W. Zhu, X. Wang, and W. H. Ma. 2014. "Synthesis of Bi2O3 Architectures in DMF–H2O Solution by Precipitation Method and Their Photocatalytic Activity." *Journal of Alloys and Compounds* 614: 353–359. http://dx.doi.org/10.1016/j.jallcom.2014.06.011

Yu, T., Z. Li, S. Chen, Y. Ding, W. Chen, X. Liu, Y. Huang, and F. Kong. 2018. "Facile Synthesis of Flower-Like Bi2MoO6 Hollow Microspheres for High-Performance Supercapacitors." *ACS Sustainable Chemistry & Engineering* 6: 7355–7361. https://doi.org/10.1021/acssuschemeng.7b04673

Yuan, D., J. Zeng, N. Kristian, Y. Wang, and X. Wang. 2009. "Bi2O3 Deposited on Highly Ordered Mesoporous Carbon for Supercapacitors." *Electrochemistry Communications* 11: 313–317. https://doi.org/10.1016/j.elecom.2008.11.041

Zheng, F. L., G. R. Li, Y. N. Ou, Z. L. Wang, C. Y. Su, and Y. X. Tong. 2010. "Synthesis of Hierarchical Rippled Bi2O3 Nanobelts for Supercapacitor Applications." *Chemical Communications* 46, no. 27: 5021–5023.. http://pubs.rsc.org/doi:10.1039/C002126A

Zhong, Y., B. Li, S. Li, S. Xu, Z. Pan, Q. Huang, and W. Li. 2018. "Bi Nanoparticles Anchored in N-doped Porous Carbon as Anode of High Energy Density Lithium Ion Battery." *Nano-micro Letters* 10, no. 4: 56. https://doi.org/10.1007/s40820-018-0209-1

Zulkifli, Z. A., K. A. Razak, W. N. W. A. Rahman, and S. Z. Abidin. 2018. "Synthesis and Characterisation of Bismuth Oxide Nanoparticles Using Hydrothermal Method: The Effect of Reactant Concentrations and Application in Radiotherapy." *InJournal of Physics: Conference Series* 1082, no. 1. https://doi.org/10.1088/1742-6596/1082/1/012103

Zuo, W., W. Zhu, D. Zhao, Y. Sun, Y. Li, J. Liu, and X. W. D. Lou. 2016. "Bismuth Oxide: A Versatile High-Capacity Electrode Material for Rechargeable Aqueous Metal-Ion Batteries." *Energy & Environmental Science* 9, no. 9: 2881–2891. https://doi.org/10.1039/c6ee01871h

20

Earth-Abundant Materials for Solar Cell Applications

Agnes Chinecherem Nkele[1], Sabastine Ezugwu[2], Mutsumi Suguyima[3], and Fabian I. Ezema[1,4,5]
[1]*Department of Physics & Astronomy, University of Nigeria, Nsukka, Enugu, Nigeria*
[2]*Department of Physics and Astronomy, The University of Western Ontario, London, Ontario, Canada*
[3]*Faculty of Science and Technology, Department of Electrical Engineering, Tokyo University of Science, Noda, Japan*
[4]*Nanosciences African Network (NANOAFNET), iThemba LABS-National Research Foundation, Somerset West, Western Cape Province, South Africa*
[5]*UNESCO-UNISA Africa Chair in Nanosciences/Nanotechnology, College of Graduate Studies, University of South Africa (UNISA), Muckleneuk Ridge, Pretoria, South Africa*

20.1 Basic Concepts of Earth-Abundant Materials

Sustainable energy creation has become an essential part of human existence due to the changing climate and increasing energy demand (Lee et al. 2009). The sun is a free, available, and abundant energy source for sustainable living (https://www.european-mrs.com/earth-abundant-next-generation-materials-solar-energy-iii-emrs 2020a). Solar cells have proved to be useful in harnessing solar energy using photovoltaics for different applications. The unique nature of solar energy requires a flexible technology for harnessing its resource for diverse usage. This technology would adopt the use of relatively cheap, available, and easy to synthesize materials for energy activities (http://inside.mines.edu/fs_home/cwolden/Wolden_Webpages/Projects/EarthAbundant_index.html 2020b). Solar cells made of silicon are very stable in humid conditions and temperature, while those made of thin films are lighter and have an increased absorption coefficient (https://energyfutureslab.blog/2019/11/28/earth-abundant-materials-for-next-generation-photovoltaics/ 2019). To greatly influence the cost of solar cells, relatively cheap, non-poisonous, and naturally available materials should be used.

Recently, researchers have delved into incorporating readily available and less toxic materials into solar cells. These materials should be non-poisonous, abundant, and electrically and optically active (Lee et al. 2009). Solar cell compounds are earth-abundant if the supply of the constituent elements is readily available. Earth-abundant materials have diverse synthesis and characterization procedures which would foster an understanding of the characteristics possessed by such materials. Earth-abundant materials should have a high coefficient of absorption and be able to convert photons to energy (Vasekar and Dhakal 2013). These materials are useful in harvesting and storing energy in solar cell devices (Maduraiveeran, Sasidharan, and Jin 2019). In a bid to accurately determine the band gap energies of the materials, computational techniques like Density Functional Theory (DFT) can be adopted. DFT carries out predictions by automating k-point selection on the crystal symmetry of compounds (Lee et al. 2009). More accurate predictions can be obtained from the dielectric screening features of such compounds.

DOI: 10.1201/9781003145585-20

The features of these materials can be engineered and improved by the addition of dopants, the formation of composites, the creation of heterojunction interfaces, etc. These engineering procedures would aid the modification and improvement of the properties of earth-abundant materials by enhancing the efficiency and stability of solar cell devices. These earth-abundant materials are applied in solar cells in forms like carrier transport layers, photovoltaics, photoelectrodes, solar fuel cells, solar cell absorbers, transparent conductors, etc. (https://www.european-mrs.com/earth-abundant-next-generation-materials-solar-energy-iii-emrs 2020c).

20.2 Some Earth-Abundant Solar Cell Materials

For wide commercialization purposes solar cells require materials that would be available at affordable rates and are found abundantly on earth (Le Donne, Trifiletti, and Binetti 2019). Some of the earth-abundant materials that find useful applications in solar cells are discussed.

20.2.1 Manganese

Manganese is a cheap and available element that exists as rhodochrosite, braunite, and pyrolusite (Le Donne, Trifiletti, and Binetti 2019). Pure manganese can be electrochemically manufactured by leaching manganese ore with sulphuric acid. Manganese can serve as a dopant to enhance the photovoltaic performance of solar cells (Jung et al. 2019; Lv et al. 2019; Marandi, Talebi, and Bayat 2019; Maziar Marandi, Talebi, and Moradi 2019). Manganese ions support increased charge transport mechanisms in crystalline silicon solar cells (Li and Van Deun 2019). Perovskite solar cells can be sandwiched with manganese ions for improved efficiency (Liu et al. 2019). Manganese states have also proved to be storage centres for energy as excited electrons get trapped in it (Maiti, Dana, and Ghosh 2019).

20.2.2 Iron

Iron is among the top five abundant elements. It can exist naturally as hematite (Fe_2O_3), siderite ($FeCO_3$), and magnetite (Fe_3O_4) (Le Donne, Trifiletti, and Binetti 2019). Other iron-derived compounds like silicate perovskite and ferropericlase can also be found in the lower part of the earth's mantle. Carbothermic reactions can be employed to commercially produce iron. Iron can react with oxygen to form oxides that would serve as barrier fillers in solar cells (Yuwawech, Wootthikanokkhan, and Tanpichai 2020) and as counter electrodes in dye-sensitized solar cells (Li et al. 2019). Incorporating iron with carbon materials can enhance the cell's stability and reduce recombination effects (Mar-Ortiz et al. 2020). Incorporating iron into silicon solar cells increases the photoconductivity of the fabricated devices (Pengerla et al.). Doping iron with nanoparticles results in the efficient output of solar cells (Neupane, Kaphle, and Hari 2019).

20.2.3 Nickel

Nickel can be obtained from garnierite and pentlandite. To process pure nickel from ores, extractive metallurgy, roasting, and reduction methods are employed (Le Donne, Trifiletti, and Binetti 2019). Nickel composites aid efficient hole extraction and transport in perovskite solar cells (Wang et al. 2014; Pitchaiya et al. 2018; Abdy et al. 2019). The stability of perovskite solar cells can be enhanced by incorporating nickel oxide as a transport material (Haider et al. 2018; Nkele et al. 2019; Nkele et al. 2020). The reaction between nickel and oxygen to form nickel(II) oxide films makes it a useful photocathode for solar cells (Nattestad et al. 2008). NiO produces low band gap energies, fill factor value of 0.44, and open circuit voltage of 0.96 V (Kawade, Chichibu, and Sugiyama 2014).

20.2.4 Sulphur

Sulfur is obtainable in its pure form from sulphate compounds like sulphides of iron, lead, mercury, zinc, antimony, potassium, calcium, aluminum, and barium (Le Donne, Trifiletti, and Binetti 2019).

Sulphur can also be obtained after the purification of fossil fuels and natural gas. Sulphur influences the rate of formation of thin-film solar cell materials (Timmo et al. 2010; Platzer-Björkman et al. 2012). Sulphur can be infused into hole transport materials to improve device performance (Chiu et al. 2020) and enhance charge transfer (Chalapathy, Jung, and Ahn 2011; Zafar, Yun, and Kim 2019). Sulphur can serve as a dopant in counter electrodes (Guai et al. 2012; Wang et al. 2018), and in tuning the band gap of solar cell materials (Xia et al. 2013).

20.2.5 Tin

Tin can react with sulphur and selenium elements at varying ratios to, respectively, form tin sulphide and tin selenide compounds for optical absorption (Nair, Barrios-Salgado, and Nair 2016). Nitrogen can also be incorporated as a doping agent to increase its rectification power and improve the interface features (Le Donne, Trifiletti, and Binetti 2019). To boost the performance of tin halide films, heterojunctions can be fabricated and the right buffer layer should be used as the absorber material. Tin oxide serves as a good electron transport layer in perovskite solar cells (Pulvirenti et al. 2018; Murugadoss et al. 2016). It can also be useful in extracting electrons in organic solar cells (Trost et al. 2015). Heterojunctions of tin sulphide yield devices with low band gap energy carriers (Sugiyama et al. 2014).

20.2.6 Barium

Barium is largely obtained from barium sulphate ($BaSO_4$) with very high purity. The mineral is washed, grounded, separated, and passed through several chemical reactions to yield barium (Le Donne, Trifiletti, and Binetti 2019). Barium is an effective cathode material for use in heterojunction cells (Gupta et al. 2013). The use of barium as a dopant makes the material useful for photovoltaic applications (Wu et al. 2018; Chan et al. 2017). Nanoparticles of barium compounds serve as working electrodes in dye-sensitized solar cells. Hydroxide form of barium enhances charge transport when incorporated into perovskite devices (Rehman et al. 2019). Barium can also be used in composite forms as absorber materials (Vismara, Isabella, and Zeman 2016). Barium enhances the stability and efficiency of perovskite solar cells (Guo, Li, and Zhang 2010; Scholtz et al. 2018; Mali, Patil, and Hong 2019).

20.2.7 Chalcogenides

Chalcogenide copper films (Cu_2S, CuS, Cu_2Se) are relatively cheap, non-poisonous, and p-type materials for solar cells (Le Donne, Trifiletti, and Binetti 2019). Cu_2S has a small band gap energy and indirect nature that is useful as absorber materials and sensitizer in DSSCs. Copper selenides have varied band gap nature and different structural phases like marcasite, umagnite, klockmannite, etc. Malekar, Gangawane, and Fulari grew copper selenide films under ambient temperatures (Malekar, Gangawane, and Fulari 2020). Results showed numerous particle distribution at the surface, crystalline peaks, low band gap energies between 12 nm and 29 nm, and an observed sharp peak at 3187 cm^{-1} as shown in Figure 20.1. Thanikaikarasan, Dhanasekaran, and Sankaranarayanan investigated the features of copper selenide films prepared via electrochemical method from copper sulphate (Thanikaikarasan, Dhanasekaran, and Sankaranarayanan 2020). They obtained polycrystalline films, smooth morphological surface, and band gap energy of 2.84 eV. Heterojunctions of copper chalcogenides yield better performance, open-circuit voltage, and fill factor values. Metal dichalcogenides are emerging materials for solar cell applications and include iron selenide ($FeSe_2$ and $FeSe$), iron sulphide (FeS_2), molybdenum sulphide (MoS_2), zinc selenide ($ZnSe$) (Ezema, Ekwealor, and Osuji 2006), etc. These 2D materials have high catalytic activities and exceptional optical and electronic features (http://inside.mines.edu/fs_home/cwolden/Wolden_Webpages/Projects/EarthAbundant_index.html 2020).

The overall electrochemical reactions that produce copper selenide from copper sulphate have been outlined in equations (20.1)–(20.4) (Thanikaikarasan, Dhanasekaran, and Sankaranarayanan 2020):

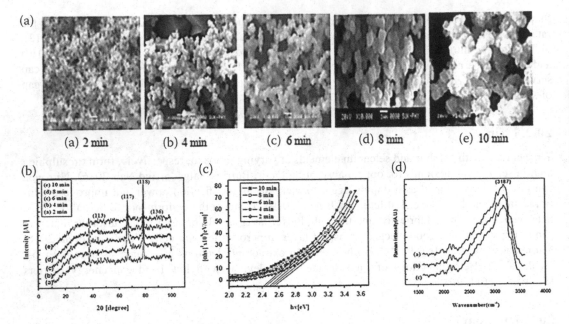

(a) 2 min   (b) 4 min   (c) 6 min   (d) 8 min   (e) 10 min

FIGURE 20.1   (Malekar, Gangawane, and Fulari 2020) (a) Surface morphology (b) Structural (c) band gap energy plot (d) Fourier transform-Raman plots of copper selenide films (Malekar, Gangawane, and Fulari 2020). Reproduced with permission Copyright 2020, Elsevier.

$$CuSO_4.5H_2O \rightarrow CuSO_4 + 5H_2O \tag{20.1}$$

$$CuSO_4 \rightarrow Cu^{2+} + SO_4^{2-} \tag{20.2}$$

$$Cu^{2+} + 2e^- \rightarrow Cu \tag{20.3}$$

$$2Cu^{2+} + Se^{2-} \rightarrow Cu_2Se \tag{20.4}$$

20.2.8 Metallic Sulphides

Sulphides of iron and tin are also efficient solar cell materials (Vasekar and Dhakal 2013). Pyrite or iron sulphide is an abundant, lustrous, and non-poisonous, yet sustainable, semiconductor. Its high specific energy, high coefficient of absorption, and low optical band gap energy enhance its usefulness in solar cells. The low open-circuit voltage of pyrite could be due to the insufficient sulphur content. A pyrite phase can have minimal defects if the ratio of iron and sulphur is greatly regulated. FeS_2 has low band gap energy that makes it applicable as a solar cell absorber and a counter electrode in dye-sensitized solar cells (Le Donne, Trifiletti, and Binetti 2019). The unique optical and electrical features make pyrites good for photovoltaic use and as performance enhancers in solar cells (Rahman and Edvinsson 2019). Paal et al. chemically deposited pyrite at different annealing temperatures (Paal et al. 2020). The films exhibited improved crystallinity and reduced band gap energies. Xie et al. studied the effect of various sources of sulphur on chemically vapour deposited iron sulphide films (Xie et al. 2020). The results in Figure 20.2 show uniformly distributed lumps, crystalline films, high impedance with the equivalent electrochemical circuit diagram. Tin sulphide, SnS, is a chalcogenide material with the cubic crystal structure, diverse preparation methods, high coefficient of absorption, and low band gap energy (Vasekar and Dhakal 2013; Sugiyama et al. 2011). It can be synthesized by sulphurizing tin sheet and sulphur powder to yield large-grained and well-adhered films (Sugiyama et al. 2008). The sulphurized films are used to fabricate heterojunction devices for solar cell application (Sugiyama et al. 2008; Reddy et al. 2011). Tin sulphide films yield efficient solar cell devices (Miles et al. 2009; Ogah et al. 2009; Nair, Garcia-Angelmo, and Nair 2016). Nanocomposites of iron sulphide and carbon nanotubes serve as electrode materials in solar cell

(a)

(b)

(c)

(d)

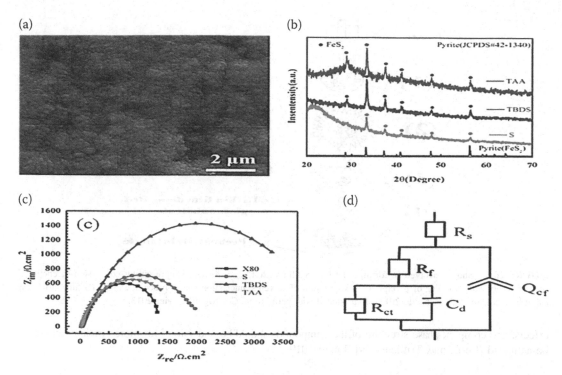

FIGURE 20.2 (a) Morphological (b) Structural (c) Nyquist and (d) Electrochemical circuit diagramplots of chemical vapour deposited pyrite film (Xie et al. 2020). Reproduced with permission. Copyright 2020, Elsevier.

devices (Yang et al. 2019). $CuSbS_2$ materials are highly conductive (due to the great copper vacancies) with reduced energy during formation, have direct band gaps, and fast mobility of carriers (Chen 2016). They have diverse synthesis methods and efficiently convert solar energy (Ezugwu, Ezema, and Asogwa 2010). Density functional theory showed reduced recombination-center defects in the forbidden gap when the sulphur content is high. It would effectively serve as a solar cell absorber.

20.2.9 Quaternary Compounds

Quaternary compounds like $Cu(In,Ga)Se_2$, Cu_2FeSnS_2, Cu_2FeSnS_4, Cu_2CoSnS_4, Cu_2MnSnS_4, Cu_2NiSnS_4, and $Cu_2ZnSn(S,Se)_4$ have increased coefficient of absorption and direct kind of band gap. Copper iron tin sulphide/CITS exhibits low band gap energies and is suitable for photovoltaic applications (Le Donne, Trifiletti, and Binetti 2019). Cu_2FeSnS_4 nanoparticles enhance optical absorption in solar cells (Deepika and Meena 2020). Copper zinc tin sulphide (CZTS) materials have low experimental efficiency, kesterite structure (Nazligul, Wang, and Choy 2020), a band gap that can be tuned, are relatively cheap, abundant, with a high coefficient of absorption, reduced energy band gaps, high stability (Ramasamy, Malik, and O'Brien 2012), recommendable structural and optoelectronic features (https://www.energy.gov/eere/solar/earth-abundant-materials 2020). It can be synthesized via several vacuum and non-vacuum methods like co-evaporation, sputtering, sulphurization, electroplating, electron beam evaporation, etc. (Shin et al. 2013). Ahmoum et al. studied the effect of preheating the deposition environment (at 150°C under nitrogen, argon, and air) on the features of spin-coated CZTS films (Ahmoum et al. 2020). The films prepared under nitrogen and air exhibited better morphological and crystalline characteristics. Olgar, Seyhan, Sarp, and Zan annealed sputtered CTZS films synthesized at varying temperatures (Olgar et al. 2020). They obtained inhomogeneous microstructures, prominent peaks, p-type conductivity, and broad peaks at about 1.36 eV to 1.37 eV. Shinde, Deokate, and Lokhande studied the features of CZTS films synthesized via spray pyrolysis as illustrated in Figure 20.3 (Shinde, Deokate, and Lokhande 2013). The obtained results showed uniform surface coverage, prominent peaks, and low optical band gap energies. Cu_2NiSnS_4 film has improved electrical conductivity and low energy band gaps, and is largely available. The defects, recombination

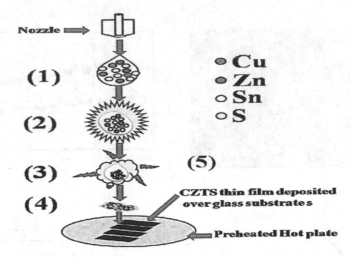

FIGURE 20.3 Stages involved in forming the CZTS film with 1, 2, 3, 4, and 5 representing conversion of chemical species to aerosols, forming of precipitates, evaporation of solvents, nucleation/growthprocess, and CZTS film formation (Shinde, Deokate, and Lokhande 2013). Reproduced with permission. Copyright Elsevier 2013.

effects, and complex phase structure of the compounds should be understood for more efficient outputs to be achieved (Le Donne, Trifiletti, and Binetti 2019).

20.3 Synthesis Methods of Earth-Abundant Materials

20.3.1 Plasma-Assisted Techniques

Plasma-assisted techniques as represented in Figure 20.4 can be employed in synthesizing chalcogenide materials as potential solar cell absorbers (https://www.energy.gov/eere/solar/earth-abundant-materials 2020a). Iron oxide, Fe_2O_3, film would be passed through hydrogen sulphide, H_2S, gas for dissociation

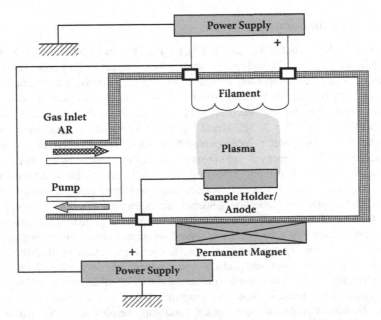

FIGURE 20.4 Schematic of a plasma-assisted technique (Ikuno et al. 2002). Reproduced with permission. Copyright Elsevier 2013.

into its constituent elements. During dissociation, oxygen atoms are replaced by sulphur atoms, thereby giving room for hematite to be converted to pyrite.

20.3.2 Chemical Vapour Deposition (CVD)

Chemical vapour deposition method uses heat-induced chemical reactions at a heated substrate surface while the reagents are supplied in gaseous form (Zhang, Sando, and Valanoor 2016). Iron sulphide layers can also be synthesized via plasma-assisted CVD by mixing hydrogen sulphide and iron cobalt, $Fe(CO)_5$ (http://inside.mines.edu/fs_home/cwolden/Wolden_Webpages/Projects/EarthAbundant_index.html 2020b). This approach yields small quantities of the material.

20.3.3 Sputtering

Sputtering is a physical method of depositing films by ejecting gaseous materials from a source on a substrate surface. Copper oxide, Cu_2O, films can be obtained from a pure copper material through direct current magnetron sputtering at constant power, controlled gas flow, base chamber pressure of 0.56×10^{-5} Pa, and changing oxygen/argon flow ratio for phase control (Lee et al. 2009). Cuprous oxide films with big grain sizes are more preferred as solar absorbers because they have increased mobility of charges and reduced charge recombination. Tin sulphide films produced via this method yield activation energy of 1.1 eV and increased crystallite size with vapour pressure (Mikami et al. 2017). Ternary alloys like $CuSbS_2$ can be synthesized using this technique. CZTS films have been observed to have deteriorating performances at longer durations and high temperatures (Vasekar and Dhakal 2013).

20.3.4 Electrochemical Deposition (ECD)

For ECD (Figure 20.5), thin adherent coatings are deposited on the surface of a conducting substrate upon electrolysis of a solution containing the desired ion. FeS_2 films electrochemically synthesized at 70°C yield an efficiency of 1.98% (Le Donne, Trifiletti, and Binetti 2019). The performance and stability of FeS_2 films can be increased by doping with cobalt ions.

20.3.5 Successive Ionic Layer Adsorption and Reaction (SILAR)

SILAR method is efficient and allows films to be deposited over large areas without vacuum involvement. Synthesizing copper iron tin sulphide film revealed high crystalline, stannite structure, and

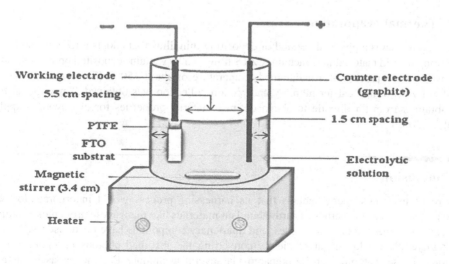

FIGURE 20.5 Setup for an electrochemical deposition technique (Ikhioya et al. 2020). Reproduced with permission. Copyright Elsevier 2020.

low band gap energy (Le Donne, Trifiletti, and Binetti 2019). Uniformly spread CZTS films can be produced by dipping the substrate into individual cation and anion solutions.

20.3.6 Chemical Synthesis

Chemical synthesis entails the production of complex materials from simpler compounds through a series of chemical reactions. Copper manganese tin sulphide, Cu_2MnSnS_4, films prepared by chemical methods exhibit small-grained crystals, high absorption coefficient, and low optical band gap energies (Le Donne, Trifiletti, and Binetti 2019).

20.3.7 Sulphurization Technique

Sulphurization is a method used to deposit high-quality films with the addition of elemental sulphur. Cu_2SnS_3 films can be sulphurized and annealed to improve the morphology and efficiency of the obtained solar cells (Sato et al. 2015). Tin sulphide films can also be sulphurized with cadmium sulphide and tin dioxide at 300°C to yield low band gap energies and highly performing devices (Sugiyama, Murata, et al. 2011).

20.3.8 Sol-Gel Method

Sol gel method is a wet chemical process used to produce materials after the solution is gradually transformed into a gel-like phase. Cu_2ZnSnS_4 material can be fabricated on soda substrates after spin coating and annealing at temperatures ranging from 300°C to 500°C. The results showed a high coefficient of absorption, reduced band gap energies, and prominent peaks whose sharpness improved with temperature (Sharmin et al. 2019).

20.3.9 Spray pyrolysis

This is a simple process for the deposition of semiconductor materials where a solution is sprayed on a heated surface. The chemical reactants are selected such that the products other than the desired compound are volatile at the deposition temperature. This method yields large-grain sized and improved carrier transport mechanisms in the films (Vasekar and Dhakal 2013). Cu_2SnS_3 materials synthesized from stannic chloride through this method yield a tetragonal crystal structure, electrical resistivity of 0.001 Ωcm, and enhanced efficiency (Sunny et al. 2017).

20.3.10 Thermal evaporation

Thermal evaporation is a physical method of depositing thin films that adopts a resistive source of heat in evaporating a solid material in a vacuum. CZTS films can be vacuum-deposited on glass substrates by thermal evaporation process at a maintained temperature of about 150°C (Shin et al. 2013). An efficiency of 8.4% with high carrier mobility and photon collection was obtained. Tin sulphide films can also be obtained from tin dioxide to obtain optimal material properties for photovoltaic applications (Miles et al. 2009).

20.4 Conclusion

The nature of the sun's energy entails that its harnessing process would incorporate low-cost and available materials. The properties of earth-abundant materials like manganese, iron, nickel, sulphur, tin, barium, chalcogenides, metallic sulphides, and quartenary compounds have been discussed. The earth-abundant materials would aim at efficiently converting the absorbed photons into energy for useful applications. These materials can be prepared via several techniques like plasma-assisted technique, chemical vapour deposition, sputtering, electrochemical deposition, SILAR, chemical synthesis, sol gel method, spray pyrolysis, thermal evaporation, amongst other methods. These methods work uniquely to boost the efficiency and performance of these abundant materials. The features of these materials can be enhanced by doping, formation of composites, and proper engineering of the heterojunction interfaces. These materials are useful in solar cell development.

REFERENCES

Abdy, Hamed, Arash Aletayeb, Mohammadreza Kolahdouz, and Ebrahim Asl Soleimani. 2019. "Investigation of Metal-Nickel Oxide Contacts Used for Perovskite Solar Cell." *AIP Advances* 9, no. 1. AIP Publishing LLC: 015216.

Ahmoum, H., P. Chelvanathan, M.S. Su'ait, M. Boughrara, G. Li, Ali HA Al-Waeli, K. Sopian, M. Kerouad, and N. Amin. 2020. "Impact of Preheating Environment on Microstructural and Optoelectronic Properties of Cu2ZnSnS4 (CZTS) Thin Films Deposited by Spin-Coating." *Superlattices and Microstructures* 140. Elsevier: 106452.

Chalapathy, R.B.V., Gwang Sun Jung, and Byung Tae Ahn. 2011. "Fabrication of Cu2ZnSnS4 Films by Sulfurization of Cu/ZnSn/Cu Precursor Layers in Sulfur Atmosphere for Solar Cells." *Solar Energy Materials and Solar Cells* 95, no. 12: 3216–3221. doi:10.1016/j.solmat.2011.07.017.

Chan, Shun-Hsiang, Ming-Chung Wu, Kun-Mu Lee, Wei-Cheng Chen, Tzu-Hao Lin, and Wei-Fang Su. 2017. "Enhancing Perovskite Solar Cell Performance and Stability by Doping Barium in Methylammonium Lead Halide." *Journal of Materials Chemistry A* 5, no. 34. The Royal Society of Chemistry: 18044–18052. doi:10.1039/C7TA05720B.

Chen, Shiyou. 2016. "Cu2ZnSnS4, Cu2ZnSnSe4, and Related Materials." In *Semiconductor Materials for Solar Photovoltaic Cells*, edited by M. Parans Paranthaman, Winnie Wong-Ng, and Raghu N. Bhattacharya, 75–103. Springer Series in Materials Science. Cham: Springer International Publishing. doi:10.1007/978-3-319-20331-7_3.

Chiu, Arlene, Eric Rong, Christianna Bambini, Yida Lin, Chengchangfeng Lu, and Susanna M. Thon. 2020. "Sulfur-Infused Hole Transport Materials to Overcome Performance-Limiting Transport in Colloidal Quantum Dot Solar Cells." *ACS Energy Letters* 5, no. 9. American Chemical Society: 2897–2904. doi:10.1021/acsenergylett.0c01586.

Deepika, R., and P. Meena. 2020. "Cu2FeSnS4 Nanoparticles: Potential Photovoltaic Absorption Materials for Solar Cell Application." *Materials Research Express* 7, no. 3. IOP Publishing: 035012.

https://www.energy.gov/eere/solar/earth-abundant-materials. "Earth-Abundant Materials." 2020a. *Energy. Gov*. Accessed October 29, 2020.

https://energyfutureslab.blog/2019/11/28/earth-abundant-materials-for-next-generation-photovoltaics/. "Earth-Abundant Materials for Next-Generation Photovoltaics." 2019.

http://inside.mines.edu/fs_home/cwolden/Wolden_Webpages/Projects/EarthAbundant_index.html. "Earth Abundant Materials for Solar Energy Conversion." 2020b. Accessed October 29, 2020.

https://www.european-mrs.com/earth-abundant-next-generation-materials-solar-energy-iii-emrs. "Earth-Abundant Next Generation Materials for Solar Energy - III | EMRS." 2020c. Accessed October 29, 2020.

Ezema, F.I., A.B.C. Ekwealor, and R.U. Osuji. 2006. "Effect of Thermal Annealing on the Band GAP and Optical Properties of Chemical Bath Deposited ZnSe Thin Films." *Turkish Journal of Physics* 30, no. 3. The Scientific and Technological Research Council of Turkey: 157–163.

Ezugwu, S.C., F.I. Ezema, and P.U. Asogwa. 2010. "Synthesis and Characterization of Ternary CuSbS 2 Thin Films: Effect of Deposition Time." *Chalcogenide Letters* 7, no. 5: 341–348.

Guai, Guan Hong, Ming Yian Leiw, Chee Mang Ng, and Chang Ming Li. 2012. "Sulfur-Doped Nickel Oxide Thin Film as an Alternative to Pt for Dye-Sensitized Solar Cell Counter Electrodes." *Advanced Energy Materials* 2, no. 3: 334–338. doi:10.1002/aenm.201100582.

Guo, Fu-an, Guoqiang Li, and Weifeng Zhang. 2010. "Barium Staminate as Semiconductor Working Electrodes for Dye-Sensitized Solar Cells." *International Journal of Photoenergy*. Hindawi. doi:https://doi.org/10.1155/2010/105878.

Gupta, Vinay, Aung Ko Ko Kyaw, Dong Hwan Wang, Suresh Chand, Guillermo C. Bazan, and Alan J. Heeger. 2013. "Barium: An Efficient Cathode Layer for Bulk-Heterojunction Solar Cells." *Scientific Reports* 3, no. 1. Nature Publishing Group: 1965. doi:10.1038/srep01965.

Haider, Mustafa, Chao Zhen, Tingting Wu, Gang Liu, and Hui-Ming Cheng. 2018. "Boosting Efficiency and Stability of Perovskite Solar Cells with Nickel Phthalocyanine as a Low-Cost Hole Transporting Layer Material." *Journal of Materials Science & Technology* 34, no. 9: 1474–1480. doi:10.1016/j.jmst.201 8.03.005.

Ikhioya, Imosobomeh L., Agnes C. Nkele, Chidimma F. Okoro, O. Obasi Chidiebere, G.M. Whyte, M. Maaza, and Fabian I. Ezema. 2020. "Effect of Temperature on the Morphological, Structural and Optical Properties of Electrodeposited Yb-Doped ZrSe2 Thin Films." *Optik* 220: 165180. Elsevier.

Ikuno, Takashi, J.-T. Ryu, T. Oyama, S. Ohkura, Y.G. Baek, Shin-ichi Honda, Mitsuhiro Katayama, T. Hirao, and K. Oura. 2002. "Characterization of Low Temperature Growth Carbon Nanofibers Synthesized by Using Plasma Enhanced Chemical Vapor Deposition." *Vacuum* 66, no. August: 341–345. doi:10.1016/ S0042-207X(02)00141-0.

Jung, Kyungeun, Jeongwon Lee, Young-Min Kim, Yun Chang Park, and Man-Jong Lee. 2019. "Effect of Manganese Dopants on Defects, Nano-Strain, and Photovoltaic Performance of Mn–CdS/CdSe Nanocomposite-Sensitized ZnO Nanowire Solar Cells." *Composites Science and Technology* 179. Elsevier: 79–87.

Kawade, Daisuke, Shigefusa F. Chichibu, and Mutsumi Sugiyama. 2014. "Experimental Determination of Band Offsets of NiO-Based Thin Film Heterojunctions." *Journal of Applied Physics* 116, no. 16. American Institute of Physics: 163108. doi:10.1063/1.4900737.

Le Donne, Alessia, Vanira Trifiletti, and Simona Binetti. 2019. "New Earth-Abundant Thin Film Solar Cells Based on Chalcogenides." *Frontiers in Chemistry* 7. Frontiers. doi:10.3389/fchem.2019.00297.

Lee, Yun Seog, Mariana Bertoni, Maria K Chan, Gerbrand Ceder, and Tonio Buonassisi. 2009. "Earth Abundant Materials for High Efficiency Heterojunction Thin Film Solar Cells," 3. IEEE.

Li, Kai, and Rik Van Deun. 2019. "Enhancing the Energy Transfer from Mn4+ to Yb3+ via a Nd3+ Bridge Role in Ca3La2W2O12:Mn4+,Nd3+,Yb3+ Phosphors for Spectral Conversion of c-Si Solar Cells." *Dyes and Pigments* 162 (March): 990–997. doi:10.1016/j.dyepig.2018.11.030.

Li, Ling, Xiaohui Wang, Zeyuan Ma, Lu Liu, Hongjie Wang, Wenming Zhang, and Xiaowei Li. 2019. "Electrospinning Synthesis and Electrocatalytic Performance of Iron Oxide/Carbon Nanofibers Composites as a Low-Cost Efficient Pt-Free Counter Electrode for Dye-Sensitized Solar Cells." *Applied Surface Science* 475. Elsevier: 109–116.

Liu, Wei, Liang Chu, Nanjing Liu, Yuhui Ma, Ruiyuan Hu, Yakui Weng, Hui Li, Jian Zhang, Xing'ao Li, and Wei Huang. 2019. "Efficient Perovskite Solar Cells Fabricated by Manganese Cations Incorporated in Hybrid Perovskites." *Journal of Materials Chemistry C* 7, no. 38. Royal Society of Chemistry: 11943–11952.

Lv, Xincong, Chenyan Hu, Jin Shang, Patrick H.-L. Sit, Frank LY Lam, and Wey Yang Teoh. 2019. "Enhanced Photoelectrochemical Charge Transfer on Mn-Doped CdS/TiO2 Nanotube Arrays: The Roles of Organic Substrates." *Catalysis Today* 335. Elsevier: 468–476.

Maduraiveeran, Govindhan, Manickam Sasidharan, and Wei Jin. 2019. "Earth-Abundant Transition Metal and Metal Oxide Nanomaterials: Synthesis and Electrochemical Applications." *Progress in Materials Science* 106 (December): 100574. doi:10.1016/j.pmatsci.2019.100574.

Maiti, Sourav, Jayanta Dana, and Hirendra N. Ghosh. 2019. "Correlating Charge-Carrier Dynamics with Efficiency in Quantum-Dot Solar Cells: Can Excitonics Lead to Highly Efficient Devices?" *Chemistry – A European Journal* 25, no. 3: 692–702. doi:10.1002/chem.201801853.

Malekar, V.P., Gangawane, S.A., and Fulari, V.J.. 2020. "Characterization and Holographic Study of Nanostructure Copper Selenide Thin Films Grown at Room Temperature." *Materials Today: Proceedings* 23. Elsevier: 202–210.

Mali, Sawanta S., Jyoti V. Patil, and Chang Kook Hong. 2019. "Hot-Air-Assisted Fully Air-Processed Barium Incorporated CsPbI2Br Perovskite Thin Films for Highly Efficient and Stable All-Inorganic Perovskite Solar Cells." *Nano Letters* 19, no. 9. American Chemical Society: 6213–6220. doi:10.1021/ acs.nanolett.9b02277.

Marandi, M., P. Talebi, and S. Bayat. 2019. "Optimization of the Doping Process and Light Scattering in CdS: Mn Quantum Dots Sensitized Solar Cells for the Efficiency Enhancement." *Journal of Materials Science: Materials in Electronics* 30, no. 4. Springer: 3820–3832.

Marandi, Maziar, Parisa Talebi, and Leila Moradi. 2019. "Co-Application of TiO2 Nanoparticles and Randomly Directed TiO2 Nanorods in the Photoelectrode of the CdS: Mn Quantum Dots Sensitized

Solar Cells and Optimization of the Doping for the Efficiency Improvement." *Optical Materials* 94. Elsevier: 224–230.

Mar-Ortiz, A.F., Jacob J. Salazar-Rábago, Manuel Sánchez-Polo, M. Rozalen, Felipe J. Cerino-Córdova, and M. Loredo-Cancino. 2020. "Photodegradation of Antihistamine Chlorpheniramine Using a Novel Iron-Incorporated Carbon Material and Solar Radiation." *Environmental Science: Water Research & Technology* 6, no. 9. Royal Society of Chemistry: 2607–2618.

Mikami, Shuntaro, Tsubasa Yokoi, Hiroki Sumi, Satoru Aihara, Ishwor Khatri, and Mutsumi Sugiyama. 2017. "Effect of Sulfur Vapor Pressure on SnS Thin Films Grown by Sulfurization." *Physica Status Solidi c* 14, no. 6: 1600160. doi: https://doi.org/10.1002/pssc.201600160.

Miles, Robert W., Ogah E. Ogah, Guillaume Zoppi, and Ian Forbes. 2009. "Thermally Evaporated Thin Films of SnS for Application in Solar Cell Devices." *Thin Solid Films*, 4th International Conference on Technological Advances of Thin Films and Surface Coatings, 517, no. 17: 4702–4705. doi: 10.1016/j.tsf.2009.03.003.

Murugadoss, Govindhasamy, Hiroyuki Kanda, Soichiro Tanaka, Hitoshi Nishino, Seigo Ito, Hiroshi Imahori, and Tomokazu Umeyama. 2016. "An Efficient Electron Transport Material of Tin Oxide for Planar Structure Perovskite Solar Cells." *Journal of Power Sources* 307 (March): 891–897. doi: 10.1016/j.jpowsour.2016.01.044.

Nair, P.K., E. Barrios-Salgado, and M.T.S. Nair. 2016. "Cubic-Structured Tin Selenide Thin Film as a Novel Solar Cell Absorber." *Physica Status Solidi (a)* 213, no. 8: 2229–2236. doi: 10.1002/pssa.201533040.

Nair, P.K., A.R. Garcia-Angelmo, and M.T.S. Nair. 2016. "Cubic and Orthorhombic SnS Thin-Film Absorbers for Tin Sulfide Solar Cells." *Physica Status Solidi (a)* 213, no. 1: 170–177. doi: 10.1002/pssa.201532426.

Nattestad, Andrew, Michael Ferguson, Robert Kerr, Yi-Bing Cheng, and Udo Bach. 2008. "Dye-Sensitized Nickel(II)Oxide Photocathodes for Tandem Solar Cell Applications." *Nanotechnology* 19, no. 29. IOP Publishing: 295304. doi: 10.1088/0957-4484/19/29/295304.

Nazligul, Ahmet Sencer, Mingqing Wang, and Kwang Leong Choy. 2020. "Recent Development in Earth-Abundant Kesterite Materials and Their Applications." *Sustainability* 12, no. 12. Multidisciplinary Digital Publishing Institute: 5138.

Neupane, Ganga R., Amrit Kaphle, and Parameswar Hari. 2019. "Microwave-Assisted Fe-Doped ZnO Nanoparticles for Enhancement of Silicon Solar Cell Efficiency." *Solar Energy Materials and Solar Cells* 201. Elsevier: 110073.

Nkele, Agnes C., Assumpta C. Nwanya, Nwankwo Uba Nwankwo, Rose U. Osuji, A.B.C. Ekwealor, Paul M. Ejikeme, Malik Maaza, and Fabian I. Ezema. 2019. "Investigating the Properties of Nano Nest-like Nickel Oxide and the NiO/Perovskite for Potential Application as a Hole Transport Material." *Advances in Natural Sciences: Nanoscience and Nanotechnology* 10, no. 4: 045009. doi: 10.1088/2043-6254/ab5102.

Nkele, Agnes C., Assumpta C. Nwanya, Nanasaheb M. Shinde, Sabastine Ezugwu, Malik Maaza, Jasmin S. Shaikh, and Fabian I. Ezema. 2020. "The Use of Nickel Oxide as a Hole Transport Material in Perovskite Solar Cell Configuration: Achieving a High Performance and Stable Device." *International Journal of Energy Research*. Wiley Online Library. doi: https://doi.org/10.1002/er.5563.

Ogah, Ogah E., Guillaume Zoppi, Ian Forbes, and R.W. Miles. 2009. "Thin Films of Tin Sulphide for Use in Thin Film Solar Cell Devices." *Thin Solid Films*, Thin Film Chalogenide Photovoltaic Materials (EMRS, Symposium L), 517, no. 7: 2485–2488. doi: 10.1016/j.tsf.2008.11.023.

Olgar, M.A., A. Seyhan, A.O. Sarp, and R. Zan. 2020. "Impact of Sulfurization Parameters on Properties of CZTS Thin Films Grown Using Quaternary Target." *Journal of Materials Science: Materials in Electronics*. Springer, 31: 1–12.

Paal, Mark, Isaac Nkrumah, Francis K. Ampong, David Ngbiche, Robert K. Nkum, and Francis Boakye. 2020. "The Effect of Deposition Time and Sulfurization Temperature on The Optical and Structural Properties of Iron Sulfide Thin Films Deposited from Acidic Chemical Baths." *Science Journal of University of Zakho* 8, no. 3: 97–104.

Pengerla, M., S. Al-Hajjawi, V. Kuruganti, J. Haunschild, N. Schüler, K. Dornich, and S. Rein. 2020. "Comparing Microwave Detected Photoconductance, Quasi Steady State Photoconductance and Photoluminescence Imaging for Iron Analysis in Silicon."

Pitchaiya, Selvakumar, Muthukumarasamy Natarajan, Agilan Santhanam, Venkatraman Madurai Ramakrishnan, Vijayshankar Asokan, Pavithrakumar Palanichamy, Balasundaraprabhu Rangasamy, Senthilarasu Sundaram, and Dhayalan Velauthapillai. 2018. "Nickel Sulphide-Carbon Composite Hole Transporting Material for (CH3NH3PbI3) Planar Heterojunction Perovskite Solar Cell." *Materials Letters* 221. Elsevier: 283–288.

Platzer-Björkman, C., J. Scragg, H. Flammersberger, T. Kubart, and M. Edoff. 2012. "Influence of Precursor Sulfur Content on Film Formation and Compositional Changes in Cu2ZnSnS4 Films and Solar Cells." *Solar Energy Materials and Solar Cells* 98 (March): 110–117. doi:10.1016/j.solmat.2011.10.019.

Pulvirenti, Federico, Berthold Wegner, Nakita K. Noel, Giulio Mazzotta, Rebecca Hill, Jay B. Patel, Laura M. Herz, et al. 2018. "Modification of the Fluorinated Tin Oxide/Electron-Transporting Material Interface by a Strong Reductant and Its Effect on Perovskite Solar Cell Efficiency." *Molecular Systems Design & Engineering* 3, no. 5. The Royal Society of Chemistry: 741–747. doi:10.1039/C8ME00031J.

Rahman, Mohammad Z., and Tomas Edvinsson. 2019. "What Is Limiting Pyrite Solar Cell Performance?" *Joule* 3, no. 10: 2290–2293. doi:10.1016/j.joule.2019.06.015.

Ramasamy, Karthik, Mohammad A. Malik, and Paul O'Brien. 2012. "Routes to Copper Zinc Tin Sulfide Cu2ZnSnS4 a Potential Material for Solar Cells." *Chemical Communications* 48, no. 46. The Royal Society of Chemistry: 5703–5714. doi:10.1039/C2CC30792H.

Reddy, KT Ramakrishna, K. Ramya, G. Sreedevi, T. Shimizu, Y. Murata, and Mutsumi Sugiyama. 2011. "Studies on the Energy Band Discontinuities in SnS/ZnMgO Thin Film Heterojunction." *Energy Procedia* 10. Elsevier: 172–176.

Rehman, Faisal, Khalid Mahmood, Arshi Khalid, Muhammad Shahzad Zafar, and Madsar Hameed. 2019. "Solution-Processed Barium Hydroxide Modified Boron-Doped ZnO Bilayer Electron Transporting Materials: Toward Stable Perovskite Solar Cells with High Efficiency of over 20.5%." *Journal of Colloid and Interface Science* 535 (February): 353–362. doi:10.1016/j.jcis.2018.10.011.

Sato, Soichi, Hiroki Sumi, Guannan Shi, and Mutsumi Sugiyama. 2015. "Investigation of the Sulfurization Process of Cu2SnS3 Thin Films and Estimation of Band Offsets of Cu2SnS3-Related Solar Cell Structure." *Physica Status Solidi c* 12, no. 6: 757–760. doi:https://doi.org/10.1002/pssc.201400294.

Scholtz, Ľubomír, Pavel Šutta, Pavel Calta, Petr Novák, Michaela Solanská, and Jarmila Müllerová. 2018. "Investigation of Barium Titanate Thin Films as Simple Antireflection Coatings for Solar Cells." *Applied Surface Science*, 5th Progress in Applied Surface, Interface and Thin Film Science and Solar Renewable Energy News, 461 (December): 249–254. doi:10.1016/j.apsusc.2018.06.226.

Sharmin, Afrina, M.S. Bashar, Samia Tabassum, and Zahid Hasan Mahmood. 2019. "Low Cost and Sol-Gel Processed Earth Abundant Cu 2 ZnSnS 4 Thin Film as an Absorber Layer for Solar Cell: Annealing without Sulfurization." *IJTFST* 8, no. 2: 65–74.

Shin, Byungha, Oki Gunawan, Yu Zhu, Nestor A. Bojarczuk, S. Jay Chey, and Supratik Guha. 2013. "Thin Film Solar Cell with 8.4% Power Conversion Efficiency Using an Earth-Abundant Cu2ZnSnS4 Absorber." *Progress in Photovoltaics: Research and Applications* 21, no. 1: 72–76. doi:10.1002/pip.1174.

Shinde, N.M., R.J. Deokate, and C.D. Lokhande. 2013. "Properties of Spray Deposited Cu2ZnSnS4 (CZTS) Thin Films." *Journal of Analytical and Applied Pyrolysis* 100 (March): 12–16. doi:10.1016/j.jaap.2012.10.018.

Sugiyama, Mutsumi, Keisuke Miyauchi, Takehiro Minemura, and Hisayuki Nakanishi. 2008. "Sulfurization Growth of SnS Films and Fabrication of CdS/SnS Heterojunction for Solar Cells." *Japanese Journal of Applied Physics* 47, no. 12R. IOP Publishing: 8723.

Sugiyama, Mutsumi, Keisuke Miyauchi, Takehiro Minemura, Kenichi Ohtsuka, Koji Noguchi, and Hisayuki Nakanishi. 2008. "Preparation of SnS Films by Sulfurization of Sn Sheet." *Japanese Journal of Applied Physics* 47, no. 6R. IOP Publishing: 4494.

Sugiyama, Mutsumi, Yoshitsuna Murata, Tsubasa Shimizu, Kottadi Ramya, Chinna Venkataiah, Tomoaki Sato, and K.T. Ramakrishna Reddy. 2011. "Sulfurization Growth of SnS Thin Films and Experimental Determination of Valence Band Discontinuity for SnS-Related Solar Cells." *Japanese Journal of Applied Physics* 50, no. 5S2. IOP Publishing: 05FH03. doi:10.1143/JJAP.50.05FH03.

Sugiyama, Mutsumi, K.T.R. Reddy, N. Revathi, Y. Shimamoto, and Y. Murata. 2011. "Band Offset of SnS Solar Cell Structure Measured by X-Ray Photoelectron Spectroscopy." *Thin Solid Films* 519, no. 21. Elsevier: 7429–7431.

Sugiyama, Mutsumi, Tsubasa Shimizu, Daisuke Kawade, Kottadi Ramya, and K.T. Ramakrishna Reddy. 2014. "Experimental Determination of Vacuum-Level Band Alignments of SnS-Based Solar Cells by Photoelectron Yield Spectroscopy." *Journal of Applied Physics* 115, no. 8. American Institute of Physics: 083508. doi:10.1063/1.4866992.

Sunny, Gincy, Titu Thomas, D.R. Deepu, C. Sudha Kartha, and K.P. Vijayakumar. 2017. "Thin Film Solar Cell Using Earth Abundant Cu2SnS3 (CTS) Fabricated through Spray Pyrolysis: Influence of Precursors." *Optik* 144. Elsevier: 263–270.

Thanikaikarasan, S., D. Dhanasekaran, and K. Sankaranarayanan. 2020. "Electrochemical, Structural, Compositional and Optical Properties of Cuprous Selenide Thin Films." *Chinese Journal of Physics* 63. Elsevier: 138–148.

Timmo, K., M. Altosaar, J. Raudoja, K. Muska, M. Pilvet, M. Kauk, T. Varema, M. Danilson, O. Volobujeva, and E. Mellikov. 2010. "Sulfur-Containing Cu2ZnSnSe4 Monograin Powders for Solar Cells." *Solar Energy Materials and Solar Cells*, Inorganic and Nanostructured Photovoltaics, 94, no. 11: 1889–1892. doi:10.1016/j.solmat.2010.06.046.

Trost, Sara, Andreas Behrendt, Tim Becker, Andreas Polywka, Patrick Görrn, and Thomas Riedl. 2015. "Tin Oxide (SnOx) as Universal 'Light-Soaking' Free Electron Extraction Material for Organic Solar Cells." *Advanced Energy Materials* 5, no. 17: 1500277. doi:10.1002/aenm.201500277.

Vasekar, Parag S., and Tara P. Dhakal. 2013. "Thin Film Solar Cells Using Earth-Abundant Materials." *Solar Cells: Research and Application Perspectives*, March. IntechOpen. doi:10.5772/51734.

Vismara, R., O. Isabella, and M. Zeman. 2016. "Organometallic Halide Perovskite/Barium Di-Silicide Thin-Film Double-Junction Solar Cells." In *Photonics for Solar Energy Systems VI*, 9898:98980J. International Society for Optics and Photonics. doi:10.1117/12.2227174.

Wang, Kuo-Chin, Jun-Yuan Jeng, Po-Shen Shen, Yu-Cheng Chang, Eric Wei-Guang Diau, Cheng-Hung Tsai, Tzu-Yang Chao, et al. 2014. "P-Type Mesoscopic Nickel Oxide/Organometallic Perovskite Heterojunction Solar Cells." *Scientific Reports* 4 (April): 4756. doi:10.1038/srep04756.

Wang, Yanfang, Yanjun Guo, Wentao Chen, Qing Luo, Wenli Lu, Peng Xu, Dongliang Chen, Xiong Yin, and Meng He. 2018. "Sulfur-Doped Reduced Graphene Oxide/MoS2 Composite with Exposed Active Sites as Efficient Pt-Free Counter Electrode for Dye-Sensitized Solar Cell." *Applied Surface Science* 452 (September): 232–238. doi:10.1016/j.apsusc.2018.04.276.

Wu, Ming-Chung, Wei-Cheng Chen, Shun-Hsiang Chan, and Wei-Fang Su. 2018. "The Effect of Strontium and Barium Doping on Perovskite-Structured Energy Materials for Photovoltaic Applications." *Applied Surface Science* 429. Elsevier: 9–15.

Xia, Congxin, Yu Jia, Meng Tao, and Qiming Zhang. 2013. "Tuning the Band Gap of Hematite α-Fe2O3 by Sulfur Doping." *Physics Letters A* 377, no. 31: 1943–1947. doi:10.1016/j.physleta.2013.05.026.

Xie, Chuang, Boyi Wang, Shujie Li, Xiangli Wen, Shikai Wei, Shuai Zhang, Min Feng, Liqiang Chen, and Shuqi Zheng. 2020. "The Dependence of Anti-Corrosion Behaviors of Iron Sulfide Films on Different Reactants." *International Journal of Hydrogen Energy*. Elsevier.

Yang, Sheng, Peng Huang, Meng Duan, Yongying Li, and Guo Gao. 2019. "Controllable Synthesis of Iron Sulfide/CNT Nanocomposites in Solvothermal System." *Crystal Research and Technology* 54, no. 7. Wiley Online Library: 1900029.

Yuwawech, Kitti, Jatuphorn Wootthikanokkhan, and Supachok Tanpichai. 2020. "Preparation and Characterization of Iron Oxide Decorated Graphene Nanoplatelets for Use as Barrier Enhancing Fillers in Polyurethane Based Solar Cell Encapsulant." *Materials Today: Proceedings*, The 6th Thailand International Nanotechnology Conference, NanoThailand 2018, 12-14 December 2018, Pathum Thani, Thailand, 23 (January): 703–711. doi:10.1016/j.matpr.2019.12.262.

Zafar, Muhammad, Ju-Young Yun, and Do-Heyoung Kim. 2019. "Improved Inverted-Organic-Solar-Cell Performance via Sulfur Doping of ZnO Films as Electron Buffer Layer." *Materials Science in Semiconductor Processing* 96 (June): 66–72. doi:10.1016/j.mssp.2019.01.046.

Zhang, Qi, Daniel Sando, and Nagarajan Valanoor. 2016. "Chemical Route Derived Bismuth Ferrite Thin Films and Nanomaterials." *Journal of Materials Chemistry C* 4 (April). doi:10.1039/C6TC00243A.

21

New Perovskite Materials for Solar Cell Applications

Agnes Chinecherem Nkele[1], Sabastine Ezugwu[2], Mutsumi Suguyima[3], and Fabian I. Ezema[1,4,5]
[1]Department of Physics & Astronomy, University of Nigeria, Nsukka, Enugu, Nigeria
[2]Department of Physics and Astronomy, The University of Western Ontario, London, OntarioN6A 3K7, Canada
[3]Faculty of Science and Technology, Department of Electrical Engineering, Tokyo University of Science, Noda, Japan
[4]Nanosciences African Network (NANOAFNET), iThemba LABS-National Research Foundation, Somerset West, Western Cape Province, South Africa
[5]UNESCO-UNISA Africa Chair in Nanosciences/Nanotechnology, College of Graduate Studies, University of South Africa (UNISA), Muckleneuk Ridge, Pretoria, South Africa

21.1 Introduction of Perovskite Solar Cells

The present worries over the harmful effects of climatic changes owing to global warming have created an urgency in the minds of researchers to generate clean energy sources that would be sustainable over long periods (Advance Science News 2020). Human society would be greatly improved with the introduction of renewable energy technologies. Presently, the solar energy community has delved into the use of perovskite technology to boost the efficiencies obtainable from solar cells. Solar cells can convert absorbed optical energy into electricity through several light reactions. Perovskite is a crystal structure with a cubic geometry that has an ABX_3 structure similar to calcium titanium oxide, where A is the alkali earth metal/methylammonium, B is tin or lead, X is the halide component as shown in Figure 21.1 (Nkele, Nwanya, et al. 2020). The cations and anions are obtained from A, B, and C, respectively. Perovskite materials have proved to have some edges over silicon solar cells in terms of their cost and efficiency. They can be made from halide/organic materials or chalcogenides. Organic perovskites are more toxic (owing to the lead component) and unstable than chalcogenide perovskite materials (Bellini 2020). Hybrid perovskites undergo quick phase changes in the presence of light, air, moisture, and high temperature. The flexible structures and compositions of perovskites create diverse perovskite properties. The unique features exhibited by perovskites are affected by the annealing temperature, the solvent used, and the fabrication technique employed. Perovskites are also desired because of their outstanding thermal, photoelectric, optical, and structural characteristics (Zhou et al. 2018). Perovskite materials enhance simultaneous mobility of charge carriers and produce devices with high open-circuit voltage, big dielectric constant values, and reduced short circuit current values (Zhou et al. 2018). An outstanding increase in the power conversion efficiency of perovskites has been steadily recorded from inception to date (Nkele, Ike, et al. 2020). The increasing device efficiencies were due to the diverse device architectures and fabrication methods.

A full perovskite system usually comprises of the perovskite component, metal contact, conducting surfaces, hole and electron transport layers (Nkele et al. 2019). Different architectures of perovskite

DOI: 10.1201/9781003145585-21

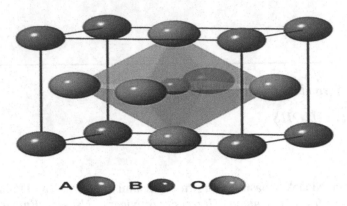

FIGURE 21.1   Diagram showing the crystal structure of perovskites (Inaguma et al. 2002). Reproduced with permission. Copyright Elsevier 2002.

solar cells like planar, inverted planar, mesoscopic, and mesoporous are geared towards improving the efficiency and stability of perovskite devices (Nkele, Nwanya, et al. 2020). Planar architectures do not have porous build-up while mesoporous structures are highly porous, very sensitive to light, and have wide specific surface areas. The components of perovskite cells have band gaps that require less exciting energy and can be tuned to allow for more light to be absorbed into the perovskite material. Upon irradiation of solar energy, photons get absorbed into the perovskite material and excitons are spontaneously generated (Zhou et al. 2018). The excitons either produce free charge carriers for electricity generation or further recombination into electron-hole pairs. The hole and electron transport layers aid minimal recombination at the interface, thereby increasing the lifetime and diffusion length of the charge carriers. Fabricating perovskite devices without hole transport layers make the device structure simple with increased stable structure and easy methods of processing. Doping perovskite materials increase their binding energy and conductivity (Zhou et al. 2018). Solvents like gamma-butyrolactone (GBL), Dimethyl Sulphoxide (DMSO), N-methyl-2-pyrrolidone (NMP), N-cyclohexyl-2-pyrollidone (CHP), and Dimethyl Formamide (DMF) can be used to easily dissolve the perovskite precursors (Seo et al. 2017). Anti-solvents like dichloromethane (DCM), chlorobenzene (CB), and toluene (TL) can be employed to eliminate remnants of solvents that may be left. Perovskite stability can be improved by doping, minimizing crystal defects, utilizing stable materials, and encapsulating the full device (Chu et al. 2019). Perovskites have simple and relatively affordable deposition methods like vapour deposition, screen printing, vapour-assisted process, and solution methods (Zhou et al. 2018). Defects associated with perovskites could be due to its scaling at the atomic level and the electrodes collecting current upon assembly of the crystal structure. Perovskite materials are useful absorber layers for effective light absorption in solar cells.

21.2 Organic-Inorganic Perovskite Materials

Hybrid solar cells comprising organic and inorganic components have been greatly studied due to the fast mobile carriers, extended life of carriers, high coefficient of extinction, and high power efficiencies (Zhou et al. 2018). It has successfully demonstrated to be a good substitute for silicon solar cells. The x-component in the composition configuration represents the halide group which could be bromide, chloride, or iodide.

21.2.1 Methylammonium Lead Halide, $CH_3NH_3PbX_3$

Methylammonium lead iodide has direct band gap nature, great electronic features, high optical absorption, small binding energy of generated excitons, and mobile charge carriers (Nkele et al. 2019).

They can serve as light sensitizers to substitute the roles of dyes in dye-sensitized solar cells. Dhamaniya et al. studied the crystal structure effect on the degrading nature of $CH_3NH_3PbX_3$ (Dhamaniya et al. 2019). They discovered that diverse crystal structures obtained from different fabrication methods and preparation routes affect the rate of degradation of perovskite devices. High transmittance and optical scintillation features are obtainable from methylammonium lead halides upon doping with bromine at low temperatures (Xu et al. 2019). The stability of these devices can be improved by incorporating interface layers that would be resistant to humid conditions and thermal stress (Ava et al. 2019). Alloys of bromine also thermally stabilize the structure of these materials (Kennard et al. 2019). Perovskites made from mixed halides have increased conductivity due to their reduction in band gap energies (Ezealigo et al. 2017). The charge transport properties of methylammonium lead iodide can be engineered by exposing it to great amounts of moisture by transforming its structural composition (Toloueinia et al. 2020). Treating lead sulphide films with polyiodide can hasten its transformation into lead iodide, although it slows down its rate of conversion in the course of yielding adhesive films (Perez et al. 2020). Wu et al. fabricated methylammonium lead iodide in an inverted architecture using phenothiazine as additive as shown in Figure 21.2 (Wu et al. 2020). The additive, phenothiazine, improved the features of the fabricated device. Crystalline films with high optical absorption and small grain boundaries with improvement in electrical features were obtained.

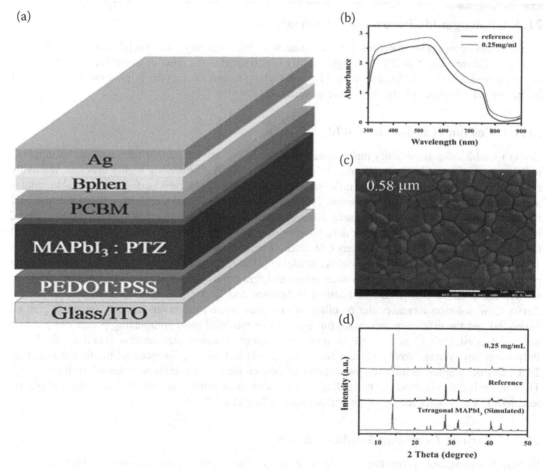

FIGURE 21.2 (a) Device architecture (b) Absorbance plot (c) SEM image (d) XRD plot of the fabricated perovskite device (Y. Wu et al. 2020). Reproduced with permission. Copyright Elsevier 2020.

21.2.2 Methylammonium Tin Halide, $CH_3NH_3SnX_3$

This perovskite material has proved to be a good substitute for lead, as it minimizes toxicity and increases device performance (Singh, Agarwal, and Agarwal 2020). It exhibits a direct band gap nature and fast mobility of charge carriers that make it vastly applied in solar cell applications. At various temperatures it exhibits several crystal phases like tetragonal, monoclinic, cubic, and orthorhombic structures. The halide element chosen determines the band gap energy of the deposited perovskite material (Hien et al. 2020). The light response, efficiency, and size of the obtained materials are affected by its band gap nature. Rahul et al. synthesized methylammonium tin chloride films from tin chloride – $SnCl_2$ and CH_3NH_3Cl solutions dissolved in dimethylformamide and stirred for 8 hours (Rahul et al. 2018). Low band gap energy, uniformly distributed crystals over the substrate surface, and monoclinic crystal structure were obtained. Ion exchange fabrication method yields efficient, no pinhole, reproducible, and stable perovskite device (Wang et al. 2020). Li et al. fabricated this device via the cation exchange method and yielded large-grained morphology, crystalline films, and power conversion efficiency of 7.13% (F. Li et al. 2019). The performance and open-circuit voltage of methylammonium tin halide can be increased by adding hydrazine monohydrobromide (Wu et al. 2020). Ion exchange method was adopted in fabricating $CH_3NH_3PbI_3$ film from CH_3NH_3Cl and SnF_2 (Wang et al. 2020). Uniform crack-free films of high crystalline and optical features were obtained with improved stability and a conversion efficiency of 7.78%.

21.3 Chalcogenide Perovskite Materials

Chalcogenide perovskite materials are less toxic with high stability and exceptional optical features (Advance Science News 2020). The high stability is attributed to the absence of the organic component in the perovskite material. Their increased light-absorbing features due to the sulphur orbitals developed in the crystal structures of the perovskite material.

21.3.1 Cesium Lead Iodide, $CsPbI_3$

Cesium lead halides have easily tuned optical features which are commonly fabricated as cesium lead iodide or cesium lead bromide (Ghorai, Midya, and Ray 2019). $CsPbI_3$ is a highly efficient, relatively affordable perovskite material with no organic component. It has a cubic structure and stable configuration (Okinawa Institute of Science and Technology (OIST) Graduate University. 2019). Cesium lead iodide could exist in alpha or beta phases. Although its alpha phase structure has high optical absorption, it can easily degrade into a form that does not absorb much light. The beta phase has better stability but less output efficiency due to the presence of cracks which cause electrons to be lost into the cell layers. The black phase is more stable and yields excellent optical features (Thomas et al. 2019). However, the energy levels can be aligned to produce better efficiency by introducing aqueous choline iodide to resolve the crack effects (Okinawa Institute of Science and Technology (OIST) Graduate University. 2019). This solution increases the mobility of electrons around the layers so that electricity can be generated and the efficiency increased. Integrating solvents with great complexing power can produce pure and crystalline films as well as increased charge transport mechanisms (Hadi et al. 2019). Polymorphism (Wang, Novendra, and Navrotsky 2019) and treating surfaces with solution (Li et al. 2020) can be adopted to improve the stability of cesium lead iodide devices as shown in Figure 21.3. The First-principles method can be studied to get a better insight into the structural properties of cesium lead iodide perovskites for optimum performance (Chen et al. 2020).

21.3.2 Barium Zirconium Sulphide, $BaZrS_3$

$BaZrS_3$ is an available perovskite material with no poisonous effect, good transport mobility of carrier charges, and high absorption of light (Bellini 2020). It also has a direct band gap nature and is chemically stable (Wei et al. 2019). It can be obtained by sulphurizing oxide films, heating to vapour $BaZrO_3$, depositing

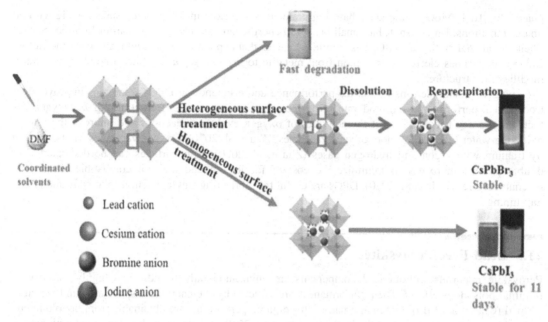

FIGURE 21.3 Formation stages for cesium lead halide upon surface treatment (D. Li et al. 2020). Reproduced with permission. Copyright American Chemical Society 2020.

the vapour on a surface, and converting the film into the final product through several chemical processes. This sulphurization process yields more conductive, high coefficient of absorption, and low band gap energy devices (Wei, Hui, and Zhao et al. 2020). Titanium can be used as a dopant to form an alloy, $BaZrTiO_3$, which would aid band gap engineering of the host compound. Barium zirconium sulphide degrades less over time because ions do not migrate quickly and there is no strong bond with water (Gupta et al. 2020). An alloy of titanium and $BaZrS_3$ produces materials with reduced band gap energies and effective transport of charge carriers (Wei, Hui, and Perera et al. 2020). Incorporating selenium as a dopant reduces the band gap energy and good transport features (Zitouni et al. 2020).

Calcium Zirconium Sulphide, $CaZrS_3$, and Strontium Zirconium Sulphide, $SrZrS_3$, are also chalcogenide perovskite materials that have excellent optical absorption. This gives room for the photogenerated holes and electrons to be collected into their respective sites for reduced recombination and increased efficiencies.

21.4 Double Perovskite Oxides (DPOs)

Double perovskite oxides have been developed to resolve toxic and stability issues posed by the presence of lead in perovskite materials. DPOs are highly stable with direct band gap nature, easily tailored characteristics, high oxygen conduction, and outstanding optical and electronic characteristics. Their high oxygen concentration increases the rate of oxygen diffusion and oxygen-ion intercalation because oxygen is the main charge carrier. Sr_2CoTaO_6 exists in different crystal structures and manifests great photocatalytic attributes (Idris et al. 2020). The concentration of lanthanum influences the thermoelectric features of $Sr_{2-x}La_xCoTiO_{6-1}$ and stabilizes its structural state. Treating Sr_2NiWO_6 or Sr_2FeWO_6 with ammonia at regulated temperatures via topochemical nitridation yields materials with antiferromagnetic properties while Pt improves its photocatalytic features (Idris et al. 2020). First-principles density functional theory explains the ferroelectric nature in strained double perovskite fluorides, Na_3ScF_6 and K_2NaScF_6. Compressive strain and increased tensile strains account for paraelectric and polar structures, respectively. $BaPrMn_{1.75}Co_{0.25}O_{5+\delta}$ are structurally reversible with enhanced redox reactions (Majee et al. 2020). Sr_2MgTeO_6 has a monoclinic structure which can transit to tetragonal phase. $Sr_{2-1}Bi_xFeMoO_6$ compounds have big grain sizes and exist in the cubic crystal structure with no phase transitions. Ln_2NiMnO_6 has small band gap energy and extended charge carrier lifetime (Sariful Sheikh et al. 2017). $Sr_2LaTiAlO_6$ has mobile vacancies that depend on the ordered pattern of the cation. $CaFeO_3$ undergoes electronic transition from metallic to insulating phase while large charge quantity modifies the structure.

Doping with suitable ions boosts the performance and efficiency of DPOs. Zinc ions improve electrochemical performance and moderate carrier concentration vacancies (Ren et al. 2020). Oxygen defects can be engineered to boost the activities of oxygen evolution reactions and improve the rate of oxidizing water by creating more oxygen vacancies (Zhu et al. 2020). Oxygen vacancies can be created by treating with argon and hydrogen gases (Zhu et al. 2020). Nanoparticles can be decorated with double perovskite oxides to minimize the cost of fabrication and make it compatible to thermo-mechanical stress (Niu et al. 2020). DPOs are useful for fabricating single junction solar cells upon band gap tuning.

21.5 Lead-Free Perovskites

Perovskite materials without lead components are environmentally friendly and highly stable with minimal poisonous effects. Their performances are affected by the charge transport and band structure formed (Chu et al. 2019). Green alternatives for organic perovskites are obtainable from double halide perovskites upon substitution of the lead ions (Chu et al. 2019). Lead ion can be substituted with ions of germanium, tin, bismuth, zinc, cobalt. For the past two years, lead-free perovskite compounds like $Cs_2NaAmCl_6$, $Cs_2NaBiCl_6$, $Cs_2AgBiCl_6$, $Cs_2AgBiBr_6$, etc., have been developed. Lead-free perovskite compounds, obtained morphologies, methods of preparation, and some energy band gaps have been listed in Table 21.1.

TABLE 21.1

Lead-Free Perovskite Materials, Morphologies, Methods of Preparation, and Some Energy Band Gaps (Chu et al. 2019)

Compounds	Morphologies	Methods of Preparation	Energy Band Gaps (eV)
$Cs_2NaAmCl_6$	Crystalline powder	Evaporate hydrochloric acid solution to dryness	–
$Cs_2AgBiCl_6$	Powder	Melt crystallization	1.05
$Cs_2AgBiBr_6$	Nanocrystal	Hot injection	2.83
$Cs_2(Ag_{1-a}Bi_{1-b})TI_xBr_6$	Single crystal	Cooling crystallization	–
$Cs_2Ag(Bi_{1-x}M_x)Br_6$ (M = In,Sb)	Single crystal	Melt crystallization	–
Cs_2AgBiI_6	Nanocrystal	Anion exchange	2.0
$Cs_2AgInCl_6$	Single crystal	Cooling crystallization	3.3
Cs_2NaBiI_6	Microcrystal	Hydrothermal technique	2.2
Cs_2SnI_6	Nanocrystal	Hot injection	1.3
Cs_2SnCl_6	Single crystal	Cooling crystallization	3.9
Cs_2PdBr_6	Nanocrystal	Anti-solvent recrystallization	1.6
Cs_2PdI_6	Nanocrystal	Anion exchange	1.69
Cs_2TiBr_6	Powder	Molten salt	2.0
$Cs_2TiI_xBr_{6-x}$	Powder	Melt crystallization	–
$(MA)_2KBiCl_6$	Powder	Evaporate hydrochloric acid solution to dryness	–
$(MA)_2TlBiBr_6$	Single crystal	Hydrothermal method	–
$(MA)_2SnI_6$	Powder	Mixed iodides	–

21.6 Conclusion and Future Perspectives

Perovskite solar cell technology has made a tremendous impact in the world of material science. This chapter has given a background on the concept of perovskite solar cells alongside different new materials useful in perovskite devices. Perovskites can be fabricated using different architectures that enhance their performance and efficiency. Emerging and new materials like methylammonium lead halide, methylammonium tin halide, cesium lead iodide, barium zirconium sulphide, calcium zirconium sulphide, strontium lead iodide, and lead-free perovskite materials have been discussed.

Researchers envisage a better future for perovskite in terms of its cost, stability, and efficiency. The lead component in perovskite materials as well as its constituent absorber materials should be replaced with non-toxic and efficient elements that would minimize degradation effects due to interaction with moisture. New designs of perovskite cells and encapsulation processes should be further studied to understand how long perovskite solar cells would work effectively on housetops. This would aid commercialization provided that the encapsulation systems are resistant to air and humid conditions. Perovskite inks and reproducible fabrication methods can be adopted to increase large area deposition and mass production of perovskite solar cells.

REFERENCES

Advanced Science News. 2020. "Almost 40% Conversion Efficiency Predicted in New Perovskite Solar Cell." *Advanced Science News*. https://www.advancedsciencenews.com/almost-40-conversion-efficiency-predicted-in-new-perovskite-solar-cell/.

Ava, Tanzila Tasnim, Abdullah Al Mamun, Sylvain Marsillac, and Gon Namkoong. 2019. "A Review: Thermal Stability of Methylammonium Lead Halide Based Perovskite Solar Cells." *Applied Sciences* 9, no. 1: 188.

Bellini, Emiliano. 2020. "A new perovskite material for solar applications." *Pv Magazine International*. Accessed November 9. https://www.pv-magazine.com/2020/01/08/a-new-perovskite-material-for-solar-applications/.

Chen, Lan, Bin Xu, Yurong Yang, and Laurent Bellaiche. 2020. "Macroscopic and Microscopic Structures of Cesium Lead Iodide Perovskite from Atomistic Simulations." *Advanced Functional Materials* 30, no. 19: 1909496.

Chu, Liang, Waqar Ahmad, Wei Liu, Jian Yang, Rui Zhang, Yan Sun, Jianping Yang, and Xing'ao Li. 2019. "Lead-Free Halide Double Perovskite Materials: A New Superstar Toward Green and Stable Optoelectronic Applications." *Nano-Micro Letters* 11, no. 1: 16. doi:10.1007/s40820-019-0244-6.

Dhamaniya, Bhanu Pratap, Priyanka Chhillar, Bart Roose, Viresh Dutta, and Sandeep K. Pathak. 2019. "Unraveling the Effect of Crystal Structure on Degradation of Methylammonium Lead Halide Perovskite." *ACS Applied Materials & Interfaces* 11, no. 25. ACS Publications: 22228–22239.

Ezealigo, Blessing N., Assumpta C. Nwanya, Sabastine Ezugwu, Solomon Offiah, Daniel Obi, Rose U. Osuji, R. Bucher, et al. 2017. "Method to Control the Optical Properties: Band Gap Energy of Mixed Halide Organolead Perovskites." *Arabian Journal of Chemistry*, September. doi:10.1016/j.arabjc.2017.09.002.

Ghorai, Arup, Anupam Midya, and Samit K. Ray. 2019. "Surfactant-Induced Anion Exchange and Morphological Evolution for Composition-Controlled Caesium Lead Halide Perovskites with Tunable Optical Properties." *ACS Omega* 4, no. 7: 12948–12954.

Gupta, Tushar, Debjit Ghoshal, Anthony Yoshimura, Swastik Basu, Philippe K. Chow, Aniruddha S. Lakhnot, Juhi Pandey, et al. "An Environmentally Stable and Lead-Free Chalcogenide Perovskite." *Advanced Functional Materials* 30, no. 23: 2001387.

Hadi, Atefe, Bradley J. Ryan, Rainie D. Nelson, Kalyan Santra, Fang-Yi Lin, Eric W. Cochran, and Matthew G. Panthani. 2019. "Improving the Stability and Monodispersity of Layered Cesium Lead Iodide Perovskite Thin Films by Tuning Crystallization Dynamics." *Chemistry of Materials* 31, no. 14: 4990–4998.

Hien, Vu Xuan, Pham Tien Hung, Jeongwoo Han, Sangwook Lee, Joon-Hyung Lee, and Young-Woo Heo. 2020. "Growth and Gas Sensing Properties of Methylammonium Tin Iodide Thin Film." *Scripta Materialia* 178: 108–113.

Idris, Ahmed Mahmoud, Taifeng Liu, Jafar Hussain Shah, Hongxian Han, and Can Li. 2020. "Sr2CoTaO6 Double Perovskite Oxide as a Novel Visible-Light-Absorbing Bifunctional Photocatalyst for Photocatalytic Oxygen and Hydrogen Evolution Reactions." *ACS Sustainable Chemistry & Engineering* 8, no. 37. ACS Publications: 14190–14197.

Idris, Ahmed Mahmoud, Taifeng Liu, Jafar Hussain Shah, Anum Shahid Malik, Dan Zhao, Hongxian Han, and Can Li. 2020. "Sr2NiWO6 Double Perovskite Oxide as a Novel Visible Light Responsive Water Oxidation Photocatalyst." *ACS Applied Materials & Interfaces* 12, no. 23:25938–25948.

Inaguma, Yoshiyuki, Tetsuhiro Katsumata, Mitsuru Itoh, and Yukio Morii. 2002. "Crystal Structure of a Lithium Ion-Conducting Perovskite La2/3−xLi3xTiO3 (X=0.05)." *Journal of Solid State Chemistry* 166, no. 1: 67–72. doi:10.1006/jssc.2002.9560.

Kennard, Rhys M., Clayton J. Dahlman, Hidenori Nakayama, Ryan A. DeCrescent, Jon A. Schuller, Ram Seshadri, Kunal Mukherjee, and Michael L. Chabinyc. 2019. "Phase Stability and Diffusion in Lateral Heterostructures of Methyl Ammonium Lead Halide Perovskites." *ACS Applied Materials & Interfaces* 11, no. 28: 25313–25321.

Li, Dan, Chang-Song Chen, Yi-Hua Wu, Zhi-Gang Zhu, Wan Y. Shih, and Wei-Heng Shih. 2020. "Improving Stability of Cesium Lead Iodide Perovskite Nanocrystals by Solution Surface Treatments." *ACS Omega* 5, no. 29: 18013–18020.

Li, Fengzhu, Chaoshen Zhang, Jin-Hua Huang, Haochen Fan, Huijia Wang, Pengcheng Wang, Chuanlang Zhan, Cai-Ming Liu, Xiangjun Li, and Lian-Ming Yang. 2019. "A Cation-Exchange Approach for the Fabrication of Efficient Methylammonium Tin Iodide Perovskite Solar Cells." *Angewandte Chemie International Edition* 58, no. 20: 6688–6692.

Majee, Rahul, Quazi Arif Islam, Surajit Mondal, and Sayan Bhattacharyya. 2020. "An Electrochemically Reversible Lattice with Redox Active A-Sites of Double Perovskite Oxide Nanosheets to Reinforce Oxygen Electrocatalysis." *Chemical Science* 11, no. 37: 10180–10189.

Niu, Bingbing, Chunling Lu, Wendi Yi, Shijing Luo, Xiangnan Li, Xiongwei Zhong, Xingzhong Zhao, and Baomin Xu. 2020. "In-Situ Growth of Nanoparticles-Decorated Double Perovskite Electrode Materials for Symmetrical Solid Oxide Cells." *Applied Catalysis B: Environmental* 270:118842.

Nkele, Agnes C., Innocent S. Ike, Sabastine Ezugwu, Malik Maaza, and Fabian I. Ezema. 2020. "An Overview of the Mathematical Modelling of Perovskite Solar Cells towards Achieving Highly Efficient Perovskite Devices." *International Journal of Energy Research* 45, no. 2:1496–1516.

Nkele, Agnes C., Assumpta C. Nwanya, Nanasaheb M. Shinde, Sabastine Ezugwu, Malik Maaza, Jasmin S. Shaikh, and Fabian I. Ezema. 2020. "The Use of Nickel Oxide as a Hole Transport Material in Perovskite Solar Cell Configuration: Achieving a High Performance and Stable Device." *International Journal of Energy Research.* doi:https://doi.org/10.1002/er.5563.

Nkele, Agnes C., Assumpta C. Nwanya, Nwankwo Uba Nwankwo, Rose U. Osuji, A. B. C. Ekwealor, Paul M. Ejikeme, Malik Maaza, and Fabian I. Ezema. 2019. "Investigating the Properties of Nano Nest-like Nickel Oxide and the NiO/Perovskite for Potential Application as a Hole Transport Material." *Advances in Natural Sciences: Nanoscience and Nanotechnology* 10, no. 4: 045009. doi:10.1088/2043-6254/ab5102.

Okinawa Institute of Science and Technology (OIST) Graduate University. 2019. "New Perovskite Material Shows Early Promise as an Alternative to Silicon." *ScienceDaily.* https://www.sciencedaily.com/releases/2019/08/190808152457.htm.

Perez, Maayan, Sa'ar Shor Peled, Tzvi Templeman, Anna Osherov, Vladimir Bulovic, Eugene A. Katz, and Yuval Golan. 2020. "A Two-Step, All Solution Process for Conversion of Lead Sulfide to Methylammonium Lead Iodide Perovskite Thin Films." *Thin Solid Films* 714: 138367.

Rahul, Pramod K. Singh, Rahul Singh, Vijay Singh, B. Bhattacharya, and Zishan H. Khan. 2018. "New Class of Lead Free Perovskite Material for Low-Cost Solar Cell Application." *Materials Research Bulletin* 97, January: 572–577. doi:10.1016/j.materresbull.2017.09.054.

Ren, Rongzheng, Zhenhua Wang, Xingguang Meng, Chunming Xu, Jinshuo Qiao, Wang Sun, and Kening Sun. 2020. "Boosting the Electrochemical Performance of Fe-Based Layered Double Perovskite Cathodes by Zn2+ Doping for Solid Oxide Fuel Cells." *ACS Applied Materials & Interfaces* 12, no. 21: 23959–23967.

Sariful Sheikh, Md., Dibyendu Ghosh, Alo Dutta, Sayan Bhattacharyya, and T. P. Sinha. 2017. "Lead Free Double Perovskite Oxides Ln2NiMnO6 (Ln=La, Eu, Dy, Lu), a New Promising Material for Photovoltaic Application." *Materials Science and Engineering: B* 226, December: 10–17. doi:10.1016/j.mseb.2017.08.027.

Seo, You-Hyun, Eun-Chong Kim, Se-Phin Cho, Seok-Soon Kim, and Seok-In Na. 2017. "High-Performance Planar Perovskite Solar Cells: Influence of Solvent upon Performance." *Applied Materials Today* 9, December: 598–604. doi:10.1016/j.apmt.2017.11.003.

Singh, Neelima, Alpana Agarwal, and Mohit Agarwal. 2020. "Numerical Analysis of Lead-Free Methyl Ammonium Tin Halide Perovskite Solar Cell." In *AIP Conference Proceedings*, 2220: 130033.

Thomas, Cherrelle J., Yangning Zhang, Adrien Guillaussier, Khaled Bdeir, Omar F. Aly, Hyun Gyung Kim, Jungchul Noh, et al. 2019. "Thermal Stability of the Black Perovskite Phase in Cesium Lead Iodide Nanocrystals Under Humid Conditions." *Chemistry of Materials* 31, no. 23: 9750–9758.

Toloueinia, Panteha, Hamidreza Khassaf, Alireza Shirazi Amin, Zachary M. Tobin, S. Pamir Alpay, and Steven L. Suib. 2020. "Moisture-Induced Structural Degradation in Methylammonium Lead Iodide Perovskite Thin Films." *ACS Applied Energy Materials* 3, no. 9: 8240–8248.

Wang, Bin, Novendra Novendra, and Alexandra Navrotsky. 2019. "Energetics, Structures, and Phase Transitions of Cubic and Orthorhombic Cesium Lead Iodide (CsPbI3) Polymorphs." *Journal of the American Chemical Society* 141, no. 37: 14501–14504.

Wang, Pengcheng, Fengzhu Li, Ke-Jian Jiang, Yanyan Zhang, Haochen Fan, Yue Zhang, Yu Miao, et al. 2020. "Ion Exchange/Insertion Reactions for Fabrication of Efficient Methylammonium Tin Iodide Perovskite Solar Cells." *Advanced Science* 7, no. 9: 1903047.

Wei, Xiucheng, Haolei Hui, Samanthe Perera, Aaron Sheng, David F. Watson, Yi-Yang Sun, Quanxi Jia, Shengbai Zhang, and Hao Zeng. 2020. "Ti-Alloying of BaZrS3 Chalcogenide Perovskite for Photovoltaics." *ArXiv Preprint ArXiv:2004.04261*.

Wei, Xiucheng, Haolei Hui, Chuan Zhao, Chenhua Deng, Mengjiao Han, Zhonghai Yu, Aaron Sheng, Pinku Roy, Aiping Chen, and Junhao Lin. 2019. "Fabrication of BaZrS3 Chalcogenide Perovskite Thin Films for Optoelectronics." *ArXiv Preprint ArXiv:1910.04978*.

Wei, Xiucheng, Haolei Hui, Chuan Zhao, Chenhua Deng, Mengjiao Han, Zhonghai Yu, Aaron Sheng, Pinku Roy, Aiping Chen, and Junhao Lin. 2020. "Realization of BaZrS3 Chalcogenide Perovskite Thin Films for Optoelectronics." *Nano Energy* 68: 104317.

Wu, Yukun, Yue He, Shiqi Li, Xiaoyan Li, Yifan Liu, Qinjun Sun, Yanxia Cui, Yuying Hao, and Yucheng Wu. 2020. "Efficient Inverted Perovskite Solar Cells with Preferential Orientation and Suppressed Defects of Methylammonium Lead Iodide by Introduction of Phenothiazine as Additive." *Journal of Alloys and Compounds* 823: 153717.

Wu, Dongdong, Pengcheng Jia, Wentao Bi, Yang Tang, Jian Zhang, Bo Song, Liang Qin, Zhidong Lou, Yufeng Hu, and Feng Teng. 2020. "Enhanced Performance of Tin Halide Perovskite Solar Cells by Addition of Hydrazine Monohydrobromide." *Organic Electronics*: 105728.

Xu, Qiang, Wenyi Shao, Jun Liu, Zhichao Zhu, Xiao Ouyang, Jiafa Cai, Bo Liu, Bo Liang, Zhengyun Wu, and Xiaoping Ouyang. 2019. "Bulk Organic–Inorganic Methylammonium Lead Halide Perovskite Single Crystals for Indirect Gamma Ray Detection." *ACS Applied Materials & Interfaces* 11, no. 50: 47485–47490.

Zhou, Di, Tiantian Zhou, Yu Tian, Xiaolong Zhu, and Yafang Tu. 2018. "Perovskite-Based Solar Cells: Materials, Methods, and Future Perspectives." *Journal of Nanomaterials* 2018: 1–15. doi:10.1155/201 8/8148072.

Zhu, Yunmin, Xiao Zhong, Shiguang Jin, Haijun Chen, Zuyun He, Qiuyu Liu, and Yan Chen. 2020. "Oxygen Defect Engineering in Double Perovskite Oxides for Effective Water Oxidation." *Journal of Materials Chemistry A* 8, no. 21: 10957–10965.

Zitouni, H., N. Tahiri, O. El Bounagui, and H. Ez-Zahraouy. 2020. "Electronic, Optical and Transport Properties of Perovskite BaZrS3 Compound Doped with Se for Photovoltaic Applications." *Chemical Physics* 538: 110923.

22

The Application of Carbon and Graphene Quantum Dots to Emerging Optoelectronic Devices

Cyril Oluchukwu Ugwuoke[1,2], Sabastine Ezugwu[3], S. L. Mammah[4], A. B. C. Ekwealor[1], and Fabian I. Ezema[1,5,6]

[1]*Department of Physics and Astronomy, University of Nigeria, Nsukka, Enugu State, Nigeria*
[2]*Science and Engineering Unit, Nigerian Young Researchers Academy, Onitsha, Anambra State, Nigeria*
[3]*Department of Physics and Astronomy, The University of Western Ontario, London, Ontario, Canada*
[4]*Department of Science Laboratory Technology, School of Applied Sciences, Rivers State Polytechnic Bori, Bori, Nigeria*
[5]*Nanosciences African Network (NANOAFNET), iThemba LABS-National Research Foundation, Somerset West, Western Cape Province, South Africa*
[6]*UNESCO-UNISA Africa Chair in Nanosciences/Nanotechnology, College of Graduate Studies, University of South Africa (UNISA), Muckleneuk Ridge, Pretoria, South Africa*

22.1 Introduction

Carbon is a non-metallic element with chemical symbol C, atomic number 6, and the average atomic weight 12 amu. In universe, carbon is the sixth most abundant element, fourth in the solar system, and seventeenth on the earth crust (Shulaker et al. 2013; Okwundu, Aniekwe, and Nwanno 2018). Apart from oxygen, carbon is the most dominant element in the human body, which adds up to 18% human weight (Pace 2001) and also contributes to 0.2% of earth total mass (Allkgre et al. 1995; Marty, Alexa, and Raymond 2013). It has the feature of forming triple, double, and single bonds, as well as bonding with each other forming straight, branched, acyclic, and cyclic chains (Nasir et al. 2018). However, due to the unique features of carbon, the interest of researchers has shifted towards its use in technological advancement. In the nanoworld (i.e. engineering, materials science, and nanoscience) nanostructured carbon has been reported to exist in various forms called allotropes, including amorphous (coal, charcoal, lampblack, carbon black, carbon fibre, coke, activated carbon) and crystalline (diamond, graphite, and carbon nanomaterial) forms (Sederberg 2009; Karthik, Himaja, and Singh 2014; Loos 2015).

22.2 Graphite

This is one of the allotropes of carbon with chemical buildup similar to that of diamond but different in structure. This made graphite to have nearly opposite properties when compared to diamond. Graphite is not transparent and has a grey colour. Unlike diamond, graphite is soft and flexible. It comprises of

DOI: 10.1201/9781003145585-22

multilayers of material (called graphene) bundled up by Van der waals forces. Graphene was never believed to exist in an isolated state until 2010, when Nobel Prize in Physics was awarded to Professor Andre Geim and Konstain Novoselove from University of Manchester for the breakthrough of graphene in an isolated form. Since then, the attention of scientific and technological researchers has been growing daily on the use of graphene and its quantum dot in nanoworld (Latif, Alwan, and Al-dujaili 2012). It has been on record that graphene and its derivatives are used in modern 5G communication technology (Chattha 2019; Khan et al. 2020). Graphene quantum dot (GQD) comprises of the graphene sheet (with zero dimension) ranging from a single layer to ten layers of size less than 10 nm (Ponomarenko et al. 2008; Recher and Trauzettel 2010; Suvarnaphaet and Pechprasarn 2017). GQD is one of the best leading materials with unique features to change the world with its limitless application. It has very good mechanical stability such as bending, rolling, or stretching (Mittal and Rhee 2018) and unique electronic performance which can be tuned perfectly. These features enhance its application in the fabrication of rigid or flexible optoelectronic devices (Zeng et al. 2014; Jang et al. 2016) such as electrochemical energy storage devices (battery, symmetric and asymmetric supercapacitors), photo-voltaic devices, and sensing devices. Zhao et al. described it as a science's darling material (Zhao et al. 2010).

Apart from GQD, there are other well-known quantum dots composed of carbon nanomaterials. They are polymer dots (PDs) and carbon nanodots (CNDs). Similar to GQD, they are also made of nano-material less than 10 nm of size. CNDs were first synthesized in 2004, during electrophoresis synthesis of single-wall carbon nanotube (CNT) (Xu et al. 2004). PDs and hydrophilic CNDs are also used in the fabrication of optoelectronic devices due to their ability to exhibit fluorescence. Collectively, GQDS, PDs, and CNDs are called *carbon quantum dots* (CDs) (Wang et al. 2019). Herein, we describe various synthesis methods of CDs and their necessity in the fabrication of optoelectronic devices such as dye-sensitized solar cells, electrochemical energy storage devices (battery, symmetric and asymmetric su-percapacitors), transistor devices, and MIMO technology for LTE and fifth-generation network tech-nology (5G) antenna.

22.3 Device Fabrication

22.3.1 Dye-sensitized Solar Cell (DSSC)

A dye-sensitized solar cell (DSSC) is a special form of photovoltaic (PV) cell that makes use of organic dye as an absorbing layer. It generates electricity by mere illumination of organic dye in the electro-chemical cell, and its efficiency is based on the material used. The first synthesized DSSC has ZnO electrode. It had a very low efficiency of power conversion because the dye monolayer was only able to absorb 1% of light incident on it (Sharma, Sharma, and Sharma 2018). The efficiency of DSSC was increased by employing a porous electrode, and this leads to the invention of Grätzel cell in 1991 with improved efficiency of up to 7% (O'Regan and Gratzel 1991). When compared to other PV cells, it is cost-effective and provides a better efficiency of conversion in low radiation (Patni, Sharma, and Pillai 2017; Jiang, Michaels, and Vlachopoulos 2019). DSSC has four major components which include the substrate, electrodes, electrolytes, and the organic dye that serves as an absorbing layer. As shown in Figure 22.1, the fabrication of dye-sensitized solar cell comprises five steps. In step one, a very good transparent glass substrate is corned and heated up to 450°C. Then, the precursor of fluorine-doped tin oxide (FTO) was spray deposited on a heated glass substrate to make it conductive (step 2). In step 3, the mixture of acetic acid and titanium isopropoxide in isopropyl alcohol was used to form colloidal TiO_2 which was sprayed on FTO followed by masking of FTO. Step 4, the FTO is unmasked by compacting and heating Ag powder forming a solid mass. The unmasking made the deposited colloidal TiO_2 to be rectangular, and it enhances the transfer of electrons to the external circuit. Step 5, the organic dye is used to coat the surface forming dye-sensitized solar cell.

Colloidal TiO_2 has been used as photoelectrode in DSSCs due to its ability to absorb radiation (Patil et al. 2019) but it has a number of drawbacks. Saedi et al. (2020) reported that TiO_2 is unfavourable in

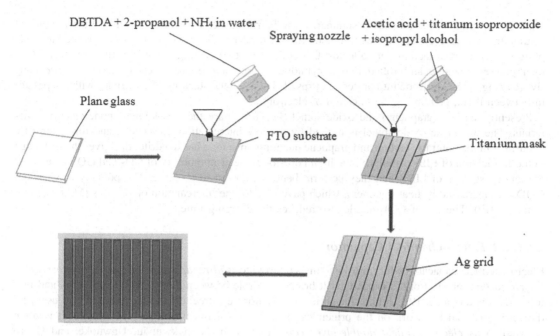

FIGURE 22.1   Step-by-step fabrication of dye-sensitizing solar cell. Adapted from Ref. Gangadharan et al. 2014.

the fabrication of DSSC because of the recombination at the metal-oxide interface. When dye absorbs a photon, electrons are produced, and before transferring the electrons from molecules of the dye to TiO_2, there is relaxation of electron-hole pair and the relaxation causes decrease of current supply. CDs and GQDs are the outstanding materials used as photoelectrode in the fabrication of DSSC due to their ability to produce a very low recombination rate at the interface, thereby increasing the electron-hole separation (Badawi 2015; Hao et al. 2016; Zhao et al. 2017). The quantum confinement effect of CDs and GQDs enhances its properties, such as good absorption of light as well as good fluorescence when illuminated. These properties give them an edge over other materials during fabrication of DSSC and other PV cells (Zhao et al. 2007).

22.3.2 Electrochemical Energy Storage System

The electrochemical energy storage systems (EESs) are energy storage devices that store electric charges in chemical form. It comprises of an electrochemical battery (rechargeable batteries) and an electrochemical capacitor (symmetric and asymmetric supercapacitor). EESs are used in automobiles, portable electronics, and renewable energy systems due to their high efficiency and low cost. The efficiency of EESs is determined by the material used in the fabrication. The unique properties of carbon and graphene quantum dots have contributed a lot in the fabrication of EES electrode.

22.3.2.1 Electrochemical Battery

After the invention of the electrochemical battery by Alessandro Volta in 1800, it has become one of the major necessities of human life. It can either be grouped as primary or secondary battery. The primary batteries undergo irreversible electrochemical reactions (e.g. zinc-carbon battery), but the secondary battery can be recharged. Sometimes it is referred as rechargeable battery, e.g. lithium-ion battery (LIB), sodium-ion battery (SIB), nickel-metal hydride battery (Ma et al. 2018). LIBs are outstanding among the secondary batteries because of their unique properties such as low self-discharge, long lifespan, and high energy density, which lead to its wide application in various automobile systems and portable electronics (Wang et al. 2009; Huang et al. 2018; Ullah, Majid, and Rani 2018). LIBs energy storage mechanism is based on the intercalation of Li^+ into the electrode (cathode and anode) materials (Lei et al. 2018). Smekens et al. (2016) summarize the fabrication of LIB into six steps (i) the mixture of conducting and active materials with the addition of binder to form electrode paste, (ii) the electrode paste is used to coat the current collector; this is done by tape casting technique, (iii) the electrode is then pressed together and rolled into a cylindrical form, a process generally called *Calendering*, (iv) resizing of the electrode into preferred strips of shape, (v) the stacking of electrodes with a separator in between them, (vi) finally the injection of electrolyte.

Presently, carbon (graphene) and oxide spinel ($Li_4Ti_5O_{12}$) are the best-known anode electrode materials. The use of an oxide spinel is limited by the low intercalation, low cycle, and formation of a dendrite. On the contrary, carbon and graphene quantum dot anode has dendrite; this gives them an edge in the fabrication of LIBs. Similarly, the high porosity or electron mobility of CDs and GQDs enhances the charge storage of LIBs, and they perform better compared to graphite. The porosity of CDs and GQDs are increased by heat treatment, which pave way for the intercalation of Li^+ ions (Mamvura and Simate 2019). The use of GQD anode also reduces the charging time.

22.3.2.2 Electrochemical Capacitor

Electrochemical capacitors which are sometimes referred to as *ultracapacitors* or *supercapacitors* are an improved type of electrolytic capacitors. it bridges the role between conventional capacitors and batteries (L. Zhou et al. 2019). Unlike conventional capacitors, electrochemical capacitors do not use solid dielectric material but work on the principles of an electrochemical double layer (EDL). It is often referred to as *electrochemical double-layer capacitors* (EDLCs) (Okwundu, Ugwuoke, and Okaro 2019). EDLCs have a unique feature which makes it desirable in the manufacturing of electric cars and this is associated with short charging time, high energy density, high surface area, long cycle lifetime, long charge/discharge cycle, high temperature tolerance, and operational safety (Shuo Zhang et al. 2018; Zou et al. 2018; Okwundu, Ugwuoke, and Okaro 2019). EDLC electrodes are composed of three major materials: the binder, active and conducting materials. The performance of EDLCs are based on the electrode material with high specific capacitance. The preferred electrode material used in the fabrication of supercapacitor is carbon-based materials such as activated carbon, carbon nanotubes, graphene, and carbon nitrites. Supercapacitors that are composed of carbon electrodes have been shown to perform poorly, and they also have low energy density because of their available narrow storage capacity (Zhang et al. 2018). On their part, CDs and GQDs have been getting more attention in recent times due to their unique properties, enhanced by quantum confinement. The high transparency, a theoretical surface area up to 2630 m^2g^{-1}, chemical inertness, and high electron mobility of GQD give it an edge in the application of electrode material in supercapacitor. A good porous electrode with improved surface area can be achieved by deposition of GQD on graphene electrode (Lee et al. 2016). GQD can also be used as electrolyte in fabrication of solid-state capacitors (Zhang et al. 2016).

Generally, supercapacitor-based carbon or carbon dot electrodes have low energy densities. Following the equation $E = \frac{1}{2}CV^2$, high energy density can only be realized by the fabrication of supercapacitor with either high capacitance or wide potential window (Jia et al. 2018). The need to fabricate electrochemical capacitors with high energy density leads to asymmetric supercapacitors. It is a blend of supercapacitor and pseudocapacitor, this means that the electrode system is made of both transition metal oxides (TMO) and carbon or carbon dot materials. GQDs have played a very good role in the fabrication of asymmetric supercapacitor electrode because of their salient features (Zhang et al. 2012). Choi and his workers were able to control the GQD band gap forming GQD/ZnO heterostructure via the establishment of Zn-O-C which result in the formation of unique properties that ZnO has never shown before (Jia et al. 2018, Son et al. 2012). Therefore, the poor conductivity of transition metal oxides is enhanced by GQD and this is advantageous in the fabrication of high performance asymmetric supercapacitors with high specific capacitance and wide potential window (Liu et al. 2013).

22.3.3 MIMO for LTE and 5G Antenna

5G technology has been the new trending communication technology for a mobile network with speed ranging from 50 Mbit/s to more than 1 gigabyte. It is termed the best and fastest mobile network so far.

Gupta, Malviya, and Charhate (2019) describe the 5G network as a mobile communication network that produces 1000 times high efficiency data compared to the one we have today and consumes high power, approximately 800 watts. The high consumption of power enhances the performance of the cell network, but sometimes maintaining its high speed download can cause overheating of mobile phones (Edmondson, Ghaffarzadeh, and Jiang 2020). The fifth-generation network technology antenna is based on the principle of multiple-input and multiple-output (MIMO) technology. MIMO is a radio frequency technology that can also be used in some modern wireless technologies such as Wi-Fi, long-term evolution (LTE), and many others. It makes use of multiple antennae which comprises the transmitter and the receiver (www.electronics-notes.com/articles/antennas-propagation/mimo/what-is-mimo-multiple-inpute-multiple-output-wireless-technology.php). The transmitter antenna transmits different radio signals with the same frequency aimed to avoid interference of the signals. The reason for using different transmission signals is to obtain a very high data rate as well as enhancing its accuracy and transmission range without using additional bandwidth (Chattha 2019). During the fabrication of MIMO for LTE or 5G antenna, a highly efficient dielectric is used with good properties such as easiness to fabricate, cost-effectiveness, mechanical strength, good permittivity, and elasticity (Choudhary, Kumar, and Gupta 2015). GQD is one of the best materials that possess such properties with the additional excellent electrical property. Sa'don et al. (2019) reported the fabrication of graphene-based antenna using a screen-printing technique where Kapton polyamide was used as a substrate and the precursor of GQD was graphene ink. The substrate was first coated with the patch and the ground plane followed meshing onto graphene ink. Then the printed antenna was heated to achieve a considerable bandwidth and thereafter the introduction of a rectangular slot that helps to achieve the required matching impedance and resonance frequency.

22.3.4 Transistor Devices

A transistor is a device that is composed of three terminals of semiconductor material which can be used to amplify or switch electronic signal and electrical power. The three terminals are the base, collector, and emitter. The base is between the emitter and collector. The emitter is usually forward biased, and collector is in reverse biased. Transistor devices work on the principle of p-n junction and it is of two forms: bipolar junction transistor (BJT) and field-effect transistor (FET). FET is a basic form of a transistor that comprises of the drain, source, and gate electrodes as the terminals. The electron enters the FET transistor through the source and drain electrodes while the gate electrode is used for control. In the semiconductor industry, silicon compounds (e.g. SiO_2) have been the leading material in the fabrication of FET transistor electrodes and the idea of increasing the number of transistors in electronic devices whilst reducing cost has been the interest of researchers, as explained by Moore's law (Lundstrom 2003; Vardi 2005; MacK 2011). In regards to Moore's law and miniaturization of Si material to achieve a flexible transistor leads to major limitations of Si in the fabrication of flexible FET because Si is rigid (Sharma and Ahn 2013). GQD has been the most outstanding material in the fabrication of flexible/wearable transistor devices. This is achieved because of its unique electrical and mechanical properties such as spring constant, tensile strength, Young modulus, mobility, and its ability to be stretched (Lee et al. 2008; Molitor et al. 2011; Song et al. 2017).

22.4 Structural, Optical, and Electronic Properties of CDs and GQDs

The sizes of CDs and GQDs are identical, but they vary in the chemical group and internal structure. They are made of carbon- based material and various oxygen-associated functional groups. CDs are sp^3 and consist of crystalline and amorphous nanoparticles resulting in low crystallinity (Li et al. 2014; Sciortino, Cannizzo, and Messina 2018) while GQDs are purely nanocrystalline particles that are composed of σ and π bonds and the distance between the carbon–carbon atom is approximately 0.142 nm. The π bond is hybridized together to form π bands and π* bands. It has a hexagonal structure which is peculiar to the three σ bonds in each of the lattices and each of the carbon atoms at the tip of the hexagonal structure is sp^2 hybridized as shown in Figure 22.2. Zhen and Zhu (2018) describe the diameter of GQD to be 0.35 nm, i.e. about 1/200,000 times that of human hair. The electrical and optical

(a)   (b)

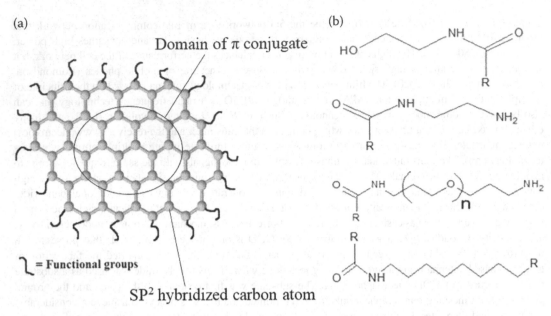

FIGURE 22.2   (a) The structure of GQD and (b) the end functional groups. Adapted from Ref. Shen et al. 2012.

properties of GQDs are peculiar to the π bond stationed vertically to the plane (Kavitha and Jaiswal 2016; Zhen and Zhu 2018) and this is determined by the arrangement of electrons in the outermost shell. A carbon atom has six electrons, two in the innermost shell, and four at the outermost shell which are used in chemical bonding. In 2D graphene, three electrons are interconnected to each other leaving an electron free in the third dimension. This free electron is referred to as π electron, and this has a great impact on the conductivity of GQD. Two-dimensional graphene has zero band gaps; this has been the major limitation on the use of graphene in optoelectronics devices. It can be confined to 0D which is GQD, to have a finite bad gap (Armstrong 2017). The quantum confinement enhances the photo-luminescence (Bharathi et al. 2017). The photoluminescence of CDs and GQDs is similar and this has brought up a lot of controversies which prompted L. Cao et al. (Cao, Meziani, and Sahu 2013) to describe it as an open debate among researchers. Although, Zhu et al. (2014) were able to explain the mechanism of photoluminescence of GQD using graphene oxide (GO) because they have the same chemical structure, it has been observed that the thickness and surface chemistry of CDs and GQDs influence their optical properties (Zhu et al. 2016). Single-layer GQD has a light absorption rate ~2.3%, i.e. 97.7 % of light transmitted through it, showing that GQDs are material with good optical trans-parency. The absorption rate of GQD is good enough that it can absorb light rays over a wide range of frequencies. The bond strength of carbon–carbon bond contributes to its unique mechanical properties such as tensile strength which is approximately 130 GPa, spring constant ~1–5 N/m, young modulus ~0.5–1 TPa, and good elasticity (Radadiya 2015; Das et al. 2018).

22.5 Characterization Technique of CDs and GQDs

The characterization of CDs and GQDs are done using X-ray diffraction (XRD) technique, Raman spectroscopy, Atomic force microscopy (AFM), transmission electron microscopy (TEM), and UV-VIS spectroscopy. Among them, TEM and AFM are used in imaging the cell internal structure and to study the mechanical properties of CDs and GQDs. Raman shift is used to study the crystalline structure of CDs and GQDs and can as well be used with XRD to determine the phase change. The characterization of CDs and GQDs has been a very challenging one because they show similar properties. Therefore, to determine the specific synthesized CDs and GQDs, different characterization techniques are used. The X-ray diffraction pattern of CDs and GQDs show either broad or small

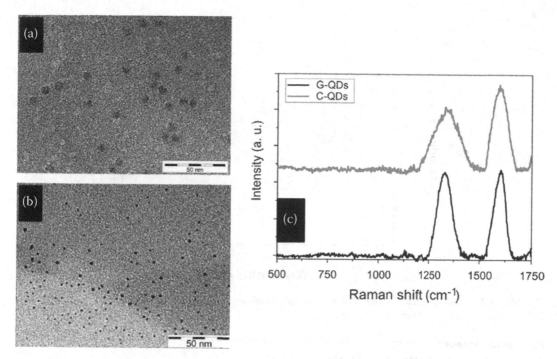

FIGURE 22.3 TEM images of (a) GQDs, (b) CDs and (c) Raman shift of CDs and GQDs (Reprinted with permission from Ref. Shan et al. 2015. Copyright © 2015, Elsevier.

diffraction peaks with 2θ falls between 20 and 26 degrees, the peaks are based on the oxygen content or defects. Dong et al. (2014) synthesized CDs and GQDs from coal, and they observed the same broad diffraction peak at 002 planes when 2θ is 25°C, this brought a lot of confusion. However, Raman spectroscopy can as well be employed to differentiate the properties of CDs and GQDs, but it does not give a clear overview. Figure 22.3c is the Raman spectra of CDs and GQDs, and they show similar peaks at ~1350 cm^{-1} and ~1560 cm^{-1}. The Raman spectra indicate the existence of structural sp^2 lattice of pristine graphene (G-band) and the basal plane defects (D-band) with the addition of higher modes, that is, 2D, D + G, and 2G (Tajik et al. 2020). Dervishi et al. (2019) observed that the intensity and frequencies of D and G bands increase with the size of GQDs. Although the clear-cut was gotten by higher modes as it increases with an increase in intensity and the I_D/I_G ratio increases when the thickness of GQDs nears 2 nm. However, the material is assumed to be crystalline when I_D/I_G ratio is less than 1. The I_D/I_G ratio of GQDs ranges from 0.5 to 1 and that of CDs is slightly close to 1 or above. The distinct feature of CDs and GQDs is obtained by characterizing them with AFM or TEM. Figure 22.3a and b show the TEM images of CDs and GQDs, indicating the size of GQD to be approximately 6 nm and CD as approximately 2 nm. Figure 22.3a shows that the TEM image of GQDs is hexagonal in shape with smooth size distribution (Kumar et al. 2017; Tajik et al. 2020). Similarly, Figure 22.3b shows the TEM image of CDs which are uniformly dispersed and are spherical in shape with a narrow size distribution (He et al. 2018). The UV-VIS spectroscopy is used to study the optical property and the composition of carbon compounds in CDs and GQDs. Figure 22.4 is the UV-VIS spectroscopy of GQDs. The analysis shows in two different peaks that arise due to the electron transition in GQDs: the strong peak and the weak peak. The strong peak occurs below 300 nm, and it is associated with the transformation of π to π* because of C=C bond structure. Similarly, the strong peak occurs within 300 nm and 390 nm, and it is related to electron transformation from *n* to π* because of the oxygen functional group (C=O) in GQDs (Wei and Alto 2014, Kundu and Pillai 2019). The reduction in the size of sp^2 cluster or conjugation length in GQDs causes oxygen functional group (C=O) to shift from 295 nm to 281 nm and C=C bond structure to shift from 242 nm to 235 nm.

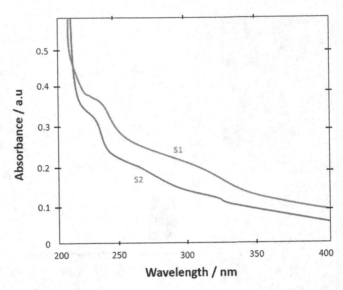

22.6 Synthesis of CDs and GQDs

CDs and CQDs are synthesized using various techniques which are categorized into two: the top-down and the bottom-up. The top-down technique involves cutting the large size of carbon-based materials (e.g. graphite, coal, carbon black, biomass, carbon fibre, etc.) into nanoparticles either by chemical or physical means. The top-down technique comprises acidic oxidation, arc discharge, laser ablation, and thermal evaporation while the bottom-up makes use of small-size precursors, and it comprises solvothermal/hydrothermal technique, combustion technique, and microwave technique (Figure 22.5).

22.6.1 Bottom-Up

22.6.1.1 Hydrothermal/Solvothermal Technique

Solvothermal method involves the preparation of crystalline materials by heating the precursor in a solution inside a sealed reactor (autoclave). Hydrothermal and solvothermal methods are very similar. The precursors used in the solvothermal process are usually non-aqueous and the solvents are organic compounds, e.g. glycerol, ethylene, etc. In hydrothermal, water is used as a solvent (Biomedicine et al. 2013). CDs and GQDs are commonly synthesized using hydrothermal/solvothermal method because it is cost-effective, eco-friendly, and easy to control the size and shapes of CDs or GQDs produced (Tajik et al. 2020). This technique involves the generation of reactant precursors and carbonization. The reactant precursor is obtained by pyrolysis of organic/polymer molecules in organic solvent or water. Then the formed precursor is moved to a sealed reactor where the carbonization takes place and it is finally converted to either CDs or GQDs. The reactant precursors not only serve as the source of carbon but enhance the doping element, size, and shapes of CDs and GQDs (Shaker, Riahifar, and Li 2020). For example, the CDs and GQDs can be doped with N_2 using melamine and urine as the source. A. kaliuri et al (2018) reported that the optimal precursor used in synthesizing GQD using hydrothermal/solvothermal is graphene sheet or nanoparticles (GNPs) due to the presence of carboxyl (– COOH) and hydroxyl (– OH) functional groups in graphene structures, which are fragile, and they can easily be involved in the reaction process. Pan et al. (2010) was the first to synthesize GQDs using the hydrothermal technique. They used graphene oxide as the starting material, and it was thermally treated to achieve a graphene sheet. The obtained graphene sheet was cut and oxidized in acid under ultrasonication; thereafter the mixture was diluted and filtered to remove the acidic

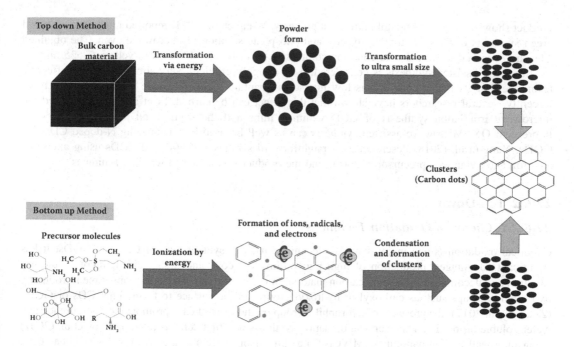

FIGURE 22.5 General synthesis of CDs (Sharma and Das 2019).

contents, and it was transferred to polytetrafluoroethylene (Teflon)-lined autoclave for carbonization, which results in the formation of colloidal GNPs solution with emission of a weak fluorescent. Finally, the colloidal GNPs were dialyzed over night to achieve a better GQD crystalline with strong fluorescent.

On the other hand, in synthesizing CDs, surfactant material is used due to the high content of carbon material. O. Kozák et al. (2013) synthesized CDs using cetylpyridinium chloride (CPC) as a source of carbon, and the autoclaving is based on alkaline compounds. The structure of CPC is composed of π conjugated carbon atoms, which not only act as a carbon source but as a stabilizing agent and as well contribute to the fluorescence of CDs. The alkaline autoclaving enhanced the photoluminescence of the formed CDs (Prikhozhdenko, Bratashov, and Mitrofanova 2018). However, the tuning and the fluorescence of hydrothermal/solvothermal synthesized CDs and GQDs are best achieved by doping with a nitrogen atom. A better controlled fluorescence of CDs and GQDs was described by Holá et al. (2017). They used the hydrothermal process to study the fluorescence of CDs and GQDs at a different wavelength by varying the amount of graphitic nitrogen. More so, Lu et al. (2018) observe that the inner structure of N-doped CDs and GQDs having biomolecules precursor that is enriched with carbon and nitrogen is perfectly tuned under hydrothermal/solvothermal condensation.

22.6.1.2 Microwave Irradiation Technique

The synthesis and commercialization CDs and GQDs are best done by microwave irradiation technique. In this method, the temperature of the precursor such as graphite and other carbonaceous materials is raised for a short time by a high electromagnetic wave. The process at which the precursor is heated can be demonstrated by either ionic conduction or dipolar polarization (Seekaew, Arayawut, and Timsorn 2019). When the microwave irradiates the samples of ionic or polar molecules, the generated electric field is sensitive to them and they aligned with electric field by rotation. During the synthesis of CDs and GQDs, a microwave heating power ranging from 500 W to 900 W is required (Chen, Yan, and Bangal 2009; Alsharaeh et al. 2016). Various liquid precursors have been used in synthesizing CDs and GQDs using hydrothermal, microwave pyrolysis, and ultrasonication. Zhu et al. (2009) described the synthesis of CD using microwave pyrolysis by mixing saccharide and poly(ethylene glycol) (PEG-200) in water forming a transparent solution, which is subsequently heated in a microwave oven. The achieved product shows a very good photoluminescent property. Wang et al. (2011) reported microwave-assisted green synthesis of CDs by mixing glycerol and phosphate solution in microwave oven. The obtained CDs are observed to show broad emission colour spectra of blue (~430 nm) to yellow (~525 nm).

Microwave-assisted synthesis is considered to be a highly efficient, very simple, and cost-effective technique, but its major limitation is low florescent quantum yield (QY) due to long reaction time. Recently, several researchers have shown that this effect can be curbed. Li et al. (2018) reported the microwave irradiation synthesis of GQD within 3 mins and the synthesized product has 35% of fluorescent QY. Microwave-assisted synthesis can as well be used in synthesizing N-doped CDs and GQDs. Yang et al. (2018) described the straightforward synthesis of N-doped CQDs using ammonia solution and xylan as a precursor at 200°C and the product was achieved within 10 minutes.

22.6.2 Top-Down

22.6.2.1 Chemical Oxidation Technique

Chemical oxidation is one major synthesis technique used in synthesizing of CDs and GQDs. It has significant advantages: production of high purity film, low cost, and high yield. It involves the exfoliation and decomposition of bulk carbon material into nanoparticles and at the same time introducing hydrophilic groups such as carboxyl or hydroxyl groups on the surface to form either CDs or GQDs (Zhang et al. 2017); the presence of hydrophilic group on the formed CDs promote the fluorescence and water-soluble properties. The chemical oxidation synthesis was first demonstrated by Zhou et al. (2007) using multiwall carbon nanotubes (MWCNTs) as precursor. Recently, the synthesis of CDs has been reported using carbon soot as precursor. The large-scale heteroatom (e.g. Se, S, and N)-doped CQD synthesized using chemical oxidation and reduction technique was first reported by Yang et al. (2014). The carbon nanoparticles obtained from Chinese ink were oxidized in the mixture of $NaClO_3$, H_2SO_4, and HNO_3. The obtained solution was neutralized with cold ammonia. Thereafter, the oxidized CQD undergoes reduction reaction in Teflon-lined autoclave in the presence of sodium selenium (NaHSe), sodium hydrosulphide (NaHS), and dimethylformamide (DMF), which serve as the source of selenium, sulphur, and nitrogen, respectively. The achieved Se-, S-, and N- doped CQD shows tunable photoluminescent property and high yield up to 61%, 69%, and 71%, respectively.

22.6.2.2 Thermal (Vacuum) Evaporation

Thermal evaporation deposition is one of the physical vapour deposition techniques used for the deposition of CD or GQD films directly on a substrate. The thermal evaporation deposition of CD and GQD films is achieved via two stages (Wei, Xu, and Wang 2009): the evaporation of the source material (usually carbon source) and the condensation of evaporated carbon atoms on a substrate. The deposition starts when the substrate is attached to the holder in a vacuum chamber and the source material (carbon or graphite) is loaded on a crucible and then heated very gently by resistive heating method, i.e. passing a large current through a tungsten filament. This heats up the source material until a high vapour pressure is attained, the atoms of the source material melt and begin to evaporate. The evaporated atom moves across the vacuum chamber and condenses directly on the surface of the substrate where it forms CD or GQD film coatings.

22.7 Conclusion

The carbon quantum dots are emerging carbon material–based quantum dots that have shown outstanding properties. Their unique characteristics are very important for fabrication and application in optoelectronic devices such as dye-sensitized solar cells, transistor devices, electrochemical double-layer capacitors, and 5G antennas. The graphene quantum dot (GQD) is best known for its high mechanical stability and excellent electronic performance. Other well-known quantum dots such as polymer dots (PDs) and carbon nanodots (CNDs) are also increasingly important and unique in their own respect. PDs and hydrophilic CNDs are now frequently used in the fabrication of optoelectronic devices due to their ability to fluorescence. As research efforts targeted at understanding these materials better are intensified, it is expected that our knowledge of their unique structural, optical, and electronic properties will increase, which will lead to better and enhanced device performances for application in optoelectronics devices.

REFERENCES

Allkgre, Claude J., Jean-paul Poirier, Eric Humler, and Albrecht W. Hofmann. 1995. "The Chemical Composition of the Earth." *Earth and Planetary Science Letters* 134: 515–526.

Alsharaeh, Edreese, Faheem Ahmed, Yazeed Aldawsari, and Majdi Khasawneh. 2016. "Novel Synthesis of Holey Reduced Graphene Oxide (HRGO) by Microwave Irradiation Method for Anode in Lithium-Ion Batteries." *Scientific Reports*, 1–13. https://doi.org/10.1038/srep29854.

Armstrong, Gordon. 2017. "Influence of Graphene Quantum Dots on Electrical Properties of Polymer Composites." *Material Research Express*. https://doi.org/10.1088/2053-1591/aa7a75.

Badawi, Ali. 2015. "Decrease of Back Recombination Rate in CdS Quantum Dots Sensitized Solar Cells Using Reduced Graphene Oxide." *Chinese Physical Society* 4: 047205–047213. https://doi.org/10.1088/1674-1056/24/4/047205.

Bharathi, Ganapathi, Devaraj Nataraj, Sellan Premkumar, and Murugaiyan Sowmiya. 2017. "Graphene Quantum Dot Solid Sheets: Strong Blue-Light-Emitting & Photocurrent-Producing Band-Gap-Opened Nanostructures." *Scientific Reports*, 1–17. https://doi.org/10.1038/s41598-017-10534-4.

Biomedicine, Nano, Minghan Xu, Xingzhong Zhu, Hao Wei, and Zhi Yang. 2013. "Hydrothermal/Solvothermal Synthesis of Graphene Quantum Dots and Their Biological Applications." *Nano Biomedicine and Engineering*, no. May 2016. https://doi.org/10.5101/nbe.v4i3.p65-71.

Cao, L. I., Mohammed J. Meziani, and Sushant Sahu. 2013. "Photoluminescence Properties of Graphene versus Other Carbon Nanomaterials." *Account of Chemical Research* 46, no. 1: 171–180. https://doi.org/10.1021/ar300128j.

Chattha, Hassan T. 2019. "4-Port 2-Element MIMO Antenna for 5G Portable Applications." *IEEE Access* PP, no. 4: 1. https://doi.org/10.1109/ACCESS.2019.2925351.

Chen, Wufeng, Lifeng Yan, and Prakriti R. Bangal. 2009. "Preparation of Graphene by the Rapid and Mild Thermal Reduction of Graphene Oxide Induced by Microwaves." *Carbon* 48, no. 4: 1146–1152. https://doi.org/10.1016/j.carbon.2009.11.037.

Choudhary, P., R. Kumar, and N. Gupta. 2015. "Dielectric Material Selection of Microstrip Patch Antenna for Wireless Communication Applications Using Ashby's Approach." *International Journal of Microwave and Wireless Technologies* 7: 579–587. https://doi.org/10.1017/S1759078714000877.

Das, Tanmoy, Bhupendra K. Sharma, Ajit K. Katiyar, and Jong-Hyun Ahn. 2018. "Graphene-Based Flexible and Wearable Electronics." *Journal of Semiconductors* 39: 1–19. https://doi.org/10.1088/1674-4926/39/1/011007.

Dervishi, Enkeleda, Zhiqiang Ji, Han Htoon, Milan Sykora, and K. Stephen Doorn. 2019. "Raman Spectroscopy of Bottom-Up Synthesized Graphene Quantum Dots: Size and Structure Dependence." *Nanoscale*, 1–27. https://doi.org/10.1039/C9NR05345J.

Dong, Yongqiang, Jianpeng Lin, Yingmei Chen, Fengfu Fu, Yuwu Chi, and Guonan Chen. 2014. "Graphene Quantum Dots, Graphene Oxide, Carbon Quantum Dots and Graphite Nanocrystals in Coals." *Nanoscale* 6: 7410–7415. https://doi.org/10.1039/c4nr01482k.

Edmondson, James, Khasha Ghaffarzadeh, and Luyun Jiang. 2020. "Thermal Management for 5G." IDTechEX. 2020. http://www.idtechex.com/fr/research-report/thermal-management-for-5g/757.

Gangadharan, Deepak Thrithamarassery, G.s. Anjusree, Amrita Vishwa Vidyapeetham, Narendra Pai, and Devika Subash. 2014. "Fabrication of a Dye-Sensitized Solar Cell Module by Spray Pyrolysis." *RSC Advances* 4: 23299–23303. https://doi.org/10.1039/C4RA01883D.

Gupta, Parul, Leeladhar Malviya, and S. V. Charhate. 2019. "5G Multi-Element/Port Antenna Design for Wireless Applications: A Review." *International Journal of Microwave and Wireless Technologies*. https://doi.org/10.1017/S1759078719000382.

Hao, Yan, Yasemin Saygili, Jiayan Cong, Anna I. K. Eriksson, Wenxing Yang, Jinbao Zhang, Enrico Polanski, et al. 2016. "A Novel Blue Organic Dye for Dye-Sensitized Solar Cells Achieving High Efficiency in Cobalt-Based Electrolytes and by Co-Sensitization." *Applied Materials and Interface*. https://doi.org/10.1021/acsami.6b09671.

He, Meiqin, Jin Zhang, Hai Wang, Yanrong Kong, Yiming Xiao, and Wen Xu. 2018. "Material and Optical Properties of Fluorescent Carbon Quantum Dots Fabricated from Lemon Juice via Hydrothermal Reaction." *Nanoscale Research Letters* 13: 175–182. https://doi.org/10.1186/s11671-018-2581-7.

Holá, Kateřina, Mária Sudolská, Sergii Kalytchuk, Dana Nachtigallová, Andrey L. Rogach, Michal Otyepka, and Radek Zbořil. 2017. "Graphitic Nitrogen Triggers Red Fluorescence in Carbon Dots." *ACS Nano* 11: 12402–12410. https://doi.org/10.1021/acsnano.7b06399.

Huang, Bin, Zhefei Pan, Xiangyu Su, and Liang An. 2018. "Recycling of Lithium-Ion Batteries: Recent Advances and Perspectives." *Journal of Power Sources* 399, June: 274–286. https://doi.org/10.1016/j.jpowsour.2018.07.116.

Jang, Houk, Yong Ju Park, Xiang Chen, Tanmoy Das, Min Seok Kim, and Jong Hyun Ahn. 2016. "Graphene-Based Flexible and Stretchable Electronics." *Advanced Materials*. https://doi.org/10.1002/adma.2015 04245.

Jia, Henan, Yifei Cai, Jinghuang Lin, Haoyan Liang, Junlei Qi, Jian Cao, and Jicai Feng. 2018. "Heterostructural Graphene Quantum Dot/MnO2 Nanosheets toward High-Potential Window Electrodes for High-Performance Supercapacitors." *Advanced Science*, 1–10. https://doi.org/10.1002/advs.201700887.

Jiang, Roger, Hannes Michaels, and Nick Vlachopoulos. 2019. "Beyond the Limitations of Dye-Sensitized Solar Cells." In *Dye-Sensitized Solar Cells*, 285–324. Elsevier Inc. https://doi.org/10.1016/B978-0-12-814541-8.00008-2.

Kalluri, Ankarao, Debika Debnath, Bhushan Dharmadhikari, and Prabir Patra. 2018. "Graphene Quantum Dots: Synthesis and Applications." In *Enzyme Nanoarchitectures: Enzymes Armored with Graphene*, 1st ed., 1–20. Elsevier Inc. https://doi.org/10.1016/bs.mie.2018.07.002.

Karthik, P. S., A. L. Himaja, and Surya Prakash Singh. 2014. "Carbon-Allotropes: Synthesis Methods, Applications and Future Perspectives." *Carbon Letters*. https://doi.org/10.5714/CL.2014.15.4.219.

Kavitha, M. K., and Manu Jaiswal. 2016. "Graphene: A Review of Optical Properties and Photonic Applications." *Asian Journal of Physics* 7, no. 25: 809–831.

Khan, Karim, Ayesha Khan Tareen, Muhammad Aslam, Renheng Wang, Yupeng Zhang, Asif Mahmood, Zhengbiao Ouyang, Han Zhang, and Zhongyi Guo. 2020. "Dimensional Materials and Their Applications." *Journal of Materials Chemistry* 8: 387–440. https://doi.org/10.1039/c9tc04187g.

Kozák, Ondřej, Kasibhatta Kumara Ramanatha Datta, Monika Greplová, Václav Ranc, Josef Kašlík, and Radek Zbořil. 2013. "Surfactant-Derived Amphiphilic Carbon Dots with Tunable Photoluminescence." *Journal of Physical Chemistry* 117: 24991–24996. https://doi.org/10.1021/jp4040166.

Kumar, Sumeet, Animesh K. Ojha, Bilal Ahmed, Ashok Kumar, and Jayanta Das. 2017. "Tunable (Violet to Green) Emission by High-Yield Graphene Quantum Dots and Exploiting Its Unique Properties towards Sun-Light-Driven Photocatalysis and Supercapacitor Electrode Materials." *Materials Today Communications Journal* 11: 76–86. https://doi.org/10.1016/j.mtcomm.2017.02.009.

Kundu, Sumana, and Vijayamohanan K. Pillai. 2019. "Synthesis and Characterization of Graphene Quantum Dot." *Physical Sciences Reviews*, 1–35. https://doi.org/10.1515/psr-2019-0013.

Latif, I., Taghreed B. Alwan, and Ammar H. Al-dujaili. 2012. "Low Frequency Dielectric Study of PAPA-PVA-GR Nanocomposites." *Nanoscience and Nanotechnology* 2, no. 6: 190–200. https://doi.org/10.5 923/j.nn.20120206.07.

Lee, Changgu, Xiaoding Wei, Jeffrey W. Kysar, and James Hone. 2008. "Measurement of the Elastic Properties and Intrinsic Strength of Monolayer Graphene." *Scientific Reports* 385: 385–388. https://doi.org/10.1126/science.1157996.

Lee, Keunsik, Hanleem Lee, Yonghun Shin, Yeoheung Yoon, Doyoung Kim, and Hyoyoung Lee. 2016. "Nano Energy Highly Transparent and Flexible Supercapacitors Using Graphene-Graphene Quantum Dots Chelate." *Nano Energy* 26: 746–754. https://doi.org/10.1016/j.nanoen.2016.06.030.

Lei, Xiang, Ke Yu, Ruijuan Qi, and Ziqiang Zhu. 2018. "Fabrication and Theoretical Investigation of MoS2-Co3S4hybrid Hollow Structure as Electrode Material for Lithium-Ion Batteries and Supercapacitors." *Chemical Engineering Journal* 347, March: 607–617. https://doi.org/10.1016/j.cej.2018.04.154.

Li, Xiaoming, Shengli Zhang, Sergei A. Kulinich, Yanli Liu, and Haibo Zeng. 2014. "Engineering Surface States of Carbon Dots to Achieve Controllable Luminescence for Solid-Luminescent Composites and Sensitive Be2+ Detection." *Scientific Reports*, 1–8. https://doi.org/10.1038/srep04976.

Li, Weitao, Ming Li, Yijian Liu, Dengyu Pan, Zhen Li, Liang Wang, and Minghong Wu. 2018. "Three-Minute Ultra-Rapid Microwave-Assisted Synthesis of Bright Fluorescent Graphene Quantum Dots for

Live Cell Staining and White LEDs Three-Minute Ultra-Rapid Microwave-Assisted Synthesis of Bright Fluorescent Graphene Quantum Dots for Live Cell Stainin." *Apllied Nano Matherials*, 1–21. https://doi.org/10.1021/acsanm.8b00114.

Liu, Wen-wen, Ya-qiang Feng, Xing-bin Yan, Jiang-tao Chen, and Qun-ji Xue. 2013. "Superior Micro-Supercapacitors Based on Graphene Quantum Dots." *Advanced Functional Materials*, 4111–4122. https://doi.org/10.1002/adfm.201203771.

Loos, Marcio. 2015. "Allotropes of Carbon and Carbon Nanotubes." In *Carbon Nanotube Reinforced Composites: CNR Polymer Science and Technology*. https://doi.org/10.1016/B978-1-4557-3195-4.00003-5.

Lu, Siyu, Laizhi Sui, Min Wu, Shoujun Zhu, Xue Yong, and Bai Yang. 2018. "Graphitic Nitrogen and High-Crystalline Triggered Strong Photoluminescence and Room-Temperature Ferromagnetism in Carbonized Polymer Dots." *Advanced Sciences* 1801192: 1–8. https://doi.org/10.1002/advs.201801192.

Lundstrom, M. 2003. "Moore's Law Forever?" *Science*. https://doi.org/10.1126/science.1079567.

Ma, Shuai, Modi Jiang, Peng Tao, Chengyi Song, Jianbo Wu, Jun Wang, Tao Deng, and Wen Shang. 2018. "Temperature Effect and Thermal Impact in Lithium-Ion Batteries: A Review." *Progress in Natural Science: Materials International*. https://doi.org/10.1016/j.pnsc.2018.11.002.

MacK, Chris A. 2011. "Fifty Years of Moore's Law." *IEEE Transactions on Semiconductor Manufacturing*. https://doi.org/10.1109/TSM.2010.2096437.

Mamvura, Tirivaviri A., and Geoffrey S. Simate. 2019. "The Potential Application of Graphene Nanotechnology for Renewable Energy Systems." In *Graphene-Based Nanotechnologies for Energy and Environmental Applications*, 59–80. Elsevier Inc. https://doi.org/10.1016/B978-0-12-815811-1.00004-1.

Marty, Bernard, Conel M. O'D. Alexa, and Sean N. Raymond. 2013. "Primordial Origins of Earth' s Carbon." *Reviews in Mineralogy & Geochemistry* 75: 149–181. https://doi.org/10.2138/rmg.2013.75.6.

Mittal, Garima, and Kyong Y. Rhee. 2018. "Role of Graphene in Flexible Electronics." *COJ Electronics & Communications*, 2–3. https://doi.org/10.31031/COJEC.2018.01.000509.

Molitor, F., J. Güttinger, C. Stampfer, S. Dröscher, A. Jacobsen, T. Ihn, and K. Ensslin. 2011. "Electronic Properties of Graphene Nanostructures." *Journal of Physics Condensed Matter*. https://doi.org/10.1088/0953-8984/23/24/243201.

Nasir, Salisu, Mohd Zobir Hussein, Zulkarnain Zainal, and Nor Azah Yusof. 2018. "Carbon-Based Nanomaterials/Allotropes: A Glimpse of Their Synthesis, Properties and Some Applications." *Materials* 11: 295–319. https://doi.org/10.3390/ma11020295.

O'Regan, B., and M. Gratzel. 1991. "A Low-Cost, High-Efficiency Solar Cell Based on Dye-Sensitized Colloidal TiO2 Films." *Nature*, 353: 737–740.

Okwundu, Onyeka Stanislaus, Emmanuel Uche Aniekwe, and Chinaza Emmanuel Nwanno. 2018. "Unlimited Potentials of Carbon: Different Structures and Uses (a Review)." *Metallurgical and Materials Engineering* 24 (3): 145–171. https://doi.org/10.30544/388.

Okwundu, Onyeka Stanislaus, Cyril Oluchukwu Ugwuoke, and Augustine Chukwujekwu Okaro. 2019. "Recent Trends in Non-Faradaic Supercapacitor Electrode Materials." *Metallurgical and Materials Engineering* 25: 105–138. https://doi.org/https://doi.org/10.30544/417.

Pace, Norman R. 2001. "The Universal Nature of Biochemistry." In Proceedings of the National Academy of Sciences of the United States of America, 98 (3): 805–808.

Pan, ByDengyu, Jingchun Zhang, Zhen Li, and Minghong Wu. 2010. "Hydrothermal Route for Cutting Graphene Sheets into *Advanced Matererials*, 734–738. https://doi.org/10.1002/adma.200902825.

Patil, Kaustubh, Soheil Rashidi, Hui Wang, and Wei Wei. 2019. "Review Article Recent Progress of Graphene-Based Photoelectrode Materials for Dye-Sensitized Solar Cells." *Hindawi International Journal of Photoenergy*. https://doi.org/10.1155/2019/1812879.

Patni, Neha, Pranjal Sharma, and Shibu G. Pillai. 2017. "Synthesis of Dye-Sensitized Solar Cells Using Polyaniline and Natural Dye Extracted from Beetroot Synthesis of Dye-Sensitized Sensitized Solar Cells Using Polyaniline and Natural Dye Extracted from Beetroot." *Research Journal of Recent Sciences* 6: 35–39.

Ponomarenko, L. A., F. Schedin, M. I. Katsnelson, R. Yang, E. W. Hill, K. S. Novoselov, and A. K. Geim. 2008. "Chaotic Dirac Billiard in Graphene Quantum Dots." *Science*. https://doi.org/10.1126/science.1154663.

Prikhozhdenko, Ekaterina S., Daniil N. Bratashov, and Anastasiya N. Mitrofanova. 2018. "Solvothermal Synthesis of Hydrophobic Carbon Dots in Reversed Micelles." *Journal of Nanoparticle Research* 20: 234. https://doi.org/10.1007/s11051-018-4336-x.

Radadiya, Tarun. 2015. "A Properties of Graphene." *European Journal of Material Sciences* 2: 6–18.

Recher, Patrik, and Björn Trauzettel. 2010. "Quantum Dots and Spin Qubits in Graphene." *Nanotechnology*. https://doi.org/10.1088/0957-4484/21/30/302001.

Sa'don, Siti Nor Hafizah, Mohd Haizal Jamaluddin, Muhammad Ramlee Kamarudin, Fauzan Ahmad, Yoshihide Yamada, Kamilia Kamardin, and Izni Husna Idris. 2019. "Analysis of Graphene Antenna Properties for 5G Applications." *Sensors* 19: 4835–4854. https://doi.org/10.3390/s19224835.

Saedi, Afsoon, Ali Mashinchian, Moradi Salimeh, Kimiagar Homayon, and Ahmad Panahi. 2020. "Efficiency Enhancement of Dye - Sensitized Solar Cells Based on Gracilaria / Ulva Using Graphene Quantum Dot." *International Journal of Environmental Research*. https://doi.org/10.1007/s41742-020-00265-2.

Sciortino, Alice, Andrea Cannizzo, and Fabrizio Messina. 2018. "Carbon Nanodots: A Review—From the Current Understanding of the Fundamental Photophysics to the Full Control of the Optical Response." *Journal of Carbon Research* 4: 67–102. https://doi.org/10.3390/c4040067.

Sederberg, D. 2009. "Allotropes of Carbon." *National Center for Learning and Teaching in Nanoscale Science and Engineering*. https://doi.org/10.1002/9783527619214.fmatter.

Seekaew, Yotsarayuth, Onsuda Arayawut, and Kriengkri Timsorn. 2019. "Applications of Graphene and Derivatives." In *Carbon-Based Nanofillers and Their Rubber Nanocomposites*, 259–284. Elsevier Inc. https://doi.org/10.1016/B978-0-12-813248-7.00009-2.

Shaker, Majid, Reza Riahifar, and Yao Li. 2020. "A Review on the Superb Contribution of Carbon and Graphene Quantum Dots to Electrochemical Capacitors Performance: Synthesis and Application." *FlatChem*, 100171. https://doi.org/10.1016/j.flatc.2020.100171.

Shan, Chee, Katerina Hola, Adriano Ambrosi, Radek Zboril, and Martin Pumera. 2015. "Graphene and Carbon Quantum Dots Electrochemistry." *Electrochemistry Communications* 52: 75–79. https://doi.org/10.1016/j.elecom.2015.01.023.

Sharma, Bhupendra K., and Jong-hyun Ahn. 2013. "Graphene Based Field Effect Transistors: Efforts Made towards Flexible Electronics." *Solid-State Electronics* 89: 177–188. https://doi.org/10.1016/j.sse.2013.08.007.

Sharma, Anirudh, and Joydeep Das. 2019. "Small Molecules Derived Carbon Dots: Synthesis and Applications in Sensing, Catalysis, Imaging, and Biomedicine." *Journal of Nanobiotechnology*, 1–24. https://doi.org/10.1186/s12951-019-0525-8.

Sharma, Khushboo, Vinay Sharma, and S. S. Sharma. 2018. "Dye-Sensitized Solar Cells: Fundamentals and Current Status." *Nanoscale Research Letters* 6. https://doi.org/10.1186/s11671-018-2760-6.

Shen, Jianhua, Yihua Zhu, Xiaoling Yang, and Chunzhong Li. 2012. "Graphene Quantum Dots: Emergent Nanolights for Bioimaging, Sensors, Catalysis and Photovoltaic Devices." *Chemical Communication* 48: 3686–3699. https://doi.org/10.1039/c2cc00110a.

Shulaker, Max M., Gage Hills, Nishant Patil, Hai Wei, Hong-yu Chen, H. Philip Wong, and Subhasish Mitra. 2013. "Carbon Nanotube Computer." *Nature* 501 (7468): 526–530. https://doi.org/10.1038/nature12502.

Smekens, Jelle, Vrije Universiteit Brussel, Rahul Gopalakrishnan, Avesta Battery, Nils Van Den Steen, Vrije Universiteit Brussel, Noshin Omar, and Avesta Battery. 2016. "Influence of Electrode Density on the Performance of Li-Ion Batteries: Experimental and Simulation Results." *Energies* 9: 104–117. https://doi.org/10.3390/en9020104.

Son, Dong Ick, Byoung Wook Kwon, Dong Hee Park, Won-seon Seo, Yeonjin Yi, Basavaraj Angadi, Chang-lyoul Lee, and Won Kook Choi. 2012. "Emissive ZnO–Graphene Quantum Dots for White-Light-Emitting Diodes." *Nature Nanotechnology*, 1–7. https://doi.org/10.1038/nnano.2012.71.

Song, Na, Dejin Jiao, Siqi Cui, Xingshuang Hou, Peng Ding, and Liyi Shi. 2017. "Highly Anisotropic Thermal Conductivity of Layer-by-Layer Assembled Nanofibrillated Cellulose/Graphene Nanosheets Hybrid Films for Thermal Management." *ACS Applied Materials and Interfaces*. https://doi.org/10.1021/acsami.6b11979.

Suvarnaphaet, Phitsini, and Suejit Pechprasarn. 2017. "Graphene-Based Materials for Biosensors: A Review." *Sensors*. https://doi.org/10.3390/s17102161.

Tajik, Somayeh, Zahra Dourandish, Kaiqiang Zhang, and Hadi Beitollahi. 2020. "Carbon and Graphene Quantum Dots: A Review on Syntheses, Characterization, Biological and Sensing Applications for Neurotransmitter Determination." *RSC Advances* 10: 15406–15429. https://doi.org/10.1039/d0ra00799d.

Ullah, Arslan, Abdul Majid, and Naema Rani. 2018. "A Review on First Principles Based Studies for Improvement of Cathode Material of Lithium Ion Batteries." *Journal of Energy Chemistry*. https://doi.org/10.1016/j.jechem.2017.09.007.

Vardi, Nathan. 2005. "Moore's Law." *Forbes*. https://doi.org/10.1201/noe0849396397.ch213.

Wang, Xiaohui, Konggang Qu, Bailu Xu, and Xiaogang Qu. 2011. "Microwave Assisted One-Step Green Synthesis of Cell-Permeable Multicolor Photoluminescent Carbon Dots without Surface Passivation Reagents." *Journal of Materials Chemistry* 21: 2445–2450. https://doi.org/10.1039/c0jm02963g.

Wang, Xiao, Yongqiang Feng, Peipei Dong, and Jianfeng Huang. 2019. "A Mini Review on Carbon Quantum Dots: Preparation, Properties, and Electrocatalytic Application." *Frontiers in Chemistry* 7: 1–9. https://doi.org/10.3389/fchem.2019.00671.

Wang, Donghai, Daiwon Choi, Juan Li, Zhenguo Yang, Zimin Nie, Rong Kou, Dehong Hu, et al. 2009. "Self-Assembled Tio2-Graphene Hybrid Nanostructures for Enhanced Li-Ion Insertion." *ACS Nano*. https://doi.org/10.1021/nn900150y.

Wei, Junhua, and Palo Alto. 2014. "Tunable Optical Properties of Graphene Quantum Dots by Centrifugation." *International Mechanical Engineering Congress & Exposition*. https://doi.org/10.1115/IMECE2013-64756.

Wei, Q., X. Xu, and Y. Wang. 2009. "Textile Surface Functionalization by Physical Vapor Deposition (PVD)." In *Surface Modification of Textiles Common*, 58–90. https://doi.org/10.1533/9781845696689.58. www.electronics-notes.com/articles/antennas-propagation/mimo/what-is-mimo-multiple-inpute-multiple-output-wireless-technology.php "What Is MIMO Wireless Technology" Accessed 15th June 2020.

Xu, Xiaoyou, Robert Ray, Yunlong Gu, Harry J. Ploehn, Latha Gearheart, Kyle Raker, and Walter A. Scrivens. 2004. "Electrophoretic Analysis and Purification of Fluorescent Single-Walled Carbon Nanotube Fragments." *Journal of American Chemical Society* 126: 12736–12737.

Yang, Pei, Ziqi Zhu, Minzhi Chen, Weimin Chen, and Xiaoyan Zhou. 2018. "Microwave-Assisted Synthesis of Xylan-Derived Carbon Quantum Dots for Tetracycline Sensing." *Optical Materials* 85, June: 329–336. https://doi.org/10.1016/j.optmat.2018.06.034.

Yang, Siwei, Jing Sun, Xiubing Li, Wei Zhou, Zhongyang Wang, Pen He, Guqiao Ding, Xiaoming Xie, Zhenhui Kang, and Mianheng Jiang. 2014. "Quantum Dots with Tunable-Photoluminescence." *Journal of Materials Chemistry* 2: 8660–8667. https://doi.org/10.1039/c4ta00860j.

Zeng, Wei, Lin Shu, Qiao Li, Song Chen, Fei Wang, and Xiao Ming Tao. 2014. "Fiber-Based Wearable Electronics: A Review of Materials, Fabrication, Devices, and Applications." *Advanced Materials*. https://doi.org/10.1002/adma.201400633.

Zhang, Zhipan, Jing Zhang, Nan Chen, and Liangti Qu. 2012. "Environmental Science Graphene Quantum Dots: An Emerging Material for Energy-Related Applications and Beyond." *Energy & Environmental Science* 5: 8869–8890. https://doi.org/10.1039/c2ee22982j.

Zhang, Qinghong, Xiaofeng Sun, Hong Ruan, Keyang Yin, and Hongguang Li. 2017. "Production of Yellow-Emitting Carbon Quantum Dots from Fullerene Carbon Soot." *Science China Materials* 60 no. 2: 141–150. https://doi.org/10.1007/s40843-016-5160-9.

Zhang, Su, Yutong Li, Huaihe Song, Xiaohong Chen, Jisheng Zhou, and Song Hong. 2016. "Graphene Quantum Dots as the Electrolyte for Solid State Supercapacitors." *Scientific Report*, 1–7. https://doi.org/10.1038/srep19292.

Zhang, Shuo, Lina Sui, Hongzhou Dong, Wenbo He, Lifeng Dong, and Liyan Yu. 2018. "High-Performance Supercapacitor of Graphene Quantum Dots with Uniform Sizes." *Applied Materials & Interfaces* 10: 12983–12991. https://doi.org/10.1021/acsami.8b00323.

Zhao, Weifeng, Ming Fang, Furong Wu, Hang Wu, Liwei Wang, and Guohua Chen. 2010. "Preparation of Graphene by Exfoliation of Graphite Using Wet Ball Milling." *Journal of Materials Chemistry* 20, no. 28: 5817–5819. https://doi.org/10.1039/c0jm01354d.

Zhao, Qidong, Tengfeng Xie, Linlin Peng, Yanhong Lin, Ping Wang, Liang Peng, and Dejun Wang. 2007. "Size- and Orientation-Dependent Photovoltaic Properties of ZnO Nanorods." *Journal of Physical Chemistry*, 17136–17145. https://doi.org/10.1021/jp075368y.

Zhao, Yan, Yue Zhang, Xiaoman Liu, Hui Kong, Yongzhi Wang, Gaofeng Qin, and Peng Cao. 2017. "Novel Carbon Quantum Dots from Egg Yolk Oil and Their Haemostatic Effects." *Scientific RepoRts* 7: 1–8. https://doi.org/10.1038/s41598-017-04073-1.

Zhen, Zhen, and Hongwei Zhu. 2018. "Structure and Properties of Graphene." In *Graphene*, 1–12. Elsevier Inc. https://doi.org/10.1016/B978-0-12-812651-6.00001-X.

Zhou, Jigang, Christina Booker, Ruying Li, Xingtai Zhou, and Tsun-kong Sham. 2007. "An Electrochemical Avenue to Blue Luminescent Nanocrystals from Multiwalled Carbon Nanotubes (MWCNTs)." *Journal of American Chemical Society* 8: 744–745. https://doi.org/10.1021/ja0669070.

Zhou, Lei, Chunyang Li, Xiang Liu, Yusong Zhu, Yuping Wu, and Teunis Van Ree. 2019. "Metal Oxides in Supercapacitors." *Metal Oxides in Energy Technologies*, 169–203. https://doi.org/10.1016/B978-0-12-811167-3.00007-9.

Zhu, Hui, Xiaolei Wang, Yali Li, Zhongjun Wang, and Xiurong Yang. 2009. "Microwave Synthesis of Fluorescent Carbon Nanoparticles with Electrochemiluminescence Properties W." *Chemical Communications*, 5118–5120. https://doi.org/10.1039/b907612c.

Zhu, Shoujun, Yubin Song, Xiaohuan Zhao, Jieren Shao, Junhu Zhang, and Bai Yang. 2014. "The Photoluminescence Mechanism in Carbon Dots (Graphene Quantum Dots, Carbon Nanodots, and Polymer Dots): Current State and Future Perspective." *Nano Research* 8, no. 2: 355–381. https://doi.org/10.1007/s12274-014-0644-3.

Zhu, Shoujun, Yubin Song, Joy Wang, Hao Wan, Yuan Zhang, Yang Ning, and Bai Yang. 2016. "Photoluminescence Mechanism in Graphene Quantum Dots: Quantum Confinement Effect and Surface/Edge State." *Nano Today*. https://doi.org/10.1016/j.nantod.2016.12.006.

Zou, Changfu, Lei Zhang, Xiaosong Hu, Zhenpo Wang, Torsten Wik, and Michael Pecht. 2018. "A Review of Fractional-Order Techniques Applied to Lithium-Ion Batteries, Lead-Acid Batteries, and Supercapacitors." *Journal of Power Sources* 390, February: 286–296. https://doi.org/10.1016/j.jpowsour.2018.04.033.

23

Solar Cell Technology: Challenges and Progress

Newayemedhin A. Tegegne and Fekadu Gashaw Hone
Department of Physics, Addis Ababa University, Addis Ababa, Ethiopia

23.1 Introduction

The main energy source that meets the energy demand of our world are fossil fuels. The use of these fossil fuels often is hazardous to the environment, due to greenhouse gas emissions, climate change, and ecosystem instability. The catastrophes occurring repeatedly in different parts of the world have increased the global interest for cleaner and sustainable sources of energy like wind and solar energy. Solar energy is considered renewable as the sun's life is between 5000 and 10,000 billion years. The amount of solar energy that reaches the surface of the earth in less than one hour is enough to satisfy the yearly energy demand of our world. In addition to its abundance, solar energy is a clean source without any harmful effects to the environment (Servaites et al. 2011). Solar energy can be harvested using photovoltaic (PV) cells for electricity generation. PV cells convert solar irradiation to electrical energy through a photovoltaic effect. The French scientist Alexandre Edmond Becquerel discovered photovoltaic effect in 1834 on a platinum electrode covered with silver chloride or silver bromide, which was then exposed to electromagnetic radiation (Petrova-Koch 2009). After 50 years, the first photoelectric effect on solid selenium was reported (Hertz 1887). The discovery led to the invention of the first photovoltaic device with a power conversion efficiency (PCE) of 6% in 1954 in the Bell Laboratories (Huo et al. 2009). Later, the first commercialized solar cell was used in a space programme in 1960 during the Cold War between Americans and Soviets (Welch and Bazan 2011).

The PV technology showed a tremendous improvement in the past four to five decades. The operation of PV cells mainly requires three processes: absorption of photons that leads to generation of charge carrier (free or bound) and finally extraction of free charge carriers by the external circuit. The semiconductor material that is used in the fabrication of PV cells determines the versatility of the technology. The first-generation solar cells use silicon. The technology is matured with a laboratory efficiency exceeding 25% (Green 2009) and stability of 20–30 years. The development of the silicon-based PV technology is limited by the sophisticated manufacturing techniques that require high energy and cost. The second-generation solar cells called thin-film solar cells use materials like gallium and indium, which are low in the crust of the earth, and cadmium which is toxic. The third generation also called *emerging solar cell technologies* includes organic photovoltaic (OPV), dye-sensitized solar cells (DSSCs), and perovskite solar cells (PSCs). The unique features of these solar cells are low cost, mechanical stability, and material abundance. In the following sections each of the three generations of solar cell technology will be discussed in detail.

23.2 First-Generation Solar Cells: Crystalline Silicon Solar Cells

The first-generation solar cells that include the mono- and polycrystalline solar cells make 90% global PV market share owing to the introduction of novel technologies after the first practical p-n junction

DOI: 10.1201/9781003145585-23

solar cell in Bell Labs. The cell design has improved to maximize PCE of the devices mainly by reducing optical losses such as reflection and electrical losses. The improvements also are selected to minimize cost of production. In the following sections, silicon solar cell designs and their progress will be discussed.

23.2.1 Back-Surface Field Solar cells

A regular solar cell consists of two layers of silicon with different electrical properties known as the *base* and the *emitter*. A strong electrical field is generated where the two layers meet, which pulls negatively charged particles (electrons) into the emitter when they reach this interface. These devices are called *back surface field (BSF) solar cells*. The electrons are generated by light entering the cell and releasing electrons from the silicon atoms. Electrons travel freely through the cell and contribute to the electrical current only if they are able to reach the interface. The device geometry of the BSF solar cells is shown in Figure 23.1. The charge generation in silicon solar cells can generally be summarized as follows.

STEP-1: Light knocks out electrons from silicon atoms
STEP-2: Electrons move freely in the silicon wafer
STEP-3: The electrons that reach the emitter-base interface moves to the emitter creating a voltage difference in the device

Fabrication of conventional solar cells includes the following processes. Single crystal of silicon (c-Si) accounting for 30% of the world PV market are grown by Czochralski and float zone methods. Float zone silicon (FZ-Si) substrates have significantly lower impurity concentration than Czochralski silicon (CZ-Si). A simpler processing technique is used to grow the multicrystalline silicon (Mc-Si). The material quality of Mc-Si is poor compared to single crystalline silicon. However, the easier processing technique made Mc-Si more preferable commercially that it is used in approximately 50% worldwide

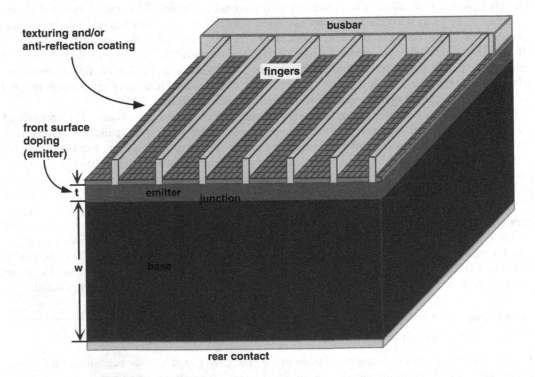

FIGURE 23.1 Device geometry of BSF cells.

PV technology. The substrates grown will be sliced using a wire saw or diamond-edge saw. The cut wafers will be cleaned in a series of chemical baths followed by lapping process to get a uniform thickness. Finally, the wafers will be polished and cleaned in a clean room. The surface of the wafers will be textured to remove damage induced during sawing. Texturing increases the light trapping within the solar cell, especially in the longer wavelength region. The p-n junction formation is the next step. The most commonly used junction formation on a silicon substrate include diffusion and ion implantation. A defined region is masked for doping. A bare silicon substrate reflects 30% of the incident light. An anti-reflection coating is applied on the surface of the silicon substrate to reduce its optical reflectivity and couple the light back to the solar cells. The final step in the solar cell fabrication is metallization in which the contacts are formed on both sides of the surfaces of the silicon for charge extraction. Screen printing of the contacts is commonly used in commercial solar cells.

23.2.2 High-Efficiency cells

The cost of Si-PV cells has reduced greatly due to the introduction of high-efficiency cells using sophisticated technologies. Large number of researchers and manufacturing companies have been involved in the design and fabrication of high-efficiency cells. The research also focused on low-cost processing technologies to increase the competitiveness of the technology with the conventional fossil fuel–based technologies. In this section, high-efficiency solar cells will be discussed.

23.2.2.1 Passivated Emitted and Rear Cell and Passivated Emitted Rear Locally Diffused Cell

The Passivated Emitted and Rear Cell (PERC) was first proposed by Martin Green in 1983 (Green et al. 1984). In a conventional solar cell, the aluminium metallization layer makes contact with the full area of the back of the cell. The back of the cell in PERC device is coated with dielectric materials that have small holes, on top of which the aluminium metallization is applied. The device structure is shown in Figure 23.2a. The rear side of the Si-wafer is polished after phosphorous diffusion. A thin dielectric layer of Al_2O_3 is deposited on the rear P-type surface and SiN_x anti-reflection coating on both front and rear side of the cell is deposited. Tiny holes are drilled in the dielectric deposited on the rear side using laser to allow the printing of Al paste. On the front side, Ag paste is screen printed. The PERC cell needs three more fabrication steps that results in an increase of PCE of 1% (from 19% to 20%).

An incident light on the surface of the solar cell generates electrons at different depths with different wavelengths. The blue part of the solar spectrum is absorbed on the front surface while the red part gets deeper to the back of the cells to generate electrons or passes through without generating current. In silicon solar cells only 70% of the infrared light that reaches the Al rear contact gets reflected back (Lorenz et al. 2010). The dielectric layer in the PERC cell reflects back any light that passes through to the rear without generating current and gives it more chance of absorption. The improved light capture at the longer wavelength region of the solar spectrum in PERC cells can increase the optical thickness of the device without the need to increase the physical thickness. The passivated rear layer should be optimized to reduce recombination. However, the optimization of the rear side is fully independent of the front side which makes improvement in PERC structure easier. An atomic layer deposited Al_2O_3 and thermally grown SiO_2 are the most preferable dielectric layers that are used for rear passivation in PERC cells. Passivation layer stacks such as Al_2O_3/SiN_x or SiO_2/SiN_x are also used in high-efficiency PERC cells. The internal quantum efficiency (IQE) and reflectance of PERC compared to Al-BSF solar cells was found to improve in the longer wavelength region above 900 nm (Dullweber et al. 2012). PCE of 19.0% and 19.4% was found in PERC cells due to improved reflectance and IQE with Al_2O_3/SiN_x and SiO_2/SiN_x passivation stacks.

A further improvement of PERC cells was achieved by a Passivated Emitter Rear Locally diffused (PERL) solar cells. Similar to the PERC cell, both the rear and the emitter cells are passivated in PERL cells. The thermally grown SiO_2 passivation layer in the rear of PERL cell is opened to allow locally born diffusion in the back P^+ surface field. The locally diffused rear layer reduces

(a)   (b)

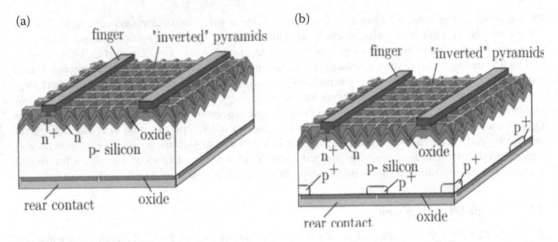

FIGURE 23.2   Device structure of (a) PERC and (b) PERL cells.

recombination by maintaining the electrical contact. The device structure of PERL and PERC cells is shown in Figure 23.2a and b. A record PCE of 22.8% was reported by Blakers et al. as early as 1989 in small area PERC solar cell (Blakers et al. 1989). The improvement in the structure of PERC cells mainly focusing on the rear passivation film has improved the PCE of the industrial device. Print-to-print technology to deposit Ag is other advancement. In print-to-print technique the number of silver fingers could be increased from the standard 50 fingers to 60 resulting in higher short circuit current. Physical vapour deposition was also used for metallization in PERC cells instead of the standard screen printing. Less shadowing and resistance were found due to the improved deposition. A high PCE of 21.1% in large-scale (125 mm^2) PERC cell based on FZ-Si was found in Fraunhofer Institute for Solar Energy System, Germany, using the sputtered contact. In the past 5 years, the PCE of PERC cells has shown tremendous improvements. In 2015, PCE of 21.7% by German manufacturer Solar World in PERC cells was achieved. In the same year, Trina Solar in China reported a PCE of 22.13% breaking the long time record in PERC cells (Ye et al. 2016). In 2016, the record PCE in PERC cells was once again broken by Trina Solar with a new PCE of 22.61% in a large-sized boron-doped CZ-Si substrate (Deng et al. 2017). In the following year, LERRI Solar reported a PCE of 23.26%. The PCE of PERC solar cells is expected to exceed 24% in the near future with the continuous improvement undergoing in the technology (Min et al. 2015).

23.2.2.2 PERT, TOPCon, and Bifacial Cells

P-type silicon wafers are the commonly used starting materials for c-Si and mc-Si-based solar cells mainly because of their high life time. Phosphorous-doped N-type silicon wafers also have a life time on the orders of milliseconds which makes them good starting material for highly efficient solar cells. However, the PN junction formation with boron diffusion of N-type Si needs high temperature of over 1000°C. The mc-Si can survive such a high temperature unlike the mc-Si which makes the N-type-based Si high-efficient solar cell compatible only with mc-Si. N-type passivated emitter rear totally diffused (PERT) solar cells can be fabricated using N-type mc-Si wafer which is textured and boron diffused. The rear of the wafer will then be phosphorus-doped to create N$^+$N structure at the back surface to reduce recombination due to the high electric field in the junction. The front and the back of the PERT cell similar to the PERC one will be passivated using thin layers of SiO$_2$ or Al$_2$O$_3$ and SiN$_x$, respectively. A thin SiN$_x$ layer is deposited on the front to serve as an anti-reflection coating. The device geometry is shown in Figure 23.3a. The PERL structure was first designed to reduce the current accumulation in the PERC cells and increase their fill factor (Wang 1992). The fill factor of small area cells (4 cm^2) fabricated using MCZ-Si substrate was improved from 81.1% in PERL to 83.5% in PERT devices which also increased the efficiency from 23.5% to high efficiency of 24.5% due to the reduced current jam in PERT cells. The n-type Si PERT cells achieved a

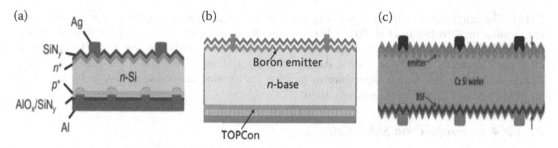

FIGURE 23.3 Device structure of (a) PERT, (b) TOPCon and (c) bifacial cells.

PCE of 21.1% and 21.9% using CZ and FZ-Si substrates, respectively. Jianhua Zhao and Aihua Wang developed an n-type PERT cell with the boron-diffused emitter at the rear surface (Zhao and Wang 2006). The development in the PERT design resulted in a significant improved PCE of 22.7% in 22 cm²-large cells.

The N-type Si wafer–based solar cells also generated high efficiency using a tunnel oxide passivated contact (TOPCon) structure. The TOPC device structure is similar to PERT cells in the front. On the rear of TOPCon cells, an ultra-thin layer of SiO_2 (1–2 nm) on which charge carrier can tunnel through is deposited as a passivation layer. An N^+-layer will then be deposited below the tunnelling layer. The device structure is shown in Figure 23.3b. In comparison to PERC cells, TOPCon cells are relatively easy to fabricate. The rear contact in the PERC cells is restricted to the openings which may limit the charge carrier transport to the shortest perpendicular distance and create accumulation of charge carriers. However, the charge carriers in TOPCon cells can easily tunnel through the ultra-thin SiO_2 layers which results in an increased carrier life time. The full-rear metal contact in TOPCon cells increases charge flow efficiency resulting in high fill factor and a high PCE of 23% at the early stage of the technology. The increase in PCE continued to improve and a world record PCE of 25.7% on TOPCon solar cells was achieved in few years (Chen et al. 2017).

A solar cell that is capable of collecting diffused light from the back surface can increase the total efficiency. Bifacial solar cells are designed to generate electricity by harnessing light both from the front and rear side of the cell. The main difference in bifacial solar cells from the standard is the structure of the rear side. A metal grid "fingers" is deposited on both sides of the cell to allow light from both sides. The device structure is shown in Figure 23.3c. For an n-type substrate a bifacial layer is prepared using a boron-diffused P^+ layer to form P^+N junction and phosphorus-doped N^+ region forming N^+N junction. The anti-reflection coat on top and bottom layers will then be covered by metal grids to allow light in both sides (Yang et al. 2011). Yang et al. reported large area (149 cm²) bifacial solar cells using CZ mc-Si wafer with PCE of 16.6% and 12.8% on the front and rear cells, respectively (Yang et al. 2011). In 2017, a PCE of 21.2% and up to 16.7% on the front and rear side of a bifacial solar cell, respectively, with five busbar Al fingers was reported (Dullweber et al. 2015).

23.2.2.3 Inter-Digitated Back Contact Cell

An optical shading from the front side of the conventional solar cell accounts for 7–10% loss. Interdigitated back contact (IBC) cells avoid this loss by moving the front metallization to the rear side. IBC cells are mainly fabricated on N-type Si wafers because of the higher carrier life time in the semiconductor that allows localization of the charge carriers to the rear side (Schwartz and Lammert 1975). The front side recombination is reduced by the phosphorus-doped N^+ and SiO_2 passivated layer. On the rear side, N^+ and P^+ narrow strips separated by insulation are deposited. This process needs a complex lithography which adds to the cost. A passivation layer of SiO_2 will be deposited on the rear through which the interdigitated contacts will be applied either by screen printing or e-beam evaporation to reduce the contact of the metal with the silicon wafer and further reduce recombination. A high efficiency of over 23% was reported in 2015 (Glunz et al. 2015). Sunpower Corp. reported a high efficiency of 24.2% based on a large area (155.1 cm²) IBC cell (Cousins et al.

2015). The efficiency of IBC cells was improved to 25.2% in large area (153.49 cm^2) by the same corporation in 2016 (Smith et al. 2016). IBC cells are becoming more interesting to manufacturers as more new fabrication methods to increase their commercial efficiency are reported. The Australian National University (ANU) adopted a damage-free laser ablation method in the place of thermal diffusion that requires high temperature to increase the commercial value of IBC cells and found an efficiency of 23.5% (Franklin et al. 2016).

23.2.2.4 Heterojunction Solar Cells

Silicon-based solar cells normally are fabricated from a single material that is diffused with different dopants in the front and rear side. The silicon heterojunction (SHJ) solar cells are a further development of the highly efficient materials composed of a-Si/c-Si. A p- or n-doped hydrogenated amorphous silicon (a-Si:H) is used as an emitter on top of the c-Si. The efficiency of these devices is generally low due to unpassivated substrate. The SHJ device geometry was further developed by inserting a thin intrinsic a-Si:H layer between the N-type c-Si wafer and the P-type a-Si:H to fabricate heterojunction with intrinsic thin layer (HIT). The thin layer passivates the dangling bonds in c-Si wafer. Tanaka et al. reported efficiency of 14.8% with a 4 nm-thick intrinsic thin layer. Further development of HIT cells was achieved by surface texturing the c-Si wafer on both sides to achieve an optimal light trapping and introducing the intrinsic a-Si:H thin layers on both sides of the textured c-Si wafer for good passivation. An N-type a-Si:H layer is deposited below the intrinsic layer at the rear to create a back-surface field. Transparent conductive oxides are deposited on both sides of the device to reduce the resistance for charge flow due to the high resistance of a-Si:H. The main advantages of the HIT cells include its low temperature processing typically below 200°C and an improved temperature coefficient compared to conventional cells that resulted in higher V_{oc}. The HIT cells achieved an efficiency of 20% as early as 1994 in 1 cm^2 area by Sanyyo. The HIT cells' performance further improved to a record efficiency of 25.1% in 2015. The device design of HIT cells was further improved by introducing interdigitated back contact to reduce optical loss and surface recombination from the front side. A record efficiency of 25.6% in HIT cells in an area of 143.7 cm^2 with interdigitated contacts was achieved in 2014 breaking the record of c-Si based solar cells. The record efficiency of 26.6% in interdigitated back-contact HIT cells is achieved later by utilizing anisotropic etching to minimize the reflection loss.

23.3 Second-Generation Solar Cells: Thin-Film Silicon Solar Cells

Second-generation solar cells are solar cells that are fabricated using thin semiconductors of thickness in the order of μm while the silicon-based solar cells have a thickness of around 350 μm (Chopra et al. 2004). Materials used for thin-film solar cells have higher extinction coefficient that enables them to absorb substantial amount of irradiation (Choubey et al. 2012; Irvine 2017; Luceño-Sánchez et al. 2019) Their thin layer could potentially provide lower cost electricity than c-Si wafer-based solar cells. In addition, thin-film solar cells are flexible and light, which can be easily integrated into buildings. The market share of this technology has increased to 20%. The main classes of thin-film solar cells, their working principle, and technology advancement will be discussed in this section.

23.3.1 Amorphous Silicon (a-Si) and Microcrystalline Silicon (mc-Si)

Amorphous silicon (a-Si) and microcrystalline silicon (*mc*-Si) are used in optoelectronic devices, such as solar cells, thin-film transistors, etc. (Hirose et al. 1979). Thin-film solar cells based on a-Si can be deposited onto a variety of flexible substrates and glass or metal using chemical vapour deposition (CVD); the device structure always features a p-doped layer on top (a side through which light enters the amorphous Si layer), while the bottom layer is n-doped (PECVD), catalytic CVD, or by sputtering (Raut et al. 2018; Reddy 2012). One of the advantages of a-Si compared to other thin-film solar cell materials is that it does not pose any health hazard. a-Si does not form ordered tetrahedral structure unlike the mono/polycrystalline silicon, which leaves some of the bonds dangling. The dangling bonds in a-Si cause anomalous electrical behaviour, poor photoconductivity, and prevents doping, which is critical to producing semiconductor properties (Collins et al. 2003; Alajlani et al. 2018). Hydrogen is incorporated in a-Si structure producing hydrogenated a-Si (a-Si:H) to passivate the dangling bonds resulting in better optical and electrical properties. The optical band gap of a-Si:H could be reduced to 1.84 eV and 2eV in a film of thickness 4 µm and below 4 µm, respectively. The electrical conductivity of a-Si is lower than c-Si; however, a-Si:H can be made thinner than c-Si wafers, which reduces the cost of production. Furthermore, a-Si:H can be deposited at temperatures as low as 75°C, which permits deposition on plastic surfaces. In the a-Si:H thin-film solar cell, large number of traps reduce the carriers' life time that limit the photovoltaic performance, especially the power conversion efficiency (Thiyagu et al. 2012). Doped a-Si, obtained by adding diborane to the gas flow for p-type material and phosphine for n-type, contains much more defect than intrinsic a-Si:H. Therefore, only intrinsic a-Si:H can be used as an absorber material (Alajlani et al. 2018). Alloys can be deposited by adding germanium (to form a-SiGe:H) or carbon (to form a-SiC:H) precursors to the gas flow, so that the band gap can be tuned to some extent (Beaucarne 2007).

Charge carrier mobility in a-Si is much lower than in c-Si. A good a-Si:H layer will show an electron mobility of around 20 cm²/Vs at best. As a result, a-Si layers cannot conduct charge carriers laterally over a long distance. The key parameters for a-Si are its dark conductivity, photo-conductivity, and its mobility (μ)/lifetime (τ) product. The conductivity of intrinsic a-Si:H in the dark is extremely low ($<10^{-10}$ S/cm) because of the low mobility, the large band gap, and the fact that charge carriers at low concentration are trapped at defects. Under illumination, however, many of the defects get filled with photogenerated carriers and are saturated. As a result, many more charge carriers are available for charge transport, and the conductivity is many orders of magnitude higher than that in the dark. The photoresponse, defined as the ratio of the illuminated conductivity to the dark conductivity, is a good indication for the suitability of the material for devices, and should be larger than 10^5. The mobility-lifetime ($\mu \times \tau$) product is the crucial parameter for the transport properties of excess charge carriers in the layer, and in device grade, a-Si is larger than 10^7 cm²/V (Beaucarne 2007; Kabir et al. 2012).

As a result of the low carrier mobility and low lifetime, collection of charge carriers in a-Si thin-film solar cells cannot take place through diffusion, which makes a strong drift field is an absolute requirement. This is achieved by sandwiching the intrinsic (i) absorber layer between doped layers of opposite doping. Most of the generation takes place close to the top surface of the i-layer. Consequently, the device structure should always feature a p-doped layer on top (a side through which light enters the amorphous Si layer), while the bottom layer is n-doped to compensate for the much lower mobility of holes compared to electrons. Most of the holes are generated close to the p-layer where they are majority carriers and no longer can recombine. Holes generated deeper in the "i"-layer have a higher probability of recombination. In contrast, electrons have a high collection probability throughout the "i"-layer. Due to the very low lifetime in doped layers, all light absorbed in the p-layer is lost for conversion. To minimize this effect, not only the p-layer is kept very thin (~10 nm), but can also be added so that the band gap increases and the absorption decreases (Beaucarne 2007). The device structure of a-Si thin-film solar cells is basically an intrinsic-a:Si layer sandwiched between p(top) and n(bottom)-doped layers. However, a rear N+a-Si layer is deposited as shown in Figure 23.4 to provide a back-surface field to reduce recombination (Ghosh et al. 2010). The main issue of a-Si solar cell is the poor and almost unstable efficiency. The cell efficiency automatically falls at PV module level. Currently, the efficiencies of commercial PV modules is in the range of 4–8%. The cells can easily be operated at elevated temperatures, and are suitable for the changing climatic conditions where sun shines for few hours (Sharma et al. 2015).

Most types of a-Si solar cells are in superstrate configuration, which means that the light enters the solar cell through the supporting substrate. This configuration requires a highly transparent substrate material and the presence of a transparent conductive oxide between the substrate and the active layer. These cells are also called p-i-n solar cells, referring to the sequence in which the different layers are deposited. The other option is to make a-Si:H cells in substrate configuration (n-i-p solar cells), which enables the use of a wide range of substrate materials (Beaucarne 2007; Ghahremani and Fathy 2016).

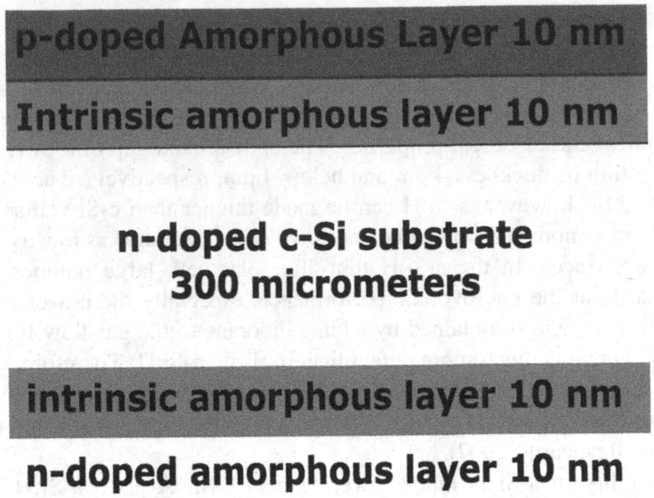

p-doped Amorphous Layer 10 nm

Intrinsic amorphous layer 10 nm

n-doped c-Si substrate
300 micrometers

intrinsic amorphous layer 10 nm

n-doped amorphous layer 10 nm

FIGURE 23.4   Device structure of thin-film solar cells.

Advance and Challenges in a-Si Thin-Film Solar Cells

The technology of a-Si:H for PV is based on two types of device design: a single-junction and multijunction p-i-n structure. Although major progress has been made in recent years in improving the deposition processes, material quality, device design, and manufacturing processes, the improvement of cell efficiency appears to have hit a bottleneck. It is generally recognized that any significant increase in efficiency can only be achieved by using multijunction devices (Deb 2000). It is indeed the case as shown by the achievement of a world-record stable efficiency of 13% (initial efficiency of 14.6%) in a triple-junction structure (Yang et al. 1997). The previous best stable efficiency was 11.8%. This was accomplished by optimization of factors such as hydrogen dilution for film growth, band gap profiling, current matching, and microcrystalline tunnel junction. Similarly, a new record in stabilized efficiency of 9.5% for 1200 cm^2 a-Si:H/a-SiGe:H has been reported recently (Cousins et al. 2015). This was possible by low-temperature (180°C) deposition of a-SiGe:H film while maintaining good optoelectronic properties. Kabir et.al fabricated single-junction a-SiC:H/a-SiC:H-buffer/a-Si:H/a-Si:H solar cells by using PECVD method, where the best initial conversion efficiency of 10.02% has been achieved for a small area cell (0.086 cm^2). The external quantum efficiency characteristic shows the cell has a better spectral response in the wavelength range of 400–650 nm, which proves it to be a potential candidate as the middle cell in a-Si-based multijunction structures (Zhao and Wang 2006).

Marsui et al. prepared highly stabilized and efficient hydrogenated amorphous silicon (a-Si:H) absorbers using triode-type PECVD method with an improved light-soaking stability and PV performance. As a result, they obtained independently confirmed stabilized efficiencies of 10.1% and 10.2% for a-Si:H single-junction solar cells (absorber thickness: t = 220–310 nm) and 12.69% for an a-Si:H (t = 350 nm) for hydrogenated c-Si:H tandem solar cell fabricated using textured SnO_2 and ZnO substrates. The relative efficiency degradations of these solar cells under 1 sun illumination at 50°C for 1000 h are 10% and 3%, in single and multijunction devices, respectively (Matsui et al. 2015). J.S. Cashmore et al. developed a world record large area 1.43 m^2 thin-film a-Si PV module with a tandem junction cell (a-Si:H/c-Si:H) with a stabilized power conversion efficiency of 2.34%.

The main technology contributions in the a-Si thin film solar cells device design for this breakthrough result that generated more than 13.2% stabilized efficiency from each equivalent 1 cm^2 of the active area of the full module are described (Cashmore et al. 2016). In addition, the intrinsic amorphous silicon oxide (i-a-SiO:H) films were used as passivation layers in crystalline silicon heterojunction (c-Si-HJ) solar cells. The effective lifetime (eff) and PV parameters were improved by controlling the CO_2/SiH_4 ratio in the i-a-SiO:H rear passivation layer. The c-Si-HJ solar cells using an i-a-SiO:H/i-a-Si:H stack passivation layer showed a high V_{oc} and compared to using a straight i-a-SiO:H passivation layer. The highest efficiency obtained for a c-Si-HJ solar cell using an i-a-SiO:H/i-a-Si:H stack passivation layer was 19.4% (V_{oc} = 715 mV, J_{sc} = 34.9 mA/cm^2, FF = 0.78) (Krajangsang et al. 2017).

23.3.2 Cadmium Telluride (CdTe) Thin-Film Solar Cells

Cadmium telluride solar cell technology has reached an unsubsidized levitated cost of electricity competitive with wind technology, slightly less than Si PV, and generally less than all other electricity generation sources. Typically, the module manufacturing is an inline process, where glass enters a factory and exits as a completed solar panel in several hours. The semiconductor layers are several microns thick, and they cost just a few pennies per watt to manufacture. Capital and energy expenditures are relatively low. In 2019, cumulative CdTe module shipments reached 25 GW (Wilson et al. 2020). Among thin- film solar cells, CdTe is one of the leading candidate for the development of cheaper, economically viable photovoltaic devices, and it is also the first PV technology at a low cost (Rathore et al. 2019). Just like other thin-film solar cells, the CdS/CdTe devices are descendants of the first CdS/Cu$_2$S solar cells. In the mid-1960s, the first experiments with tellurides were performed. Efficiencies between 5% and 6% were obtained for CdTe/CuTe devices (Bonnet 1992). Since Cu diffusion led to instabilities in these devices, instead, CdS and CdTe were combined to form a p-n heterojunction with efficiencies around 6% (Bonnet and Rabenhorst 1972).

Thirty years later, the efficiency has increased to more than 21% (Bosio et al. 2020). The continuous increase in efficiency is largely due to the optimization of the layers making up the anti-reflection coating, the transparent electrical contact, and the window layer, as well as a careful choice of the glass substrate. This led to a substantial increase in the photocurrent, going from 26.1 mA/cm^2 for a cell with 17.6% efficiency (Wu et al. 2001) to 28.59 mA/cm^2 photocurrent for a cell with 19.6% efficiency, corresponding to a rather modest increase in photovoltage (Green et al. 2019). In addition, CdTe solar modules represent by far the most successful photovoltaic thin-film technology with a share of about 10% in the global PV market in 2014, and commercial modules with an efficiency of 18.2% are now available on the market (available at: http://www.firstsolar.com/-/media/First-Solar/Technical-Documents/Series-6-Datasheets/Series-6-Datasheet.ashx). The major challenge for reliable manufacture of efficient devices is to produce a stable and ohmic back contact to the CdTe absorber with its high electron affinity. Often, back contacts are made with materials that contain Cu, such as Cu$_2$Te, ZnTe:Cu, or HgTe:Cu, enabling a relatively low contact resistance. However, Cu diffusion in CdTe is fast and extends deeply into the absorber, thereby affecting considerably the stability of the device (Dobson et al. 2000). Cu-free alternative contact materials include embrace, for example, Sb$_2$Te$_3$ (Romeo et al. 2004).

As a polycrystalline CdTe has a high light absorption level only about a micrometre thick layer can absorb 90% of the solar spectrum. It has a direct optical band gap about 1.45 eV, which makes it a good candidate for converting sunlight into electricity in single-junction cells (Luceño-Sánchez et al. 2019). A CdTe-based solar cell is typically realized with very thin stacked layers, arranged in such a way as to form a high-quality heterojunction (See Figure 23.5). When CSS or CSVT techniques are used, CdTe thin films naturally grow as p-type, and are forced to select a n-type partner to make the p-n junction. The architecture of the heterojunction allows sunlight to pass through the n-type layer to reach CdTe where photogeneration takes place. To date, the best-found configuration is the CdS/CdTe system, where CdS is the window, while CdTe is the absorber. For this reason, the thickness of CdS film is never more than 100 nm to ensure excellent transparency. The free carrier concentration of both the window and absorber films ensures that the electric field mostly falls into the absorber material, so that all the photogenerated electron-hole pairs can be separated and pushed towards the external electrical contacts. These auxiliary layers complete the device, ensuring the passage of the photocurrent. High-efficiency CdS/CdTe solar cells are produced in a superstrate configuration, which means light passes through the substrate and the front contact is as transparent and conductive as possible. Instead, the back contact,

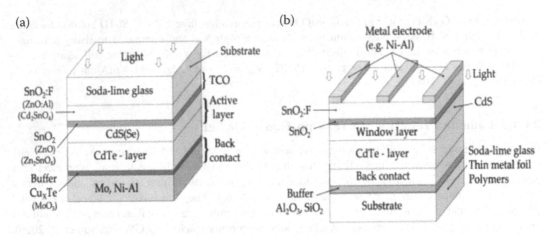

FIGURE 23.5 The CdTe/CdS solar cell; (a) superstrate configuration (light enters through the substrate), (b) substrate configuration (substrates could be opaque).

which is generally opaque, has to ensure the ohmicity with p-type CdTe in order to efficiently harvest all the photocurrents (Figure 23.5a).

In substrate configuration, the manufacturing sequence starts with the back contact, followed by the CdTe/CdS films, ending with the transparent electrical contact (Figure 23.5b). This configuration is particularly intriguing since it allows devices to be made on opaque substrates such as metal and polymeric foils, opening a means for the production of flexible, light, and cost-effective modules. By using Te/MoO$_3$ and CuxTe as buffer layers, coupled with a Mo metal contact, solar cells exhibiting efficiencies in the range 13.6–11.3% are obtained (Gretener et al. 2013). M.A. Islam et al. prepared CdTe (up to 2.0 μm thick) and oxygenated CdS (CdS:O) (up to 100 nm thick) films by magnetron sputtering, and optimum conditions of film growth were investigated for CdS:O/CdTe solar cells. The complete cell was fabricated with the screen-printed copper-doped graphite paste and silver metal (C:Cu/Ag) back contact with the configuration of FTO/ZnO:Sn/CdS:O/CdTe/Cu:C/Ag on glass. The J-V characteristics and external quantum efficiency were measured for the solar cells under the illumination of AM 1.5G, and the highest efficiency of 10.3% was achieved for the optimized CdTe growth rate of 5.4 s^{-1}, while CdS:O growth rate was 0.25 s^{-1} (Islam et al. 2017). The CdTe solar cell is made from a heterojunction between the wider bandgap CdS and the CdTe absorber layer. CdS is often referred to as the window layer which exhibits n-type conduction due to stoichiometry defects, such as sulphur vacancies, which are formed during the film growth. With an energy gap of 2.42 eV, it allows sunlight to pass up to a wavelength of 512 nm, inhibiting the passage of the near ultraviolet light to which SLG substrate and TCO layers are still transparent. Moreover, self-doping is effective in obtaining resistivity of the order of 10^6–10^7 Ωcm, the same order of magnitude as was obtained with CdTe films (Bosio et al. 2020). The most common method for depositing CdS is by chemical bath deposition (CBD). The layer is n-type, forming a junction with the p-type CdTe. The thickness is kept to a minimum (around 100–200 nm) in order to minimize absorption at the blue end of the spectrum.

23.3.2.1 Advances and Challenges in CdTe Thin-Film Solar Cells

A near-term focus is to increase cell efficiency to 25%. Photocurrent and fill factor have nearly been maximized for current technology. Consequently, the greatest challenge is to increase photovoltage by overcoming fundamental material issues such as recombination and low carrier concentration. GaAs solar cells have achieved open-circuit voltage greater than 1.1 V with a band gap of 1.4 eV, whereas production of CdTe solar cells with a similar band gap have been limited to a V$_{oc}$ of 800–900 mV (Wilson et al. 2020; Burst et al. 2016). Once V$_{oc}$ is increased, further gains in fill factor are possible. Recent efforts in doping single-crystal CdTe with Group-V elements such as P, As, and Sb demonstrated that CdTe is capable of 10^{16}–10^{17} cm^{-3} hole density with radiatively limited life times and

V_{oc} exceeding 1 V for the first time (Sharma et al. 2015). The V_{oc} of Group-V-doped CdTe devices is less than expected from absorber hole density and life time measurements. This can be reconciled by front interface recombination consistent with V_{oc} data, other experimental observations, and computational modelling (Metzger et al. 2019). Simultaneously optimizing the p-n junction interface and carrier activation remains a challenge to increase performance. CdTe solar technology using Cu has achieved state-of-the-art long-term degradation less than 0.5% per year as well as excellent field performance. However, there is room to improve. Cu migration has been a primary limiting factor of cell stability.

Early results with Group-V doping and no Cu show far superior cell stability and energy yield compared to their Cu counterpart, thus providing a path to increase energy yield and product life time beyond today's technologies. Theory and experiment need to describe the interaction of dopants with impurities such as Cl, and novel synthesis methods to increase activation are critical. Understanding the science of the p-n junction interface and carefully manipulating its properties is another frontier. There are a number of next-generation technology possibilities. These include but are not limited to seeding large grain growth with low-cost methods, achieving nucleation that removes the need for the $CdCl_2$ anneal, establishing n-type absorbers, creating new cell architectures such as homojunctions, transparent emitters with minimal band offsets and implementing interdigitated back contacts.

23.3.3 Copper-Indium-Gallium-Diselenide (CIGS)

Quaternary chalcopyrite semiconductor alloy $Cu(In,Ga)Se_2$ (CIGS) represents one of the most suitable materials to produce low-cost and high-efficiency photovoltaic modules and hence can be considered as an appropriate alternative to the silicon technology. In the CIGS crystal structure, some indium is replaced by Ga in the $CuInSe_2$ film and consequently the CIGS band gap increases from 1.04 eV to about 1.7 eV, if all indium is replaced by Ga. The advantages of CIGS-based solar cells compared to CIS- based solar cells are: (i) the optical band gap can be tuned by varying the Ga/In ratio to match the solar spectrum, (ii) an improved V_{oc} can be obtained due to Ga incorporation since $V_{oc} \sim E_g/2$, and (iii) a single-phase CIGS absorber can be made more easily than a single-phase CIS absorber (Ramanujam and Singh 2017). The tetragonal CIGS unit cell is shown in Figure 23.6. The lattice parameter ratio (c/a) is close to 225 and any deviation from this value is known as tetragonal distortion which comes from CuSe, InSe, or GaS bonds. In the chalcopyrite $Cu(In,Ga,Al)(Se,S)2$ alloy system, the band gap can varies from 1.04 eV (CuInSe2) to about 3.5 eV (CuAlS2), almost covering the visible spectrum.

Figure 23.6 CIGS-based solar cells have emerged as one of the most promising candidates for high-efficiency low-cost thin-film solar cells, and high efficiencies of as high as $h = 20.0\%$ have been

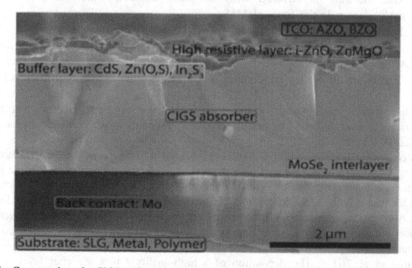

FIGURE 23.6 Cross-section of a CIGS cell taken from Ref. (Krajangsang et al. 2017).

reported (Niki et al. 2010). Solar cells with an absorber layer made from $Cu(In,Ga)Se_2$ are currently the state-of-the-art of the evolution of Cu-based chalcopyrites for use as solar cells. Heterojunctions between CdS and Cu_2S were the basis for first approaches for thin-film solar cells since the 1950s. In 1974, first work on the light emission and light absorption of $CdS/CuInSe_2$ diodes was published. While $CuInSe_2$ was not further considered for applications as a near-infrared light-emitting diode, its high absorption coefficient and its electronically rather passive defects make it a perfect choice for use as a microcrystalline absorber material. Inclusion of Ga atoms on the In lattice site such that the ratio of [Ga]/([Ga]+[In]) becomes around 20% shifts the band gap from 1.04 eV to around 1.15 eV, which is nearly perfect for a single-junction cell. Today, thin-film solar cells with a $Cu(In,Ga)Se_2$ absorber layer are the most efficient thin-film technology with laboratory efficiencies up to 21.7% (Jackson et al. 2015). The use of $Cu(In, Ga)(Se, S)_2$, together with a high temperature heavy alkali (Cs) post-treatment of the absorber layer, results in a reduced defect density. The enhanced quality of the $Cu(In, Ga)(Se, S)_2$ allows the opportunity to benefit from the effects of a wider absorber band gap. Consequently, the reverse saturation current density decreases, producing an important enhancement in photovoltage and fill factor leading to the world record efficiency result. Recently, a CIGS-based solar cell world record efficiency of 23.35% was achieved by Solar Frontier (Nakamura et al. 2019) quite similar to the value obtained by crystalline silicon cells, that is, around 25%. Specifically, thin-film CIGS solar cells have emerged as a technology that could challenge the current dominance of silicon solar cells. This is possible thanks to the peculiar optical and structural properties of CIGS cells, which possess an extraordinary stability under a wide range of operating conditions. Today, CIGS technology, with an annual production of around 1.8 GW power, covers a market share of 1.9%, which is principally supported by three producers: Miasol, Solibro, and Solar Frontier (Bosio et al. 2020).

In common with other thin-film PV technologies, these films can also be deposited onto cheap substrates at relatively low temperature. Currently, four main categories of deposition techniques are employed when fabricating CIGS films: (1) metal precursor deposition followed by sulpho-selenization; (2) reactive co-deposition; (3) electrodeposition; and (4) solution processing. All recent world records and the most commercially successful have utilized two-step sulpho-selenization of metal precursors or reactive codeposition (Choubey et al. 2012). The absorber layer is based on the chalcopyrite-phase CuInSe (CIS) with a band gap of 1.04 eV, which is lower than the band gap for optimum efficiency in the Shockley Queisser limit (Kasap and Capper 2017). CIGS cells are usually manufactured following five steps:

STEP-1: A substrate such as Na_2CO_3CaO, a metal, a ceramic or a polymer sheet is placed to support the rest of the cell.

STEP-2: The substrate is covered with the back contact, which is usually pulverized molybdenum in the form of $MoSe_2$.

STEP-3: The CIGS layer (p-type) is grown by a co-evaporation process.

STEP-4: A buffer layer (n-type), currently formed by a TCO such as zinc oxide (ZnO) with or without doping is deposited.

STEP-5: Finally, an anti-reflective coating is applied to improve the cell efficiency.

However the steps are not binding, for instance, the third step can be replaced by a spraying of the CIGS precursor elements followed by selenization and sulphuration, which results in laboratory efficiency of 22.3% (Feurer et al. 2017).

The classical layer stack for this type of solar cell is shown in the cross-section structure of a CIGS cell in Figure 23.6. CIGS films prepared by a three-stage co-evaporation technique result in a double-graded conduction band profile, with a deep notch point closer to the absorber front surface. A wide band gap is located at both sides (due to high Ga percentage) and a narrow band gap is located at the notch point. A graded band gap composition is used to improve the electronic properties such as reduced recombination losses etc. Front grading improves V_{oc}; back grading increases carrier collection and J_{sc}, and decreases bulk/surface recombination at the back contact interface due to the BSF created by Ga grading (Parisi et al. 2015). The deposition of a high-quality CIGS absorber layer is the crucial processing step and thus far, the PVD technique appears to be the preferred method. Although PVD is the preferred method for high-efficiency cell fabrication, recent results suggest that a wide variety of techniques, such as sputtering, spray pyrolysis, closed space sublimation (CSS), molecular-beam epitaxy (MBE), and electrodeposition, are currently being pursued. CIGS solar cells are an alternative to Si solar cells and less susceptible to damage than Si solar cells. Different layers used in CIGS cells help to reduce J_{sc} losses. CIGS technology is highly competitive and involves less raw materials, cost, and time. Module production requires a lower thermal budget (550^0C) than c-Si (1100^0C), and CIGS cells can be monolithically integrated during cell processing, whereas in Si solar cells, monolithic integration is not possible. The lab-scale (2 cm x 2cm) solar cell efficiencies of these technologies are closer (CIGS 22.8%, c-Si 25.6%); CIGS solar panels are heavier than Si panels; two panes of glass are used in CIGS modules whereas in Si modules only one pane of glass is used (Ramanujam and Singh 2017).

23.3.3.1 Advances and Challenges of CIGS Thin-Film Solar Cells

The greatest challenge to increasing CIGS device efficiency is reducing its voltage deficit ($V_{OC,def}$) towards the AM1.5G detailed balance limit of solar cell parameters. The voltage deficit is defined as: $V_{OC,def} = V_{OC,DBL} - V_{OC}$, where $V_{OC,DBL}$ is the detailed balance limit open-circuit voltage which is a function of the band gap of the cell (R¨uhle 2016). Optimizing processes to increase intergrain homogeneity is a promising strategy. World-record devices that utilize sulphur substitution on the absorber Group VI lattice site showed reduced voltage deficit with Zn(O,S,OH) relative to CdS buffers, attributed to reduced space-charge region recombination (Nakamura et al. 2019). Complete understanding of issues, including secondary phases, doping, and band gap alignment, that are deleterious to the buffer/CIGS interface, would provide opportunities to better match these layers based on their processing and resulting properties. In addition of the above cases, several challenges still need to be addressed as emerging and new groups develop CIGS thin-film PV technologies. The following five challenges are critical for developing low-cost and reliable CIGS products:

1. standardization of equipment for the growth of the CIGS absorber films,
2. higher module conversion efficiencies,
3. improved processing for CIGS deposited by alternative process for high-efficiency cells and modules,
4. thinner absorber layers of less than 1 μm or less, and
5. CIGS absorber film stoichiometry, and junction and film uniformity over large areas for power module fabrication (Ullal 2008).

Furthermore, continuing developments promise increased profitability through advances and optimization in production technology. The production of flexible CIGS solar modules in roll-to-roll systems, once optimized, will provide a significant breakthrough in the reduction of production costs, and will significantly simplify the transportation and distribution of the flexible, light-weight modules. Several manufacturers are already in different phases of realizing remarkable expansion plans for the production of both glass-based and flexible CIGS solar modules. With exponential growth figures, CIGS solar module production is truly flourishing from time to time.

23.4 Third-Generation Solar Cells: Emerging Solar Cell Technologies

The PV technology available commercially is dominated by the well-developed inorganic solar cell that has an energy conversion efficiency between 15–20% and a life time of 20–30 years. The cost related to the production of these PV technologies and the limited resource of the materials limited the wide spread of the technology to the electricity generation of the globe. The policy-makers and scientists proposed a research and development direction on how to reduce the cost of the production of this PV technology.

23.4.1 Polymer Solar Cells

Semiconductors, mainly inorganic ones, are used in the fabrication of PV cells for more than five decades. The discovery of electrical conductivity in polyacetylene when it was partially oxidized in 1970s opened a cheaper alternative to the Si-based PV technology. In 2000, Alan J. Heeger, Alan G. MacDiarmid, and Hideki Shirakawa won the Noble Prize in Chemistry for their discovery of conductivity in organic materials. Organic semiconductors have a unique semiconducting property, with a chemical composition of plastic, that opens an endless opportunity to tailor their optical, electrical, and mechanical properties. Organic semiconductors found a wide variety of applications after Tang demonstrated the first organic light emitting diode in the 1970s. Organic semiconductors have a high extinction coefficient (above 10^5/cm) that enables efficient absorption of light in a layer of thickness of few hundreds of nanometres. The mass production of organic solar cells can thus be done with simple roll-to-roll printing or ink-jet printing. The possibility of tailoring their optical property is also another eye-catcher for the organic electronics technology that adds an aesthetic value to their application. A recent report of self-powered electronic eye glass with integrated organic solar cells showed the advantage of tuning their colours and mechanical flexibility (Landerer et al. 2017).

Organic semiconductors are mainly carbon-based molecules with very few heteroatoms like sulphur, oxygen, and nitrogen included. The backbone of this class of materials is formed by alternating single and double bonds of carbon atoms which are the origin of de-localized electrons that are responsible for their semiconducting property. The dielectric constant of these materials (ε_r) is low compared to inorganic semiconductors like Si ($\varepsilon_r = 11$). The low dielectric constant of organic semiconductors is not enough to screen out the bonding energy between an electron and a hole in such molecules which results in a tightly bound electron-hole pair upon excitation of the electron from the ground to the excited state. The excitations in organic semiconductors are therefore electron-hole pairs called excitons unlike free charge carriers as in the inorganic semiconductors. Since the binding energy of exciton is high (0.5–1 eV) (Clarke and Durrant 2010) compared to thermal energy at room temperature (kT = 0.025 eV), thermal excitation is negligible in these kind of materials. The conducting nature of organic semiconductors, unlike inorganic semiconductors, is mainly extrinsic, which is either the result of charge injection by electrodes or dissociation of photoexcited electron-hole pairs.

23.4.1.1 Origin of the Electrical Conductivity and Band Gap in Conjugated Polymers

Organic semiconductors are broadly classified as small molecules and conjugated polymers based on the number of repeating units called monomers. Small molecules also called *oligomers* are composed of less than 100 monomers while the conjugated polymers are a chain of more than 100 repeating units. Carbon is the main element in polymers and its electronic configuration in the backbone of the material determines the electrical, optical, and mechanical properties. The four valence electrons in C atom are arranged as: $2s^2$, $2p^1_x$, $2p^1_y$, $2p^0_z$. Hybridization of these atomic orbital creates a newly hybridized orbital according to the molecular orbital theory. The three possible hybridized orbitals in C atom are: sp, sp^2, and sp^3. Bond formed between sp^2 hybridized C atoms form three strong "σ" bonds that keep a molecule intact while the remaining p_z orbital of the sp^2 hybridized carbon atoms form π-bonds with the neighbouring carbon atoms in the chain as shown in Figure 23.7. The overlap of the p_z orbitals along the chain leads to alternating single and double bonds due to Peirels effect. Since the alternating single and double bonds are inherently isomeric, the π-electrons are de-localized in nature resulting in a considerable electrical conductivity in the material.

The band gap of a material determines its electrical property. The molecular orbitals in organic semiconductors evolve from the linear combination of the atomic orbital. In this regard, it is important to see each of the possible atomic orbital overlaps and the resultant superposition to understand the energy levels of organic semiconductors. As mentioned above, the sp^2 hybridized orbitals of each C atom overlap to form a σ-bond. The overlap of these orbitals is along the internuclear axes of the C atoms which results in a charge density along the axes. The overlap results either in an enhanced or reduced charge density between the nuclei that leads to a σ- bonding and σ*-anti-bonding molecular

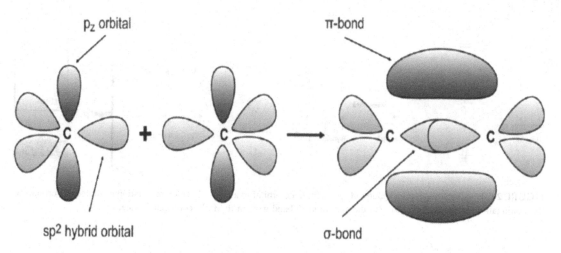

FIGURE 23.7 Simplified representation of sp^2 hybridized orbital in a C_2H_2 (left), representation of molecular orbitals as the conjugation increases (centre) and the evolution of band gap in the fully conjugated system.

orbitals, respectively. Since σ and σ^* are close to each other, the resonance between these molecular orbitals is strong that leads to a larger gap between them. The remaining p_z orbitals of the C atoms overlap far from the nuclei, which results in a lower interaction between the π bonding and π^*- anti-bonding molecular orbitals. The separation between the π and π^* will be smaller, thus lies in between the σ and σ^* molecular orbitals. The number of the molecular orbitals increases as the length of the chain is extended. The highest unoccupied molecular orbital (HOMO) is the π-orbital and the next lowest unoccupied molecular orbital (LUMO) is π^*-orbital. The gap between HOMO and LUMO levels of conjugated polymers lies in the UV-Visible part of the solar spectrum that made them good candidates for optoelectronic application.

The band gap of conjugated polymers can be effectively reduced by increasing the chain length. As more and more sp^2 hybridized carbon atoms are covalently bonded and the p_z orbitals overlap sufficiently, the π-electrons become de-localized in the extended system thereby effectively reducing the gap between the HOMO and LUMO. The narrowing of band gap of conjugated polymers by increasing the number of monomers in the chain is shown in Figure 23.8.

23.4.1.2 Working Principle of Organic Solar Cells and Efficiency Limiting Factors

Photogeneration in organic solar cells (OSCs) is a multistep electronic process as opposed to thermal process. It can be classified into four important consecutive steps as follows and is summarized in Figure 23.9.

- **Absorption of Photons:** The first step in any solar cell technology is absorption of photons. Absorption of photons of enough energy promotes the electrons in the ground state (S_0) to higher energy states (S_n) leaving a hole behind and leads to a formation of electron-hole pairs – excitons. Extinction coefficient of conjugated polymers ($\sim 10^5$/cm) is relatively higher compared to inorganic semiconductors like Si. A thin film of thickness $\sim 1\mu m$ can absorb most of the solar irradiance within its absorption width. The absorption efficiency of an organic material is limited by the thickness of the device, as it has to be comparable to the shorter charge carrier diffusion length (typically ~ 100 nm). The other limiting factor for efficient absorption is the spectral overlap of the conjugated polymers with the solar irradiance. To maximize absorption of photons, the polymer should be able to harvest the photon in longer wavelength region. Low band gap materials can harvest substantial amount of solar irradiance. But organic solar cells (OSCs) fabricated using small band gap materials suffer from smaller photovoltage. A trade-off between the solar harvest and photovoltage has to be considered when using low band gap organic semiconductors.

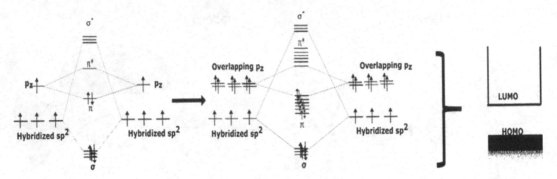

FIGURE 23.8 Simplified representation of sp^2 hybridized orbital in a C_2H_2 (left), representation of molecular orbitals as the conjugation increases (middle), and the evolution of band gaps in the fully conjugated system.

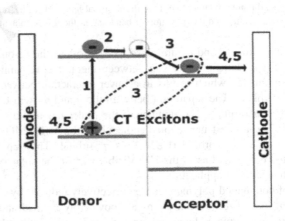

FIGURE 23.9 Sketch of working principle of organic solar cells (1) exciton generation (2) exciton diffusion (3) exciton dissociation and CT formation (4,5) charge separation and transport/collection.

- **Exciton Diffusion:** The excitons generated by absorption of photons are tightly bound and remain localized on the molecule or a unit in the conjugated polymer. The excitons drift along the conjugated chain and hop to another conjugated chain, once they reach the end. Exciton diffusion efficiency is mainly limited by its short diffusion length and life time due to the large electrostatic force (Tegegne et al. 2020). A number of researchers are working on increasing the diffusion length of exciton by tailoring the molecular structures of the polymers, while others are focusing on decreasing the distance the exciton has to diffuse before dissociation (Tegegne et al. 2020; Dennler et al. 2009; Tegegne et al. 2018).

- **Exciton Dissociation:** The tightly bound exciton can dissociate by a driving force created by the difference between the LUMOs of the donor (D) and acceptor (A) materials. The exciton that reached the D/A interface dissociates by transferring its electron to the LUMO of the A material leaving behind a hole in the D's HOMO. Exciton dissociation efficiency is mainly limited by the D/A interface within the active layer. Bulk heterojunction of D and A materials in the active layer can efficiently dissociate exciton close to 100% efficiency provided there is enough driving force for dissociation. A LUMO offset of 0.3 eV between the D and A materials is empirically suggested for efficient dissociation (Brédas et al. 2004; Mola and Abera 2014).

- **Free Charge Generation:** Exciton dissociation at the D/A interface will create electron in the A LUMO and hole in D HOMO which are still bound and are called charge transfer (CT) excitons. Since the charge transfer in exciton dissociation occurs in a close proximity, the electrostatic attraction in CT exciton cannot be screened out due to the low dielectric constant of the D and A materials. Free charge generation therefore is a competitive process between the recombination of CT exciton and separated charges generation. Once the charges are separated, the electron will be in the A material while the hole is in the D material and are free to move.

- **Charge Transport and Collection:** The free charge carriers will be moving through their respective transport materials for collection in the external circuit. The charge transport within the materials is determined by the charge mobility and percolation paths in the active layer. The active layer morphology plays an important role in this process. The charge carriers that reached the electrodes can be collected easily if the barrier height between the electrodes and the active layer is small enough for charge collection.

23.4.1.3 The Bulk Heterojunction Concept

The charge photogeneration in OSCs necessitates optimization of each charge generation step for increasing their performance. The architecture of OSCs has developed in the past three decades after the first OSC was reported by Weinberger et al. in 1982. This device is the simplest architecture in which an organic semiconductor is sandwiched between two asymmetry electrodes usually indium tin oxide (ITO) (a transparent conducting glass to allow light in) and aluminium (Al). The incoming light generates exciton in the active layer that will be dissociated by the electric field that is generated in the active layer by the work function difference of the electrodes. However, exciton dissociation is not efficient in the single-layered OSCs because their PCE is poor (<0.1%).

The second-generation, bilayer OSCs, in which two layers of organic semiconductor with different electron affinity and ionization energy are deposited on top of each other, were brought forth to enhance exciton dissociation in the active layer of the OSC. Tang et al. reported the first bilayer OSC using a vacuum deposited CuPc/Perlene derivative as donor/acceptor, respectively. In this device geometry, the excitons generated in the donor material dissociate if they reach the donor/acceptor (D/A) interface, but those that are generated further than their diffusion lengths from the interface will relax readily. The limitation in this device is the short exciton diffusion length, typically 5–15 nm, which imposes very thin absorber which will drastically reduce the absorption of the device. This inherently reduces the efficiency of such devices and their efficiency never exceeded 1% (Halls et al. 1996).

Exciton dissociation can be enhanced by increasing the D/A interface in the active layer of OSC. Bulk heterojunction (BHJ) device structure provides an interpenetrating network of D and A materials in the active layer of the OSC device, which significantly increases the D/A interface compared to bilayer device structure. The exciton generated in D material can easily get a D/A interface nearby for dissociation due to the intermixed domains in the active layer. The electron and hole will then find their way to the electrodes. The power conversion efficiency of BHJ OSC has increased from 2.9% (Yu et al. 1995) of the first BHJ solar cell to over 16% (Xiong et al. 2019). BHJ device architecture is a well-developed and accepted device structure for OSCs. The main challenge in this device structure is the morphology of the active layer, which will be discussed in the next section. The generation of OSCs is briefly summarized in Figure 23.10.

23.4.1.4 Morphology of Active Layer of BHJ Organic Solar Cells

The active layer of a BHJ organic solar cell is an intermixed D and A phases in a nanoscale which plays a crucial role in the charge generation of the device. Charge photogeneration in organic solar cell starts with generation of tightly bound excitons that will dissociate at the D/A interface. After dissociation, the free charges need uninterrupted pathway to the electrodes to minimize recombination. The active layer morphology should favour both exciton dissociation by intermixing D/A within the diffusion of excitons

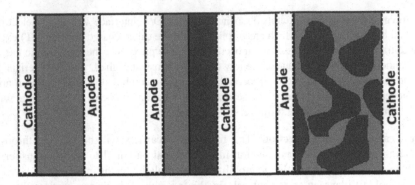

FIGURE 23.10   Generation of OPV from left to right: Single layer, Bilayer, Bulk heterojunction.

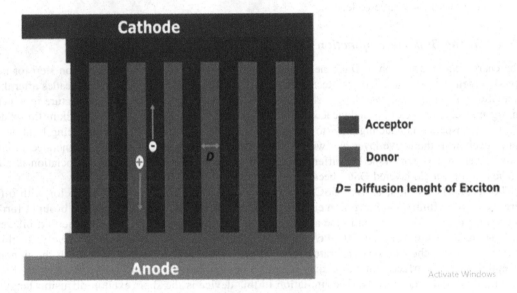

FIGURE 23.11   Ideal morphology for high-performance organic solar cell.

and uninterrupted percolation path for free charge carriers. The ideal morphology is therefore a pillar of one of the materials (such as the D) with thickness of exciton diffusion length embedded in the matrix of the other material (such as A) as shown in Figure 23.11. In this ideal morphology, excitons generated in the D materials can diffuse to the D/A interface which is its diffusion length away and dissociate. The free charges in the D and A phases have percolation paths to their respective electrode that enables efficient charge transport and collection.

The morphology of the active layer of OSCs affects all the four photovoltaic parameters the device. Unoptimized morphology reduces the amount of current that can be extracted in a BHJ OSC due to poor exciton dissociation and/or charge transport and collection. The open circuit voltage and the fill factor of the OSC is also affected by the nanomorphology if the active layer/electrode interface induces extraction barrier due to accumulation of undesired phases. As shown in the ideal morphology (Figure 23.11), the top and bottom of the cell is covered with either the D or A material to enhance charge collection. Therefore, morphology optimization needs to be carefully done to enhance the performance of an OSC. A number of ways including solvent choice, annealing, material design, and deposition method can be used to control the morphology of the BHJ active layer.

In solution-processed materials for OSC application, the D and A materials form a homogeneous mixture in solution but during casting as the solvent evaporates the D and A materials will start to phase separate into domains which are rich in one of the materials. The drying kinetics of the spin-coated film determines the morphology of the active layer. Fast drying solvents like chloroform do not give the film enough time to grow large domains of each of the phases (Hoppe et al. 2004). Besides the drying kinetics, the solubility of the D and A materials in the solvent has an effect on morphology. This can be improved by choosing solvent or solvent additives that have high boiling point and can selectively dissolve one of the materials more than the other. The phase separation in such solvents start earlier which gives more time for the phase separation and formation of large domains. The power conversion efficiency of OSCs fabricated using diketopyrrolopyrrole-quinquethiophene (PDPP5T) copolymer as donor and phenyl-C71-butric acid methyl ester (PCBM70) as fullerene acceptor increased from 2% to more than 5% by using co-solvents instead of a single solvent (Lucěno-Sánchez et al. 2019). High boiling point solvent additives treatments like diiodooctane (DIO) are commonly used to improve the morphology of BHJ solar cell. Addition of DIO in PTB7:PC70BM blends films prevents segregation of PC70BM by decreasing the PC70BM size in the polymer-rich matrix which is beneficial to improve the morphology for increasing the PCE of the OSC (Collins et al. 2013). The effect of solvent additives on morphology is different for different blends in the active layer. Thermal annealing can also induce phase separation and modify the morphology of the active layer. P3HT:PCBM-based solar cells are common examples for thermal annealing–induced morphology improvement. Padinger et al. reported a PCE increase from 0.4% to 2.5% of P3HT:PCBM-based solar cells by thermal annealing (Padinger et al. 2003) which the authors attribute to enhanced charge carrier transport. Different authors demonstrated thermal annealing improves the morphology of the P3HT:PCBM BHJ layer by increasing the size of P3HT domain and crystallinity of P3HT (Erb et al. 2005). Though thermal annealing has been one of the important tools, it is not always effective as the active layer might be damaged by the heat (Park et al. 2011). Solvent vapour annealing, which means confining the photoactive layer in the solvent vapour in closed system, has also been used to modify the morphology of the active layers like P3HT:PCBM and improve the PCE to 4.4% (Wu et al. 2013).

The ratio of donor to acceptor materials in the BHJ active layer is also another factor that determines the morphology of the active layer. The optimized D:A ratio favours optimized carrier dissociation, transport, and extraction. A small amount of PC70BM in the PTB7:PC70BM blend tends to dock the fullerene acceptor in the PTB7 molecule that will make the fullerene as an electron trap. An increased amount of PC70BM dramatically increased the electron mobility in the PTB7:PC70BM film (Ho et al. 2017). A sharp increase in photocurrent and FF of MDMO-PPV:PC60BM-based OSC was recorded as the PC60BM concentration increases to 67% (van Duren et al. 2004). On the contrary for P3HT:PC60BM-based OSC the optimized morphology is for 65 wt.% of P3HT. The optimized D:A ratio should favour high crystallinity of the components, bi-continuous percolation, and interconnected pathways.

Materials designing is one of the effective ways to control morphology of a BHJ active layer. The molecular structure of the donor and acceptor material determines the miscibility of D/A in the BHJ (Ghasemi et al. 2017). Solubility of the D/A materials is also determined by the structure of the organic materials which is a crucial factor to optimize the morphology of OSC. Solubility of the organic materials can be enhanced by appending different side chains at different positions but bulky side chains hinder molecular packing (Tegegne et al. 2020). The molecular weight of the polymer also is one parameter to optimize the solubility. Long polymer chains impede the freedom of movement of the polymer and thus reduce the solubility. Besides an intermixed D/A phase produced by increasing the solubility of the D and A materials, some degree of self-aggregated phases is also important for the percolation of free charges in an optimized morphology. Therefore, molecular designing should be done taking into account both the miscibility and self-aggregation of the organic materials. Side-chain engineering (Tegegne et al. 2020), controlling the molecular structure of the backbone repeating units (Osaka et al. 2012) and end groups, can be used to control conformation, molecular ordering, and optoelectric properties of the organic materials, which are essential for the photovoltaic performance of an OSC.

23.4.1.5 Advances and Challenges in Organic Solar Cells

In the past three decades, the organic solar cell research has shown a tremendous improvement in PCE. A record PCE exceeding 18% (Cui et al. 2020) in single junction and 17% (Meng et al. 2018) in multijunction OSCs devices are reported. The main contribution to this leap in performance includes the synthesis of low band gap polymer semiconductors and non-fullerene acceptors. In this section, the recent advances in OSCs technology and the main challenges the research community are working on will be discussed.

• **Advances in Low Band Gap Polymers and Non-Fullerene Acceptors:** A band gap of 1.1 eV is capable of harvesting 77% of the Am 1.5 solar spectrum. Organic semiconductors usually have a band gap above 2 eV which limits their absorption to around 33%. The solar harvest of OSCs can be improved by the synthesis of low band gap polymers that have a band gap of 1–2 eV. Low band gap polymers are usually blended with fullerene-based acceptors like PC60BM and PC70BM.

• **Synthetic Principles of Low Band Gap Polymers for High Performance OSCs:** Fused heterocycles are one of the important building blocks of low band gap polymers. The first low band gap polymer known was polyisothianaphtalene (PITN) in which a quinoid mesomeric structure is favoured in the backbone to maintain the benzene aromaticity. The resulting polymer has a band gap as low as 1 eV (Wudl et al. 1984). Fused heterocyles are also commonly used as building blocks of donor-acceptor copolymers in which the electron-deficient and electron-rich units are coupled in an alternating fashion. The intramolecular charge transfer between the donor and acceptor units in the copolymer results in a reduced band gap than the respective individual units (Tegegne et al. 2020; Tegegne et al. 2018). Thienothiophene (TT) is the most successful fused heterocycle as a building block of low band gap polymers. The most studied TT-based copolymer is synthesized by alternating TT and the fused benzodithiophene (BDT) moieties called *PTB polymers*. The band gaps of PTB polymers can go as low as 1.6 eV and power conversion efficiency over 5% was achieved when these polymers were blended with PC70BM acceptor in BHJ OSC (Invernale et al. 2009; Liang et al. 2009). Yu et.al synthesized a series of TT containing BDT donor-acceptor copolymers by changing their side chains as shown in Figure 23.12 (Invernale et al. 2009). The photovoltaic performance of these series of copolymers increased from 5.3% (P1) (Invernale et al. 2009) to over 9% (P7) (Lee et al. 2008). However, the band gap of the copolymers was almost similar. The main reason for this enhancement in PCE was the

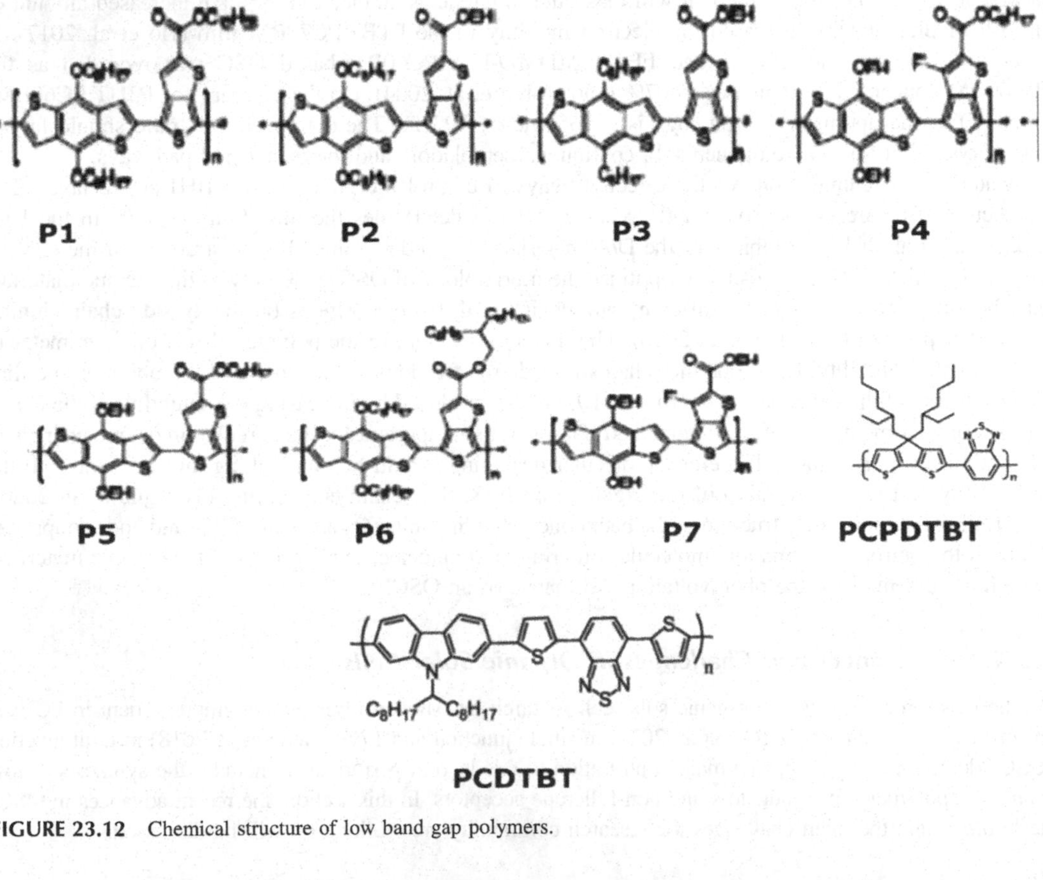

FIGURE 23.12 Chemical structure of low band gap polymers.

decrease in the HOMO level which gave rise to an improved V_{oc}. The other commonly used monomers in low band gap copolymers benzothiadiazole (BT) and pyridalthiadiazole include PCDTBT and PCPDTBT (structure is shown in Figure 23.12). The PCE of such copolymer was 5-7% (Beaujuge et al. 2009; Chen et al. 2010; Hou et al. 2008; Coffin et al. 2009; Henson et al. 2012; Ying et al. 2011). Fluorination is also another common practice in synthesis chemistry to lower the band gap of polymers. Fluorine has a size of hydrogen atom but with much higher electronegativity (with a Pauling electronegativity of 4.0). The replacement of hydrogen atom with fluorine will increase the electronegativity of the polymer, which significantly lowers the HOMO level without introducing steric hindrance in the backbone. The TT-based polymers shown in Figure 23.12 have shown an increase of PCE from 5.3% (P1) (Invernale et al. 2009) to over 9% (P7) (Lee et al. 2008) due to the fluorine substitution in polymers that significantly lowered the HOMO level from -4.90 eV to -5.15 eV. A similar effect was reported with TT and BDT-based copolymers upon fluorination. The most studied fluorinated copolymer, PTB7, was the first polymer that broke the 7% PCE in 2010 (Liang et al. 2010). Substitution of the side chains on the backbone of the copolymers is also another strategy to tune their energy levels in addition to their solubility. A thiophene substitution in PTB7 to synthesize PTB7-Th increased the PCE to over 10% when blended with a fullerene acceptor (He et al. 2015).

- **Non-Fullerene Acceptors:** Fullerene-based acceptors such as PC60BM, PC70BM, and ICBA (indene-C60-bisadduct) have long been used in bulk heterojunction OSCs. A promising PCE over 10% was achieved using low band gap polymers as donors and fullerene-based acceptors (He et al. 2015; Liu et al. 2014). The 3D surface cages of C60 and C70 gave the fullerene acceptors efficient and isotropic electron transport. However, the 3D isotropic surface of the fullerene is also responsible for the asymmetric nature of their wave function. Consequently, light absorption of these acceptors in the UV-Visible region of the solar spectrum is limited, which made their contribution to the photogenerated current minimal. In addition, because the 3D isotropic surface of the C60 and C70 in the LUMO is delocalized, the chemical modification of the structure becomes difficult. The inability to change the LUMO level of this acceptors limits the enhancement of the V_{oc} of OSCs. Non-fullerene acceptors (NFA) were introduced to overcome the poor absorbance of fullerene-based acceptors while keeping their advantage of highly efficient electron transport. The emergence of these NFAs has revitalized the organic solar cell research and the PCE of the devices has exceeded 18% in single junction (Cui et al. 2020). The structure of these acceptors can easily be tuned so that V_{oc} of 1.18 V was obtained using P3HT as a donor and a NFA (Holliday et al. 2015). The total fraction of photon that can be absorbed in NFA-based OSCs is maximized by choosing donor and NFA materials that absorb in different regions of the solar spectrum.

NFA can be small molecules or polymers. The calamitic small molecules synthesized by acceptor-donor-acceptor (A-D-A) copolymerization gained good attention in organic solar cells as NFAs. The flourene and carbazole donor unit-based NFAs are among these calamatic A-D-A NFAs that were first introduced. An impressive PCE over 10% was achieved using FDICTF non-fullerene acceptor and PBDB-T donor (Qiu et al. 2017). The main drawbacks in this class of NFAs are that they have very small absorbance which leads to low photocurrent generation and the amorphous nature of the acceptors due to the twist in the backbone structure. The morphology of the active layer prepared using these NFAs is too mixed that inhibits the formation of aggregation to facilitate percolation of charge carriers in the device. The indacenodithipohene and indacenodithinethiophe-based AD-A NFA were introduced later to overcome the lower photon absorption in fluorine and carabazole-based NFA. IEIC was among the first A-D-A acceptor which contains IDT core flanked by thiophene spacer units and DCI end groups. Devices fabricated using IEIC as an acceptor and PTB7-Th as donor resulted in a moderate JSC of 13.55 mA cm^{-2} due to poor spectra coverage. A small structural change by substituting the alkyl side chains by aloxy on the thiophene units to synthesize IEICO resulted in a lower band gap of 1.34 eV. Consequently, an improved photon harvest increased the JSC into 17.7 mA cm^{-2} (Yao et al. 2016). ITIC was later developed based on IDTT as the core instead of IDT and without a *p*-conjugated spacer. The energy levels of ITIC allow for good alignment with low band gap polymers, resulting in enhanced charge separation efficiency and reduced energy loss. The highest PCE (11.34%) of OSCs based on ITIC was achieved when it is blended with the wide gap polymer PBQ-4F (Zheng et al. 2017). Further improvement of ITIC acceptors by substituting the methyl units in DCI unit by four fluorine atoms resulted in a lower band gap of IT-4F. The record power conversion efficiency in 2019 of 13.1% was attained using IT-4F acceptor and PBDB-T-SF due to the high JSC of 20.88 mA cm^{-2} (Zhao et al. 2017). The progress in OSCs continued with the modification of the chemical structures of NFA. An electron-deficiency ladder type fused ring (dithienothiophen[3.2-b]- pyrrolobenzothia-diazole) with a benzothiadiazole (BT) core was employed to synthesize Y6 which resulted in a PCE over 15% (Yuan et al. 2019). Ciu et al. reported a modified NFA (BTP-4Cl) by replacing the F in Y6 by Cl which resulted in higher dipole moment. An improved charge separation in the donor-acceptor blends can be obtained due to the increased dipole moment of the chlorinated NFA which resulted in higher FF over 75%. The OSC device based on PBDB-TF:BTP-4Cl attained a PCE of 16.5% (Cui et al. 2019). This PCE was a record efficiency until the recently record efficiency of 18.22% using the Y6 acceptor blended with a wide band gap polymer D18 (Liu et al. 2020).

23.4.1.6 Stability: Challenges of Organic Solar Cells

A tremendous effort to improve PCE of OSC has been employed to reach over 18% which includes designing of electron donor and acceptor organic semiconductors and device structure (Cui et al. 2019; Liu et al. 2020). The record PCE of OSCs obtained has reached the efficiency of commercialized thin-film solar cells (Liu et al. 2020). However, the industrialization of the technology faces many problems of which stability of the devices is the main factor. An economic assessment done on OSCs and silicon solar cells showed that OSCs can be economically competitive to silicon-based solar cells if their PCE reached 7% in large area with life time of at least 5 years (Zuo et al. 2015). In this section, factors affecting stability of OSCs and strategies to improve their stability will be discussed.

23.4.1.6.1 Factors Affecting Stability of OSCs

The factors that limit stability of organic solar cells can be broadly divided into extrinsic and intrinsic factors. The extrinsic factors are mainly due to environmental condition that includes oxygen and water ingression, irradiation, heating and mechanical stress. The intrinsic factors are mainly the metastable morphology and diffusion of electrodes and buffer layers. Oxygen reduces the stability of OSCs in many aspects. The oxidation of low work function electrodes and an insulating layer that is formed between the electrodes and the buffer layer will create a charge collection barrier. In addition to the electrodes, the active layer materials (donor and/or acceptor) can be oxidized and change their chemical structure that can affect both charge mobility and absorption. Oxygen doping is also another process that can occur due to its reaction with the active layer materials. The increase in hole concertation due to the trapping of electrons by oxygen will lead to unbalanced charge transport resulting in loss in FF and V_{oc} (Schafferhans et al. 2010; Seemann et al. 2011). Water ingression in an OSC can oxidize the low work-function electrode and at the same time create pinholes that can serve as a pathway for more water ingression. The active layer morphology can be tempered due to the excessive aggregation of PC60BM due to water ingression. Consequently, the exciton dissociation will be limited in the active layer. A solar cell works under sunlight, which makes stability under irradiation a paramount importance. The main degradation processes due to irradiation are photooxidation of the active layer materials (Tegegne et al. 2020) and photo-physical degradation of electrodes and buffer layers. Photooxidation of the active layer material impairs its absorption which will result in lower exciton generation. The oxidized active layer materials can also have increased sub-band gap states that increases recombination (Peters et al. 2012; Clarke et al. 2013).

Photo-induced charge accumulation in the active layer can lead to photo-physical degradation that modifies the energetics in the device. As a result, the open circuit voltage of the device is reduced. The continuous irradiation of solar cells increases the temperature of the device above the environmental temperature. Heating is therefore one of the extrinsic degradation factors. The glass transition temperature of many conjugated polymers is high that makes heat-induced decomposition far above the heating of the OSCs working under illumination. Therefore, heat-induced degradation in conjugated polymers is physical degradation that includes mobility of the active layer materials. The mobility of the materials in the active layer due to heating is mostly seen when the temperature exceeds their glass transition temperature. Consequently, the active layer morphology will be changed. Aggregation/ crystallization of fullerenes is also another heat-induced morphological change that leads to loss of power. Nucleation of fullerenes in the active layer leads to growth of large domains that reduces the donor/acceptor interface resulting in the exciton dissociation efficiency loss (Lindqvist et al. 2013). In addition to the active layer, heating-induced degradation can occur at the interface between the active layer/electrodes and buffer layers (Sachs-Quintana et al. 2014).

Mechanical stability is important to organic solar cells because of the need to manufacture them using roll-to-roll technique. Mechanical stress can cause fracture on the active layer, the electrodes, and the buffer layers. In addition, de-cohesion of the active layer can also be caused due to mechanical stress which leads to ingression of water. Metastable morphology is also another factor that causes degradation of OSCs. The morphology of an OSC can become unstable due to the movement of the two or three phases (donor and acceptor and/or a third component) in the active layer. In addition, the electrodes also show some degree of mobility that impairs the charge photogeneration and/or collection in the OSC. ITO diffusion to PEDOT:PSS layer and to the active layer (Sharma et al. 2011) is one of the cause for unstable morphology in the active layer. The PEDOT: PSS layer can also directly diffuse to the active layer reducing the stability of the device.

23.4.1.6.2 *Mechanism to Improve Stability of OSCs*

There is a tremendous effort to improve the stability of organic solar cells. Some of the mechanisms that have proved to be working will be discussed in this section. Rational molecular structure design of donor materials plays an important role in the stability of the active layer materials. Substitution of oxidizable units of the donor materials by chemically stable units is one of the design rules for stable materials (Manceau et al. 2011). These include avoiding the excocyclic double bonds and readily cleavable bonds and quaternary carbon atoms, and including aromatic polycyclic units in the backbone of the donor material is recommended by Krebs et al. (Carlé et al. 2014). The crystallinity of the donor polymer is the other parameter that can be tailored to enhance their photochemistry stability. Electrons and holes in a crystalline polymer are spatially separated which decreases trap assisted recombination unlike amorphous polymers. McGhee et al. reported a PCE loss of 16% in the amorphous PCDTBT-based device while the loss in the crystalline KP115 was only 5% after 60 h of aging (Heumueller et al. 2014). On the other hand, the mechanical stability of polymers increases when they are robust and amorphous. Increasing the degree of polymerization could eliminate trap-assisted recombination in PCDTBT/PCBM70-based solar cells (Kong et al. 2014). In addition, the glass transition temperature, T_g, of the material can be increased with increasing degree of polymerization (Papadopoulos et al. 2004). Hence, degree of polymerization can be used to control the photochemistry and thermal stability of donor polymers. Side-chains tailoring is also one of the synthesis parameter to control the stability of donors. Side chains play an important role in the solubility of the donor material. However, once the film is deposited these side chains are useless and are also source of degradation. Cleavable side chains are employed to increase the stability of donors. The acceptor materials in the active layer of the OSC are also source of degradation. Fullerene-based acceptors (C60, C70, PC60BM, PC70BM, and ICBA) are well developed acceptor materials with favourable nanoscale network with donors and high electron mobility. These acceptors undergo dimerization upon irradiation and form large aggregates upon heating which are both detrimental to charge photogeneration in OSCs. Changing side chains of the fullerene-based acceptors, such as increasing the size of the side chain, broke the high symmetry of PC60BM and decreased crystallization, resulting in an improved stability of OSCs.

23.4.2 Perovskite Solar Cells

The need for low-cost material and easy processing techniques in solar cells brought generations of solar cells. Perovskite solar cells (PCSs) are one of the promising technologies that combine high PCE with easy solution processing technique. The name of the absorber, perovskite, came from the Russian mineralogist Lev Perovski, who discovered $CaTiO_3$. Perovskite materials are light absorbers that has a chemical structure of ABX_3 with A being monovalent cations while B and X are divalent metal ions such as Pb^{2+} and Sn^{2+} and halogens such as I^{2-}, Br^{2-}, and Cl^{2-}, respectively. The crystallographic

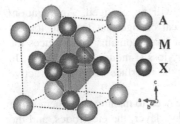

FIGURE 23.13 Perovskite materials crystal structure.

structure of perovskites is shown in Figure 23.13. The most studied organic-inorganic lead halide perovskite for PSCs are MAPbI$_3$ (MA = methylamonium) and FAPbI$_3$ (FA = HC(NH$_2$)$^{2+}$; formamidium). The PCE of organic-inorganic lead halide perovskite-based PSCs has increased from 3.8% in 2009 to over 25% in 2019 (Kojima et al. 2009; Green et al. 2020). The PCE of PSCs improved from 6.5% to over 20% in three years mainly due to the advancement in device preparation techniques (Im et al. 2011; Green et al. 2013). The development to the current champion PCE of 25.1% was recorded after 4 years (Green et al. 2020). The total leap in this high PCE in PCS occurred in a much shorter time than the matured technologies like silicon or thin-film solar cells. The PCE of PSCs has already surpassed the commercialized thin-film solar cells based on CdTe, GaSe$_2$, and amorphous Si and is very close to the crystalline Si-based solar cell (PCE = 26.6%) (Meng et al. 2018). In this section, the progress in device structure followed by the evolution of the processing techniques will be discussed. Finally, the challenges PSC device technology face for commercialization will be discussed.

23.4.2.1 Evolution of Perovskite Solar Cells Device Structure

23.4.2.1.1 Liquid Electrolyte Dye-Sensitized Solar Cells

Dye-sensitized solar cells (DSSCs) are the forerunners to PCSs. The DSSCs are fabricated on a transparent conductive oxide, commonly fluorine doped titanium-oxide (FTO)-coated glass. A thin compact TiO$_2$ (c-TiO$_2$) layer is deposited on FTO to serve as an electron blocking layer followed by thick mesoporous TiO$_2$ (mp-TiO$_2$) layer sensitized with dye. The DSSCs will then be capped with a liquid electrolyte containing redox couples and counter electrode (Pt/TCO/glass). The device geometry is shown in Figure 23.14a.

Charge photogeneration in DSSCs occurs in the following steps. The first step is the absorption of photons by the sensitizer exciting the dye. The excited dye will oxidize itself by injecting electron to the semiconductor. The electron will percolate in the semiconductor layer to the back electrode which will then be transported through the external load to the counter electrode. The electrons will be injected to the redox which will regenerate the oxidized dye and complete the circuit. The working principle of DSSCs is shown in Figure 23.14b.

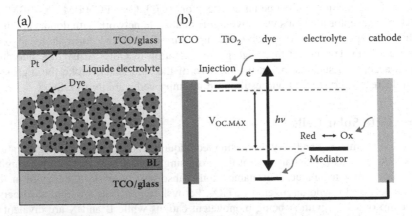

FIGURE 23.14 (a) Device structure of liquid electrolyte DSC and (b) their working principle.

Kojima et al. reported the first perovskite-sensitized solar cells using MAPBI$_3$ sensitizer in iodide/triiodide redox couple with PCE of 3.5% (Kojima et al. 2009). Although this work showed the potential of perovskite materials in PV application, it did not get much attention due to the low efficiency and extremely poor life time of few minutes. Two years later the efficiency of PSCs was improved to 6.5% by optimizing the mp-TiO$_2$ surface morphology and using perovskite quantum dots (Im et al. 2011). The dissolution of the perovskite layer in the liquid electrolyte put a stumbling block for the development of liquid electrolyte perovskite-sensitized solar cells. The need to replace the liquid electrolyte with solid-state hole transport layer brought forth the birth of solid-state perovskite-sensitized solar cells.

23.4.2.1.2 *Solid-State PSCs with Mesoporous TiO$_2$ Scaffold*

The first solid-state PSC was fabricated by using a solid hole transport material 2',2',7',7'-tetrakis-(N,N-di-p-methoxyphenyl-amine)-9,9'-spirobifluorene (Spiro-MeOTAD) (Kim et al. 2012). The perovskite layer was synthesized by spin coating equimolar γ-butyrolacetone solution containing MAI and PbI$_2$. The thickness of the mp-TiO$_2$ layer was reduced from 3 μm in liquid electrolyte to ~0.6 μm in solid hole transport materials (HTM)-based PCS. The HTM penetrates through the pore of the mp-TiO$_2$ to make contact with the perovskite absorber with some part of it forming a dense capping layer on top of the mp-TiO$_2$ electron transport material (ETM) to reduce shunts. The device geometry is shown in Figure 23.15a. The device achieved a PCE of 9.7% with V$_{oc}$ of 888 mV and fill factor of 0.62. The poor fill factor is indicative of poor interfaces within the device. Unlike the liquid electrolyte PSCs, hole transport in solid-state PSCs was efficient. This is due to a better alignment of the HOMO of the Spiro-OMeTAD with the valence band of the perovskite layer (Kim et al. 2012). A thicker mp-TiO$_2$ of 1.6 μm decreased the performance of the MAPbI$_3$-based PSC to 9.4%. The increased thickness of the ETM reduced the pore filling of the HTM. Br-I mixed perovskite MAPbI$_x$Br$_{2-x}$ in mp-TiO$_2$ is another highly studied absorber. Polymer HTM like PCBTDPP and P3HT are also used in MAPbBr$_3$ absorber in mp-TiO2 PSCs. The V$_{OC}$ in PSCs fabricated using PCDTDPP (HOMO = -5.2 eV) and P3HT (HOMO = -5.4 eV) as HTM were 1.2 V and 0.5 V, respectively. This large difference in V$_{oc}$ with a small HOMO difference of 0.2 eV is due to the chemical interaction of the HTM with the perovskite absorber. The results show that the V$_{oc}$ of the PSCs depends on other properties of the HTM besides their HOMO level. A record PCE of 25.2% is recorded in PSCs containing mp-TiO$_2$ layer (Green et al. 2013).

23.4.2.1.3 *Meso-Superstructured PCSs Based on Non-Injecting Oxides*

The mp-TiO$_2$ was replaced by the insulating Al$_2$O$_3$ in meso-superstructured PSCs. The device geometry is similar to the mesoporous structure as shown in Figure 23.15b. A meso-superstructured PSC using MAPbI$_2$Cl mixed perovskite achieved a PCE of 10.9% (Leijtens et al. 2014). A control device fabricated using mp-TiO$_2$ demonstrated an inferior PCE of 7.6% with a lower V$_{oc}$. Photo-induced spectroscopy study shows that electrons from the perovskite absorber are not injected to Al$_2$O$_3$.

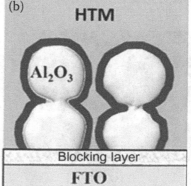

FIGURE 23.15 Device structure of PSCs using (a) mp-TiO2 (b) mp-Al2O3. Ref. Kim et al. 2014.

This resulted in low non-radiative recombination generating higher V_{oc} in meso-superstructured cells. The electron from the perovskite absorber cannot be injected to the Al_2O_3 for two reasons. One is the conduction band of $Al_2O_3$is high and electron injection from the perovskite is energetically unfavourable. The second reason is that Al_2O_3 is an insulator. Therefore, Al_2O_3 in the meso-superstructured cell serves as a scaffold. The electron from the perovskite will then be injected to the c-TiO_2 layer due to the long carrier diffusion length in the absorber.

The meso-superstructured device geometry was also tried using ZrO_2 scaffold and a high efficiency of 10.8% using MAPbI3 absorber with V_{oc} of 1.07 V was achieved. The success of meso-superstructured cells mainly lies on the excellent optical and electrical properties of the absorber perovskite layer. The mp-Al_2O_3 layer can therefore be left out in the fabrication of PSCs in the so-called planar structure. Snaith et al. reported a PCE of 1.8% in a preliminary planar device structure (Lee et al. 2012). The PCE of PSCs with the meso-superstructured Al_2O_3 has reached 10.9% (Green et al. 2013).

23.4.2.1.4 Planar Heterojunction

The thickness of the mesoporous layer was reduced further to fabricate planar heterojunction PSCs in n-i-p structure. Due to the ambipolar charge transport property, low exciton binding energy, and high carrier diffusion length in perovskite, the mesoporous layer might not be needed. A transparent conductive oxide is coated with a n-type ETM. The intrinsic perovskite layer and the p-type HTM will then be deposited successively. The device geometry is shown in Figure 23.16a. The first planar heterojunction PSC was fabricated in 2012 and generated a performance of 1.8%. The poor performance of the planar PSC device was due to the poor film coverage that created shunts in the device. The performance of PSCs in this device geometry increased to 15.4% by optimizing the device processing conditions and methods (Bi et al. 2013). The high efficiency PSCs that are reported in recent years exceeding 20% are in planar geometry with improved film deposition techniques (Anaraki et al. 2016).

Inverted planar heterojunction PSCs have gained a lot of attention. This device geometry is constructed based on a p-i-n structure. The TCO is coated with a p-type HTM followed by the intrinsic perovskite layer. The device is capped with a n-type ETM/Anode. The device geometry is shown in Figure 23.16b. Among the first inverted planar heterojunction PSCs was the device fabricated using ITO as a transparent oxide. The structure of the device is ITO/PEDOT:PSS/$MAPbI_{3-x}Cl_x$/PCBM/Al and yields a PCE of 3.9% (Liu et al. 2013). The p-i-n planar heterojunction solar cells showed a tremendous improvement that in only few years a PCE of 17.7% is achieved by optimizing the deposition technique for the perovskite layer and the E/HTMs. In 2016, a PCE of 16.5% with J_{SC} = 22.6 mA/cm^2, V_{oc} = 1.07 V, and FF = 88.6% was reported (Bai et al. 2016). The performance of planar devices was improved to 23.7% by using $FA_{0.92}MA_{0.06}PbI_3$ perovskite absorber and SnO_2 ETM (Kim et al. 2019).

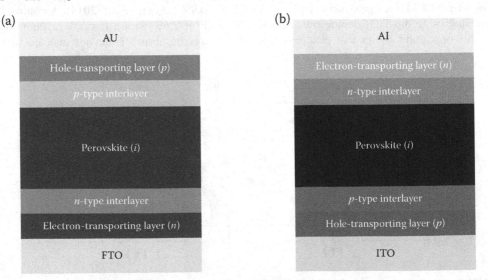

FIGURE 23.16 Device geometry of (a) normal planar (b) inverted planar PSCs.

23.4.2.2 Progress in Fabrication Techniques and Stability Of PSCs

The development in the deposition techniques of the perovskite layer played a huge role in the drastic improvement of PCE of PSCs. The material properties of the perovskite that includes its crystallographic phase and grain structure can be controlled by the processing methods. In this section, the progress in perovskite film depositions that led to high PCE in PSCs will be discussed briefly.

- **Single-Step Solution Process:** A single-step process is the simplest processing technique to prepare perovskite films. The precursor solution will be spin coated on a substrate and will be annealed. The precursor solution is prepared by mixing MAI and PbI_2 in polar solvent like γ-butyrolactone (GBL), dimethyl formamide (DMF), dimethyl sulphoxide (DMSO), N-2-methyl pyrrolidone, or a mixture of them. The ratios of the organic halide to the metal-halide, the solvents and additives and annealing temperatures have a profound impact on the final film. In addition, the environmental condition in which the spin coating is done affects the quality of the film.

- **Two-Step Sequential Deposition:** The poor film formation of single-step processed perovskite layers could be improved by two-step sequential deposition. The PbI_2 will first be spin coated on a substrate followed by deposition of MAI either by dipping or spin coating to form $MAPbI_3$ layer. The perovskite layer deposited in a two-step method by far is more compact, uniform, and reproducible. PSCs with this advanced deposition technique have exceeded PCE of 20% (Yang et al. 2015).

- **Vapour-Assisted Deposition:** Vapour assisted deposition is a modified two-step processing method. The PbI_2 seed layer is spin coated on a substrate which will be exposed to MAI vapour to form the $MAPbI_3$ layer. The vapour-assisted deposition method combines solution processing and low temperature vapour deposition that resulted in compact, pinhole-free perovskite films (Hu et al. 2014). However, the formation of the MAPbI3 takes longer and hence the deposition technique is not favoured for practical applications.

23.4.2.3 Stability of Perovskite Solar Cells: Challenge to Commercialization

The commercialization of PSCs depends on the two most important factors; efficiency and stability. The success story of PSC in high efficiency exceeding 25% is out due to the low stability of the device. The stability of PSCs was only several seconds in early 2009. Later, it was improved to a life time of 1 year (Grancini et al. 2017). However, it is still far behind the 25 years life time of the commercialized PV technologies. Degradation in PSC stacks can occur at any of the layers that includes the perovskite absorber, the charge transport layer, electrodes, and interfaces. The exposure to humidity, oxygen, UV-radiation, and elevated temperatures leads to deterioration of the PSCs.

Perovskite materials used in PSCs, mainly the organic-inorganic hybrids, exhibit soft ionic lattice with unstable weak interactions. This leads to easy ion migration affecting the perovskite film drastically when exposed to illumination. Photochemical reaction that leads to photodecomposition is also one of the degradation mechanisms in PSCs. The photodecomposition in the absorber is facilitated with the presence of oxygen and water. A humidity level of 55% has proved to degrade the perovskite film changing its colour from dark brown to yellow. Introducing Al_2O_3 scaffold and changing the perovskite absorber to $MAPbBr_3$ have proved to increase the stability of the PSCs (Ma et al. 2009). Encapsulation and preheating the substrate before deposition are other proven techniques to improve stability of PSCs. The working PSCs under illumination also face thermal degradation due to the phase transitions of $MAPbI_2$ from tetragonal to cubic structure at $56^\circ C$ (Poglitsch and Weber 1987). Thermal stability of the perovskite films can be improved by optimizing precursors.

The electron transport layer plays an important role in the high efficiency PSCs; thus, it also needs to be stable. Metal oxides such as TiO_2, SnO_2, ZnO, and $BaSnO_3$ and fullerene derivatives are the mostly used ETM in PSCs. Photo-induced desorption of O_2 in metal oxides lead to trap-induced degradation in the device. Chlorine-sapped TiO_2 (TiO_2-Cl) was found to reduce trap formation leading to reduced recombination (Tan et al. 2017). PSCs with TiO_2-Cl HTM achieved a PCE of 20.1% and kept 90% of its performance after 500 h 1-sun light illumination. Fullerenes are also not stable under illumination. N-doped C60 with 4-(1,3-dimethyl-2,3-dihydro-1H-benzimidazol-2-yl)-N,N-diphenylaniline (N-DPBI) increased the stability of PSCs that the device fabricated using the new electron transport material kept 76% of its initial PCE while the undoped ETM kept only 56% after 2000 h exposure to 1-sun illumination (Armstrong et al. 2009). In general, light-induced degradation in ETM of PSCs is a huge challenge. The other part of the perovskite stack is the HTM. The commonly used hole transport materials include organic semiconductors like spiro-OMeTaD, PTTA, P3HT, etc., and inorganic semiconductors like NiO_x, CuSCn, Cul. The low conductivity of organic HTM can be improved by doping with salts like Li-TFSI. The organic HTM tends to crystallize and/or decompose at high temperatures reducing the stability of PSCs. Solvent optimization and introduction of poly(4-vinylpyridine)(P4VP) in spiro-OMeTAD during film formation could increase its stability against humidity (Bella et al. 2016).

In summary, PSCs can be used for electric generation given their life time is improved. Degradation channels both in operation and shelf life need to be well understood to further improve their stability.

23.5 Future Outlooks

The well-developed crystalline solar cell technology has covered more than 90% of the solar cell market. The price of solar cell technology has been reduced greatly with introduction of low processing techniques and new device geometry. Further, research to reduce the cost of production is needed for the technology to be competitive with fossil fuels. On the other hand the thin-film solar cells suffer with low power conversion deficiency as well as life time. The toxicity of the materials used in the technology is also another bottleneck for commercialization. Introduction of new materials that can substitute the existing semiconductors for thin-film solar cell is important. The emerging solar cell technologies, both the organic and perovskite solar cells, have attained commercial PCE. The energy payback time in these devices is reduced to the order of days as opposed to a year in silicon solar cells. However, these low-cost materials are stable both in working conditions and their shelf life. The basic understanding of the device degradation mechanisms will undoubtedly contribute to the maturing of the technology for commercialization.

REFERENCES

Alajlani, Y., A. Alaswad, F. Placido, D. Gibson and A. Diyaf. 2018. "Inorganic Thin Film Materials for Solar Cell Applications." In *Reference Module in Materials Science and Materials Engineering.* Amstedam: Elsevier BV.

Anaraki, E.H., A. Kermanpur, L. Steier, K. Domanski, T. Matsui, W. Tress, M. Saliba, et al. 2016. "Highly Efficient and Stable Planar perovskite Solar Cells by Solution-Processed Tin Oxide." *Energy & Environmental Science* 9, no. 10: 3128–3134.

Armstrong, N.R., P.A. Veneman, E. Ratcliff, D. Placencia and M. Brumbach. 2009. "Oxide Contacts in Organic Photovoltaics: Characterization and Control of Near-Surface Composition in Indium-Tin Oxide (ito) Electrodes." *Accounts of chemical research* 42, no. 11: 1748–1757.

Bai, Y., Q. Dong, Y. Shao, Y. Deng, Wang Q., L. Shen, D. Wang, W. Wei and J. Huang. 2016. "Enhancing Stability and Efficiency of perovskite Solar Cells with Crosslinkable Silanefunctionalized and Doped Fullerene." *Nature Communications* 7, no. 1: 1–9.

Beaucarne, G. 2007. "Silicon Thin-film Solar Cells." *Advances in OptoElectronics* 2007, 35383.

Beaujuge, P.M., W. Pisula, H.N. Tsao, S. Ellinger, K. Mullen and J.R. Reynolds. 2009. -"Tailoring Structure-Property 99Relationships in Dithienosilole- Benzothiadiazole Donor- Acceptor Copolymers." *Journal of the American Chemical Society* 131, no. 22: 7514–7515.

Bella, F., G. Griffini, J.-P. Correa-Baena, G. Saracco, M. Gr¨atzel, A., Hagfeldt, Turri S. and C. Gerbaldi. 2016. "Improving Efficiency and Stability of perovskite Solar Cells with Photocurable Fluoropolymers." *Science* 354, no. 6309: 203–206.

Bi, D., S.-J. Moon, L. Häggman, G. Boschloo, L. Yang, E.M. Johansson, M.K. Nazeeruddin, M. Grätzel and A. Hagfeldt. 2013. "Using a Two-Step Deposition Technique to Prepare perovskite (ch 3 nh 3 pbi 3) for Thin Film Solar Cells Based on ZrO2 and TiO2 Mesostructures." *Rsc Advances* 3, no. 41: 18762–18766.

Blakers, A.W., A. Wang, A.M. Milne, J. Zhao and M.A. Green. 1989. "22.8% Efficient Silicon Solar Cell." *Applied Physics Letters* 55, no. 13: 1363–1365.

Bonnet, D. 1992. "The CdTe Thin Film Solar Cell: An overview." *International Journal of Solar Energy* 12, no. 1-4: 1–14.

Bonnet, D. and H. Rabenhorst. 1972. "New Results on the Development of a Thin-film P-CdTe-NCdS Heterojunction Solar Cell." In Photovoltaic Specialists Conference, 9th, Silver Spring, Md, pp. 129–132.

Bosio, A., Pasini, S. and Romeo, N. 2020. "The History of Photovoltaics with Emphasis on CdTe Solar Cells and Modules." *Coatings* 10, no. 4: 344.

Brédas, J.-L., D. Beljonne, V. Coropceanu and J. Cornil. 2004. "Charge-Transfer and Energy Transfer Processes in *p*-conjugated Oligomers and Polymers: A Molecular Picture." *Chemical reviews* 104, no. 11: 4971–5004.

Burst, J.M., J.N. Duenow, D.S. Albin, E. Colegrove, M.O. Reese, J.A. Aguiar, C.-S. Jiang, M. Patel, M.M. Al-Jassim, D. Kuciauskas, et al. 2016. "Cdte Solar Cells with Open-circuit Voltage Breaking the 1 v Barrier." *Nature Energy* 1, no. 3: 1–8.

Carlé, J.E., M. Helgesen, M.V. Madsen, E. Bundgaard and F.C. Krebs. 2014. "Upscaling from Single Cells to Modules–Fabrication of Vacuum-and ito-Free Polymer Solar Cells on Flexible Substrates with Long Lifetime." *Journal of Materials Chemistry C* 2, no. 7: 1290–1297.

Cashmore, J., M. Apolloni, A. Braga, O. Caglar, Cervetto, V., Fenner Y., S. Goldbach-Aschemann, et al. 2016. "Improved Conversion Efficiencies of Thin-film Silicon Tandem (Micromorph) Photovoltaic Modules." *Solar Energy Materials and Solar Cells* 144: 84–95.

Chen, H.-Y., J. Hou, A.E. Hayden, H. Yang, K. Houk and Y. Yang. 2010. "Silicon atom Substitution Enhances Interchain Packing in a Thiophene-based Polymer System." *Advanced materials* 22, no. 3: 371–375.

Chen, C.-W., M. Hermle, J. Benick, Y. Tao, Y.-W. Ok, A. Upadhyaya, A.M. Tam and A. Rohatgi. 2017. "Modeling the Potential of Screen Printed Front Junction Cz Silicon Solar Cell with Tunnel Oxide Passivated Back Contact." *Progress in Photovoltaics: Research and Applications* 25, no. 1: 49–57.

Chopra, K., P. Paulson and V. Dutta. 2004. "Thin-Film Solar Cells: An Overview." *Progress in Photovoltaics: Research and Applications* 12no. 2-3: 69–92.

Choubey, P., A. Oudhia and R. Dewangan. 2012. "A Review: Solar Cell Current Scenario and Future Trends." *Recent Research in Science and Technology* 4, no. 8: 99–101.

Clarke, T.M. and J.R. Durrant 2010. "Charge Photogeneration in Organic Solar Cells." *Chemical Reviews* 110, no. 11: 6736–6767.

Clarke, T.M., C. Lungenschmied, J. Peet, N. Drolet, K. Sunahara, A. Furube and A.J. Mozer. 2013. "Photo Degradation in Encapsulated Silole-Based Polymer: Pcbm Solar Cells Investigated Using Transient Absorption Spectroscopy and Charge Extraction Measurements." *Advanced Energy Materials* 3, no. 11: 1473–1483.

Coffin, R.C., J. Peet, J. Rogers and G.C. Bazan. 2009. "Streamlined Microwave-assisted Preparation of Narrow-bandgap Conjugated Polymers For High-performance Bulk Heterojunction Solar Cells." *Nature Chemistry* 1, no. 8: 657–661.

Collins, B.A., Z. Li, J.R. Tumbleston, E. Gann, C.R. McNeill and H. Ade. 2013. "Absolute Measurement of Domain Composition and Nanoscale Size Distribution Explains Performance in ptb7: Pc71bm Solar Cells." *Advanced Energy Materials* 3, no. 1: 65–74.

Collins, R., A. Ferlauto, G. Ferreira, C. Chen, J. Koh, R. Koval, Y. Lee, 2003. "Evolution of Microstructure and Phase in Amorphous, Protocrystalline, and Microcrystalline Silicon Studied by Real Time Spectroscopic Ellipsometry." *Solar Energy Materials and Solar Cells* 78, no. 1-4: 143–180.

Cousins, P.J., D. D Smith, Hsin-Chiao Luan, Jane Manning, Tim D. Dennis, Ann Waldhauer, Karen E. Wilson, Gabriel Harley and G.P. Mulligan. 2015. "Gen iii: Improved Performance at Lower Cost." In Proceedings of the 35th IEEE Photovoltaic Specialists Conference, Honolulu, Hawaii,-2010, pp. 823–826.

Cui, Y., H. Yao, J. Zhang, T. Zhang,Y. Wang, L. Hong, K. Xian, et al. 2019. "Over 16% Efficiency Organic Photovoltaic Cells Enabled by a Chlorinated Acceptor with Increased Open-circuit Voltages." *Nature Communications* 10, no. 1: 1–8.

Cui, Y., H. Yao, J. Zhang, K. Xian, T. Zhang, L. Hong, Y. Wang, et al. 2020. "Single-Junction Organic Photovoltaic Cells with Approaching 18% Efficiency." *Advanced Materials* 32, no. 19: 1908205.

Deb, S.K. 2000. "Recent Developments in High-Efficiency PV Cells." In *World Renewable Energy Congress VI*, 2658–2663. Brighton, U.K: Elsevier.

Deng, W., F. Ye, R. Liu, Y. Li, H. Chen, Z. Xiong, Y. Yang, Y. Chen, Y., Wang, P.P. Altermatt et al. 2017. "22.61% Efficient Fully Screen Printed Perc Solar Cell." In *2017 IEEE 44th Photovoltaic Specialist Conference (PVSC)*, pp. 2220–2226. IEEE.

Dennler, G., M.C. Scharber and C.J. Brabec. 2009. "Polymer-Fullerene Bulk-Heterojunction Solar Cells." *Advanced Materials* 21, no. 13: 1323–1338.

Dobson K.D., I. Visoly-Fisher, G. Hodes and D. Cahen. 2000. "Stability of CdTe/CdS Thin-Film Solar Cells." *Solar Energy Materials and Solar Cells* 62, no. 3: 295–325.

Dullweber, T., S. Gatz, H. Hannebauer, T. Falcon, R. Hesse, J. Schmidt and R. Brendel. 2012. "Towards 20% Efficient Large-Area Screen-Printed Rear-Passivated Silicon Solar Cells." *Progress in Photovoltaics: Research and Applications* 20, no. 6: 630–638.

Dullweber, T., C. Kranz, R. Peibst, U. Baumann, H. Hannebauer, A. F¨ulle, S. Steckemetz, T. Weber, M. Kutzer, M. M¨uller et al. 2015. "The perc+ cell: A 21%-Efficient Industrial Bifacial Perc Solar Cell." In Proceedings of 31st European Photovoltaic Solar Energy Conference, pp. 341–350.

Erb, T., U. Zhokhavets, G. Gobsch, S. Raleva, B. St¨uhn, P. Schilinsky, C. Waldaufand C.J. Brabec. 2005. "Correlation between Structural and Optical Properties of Composite Polymer/ Fullerene Films for Organic Solar Cells." *Advanced Functional Materials* 15, no. 7: 1193–1196.

Feurer, T., P. Reinhard, E. Avancini, B. Bissig, J. L¨ockinger, P. Fuchs, R. Carron, et al. 2017. "Progress in Thin Film cigs Photovoltaics–Research and Development, Manufacturing, and Applications." *Progress in Photovoltaics: Research and Applications* 25, no. 7: 645–667.

First Solar. "First Solar Series 6™ Advanced Thin Film Solar Technology." 2021 http://www.firstsolar.com/-/media/First-Solar/Technical-Documents/Series-6.-Datasheets/Series-6-Datasheet.ashx

Franklin, E., K. Fong, K. McIntosh, A. Fell, A. Blakers, T. Kho, D., Walter, et al. 2016. "Design, Fabrication and Characterisation of a 24.4% Efficient Interdigitated Back Contact Solar Cell." *Progress in Photovoltaics: Research and Applications* 24, no. 4: 411–427.

Ghahremani, A. and A.E. Fathy. 2016. "High Efficiency Thin-Film Amorphous Silicon Solar Cells." *Energy Science & Engineering* 4, no. 5: 334–343.

Ghasemi, M., L. Ye, Q. Zhang, L. Yan, J.-H. Kim, O. Awartani, W. You, A. Gadisa and H., Ade. 2017. "Panchromatic Sequentially Cast Ternary Polymer Solar Cells." *Advanced Materials* 29, no. 4: 1604603.

Ghosh, K., C.J. Tracy, S. Herasimenka, C. Honsberg and S. Bowden. 2010. "Explanation of the Device Operation Principle of Amorphous Silicon/Crystalline Silicon Heterojunction Solar Cell and Role of the Inversion of Crystalline Silicon Surface." In *2010 35th IEEE Photovoltaic Specialists Conference*, pp. 001383–001386. IEEE.

Glunz, S., F. Feldmann, A. Richter, M. Bivour, C. Reichel, H. Steinkemper, J. Benick and M. Hermle. 2015. "The Irresistible Charm of a Simple Current Flow Pattern–25% with a Solar Cell Featuring a Full-Area Back Contact." In Proceedings of the 31st European Photovoltaic Solar Energy Conference and Exhibition, pp. 259–263.

Grancini, G., C. Rold´an-Carmona, I. Zimmermann, E. Mosconi, X. Lee, D. Martineau, S. Narbey, et al. 2017. "One-Year Stable perovskite Solar Cells by 2d/3d Interface Engineering." *Nature Communications* 8, no. 1: 1–8.

Green, M.A. 2009. "The Path to 25% Silicon Solar Cell Efficiency: History of Silicon Cell Evolution." *Progress in Photovoltaics: Research and Applications* 17, no. 3: 183–189.

Green, M.A., K. Emery, Y. Hishikawa, W. Warta and E.D. Dunlop. 2013. "Solar Cell Efficiency Tables (Version 42)." *Progress in Photovoltaics: Research and Applications* 21, no. 5: 827–837.

Green, M.A., E.D. Dunlop, D.H. Levi, J. Hohl-Ebinger, M. Yoshita and A.W. Ho-Baillie. 2019. "Solar Cell Efficiency Tables (version 54)." *Progress in Photovoltaics: Research and Applications* 27, no. 7: 565–575.

Green, M.A., E.D. Dunlop, J. Hohl-Ebinger, M. Yoshita, N. Kopidakis and X. Hao. 2020. "Solar Cell Efficiency Tables (Version 56)." *Progress in Photovoltaics: Research and Applications* 28, no. 7: 629–638.

Green, M., A. Blakers, J. Kurianski, S. Narayanan, J. Shi, T. Szpitalak, M. Taouk, S. Wenham and M. Willison. 1984. "Ultimate Performance Silicon Solar Cells." *Final Report, NERDDP Project* 81, no. 1264: 83.

Gretener, C., J. Perrenoud, L., Kranz, L. Kneer, R. Schmitt, S. Buecheler and A.N. Tiwari. 2013. "Cdte/cds Thin Film Solar Cells Grown in Substrate Configuration." *Progress in Photovoltaics: Research and Applications* 21, no. 8: 1580–1586.

Halls, J.J., K. Pichler, R.H. Friend, S. Moratti and A. Holmes. 1996. "Exciton Diffusion and Dissociation in a Poly (p-phenylenevinylene)/c60 Heterojunction Photovoltaic Cell." *Applied Physics Letters* 68, no. 22: 3120–3122.

He, Z., B. Xiao, F. Liu, H. Wu, Y. Yang, S. Xiao, C. Wang, T.P. Russell and Y. Cao. 2015. "Single-Junction Polymer Solar Cells with High Efficiency and Photo Voltage." *Nature Photonics* 9, no. 3: 174–179.

Henson, Z.B., G.C. Welch, T. van der Poll and G.C. Bazan. 2012. "Pyridalthiadiazole-Based Narrow Band Gap Chromophores." *Journal of the American Chemical Society* 134, no. 8: 3766–3779.

Hertz, H. 1887. "U¨ber einen einfluss des ultravioletten lichts auf die elekrische entladung." *Sitzungsberichte der K¨oniglich Preußischen Akademie der Berliner Akademie der Wissenschaft* 2: 487–490.

Heumueller, T., W.R. Mateker, I. Sachs-Quintana, K. Vandewal, J.A. Bartelt, T.M. Burke, T. Ameri, C.J. Brabec and M.D. McGehee. 2014. "Reducing Burn-in Voltage Loss in Polymer Solar Cells by Increasing the Polymer Crystallinity." *Energy & Environmental Science* 7, no. 9: 2974–2980.

Hirose, M., T. Suzuki and G. D¨ohler. 1979. "Electronic Density of States in Discharge-Produced Amorphous Silicon." *Applied Physics Letters* 34, no. 3: 234–236.

Ho, C.H.Y., S.H. Cheung, H.-W. Li, K.L. Chiu, Y. Cheng, H. Yin, Chan M.H., F. So, S.-W. Tsang and S.K. So. 2017. "Using Ultralow Dosages of Electron Acceptor to Reveal the Early Stage Donor–Acceptor Electronic Interactions in Bulk Heterojunction Blends." *Advanced Energy Materials* 7, no. 12: 1602360.

Holliday, S., R.S. Ashraf, C.B. Nielsen, M. Kirkus, J.A. Rohr, C.-H. Tan, E. Collado-Fregoso, et al. 2015. "A Rhodanine Flanked Nonfullerene Acceptor for Solution-processed Organic Photovoltaics." *Journal of the American Chemical Society* 137, no. 2: 898–904.

Hoppe, H., M., Niggemann, C. Winder, J. Kraut, R. Hiesgen, A. Hinsch, D. Meissner and N.S. Sariciftci. 2004. "Nanoscale Morphology of Conjugated Polymer/Fullerene-Based Bulk Heterojunction Solar Cells." *Advanced Functional Materials* 14, no. 10: 1005–1011.

Hou, J., H.-Y. Chen, S. Zhang, G. Li and Y. Yang. 2008. "Synthesis, Characterization, and Photovoltaic Properties of a Low Band Gap Polymer Based on Silole-Containing Polythiophenes And 2, 1, 3-benzothiadiazole." *Journal of the American Chemical Society* 130, no. 48: 16144–16145.

Hu, H., D. Wang, Y. Zhou, J. Zhang, S. Lv, S. Pang, X. Chen, et al.2014. "Vapour-Based Processing of Hole-conductor-free ch 3 nh 3 pbi 3 perovskite/c 60 Fullerene Planar Solar Cells." *RSC Advances* 4, no. 55: 28964–28967.

Huo, L., H.-Y. Chen, J. Hou, T.L. Chen and Y. Yang. 2009. "Low Band Gap Dithieno [3, 2-b: 2, 3- d] Silole-containing Polymers, Synthesis, Characterization and Photovoltaic Application." *Chemical Communications*, 37: 5570–5572.

Im, J.-H., C.-R. Lee, J.-W. Lee, S.-W. Park and N.-G. Park. 2011. "6.5% Efficient perovskite Quantum-Dot-Sensitized Solar Cell." *Nanoscale* 3, no. 10: 4088–4093.

Invernale, M.A., V. Seshadri, D.M.D. Mamangun, Y. Ding, J. Filloramo and G.A. Sotzing. 2009. "Polythieno [3, 4-b] Thiophene as an Optically Transparent Ion-Storage Layer." *Chemistry of Materials* 21, no. 14: 3332–3336.

Irvine, S. 2017. "Solar Cells and Photovoltaics." In *Springer Handbook of Electronic and Photonic Materials*, pp. 1–1. Springer, Cham.

Islam, M., K. Rahman, K. Sobayel, T. Enam, A. Ali, M. Zaman, M. Akhtaruzzaman, and N. Amin. 2017. "Fabrication of High Efficiency Sputtered Cds: O/cdte Thin Film Solar Cells from Window/absorber Layer Growth Optimization in Magnetron Sputtering." *Solar Energy Materials and Solar Cells* 172: 384–393.

Jackson, P., Hariskos, D., Wuerz, R., Kiowski, O., Bauer, A., Friedlmeier, T.M. and Powalla, M. 2015. "Properties of cu (in, ga) se2 Solar Cells with New Record Efficiencies up to 21.7%." *physica status solidi (RRL)–Rapid Research Letters* 9, no. 1: 28–31.

Kabir, M.I., S.A. Shahahmadi, V. Lim, S. Zaidi, K. Sopian and N. Amin. 2012. "Amorphous Silicon Single-Junction Thin-film Solar Cell Exceeding 10% Efficiency by Design Optimization." *International Journal of Photoenergy* 2012. https://doi.org/10.1155/2012/460919

Kasap, S. and Capper, P. 2017. *Springer Handbook of Electronic and Photonic Materials*. New York: Springer.

Kim, H.-S., S.H. Im and N.-G. Park. 2014. "Organolead Halide perovskite: New Horizons in Solar Cell Research." *The Journal of Physical Chemistry C* 118, no. 11: 5615–5625.

Kim, H.-S., A. Hagfeldt and N.-G. Park. 2019. "Morphological and Compositional Progress in Halide perovskite Solar Cells." *Chemical Communications* 55, no. 9: 1192–1200.

Kim, H.-S., C.-R. Lee, J.-H. Im, K.-B. Lee, T. Moehl, A. Marchioro, S.-J. Moon, et al. 2012. "Lead Iodide perovskite Sensitized All Solid-State Submicron Thin Film Mesoscopic Solar Cell with Efficiency Exceeding 9%." *Scientific Reports* 2, no. 1: 1–7.

Kojima, A., K. Teshima, Y. Shirai and T. Miyasaka. 2009. "Organometal Halide perovskites as Visible-Light Sensitizers for Photovoltaic Cells." *Journal of the American Chemical Society* 131, no. 17: 6050–6051.

Kong, J., S., Song, M. Yoo, G. Lee, O. Kwon, J. Park, H. Back, et al. 2014. "Long-Term Stable Polymer Solar Cells with Significantly Reduced Burn-in Loss." *Nature Communications* 5, no. 1: 1–8.

Krajangsang, T., S. Inthisang, J. Sritharathikhun, A. Hongsingthong, A. Limmanee, S. Kittisontirak, P. Chinnavornrungsee, R. Phatthanakun and K. Sriprapha. 2017. "An Intrinsic Amorphous Silicon Oxide and Amorphous Silicon Stack Passivation Layer for Crystalline Silicon Heterojunction Solar Cells." *Thin Solid Films* 628: 107–111.

Landerer, D., D. Bahro, H. R"ohm, M. Koppitz, A. Mertens, F. Manger, F. Denk, et al. 2017. "Solar Glasses: A Case Study on Semitransparent Organic Solar Cells for Self-powered, Smart, Wearable Devices." *Energy Technology* 5, no.11: 1936–1945.

Lee, M.M., J. Teuscher, T. Miyasaka, T.N. Murakami and H.J. Snaith. 2012. "Efficient Hybrid Solar Cells Based on Meso-superstructured Organometal Halide perovskites." *Science* 338, no. 6107: 643–647.

Lee, J.K., W.L. Ma, C.J. Brabec, J. Yuen, J.S. Moon, J.Y. Kim, Lee K., G.C. Bazan and A.J. Heeger. 2008. "Processing Additives for Improved Efficiency from Bulk Heterojunction Solar Cells." *Journal of the American Chemical Society* 130, no. 11: 3619–3623.

Leijtens, T., S.D. Stranks, G.E. Eperon, R. Lindblad, E.M. Johansson, I.J. McPherson, Rensmo H., et al.2014. "Electronic Properties of Mesosuperstructured and Planar Organometal Halide perovskite Films: Charge Trapping, Photodoping, and Carrier Mobility." *ACS Nano* 8, no. 7: 7147–7155.

Liang, Y., Y. Wu, D. Feng, S.-T. Tsai, H.-J. Son, G. Li and L. Yu. 2009. "Development of New Semiconducting Polymers for High Performance Solar Cells." *Journal of the American Chemical Society* 131, no. 1: 56–57.

Liang, Y., Z. Xu, J. Xia, S.-T. Tsai, Y. Wu, G. Li, C. Ray and L. Yu. 2010. "For the Bright Futurebulk Heterojunction Polymer Solar Cells with Power Conversion Efficiency of 7.4%." *Advanced materials* 22, no. 20: E135–E138.

Lindqvist, C., A. Sanz-Velasco, E. Wang, O. B"acke, S. Gustafsson, E. Olsson, M.R. Andersson and C. Müller. 2013. "Nucleation-Limited Fullerene Crystallisation in a Polymer–Fullerene Bulk-Heterojunction Blend." *Journal of Materials Chemistry A* 1, no. 24: 7174–7180.

Liu, M., M.B. Johnston and H.J. Snaith. 2013. "Efficient Planar Heterojunction perovskite Solar Cells By Vapour Deposition." *Nature* 501, no. 7467: 395–398.

Liu, Y., J. Zhao, Z. Li, C. Mu, W. Ma, H. Hu, K. Jiang, et al. 2014. "Aggregation and Morphology Control Enables Multiple Cases of High-efficiency Polymer Solar Cells." *Nature communications* 5, no. 1: 1–8.

Liu, Q., Y. Jiang, K. Jin, J. Qin, J. Xu, W. Li, J. Xiong, et al. 2020. "18% Efficiency Organic Solar Cells." *Science Bulletin* 65, no. 4: 272–275.

Lorenz, A., J. John, B. Vermang and J. Poortmans. 2010. "Influence of Surface Conditioning and Passivation Schemes on the Internal Rear Reflectance of Bulk Silicon Solar Cells." In Proceedings of the 25th European Photovoltaic Solar Energy Conference and Exhibition-EPVSEC, pp. 2059–2061.

Luceño-Sánchez, J.A., A.M. Díez-Pascual and R. Peña Capilla. 2019. "Materials for Photovoltaics: State of Art and Recent Developments." *International Journal of Molecular Sciences* 20, no. 4: 976.

Ma, B., R. Gao, L. Wang, F. Luo, C. Zhan, J. Li and Y. Qiu. 2009. "Alternating Assembly Structure of the Same Dye and Modification Material in Quasi-solid State Dye-sensitized Solar Cell." *Journal of Photochemistry and Photobiology A: Chemistry* 202, no. 1: 33–38.

Manceau, M., E. Bundgaard, J.E. Carl´e, O. Hagemann, M. Helgesen, R. Søndergaard, M. Jørgensen and F.C. Krebs. 2011. "Photochemical Stability of *p*-Conjugated Polymers for Polymer Solar Cells: A Rule of Thumb." *Journal of Materials Chemistry* 21, no. 12: 4132–4141.

Matsui, T., K. Maejima, A. Bidiville, H. Sai, T. Koida, T. Suezaki, M. Matsumoto, et al. 2015. "High-Efficiency Thin-Film Silicon Solar Cells Realized by Integrating Stable a-si: H Absorbers into Improved Device Design." *Japanese Journal of Applied Physics* 54, no. 8S1: 08KB10.

Meng, L., J. You and Y. Yang. 2018. "Addressing the Stability Issue of perovskite Solar Cells for Commercial Applications." *Nature communications* 9, no. 1: 1–4.

Meng, L., Y. Zhang, X. Wan, C. Li, X. Zhang, Y. Wang, X. Ke, et al. 2018. "Organic and Solution-Processed Tandem Solar Cells with 17.3% Efficiency." *Science* 361, no. 6407: 1094–1098.

Metzger, W.K., S. Grover, D. Lu, E. Colegrove, J. Moseley, C. Perkins, X. Li, et al. 2019. "Exceeding 20% Efficiency with In Situ Group v Doping in Polycrystalline cdte Solar Cells." *Nature Energy* 4, no. 10: 837–845.

Min, B., H. Wagner, M¨uller M., H. Neuhaus, R. Brendel and P. Altermatt. 2015. "Incremental Efficiency Improvements of Mass-Produced Perc Cells up to 24%, Predicted Solely with Continuous Development of Existing Technologies and Wafer Materials." In 31st European Photovoltaic Solar Energy Conference and Exhibition, pp. 473–476.

Mola, G.T. and N. Abera. 2014. "Correlation between Lumo Offset of Donor/Acceptor Molecules to an Open Circuit Voltage in Bulk Heterojunction Solar Cell." *Physica B: Condensed Matter* 445: 56–59.

Nakamura, M., K. Yamaguchi, Y. Kimoto, Y. Yasaki, T. Kato and H. Sugimoto. 2019. "Cd-Free cu (in, ga) (se, s) 2 Thin-Film Solar Cell with Record Efficiency of 23.35%." *IEEE Journal of Photovoltaics* 9, no. 6: 1863–1867.

Niki, S., M. Contreras, I. Repins, M. Powalla, K. Kushiya, S. Ishizuka and K. Matsubara. 2010. "Cigs Absorbers and Processes." *Progress in Photovoltaics: Research and Applications* 18, no. 6: 453–466.

Osaka, I., M. Shimawaki, H. Mori, I. Doi, E. Miyazaki, T. Koganezawa and K. Takimiya. 2012. "Synthesis, Characterization, and Transistor and Solar Cell Applications of a Naphthobisthiadiazole-Based Semiconducting Polymer." *Journal of the American Chemical Society* 134, no. 7: 3498–3507.

Padinger, F., R.S. Rittberger and N.S. Sariciftci. 2003. "Effects of Postproduction Treatment on Plastic Solar Cells." *Advanced Functional Materials* 13, no. 1: 85–88.

Papadopoulos, P., G. Floudas, C. Chi and G. Wegner. 2004. "Molecular Dynamics of Oligofluorenes: A Dielectric Spectroscopy Investigation." *The Journal of Chemical Physics* 120, no. 5: 2368–2374.

Parisi, A., R. Pernice, V. Rocca, L. Curcio, S. Stivala, A.C. Cino, G. Cipriani, et al. 2015. "Graded Carrier Concentration Absorber Profile for High Efficiency Cigs Solar Cells." *International Journal of Photoenergy* 2015, 410549.

Park, J.K., J. Jo, J.H. Seo, J.S. Moon, Y.D. Park, K. Lee, A.J. Heeger and G.C. Bazan. 2011. "End-capping Effect of a Narrow Bandgap Conjugated Polymer on Bulk Heterojunction Solar Cells." *Advanced Materials* 23, no. 21: 2430–2435.

Peters, C.H., I. Sachs-Quintana, W.R. Mateker, T. Heumueller, J. Rivnay, R. Noriega, Z.M. Beiley, et al. 2012. "The Mechanism of Burn-in Loss in a High Efficiency Polymer Solar Cell." *Advanced Materials* 24, no. 5: 663–668.

Petrova-Koch, V. 2009. "Milestones of Solar Conversion and Photovoltaics." In *High-Efficient Low-Cost Photovoltaics*, 1–5. Springer, Cham.

Poglitsch, A. and D. Weber. 1987. "Dynamic Disorder in Methylammoniumtrihalogenoplumbates (ii) Observed by Millimeter-Wave Spectroscopy." *The Journal of Chemical Physics* 87, no. 11: 6373–6378.

Qiu, N., H. Zhang, X. Wan, C. Li, X. Ke, H. Feng, B. Kan, et al. 2017. "A New Nonfullerene Electron Acceptor with a Ladder Type Backbone for High-performance Organic Solar Cells." *Advanced Materials* 29, no. 6: 1604964.

R¨uhle, S. 2016. "Tabulated Values of the Shockley–Queisser Limit for Junction Solar Cells." *Solar Energy* 130: 139–147.

Ramanujam, J. and U.P. Singh. 2017. "Copper Indium Gallium Selenide Based Solar Cells–A Review." *Energy & Environmental Science* 10, no. 6: 1306–1319.

Rathore, N., N.L. Panwar, F. Yettou and A. Gama. 2019. "A Comprehensive Review of Different Types of Solar Photovoltaic Cells and Their Applications." *International Journal of Ambient Energy*, 12, no. 10: 1200–1217.

Raut, K.H., H.N. Chopde and D.W. Deshmukh. 2018. "A Review on Comparative Studies of Diverse Generation in Solar Cell." *International Journal of Electrical Engineering and Ethics* 1: 1–9.

Reddy, P.J. 2012. *Solar Power Generation: Technology, New Concepts & Policy.* Taylor and Francis, UK: CRC Press.

Romeo, N., A. Bosio, V. Canevari and A. Podesta. 2004. "Recent Progress on CdTe/CdS Thin Film Solar Cells." *Solar Energy* 77, no. 6: 795–801.

Sachs-Quintana, I., T. Heum¨ uller, W.R. Mateker, D.E. Orozco, R. Cheacharoen, S. Sweetnam, C.J. Brabec and M.D. McGehee. 2014. "Electron Barrier Formation at the Organic-back Contact Interface Is the First Step in Thermal Degradation of Polymer Solar Cells." *Advanced Functional Materials* 24, no. 25: 3978–3985.

Schafferhans, J., A. Baumann, A. Wagenpfahl, C. Deibel and V. Dyakonov. 2010. "Oxygen Doping of p3ht: Pcbm Blends: Influence on Trap States, Charge Carrier Mobility and Solar Cell Performance." *Organic Electronics* 11, no. 10: 1693–1700.

Schwartz, R. and M. Lammert. 1975. "Silicon Solar Cells for High Concentration Applications." In 1975 International Electron Devices Meeting, pp. 350–352. IEEE.

Seemann, A., T. Sauermann, C. Lungenschmied, O. Armbruster, S. Bauer, H.-J. Egelhaaf and J. Hauch. 2011. "Reversible and Irreversible Degradation of Organic Solar Cell Performance by Oxygen." *Solar Energy* 85, no. 6: 1238–1249.

Servaites, J.D., M.A. Ratner and T.J. Marks 2011. "Organic Solar Cells: A New Look at Traditional Models." *Energy & Environmental Science* 4, no. 11: 4410–4422.

Sharma, A., G. Andersson and D.A. Lewis. 2011. "Role of Humidity on Indium and Tin Migration in Organic Photovoltaic Devices." *Physical Chemistry Chemical Physics* 13(10): 4381–4387.

Sharma, S., K.K. Jain, A. Sharma. 2015. "Solar Cells: In Research and Applications Review." *Materials Sciences and Applications* 6, no. 12: 1145.

Smith, D.D., G. Reich, M. Baldrias, M. Reich, N. Boitnott and G. Bunea. 2016. "Silicon Solar Cells with Total Area Efficiency above 25%." In 2016 IEEE 43rd Photovoltaic Specialists Conference (PVSC), pp. 3351–3355. IEEE.

Tan, H., A. Jain, O. Voznyy, X. Lan, F.P.G. De Arquer, J.Z. Fan, R. Quintero-Bermudez, et al. 2017. "Efficient and Stable Solution-processed Planar perovskite Solar Cells via Contact Passivation." *Science* 355, no. 6326: 722–726.

Tegegne, N.A., Z. Abdissa, W. Mammo, M.R. Andersson, D. Schlettwein and H. Schwo erer. 2018. "Ultrafast Excited State Dynamics of a Bithiophene-Isoindigo Copolymer Obtained by Direct Arylation Polycondensation and Its Application in Indium Tin Oxide-Free Solar Cells." *Journal of Polymer Science Part B: Polymer Physics* 56, no. 21: 1475–1483.

Tegegne, N.A., Z. Abdissa, W. Mammo, T. Uchiyama, Y. Okada-Shudo, F. Galeotti, W. Porzio, et al. 2020. "Effect of Alkyl Side Chain Length on Intra-and Intermolecular Interactions of Terthiophene–isoindigo Copolymers." *The Journal of Physical Chemistry C* 124, no. 18: 9644–9655.

Tegegne, N.A., H. Wendimu, Z. Abdissa, W. Mammo, M.R. Andersson, F.G. Hone, D.M. Andoshee, O. Olaoye and G. Bosman. 2020. "Light-Induced Degradation of a Push–Pull Copolymer for ito-Free Organic Solar Cell Application." *Journal of Materials Science: Materials in Electronics* 31: 21303–21315.

Thiyagu, S., Z. Pei and M.-S. Jhong. 2012. "Amorphous Silicon Nanocone Array Solar Cell." *Nanoscale research letters* 7, no. 1: 1–6.

Ullal, H.S. 2008. *Overview and Challenges of Thin Film Solar Electric Technologies.* Tech. Rep., National Renewable Energy Lab.(NREL), Golden, CO (United States).

van Duren, J.K., X. Yang, J. Loos, C.W. Bulle-Lieuwma, A.B. Sieval, J.C. Hummelen and R.A. Janssen. 2004. "Relating the Morphology of Poly (p-phenylene vinylene)/ ethanofullerene Blends to Solar-Cell Performance." *Advanced Functional Materials* 14, no. 5: 425–434.

Wang, A. 1992. "High Efficiency PERC and PERL Silicon Solar Cells." Ph.D. thesis, University of New South Wales.

Welch, G.C. and G.C. Bazan. 2011. "Lewis Acid Adducts of Narrow Band Gap Conjugated Polymers." *Journal of the American Chemical Society* 133, no. 12: 4632–4644.

Wilson, G., M.M. Al-Jassim, W. Metzger, S.W. Glunz, P. Verlinden, X. Gang, L. Mansfield, et al. 2020. "The 2020 Photovoltaic Technologies Roadmap." *Journal of Physics D: Applied Physics*, 52, no. 49: 493001.

Wu, F.-C., S.-W. Hsu, H.-L. Cheng, W.-Y. Chou and F.-C. Tang. 2013. "Effects of Soft Insulating Polymer Doping on the Photovoltaic Properties of Polymer–Fullerene Blend Solar Cells." *The Journal of Physical Chemistry C* 117, no. 17: 8691–8696.

Wu, X., J. Keane, R. Dhere, C. DeHart, A. Albin, A. Duda, T. Gessert, et al. 2001. "16.5%-Efficient CdS/CdTe Polycrystalline Thin-Film Solar Cell." In Proceedings of the 17th European Photovoltaic Solar Energy Conference, vol. 995. James & James Ltd.: London.

Wudl, F., M. Kobayashi and A. Heeger. 1984. "Poly (isothianaphthene)." *The Journal of Organic Chemistry* 49, no. 18: 3382–3384.

Xiong, J., K. Jin, Y. Jiang, J. Qin, T. Wang, J. Liu, Q. Liu, et al. 2019. "Thiolactone Copolymer Donor Gifts Organic Solar Cells a 16.72% Efficiency." *Science Bulletin* 64, no. 21: 1573–1576.

Yang, J., A. Banerjee, T. Glatfelter, S. Sugiyama and S. Guha. 1997. "Recent Progress in Amorphous Silicon Alloy Leading to 13% Stable Cell Efficiency." In Conference Record of the Twenty Sixth IEEE Photovoltaic Specialists Conference-1997, pp. 563–568. IEEE.

Yang, L., Q. Ye, A. Ebong, W. Song, G. Zhang, J. Wang and Y. Ma. 2011. "High Efficiency Screen Printed Bifacial Solar Cells on Monocrystalline Cz Silicon." *Progress in Photovoltaics: Research and Applications* 19, no. 3: 275–279.

Yang, W.S., J.H. Noh, N.J. Jeon, Y.C. Kim, S. Ryu, J. Seo and S.I. Seok. 2015. "Highperformance Photovoltaic perovskite Layers Fabricated through Intramolecular Exchange." *Science* 348, no. 6240: 1234–1237.

Yao, H., Y. Chen, Y. Qin, R. Yu, Y. Cui, B. Yang, S. Li, K. Zhang and J. Hou. 2016. "Design and Synthesis of a Low Band Gap Small Molecule Acceptor for Efficient Polymer Solar Cells." *Advanced Materials* 28, no. 37: 8283–8287.

Ye, F., W. Deng, W. Guo, R. Liu, D. Chen, Y. Chen, Y. Yang, N. Yuan, J. Ding, Z. Feng et al. 2016. "22.13% Efficient Industrial p-Type Mono Perc Solar Cell." In 2016 IEEE 43rd Photovoltaic Specialists Conference (PVSC), pp. 3360–3365. IEEE.

Ying, L., B.B. Hsu, H. Zhan, G.C. Welch, P. Zalar, L.A. Perez, E.J. Kramer, et al. 2011. "Regioregular Pyridal [2, 1, 3] Thiadiazole *p*-Conjugated Copolymers." *Journal of the American Chemical Society* 133, no. 46: 18538–18541.

Yu, G., J. Gao, J.C. Hummelen, F. Wudl and A.J. Heeger. 1995. "Polymer Photovoltaic Cells: Enhanced Efficiencies via a Network of Internal Donor-Acceptor Heterojunctions." *Science* 270, no. 5243: 1789–1791.

Yuan, J., Y. Zhang, L., Zhou, G. Zhang, H.-L. Yip, T.-K. Lau, X. Lu 2019. "Single-Junction @@Organic Solar Cell with over 15% Efficiency Using Fused-ring Acceptor with Electron-deficient Core." *Joule* 3, no. 4: 1140–1151.

Zhao, J. and A. Wang 2006. "Rear Emitter n-Type Passivated Emitter, Rear Totally Diffused Silicon Solar Cell Structure." *Applied Physics Letters* 88, no. 24: 242102.

Zhao, W., S. Li, H. Yao, S. Zhang, Y. Zhang, B. Yang and J. Hou. 2017. "Molecular Optimization Enables over 13% Efficiency in Organic Solar Cells." *Journal of the American Chemical Society* 139, no. 21: 7148–7151.

Zheng, Z., O.M. Awartani, B. Gautam, D. Liu, Y. Qin, W. Li, A. Bataller, et al. 2017. "Efficient Charge Transfer and Fine-Tuned Energy Level Alignment in a ThF-Processed Fullerene-Free Organic Solar Cell with 11.3% Efficiency." *Advanced Materials* 29, no. 5: 1604241.

Zuo, L., S. Zhang, H. Li and H. Chen. 2015. "Toward Highly Efficient Large-area ito-Free Organic Solar Cells with a Conductance-gradient Transparent Electrode." *Advanced Materials* 27, no. 43: 6983–6989.

24

Stannate Materials for Solar Energy Applications

Calister N. Eze[1], Raphael M. Obodo[2,3,4], Fabian. I. Ezema[2,5,6], and Mesfin A. Kebede[7,8]
[1]*Department of Physics, Federal University of Technology, Minna, Niger State, Nigeria*
[2]*Department of Physics and Astronomy, University of Nigeria, Nsukka, Enugu State, Nigeria*
[3]*National Center for Physics, Islamabad, Pakistan*
[4]*NPU-NCP Joint International Research Center on Advanced Nanomaterials and Defects Engineering, Northwestern Polytechnical University, Xi'an, China*
[5]*Nanosciences African Network (NANOAFNET) iThemba LABS-National Research Foundation, Somerset West, Western Cape Province, South Africa*
[6]*UNESCO-UNISA Africa Chair in Nanosciences/Nanotechnology, College of Graduate Studies, University of South Africa (UNISA), Muckleneuk Ridge, Pretoria, South Africa*
[7]*Energy Centre, Council for Scientific & Industrial Research, Pretoria, South Africa*
[8]*Department of Physics, Sefako Makgatho Health Science University, Medunsa, South Africa*

24.1 Introduction

The amount of energy used up by humanity is greatly rising. Increase in population and advancement in human progress determine this rise across the globe. The global energy request hinges on earth natural resources, which are inadequate. Research has shown that of the 7 billion people on earth, about 2 billion people live without electricity, and about 2.5 billion use wood, charcoal, dung, etc., as major sources of energy. This is chiefly used by non-advanced nations and is not a clean form of energy. It causes pollution, health problems, etc. Fossil fuels generally cause universal heating and climatic changes, which results in increase in normal global temperatures (Gaurav et al. 2012).

It is obvious that global stability and sustainability will be continuously affected by the utilization of fossil fuels. The escalation of global population, speedy technological expansion, and rising energy needs will further complicate this matter. In the existing position of energy crunch and global warming, there is the necessity to build up energy production processes that protract the necessities of sustainability and renewability. There is the need for the vast upgrading in the invention and utilization of renewable energy sources for the provision of adequate energy for people and the safeguarding of our immediate environment.

The adoption of the usage of a cleaner form of energy like electricity is the aim of the backward nations.

No advanced countries should reduce their regular way of life but increase energy production without harming environments and humanity. They should function very efficiently but should hold onto less harmful energy sources. Renewable energy sources, which are environmentally harmless and tie up natural processes, should be encouraged, such are tidal, wave, solar, biomass, geothermal, hydroelectric, hydrogen fusion, etc. In opposition, each of these has a major shortcoming and cannot be relied on currently as a way out of the approaching energy crunch. Amid these stated seemingly unfailing

DOI: 10.1201/9781003145585-24

renewable energy sources, solar power draws greater awareness because it is plenty, clean, and without payments.

Solar energy is a less harmful, renewable, and serviceable energy source, which could be produced from the sun's rays. It requires accurate harvesting process to get high quality and quantity of solar energy needed by the world's populace. This means that solar energy could be technically collected and utilized, but there could still be an improved method of collecting the solar energy for more advanced utilization called *solar energy harvesting*. Achieving energy harvesting success involves the process of doping metal oxide thin films with impurities like carbon derivatives, organic synthetic dyes, etc. This will harness dopant characteristics for optimal performance. The doped materials help in controlling the composition and structure of dopants, which enhance their performance.

Structures of nanomaterials are different from the bulk materials. The changes made when doping them signify some hopeful electronic characteristics, such as enhanced light absorption and speedy light response. Among all forms of nanomaterials, one-dimensional materials have captivated more responsiveness in electronic device function, since they are simple to produce into building blocks and simple to test with two terminals at two ends. Thus, metal oxide nanorods, nanowires, nanoneedles, and nanotubes are deposited by varieties of processes, as they are proper for a big range of functions in biosensors, smart windows, solar cells, supercapacitors, photodetectors, light-emitting diodes, and field effect transistors.

In this chapter, typical consideration is given to stannates like barium stannate ($BaSnO_3$), strontium stannate ($SrSnO_3$), and zinc stannate ($ZnSnO_4$) thin films deposited via solid-state reactions and hydrothermal methods. However, other numerous available methods can be used as listed in Section 24.9. Multication oxides possess liberty to modify their materials' chemical and physical properties by varying their compositions. Doping stannates with appropriate materials like stibium (Sb), lanthanum (La), etc., enhances their applications in diverse areas.

24.2 Solar Energy Harvesting

It has been proved that energy and environment are top among the list of critical troubles of people as well as civilization even in the next 50 years. The tremendous rise in worldwide energy utilization has amplified the exhaustion of fossil fuels. Equally, the burning of fossil fuels results to adverse environmental effects. Mutually, energy and environmental harms prompt worry on the pressing call for seeking substitute renewable and green energy. Renewable energy means any energy supply which can be replaced fast through natural procedure and is practically endless, like sunshine, that is solar energy. Other forms of renewable energy include tidal power, wave power, flowing water that is hydropower, biological processes, biofuel, biomass, and geothermal heat flows (Sukhatme 2008). Up until now, humans have tied these types of renewable energy together. Amid these stated renewable energies, solar power draws greater awareness. It is this need for solar energy as a better substitute to fossil fuel that led to fast expansion of solar industry increasing market share for photovoltaic (PV) modules, which has been estimated to increase even the more between year 2000 and 2015. If solar cell of 10% efficiency can cover 0.1% of the earth's shell, then the world power need can be supplied (Shruti et al. 2015). The figure below explains the above narration by Nobel Laureate Richard Smally (Figure 24.1).

The solar energy needs to be captured, harvested, and utilized by processing it into useful power applications, see Figure 24.2 below. Solar energy collecting is the method of netting and packing solar energy that was emitted from the sun. It will then be changed from light or heat energy to electrical energy by using appropriate technique such as photovoltaic cells (photovoltaic modules): This is habitually said to mean solar panels, and they are chiefly omnipresent solar power's harvesting technology (Wang et al. 2010) as shown in Figure 24.2(b). Solar thermal collectors can be called *concentrating solar power* (CSP). It is however not a well-known method of extracting solar energy.

Solar energy harvesting could be achieved by the process of doping metal oxide thin-film semiconductors with impurities like transition metal oxides, carbon derivatives, synthetic and nonsynthetic dyes, etc. This will present a valuable technique to achieve solar energy harvesting for improved enhancement. This process will require harnessing dopant-induced characteristics and

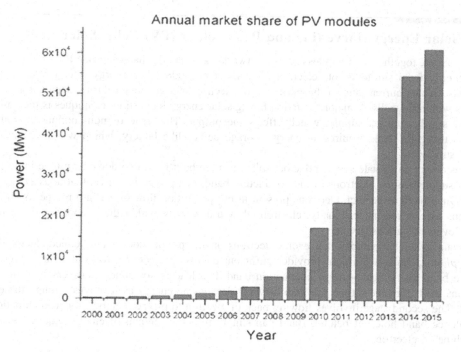

FIGURE 24.1 Increasing market share for PV modules from year 2000 to 2015 (Shruti et al. 2015). Reprinted with permission from Ref. Shruti et al. (2015), copyright 2015, Scientific Research publishing Inc.

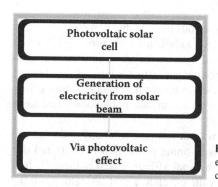

FIGURE 24.2 Production of electricity using photovoltaic cells (Wang et al. 2010). Reprinted with permission from Ref. Wang et al. (2010), copyright 2010, Elsevier.

paying attention to control of composition and structure of dopants (Wang et al. 2012). On the other hand, deposition of doped nanocrystals with specific control over composition and structure is a considerable challenge.

Energy harvesting is helpful because it offers techniques of powering electronics when accustomed power sources are deficient. It moreover stops the requirement for changing batteries regularly and connecting wires to end applications. It is appallingly a considerable advance for valuable energy utilization. This is because solar energy is a superior option for substituting fossil fuel as the core energy provider. It also follows that solar power is renewable with utterly no payments made for its use. These renewable energy sources can be used unswervingly to generate additionally other suitable types of energy. Some examples of unswerving utilization are solar ovens, geothermal heating, windmills, etc. The list of indirect exercise that calls for energy harvesting include electricity production using wind turbines, photovoltaic cells, or fabrication of fuels like ethanol from biomass (Affrin 2018; Patel 2016). Nevertheless, a major shortcoming or more affects every one of the above-mentioned methods. This makes it unreliable as a solution to forthcoming energy crisis.

24.3 Solar Energy Harvesting and Photovoltaic (PV) Cells (Solar Cells)

Most systems, together with wireless sensor networks solar energy harvester use PV cells in changing solar energy from sun to usable electricity (Arms et al. 2008). The energy released from liberated electrons creates current due to the existence of electric field generated during separation of charge carriers within the cells. A major limitation facing solar energy harvesting techniques is the absence of 24-hour steady sunlight supply, which affects the output. The issue of non-continuous sunlight in systems using PV cells requires an energy storage device like battery, which will guarantee steady energy supply.

In customary solid-state semiconductors solar cells are built from two dopants: n-type impurity, that creates supplementary electrons in the conduction band, and p-type impurity, that adds extra electron holes. Once fixed in contact, electrons present in n-type portion flow direct into p-type to fill the lost electrons as electron holes. Finally, electrons flow transversely within the boundaries to balance the Fermi levels of various devices.

The sun generates photons that excite electrons on the p-type side of the semiconductor devices, called *photoexcitation*. Sunlight provides sufficient energy to eject electron out of its lower energy valence band to its higher-energy conduction band. If a load is positioned transversely with the cell, electrons drift from the p-type side into n-type side, loosing energy as it moves through the external circuit. The second stage involves flowing back of electron to p-type material, for recombination with the valence band hole left behind. Band gap simply refers to amount of energy required to eject or recombine an electron.

24.4 Current Technology

Stannate perovskite oxides for solar cells have similarly enticed serious interest due to their exceptional properties. Once a stannate material is irradiated with sunlight electron/hole pairs are produced. Various photogenerated transporters are separated and compelled towards electrodes using polarized and induced inner electric field, producing a photovoltaic effect. Photovoltaic effect witnessed in stannate compounds is a bulky-based effect, which varies from p-n junction-centered semiconductors photovoltaic effect. The internal electric field covers the entire region in a stannate; photovoltaic responses can be creäted without establishing complex junction structures.

The solar cell, otherwise called *photovoltaic (PV) cell*, was developed to change sunlight to electricity via PV effect procedures. In any case, its business in home utilization is restricted due to the expensive and complicated production technique for a usual silicon-based solar cell. It is the bid to cut down this high cost and at the same time retain the efficiency that brought about the production of dye-sensitized solar cells (DSSCs), which are third-generation solar cells, and other types of solar cells. The 15% or more efficiency achieved by DSSC was when perovskite was used as sensitizer.

24.5 Types of Solar Cells

24.5.1 Semiconductor Solar Cells

In semiconductor materials, the mainstream charge carriers are electrons and the Fermi levels lie closer to the conduction band of the semiconductor. A variety of such materials such as TiO_2 (Nowotny et al. 2008; Park et al. 2000; Barbe' et al. 1997; Wong et al. 2006) ZnO, Fe_2O_3, Nb_2O_5, CeO_2, Zn_2SnO_4, $SrTiO_3$, and $BaSnO_3$ have been tested as photo anodes in DSCs/DSSCs.

24.5.2 Dye-Sensitized Solar Cells

This is a third-generation solar cell. Its history dates back to 1972 when ZnO (zinc oxide) photo-anode sensitized with chlorophyll was used. It was discovered that the cell effectiveness was small because of inadequate surface area available for light harvesting. It was in 1991 that nanostructured titanium dioxide (TiO_2) film was covered with a monolayer of ruthenium dye as photoanode and was achieving self-efficiency of 7.1–7.9%. The nanostructured TiO_2 film extensively improved the obtainable surface area by a factor greater than a thousand, which was necessary for competent dye loading. This advance enhanced the light absorption and efficiency of DSSC considerably, which made DSSC to be looked at as a strong competitor to their obtainable solar cell know-how. From there, the efficiency and stability of DSSC have continuously improved out of research. Nazeerudeen et al. reported DSSC efficiency of 10.4%, Chiba et al. reported that of 11.1%, Mathew et al. reported 13%, while 15% obtained is based on perovskite sensitizer. To get a high-efficiency DSSC, it is imperative to comprehend the kinetics of photo-excited electrons in the photo electrochemical cell throughout its operation. It is important to know the performance of photo-excited electrons in DSSC. The photo-anode plays a significant role in governing the collection and transportation of photo-excited electrons. The variations of photo-anode in terms of morphology, doping, and film thickness have important control in the PV performance of DSSCs.

24.5.3 Perovskite Solar Cells (PSCs)

Perovskite solar cells grounded on organometallic halides symbolize an evolving photovoltaic technology. When perovskite sensitizer was used in DSSC, an efficiency of 15% and even more was achieved (Lee et al. 2017).

The crossbreed organic-inorganic metal halide perovskite gives about the best solar cell operation like methylammonium lead iodide ($CH_3NH_3PbI_3$) or methylammonium lead bromide ($CH_3NH_3PbI_3$). The advancement in perovskite solar cells shifted from methylammonium lead halide ($CH_3NH_3PbX_3$) (X = Br, I) sensitized liquid solar cell at 3.8% to over 20% of formamidinium lead iodide iodide ($NH_2CHNH_2PbI_3$) solar cells from 2009 to 2015 (Park 2014). Qin et al. created an apparatus configuration involving titanium dioxide as a scaffold as well as electron collector, lead halide perovskite as the light harvester, and CuSCN as the hold transport material (HTM) with high power conversion efficiency of about 12.4% under full sun brightness.

24.5.4 Spinel Oxide Solar Cells

Numerous spinel oxides with structure of AB_2O_4 arrangements are seen as light absorbers for all-oxide thin film photovoltaic cells due to their almost perfect optical band gap of about 1.5 eV, that can get to 40% of the theoretical conversion efficiency. Some spinel oxides were seen as alternative materials for photo anodes in thin film photovoltaic cells due to their wide optical band gap of about 3.6 eV. Similarly, heterojunctions of spinel oxides for inorganic solar cells like TiO_2/Co_3O_4 where Co_3O_4 is a p-type semiconductor, which has two obvious direct transitions in the visible range with band gaps energy of about 1.45 eV and 2.26 eV.

All-oxide photovoltaics could be the next generation of photovoltaic cell.

24.6 Crystal Structures of Spinels and Perovskites Stannates

24.6.1 Crystal Structures: Spinel

Here we consider zinc stannate spinel structure belonging to space group $Pm\bar{3}m$. The spinel structure is formulated as $MM^1_2X_4$, where M and M^1 are tetrahedral and octahedral. The structure is named after the mineral $MgAlO_4$. The compounds from this family can be represented by a different general formula AB_2O_4, where A and B are divalent (+2) and trivalent (+3) or tetravalent (+4) and divalent cations, respectively. The crystal structure could be determined independently (Bragg 1915) (Figures 24.3 and 24.4).

Cubic spinels have been assigned to the space group $Fd\bar{3}m$. the unit cell of a spinel consists of eight formula units with the oxygen ions forming a close-packed face centred structure.

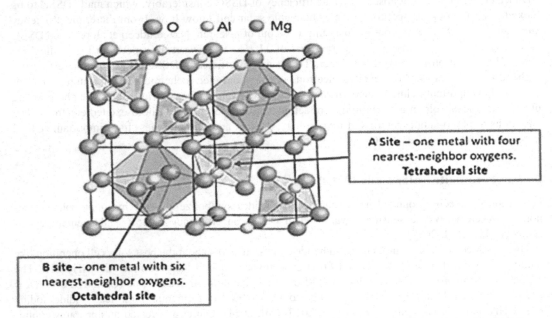

FIGURE 24.3 Schematic of the spinel structure of $MgAl_2O_4$ (Bragg 1915). Reprinted with permission from Ref. Bragg (1915), copyright 1915, Taylor & Francis.

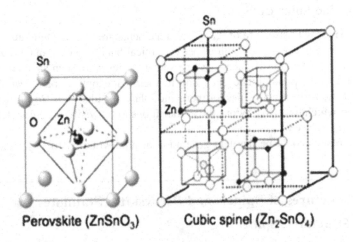

FIGURE 24.4 Crystal structures of zinc stannate: perovskite structure for zinc metastannate ($ZnSnO_3$) and cubic spinnel structure of zinc orthostannate ($ZnSnO_4$). A single atom of each element is labelled in the structural diagrams. Bragg (1915). Reprinted with permission from Ref. Bragg (1915), copyright 1915, Taylor & Francis.

24.6.2 Crystal Structure: Perovskite

A perovskite is any substance with a crystal structure like that of a mineral known as perovskite. This has calcium titanate oxide ($CaTiO_3$) structure. The perfect cubic structure has the B cation in 6 folds coordination surrounded by an octahedron of anions and the A cations in 12 folds cuboctahedral coordination (Yang et al. 2015).

The perovskite structure has simple cubic symmetry, but is related to the fcc lattice in the sense that the A site cations and the three O atoms comprise a fcc lattice. The B-site cations fill one-fourth of the octahedral holes and are surrounded by six oxide anions.

24.6.3 Band Structure

The conduction band structure of cubic $SrSnO_3$ is qualitatively related to that of $BaSnO_3$.

The optoelectronic properties of spinel oxides are directly related to their band structures, as most of transition metal oxides. In general the valence band (VB) is filled with oxygen orbitals (orbitals $2p^6$–full) while the conduction band (CB) consists of cations orbitals (orbital d-empty). Depending on the value of the band gap, the oxide presents a semiconducting or insulating behaviour. The light absorption characteristics of spinel oxides are associated to the charge (electron) transfer in material. There are two key groups of electronic transfers:

In spinel-type oxides, the charge transfers of cation-oxygen (band theory) give the highest absorption intensities. When the energy band gap value of an oxide compound is in the visible range, a transfer of electrons explains the absorption phenomenon from the orbital $2p^6$ of oxygen (VB) to the empty orbital $3d^1$ of metal cation (CB).

Some transition metal oxides with wide band gap may appear transparent, while some with wide band gap appear coloured when doped with an electron acceptor element or electron donor element in the form of impurities. Other absorber properties of transition metal oxides are related to interatomic charge transfer between cations, which is called an *intervalence charge transfer*, representing an internal oxidation-reduction process. Such transfer may be direct (in the case of orbital overlap cations) or indirect (via transfer oxygen). It occurs between two different oxidation states of cations (Liu et al. 2014).

24.7 Doped Stannates

Doping is the deliberate addition of impurities into a material to be able to manage the qualities of the materials. It improves the uniqueness of the material doped (for example, metal oxide nanoparticles/thin films) and enhances its appropriateness for diverse fields. Doping nanomaterials gives a stretchy way to change the qualities of the materials whilst keeping their high surface areas. It is an extensively utilized method for the alteration of nanoparticles to boost their electrical, optical, and biological activities. It has been evidenced that doping may enhance the antimicrobial effect.

It actually depends on the class of nanomaterial that will comprise the medium. The doping process will vary in case of metallic nanoparticles, oxides, or semiconductors (such as chalcogenides). Nevertheless, the common trend is to add the doping precursor at the beginning of deposition, before the nucleation of NPs, in the preferred molar ratio, and a yield of about 100% is predicted. For metals, the precursors are normally metal salts. For oxides, it might be an alcoxide, an organometallic compound, or a complex of the transition metal like acetylacetonate. The mechanism of nucleation+growth for the material is a reduction method (like for metal NPs), a sol gel process (for oxides), or a thermal decomposition (for some chalcogenides) (Rao and Vanaja 2015).

Doping diverse elements helps to change and direct the material towards required relaxation properties like electronic, optical, photochemical, photo electrochemical, photocatalytic, and photo excited properties. The substances can be engineered towards particular usages by cautious choosing of the dopants. Doped nanoparticles give the anticipation that doped materials will find applications in optical storage, radiation detection, infrared detection, and dosimetry (Karunakaramoorthy and Suresh 2014).

The energy structure and physical characteristics of doped nanomaterials plus nano devices are being affected by quantum size confinement. Near-infrared and up conversion luminescence nanoparticles are particularly hopeful for biological imaging because auto-fluorescence can be overcome and higher imaging resolutions can be obtained. Doped insulator nanomaterials as well as carbon nanotubes signify a new dimension. As a novel kind of biological labelling agent, insulator nanoparticles are less poisonous than semiconductor nanoparticles and are skilled for cancer detection, diagnosis, and treatment. Doped nanomaterials are likely to make central assistance to nanotechnology for practical usages in the fields of electronics, photonics, optics, homeland security, and medical sciences.

Doping of metal oxide thin-film semiconductors with transition metals presents amazingly valuable technique to achieve solar energy harvesting for improved enhancement. Doped oxide nanocrystals embrace pledge for a broad range of use if dopant-induced characteristic properties can be suitably harnessed. On the other hand, deposition of doped nanocrystals with specific control over composition and structure poses a considerable challenge. Nanocrystal composition is difficult to manage owing to the contrary reactivity of dopant and host precursors. The multiplicity of dopant atoms, which could be included, is restricted with effectiveness of dopant atom inclusion varying (Zhang et al. 2012).

Moreover, the enlarged electron density resulting from the doped resources enhances the fill factor of the solar cell. The enlarged electron density speeds up the transfer rate of electrons in the doped resources of metal oxide thin films in contrast with undoped films. This could be established using intensity-modulated photocurrent spectroscopy measurements.

The above analogies are applicable to doped stannates.

Stibium (Sb), lanthanum (La), etc., can be used to dope stannates like $BaSnO_3$ (BSO) to a highly conductive status. When doped with La, it is called Barium Lanthanum Stannic Oxide (BLSO), which has tremendous metallic conduction.

24.8 Peculiarities/Properties of the Stannates

The term *stannate* means compounds of tin (Sn). They are ternary metal oxide semiconductors.

The stannates under consideration in this chapter are barium stannate (Ba_3SnO), strontium stannate ($SrSnO_3$), and zinc stannate ($ZnSnO_4$). The $BaSnO_3$ and $SrSnO_3$ have perovskite structures while $ZnSnO_4$ has spinel structure. The common formula for the stannate-based perovskite type is $ASnO_3$ while that for spinel type is $ASnO_4$ (Raghavan et al. 2016).

Alkaline earth metal is any of the six chemical elements that comprise Group 2 (IIa) of the periodic table. The elements are beryllium (Be), magnesium (Mg), calcium (Ca), strontium (Sr), barium (Ba), and radium (Ra). Zinc stannate (ZSO) which is of spinel nature equally belongs to alkaline earth stannate class like the perovskite-natured stannates. It is not a perovskite type but can be an alternative to perovskite substance. It has a non-cubic structure and belongs to R3c space class. Its lattice constant is $\alpha = 7.758$ Å. Barium stannate (BSO) is among the $Pm\bar{3}m$ space group. It has $\alpha = 4.139$ Å as its lattice constant. Equally, strontium stannate (SSO) is a member of the Pbnm space class. It has average size of 45 nm. It has lattice constant $\alpha = 5.709$ Å, which is pseudocubic.

The alkali earth stannates ($MSnO_3$ or ABO_3 where M or A = Sr, Ba, etc.) are distinctive for their good stability at high temperatures and outstanding dielectric and sensing properties and are widely used in capacitors, moisture sensing, solar cells, etc.

The term *stannate* is also used in naming conventions, as a suffix, for example, the hexachlorostannate ion is $SnCl_6^{2-}$.

One of the substances belongs to the set of alkaline earth stannates where A site is to be occupied by either Ba or Sr with the ionic radius of 135 pm and 118 pm. Even though they have wide optical band gaps, the electron conductivity is outstandingly better. In electronic industries, they are equally largely applied. This is as a result of their unbelievable dielectric and gas sensing characteristics. Alkaline earth stannates are equally applied in conduction, and as a result are drawing attention in the field of technology. In production of transparent electrodes involving utilization in photovoltaic cells and organic light-emitting diodes, stannates play a wonderful role. Transparent conductors are materials that have DC electrical conductivity but are transparent to light over a wavelength range of interest making them essential in solar photovoltaics, displays, and other technologies.

The phase purity, crystallization idiosyncrasies, and microstructural evolution, etc., of derived alkaline earth metal stannate powders could be studied by XRD and SEM measurements, etc. The photovoltaic performance of $SrSnO_3$ nanoparticles is used as electrode material in DSSCs. Substances that have perovskite structure have shown excessive fascinating physical characteristics like great photovoltaic effects, superconductivity, etc.

24.8.1 Barium Stannate (Barium Stannic Oxide), Barium Tin Oxide $BaSnO_3$ (or BSO)

The barium stannate is an n-type substance, which has cubic perovskite structure with band gap of about 3.4 eV. It maintains stability even at 1000°C (John et al. 2017). It is among the Pm$\bar{3}$m space group. It has α = 4.139 Å as its lattice constant. Its applications are numerous, they are used as thermally stable capacitors, humidity sensors, gas sensors, etc. it is also used in optical applications, capacitors, ceramic boundary layers, and equally a promising material to produce gas phase sensors for the detection of carbon monoxide and carbon dioxide (Li et al. 2017). The figure below shows the cubic configuration of barium stannate of Pm$\bar{3}$m class (Figure 24.5).

Stibium (Sb) and Lanthanum and other dopants can be used to dope BSO to a highly conductive status. When doped with La, it is called barium lanthanum stannic oxide (BLSO), which has tremendous metallic conduction as depicted in Table 24.1 below.

Apart from working better due to high mobility, BLSO equally has good stable state for sustenance at elevated thermal conduction (Lee et al. 2017). However, BSO in its epitaxial form has lesser mobility, which actually is contradicting to the very high mobility of BLSO. This is as a result of grain boundaries and dislocation that is causing charge traps and scattering and consequently bringing down the carrier density and mobility at the same time.

When lattice-matched substrate is selected, it will help to lessen dislocation as well as grain boundaries thereby improving dislocation in thin films (Zang et al. 2017).

Barium stannate, when prepared by solid-state reaction, served as starting material for further heat treatment. Suitable quantities of $BaCO_3$ as well as dried SnO_2 mutually of purity greater than 99.5 %

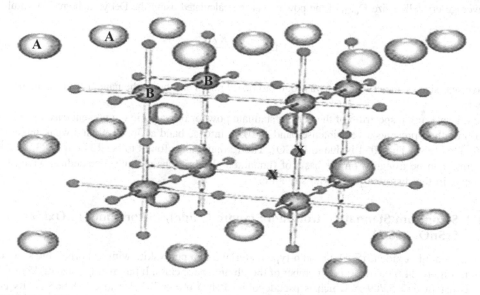

FIGURE 24.5 ABX_3 perovskite structure. Here, A, B, and X are big, small, and smallest, respectively (Li et al. 2017). Reprinted with permission from Ref. Li et al. 2017, copyright 2017, Royal Society of Chemistry.

TABLE 24.1

The BSO Performance

Dopant	Band Gap (eV)	Resistivity (Ω/cm)	Conductivity (S/cm)	Mobility ($cm^2V^{-1}s^{-1}$)	Trans (%)	Deposition Method
Lanthanum	4.05	5.9×10^{-4}	1695	103	–	Solid state reaction
Lanthanum	–	10×10^{-3}	100	320	–	Single crystal
Lanthanum	–	–	–	70	–	Pulse layer depositn (epitaxial films)
Lanthanum	–	–	–	150	–	Modified MBE
Lanthanum	3.95	–	–	–	>80	Sol-gel

were homogenized in agate motor for 6 hrs as the mixing medium. The homogeneous mixture was heated in an alumina crucible up to temperature 1000 °C. It was maintained at that temperature for about 8 hrs. There was addition of polyvinyl alcohol into the powder, which was calcined for the synthesis of pellets. This was burnt out via elevated temperature sintering. Five tons were utilized for the synthesis of pellets of circular disc shape. The pellets were ground and sintered in a platinum crucible up to a very high temperature of about 1250 °C for 6 hours.

The structural, morphological, and optical properties of the materials could be examined extensively. The crystalline quality and crystallographic orientation could be studied by X-ray diffraction analysis using wavelength of 1.54060 Å in 2θ range of 10– 100°. The vibrational spectrum could be carried out using micro-Raman spectrometer applying a laser radiation of wavelength 514 nm from an argon ion laser. The spectrum could be carried out using spectral resolution of 1 cm^{-1}. The morphology and elemental analysis of the compound could be studied using Energy Dispersive X-ray spectrometer. Both absorbance and reflectance spectra of the material in the spectral range of 200–900 nm could be carried out using UV-visible double beam spectrometer. The TEM of the image could be studied using electron microscope working at 200 KeV.

The XRD pattern of BaSnO$_3$ powder was prepared by solid-state reaction method, and it presents a polycrystalline nature. X-ray diffraction pattern of the prepared sample presents all the characteristic peaks of cubic phase of BaSnO$_3$. The inter-planar distance "d" of the powder is calculated using Bragg's relation 2dsinθ = nλ where λ is the wavelength of the X-ray radiation and θ is the diffraction angle. The lattice constant for the compound was calculated and found to be 4.101 Å, which is in agreement with the reported value (4.112Å). The average crystallite size D$_{(hkl)}$ of the powder can be calculated using the Debye Scherrer's formula.

$$D_{(hkl)} = \frac{K\lambda}{\beta cos\theta} \tag{24.1}$$

The average size of the crystallites was found to be 49 nm. This shows the nanocrystalline nature of the powder.

The micro-Raman spectrum of the barium stannate powder prepared in the present case by solid-state reaction method presents a very intense band at 1054, intense band at 560, 139, and weak intense band at 686, 336, 1130. The 1054 is due to BaCO$_3$. The Raman bands found to be 139 cm^{-1}, 833 cm^{-1}, and 1130 cm^{-1} can be assigned on the basis of fundamental vibrations of SnO$_6$ octahedron which has O$_h$ symmetry, in the distorted perovskite structure.

24.8.2 Strontium Stannate (Strontium Stannic Oxide), (Strontium Tin Oxide), SrSnO$_3$ (SSO)

Strontium stannic oxide (SrSnO$_3$) is an n-type material and a perovskite with orthorhombic structure as shown in Figure 24.6 below. It is a member of the Pb nm space class. It has average size of 45 nm. It has lattice constant α = 5.709 Å which is pseudocubic. It has b = 5.703 Å and c = 8.065 Å. Its energy band gap is large, about 4.27 eV. The band gap transition is indirect (m = 2). Band gap energy of semiconductor nanoparticles increases with a decrease in the grain size.

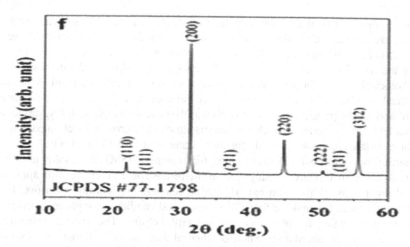

FIGURE 24.6 XRD pattern of the $SrSnO_3$ (Ong et al. 2015). Reprinted with permission from Ref. Ong et al. (2015), copyright 2015, AIP publishing LLC.

It is synthesized using solid-state reaction, chemical bath, simple wet chemical reaction, electro-chemical methods, etc. In our work, solid-state reaction technique is considered.

Commonly, $SrSnO_3$ powder (for industrial application) is synthesized using solid-state reaction. This technique is simple and cheap. Strontium carbonate and titanium dioxde or strontium carbonate ($SrCO_3$) and tin oxide (SnO_2) which are of high purity, up to 99% are used as starting marterials. Over 900°C calcination temperature and a long time is needed in this method. This however has many disadvantages including big particle size with big size distribution and big degree of particle agglomeration. When they are mixed stoichiometrically by ball milling with zirconia in ethanol for 24 h and later calcined at 1100°C for 6 h in air, the process produces a uniform, facile, small-sized $SrSnO_3$ powder. The starting materials could be strontium carbonate ($SrCO_3$) and tin oxide (SnO_2), which are of high purity, up to 99% as stated above.

The $SrSnO_3$ could be doped with impurities such as La, Ne, etc. As doping increases, the distortion in the orthorhombic pattern of the perovskite configuration in SSO results in absorption in the visible. Smaller lattice parameters are better well-matched with ordinary oxide electronic substrates. The mobility performance of SSO is lower than that of BSO perovskite. Liu et al. were able to get a mobility of $18.5 \ cm^2V^{-1}s^{-1}$ when they introduced tantalum (Ta) as dopant. This can be seen in Table 24.2 below. The potential barrier is reduced when the carrier concentration and ionization efficiency become higher as a result of raising the Ta content. Due to that, mobility is tremendously enhanced by half a 100 folds compared with the previous achievements.

The use of $SrSnO_3$ is visible in the area of photocatalysis. This process has helped in solving the trouble of environmental pollution and energy collapse by assisting in disintegration of contaminants and water

TABLE 24.2

SSO Performance

Dopant	Band Gap (eV)	Resistivity (Ω/cm)	Conductivity (S/cm)	Mobility ($cm^2V^{-1}s^{-1}$)	Trans (%)	Synthesis Method
Erbium	–	$0/1 \times 10^{-3}$	1×10^6	–	–	Sol gel
Stibium	4.53	23×10^{-3}	43.48	0.329	>90	Pulsed laser deposition
Neodymium	–	21×10^{-3}	47.62	0.104	>90	Pulsed laser deposition
Tantalum	4.63	3.33×10^{-3}	300	18.5	>90	Pulsed laser deposition
Ferum	–	400×10^9	2.5×10^{-12}	–		Solid-state synthesis
	4.23	–	–	–	50–90	Pulsed laser deposition
Chromium	3.8	–	–	–	–	Chemical precipitation

splitting. Usually materials used as photocatalysts need to have elevated optical absorption properties, elevated surface area, elevated crystallinity, elevated chemical stability, etc. Considering all these properties, high surface area is the more important feature upsetting photocatalytic activity. This problem can be solved by controlling the size, shape, and morphology of the photocatalyst (Chen and Umezawa 2014).

Usually, photocatalytic reaction has three procedures namely, optical absorption, migration of charge carriers, and catalytic reaction at the crystalline surface structure. Consequently, there are many important features for influencing the photocatalytic activity such as the optical absorption ability, surface area, particle size, and crystallinity. The generations of charge carriers are greatly improved by elevated optical absorption capability. Huge surface area makes it likely for more water molecules to be adsorbed.

Alkaline earth stannates of perovskite structure, for example, $SrSnO_3$ are greatly used in energy and environment. Apart from this area of application, there are some other areas of application like in dye-sensitized solar cells, lithium-ion batteries, high temperature humidity sensors. It is this high photocatalytic activity that is making $SrSnO_3$ to be a good candidate for photocatalysis. The spatial structure of $SrSnO_3$ makes its photocatalytic activity helpful. The spatial structure has three-dimensional network of corner-sharing SnO_6 octahedral that can help charge carriers to move more easily, and the octahedral tilting distortion has an optimistic outcome on local charge separation.

The photovoltaic performance of $SrSnO_3$ nanoparticles is used as electrode material in DSSCs (Ong et al. 2015).

In characterizing $SrSnO_3$, there was discovery of the crystal structure of the synthesized powder by using XRD shown in Figure 24.7a. Likewise, the study of microstructures and morphologies was done

FIGURE 24.7   (a) FESEM, (b) TEM, (c) SAED, and (d and e) HRTEM images of SrSnO3 (Ong et al. 2015). Reprinted with permission from Ref. Ong et al. (2015), copyright 2015, AIP publishing LLC.

using field-emission scanning microscopy (FESEM) as shown in Figure 24.7b. The TEM was taken at a highly increasing voltage of 300 Kv as shown in Figure 24.7c. The UV-vis diffuse reflectance spectra were projected via a UV-vis spectrometer. Besides, the use of Raman spectroscopy and X-ray photo-electron spectroscoscopy (XPS) provides the structural idea. The selected area electron diffraction (SAED) pattern shows that multiple particles are put into aperture. Figure 24.7 (d and e) shows this SAED pattern. This is orthorhombic phase of $SrSnO_3$.

24.8.3 Zinc Stannate or Zinc Stannic Oxide (ZSO) or Zinc Tin Oxide (ZTO)

ZSO is of spinel structure belonging to alkaline earth stannate class. It is not a perovskite type but can be an alternative to perovskite substance. It has a non-cubic spinel structure, which belongs to R3c space class. Its lattice constant is $\alpha = 7.758$ Å.

ZTO is an n-type material with an exceptional electronic and optical property. It has spinel non-cubic structure. ZTO has large 3.6 eV and 2 values of band gap energy and refractive index, respectively. Its larger band gap makes it to have better photo stability against UV light than even binary and other stannate components. It is of high-speed charge injection with fast electron flow. It also has soaring electron mobility ($10-15Cm^2V^{-1}S^{-1}$) and finally first-class light-harvesting properties. ZTO metal oxide semiconductor particles are identified for having steady properties under severe situation. They have their use generally in making photo anodes given that they have large surface area and elevated porous structure. This helps in absorbing dye into semiconductor material.

The challenge led to a recent proposal of many new materials, such as n-type cubic perovskite $BaSnO_3$. These materials have number of benefits such as abundance, non-toxicity, environmental friendliness, etc. They have many promising applications in gas sensors, solar cells, DSSCs, photo catalysts, and electrodes for rechargeable lithium-ion batteries (Sharma et al. 2018).

Even though ZSO is of non-cubic structure, it can maintain admirable transparency when doped with an appropriate dopant. This makes it a good material for transparent conductor. Considering it in R3c group, it has inborn ferroelectric properties with elevated dielectric constants. When these are modified, it will increase the mobility when it is doped with dupable/promising dopants.

Researchers have established that ZSO is one of those prospective transparent conductive materials for few decades now. But of late there is generated and regained interest in the understanding of its use in high mobility field-effect transistors (Tan et al. 2007) (Table 24.3).

ZSO is a transparent conducting oxide. It is widely used in DSSC. It is used as electrode material in DSSC due to its stability against acid, making it better than its binary counterparts. To achieve this application in DSSC, there should be a control of the particle size to guarantee large surface area for dye absorption (Tan et al. 2007).

The ZSO nanoparticles are achieved by hydrothermal technique. This is by the decomposition of combination of zinc and tin tert-butylamine complexes. There were five solar cell dyes of the same thickness (4.3 μm) with different sensitization times used in this technique. The adsorption increases fast with sensitization time and later becomes saturated. The short circuit current density (J_{sc}) equally increases fast as sensitization time increases and later becomes saturated. The open circuit voltage (V_{oc}) and fill factor (ff) are stable with diverse sensitization times (Das et al. 2016). This shows that ZSO is stable against acidic dye molecules. All films fabricated from ZSO were transparent. The highest energy conversion efficiency got from a ZSO cell is 3.8% with film thickness of 5.6 μm, which is near to the highest reported efficiency ever achieved of about 4.5%. However, the major setback for ZSO cell is the short electron diffusion length with increased film thickness. A higher dye loading on a ZSO film, the electron injection, as well as transport ought to be poor. It is expected to have low open-circuit voltage.

TABLE 24.3

ZTO Performance

Dopant	Band Gap (eV)	Resistivity (Ω/cm)	Conductivity (S/cm)	Mobility (cm^{2}V^{-1}s^{-1})	Trans (%)	Synthesis Method
Undoped	3.84	–	–	–	–	Wet chemical deposition
Undoped	–	–	–	45	–	Molecular beam epitaxy
Undoped	–	4×10^{-3}	–	–	>80	Magnetic sputtering

24.9 Methods of Synthesis

Stannate compounds could be fabricated via environmentally friendly methods whether or not in monophasic state. The character of alkaline earth metal source on the phase purity of various metal stannates is of immense influence.

Alkaline earth orthostannates (A_2SnO_4) have created great interest in researchers for novel phosphorus resulting in their stable crystalline structure as well as elevated physical and chemical stability. Hence the need to fabricate them.

There are numerous methods used in the fabrication of stannates such as sol gel, high temperature solid-state reaction, sol gel combustion, hydrothermal, polymer precursor, spray pyrolysis, hydrothermal, wet chemical, chemical precipitation and reverse micelle, magnetic sputtering, etc. Out of all these techniques, solid-state reaction fabricating means is used in fabrication of stannate compounds together with hydrothermal method. This is because it is a better and a convenient technique due to the fact that it is simple, has better mixing of starting materials, has relatively little reaction temperature, and simple control of the chemical composition of the end product.

Effective deposition of stannates nanopowders via a chemical or physical route depicts good structure for doping.

The metal oxide semiconductor thin films are synthesized via varieties of processes on working active layer materials with stable interfaces.

24.9.1 Thin Films

Substances that are made to be thin films are easily integrated into different devices. The uniqueness of such substances significantly varies when they are examined as thin films. The bulk of the working substances are rather important in thin film structure due to some particular properties such as electrical, magnetic, and optical characteristics or wear resistance. In the technology of thin film, there is the utilization of the idea that the properties can habitually be checked by the thickness factor. Through deposition processes, thin films are made either by physical or chemical techniques. When thin films are synthesized, they have applications in microelectronic and optoelectronics appliances (these devices belong to the supreme hi-tech movers of our economy). They are equally applied in magnetic thin films in recording devices, creation and maintenance of energy, display and storage applications, magnetic sensors, gas sensors, coating of all kinds, biotechnology, photoconductors, IR detectors, interference filters, solar cells, polarizers, temperature controllers in satellites, superconducting films, communication and information processing, anti-corrosive and decorative coatings. Thin-film practices have a fixed place in material science and engineering.

24.9.2 Metal Oxide Thin Films

Metal oxides are made up of positive metallic and negative oxygen ions (Obodo et al. 2020a). Strong ionic bond is attained when the metallic and oxygen ions are bound by electrostatic interaction (Obodo et al. 2019a). The majority of the metal oxides confirmed outstanding thermal and chemical steadiness. Their several outstanding characteristics resulting from partially or fully filled *d*-shells help in all kinds of electronic device usages such as solar cells, photo electrochemical cells, thin-film transistors, photocatalysts, and sensors (Obodo et al. 2019b).

They played a big role in the enhancement of modern discovery of optics, sensors, photocatalysis, electrocatalysis, anti-corrosion, electronic and magnetic utilizations (Obodo et al. 2020b). The interesting physical distinctiveness observed in nanoscale metals as well as semiconductors are related to alterations in electronic structure as well as the vast surface-to-volume ratios, which make them distinctly different from comparable bulk solids. Breathtaking power over material dimension, figure, as well as inconsistency in the nanometre length scale are being permitted the enhancement of synthetic measures resulting from their extraordinary characteristics. There is construction of nanoparticle assemblies in the solid state given that the practice expected and devices that are projected to be created cannot be solution-based or single particle-based. Consequently, development of measures for assembling of nanoparticles into solid-state structures, as keeping their characteristic physical traits, becomes a noteworthy setback (Obodo et al. 2020c).

The synthesis and study of nanosized materials are important because of their characterization interest; their interesting catalytic, thermal, magnetic, electrical, and optical characteristics; and the range of applications associated with them. Research into oxide thin films is being importunate since 1960s; however, it was the discovery of soaring temperature superconductivity in 1986 that gave a key motion to the research in the area of multicomponent complex oxide thin films (Obodo et al. 2020d; Obodo et al. 2020e).

24.10 Conclusion

The stannates can be prepared using a variety of ways, but here the predominant process are solid-state reaction and hydrothermal processes. The stannates, typically perovskite-type or spinel-type, are utilized in solar cells (alternative materials, photoanodes, future transparent conductive tool, etc.), photocatalysts, etc. Its synthesis improvement is a pointer that its optical as well as electrical performance could be advanced. Doping stannates (which have revealed a significant performance as hopeful) offers a further alternative when doped with the right material. If doped with the right dopants, high electron mobility can be achieved, making the resistivity which is a reflection of conductivity to be low with high band gap. The incorporation of doped stannates into solar energy harvesting devices helps in solving world energy problems by making the solar cells have a very high functional performance.

REFERENCES

Affrin, I. 2018. Solar Energy Harvesting and Its Applications, Ladydoak college, Madurai.

Arms, S., C. Townsend and D. e. a. Churchill. 2008. "Energy Harvesting, Wireless, Structural Health Monitoring and Reporting System." in *Preceedings of the 2nd Asia Pacific Workshop on SHM*, Melboune, Australia., December.

Bragg, W. 1915. "The Structure of the Spinel Group of Crystals." *Philosophical Magazine Series 6* 30: 305–315.

Chen, H. and N. Umezawa. 2014. "Sensitization of perovskite Strontium Stannate SrSnO3 towards Visible-light Absorption by Doping." Photocatalysis and Photoelectrochemistry for Solar Fuels.

D. Chen and J. Ye. 2007. "SrSnO3 nanostructures: synthesis, characterization, and photocatalytic properties." *Chemistry of Materials* 19, no. 18: 4585–4591.

Das, P., A. Roy and P. Devi. 2016. "Zn2SnO4 as an Alternative Photoanode for Dye Sensitized Solar Cells: Current Status and Future Scope." *Transactions: Indian Ceramic Society* 75, no. 3: 1–8.

Gaurav, A., S. Madhungiri and F. Karale. 2012. "High Solar Energy Concentration with a Fresnel Lens." *A Review* 2, no. 3: 1381–1385.

John, J., V. Pillai, A. Thomas, R. Joseph, S. Muthunatesan and R. Prabhu. 2017. "Synthesis, Structural and Morphological Property of BaSnO3 Nanopowder Prepared by Solid State Ceramic Method." *Material Science and Engineering* 195: 012007.

Karunakaramoorthy, K. and G. Suresh. 2014. "Synthesis and Characterization of the Al- Doped and Al-Mn Co-doped Zno Nanoparticles by Sol-gel Method." *555Xresearch2* 4, no. 7, July.

Lee, W., H. Kim, J. Kang, D. Jang, T. Kim and J. Lee. 2017. "Transparent perovskite barium Stannate with High Electron Mobility and Thermal Stability." *Annual Review of Materials Research* 47: 391–423.

Lee, C., D. Kim, I. Cho, S. Park, S. Shin, S. Seo and K. Hong. 2012. "Simple Synthesis and Characterization of SrSnO3 Nanoparticles with Enhanced Photocatalytic Activity." *International Journal of Hydrogen Energy* 37: 10557–10563.

Lee, W., H. Kim, J. Kang, D. Jang, T. Kim, J. Lee and K. Keem. 2017. "Transparent perovskite Barium Stannate with High Electron Mobility and Thermal Stability." *Annual Reviews of Materials Research* 47: 391–423.

Li, J., Z. Ma and K. Wu. 2017. "Improved Thermoelectric Power Factor and Conversion Efficiency of perovskite Barium Stannate." *RSC Advances* 7, no. 52: 32703–32709.

Liu, Q., H. Li, B. Li, W. Wang, L.Q.Y. Zhang and J. Dai. 2014. "Structure and Band Gap Engineering of Fe-doped SrSnO3 Epitaxial Films." *EPL (Europhysics Letters)* 108, no. 3: 37003.

Minami, T., H. Sonohara, S. Takata and H. Sato. 1994. "High Transparent and Conductive Zinc-stannate Thin Films Prepared by Rf Magnetron Sputtering." *Japanese Journal of Applied Physics* 33, no. 12A: L1693.

Obodo, R.M., A. Ahmad, G.H. Jain, I. Ahmad, M. Maaza and F.I. Ezema. 2020. "8.0 MeV copper ion (Cu^{++}) Irradiation-Induced Effects on Structural, Electrical, Optical and Electrochemical Properties of Co_3O_4-NiO-ZnO/GO Nanowires." *Materials Science for Energy Technologies* 3: 193–200.

Obodo, R.M., A.C. Nwanya, A.B.C. Ekwealor, I. Ahmad, T. Zhao, M. Maaza and F.I. Ezema. 2019. "Influence of pH and Annealing on the Optical and Electrochemical Properties of Cobalt (III) Oxide (Co_3O_4) Thin Films." Surfaces and Interfaces 16: 114–119.

Obodo, Raphael. M., Assumpta C. Nwanya, Tabassum Hassina, Mesfin Kebede, Ishaq Ahmad, M. Maaza and Fabian I. Ezema. 2019. "Transition Metal Oxide-Based Nanomaterials for High Energy and Power Density Supercapacitor." In F.I. Ezema and Mesfin Kebede (Eds.), *Electrochemical Devices for Energy Storage Applications*, 131–150. London: Taylor & Francis Group, CRC Press.

Obodo, R.M., N.M. Shinde, U.K. Chime, S. Ezugwu, A.C. Nwanya, I. Ahmad, M. Maaza and F.I. Ezema. 2020. "Recent Advances in Metal Oxide/Hydroxide on Three-Dimensional Nickel Foam Substrate for High Performance Pseudocapacitive Electrodes." *Current Opinion in Electrochemistry* 21: 242–249.

Obodo, R.M., A.C. Nwanya, C. Iroegbu, B.A. Ezekoye, A.B.C. Ekwealor, I. Ahmad, M. Maaza and F.I. Ezema. 2020. "Effects of Swift Copper (Cu^{2+}) Ion Irradiation on Structural, Optical and Electrochemical Properties of Co_3O_4-CuO-MnO_2/GO Nanocomposites Powder." *Advanced Powder Technology*, 1 (4): 1728–1735.

Obodo, R.M., A.C. Nwanya, M. Arshad, C. Iroegbu, I. Ahmad, R.U. Osuji, M. Maaza and F.I. Ezema. 2020. "Conjugated NiO-ZnO/GO Nanocomposite Powder for Applications in Supercapacitor Electrodes Material." *International Journal of Energy Resources* 44: 3192 – 3202.

Obodo, R.M., E.O. Onah, H.E. Nsude, A. Agbogu, A.C. Nwanya, I. Ahmad, T. Zhao, P.M. Ejikeme, M. Maaza and F.I. Ezema. 2020. "Performance Evaluation of Graphene Oxide Based Co_3O_4@GO, MnO_2@GO and Co_3O_4/MnO_2@GO Electrodes for Supercapacitors." Electroanalysis 32: 1–10.

Ong, K., X. Fan, A. Subedi, M. Sulivan and D. Singh. 2015. "Transparent Conducting Properties of SrSnO3 and ZnSnO3." *Applied Physics Letters* 3, no. 6. https://doi.org/10.1063/1.4919564

Park. 2014. "perovskite Solar Cells: An Emerging Photovoltaic Technique." *Materials Today* 18, no. 2: 65–72.

Patel, M. 2016. *Wind and Solar Power Plants*. CRC Press.

Raghavan, S., T. Schumann, H. Kim, C.T. Zhang and S. Stemmer. 2016. "High Mobility BaSnO3 Grown by Oxide Molecular Beam Epitaxy." *APL Materials* 4, no. 1: 016106.

Rao, K. and Vanaja. 2015. "Influence of Transition Metal (Cu,Al) Ions Doping on Structural and Optical Properties of ZnO Nanopowders." in *Materials Today: proceedings*, 2, 3743–3749.

Sharma, A., Y. Kumar and P. Shirage. 2018. "Structural, Optical and Excellent Humidity Sensing Behaviour of ZnSnO3 Nanoparticles: Effect of Annealing." *Journal of Materials Science: Materials in Electronics* 1–15.

Shruti, S., K. Kamlosh and S. Asutosh. 2015. "Solar Cells in Research and Applications." *Material sciences and applications* 612113: 1145–1155, December.

Sukhatme, S.P. 2008. *Solar Energy*. Tata McGraw Hill publication.

Tan, B., E. Toman, Y. Li and Y. Wu. 2007. "Zinc Stannate Dye-Sensitized Solar Cells." *JACS*, 129 (14): 4162.

Tan, B., E. Toman, Y. Li and Y. Wu. 2007. "Zinc Stannates (Zn2SnO4) Dye- Sensitized Solar Cells." *Journal of American Chemical Society* 129: 4162–4163.

Wang, L., Y. Zhang, X. Li, Z. Liu, E. Guo, X. Yi, J. Wang, H. Zhu and G. Wang. 2012. "Interface and transport properties of GaN/graphene junction in GaN-based LEDs." *Journal of Physics D: Applied Physics*, 45 (50): 505102.

Wang, L., X. Fang, and Z. Zhang 2010. "Design Methods for Large Scale Dye-sensitized Solar Modules and the progress of Stability Research." *Renewable and Sustainabl Energy Reviews* 14, no. 9: 3178–3184.

Yang, W., J. Noh, N. Joen, Y. Kim, S. Ryu, J. Seo and S. Seok. 2015. "High Performance Photovoltaic perovskite lAyers Fabricated through Intramolecular Exchange." *Science* 348, no. 6240: 1234–1237.

Zang, Y., J. Wang, M. Sahoo and T.K.T. Shimada. 2017 "Strain Induced Ferroelectricity and Lattice Coupling in BaSnO3 and SrSnO3." *Physical Chemistry Chemical Physics* 19, no. 38: 26047–26055.

Zhang, Y.D., X.M. Huang, D.M. Li, Y.H. Luo and Q.B. Meng, 2012. "How to Improve the Performance of Dye-sensitized Solar Cell Modules by Light Collection." *Solar Energy Materials and Solar Cells* 98: 417–423.

Index

Printed in the United States
by Baker & Taylor Publisher Services